Challenges and Advances in Computational Chemistry and Physics

Volume 29

Series editor

Jerzy Leszczynski
Department of Chemistry and Biochemistry
Jackson State University, Jackson, MS, USA

This book series provides reviews on the most recent developments in computational chemistry and physics. It covers both the method developments and their applications. Each volume consists of chapters devoted to the one research area. The series highlights the most notable advances in applications of the computational methods. The volumes include nanotechnology, material sciences, molecular biology, structures and bonding in molecular complexes, and atmospheric chemistry. The authors are recruited from among the most prominent researchers in their research areas. As computational chemistry and physics is one of the most rapidly advancing scientific areas such timely overviews are desired by chemists, physicists, molecular biologists and material scientists. The books are intended for graduate students and researchers.

All contributions to edited volumes should undergo standard peer review to ensure high scientific quality, while monographs should be reviewed by at least two experts in the field. Submitted manuscripts will be reviewed and decided by the series editor, Prof. Jerzy Leszczynski.

More information about this series at http://www.springer.com/series/6918

Ewa Broclawik · Tomasz Borowski ·
Mariusz Radoń
Editors

Transition Metals in Coordination Environments

Computational Chemistry and Catalysis Viewpoints

Springer

Editors
Ewa Broclawik
Jerzy Haber Institute of Catalysis
and Surface Chemistry
Polish Academy of Sciences
Kraków, Poland

Tomasz Borowski
Jerzy Haber Institute of Catalysis
and Surface Chemistry
Polish Academy of Sciences
Kraków, Poland

Mariusz Radoń
Faculty of Chemistry
Jagiellonian University
Kraków, Poland

ISSN 2542-4491 ISSN 2542-4483 (electronic)
Challenges and Advances in Computational Chemistry and Physics
ISBN 978-3-030-11713-9 ISBN 978-3-030-11714-6 (eBook)
https://doi.org/10.1007/978-3-030-11714-6

Library of Congress Control Number: 2018967746

This Springer imprint is published by the registered company Springer Nature Switzerland AG
The registered company address is: Gewerbestrasse 11, 6330 Cham, Switzerland

Preface

From its very foundation, quantum chemistry has been closely accompanying experiment in search of properties of a coordinate bond and structures of coordination compounds. The impact of the molecular orbital concept exerted on the model crystal field theory converted it to the ligand field theory which paved the way to investigate transition metal (TM) systems from first principles. Already early and of necessity simplified wave function methods were applied to study structures and other properties of transition metal complexes, but only after the advent of robust density functional-based methods, the number of quantitative applications dramatically increased. The number of theoretical works devoted to transition metal coordination compounds, dealing with their intricate electronic structures, spectral and magnetic properties, and related catalytic activity has already exceeded ten thousand, with recent contributions reaching over 500 papers each year. This is an obvious direction since the electronic properties of coordinated transition metals are largely responsible for unique activity of TM sites in materials and bio- or inorganic catalysis, and fine control of an intricate TM chemistry is vital for the progress of society and environmental safety.

Wide demand for theory-based practical guidelines for designing modern materials in the fields of medicine, electronics, and sustainable technology compelled the studies on transition metal compounds to verge upon the limits. However, it was soon recognized that the exact description of advanced electronic and magnetic properties of TM sites in materials required involvement of high-level theoretical approaches, more demanding than standard density functional theory (DFT). This prompted us to design this book as a collection of chapters, dealing with both the advancements in correlated wave function theory, making it still more accurate but less costly, and summarizing the attempts to upgrade DFT methods to make them sufficiently accurate to become robust and reliable tools for quantitative predictions. On the other hand, the quest for exact description of electron density-based properties of coordinated transition metals is challenging to both theory and experiment; thus, the interplay between these two areas remains in the focus of consecutive chapters devoted to selected subfields, here spin-related phenomena, spectroscopic, electrochemical, and catalytic properties of TM in various coordination environments. In this spirit, the following chapters are not strictly ordered either according to the level of theory or along formal distinction with respect to the branch of chemistry or physics, but cover rather related fields, combining various levels of theory and experiment.

The initial batch of five chapters deals with magnetic interactions, untypical systems, and related dedicated theoretical approaches. Chapter "The Electronic Determinants of Spin Crossover Described by Density Functional Theory" by Kasper P. Kepp deals with spin crossover (SCO) phenomenon which plays a vital role in living systems and in many emerging technologies. In line with the introductory paragraphs, the accurate prediction and design of SCO systems are of high current priority, while SCO tendency is extremely sensitive not only to the level of theory but also to its ability to cover such physical effects as dispersion, relativistic effects, and vibrational entropy for large molecular systems. It is shown that the density functional theory (DFT) is the tool of choice that can be predictive for this purpose only if the study is carried out scrupulously. A similar approach is taken in Chapter "Anisotropic Magnetic Spin Interactions of Transition Metal Complexes and Metalloenzymes from Spectroscopy and Quantum Chemistry" where Matthias Stein on example of transition metal-containing model complexes and metalloenzymes discusses the advancement of DFT computational approaches to calculate the parameters of the effective Spin Hamiltonian such as the electronic g- and hyperfine tensors in order to support the analysis and interpretation of complex magnetic resonance spectra. Chapter "Non-covalent Interactions in Selected Transition Metal Complexes" by Filip Sagan and Mariusz P. Mitoraj discusses how DFT is able to describe non-covalent chemical interactions in transition metal complexes where routine DFT methods have been known to occasionally fail. The authors illustrate on several examples of nonstandard bonds that good separation of donation (ligand to metal) and back-donation (metal to ligand) charge transfer processes, which govern this type of bonding, may be done by proper analysis of the deformation density.

Different approach is taken by Vera Krewald and Dimitrios A. Pantazis (Chapter "Applications of the Density Matrix Renormalization Group to Exchange-Coupled Transition Metal Systems"). This chapter deals with inherently multireference problem, like oligonuclear transition metal complexes containing magnetically coupled open-shell ions, not always tractable by broken-symmetry DFT. In such cases, the use of multireference methods remains the deal to treat the exchange coupling. However, the applicability of these methods has been severely restricted due to their computational cost, and only in recent years, the introduction of the density matrix renormalization group (DMRG) to quantum chemistry has enabled the multireference treatment of exchange-coupled transition metal systems. The first detailed applications of DMRG-based methods to exchange-coupled systems are reviewed, and the lessons learned so far regarding the applicability, apparent limitations, and future promise of this approach are discussed. In the same spirit, Chapter "New Strategies in Modeling Electronic Structures and Properties with Applications to Actinides" is written by Aleksandra Leszczyk, Paweł Tecmer, and Katharina Boguslawski. After a short overview of relativistic Hamiltonians, frequently applied to account for relativistic effects, the authors review conventional and unconventional single- and multireference approaches, developed to model the multireference nature of actinide compounds and their ground- and excited-state electronic structures, finishing with geminal-based approaches. In addition, concepts of quantum information theory are introduced, providing a qualitative understanding of complex electronic structures of actinide compounds using the picture of interacting orbitals.

The set of next four chapters is generally devoted to spectroscopic issues where the interplay between experiment and theory is frequently indispensable for the interpretation of the spectra and translates the gained information into chemical insights. In Chapter "Computational Versus Experimental Spectroscopy for Transition Metals" Maja Gruden, Wesley R. Browne, Marcel Swart, and Carole Duboc discuss a variety of examples where different spectroscopy techniques aided by computations allow to determine intricate and elusive properties, like the oxidation state, spin state, or coordination environment around redox-active metal ions such as iron, manganese, or nickel.

Marcus Lundberg and Mickaël G. Delcey show in Chapter "Multiconfigurational Approach to X-ray Spectroscopy of Transition Metal Complexes" how close correlation between theoretical modeling and X-ray experiment allows for the identification of the electronic and geometric structure of transition metal system through their spectral fingerprint from the core excitation energies. Compared to ground state calculations, modeling of X-ray spectra is complicated by the presence of the core hole, which typically leads to multiple open shells and large effects of spin–orbit coupling. Thus, reliable fingerprinting requires a theoretical model that is accurate enough, and the authors show that multiconfigurational wave function approaches, recently extended to model a number of X-ray processes of transition metal complexes, are suitable for that purpose. Chapter "Assessing Electronically Excited States of Cobalamins via Absorption Spectroscopy and Time-Dependent Density Functional Theory" by Megan J. Toda, Pawel M. Kozlowski, and Tadeusz Andruniów is specifically devoted to one type of systems, B12 chemistry. Due to the complexity and the size of the cobalamins, the computational analysis is almost exclusively represented by DFT and time-dependent DFT (TD-DFT) methods; thus, the proper choice of exchange-correlation functional discussed by authors is of paramount importance in predicting electronic transitions and simulating the full spectrum reliably. Chapter "Photodeactivation Channels of Transition Metal Complexes: A Computational Chemistry Perspective" by Daniel Escudero which concludes this section deals with the fate of the excited states in a transition metal compound, deactivating via a plethora of interconnected relaxation processes, competing with each other and controlled by the subtle interplay of electronic and geometrical rearrangements. The author provides critical overview of the state-of-the-art quantum chemical and reaction dynamic methods to study the photodeactivation dynamics in transition metal compounds and illustrates the progress and challenges in this field with recent examples on a variety of excited states in photoactive iridium and ruthenium complexes.

Finally, the third section of the book is devoted to computational approaches to catalysis by transition metal compounds, with enzymatic and biomimetic systems allocated as the last, separate part. The section opens with Chapter "Mechanism and Kinetics in Homogeneous Catalysis: A Computational Viewpoint" where Jeremy N. Harvey critically reviews the use of computational methods based on electronic structure theory and statistical mechanics to study reaction mechanisms and kinetics in homogeneous catalysis, especially organometallic catalysis and organocatalysis. The chapter is based on suitably selected examples from the authors' own group, which are discussed in the perspective of progress and still open challenges for computational chemistry to model actual chemistry. A careful reading of this chapter may be especially recommended as a warning to researches pursuing this domain of computational modeling because it draws their attention to still existing (and occasionally even growing) number of doubts and pitfalls awaiting inexperienced users of "user-friendly" computational, half-automated tools. The main line set in the introductory chapter for this part is actually followed in consecutive three chapters which are devoted to complicated reaction patterns found in catalytic systems and summarize attempts to allow various extended environments into the modeling.

Chapter "Computational Modelling of Structure and Catalytic Properties of Silica-Supported Group VI Transition Metal Oxide Species" by Jarosław Handzlik deals with chromium, molybdenum, and tungsten oxides supported on amorphous silica, the catalysts for many reactions, including large-scale industrial processes. Although these systems have been extensively studied for many years, there are still unresolved issues, concerning mainly the nature of the active sites and mechanisms of their formation. Computational studies, using carefully selected cluster or periodic models to represent the catalyst surface, are helpful in interpretation of spectroscopic data and can provide complementary information about the catalytic process and lead to complex structure–activity relationships. Nevertheless, even if a great progress has been achieved in modeling of these systems, theoretical determining of complex reaction mechanisms using surface models with representative distribution of metal sites is still a challenge for computational catalysis. Chapter "Catalytic Properties of Selected Transition Metal Oxides—Computational Studies" by Witold Piskorz and Filip Zasada constitutes the review of computational methods applied specifically to transition metal oxides (most abundant in heterogeneous catalysis, used as bulk or supported oxides) and is focused on the influence of the environment on the transition metal cation properties. The shortcomings of the DFT level of theory, most commonly used for modeling, are discussed, and its extensions toward more realistic environment are presented. The modern reactive force field methods are also mentioned, and the embedding schemes most commonly found in the quantum chemical or classical description of the heterogeneous processes are discussed. It is shown that in all discussed systems, the appropriately applied Hubbard DFT GGA+U methods are successful provided that the Hubbard correction is carefully tuned for modeling of particular properties. In turn, Chapter "Molecular Electrochemistry of Coordination Compounds—A Correlation Between Quantum Chemical Calculations and Experiment" by

Piotr P. Romańczyk and Stefan S. Kurek describes the progress in efficient solvation models that helped to develop effective computational protocols allowing for accurate reproduction of experimental redox potentials of mono- and dinuclear complexes, including electrocatalytically relevant systems and mixed-valence compounds. Examples of such computational protocols that use DFT combined with continuum solvent models, as well as a mixed, discrete-continuum approach, are shown in this contribution. The ferrocenium/ferrocene system, widely used as an internal standard, is discussed, followed by the presentation of intriguing properties of mono- and bimetallic Mo/W scorpionates, in particular electrochemical communication between metal centers and a baffling dehalogenation, the mechanism of which was elucidated only owing to the application of DFT-D3 calculations.

The end section of the book entails four chapters and discusses enzymatic and biomimetic systems. The topic is singled out even if problems intrinsic to very large systems incorporating transition metals are generally common; nevertheless, transition metals in bio- or bioinorganic complexes share several specific properties and specific techniques common among them but distinctive from inorganic transition metal complexes. This part opens with Chapter "The Quest for Accurate Theoretical Models of Metalloenzymes: An Aid to Experiment" by Matthew G. Quesne and Sam P. de Visser. The chapter reviews two key computational approaches to metalloenzymes, namely quantum mechanics/molecular mechanics (QM/MM) on complete enzyme structures and QM cluster models on active site structures only. The former take the structure of the full enzyme with a solvent layer into consideration, whereas the latter only include key features of the first and second coordination sphere. The examples are discussed where the QM cluster approach worked well; however, for systems where substrate binding is tight and or a network of hydrogen-bonding interactions exists, a complete QM/MM approach may be more appropriate. The following Chapter "Applications of Computational Chemistry to Selected Problems of Transition-Metal Catalysis in Biological and Nonbiological Systems" by Hajime Hirao describes as well recent attempts to study the structure and catalytic properties of transition metal-containing systems of different sizes, including metalloenzymes but also metal–organic frameworks (MOFs). Similar techniques (DFT and hybrid techniques for embedding) are used, but examples are selected specifically and substantially broaden the spectrum of applications, increasing the pool for critical analyses and benchmarking. The same concerns Chapter "How Metal Coordination in the Ca-, Ce-, and Eu-Containing Methanol Dehydrogenase Enzymes can Influence the Catalysis: A Theoretical Point of View" written by Tiziana Marino, Mario Prejanò, and Nino Russo, where the pool of examples is farther enriched with studies on lanthanide-containing enzymes where relativistic effects played a significant role.

The section is finalized by Tomasz Borowski and Maciej Szaleniec in Chapter "Challenges in Modelling Metalloenzymes" which gives a critical summary of the entire process of constructing a reliable computational model for metalloenzymes. This contribution, complementary to preceding chapters, nicely illustrates and validates the key decisions and steps one has to take in such projects: validating

X-ray structures, re-refining metal sites, determining residues protonation state, modeling missing loop regions or whole proteins by comparative modeling, performing MD simulations, choosing representative snapshots and constructing QM models. While this list is by no means exhaustive, the content of this chapter, combined with the material covered in other chapters of this book offers a more comprehensive view on modeling metalloenzymes and may supply additional suggestions for modeling other systems related to modern materials.

In the closing paragraph of this Preface, the reader deserves some remarks justifying the need for publishing (and reading!) still new book on modeling transition metals in various coordination environments despite that many works on the topic have already appeared. We hope we succeed in constructing the book as a collection of chapters well obeying the balance between methodological and practical information. The first nine chapters deal with advancements in correlated wave function and DFT methods making them applicable to nonstandard problems, whereas the next eight contributions bring as well instructive examples of their applications as practical guidance (and warnings!) to prospective users of modeling protocols in related disciplines.

We are very grateful to all authors for their contributions shaping this book and for their patience.

Kraków, Poland

Ewa Broclawik
Tomasz Borowski
Mariusz Radoń

Contents

Contributors

Tadeusz Andruniów Advanced Materials Engineering and Modelling Group, Department of Chemistry, Wroclaw University of Science and Technology, Wrocław, Poland

Katharina Boguslawski Faculty of Physics, Astronomy and Informatics, Institute of Physics, Faculty of Chemistry, Nicolaus Copernicus University in Torun, Toruń, Poland

Tomasz Borowski Jerzy Haber Institute of Catalysis and Surface Chemistry, Polish Academy of Sciences, Kraków, Poland

Wesley R. Browne Faculty of Science and Engineering, Stratingh Institute for Chemistry, Groningen, The Netherlands

Sam P. de Visser The Manchester Institute of Biotechnology and School of Chemical Engineering and Analytical Science, The University of Manchester, Manchester, UK

Mickaël G. Delcey Department of Chemistry - Ångström Laboratory, Uppsala University, Uppsala, Sweden

Carole Duboc Departement de Chimie Moleculaire, University Grenoble Alpes, CNRS, DCM, Grenoble, France

Daniel Escudero Department of Chemistry, KU Leuven, Leuven, Belgium

Maja Gruden Faculty of Chemistry, University of Belgrade, Belgrade, Serbia

Jarosław Handzlik Faculty of Chemical Engineering and Technology, Cracow University of Technology, Kraków, Poland

Jeremy N. Harvey Department of Chemistry and Division of Quantum Chemistry and Physical Chemistry, KU Leuven, Leuven, Belgium

Hajime Hirao Department of Chemistry, City University of Hong Kong, Hong Kong, China

Kasper Planeta Kepp DTU Chemistry, Technical University of Denmark, Kongens Lyngby, Denmark

Pawel M. Kozlowski Department of Chemistry, University of Louisville, Louisville, KY, USA

Vera Krewald Fachbereich Chemie, Technische Universität Darmstadt, Darmstadt, Germany

Stefan S. Kurek Faculty of Chemical Engineering and Technology, Molecular Electrochemistry Group, Cracow University of Technology, Kraków, Poland

Aleksandra Leszczyk Faculty of Physics, Astronomy and Informatics, Institute of Physics, Nicolaus Copernicus University in Torun, Toruń, Poland

Marcus Lundberg Department of Chemistry - Ångström Laboratory, Uppsala University, Uppsala, Sweden

Tiziana Marino Dipartimento di Chimica e Tecnologie Chimiche, Università della Calabria, Rende, Italy

Mariusz P. Mitoraj Department of Theoretical Chemistry, Faculty of Chemistry, Jagiellonian University, Kraków, Poland

Dimitrios A. Pantazis Max-Planck-Institut Für Kohlenforschung, Mülheim an der Ruhr, Germany

Witold Piskorz Faculty of Chemistry, Jagiellonian University, Kraków, Poland

Mario Prejanò Dipartimento di Chimica e Tecnologie Chimiche, Università della Calabria, Rende, Italy

Matthew G. Quesne School of Chemistry, Cardiff University, Cardiff, UK

Piotr P. Romańczyk Faculty of Chemical Engineering and Technology, Molecular Electrochemistry Group, Cracow University of Technology, Kraków, Poland

Nino Russo Dipartimento di Chimica e Tecnologie Chimiche, Università della Calabria, Rende, Italy

Filip Sagan Department of Theoretical Chemistry, Faculty of Chemistry, Jagiellonian University, Kraków, Poland

Matthias Stein Molecular Simulations and Design Group, Max Planck Institute for Dynamics of Complex Technical Systems, Magdeburg, Germany

Marcel Swart ICREA, Barcelona, Spain;
IQCC, Universitat de Girona, Girona, Spain

Maciej Szaleniec Jerzy Haber Institute of Catalysis and Surface Chemistry, Polish Academy of Sciences, Kraków, Poland

Paweł Tecmer Faculty of Physics, Astronomy and Informatics, Institute of Physics, Nicolaus Copernicus University in Torun, Toruń, Poland

Megan J. Toda Department of Chemistry, University of Louisville, Louisville, KY, USA

Filip Zasada Faculty of Chemistry, Jagiellonian University, Kraków, Poland

The Electronic Determinants of Spin Crossover Described by Density Functional Theory

Kasper Planeta Kepp

Abstract Spin crossover (SCO) plays a vital role in living systems and in many emerging technologies, and the accurate prediction and design of SCO systems is of high current priority. Density functional theory (DFT) is the state-of-the-art tool for this purpose due to its ability to describe large molecular electronic systems with an accuracy that can be predictive if carried out correctly. However, the SCO tendency, i.e., the free-energy balance of high- and low-spin states, is extremely sensitive to the theoretical description and physical effects such as dispersion, relativistic effects, and vibrational entropy. This chapter summarizes the recent fundamental insight into SCO gained from DFT and efforts that approach the accuracy needed (~10 kJ/mol) for rational design of SCO to become reality.

1 Introduction

Spin crossover (SCO) is the process, whereby two electronic states of different quantum-mechanical spins interconvert upon perturbation such as chemical bonding, temperature, light, or applied pressure [1–9]. Since its discovery in coordination complexes [10], it has emerged as one of the most important chemical processes [1–3, 11]. If this importance is not immediately appreciated, please consider that without SCO, the reader would suffocate within 2–3 min due to the absence of O_2 binding to hemoglobin within the honorable reader's lung arteria.

Some chemical systems can undergo transition between the two spin states without any change of chemical composition, whereas others only do so upon interaction with other molecules, such as ligand binding to a metal. The term "SCO system" or "SCO complex" is mainly used if the transition occurs relatively near standard conditions of temperature and pressure. Thus, thermal SCO systems are particularly notable as those where SCO occurs close to atmospheric pressure and within the range of room temperature by a few hundred Kelvin. Such systems are central to life

K. P. Kepp (✉)
DTU Chemistry, Technical University of Denmark, 2800 Kongens Lyngby, Denmark
e-mail: kpj@kemi.dtu.dk

E. Broclawik et al. (eds.), *Transition Metals in Coordination Environments*,
Challenges and Advances in Computational Chemistry and Physics 29,
https://doi.org/10.1007/978-3-030-11714-6_1

processes via the management of $^3O_2$ by hemes [12–14]. They are also important to many emerging technologies [1, 3], as they manifest as two distinct quantum states that can be interconverted by external stimuli. This makes them suitable for, e.g., molecular electronics [3, 15–17], sensors [18–20], and nano-mechanical behavior such as molecular motors [3, 7, 21–27].

For convenience, we will write the SCO process as a conversion from a low-spin state (LS) to a high-spin state (HS),

$$\text{LS} \rightleftharpoons \text{HS} \tag{1}$$

where the HS state has the highest spin quantum number or, if this number is not well-defined, the largest magnetic moment due to parallel alignment of electron spins. The fundamental requirement of SCO is then that the free energy difference of the two electronic spin states approaches zero [2, 28, 29]:

$$\Delta G_{\text{SCO}} = \Delta H_{\text{SCO}} - T\,\Delta S_{\text{SCO}} \approx 0 \tag{2}$$

The enthalpy ΔH_{SCO} largely derives from changes in the ground-state electronic structure during SCO. These effects can be obtained from electronic energy calculations using standard quantum-mechanical programs and a suitable Hamiltonian, but it includes various energy terms not always considered that systematically contribute to ΔH_{SCO}, most notably the differential zero-point energy, dispersion forces, and relativistic effects of the two states [28, 30]. In contrast, the entropy of the process ΔS_{SCO} to a large extent (but not completely) arises from changes in the vibrational state functions [2, 30, 31] and favors the weaker and longer M–L bonds of the HS state [24, 28, 32, 33]. The electronic degeneracy contribution to this entropy is somewhat less important than the vibrational entropy of the involved chemical bonds [28, 30, 34, 35], as first recognized by Sorai and coworkers [24, 32].

Written as in (1) and (2), because ΔS_{SCO} is positive, higher temperature will favor the right-side HS state via $-T\Delta S_{\text{SCO}}$ of (2), and thus conversion from an initial LS state to HS is induced by raising the temperature [24, 29]. It turns out by inspection of experimental data for iron SCO systems, but it remains to be confirmed as a general law, that the entropy and enthalpy terms of (2) tend to compensate each other substantially, as also seen in some other processes [36]. This would suggest that SCO may be a true entropy–enthalpy compensation process not just with entropy favoring reaction toward the right, but with the two terms canceling over a broad range of enthalpies and entropies; this possibility is explored further in the present chapter as it has direct implications for accurate prediction of the SCO tendency.

Transition metal ions of the middle of the first row of the d-block, specifically Mn, Fe, and Co, are particularly common in SCO systems: SCO has been observed in d^4, d^5, d^6, d^7 and arguably in some Ni(II) d^8 systems [37]. This is partly because the ligand field stabilization energy has a magnitude that makes the vibrational entropy cancel the energy terms almost perfectly. However, the balance is a compromise between the metal ion and ligand field strength, as both contribute to the SCO tendency. This is illustrated by the spectrochemical series of the ligands and the corresponding series of the metal ions [38, 39]. Thus, for example, most SCO systems contain moderate-field nitrogen-donor ligands combined with Fe(II), as exemplified by the much studied, first synthetic iron(II) SCO system $Fe(Phen)_2SCN_2$ [40–43]. However, Fe(III) and Co(II) are also relatively common SCO metal ions, with the first-reported Fe(III) SCO systems being those of Cambi et al. [10] and the first Co(II) SCO system being that of Stoufer et al. [44].

In this chapter, the basis for describing SCO accurately by chemical–physical principles and the role of various contributions to the SCO tendency will be discussed. These include important systematic energy terms, i.e., the zero-point vibration energy, relativistic contributions to SCO, and dispersion forces that modulate the HS–LS equilibrium already at the single-molecule level. The importance of modeling the vibrational entropy contribution in the theoretical study of SCO systems is emphasized. The difference between the spectrochemical series and the "thermochemical series" of spin-state propensities are discussed. The performance of DFT and various ingredients of the functionals that affect the accuracy are analyzed.

2 Fundamentals of Spin Crossover

2.1 The Dilemma and Choice Between LS and HS

As taught in basis inorganic chemistry, when ligands are placed around a metal ion, the energies of the d-orbitals split into several energy levels due to the symmetry breaking, i.e., the d-orbitals experience different environments. If the ligand field is octahedral (O_h symmetry), two levels occur: The two high-lying degenerate e_g orbitals are aligned toward the ligands and thus experience more electronic repulsion, and the threefold degenerate low-lying t_{2g} orbital level becomes less repelled as these orbitals (originating from d_{xy}, d_{xz}, and d_{yz}) distribute further from the ligands.

Depending on the energy splitting Δ_o between the two levels, the electrons face a dilemma after occupying the three t_{2g} orbitals by one electron according to Hund's Rule: Either the additional electrons distribute in the normal fashion by pairing with the three first t_{2g} electrons or, if the energy distance is small, they may in fact move to the next level, e_g. The solution to this dilemma partly (but not completely, as discussed below) lies in resolving the relative magnitude of the penalty of moving up to the e_g orbitals, i.e., Δ_o, versus the penalty of occupying a t_{2g} orbital where another electron is already residing close in space, i.e., the spin-pairing energy penalty P. This situation is also entropically unfavorable, to be discussed below. If the fourth and fifth electron decide to move to e_g, the system will experience more aligned electron spins; this state is the HS state. If it is more favorable to pair with the t_{2g} electrons first, the resulting spin and magnetic moment becomes smaller; this is the LS state.

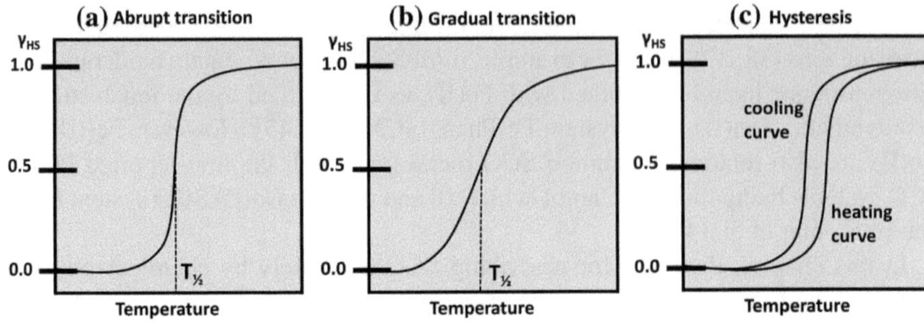

Fig. 1 Schematic representation of the transition from a low-spin electronic state with zero or little magnetism, which dominates at low temperature, to a more magnetic high-spin state, which dominates at higher temperature. **a** Abrupt transitions increase magnetism quickly near the transition temperature; **b** gradual transitions display smaller temperature gradients of the magnetism near transition; **c** hysteresis involves different transition temperatures upon heating and cooling

The question now arises: What happens if this dilemma remains unsolved, in other words, if the energy cost of pairing in the same orbitals and the entropy loss associated with this more compact LS state is almost perfectly outweighing the benefits of the lower orbital energies? If this is the case, the two possible occupations may be realized not far from standard conditions, as speculated by Pauling in his third paper in the series on the chemical bond, where he discussed the magnetic criterion for transition between HS and LS states [45]; these systems are the SCO systems.

The temperature at which the conversion in (1) takes place is referred to as the transition temperature, $T_{½}$, the temperature at which half of the system is in the HS state (the most magnetic), written as $\gamma_{\mathrm{HS}} = ½$, and the other half is in the LS state (the least magnetic or even diamagnetic state, as in, e.g., Fe(II) LS), written as $\gamma_{\mathrm{LS}} = ½$. Accordingly, at higher temperature, the fraction of HS, γ_{HS}, exceeds ½. This situation is shown schematically in Fig. 1a for an abrupt transition. The process can also be considerably more gradual, as shown schematically in Fig. 1b, characterized by a smaller magnetic susceptibility gradient at $T_{½}$. For abrupt processes, hysteresis is commonly observed (Fig. 1c), as discussed in detail in this chapter. At $T_{½}$, the isobaric heat capacity C_{p} displays a major peak reflecting the transition, being either narrow and steep or broader depending on whether the transition is abrupt or gradual [24].

Under actual equilibrium conditions, which are rarely realized in practice, the equilibrium constant K_{SCO} is equal to unity and the free energy of the process is then $\Delta G_{\mathrm{SCO}} = 0$. However, due to the nonequilibrium nature of the actual transition, one can hardly consider this definition exact. Still it is theoretically meaningful to separate contributions to the free energy and transition temperature, as the following discussion shows. Simply put, the core premise of theoretical studies of SCO is that if the relationship holds, we should be able to predict $T_{½}$ from an estimate of $\Delta G_{\mathrm{SCO}} = 0$, determined by electronic structure calculations. Under such conditions,

$T_{½} = \Delta H_{SCO}/\Delta S_{SCO}$, which can, as discussed below, be obtained from quantum-chemical computations at variable accuracy.

2.2 *The Spectrochemical Series*

In text books, the spectrochemical series [38, 46, 47] is traditionally used to estimate the preference for either HS or LS in a given coordination complex. This series is based on spectroscopic measurements of the absorption peak for the d–d transitions of cobalt(III) complexes and ranks common coordinating ligands according to Δ_o in an octahedral field. A rough rule-of-thumb order is:

$$I^- < Br^- < Cl^- < F^- < \text{S } \sigma\text{-donor} < \text{O } \sigma\text{-donor} < \text{N } \sigma\text{-donor} < \pi\text{-acceptors}$$

Although sometimes forgotten, this series is mainly based on Co(III), and the series, more specifically, the ligand field stabilization energy calculated from Δ_o, estimates the relative preference for HS versus LS *if this preference was only due to electronic energy* as measured by the absorption spectroscopy. The series, moreover, reflects a non-thermal electronic excitation, whereas the SCO systems of interest involve the thermal excitation of typically two electrons. Although widely used and displayed in textbooks, the estimates based on the spectrochemical series thus miss vibrational relaxation, spin pairing, and entropic effects, and they do not necessarily accurately convey the spin-state preference in a real chemical system at thermal equilibrium. Still, because the energy described by absorption maximum is a large part of the typical thermodynamic preference between the spin states, it is often accurate when applied to trend predictions, which largely explains its success [48–51].

2.3 *The Thermochemical Spin Series*

The real thermochemical spin-state preference and thus the adequate tool for rationalizing and predicting SCO can be argued to be a "thermochemical series" of spin-state propensity [52]. This series takes into account ground-state geometry relaxation of the HS state and entropy terms that also favor HS [31, 52, 53]. This series is straightforward achievable from DFT computations of the fully relaxed ground-state geometry of the HS state, which corrects the spectrochemical series based on electronic transitions in which the HS state features as an excited state. Furthermore, DFT can compute the vibrational entropy term with decent accuracy [36, 52, 54] so that the real preference as given by the free energy in (2) is honored. The series is importantly independent on the functional used [52], because the trend of interest involves cancellation of the major systematic errors in DFT that are discussed below.

The resulting thermochemical series resembles the spectrochemical series, but notably differs in several aspects. One of the most interesting differences is that Cl^- and Br^- have similar spin-state propensity once the thermochemical, spin-pairing, and vibrational-structural corrections are accounted for [52]. In the spectrochemical series, they separate clearly in the halide trend. Another difference is the preference for negatively charged versus neutral π-acceptors such as CN^- and CO, which change place in the thermochemical series depending on the oxidation state of the metal ion. A third difference relates to coordination isomers such as $\underline{S}CN^-$ and $\underline{N}CS^-$ that also change relative position depending on metal oxidation state [52], of relevance to the much studied SCO system $[Fe(SCN)_2(Phen)_2]$ and its derivatives.

The thermochemical series is quantitative and includes vibrational relaxation and entropy terms directly [52]. Thus, one can predict that Mn(III) systems will have relatively similar ligand preferences as Fe(II) in order to induce SCO, whereas Fe(III) should have a slightly weaker total ligand field; Co(II) SCO systems are predicted to be realized with stronger ligand fields than for Mn(III), Fe(III), and Fe(II), and even stronger ligand fields are required for Mn(II) such that even CO and CN^- become relevant. In contrast, Co(III) requries very weak ligands due to its strong LS preference, with SCO most likely occuring between F^- and weak O-donor ligands such as water, ethers, and alcohols [52]. Examples of SCO systems that illustrate these preferences are shown in Fig. 2, with 6N coordination for Fe(II) (Fig. 2a), weaker 4N2O coordination for Fe(III) (Fig. 2b), tunable ligand fields for porphyrins with Fe(II)/Fe(III) (Fig. 2c), 6N coordination for Co(II) (Fig. 2d), and correspondingly weaker 6O ligand field for Co(III) (Fig. 2e).

2.4 *The Oxidation State on the Central Metal Ion*

The metal ion's oxidation state plays a major role in defining the spin-state propensity, with higher oxidation state favoring LS. This effect ("the spectrochemical series of the central ions") was originally formulated by Jørgensen on the basis of spectroscopic data [39], but its fundamental truth is easily recovered and even quantified using modern DFT [52]. Specifically, a difference in LS preference of Fe(II) and Fe(III) of ~50 kJ/mol has been estimated [52]; for Co(III) versus Co(II), the difference is even higher and can reach 100 kJ/mol [52]. This effect clearly needs to be considered if SCO systems of variable metal oxidation states are to be developed by rational screening.

The fact that Fe(III) favors LS more than Fe(II) is easily understood from the stronger and shorter metal–ligand bonds formed in the higher oxidation states, which increases the σ-donation and thus the ligand field splitting and tendency toward low spin. Because of this, the most common coordination structure for Fe(II) SCO systems is 6N [55], whereas for Fe(III) SCO systems, the 4N2O coordination structure is common [6], where two nitrogen donors have been changed into weaker oxygen donor atoms. There are many exceptions to this preference: For example, some 4N2O iron(II) systems have been reported [56], and an Fe(II) 5N1S system has been made

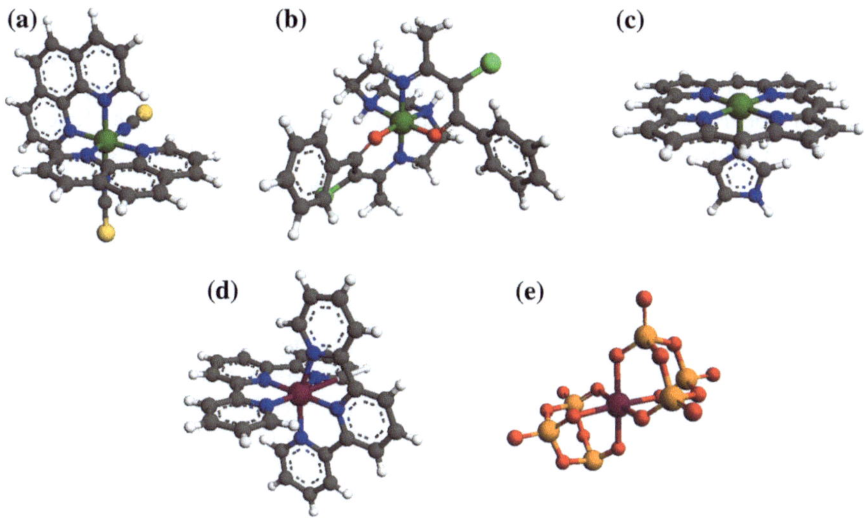

Fig. 2 Some examples of mononuclear SCO coordination complexes: **a** the iron(II) compound $[Fe(SCN)_2(Phen)_2]$; **b** the iron(III) compound $[Fe(bzacCl)_2trien]^+$ (bzac = benzoylacetonate-triethylenetetramine); **c** deoxyiron(II)porphine as a generalized representation of porphyrins and hemes; **d** the cobalt(II) compound $[Co(terpy)_2]^{2+}$; **e** the cobalt(III) compound $[Co(P_3O_9)_2]^{3-}$

[57]. This is possible because the constraints imposed by the thermochemical spin preference of the monodentate ligands can be broken in more complex coordination environments where the σ-donation is modulated electronically either by induction effects to the donor atom or by strained geometries, as commonly seen in SCO systems using multidentate ligands.

2.5 *Homoleptic SCO Complexes and the Case of Co^{3+}(aq)*

Very few, if any, real SCO systems possess O_h symmetry (one candidate is $[NiF_6]^{3-}$). Even if the complex is homoleptic (sharing chemically identical ligand donor atoms) as is reported in few cases [58, 59], Jahn–Teller distortion will cause the t_{2g} and e_g levels to split for the d^4 HS and LS configurations, for the d^6 HS configuration, and for the d^5 and d^7 LS configurations. Thus, the O_h symmetry is broken in almost all real cases to a variable extent. In fact, it would be an interesting academic challenge to identify a homoleptic SCO system that possesses almost perfect, unstrained O_h symmetry without Jahn–Teller distortion in one of its spin states, i.e., d^5 HS or d^6 LS. This state would represent an intrinsic, unstrained fit of the ligand and metal to enable SCO, something that puts major restriction on the exact ligand fields' strength of the six identical donor atoms.

Homoleptic coordination complexes for a range of different simple monodentate ligands with these d-electron configurations for Mn(II), Mn(III), Fe(II), Fe(III), Co(II) and Co(III) have been studied by DFT [52] and can help to suggest how such an "ideal" symmetric homoleptic system may be realized. Using the functionals known to be more accurate for the purpose (e.g., B3LYP* [60, 61] or TPSSh [62], e.g., as shown previously [63, 64]), one sees that some combinations of metal ions and ligands can bring a homoleptic complex very close to SCO. Co(III) has the strongest LS preference of all first-row d-block metal ions up to the III oxidation state as it has maximal ligand field stabilization energy in LS due to its t_{2g}^6 configuration and has higher charge than the iso-electronic Fe(II) LS. Thus, Co(III) SCO systems are rare and require very weak total ligand field strengths to reach SCO; the 6O coordination structures by Kläui and associates are notable in this regard [65–67]. Examples of Co(III) SCO may also exist in mixed-metal oxides [68], and a new example of a bistrimetaphosphate Co(III) complex with probable SCO properties was reported recently (Fig. 2e) [69]. In contrast, Co(II) SCO systems are relatively common and often feature 6N-coordination structures, e.g., bis-terpyridines (Fig. 2d) [70], consistent with the discussion above [71–73].

This raises an interesting and important question, namely how far above the LS state is the HS state of Co(III)(aq)? All other M(aq) systems of the first row of the d-block are HS due to water's weak ligand field, and their spin states were recently studied by CASPT2 [74]. Standing out alone, Co(III) is known to be LS under typical conditions studied (which are very acidic, because Co(III)(aq) readily undergoes reductive hydrolysis to Co(II) at neutral pH). Moreover, $[CoF_6]^{3-}$ is known to be HS, and this makes the range to HS very small since H_2O is close-by in the spectrochemical series [38]. The Co(III)(aq) is assumed (and has been shown in older data) to feature LS, but this produces several anomalies such as a much faster self-exchange electron transfer rate and faster ligand substitution than expected. Recent DFT computations [75] of the relative self-exchange rates of hydrated transition metal ions accurately recover the experimental 10^5 anomaly of Co(II)/Co(III) when plotting the trend in reorganization energies versus experimental rate constants and using LS Co(III) as commonly assumed. When doing the same correlation for HS Co(III)(aq), the anomaly disappears almost completely. DFT can also be used to correct previous spectroscopic estimates of the HS–LS gap with entropy and vibrational geometry relaxation showing that Co(III)(aq) is very close to SCO. In conclusion, this analysis converges on the view that the HS state is probably active during much of the chemistry of the hydrated Co(III), in stark contrast to text book consensus based on early NMR and absorption spectroscopy measurements in strong acid [76–78], but explaining the anomalous high ligand substitution and electron transfer capabilities of Co(III)(aq) [79, 80]. Indeed, several SCO systems of Co(III) with O-donor ligands are known [66, 68]. Future exploration of the SCO properties of Co(III) in water-like coordination environments should therefore be of interest.

2.6 Geometry Preferences and Changes During SCO

In the HS state, e_g occupation and associated ligand d-electron repulsion expands the metal–ligand bond lengths [52, 81]. Accordingly, LS states are generally more compact than HS states, and the system tends to expand upon SCO to the HS state [2], as schematically shown in Fig. 3a. However, despite the increased molar volume, the crystal symmetry is typically unaffected [24]. The longer, weaker, and more entropic metal–ligand bonds largely explain why HS is favored by temperature, viz the $T\Delta S$ term in (2). It is also the main reason why applied pressure tends to often favor the more compact LS state. In a crystal state, expansion of the core system leads to a larger unit cell and to a change in the intermolecular crystal packing forces. In a solution state, the expansion can affect the solvation energy of the two states differently, since the HS state will tend to be a slightly larger solute [28]. Not only the bond lengths but also the bite angles and the distances between the N-donor atoms of multidentate ligands may change accordingly, and the different geometric preferences of the two electronic states can thus sometimes be used to predict spin-state preferences by simple geometric inspections [58].

To put approximate numbers to this geometry effect, one can again turn to systematic studies of homoleptic mononuclear octahedral coordination complexes [52]. A summary of this analysis is shown in Fig. 3b. It turns out that the geometry changes upon the conversion of spin state are very dependent on the involved ligands and metal ion. Importantly, the geometry change scales almost monotonically with the ligand field strength such that weak-field halides give small geometry changes of the order of 0.02–0.12 Å (depending on metal ion), whereas larger changes of 0.15–0.40 Å occur for strong-field π-acceptor ligands [52]. Furthermore, the *variation* in geometry relaxation also scales with ligand field strength. This implies that the metal ion effect on HS–LS geometric relaxation upon SCO is much larger for strong-field

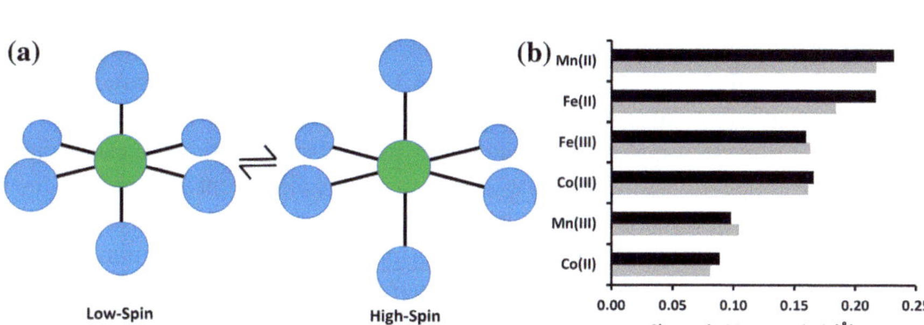

Fig. 3 **a** Schematic representation of the change in geometry associated with a transition from a LS state to a HS state in a mononuclear coordination complex; **b** DFT-computed changes in average metal–ligand bond lengths for d^4–d^7 configuration systems. The present figure is made from previously published data [52]: black color shows results for hexamine complexes, whereas gray color shows the average over a range of ligand types (halides, O-, and N-donor, and strong π-acceptors)

ligands than for weak-field ligands. The largest structural effects are, interestingly, seen for Mn(II) and Fe(II). A change of 0.2 Å or 0.3 Å is considerable; it occurs on average for all six bonds in a six-coordinate complex. For Fe(II) SCO systems with N-donor ligands, one can expect a typical average increase in Fe–N bond length of ~0.2 Å, as shown in black color for Fe(II) in Fig. 3b [52].

As discussed below, metal ions with large structural relaxation upon SCO are expected to also show more abrupt transitions with hysteresis. Accordingly, DFT as summarized in Fig. 3b largely explains why Fe(II) SCO systems commonly display high hysteresis but also predicts that Mn(II) systems should have similar or even larger ability to do so, of course under a modulating influence of other effects outside the first coordination sphere.

Since the geometric and environmental effects required to cause SCO can be subtle, SCO may be induced by adsorption of one molecular system to another, as the adsorbed state affects the molecular environment of the SCO system. The classic example is the host–guest systems [82], such as that of Halder and Kepert et al. [83] Supramolecular cages with SCO properties constitute one class of systems [84], whereas a recent example of "on-surface" SCO is that of Kumar et al. [85] Hemes, which are well known to change spin state upon changes in coordination environment, also seem to undergo on-surface SCO upon relatively weak adsorption to, e.g., gold surfaces [86], although this finding needs stronger experimental bearing.

2.7 *The Nature of the SCO Transition*

The actual transition from one spin state to the other can occur either gradually, or abruptly, and be subject to small or large hysteresis, as shown in Fig. 1. It is also possible, although not shown in Fig. 1, to have a multistep transition, and two-step processes have recently been studied by DFT [27]. Such multistep processes typically arise from a heterogeneous SCO sample, i.e., the presence of two or more the individual and distinct sites undergoing SCO, or possibly from restructuring of the system (including solvent) near the transition temperature.

The SCO transition curve has a sigmoidal form characteristic of a cooperative process. The cooperativity can be partly due to magnetic alignment as seen in an Ising model, and partly to structural phase transitions occurring locally: If the molecules interact closely, the intermolecular interactions produce free energy minima distinct to the larger HS and smaller LS volume. Accordingly, the conversion into a given spin state of one molecule makes it more favorable for neighbor molecules to attain the same spin state [27]. The extent of cooperativity, and accordingly the abruptness of the transition, thus depends greatly on the surroundings of the single molecule. In a solid, each magnetic center has contact with several neighbors, and thus the geometric spin-state preference induces a friction in the tendency to change spin state, which is greatly influenced by intermolecular interactions. In a solvent, cooperative interactions can be modulated by the presence of counter ions and solvent molecules, and the solvent can separate the SCO solutes so well that the spin transition becomes gradual. Accordingly, the transition behavior is typically very different in solid and solution [2, 3], and largely influenced by electrostatic interactions of molecules [87]. Covalent linkers and hydrogen bonds can be introduced to enhance the cooperativity [2, 3, 25, 88].

Hysteresis, defined as magnetization curves that differ upon heating and cooling *for the same molecular composition*, is shown schematically in Fig. 1c. Hysteresis is a priori expected during SCO because of the change in molar volume associated with the more expanded HS state. In case of hysteresis, one has to distinguish two transition temperatures, $T_{½}$ (↑) for heating and $T_{½}$ (↓) for cooling. Because of the "friction" due to the distinct cooperative interactions in each state of distinct molar volumes, it generally holds that $T_{½}(\downarrow) < T_{½}(\uparrow)$ (Fig. 1c). The difference in these two values,

$$\Delta T_{½} = T_{½}(\uparrow) - T_{½}(\downarrow) \tag{3}$$

defines the extend of hysteresis. For the purpose of theoretical modeling, the *single* molecule $T_{½}$ can be assumed to be

$$T_{½} \approx (T_{½}(\uparrow) + T_{½}(\downarrow)) / 2 \tag{4}$$

The *intermolecular* correlations causing $\Delta T_{½}$ should then be modeled separately [89]. Hysteresis may be utilized technologically because it produces a molecular "memory" to the system near the transition region [88]. Accordingly, large hysteresis is rare but desired for the purpose of switchable materials [4, 90].

2.8 True Hysteresis and Intrinsic Hysteresis

True hysteresis is defined only for systems where the composition is identical before and after transition. However, many SCO systems are prepared as hydrates, or with other co-crystallized molecules. Many transitions of interest occur at temperatures where these molecules begin to evaporate from the complexes, and this process is irreversible. Accordingly, if $T_{½}$ provides thermal energy enough to release these molecules, a large separation in the first heating and subsequent cooling curves will be observed which is *not* hysteresis, but simply reflects two different molecular systems being studied. Thus, several of the SCO systems reported to have large hysteresis may in fact reflect different molecular systems rather than true hysteresis [24]. Any technological application of hysteresis obviously requires microscopic reversibility, which is only obtained with stable systems of the same composition. This requirement substantially narrows down the number of observed cases of very large "true" hysteresis.

Inspection of $\Delta T_{½}$ values suggests a natural or *intrinsic* hysteresis of 0–20 K for many studied cases, which for some SCO systems is augmented by additional hysteresis. Although not discussed there, this can be inferred from the data compiled by Sorai [24]. The intrinsic hysteresis probably reflects the simple volume reordering effects of the first coordination sphere, which is approximately similar for all SCO systems (although the expansion depends on ligand and metal type as discussed above). Additional contributions to $\Delta T_{½}$ can arise from larger reorganizations due to bulky groups or intermolecular interactions beyond the local volume changes of the first coordination sphere that is generic to all SCO systems; this distinction between two parts of the hysteresis (which is, as a disclaimer, only the author's view) probably warrants further exploration.

As shown in Fig. 3b, Fe(II) systems exhibit some of the largest geometric changes upon SCO among d^4–d^7 systems [52]. Consistent with the volume-friction interpretation discussed above, this probably explains why Fe(II) more commonly displays hysteresis in comparison with other SCO systems [2]. This observation seems to confirm that hysteresis at least partly arises from the geometric friction of the heating and cooling processes caused by the different free energy minima at different molar volumes for HS and LS. The intermolecular interactions that define the transition are diverse and harder to systematize than the electronic structure of the molecule itself, but the intrinsic contribution from the first coordination sphere, as discussed above, seems to be predictable by DFT, which computes geometric changes with good accuracy.

3 Important Contributions to Single-Molecule SCO

3.1 Zero-Point Vibrational Energy

The vibrational zero-point energy (ZPE) is one of the electronic effects that always contribute to the SCO tendency regardless of the environment. It has been known for a long time, and was described clearly in the pioneering DFT work on SCO by Paulsen et al. [34], that the differential ZPE of the HS and LS states is an important contribution to the SCO process. This paper also reported the strong bias toward LS of the non-hybrid GGA functionals and the preference for HS for the hybrid B3LYP functional, an important observation that, for example, motivated the later development of the B3LYP* functional by Reiher et al. with a smaller 15% HF exchange [60].

The importance of ZPE lies both in the fact that its magnitude is of the order of 10 kJ/mol [34, 64], similar to the typical values of the full ΔH_{SCO} [36], and it is systematic, as it almost exclusively favors the HS state. This favoring follows directly from the longer and weaker M–L bonds of the HS state, which accordingly have smaller ZPEs than the LS states. Not only the SCO process but also a general chemical process involving multiple spin states will experience this effect. On an energy profile of competing spin states, neglect of ZPE will tend to provide an artificial bias in favor of the LS states in the reaction coordinate diagram.

More systematic studies of the ZPE contribution indicate that the differential ZPE is very dependent on the type ligand and to a lesser extent the type of metal ion [52]: Typical differential ZPEs in favor of HS range from 5 to 25 kJ/mol and grow more or less monotonically with the ligand field strength. Accordingly, for weak ligands such as halides, the differential ZPE can be almost neglected, whereas for strong-field π-acceptor ligands such as CN^{-} and CO, the ZPE dramatically favors HS by more than 20 kJ/mol. Thus, the importance of remembering the ZPE correction depends very much on the system of interest. This follows trivially from the fact that the ZPE scales monotonically with the overall strength of the M–L bond, which again scales with the ligand field strength. In the middle range, typical ligands relevant to SCO systems have differential ZPEs of 10–15 kJ/mol [52]. The ZPE can also vary by >10 kJ/mol due to metal ion and d^q configuration. This also affects the bond strengths of the M–L bonds in the two spin states, but since these d^q configurations vary substantially in terms of electronic structure and some, such as HS Mn(III) and LS Co(II) induce strong Jahn–Teller distortions, this metal effect is not trivial to interpret.

3.2 Dispersion Contributions to the Spin Crossover Equilibrium

Dispersion effects are the second-order interaction of instantaneously induced dipole moments of electron densities that affect bonding in all systems. Dispersion is not intrinsically included in most modern density functionals, and the most popular way to do so is by using an empirically parameterized correction to the electronic energies computed by the functional, such as the popular D3 correction by Grimme and coworkers [91]. Within the last decade, dispersion corrections have become increasingly mandatory in DFT calculations of chemical structure and reactivity.

Since the dispersion forces contribute substantially to the intermolecular interactions, their inclusion is important if one wants to understand the transition behavior [92, 93]. Dispersion interactions also affect $T_{1/2}$ itself, by favoring either HS or LS. Depending on the nature of the ligands and their packing, dispersion forces will either compress or expand the first coordination sphere, shifting the potential energy surface either toward longer M–L bonds to favor HS or to shorter M–L bonds favoring LS. The favoring of HS or LS may be very dependent on the intermolecular ligand–ligand interactions.

A priori, dispersion effects might not be expected to contribute to the SCO tendency of *single* molecules in solution with simple monodentate ligands having no systematic steric strain. Dispersion would be expected to mainly affect closely interacting parts of different molecules, or bulky parts of the same molecule. With the advent of empirical dispersion corrections of DFT, one could explore whether dispersion forces also contribute to this important process *even* at the single-molecule level, i.e., if there is a generic first coordination sphere contribution from dispersion to the SCO thermodynamics. This question can be directly addressed by dispersion-corrected DFT [91], because the dispersion energy correction is calculated explicitly and is separated from the remaining electronic energy of the system [54].

Strained five-coordinate iron(III)porphyrins with variable substituted axial ligands have been studied with and without dispersion forces included [54]. It was found that dispersion forces contribute already for the single molecule by affecting the free energy gap of (2) by often 10 kJ/mol or more. Considering that the total ΔH_{SCO} is in the order of 5–20 kJ/mol, this makes account of the dispersion forces critical. Four of the porphines had axial phenyl ligands attached directly by Fe–C bonds with 3–5 fluorides as phenyl substituents. This produces unusual short-range interactions and clashes between the fluorine and hydrogen atoms and the porphyrin ring in these particular systems.

However, with the advent of computational dispersion corrections to DFT, it was discovered that also in most other, unstrained single molecules, intramolecular dispersion tends to favor LS due to the stronger electronic stabilization of the more compact LS state [52]. This suggests that there is a generic, intrinsic contribution of dispersion interactions to the SCO tendency arising for the first coordination sphere of any complex of typically 5–15 kJ/mol which contributes to the real, observed $T_{½}$ [36, 52]. This contribution may then be compensated or increased by other ligand–ligand interactions. In bulky systems with ligand–ligand strain from close contacts, which tend to expand the first coordination sphere and favor HS, dispersion will further remedy some of the strain and reduce the expansion, and thus by itself favor LS.

This discovery of a generic first coordination sphere dispersion contribution to SCO arises because the attractive close-range dispersion energy favors the more compact LS state more than the HS state. As an example of consequence, if dispersion is included in B3LYP (as in B3LYP-D3), the SCO prediction becomes better because the intrinsic first coordination sphere contribution to SCO is included and counteracts the HS bias, and thus B3LYP-D3 is generally more accurate than B3LYP [36, 94].

It has now been found repeatedly that the dispersion forces of the SCO process work to favor the LS state of the single-molecule first coordination sphere [36, 95, 96]. The effect can easily reach 15–20 kJ/mol and averages 10 kJ/mol for the 30SCOFE database [36]. Thus, in order to model and predict the relative thermodynamics, SCO tendency, and $T_{½}$ of a series of compounds, dispersion forces need to be explicitly included.

3.3 Relativistic Stabilization of LS

Most studies of first-row transition metal systems do not include relativistic contributions to the energy. This is probably because relativistic effects are relatively less important for Sc–Zn, and partly because relativistic computations can be very demanding in terms of computational resources. However, relativistic effects can be substantial already for the first row of the d-transition series [97], and, e.g., for

M–L bond enthalpies, relativistic effects typically surpass 5 kJ/mol [98]. It is of fundamental scientific interest to understand whether relativistic effects contribute to SCO. Also, from the point of view of theoretical prediction of SCO systems, comparing nonrelativistic energy estimates directly to the experimental energy gaps could cause an error in the conclusion on the quality of the applied nonrelativistic method, whether it be CASPT2, CCSD(T), or DFT.

It has been shown [36] that scalar-relativistic corrections to the HS–LS energy gap accurately reproduce relativistic effects computed using both second- and fourth-order Douglas–Kroll–Hess energies [99], which simplify the Dirac equation by separating the positive and negative energy states [100]. The success of scalar-relativistic estimates arises from the small spin-orbit coupling of the light transition metal ions (Mn, Fe, Co) that undergo SCO (~1 kJ/mol [101]), although spin-orbit coupling is formally required for transition to occur in the first place and plays a qualitative role in the process as seen, e.g., for light-induced SCO [102]. The Douglas–Kroll–Hess two-component formalism with and without spin-orbit coupling only changes the scalar-relativistic energies by typically ~1 kJ/mol. Order 4 and 2 give similar results within ~1 kJ/mol, and the spin-orbit coupling corrections are 0–3 kJ/mol for the HS–LS gap, justifying the use of scalar-relativistic corrections which can recover most of the real relativistic LS stabilization by fast computation [36].

It turns out that there are significant relativistic contributions to SCO [36, 103, 104]. Interestingly, the relativistic energies tend to generally favor the LS state and are quite systematic and not very variable, because they tend to be localized to the metal center rather than other lighter atoms of the SCO system. The simplest explanation for this relativistic SCO effect is that the LS state is more compact with lower spin and angular momentum, and thus features stronger stabilization (reduction in inter-shell electron repulsion) once the 1s-orbital on iron is relativistic stabilized and contracted. In contrast, the reduced effective nuclear charge resulting from relativistic contraction mainly destabilizes the diffuse higher-angular momentum d-orbitals, in particular, the e_g-type d-orbitals of the HS state [36]. This explanation of the relativistic SCO effect follows closely the standard principles seen for other observables as outlined and discussed by Pyykkö [105].

The relativistic LS stabilization averages to 9 kJ/mol for iron SCO systems [36]. This is, remarkably, of the same magnitude as the dispersion and ZPE corrections. In other words, ZPE, dispersion, and relativistic effects work together to affect the energy difference between the HS and LS states, which is of a net magnitude of 5–20 kJ/mol, and they all are of similar importance, on average ~10 kJ/mol or so [36]. These three energy terms are systematic, i.e., they tend to favor one spin state consistently. Accordingly, they need to be included if one strives toward quantitative accuracy. To summarize this important conclusion, Fig. 4 displays the impact on a hypothetical transition curve of the three energetic contributions discussed above, i.e., ZPE (Fig. 4a), relativistic stabilization of LS (Fig. 4b), and the single-molecule component of the dispersion forces (Fig. 4c). The systematic behavior of these terms may aid us in the future rational design of powerful SCO systems with the exact energy terms desired to contribute to ΔH_{SCO} of (2).

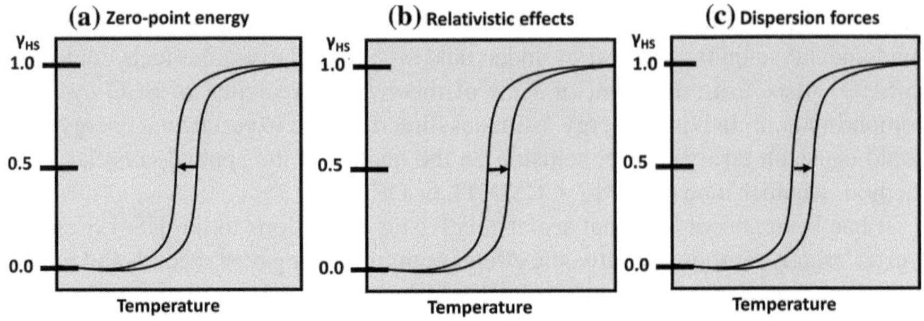

Fig. 4 Schematic representation of the theoretically expected typical impact of **a** zero-point vibrational energy, **b** relativistic effects, and **c** single-molecule dispersion forces on the transition temperature $T_{½}$, with all intermolecular effects ignored

3.4 Vibrational Entropy

The previous three sections discussed three physical effects that contribute to the energy of SCO, as measured by ΔH_{SCO}. With these three corrections to the electronic energy computed by a quantum-chemical method, one can obtain a decent estimate of how well the method performs in comparison with the experimental enthalpy of SCO. However, in order to understand and rationally predict SCO, and in particular the actual transition temperature $T_{½}$, one needs to account for the entropy term, $T\Delta S_{SCO}$ of (2), which is largely responsible for the transition to HS as the temperature is increased [24, 32, 106]. As mentioned above, the HS state contains more entropy in its longer and weaker metal–ligand bonds, due to the occupation of the e_g-type orbitals, and is accordingly favored by higher temperature because this entropy scales with T.

Paulsen et al. [34] first included vibrational entropy in the computational estimate of SCO tendency. The entropy change during SCO arises partly from the increased electronic partition function from the additional occupied orbitals (the electron configurational entropy), which provides a few kJ/mol of $T\Delta S$ in favor of the HS state near room temperature, and the vibrational entropy arising from changes in molecular geometry, which accounts for most (typically, 2/3–3/4) of the total entropy effect [24, 107]. In essentially all real SCO systems, symmetry breaks down to C_1, and the electronic degeneracy factor is no longer exactly applicable. Still the larger density of close-lying configurations prevails in the HS state and a simple estimate of the electronic degeneracy factor such as $\Delta S \sim k_B \ln \Omega$ from the Boltzmann formula gives an approximate idea of this contribution.

While Sorai and coworkers showed the importance of vibrational entropy in driving SCO [32, 107], the importance goes further: Correlation of experimental data for iron SCO systems [36] suggests that for the single-molecule first coordination sphere, the magnitude of this entropy directly relates to the enthalpy of the process, with entropy–enthalpy compensation across the range of ΔH_{SCO} and ΔS_{SCO} values for quite diverse ligand systems. Whether this is a general law remains to be

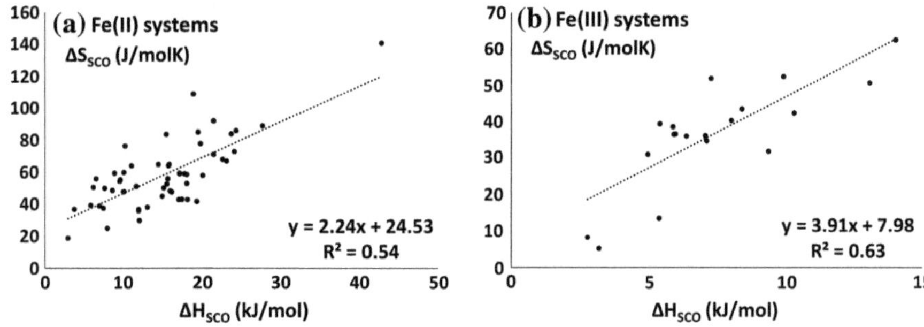

Fig. 5 Evidence for entropy–enthalpy compensation during SCO: **a** For Fe(II) systems (data were compiled from Toftlund [29], Chum et al. [109], Letard et al. [110], Strauss et al. [111], Sorai [107], Kulshreshtha et al. [35], Boča et al. [112], Nakamoto et al. [106, 113], Bartel et al. [114], and Lemercier et al. [115]); **b** for Fe(III) systems (data were compiled from Sorai [107] and Dose et al. [116])

established. However, for the purpose of this chapter, the author collected additional experimental data for ΔH_{SCO} and ΔS_{SCO} from the literature. Figure 5a shows the plot of ΔH_{SCO} and ΔS_{SCO} for a compiled data set of 62 iron(II) systems, and Fig. 5b shows this for 20 iron(III) systems for which data are available in the literature. For iron(II) systems, one set of values for $[Fe(bzimpy)_2]^{2+}$ is particular high [108] (bzimpy = 2,6-bis(benzimidazol-2′-yl)pyridine). This data point should probably be deemphasized. If correct, it doubles the range of possible ΔH_{SCO} and ΔS_{SCO} values which would be interesting. Even without this outlier, the correlation coefficient R^2 is 0.39 and remains highly significant. Figure 5 clearly shows evidence of very strong entropy–enthalpy compensation across both iron(II) and iron(III) systems during SCO and thus confirms the previous discovery [36]. Thus, not only does entropy drive thermal SCO as discovered by Sorai and coworkers [24, 32], *it also does so in proportion to the enthalpy of the same process.*

The entropy–enthalpy compensation of SCO needs to be considered when actively searching for new SCO systems and understanding their behavior. For example, an applied increased ligand field strength that increases the enthalpy of SCO in favor of LS will remarkably also tend to increase the entropy of the corresponding HS state to largely counteract the effect intended by the scientist. This compensation effect will obviously complicate rational design unless the effects are clearly separated. Understanding when the entropy–enthalpy compensation of SCO can be circumvented will thus be of particular interest.

The compensation effect also has implications for studies that estimate SCO tendency purely based on energies or proxies thereof, as has been and is still relatively common [117–119]. Most importantly, one cannot predict the $T_{1/2}$ or other real conditions of SCO without including the entropy because it largely counteracts the energy terms derived from standard electronic structure computations. Neglect of entropy is relatively common in studies of transition metal catalysis, metalloenzymes, and organometallic chemistry. Many systems have intermediates with close-lying spin states that often play an important role in the chemical process [13, 120, 121]. It makes a substantial difference if one does not include a systematic effect that favors one of these spin states consistently. Examples include hydrogenases and heme proteins, where the spin states are close in energy and important for the mechanism [14, 95, 121]. The HS state is generally more entropic than the LS state, and one can expect errors of 10–20 kJ/mol [36] *per metal site* systematically underestimating the importance of the HS state if entropy is neglected. Neglect of vibrational entropy can lead to the erroneous assignment of a LS state as the ground state. Similar errors will occur in estimates of the best theoretical method based on comparing energies without entropy directly to experimental spin states, which always represent free energies that include the entropy effect.

The question then arises whether one can model this vibrational entropy with decent accuracy. Standard approaches involve the computation of the harmonic vibration frequencies of the molecule, which is already required to obtain the ZPE, which, incidentally, also favors the HS state's longer and weaker bonds. Once this calculation has been carried out, it is straightforward to estimate the vibrational entropy by using thermodynamic state functions and the calculation of the vibration partition function Q_{vib}. Most quantum-chemistry programs can routinely perform this computation. For single molecules, the vibrational entropies correlate decently with experimental ΔS_{SCO} with errors translating into typically 5 kJ/mol for $T\Delta S_{\mathrm{SCO}}$, partly because the electronic configurational entropy is relatively similar for the systems [36]. The estimates neglect differential entropy contributions from solvent–solute and crystal packing, i.e., they represent only the contribution from single SCO molecules. In terms of intermolecular contributions, both high-frequency and low-frequency modes contribute to the entropy [122–124]. One can expect the soft vibrational modes to be associated with large *relative* errors for the computed estimates. However, importantly, the high-frequency (M–L stretch) frequencies of the first coordination sphere of the single molecule dominates the entropy effect as shown by Raman spectroscopy [122], and these are well modeled by DFT [36, 92, 125]. Thus including entropy estimates for the first coordination sphere is much better than omitting them, as they improve the $\Delta G_{\mathrm{SCO}}(T)$ and thus $T_{½}$ substantially. However, given the current limitations in the accuracy of these calculations, simply adding a constant contribution of $T\Delta S$ [52] may be a reasonable approach for many transition metal systems.

The entropy contribution to the balance between the spin states depends greatly on the nature of the ligand and the metal ion, with stronger ligands showing much larger entropy effects than weak ligands [52]. Typical $T\Delta S$ contributions of 5–30 kJ/mol are estimated for mononuclear complexes at room temperature [52]. This range is similar to the experimental range seen for Fe(II) and Fe(III) SCO systems [36]. The small effects are typical of weak field or weakly bound ligands. For nitrogen-donor ligands as are commonly found in SCO systems, the entropy contribution ranges typically from 10 to 25 kJ/mol [36, 52]. The entropy contribution is relatively insensitive to the theoretical method used as long as the geometry and vibrational frequencies are reasonable. This makes the entropy contribution more straightforward to estimate than the electronic energy contribution, which is discussed in more detail below.

4 Performance of DFT for Describing SCO

4.1 The Massive Role of HF Exchange Favoring HS

In order to fully understand and predict SCO tendencies of molecular systems, the various systematic effects discussed above need to be considered and added to the electronic energy of the HS and LS states. The enthalpy of the process can be written as:

$$\Delta H_{\mathrm{SCO}} = \Delta E_{\mathrm{SCO}} + \Delta \mathrm{PV}_{\mathrm{SCO}} \tag{5}$$

The last term is small (of the order of RT) for thermal SCO, and the energy term is:

$$\Delta E_{\mathrm{SCO}} = \Delta E_{\mathrm{el,SCO}} + \Delta E_{\mathrm{rel,SCO}} + \Delta E_{\mathrm{ZPE,SCO}} + \Delta E_{\mathrm{disp,SCO}} \tag{6}$$

where $\Delta E_{\mathrm{el,SCO}} = E_{\mathrm{el}}(\mathrm{HS}) - E_{\mathrm{el}}(\mathrm{LS})$ is the direct nonrelativistic energy gap of the HS and LS states computed by a density functional without dispersion included, $\Delta E_{\mathrm{rel,SCO}} = E_{\mathrm{rel}}(\mathrm{HS}) - E_{\mathrm{rel}}(\mathrm{LS})$ is the relativistic contribution to the HS–LS energy gap (typically 5–10 kJ/mol in favor of LS and very constant), $\Delta E_{\mathrm{ZPE,SCO}}$ is the differential ZPE (typically 10 kJ/mol in favor of HS for SCO systems but very dependent on metal and ligand type), and $\Delta E_{\mathrm{disp,SCO}}$ is the differential dispersion effect on SCO (typically 10 kJ/mol in favor of LS for single molecules, but augmented with a variable contribution depending on intermolecular interactions). Commonly, these three terms sum up to a correction of 0–20 kJ/mol in favor of LS. Once the systematic effects of (5) are accounted for, it enables us to estimate the accuracy of a theoretical method toward SCO and to identify truly spin-state-balanced density functionals.

However, it turns out that the electronic Hamiltonian used to obtain the electronic energies of the states, $E_{\mathrm{el}}(\mathrm{HS})$ and $E_{\mathrm{el}}(\mathrm{LS})$ of (6), is a major problem in itself. In the world of DFT, there are hundreds of functionals with distinct acronyms to choose from, and this diversity can easily overwhelm young researchers unless their supervisors have very strong adherence to certain functionals. So which density functionals produce accurate $\Delta E_{\mathrm{el,SCO}}$?

The use of hybrid functionals, in particular B3LYP [126–128], greatly improved the accuracy of computational main-group chemistry and have accordingly also been widely applied to study inorganic chemistry. Paulsen et al. computed the energy gap between HS and LS states for nine iron complexes using B3LYP and for some of them also the non-hybrid GGA functionals PW91 and BLYP [34]. They observed that the non-hybrid functionals produce energies much in favor of LS (by up to 104 kJ/mol for PW91), whereas B3LYP favors HS. This observation that the 20% HF exchange hybrid B3LYP favors HS and that 0% HF exchange favors LS probably inspired the development of the B3LYP* functional by Reiher a year later [60, 129].

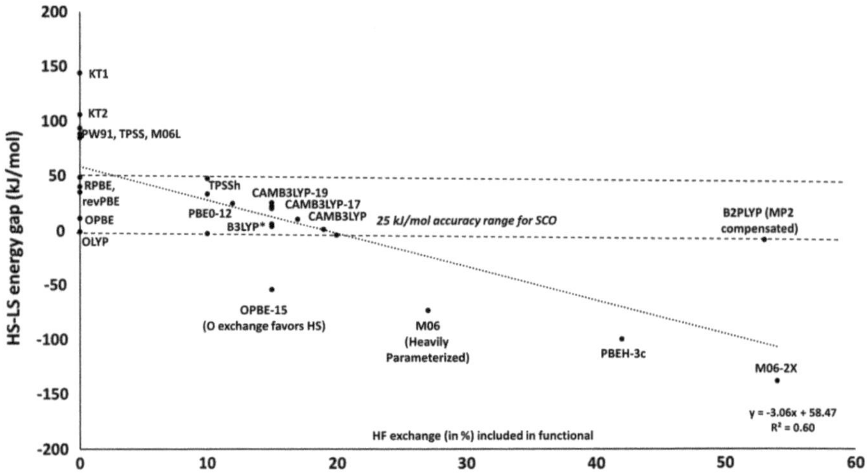

Fig. 6 The performance of various density functionals for modeling the average HS–LS energy gap of Fe(II) and Fe(III) SCO systems. The dashed line reflects a generous estimate of an acceptable result within 25 ± 25 kJ/mol of the fully corrected HS–LS gap. The figure was made using data previously published [94]

Since then it has consistently emerged that non-hybrid functionals are commonly (but with notable exceptions) not capable of describing $\Delta E_{\mathrm{el,SCO}}$ of (6) accurately [64, 89, 130–132], i.e., some inclusion of HF exchange is needed in a hybrid as is also the experience for main-group thermochemistry [133–135]. The HF state represents an artificial situation where the spin-aligned electrons are completely correlated by exchange, but the electrons of opposite spins are not correlated at all. HF exchange selectively favors HS because the exchange integrals of the Kohn–Sham determinant explicitly count only the parallel-spin electron interactions, which are more abundant in the HS state, and this exchange energy is always favorable [28, 136].

The amount of HF exchange is accordingly the single most important feature affecting $\Delta E_{\mathrm{el,SCO}}$ of a hybrid GGA functional, and $\Delta E_{\mathrm{el,SCO}}$ increases linearly with the included HF exchange [60, 61, 129, 131]. The effect depends very much on the bonding character of the t_{2g} orbitals, as recently analyzed [137]. A benchmark [36] accounting for the systematic effects of (6) concluded that B3LYP* remains one of the most accurate functionals even in competition with newer and more advanced functionals. B2PLYP [138] and TPSSh [62] also performed well. Many functionals can be tuned to perform well for SCO if they are made into hybrids with 10–20% HF exchange, suggesting that this range is perhaps generic among standard GGA functionals [94]. However, some functionals break this rule for special reasons and it is instructive to understand why this occurs. To understand these other underlying determinants of the HS–LS energy difference in more detail, the HS–LS gap for different functionals is shown in Fig. 6.

Several observations are notable: For example, double hybrids such as B2PLYP [138] perform much better than their high HF exchange fractions would suggest, because the exact exchange integrals are compensated by explicit pure non-exchange correlation terms via second-order perturbation theory [36, 94]. Also, the optimized exchange functional in the form of, e.g., OPBE [139, 140] or OLYP [141, 142], which does not include exact HF exchange, performs much better and more like a hybrid functional than other GGAs [132, 143]. The O exchange functional has been estimated to have an effect that corresponds to ~15% HF exchange [94]. The accuracy of OPBE supports previous findings by Swart [143, 144]. The O exchange functional was made from B88 by parameterization toward HF unrestricted energies of atoms of the first and second periods, and thus, this parameterization by design includes HF-like energetics. It is very interesting that re-parametrization toward HF energies or, as analyzed by Swart et al. [132] a leading s^4 term in the exchange functional, can cause a GGA non-hybrid exchange functional to behave similarly to a hybrid with 15% HF exchange because it tells us that HF exchange is not a "universal" feature by itself, but a pragmatic solution to a major problem of accuracy [127].

4.2 The Role of the Correlation Functional

Inspection of the original paper by Paulsen et al. [34] reveals that BLYP has the same ~17 kJ/mol smaller bias toward LS than PW91 for two distinct systems. This consistent difference could be coincidental, and even if not, it could be due to many features of the two functionals. Systematic comparison of functional types such as, e.g., BLYP versus BP86 (which use the same exchange functional) shows that the correlation functional, perhaps surprisingly, also contributes systematically to the spin-state balance [132]. Thus, for example, whereas BP86 and PBE give very similar results for SCO energetics and can be considered to have the same spin-state balance, the bias toward LS is reduced by typically 10–15 kJ/mol when using BLYP, and this effect is thus explicitly due to the LYP correlation functional [36].

It is interesting here to comment on the analogy between the performance of DFT applied to SCO and to the modeling of chemical bond strengths. The dissociation energy of a chemical bond is arguably the most fundamental energy of chemistry, as most chemical processes involve breaking and forming bonds with a net effect resembling the involved BDEs. It has been shown many years ago that HF exchange weakens the BDE of bonds, and correspondingly, that the LYP functional also lowers bond energies relative to other correlation functionals [145]. This has been seen repeatedly and is true for bonds involving strictly main-group elements [146] as well as bonds involving transition metals [98, 147]. Thus, the experience with modeling chemical bond strengths and spin-state energetics is intriguingly similar. The reason for this similarity has been proposed [28] to be due to HF exchange generally favoring the looser electron densities and higher spin quantum numbers reminiscent of both the dissociated states upon bond breaking and the high-spin electronic state.

Finally, it is relevant to mention the recent observation by Kulik and coworkers that also the additionally included gradient terms of the metafunctional, as shown for the TPSS functional, contribute to the spin-state balance [52, 148].

4.3 *The Use of Quantum-Chemical Benchmarks and the Post-HF Bias*

Two very important requirements for further progress in the accurate description of spin states is the use of adequate and reliable benchmark data either from high-level quantum-chemical computations or experimental data, and the proper account of systematic corrections to ensure that the actual property calculated corresponds to the experimental observable. For many SCO systems, the ΔH_{SCO} is in fact available as recently compiled in the SCOFE30 database [36]; these enthalpies can be accurately computed by DFT once ZPE, dispersion, relativistic, and solvent effects are accounted for.

However, more generally, one has to invoke adequate high-level quantum-mechanical benchmarks. One such type of calculation is CCSD(T); another is CASPT2; and a third is diffusion quantum Monte Carlo techniques (DMC). One of the main worries of these benchmarks is that they depend on a single HF reference which includes *all* of the exact exchange energy in a one-determinant basis but *none* of the compensating correlation energy. Whereas this is probably not a problem for the LS state, for the HS state, this single-determinant reference is heavily influenced by the impact of exact HF exchange and the orbitals and electron density must reflect this. A valid question is thus whether the correlated method is truly capable of bringing this overly spin-polarized reference state into complete spin balance by affording most of the compensating correlation energy.

Knowing the HS–LS biases of these benchmark methods is obviously extremely important in order to avoid false conclusions on the performance of functionals versus such a method. Notably, CASPT2 is biased toward configuration state functions with more exchange integrals (i.e., higher spin states), as they have favorable interactions with the HF reference. A modified shifted reference state was introduced into CASPT2 in 2004 [149] but a bias toward HS remained thereafter as shown from a low-lying triplet in the first CASPT2 study of O_2-binding to heme [150] or from the study of other hemes [135]; this bias can be partly remedied by using CC computation of the 3s3p correlation effects in combination with CASPT2 [151].

The CASPT2 example illustrates well a principle that may be a priori true but depends in practice on the implemented correlation method: *A single-reference post-HF method that is not perfectly correlated (i.e., is not full-CI) will not compensate completely the HF exchange of the reference state and will thus carry some bias toward this state.* Due to this "post-HF-bias", such methods will tend to favor HS too much and underbind metal–ligand bonds where more exchange integrals are presented for the dissociated states than for the bound state. A bias with this type of effect was in fact reported in the original paper on the shifted CASPT2 zeroth-order Hamiltonian [149].

The simple mononuclear nitrogen-donor octahedral complex, $[Fe(NH_3)_6]^{2+}$, is ideally suited for comparing method performance as most methods can be applied to this small system [152]. For this system with adequately optimized HS and LS geometries, the adiabatic energy difference for B3LYP is roughly 60 kJ/mol in favor of HS; this has been shown in multiple studies [52, 143, 153]. From Swart's study, OPBE gives ~80 kJ/mol in favor of HS [143]. Notably, OPBE is known to do well for SCO systems also from the same study [143], and confirmed by us [94]. Thus, one cannot expect the 80 kJ/mol estimate for the relaxed ΔH_{SCO} to be in much error, even when correcting for the systematic dispersion, relativistic, and ZPE effects described above. This value of ~80 kJ/mol is quite similar to the CASPT2 value obtained by Pierloot and Vancoille [152]. From all of these studies, it thus emerges that B3LYP at 60 kJ/mol has an error probably not larger than ~20 kJ/mol for $[Fe(NH_3)_6]^{2+}$.

In stark contrast to these findings is a new benchmark using diffusion Monte Carlo (DMC) based on HF nodal surfaces, which suggests that B3LYP favors LS too much by more than 60 kJ/mol for $[Fe(NH_3)_6]^{2+}$, and TPPSh by almost 100 kJ/mol [153]. DMC in that study suggests that HS is 120 kJ/mol below LS, which seems too much and is 40 kJ/mol more in favor of HS than CASPT2. Similar surprising results with DMC can also be deduced by comparison of $Fe(NCH)_6]^{2+}$ of the new study [153] to the study by Lawson Daku et al. [154] and Kepenekian et al. [155].

One explanation could be that DMC applied to spin-state energetics is very sensitive to the use of the HF nodes for the highly spin-polarized HS state, but this remains to be further investigated. If true, it again illustrates a post-HF bias but this time via the applied fixed node used in DMC. This suspicion is enhanced by a study [156] that investigated the use of different orbitals for the Slater–Jastrow trial wave function with DMC: CASSCF and HF orbitals give similar results and much higher absolute energies by 0.01 a.u. for the HS state and 0.03 a.u. for the LS state than the DFT Kohn–Sham orbitals, showing that the latter orbitals become better correlated during the full computation, as expected from the considerations above (i.e., the removal of the initial HF bias is very difficult). Confirming the suspicion further, the difference amounts to 0.02 a.u. in the computed HS–LS gap, or ~50 kJ/mol, very similar to the hypothesized error in DMC(HF) that can be deduced from the work discussed above [153].

This discussion illustrates (1) the effect of correlation on DMC using DFT rather than HF orbitals for the fixed node approximation, (2) that any quantum-mechanical method that starts from the HF picture will keep some bias toward this state unless fully correlated, which is in practice very hard; and (3) that extreme care should be applied when using supposedly high-level quantum-chemical methods as direct benchmark, rather than experimental data. CASPT2, rather than CCSD(T) which cannot describe non-dynamic correlation as well, is arguably the current "golden standard" of computational spin crossover and may stay so as the use of larger more appropriate active spaces and basis sets become computationally tractable.

4.4 Toward Spin-State-Balanced Density Functionals

The goal of current efforts in theoretical chemistry is to achieve a state where theory becomes truly predictive and thus, accurate enough to explain and design new systems of interest. For DFT, this would mean that a functional can be applied to chemistry broadly, and to SCO specifically, with predictive accuracy, making the functional "universal" [157–159]. The discussion above suggests that we must continue to improve the quality of density functionals, which will ultimately be used for predicting SCO behavior in larger systems and in batches of many systems, where other quantum-mechanical methods are too computationally slow. However, we should also carefully understand the biases in the quantum-mechanical benchmark methods themselves. Finally, in order to ensure that this process works, we should use the experimental data available for smaller systems where several quantum-chemical methods can be applied.

As has been widely discussed in the literature, different density functionals produce very different HS–LS gaps [28, 30, 36, 49, 52, 61, 64, 89, 118, 129, 132, 143, 160–163]. From these many studies, some consensus is, however, starting to emerge. In this author's humble view, functionals emerging as accurate for SCO include B3LYP* [61] (B3LYP [127, 128, 142] with 15% HF exchange), TPSSh [62, 164] (a meta hybrid with 10% HF exchange), and the double-hybrid B2PLYP [91]. Among GGA functionals, OLYP and OPBE are promising [94, 143]; it uses Handy and Cohen's optimized exchange [139, 141] which favors HS more than other non-hybrid GGA exchange functionals. If combined with LYP [142, 165] it gives even more HS stabilization because LYP is a HS-favoring correlation functional compared to, e.g., PBE, PW91, and P86 [36]. Particularly, encouraging is also the SSB functional which switches between PBE and OPBE [144], utilizing the high accuracy of the O exchange functional for spin-state energetics.

Even within iron SCO, which would supposedly be considered one type of systems for one type of purpose, there are system dependencies in method performance relating to oxidation state, i.e., even for the subset of iron SCO, there is no "universal" functional. It turns out that typical density functionals produce different errors in the spin-state balance for Fe(III) and Fe(II) SCO systems. Thus, for example, B3LYP* is not equally accurate for Fe(III) and Fe(II) systems, and in fact tends to produce too much high spin in Fe(II) systems and too much LS in Fe(III) systems [36]. Hybrid functionals tend to not only favor HS as explained above but also favor HS too much in Fe(II) compared to Fe(III), which has been called the "Fe(II)–Fe(III) bias" of DFT and which readily grows to 20 kJ/mol [94].

One of the more promising, recent avenues is the use of range-separated hybrids to study SCO processes, because these functionals can have several other advantages such as a small self-interaction error and more accurate transition-state energetics. One example of such a range-separated hybrid functional for use in SCO is CAMB3LYP [166]. This functional works surprisingly well by itself, but the 2018-customized versions with slightly less HF exchange (15–17%) are even more accurate [94]. Another example, also from 2018, is the optimally tuned range-separated hybrid (OT-RSH) studied by Prokopiou and Kronik [167], who also found that the short-range HF percentage is the most important parameter for achieving spin-state balance. Thus, for the broader study of transition metal catalysis where transition states are involved of variable spin states, the CAMB3LYP functional is probably among the best currently available, although this remains to be tested by additional benchmark studies.

Many of the catalytic processes that are of fundamental and technological interest involve changes in the spin states. The widely used RPBE functional [168] has a substantially better spin-state balance [94] than the original PBE functional [140] from which it was made, because it is less biased toward LS; this tendency as usual agrees well with the reduced over-binding tendency of RPBE, since HS bias and under-binding goes together [28].

5 Conclusions

This chapter has outlined the basic machinery of thermal SCO in single molecules with a particular focus on the achievements in modeling this process using DFT. The contribution of dispersion forces already in a single molecule undergoing SCO has been discussed; the generic dispersion effect on SCO arising from the first coordination sphere expansion favors the LS state by typical 10 kJ/mol; this generic dispersion term is supplemented by additional contributions from the type and bulkiness of the ligands that can favor either LS or HS. There is also a generic relativistic SCO effect with a surprisingly large contribution to the spin-state balance, typically also 10 kJ/mol in favor of LS but much less variable because the relativistic contribution is dominated by the metal ion and not the different ligands of SCO systems. Considering that typical SCO energies are of the order of 10–20 kJ/mol in favor of LS, these two effects combined favor LS more than the total energy gap. Thus, any conclusion on the accuracy of a theoretical method compared to the "observed ground-state spin" or the experimental enthalpy of SCO should consider these terms. Put another way, a functional without these two terms that gives 20 kJ/mol too much HS compared to experimental ΔH_{SCO} is excellent. If the calculation is compared to observed ground states, which reflect free energies, then one cannot ignore vibrational entropy, which largely determines the spin transition and SCO process. This entropy can be estimated decently but not very precisely from frequency analysis using standard quantum-chemical programs and is less sensitive to DFT functional used but more sensitive to the intermolecular interactions and explicit solvent effects not generally accounted for in current theoretical models of DFT. Once we master the ability to predict quantitatively the single-molecule SCO energetics, which requires an accuracy of about 10 kJ/mol, we can hope to continue to these challenges of real systems in condense states in the near future.

References

1. Halcrow MA (2013) Spin-crossover materials: properties and applications. Wiley
2. Gütlich P, Goodwin HA (2004) Spin crossover—an overall perspective. In: Spin crossover in transition metal compounds I. Springer, pp 1–47
3. Létard J-F, Guionneau P, Goux-Capes L (2004) Towards spin crossover applications. Spin Crossover Transit Met Compd III 1–19

4. Brooker S (2015) Spin crossover with thermal hysteresis: practicalities and lessons learnt. Chem Soc Rev 44:2880–2892
5. Gütlich P, Garcia Y, Goodwin HA (2000) Spin crossover phenomena in Fe(II) complexes. Chem Soc Rev 29:419–427
6. Harding DJ, Harding P, Phonsri W (2016) Spin crossover in iron (III) complexes. Coord Chem Rev 313:38–61
7. Kumar KS, Ruben M (2017) Emerging trends in spin crossover (SCO) based functional materials and devices. Coord Chem Rev 346:176–205
8. Guionneau P (2014) Crystallography and spin-crossover. A view of breathing materials. Dalt Trans 43:382–393
9. Ruiz E (2014) Charge transport properties of spin crossover systems. Phys Chem Chem Phys 16:14–22
10. Cambi L, Szegö L (1931) Über die magnetische Susceptibilität der komplexen Verbindungen. Berichte der Dtsch Chem Gesellschaft A B Ser 64:2591–2598
11. Šalitroš I, Madhu NT, Boča R, Pavlik J, Ruben M (2009) Room-temperature spin-transition iron compounds. Chem Mon 140:695–733
12. Kepp KP (2017) Heme: from quantum spin crossover to oxygen manager of life. Coord Chem Rev 344:363–374
13. Scheidt WR, Reed CA (1981) Spin-state/stereochemical relationships in iron porphyrins: implications for the hemoproteins. Chem Rev 81:543–555
14. Jensen KP, Ryde U (2004) How O_2 binds to heme: reasons for rapid binding and spin inversion. J Biol Chem 279:14561–14569
15. Gaspar AB, Muñoz MC, Real JA (2006) Dinuclear iron (II) spin crossover compounds: singular molecular materials for electronics. J Mater Chem 16:2522–2533
16. Bousseksou A, Molnár G, Demont P, Menegotto J (2003) Observation of a thermal hysteresis loop in the dielectric constant of spin crossover complexes: towards molecular memory devices. J Mater Chem 13:2069–2071
17. Miyamachi T, Gruber M, Davesne V, Bowen M, Boukari S, Joly L, Scheurer F, Rogez G, Yamada TK, Ohresser P (2012) Robust spin crossover and memristance across a single molecule. Nat Commun 3:938
18. Linares J, Codjovi E, Garcia Y (2012) Pressure and temperature spin crossover sensors with optical detection. Sensors 12:4479–4492
19. Salmon L, Molnár G, Zitouni D, Quintero C, Bergaud C, Micheau J-C, Bousseksou A (2010) A novel approach for fluorescent thermometry and thermal imaging purposes using spin crossover nanoparticles. J Mater Chem 20:5499–5503
20. Bartual-Murgui C, Akou A, Thibault C, Molnár G, Vieu C, Salmon L, Bousseksou A (2015) Spin-crossover metal–organic frameworks: promising materials for designing gas sensors. J Mater Chem C 3:1277–1285
21. Manrique-Juarez MD, Mathieu F, Shalabaeva V, Cacheux J, Rat S, Nicu L, Leïchlé T, Salmon L, Molnár G, Bousseksou A (2017) A bistable microelectromechanical system actuated by spin-crossover molecules. Angew Chem 129:8186–8190
22. Shepherd HJ, Il'ya A, Quintero CM, Tricard S, Salmon L, Molnár G, Bousseksou A (2013) Molecular actuators driven by cooperative spin-state switching. Nat Commun 4:2607
23. Harvey JN, Poli R, Smith KM (2003) Understanding the reactivity of transition metal complexes involving multiple spin states. Coord Chem Rev 238:347–361
24. Sorai M, Nakano M, Miyazaki Y (2006) Calorimetric investigation of phase transitions occurring in molecule-based magnets. Chem Rev 106:976–1031
25. Molnár G, Salmon L, Nicolazzi W, Terki F, Bousseksou A (2014) Emerging properties and applications of spin crossover nanomaterials. J Mater Chem C 2:1360–1366
26. Liu T, Zheng H, Kang S, Shiota Y, Hayami S, Mito M, Sato O, Yoshizawa K, Kanegawa S, Duan C (2013) A light-induced spin crossover actuated single-chain magnet. Nat. Commun. 4:2826
27. Cirera J, Ruiz E (2015) Theoretical modeling of two-step spin-crossover transitions in Fe^{II} dinuclear systems. J Mater Chem C 3:7954–7961

28. Kepp KP (2013) Consistent descriptions of metal–ligand bonds and spin-crossover in inorganic chemistry. Coord Chem Rev 257:196–209
29. Toftlund H (2001) Spin equilibrium in solutions. Chem Mon 132:1269–1277
30. Paulsen H, Schünemann V, Wolny JA (2013) Progress in electronic structure calculations on spin-crossover complexes. Eur J Inorg Chem 2013:628–641
31. Saito Y, Takemoto J, Hutchinson B, Nakamoto K (1972) Infrared studies of coordination compounds containing low-oxidation-state metals. I. Tris (2,2′-bipyridine) and tris (1,10-phenanthroline) complexes. Inorg Chem 11:2003–2011
32. Sorai M, Seki S (1974) Phonon coupled cooperative low-spin $^1A_1$ high-spin $^5T_2$ transition in $[Fe(phen)_2(NCS)_2]$ and $[Fe(phen)_2(NCSe)_2]$ crystals. J Phys Chem Solids 35:555–570
33. Kershaw Cook LJ, Kulmaczewski R, Mohammed R, Dudley S, Barrett SA, Little MA, Deeth RJ, Halcrow MA (2016) A unified treatment of the relationship between ligand substituents and spin state in a family of iron (II) complexes. Angew Chem Int Ed 55:4327–4331
34. Paulsen H, Duelund L, Winkler H, Toftlund H, Trautwein AX (2001) Free energy of spin-crossover complexes calculated with density functional methods. Inorg Chem 40:2201–2203
35. Kulshreshtha SK, Sasikala R, König E (1986) Calorimetric investigations of the low-spin ($^1A_1$) $\rightleftarrows$ high-spin ($^5T_2$) transition in solid dithiocyanatobis(2,2′-BI-2-thiazoline iron(III). Chem Phys Lett 123:215–217
36. Kepp KP (2016) Theoretical study of spin crossover in 30 iron complexes. Inorg Chem 55:2717–2727
37. Homma Y, Ishida T (2018) A new $S = 0 \rightleftarrows S = 2$ "Spin-crossover" scenario found in a Nickel (II) Bis (nitroxide) system. Chem Mater 30:1835–1838
38. Shimura Y, Tsuchida R (1956) Absorption spectra of Co(III) complexes. II. Redetermination of the spectrochemical series. Bull Chem Soc Jpn 29:311–316
39. Jørgensen CK (1959) Electron transfer spectra of hexahalide complexes. Mol Phys 2:309–332
40. Baker WA Jr, Bobonich HM (1964) Magnetic properties of some high-spin complexes of iron (II). Inorg Chem 3:1184–1188
41. König E, Madeja K (1966) Unusual magnetic behaviour of some iron (II)–bis-(1, 10-phenanthroline) complexes. Chem Commun 61–62
42. Koenig E, Madeja K (1967) $^5T_2$-$^1A_1$ Equilibriums in some iron (II)-bis (1, 10-phenanthroline) complexes. Inorg Chem 6:48–55
43. König E, Madeja K (1967) Infra-red spectra at the $^5T_2$-$^1A_1$ cross-over in iron (II) complexes. Spectrochim Acta Part A Mol Spectrosc 23:45–54
44. Stoufer RC, Busch DH, Hadley WB (1961) Unusual magnetic properties of some six-coordinate cobalt(II) complexes—electronic isomers. J Am Chem Soc 83:3732–3734
45. Pauling L (1932) The nature of the chemical bond. III. The transition from one extreme bond type to another. J Am Chem Soc 54:988–1003
46. Tsuchida R (1938) Absorption spectra of co-ordination compounds. I. Bull Chem Soc Jpn 13:388–400
47. Fajans K (1923) Struktur und Deformation der Elektronenhüllen in ihrer Bedeutung für die chemischen und optischen Eigenschaften anorganischer Verbindungen. Naturwissenschaften 11:165–172
48. König E (1968) Some aspects of the chemistry of bis (2,2′-dipyridyl) and bis (1,10-phenanthroline) complexes of iron (II). Coord Chem Rev 3:471–495
49. Houghton BJ, Deeth RJ (2014) Spin-state energetics of FeII complexes-the continuing voyage through the density functional minefield. Eur J Inorg Chem 2014:4573–4580
50. Deeth RJ, Anastasi AE, Wilcockson MJ (2010) An in silico design tool for Fe(II) spin crossover and light-induced excited spin state-trapped complexes. J Am Chem Soc 132:6876–6877
51. Deeth RJ, Halcrow MA, Kershaw Cook LJ, Raithby PR (2018) Ab initio ligand field molecular mechanics and the nature of metal-ligand π-bonding in Fe(II) 2,6-di(pyrazol-1-yl)pyridine spin crossover complexes. Chem Eur J 24:5204–5212
52. Mortensen SR, Kepp KP (2015) Spin propensities of octahedral complexes from density functional theory. J Phys Chem A 119:4041–4050

53. Takemoto JH, Hutchinson B (1973) Low-frequency infrared spectra of complexes which exhibit magnetic crossover. I. Iron(II) complexes of 1,10-phenanthroline and 2,2′-bipyridine. Inorg Chem 12:705–708
54. Kepp KP (2011) The ground states of iron(III) porphines: Role of entropy-enthalpy compensation, Fermi correlation, dispersion, and zero-point energies. J Inorg Biochem 105:1286–1292
55. Ashley DC, Jakubikova E (2017) Ironing out the photochemical and spin-crossover behavior of Fe (II) coordination compounds with computational chemistry. Coord Chem, Rev
56. Phonsri W, Davies CG, Jameson GNL, Moubaraki B, Ward JS, Kruger PE, Chastanet G, Murray KS (2017) Symmetry breaking above room temperature in an Fe (ii) spin crossover complex with an N4O2 donor set. Chem Commun 53:1374–1377
57. Yamasaki M, Ishida T (2015) First Iron(II) spin-crossover complex with an N5S coordination sphere. Chem Lett 44:920–921
58. Phan H, Hrudka JJ, Igimbayeva D, Lawson Daku LM, Shatruk M (2017) A simple approach for predicting the spin state of homoleptic Fe(II) tris-diimine complexes. J Am Chem Soc 139:6437–6447
59. Scott HS, Staniland RW, Kruger PE (2018) Spin crossover in homoleptic Fe(II) imidazolylimine complexes. Coord Chem Rev 362:24–43
60. Reiher M, Salomon O, Hess BA (2001) Reparameterization of hybrid functionals based on energy differences of states of different multiplicity. Theor Chem Acc 107:48–55
61. Salomon O, Reiher M, Hess BA (2002) Assertion and validation of the performance of the B3LYP* functional for the first transition metal row and the G2 test set. J Chem Phys 117:4729–4737
62. Tao J, Perdew JP, Staroverov VN, Scuseria GE (2003) Climbing the density functional ladder: nonempirical meta generalized gradient approximation designed for molecules and solids. Phys Rev Lett 91:146401
63. Jensen KP (2008) Bioinorganic chemistry modeled with the TPSSh density functional. Inorg Chem 47:10357–10365
64. Jensen KP, Cirera J (2009) Accurate computed enthalpies of spin crossover in iron and cobalt complexes. J Phys Chem A 113:10033–10039
65. Gütlich P, McGarvey BR, Kläui W (1980) Temperature-dependent $^5T_2$ (O_h) $\rightleftharpoons$ $^1A_1$ (O_h) spin equilibrium in a six-coordinate cobalt (III) complex. Investigation by phosphorus-31 NMR in solution. Inorg Chem 19:3704–3706
66. Kläui W (1980) High spin-low spin equilibrium in six-coordinate cobalt(III) complexes. Inorganica Chim Acta 40:X22–X23
67. Kläui W, Eberspach W, Guetlich P (1987) Spin-crossover cobalt(III) complexes: steric and electronic control of spin state. Inorg Chem 26:3977–3982
68. Chen J-M, Chin Y-Y, Valldor M, Hu Z, Lee J-M, Haw S-C, Hiraoka N, Ishii H, Pao C-W, Tsuei K-D (2014) A complete high-to-low spin state transition of trivalent cobalt ion in octahedral symmetry in $SrCo_{0.5}Ru_{0.5}O_{3-\delta}$. J Am Chem Soc 136:1514–1519
69. Stauber JM, Zhang S, Gvozdik N, Jiang Y, Avena L, Stevenson KJ, Cummins CC (2018) Cobalt and vanadium trimetaphosphate polyanions: synthesis, characterization, and electrochemical evaluation for non-aqueous redox-flow battery applications. J Am Chem Soc 140:538–541
70. Hayami S, Nakaya M, Ohmagari H, Alao AS, Nakamura M, Ohtani R, Yamaguchi R, Kuroda-Sowa T, Clegg JK (2015) Spin-crossover behaviors in solvated cobalt (II) compounds. Dalt Trans 44:9345–9348
71. Hayami S, Komatsu Y, Shimizu T, Kamihata H, Lee YH (2011) Spin-crossover in cobalt (II) compounds containing terpyridine and its derivatives. Coord Chem Rev 255:1981–1990
72. Krivokapic I, Zerara M, Daku ML, Vargas A, Enachescu C, Ambrus C, Tregenna-Piggott P, Amstutz N, Krausz E, Hauser A (2007) Spin-crossover in cobalt (II) imine complexes. Coord Chem Rev 251:364–378
73. Guo Y, Yang X-L, Wei R-J, Zheng L-S, Tao J (2015) Spin transition and structural transformation in a mononuclear Cobalt(II) complex. Inorg Chem 54:7670–7672
74. Radon M, Drabik G (2018) Spin states and other ligand–field states of aqua complexes revisited with multireference ab initio calculations including solvation effects. J Chem Theory Comput

75. Nielsen MT, Moltved KA, Kepp KP (2018) Electron transfer of hydrated transition-metal ions and the electronic state of Co^{3+}(aq). Inorg Chem 57:7914–7924
76. Johnson DA, Sharpe AG (1964) Reactions of cobalt(III) compounds: magnitude of Cobalt(III)/Cobalt(II) standard potential in aqueous solution. J Chem Soc 3490–3492
77. Winkler JR, Rice SF, Gray HB (1981) On the role of the high-spin state in the water exchange reaction of Hexaaquocobalt(III). Comments Inorg Chem 1:47–51
78. Navon G (1981) A search for the thermally populated high-spin excited state of hexaaquocobalt(3+) by coblt NMR. J Phys Chem 85:3547–3549
79. Habib HS, Hunt JP (1966) Electron-transfer reactions between aqueous cobaltous and cobaltic ions. J Am Chem Soc 88:1668–1671
80. Chou M, Creutz C, Sutin N (1977) Rate constants and activation parameters for outer-sphere electron-transfer reactions and comparisons with the predictions of Marcus theory. J Am Chem Soc 99:5615–5623
81. Shatruk M, Phan H, Chrisostomo BA, Suleimenova A (2015) Symmetry-breaking structural phase transitions in spin crossover complexes. Coord Chem Rev 289–290:62–73
82. Ni Z-P, Liu J-L, Hoque MN, Liu W, Li J-Y, Chen Y-C, Tong M-L (2017) Recent advances in guest effects on spin-crossover behavior in Hofmann-type metal-organic frameworks. Coord Chem Rev 335:28–43
83. Halder GJ, Kepert CJ, Moubaraki B, Murray KS, Cashion JD (2002) Guest-dependent spin crossover in a nanoporous molecular framework material. Science (80–) 298:1762–1765
84. Struch N, Bannwarth C, Ronson TK, Lorenz Y, Mienert B, Wagner N, Engeser M, Bill E, Puttreddy R, Rissanen K (2017) An octanuclear metallosupramolecular cage designed to exhibit spin-crossover behavior. Angew Chem Int Ed 56:4930–4935
85. Kumar KS, Studniarek M, Heinrich B, Arabski J, Schmerber G, Bowen M, Boukari S, Beaurepaire E, Dreiser J, Ruben M (2018) Engineering on-surface spin crossover: spin-state switching in a self-assembled film of vacuum-sublimable functional molecule. Adv Mater 30:1705416
86. Zhang L, Kepp KP, Ulstrup J, Zhang J (2018) Redox potentials and electronic states of iron porphyrin IX adsorbed on single crystal gold electrode surfaces. Langmuir 34:3610–3618
87. Kepenekian M, Le Guennic B, Robert V (2009) Primary role of the electrostatic contributions in a rational growth of hysteresis loop in spin-crossover Fe (II) complexes. J Am Chem Soc 131:11498–11502
88. Kahn O, Martinez CJ (1998) Spin-transition polymers: from molecular materials toward memory devices. Science 279:44–48
89. Cirera J, Ruiz E (2016) Theoretical modeling of the ligand-tuning effect over the transition temperature in four-coordinated FeII molecules. Inorg Chem 55:1657–1663
90. Tailleur E, Marchivie M, Daro N, Chastanet G, Guionneau P (2017) Thermal spin-crossover with a large hysteresis spanning room temperature in a mononuclear complex. Chem Commun 53:4763–4766
91. Grimme S, Antony J, Ehrlich S, Krieg H (2010) A consistent and accurate ab initio parametrization of density functional dispersion correction (DFT-D) for the 94 elements H-Pu. J Chem Phys 132:154104
92. Bučko T, Hafner J, Lebègue S, Ángyán JG (2012) Spin crossover transition of $Fe(phen)_2(NCS)_2$: periodic dispersion-corrected density-functional study. Phys Chem Chem Phys 14:5389–5396
93. Timken MD, Wilson SR, Hendrickson DN (1985) Dynamics of spin-state interconversion and cooperativity for ferric spin-crossover complexes in the solid state. 4. Pyruvic acid thiosemicarbazone complex. Inorg Chem 24:3450–3457
94. Siig OS, Kepp KP (2018) Iron (II) and Iron (III) spin crossover: toward an optimal density functional. J Phys Chem A 122:4208–4217
95. Kepp KP, Dasmeh P (2013) Effect of distal interactions on O_2 binding to heme. J Phys Chem B 117:3755–3770
96. Kepp KP (2017) Heme isomers substantially affect heme's electronic structure and function. Phys Chem Chem Phys 19:22355–22362

97. Pyykkö P (2012) Relativistic effects in chemistry: more common than you thought. Annu Rev Phys Chem 63:45–64
98. Jensen KP, Roos BO, Ryde U (2007) Performance of density functionals for first row transition metal systems. J Chem Phys 126:14103
99. Hess BA (1986) Relativistic electronic-structure calculations employing a two-component no-pair formalism with external-field projection operators. Phys Rev A 33:3742
100. Reiher M, Wolf A (2004) Exact decoupling of the Dirac Hamiltonian. I. General theory. J Chem Phys 121:2037–2047
101. Moore CE (1971) Atomic energy levels. Atomic energy levels. United States Department of Commerce & National Bureau of Standards, Washington, DC, pp 56–57
102. Sousa C, Domingo A, de Graaf C (2018) Effect of second-order spin-orbit coupling on the interaction between spin states in spin-crossover systems. Chem Eur J 24:5146–5152
103. de Graaf C, Sousa C (2010) Study of the light-induced spin crossover process of the $[Fe^{II}(bpy)_3]^{2+}$ complex. Chem Eur J 16:4550–4556
104. Paulsen H, Trautwein AX (2004) Density functional theory calculations for spin crossover complexes. In: Spin crossover in transition metal compounds III. Springer, pp 197–219
105. Pyykko P (1988) Relativistic effects in structural chemistry. Chem Rev 88:563–594
106. Nakamoto T, Tan Z-C, Sorai M (2001) Heat capacity of the spin crossover complex $[Fe(2\text{-}pic)_3]Cl_2{\cdot}MeOH$: a spin crossover phenomenon with weak cooperativity in the solid state. Inorg Chem 40:3805–3809
107. Sorai M (2001) Calorimetric investigations of phase transitions occurring in molecule-based materials in which electrons are directly involved. Bull Chem Soc Jpn 74:2223–2253
108. Addison AW, Burman S, Wahlgren CG, Rajan OA, Rowe TM, Sinn E (1987) New iron (II) spin-crossover complexes with heterocyclic amine-derived ligands and STEPS experiments on photogenerated metastable high-spin states. J Chem Soc Dalt Trans 2621–2630
109. Chum HL, Vanin JA, Holanda MID (1982) Tris (2-(aminomethyl) pyridine)iron(II): a new spin-state equilibrium in solution. Inorg Chem 21:1146–1152
110. Létard J-F, Guionneau P, Rabardel L, Howard JAK, Goeta AE, Chasseau D, Kahn O (1998) Structural, magnetic, and photomagnetic studies of a mononuclear iron (II) derivative exhibiting an exceptionally abrupt spin transition. Light-induced thermal hysteresis phenomenon. Inorg Chem 37:4432–4441
111. Strauß B, Linert W, Gutmann V, Jameson RF (1992) Spin-crossover complexes in solution, I. Substitutional lability of $[Fe(bzimpy)_2](ClO_4)_2$. Chem Mon 123:537–546
112. Boča R, Boča M, Dlháň L, Falk K, Fuess H, Haase W, Jaroščiak R, Papánková B, Renz F, Vrbová M, Werner R (2001) Strong cooperativeness in the mononuclear iron(II) derivative exhibiting an abrupt spin transition above 400 K. Inorg Chem 40:3025–3033
113. Nakamoto T, Bhattacharjee A, Sorai M (2004) Cause for unusually large thermal hysteresis of spin crossover in $[Fe(2\text{-}pic)_3]Cl_2{\cdot}$ H2O. Bull Chem Soc Jpn 77:921–932
114. Bartel M, Absmeier A, Jameson GNL, Werner F, Kato K, Takata M, Boca R, Hasegawa M, Mereiter K, Caneschi A (2007) Modification of spin crossover behavior through solvent assisted formation and solvent inclusion in a triply interpenetrating three-dimensional network. Inorg Chem 46:4220–4229
115. Lemercier G, Bousseksou A, Verelst M, Varret F, Tuchagues JP (1995) Dynamic spin-crossover in $[FeII(TRIM)_2]Cl_2$ investigated by Mössbauer spectroscopy and magnetic measurements. J Magn Magn Mater 150:227–230
116. Dose EV, Murphy KMM, Wilson LJ (1976) Synthesis and spin-state studies in solution of γ-substituted tris (β-diketonato) iron(III) complexes and their spin-equilibrium. β-ketoimine analogues derived from triethylenetetramine. Inorg Chem 15:2622–2630
117. Ye S, Neese F (2010) Accurate modeling of spin-state energetics in spin-crossover systems with modern density functional theory. Inorg Chem 49:772–774
118. Conradie J, Ghosh A (2007) DFT calculations on the spin-crossover complex Fe (salen)(NO): a quest for the best functional. J Phys Chem B 111:12621–12624
119. Li Z, Dai J, Shiota Y, Yoshizawa K, Kanegawa S, Sato O (2013) Multi-step spin crossover accompanied by symmetry breaking in an FeIII complex: crystallographic evidence and DFT studies. Chem Eur J 19:12948–12952

120. de Visser SP (2005) What affects the quartet-doublet energy splitting in peroxidase enzymes? J Phys Chem A 109:11050–11057
121. Franzen S (2002) Spin-dependent mechanism for diatomic ligand binding to heme. Proc Natl Acad Sci USA 99:16754–16759
122. Boussekou A, McGarvey JJ, Varret F, Real JA, Tuchagues J-P, Dennis AC, Boillot ML (2000) Raman spectroscopy of the high-and low-spin states of the spin crossover complex $Fe(phen)_2(NCS)_2$: an initial approach to estimation of vibrational contributions to the associated entropy change. Chem Phys Lett 318:409–416
123. Molnár G, Niel V, Gaspar AB, Real J-A, Zwick A, Bousseksou A, McGarvey JJ (2002) Vibrational spectroscopy of cyanide-bridged, iron (II) spin-crossover coordination polymers: estimation of vibrational contributions to the entropy change associated with the spin transition. J Phys Chem B 106:9701–9707
124. Ronayne KL, Paulsen H, Höfer A, Dennis AC, Wolny JA, Chumakov AI, Schünemann V, Winkler H, Spiering H, Bousseksou A (2006) Vibrational spectrum of the spin crossover complex $[Fe(phen)_2(NCS)_2]$ studied by IR and Raman spectroscopy, nuclear inelastic scattering and DFT calculations. Phys Chem Chem Phys 8:4685–4693
125. Brehm G, Reiher M, Schneider S (2002) Estimation of the vibrational contribution to the entropy change associated with the low-to high-spin transition in $Fe(phen)_2(NCS)_2$ complexes: results obtained by IR and Raman spectroscopy and DFT calculations. J Phys Chem A 106:12024–12034
126. Becke AD (1993) A new mixing of Hartree-Fock and local density-functional theories. J Chem Phys 98:1372–1377
127. Becke AD (1993) Density-functional thermochemistry. III. The role of exact exchange. J Chem Phys 98:5648–5652
128. Stephens PJ, Devlin FJ, Chabalowski CF, Frisch MJ (1994) Ab initio calculation of vibrational absorption and circular dichroism spectra using density functional force fields. J Phys Chem 98:11623–11627
129. Reiher M (2002) Theoretical study of the $Fe(phen)_2(NCS)_2$ spin-crossover complex with reparametrized density functionals. Inorg Chem 41:6928–6935
130. Cirera J, Paesani F (2012) Theoretical prediction of spin-crossover temperatures in ligand-driven light-induced spin change systems. Inorg Chem 51:8194–8201
131. Gani TZH, Kulik HJ (2017) Unifying exchange sensitivity in transition-metal spin-state ordering and catalysis through bond valence metrics. J Chem Theory Comput 13:5443–5457
132. Swart M, Groenhof AR, Ehlers AW, Lammertsma K (2004) Validation of exchange-correlation functionals for spin states of iron complexes. J Phys Chem A 108:5479–5483
133. Harvey JN (2004) DFT computation of relative spin-state energetics of transition metal compounds. In: Principles and applications of density functional theory in inorganic chemistry I. Springer, pp 151–184
134. Ghosh A (2006) Transition metal spin state energetics and noninnocent systems: challenges for DFT in the bioinorganic arena. J Biol Inorg Chem 11:712–724
135. Vancoillie S, Zhao H, Radon M, Pierloot K (2010) Performance of CASPT2 and DFT for relative spin-state energetics of heme models. J Chem Theory Comput 6:576–582
136. Hughes TF, Friesner RA (2011) Correcting systematic errors in DFT spin-splitting energetics for transition metal complexes. J Chem Theory Comput 7:19–32
137. Pinter B, Chankisjijev A, Geerlings P, Harvey JN, De Proft F (2018) Conceptual insights into DFT spin-state energetics of octahedral transition-metal complexes through a density difference analysis. Chem Eur J 24:5281–5292
138. Grimme S (2006) Semiempirical hybrid density functional with perturbative second-order correlation. J Chem Phys 124:34108
139. Cohen AJ, Handy NC (2000) Assessment of exchange correlation functionals. Chem Phys Lett 316:160–166
140. Perdew JP, Burke K, Ernzerhof M (1996) Generalized gradient approximation made simple. Phys Rev Lett 77:3865
141. Handy NC, Cohen AJ (2001) Left-right correlation energy. Mol Phys 99:403–412

142. Lee C, Yang W, Parr RG (1988) Development of the Colle-Salvetti correlation-energy formula into a functional of the electron density. Phys Rev B 37:785–789
143. Swart M (2008) Accurate spin-state energies for iron complexes. J Chem Theory Comput 4:2057–2066
144. Swart M, Solà M, Bickelhaupt FM (2009) A new all-round density functional based on spin states and S_N2 barriers. J Chem Phys 131:94103
145. Jensen K, Ryde U (2003) Theoretical prediction of the Co-C bond strength in cobalamins. J Phys Chem A 155:7539–7545
146. Kepp KP (2017) Trends in strong chemical bonding in C_2, CN, CN^-, CO, N_2, NO, NO^+, and O_2. J Phys Chem A 121:9092–9098
147. Xu X, Zhang W, Tang M, Truhlar DG (2015) Do practical standard coupled cluster calculations agree better than Kohn-Sham calculations with currently available functionals when compared to the best available experimental data for dissociation energies of bonds to 3d transition metals? J Chem Theory Comput 11:2036–2052
148. Ioannidis EI, Kulik HJ (2017) Ligand-field-dependent behavior of meta-GGA exchange in transition-metal complex spin-state ordering. J Phys Chem A 121:874–884
149. Ghigo G, Roos BO, Malmqvist P-Å (2004) A modified definition of the zeroth-order Hamiltonian in multiconfigurational perturbation theory (CASPT2). Chem Phys Lett 396:142–149
150. Jensen KP, Roos B, Ryde U (2005) O_2-binding to heme: electronic structure and spectrum of oxyheme, studied by multiconfigurational methods. J Inorg Biochem 99:45–54
151. Phung QM, Feldt M, Harvey JN, Pierloot K (2018) Toward highly accurate spin state energetics in first-row transition metal complexes: a combined CASPT2/CC approach. J Chem Theory Comput 14:2446–2455
152. Pierloot K, Vancoillie S (2006) Relative energy of the high-($^5T_{2g}$) and low-($^1A_{1g}$) spin states of $[Fe(H_2O)_6]^{2+}$,$[Fe(NH_3)_6]^{2+}$, and $[Fe(bpy)_3]^{2+}$: CASPT2 versus density functional theory. J Chem Phys 125:124303
153. Song S, Kim M-C, Sim E, Benali A, Heinonen O, Burke K (2018) Benchmarks and reliable DFT results for spin gaps of small ligand Fe(II) complexes. J Chem Theory Comput 14:2304–2311
154. Lawson Daku LM, Aquilante F, Robinson TW, Hauser A (2012) Accurate spin-state energetics of transition metal complexes. 1. CCSD(T), CASPT2, and DFT study of $[M(NCH)_6]^{2+}$ (M = Fe, Co). J Chem Theory Comput 8:4216–4231
155. Kepenekian M, Robert V, Le Guennic B, De Graaf C (2009) Energetics of $[Fe(NCH)_6]^{2+}$ via CASPT2 calculations: a spin-crossover perspective. J Comput Chem 30:2327–2333
156. Fumanal M, Wagner LK, Sanvito S, Droghetti A (2016) Diffusion Monte Carlo perspective on the spin-state energetics of $[Fe(NCH)_6]^{2+}$. J Chem Theory Comput 12:4233–4241
157. Peverati R, Truhlar DG (2014) Quest for a universal density functional: the accuracy of density functionals across a broad spectrum of databases in chemistry and physics. Philos Trans R Soc Lond A Math Phys Eng Sci 372:20120476
158. Medvedev MG, Bushmarinov IS, Sun J, Perdew JP, Lyssenko KA (2017) Density functional theory is straying from the path toward the exact functional. Science 355:49–52
159. Kepp KP (2017) Comment on "Density functional theory is straying from the path toward the exact functional". Science 356:496–497
160. Isley WC III, Zarra S, Carlson RK, Bilbeisi RA, Ronson TK, Nitschke JR, Gagliardi L, Cramer CJ (2014) Predicting paramagnetic ^1H NMR chemical shifts and state-energy separations in spin-crossover host–guest systems. Phys Chem Chem Phys 16:10620–10628
161. Ioannidis EI, Kulik HJ (2015) Towards quantifying the role of exact exchange in predictions of transition metal complex properties. J Chem Phys 143:34104
162. Coskun D, Jerome SV, Friesner RA (2016) Evaluation of the performance of the B3LYP, PBE0, and M06 DFT functionals, and DBLOC-corrected versions, in the calculation of redox potentials and spin splittings for transition metal containing systems. J Chem Theory Comput 12:1121–1128
163. Verma P, Varga Z, Klein JEMN, Cramer CJ, Que L, Truhlar DG (2017) Assessment of electronic structure methods for the determination of the ground spin states of Fe(II), Fe(III) and Fe(IV) complexes. Phys Chem Chem Phys 19:13049–13069

164. Perdew JP, Tao J, Staroverov VN, Scuseria GE (2004) Meta-generalized gradient approximation: explanation of a realistic nonempirical density functional. J Chem Phys 120
165. Colle R, Salvetti O (1975) Approximate calculation of the correlation energy for the closed shells. Theor Chim Acta 37:329–334
166. Yanai T, Tew DP, Handy NC (2004) A new hybrid exchange–correlation functional using the Coulomb-attenuating method (CAM-B3LYP). Chem Phys Lett 393:51–57
167. Prokopiou G, Kronik L (2018) Spin-state energetics of Fe complexes from an optimally tuned range-separated hybrid functional. Chem Eur J 24:5173–5182
168. Hammer B, Hansen LB, Nørskov JK (1999) Improved adsorption energetics within density-functional theory using revised Perdew-Burke-Ernzerhof functionals. Phys Rev B 59:7413

Anisotropic Magnetic Spin Interactions of Transition Metal Complexes and Metalloenzymes from Spectroscopy and Quantum Chemistry

Matthias Stein

Abstract Spectroscopic investigations of the interaction of a spin magnetic moment with an external magnetic field reveal insight into the electronic structure, e.g. the composition of the occupied molecular orbitals of the system, the oxidation state of a possible transition metal and its coordination environment. For paramagnetic systems, electron spin resonance (ESR) and related techniques probe the interaction between electron and nuclear spins, provide information about the spatial distribution of the spin density and allow identifying binding partners which are often not resolved structurally, for example hydrogen atoms. In particular, diagonalization of the electron Zeeman and electron-nuclear hyperfine interaction matrices does not only give their principal values but also their magnetic principal axes and allows making statements about the spatial arrangement of coordinating atoms and ligands. The advancement of computational approaches to calculate the parameters of the effective Spin Hamiltonian such as the electronic $\boldsymbol{g}$-tensors and hyperfine tensors and their comparison with experiment supports the analysis and interpretation of complex magnetic resonance spectra. This is discussed here for **g**- and hyperfine tensors and zero-field splitting tensors for selected examples including transition metal containing model complexes and metalloenzymes.

1 Introduction

Modern spectroscopic techniques from spin magnetic resonance experiments (either nuclear spin (NMR) or electron spin (EPR)) are able to give high resolution spectra and very detailed insight into the electronic and coordination structures of small molecules and proteins. Their analysis and interpretation rely on the concept of an 'effective Spin Hamiltonian' that phenomenologically describes the interactions of the spin (electron or nuclear) with an external magnetic field and introduces spec-

M. Stein (✉)
Molecular Simulations and Design Group, Max Planck Institute for Dynamics of Complex Technical Systems, Sandtorstrasse 1, 39106 Magdeburg, Germany
e-mail: matthias.stein@mpi-magdeburg.mpg.de

E. Broclawik et al. (eds.), *Transition Metals in Coordination Environments*, Challenges and Advances in Computational Chemistry and Physics 29,
https://doi.org/10.1007/978-3-030-11714-6_2

troscopic observables (such as chemical shift, **g**-values, spin–spin interactions and zero-field splitting) [1, 2]. Here, the concept of the descriptive Spin Hamiltonian, which is used in the analysis of spectra from magnetic resonance spectroscopy, is linked to computationally accessible spectroscopic parameters from quantum chemistry.

The focus in this chapter is on Electron Paramagnetic Resonance (EPR) spectroscopy but its fundamentals are also valid and can be formulated in an analogous way for Nuclear Magnetic Resonance (NMR). The interpretation of magnetic resonance data in the context of structural and chemical biology of paramagnetic metallic cofactors does require knowledge not only of the magnitude and strength but also the directionality and orientation of these magnetic interactions with respect to the compound or protein structural frame. Here, systems containing transition metal ions are being discussed since they are challenging to study experimentally but also computationally demanding in terms of model size and accuracy of results.

Transition metals are in particular interesting and provide the following challenges and difficulties:

(i) A multitude of oxidation states
(ii) Different electronic ground states (low-spin, intermediate spin, high spin) for a metal in a particular oxidation state
(iii) Redox activity of transition metal during catalytic cycle
(iv) Changes in spin state during a reaction ('spin state catalysis').

Thus, spectra of transition metal-containing complexes or proteins are particularly cumbersome to analyze and interpret in terms of molecular structures. The EPR **g**-values for transition metal complexes vary over a wide spectral range due to spin–orbit coupling and zero-field splitting, depending on the geometry of the complex and this variability in nature and number of ligands makes it impossible to establish an empirical approach to predict **g**-values of most transition metal systems.

As a subset, selected examples from nickel, molybdenum and manganese containing transition metal complexes and enzymatic systems are discussed.

Nickel occurs in coordination complexes and metalloenzymes in different oxidation states which may be para- or diamagnetic depending on the ligand environment. Nickel enzymes catalyze a wide range of chemical reactions using mononuclear, homo- or heterodinuclear active sites and are present among others in urease, superoxide dismutase, carbon monoxide dehydrogenase, acetyl-coenzyme A synthase/decarbonylase and methyl-coenzyme M reductase [3, 4]. [NiFe]-hydrogenase enzymes possess a heterobimetallic active site and the nickel ion occurs in the +I, +II, and +III oxidation states during the heterolytic splitting of molecular hydrogen (H_2). It is covered here as one example of the application of EPR spectroscopy in solution and in protein single crystals (for reviews see [5, 6]) hand in hand with computational means to identify the intermediates by calculating spectroscopic observables and elucidate the enzymatic reaction mechanism [7, 8].

Molybdoenzymes are widespread in eukaryotic and prokaryotic organisms where they have crucial functions in detoxification reactions in the metabolism of humans and bacteria [9]. In human, molybdenum xanthine oxidoreductase (XO), aldehyde oxidase (AO) and the mitochondrial amidoxime-reducing enzymes called mARC1 and 2 are present, the latter of which were only recently discovered. In bacteria, the two-electron reduction of molecular nitrogen (N_2) to ammonia (NH_3) by the heterotetrameric MoFe protein nitrogenase from cyanobacteria is an essential process to sustain life on Earth [10] and has inspired the design and characterization of many biomimetic complexes [11, 12]. The structure and function of mono-molybdenum enzymes have been reviewed in [13, 14].

Mn superoxide dismutase is present in eukaryotic mitochondria, and also in most bacteria. The Mn-SOD enzyme is the most ancient enzyme living in the presence of oxygen and being able deal with the toxic effects of superoxide (O_2^-) [15, 16]. The oxygen-evolving complex (OEC) of photosystem II contains a Mn_4CaO_5 cluster plus two chloride ions with Mn oxidation states ranging from +2 to +4. Five oxygen atoms serve as oxo bridges linking the five metal atoms and four water molecules bound to the cluster which were structurally resolved, see reviews [17, 18] for PSII and also biomimetic systems. Quantum chemical calculations for PSII are summarized in reference [19] for example.

Whereas most computational benchmarks and comparison with experiment have focused on calculating the principal values of the **g**- and hyperfine tensors, it is shown here that the non-uniform distribution of electron spin and the anisotropy of magnetic interactions provide additional valuable information such as the assignment of the magnetic axes into a molecular or protein structure and the resolution of conformational states of ligands close to the spin center. This information is complementary to the one that can be obtained from crystallography.

2 Electron Spin Interactions with an External Magnetic Field

In the following, the discussion will be limited to a simple paramagnetic center with one unpaired electron spin ($S = ½$, $M_s = \pm ½$) and its interaction with nuclear spins of the same spin state ($I = ½$, $M_I = \pm ½$). For a discussion of more elaborate magnetic systems, the reader is referred to textbooks about EPR and d-transition metals, such as [1, 2].

In the absence of an external magnetic field, the electron magnetic moment is randomly oriented and the two energy levels $M_s = \pm ½$ are energetically degenerate. This degeneracy of α- and β-spin orbitals is removed in the presence of an external magnetic field $\mathbf{B_0}$ due to the electron Zeeman splitting:

$$H = \hbar \mu_B \mathbf{S} \, \mathbf{g}_e \, \mathbf{B_0} \quad (1)$$

where g_e is the g-factor of the free electron ($g_e = 2.0023$), μ_B is the Bohr magneton (9.274×10^{-24} J T^{-1}), **S** is the spin operator and $\mathbf{B_0}$ is the magnitude of the external magnetic field in Tesla.

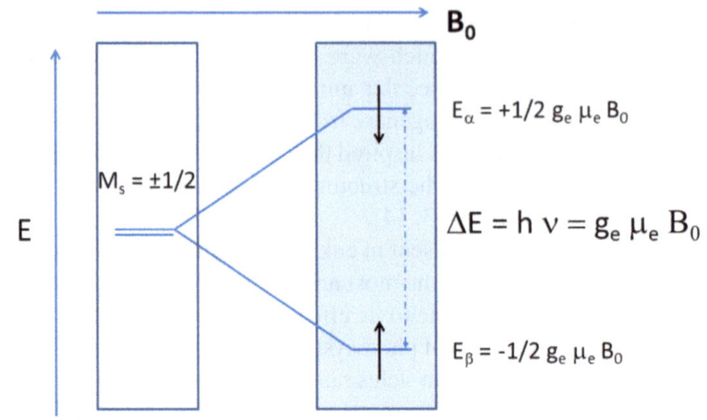

Fig. 1 Electron Zeeman splitting. The two energetically degenerate spin states split into α ($M_s = +½$) and β ($M_s = -½$) in the presence of a magnetic field $\mathbf{B_0}$

The energy splitting ΔE of α- and β-spin states is the resonance condition for the absorption of a microwave frequency (see Fig. 1).

2.1 The Concept of an Effective Spin Hamiltonian

An important ingredient to the interpretation of most EPR experiments is the concept of an effective 'Spin Hamiltonian'.

The analysis of magnetic resonance experiments makes use of such a parametrized, phenomenological 'Spin Hamiltonian' which allows experimental data to be summarized in terms of a small number of parameters (the **g**-tensor, the hyperfine tensor **A** or the quadrupole tensor **D** are being introduced).

The effective, Spin Hamiltonian '(introduced by Griffith [20] and Abragam and Pryce [21]) for an EPR transition

$$H_{\text{spin}}(\text{EPR}) = \hbar^{-1}\mu_B\, \mathbf{B_0 g S} + \hbar^{-2}\sum_{\text{N}} \mathbf{S}\,\mathbf{A}_{\text{N}}\,\mathbf{I}_{\text{N}} + \hbar^{-2}\mathbf{S\,D\,S} \tag{2}$$

contains (i) the electron Zeeman term from an interaction of an effective electron spin magnetic moment **S** with an external magnetic field $\mathbf{B_0}$, (ii) the sum of hyperfine splittings $\mathbf{A_N}$ of all N nuclei which describes the interaction of electron–spin **S** with the magnetic field from a nuclear spin $\mathbf{I_N}$ and (iii) the zero field splitting caused by electron-electron dipolar interaction and from spin–orbit coupling. This phenomenological description is closely related to the description of the chemical shift and spin–spin coupling in NMR spectra.

The concept of an effective Spin Hamiltonian establishes the connection between measured energy level differences and molecular interactions. The structural interpretation of measured EPR spectra is complicated and not straightforward (see below). It allows the identification of the type of radical and an assignment of an oxidation state but the electron-nuclear hyperfine interactions are richer in nature.

2.1.1 Isotropic and Anisotropic Electron Zeeman Interactions

In liquid solution, the rotational and translational motions of a paramagnetic molecule average out and only the isotropic average of each tensor is relevant if there is no hyperfine interaction with a nucleus of non-zero spin. Thus, the EPR spectrum appears 'isotropic'. In frozen solution, glassy environments or polycrystalline samples, the spectra are superpositions of many single-crystal-like spectra and their analysis requires the introduction of anisotropic parameters. In single crystal EPR, the various tensor components can be obtained separately.

The deviation of the **g**-values from that of the free electron g_e is dependent on the orbital angular momentum L, electron spin S, and the spin–orbit coupling (SOC). This interaction is direction-dependent and thus the spectrum becomes anisotropic.

The orientation-dependent Zeeman interaction is expressed by a (3 × 3) **g** interaction matrix which contains information about the symmetry of inner field interactions and is diagonal in its principal axes system to give its principal values g_x, g_y, g_z:

$$g = \begin{bmatrix} g_x & & \\ & g_y & \\ & & g_z \end{bmatrix} \tag{3}$$

Deviations of **g**-values from that of the free electron g_e depend on the orbital angular momentum of the electronic ground state and the nature of coordinating ligands.

$$\mathbf{g} = g_e \mathbf{1} + \Delta\mathbf{g} \tag{4}$$

The chemical bond covalency, ligand spin–orbit coupling, and charge transfer states manifest themselves in the EPR parameters.

For typical organic free radicals from first row atoms, **g** is in the range of 2.003–2.005. Because of the critical role of spin–orbit interaction in the deviation **g**-values from g_e, and the strong dependence of this SOC on the nuclear charge, the shift is sensitive to the Z value of atom in the paramagnetic species.

For transition metal ions in the condensed phase, the interactions responsible for the large g-value splitting are the crystal-field splitting and the spin–orbit coupling. In the absence of an axial crystal field symmetry, this is the origin of rhombic **g**-tensors with $g_x \neq g_y \neq g_z$ due to anisotropic interactions of the x-, y-, and z-directions of the molecule with the magnetic field (see Fig. 2) as often found for d-orbital dominated singly occupied molecular orbitals (SOMOs).

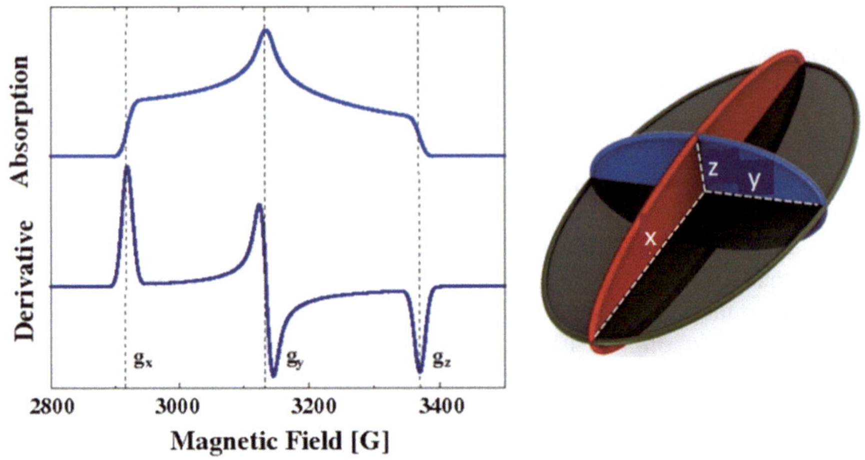

Fig. 2 Left: Simulation of a typical rhombic EPR spectrum in absorption mode (top) and its first derivative (bottom). The simulation was done with parameters characteristic for the Ni(III)-B state of the metalloenzyme [NiFe]-hydrogenase: $g_x = 2.32$, $g_y = 2.16$, $g_z = 2.01$. Right: Ellipsoidal shape associated with a rhombic **g**-tensor

The **g**-tensor can be diagonalized to give its principal values and principal axes system which is related to the orbital axes system and thus to ligand coordination and the molecular bonding situation of the metal.

The EPR spectrum will give first insight into the type of paramagnetic species (organic vs. transition metal; high-spin or low-spin). For an $S = ½$ system, the spin–orbit coupling is usually positive for a system with less than half filled valence orbitals (which leads to **g**-values < g_e), and negative for metals with a more than half filled outer shell (**g**-values > g_e). However, the assignment of **g**-values g_1, g_2 and g_3 to the magnetic principal axes x, y, z cannot be predicted a priori and depends on the individual compound. In more complicated cases, further spectral properties originate from the presence of more than one unpaired electron, for example transition metal ions with several unpaired d-electrons (high-spin Mn^{2+} or Fe^{3+}) or organic molecules in triplet states. A hyperfine interaction might be present that gives a hint as to the identity of the metal ion but can also originate from coupling to a ligand with a nuclear spin (see below).

2.1.2 Electron and Nuclear Spin Interactions

The information from an EPR spectrum is richer in the presence of nuclei with a non-zero nuclear spin quantum number in the vicinity of the electron spin, such as ^1H or ^{14}N with $I = ½$ and $I = 1$, respectively. The electron spin will then sense not only the external magnetic field $\mathbf{B_0}$ but also the local magnetic fields built up by the nuclei close to the electron spin. This interaction of the electron spin with

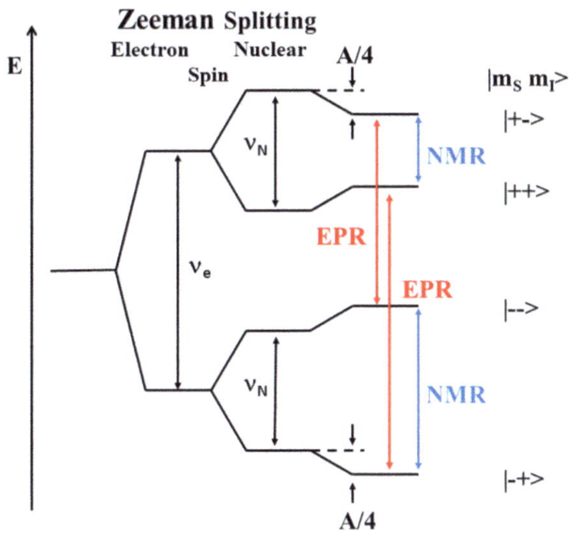

Fig. 3 Four level diagram of an S = ½ and I = ½ system. Schematic drawing of the energy level splitting in an external magnetic field with electron, nuclear Zeeman and hyperfine interactions; valid when $A/h > 0$, $2\ \nu_N > A/h$ and $g_n > 0$. The allowed EPR (red) and NMR (blue) transitions are marked

these nuclear magnetic moments is termed 'hyperfine interaction'. Note: chemical shift and nuclear spin–spin coupling interactions are typically neglected since they appear on a different energy scale compared to those of the electronic Zeeman or hyperfine interactions (by a factor 1/2000). When the hyperfine splitting is smaller than the EPR linewidth, this hyperfine interaction between electron and nuclear spin magnetic fields cannot be observed in the EPR spectrum. Electron-nuclear double resonance (ENDOR) spectroscopy allows the measurement of small hyperfine interactions by means of a double resonance experiment. The effect of a second swept radio frequency on the microwave absorption of an EPR transition is detected. When the radio frequency matches one of the NMR transitions in Fig. 3 with $\Delta M_S = 0$ and $\Delta M_I = 1$, the EPR absorption also changes and can then be detected. Pulse EPR methods, such as electron spin echo modulation (ESEEM) or hyperfine sub-level correlation spectroscopy (HYSCORE) can also be used to measure small hyperfine interactions by a transfer of spin magnetization. ENDOR thus effectively combines the resolution and nuclear selectivity of NMR with the inherent sensitivity of EPR spectroscopy which allows the very detailed resolution and characterization of structural and electronic properties.

2.1.3 Interpretation of Hyperfine Couplings

The two common mechanisms by which electrons and nuclei interact are: the Fermi contact interaction and the dipolar interaction. The isotropic coupling (or Fermi contact term) arises from the probability of finding the unpaired electron (or a fractional spin density) at the position of the nucleus (the square of the wavefunction at r = 0):

$$a_{iso} = \frac{2}{3} g_e \mu_B \, g_N \mu_N \, |\psi(O)|^2 \tag{5}$$

where μ_e is the Bohr magneton and μ_N is the nuclear magneton, g_e and g_N are the g-factors of electron and nucleus, respectively, and the atomic spin density at the nucleus specified (given by $|\Psi(0)|^2$). This interaction is due to an appearance of spin density in atomic s-orbitals from direct s-orbital occupancy of the unpaired electron or by spin polarization and is independent of sample orientation in the magnetic field.

The dipole–dipole interaction between electron and nuclear spins is directed and depends on the relative orientation of nuclear spins of atoms *i* and *j*:

$$\mathbf{A}_{\mathbf{dip,\,ij}} = \frac{\mu_B \, g_e \, \mu_N \, g_N}{h} \left\langle \frac{3r_i r_j - \delta_{ij} r^2}{r^5} \right\rangle \tag{6}$$

with the angular brackets indicating the integration over the electron wavefunction in order to remove the explicit spatial dependence of the angle between the magnetic field $\mathbf{B_0}$ and the **r**-vector.

For sufficiently large distances r between electron spin and nucleus, the spatial distribution of the electron density can be neglected and $\mathbf{A_{dip}}$ reduces to the point-dipole model

$$\mathbf{A_{dip}} = \frac{\mu_B \, g_e \, \mu_N \, g_N}{h} \cdot \frac{\rho}{r^3} \tag{7}$$

which can then be used as a direct measure of the distance between electron and coupling nuclear spins due to its $1/r^3$ dependence, for example between a transition metal ion with an unpaired electron and a nucleus from a ligand.

In transition metal systems, there is an additional third term contribution to the hyperfine interaction due to spin–orbit coupling between the nuclear spin **I** and the electron spin **S** ($\mathbf{A_{SO}}$). For light nuclei, the SOC correction is usually negligible.

The full hyperfine tensor $\mathbf{A_{tot}} = a_{iso}\,\mathbf{1} + \mathbf{A_{aniso}}$ is the sum of isotropic, dipolar plus second-order spin–orbit hyperfine interactions and can, once more, be diagonalized to yield the hyperfine tensor principal axes system for each interacting nucleus.

The 'Spin Hamiltonian' only relates to the concept of an 'effective spin'. Magnetic-resonance parameters that can be extracted from EPR spectra, electronic **g**-tensors, hyperfine coupling tensors, or zero-field splitting (see below) contain only indirect information about the metal binding site. Thus, it is often difficult or even impossible to directly relate these Spin Hamiltonian EPR parameters to structural information.

3 Quantum Chemical Calculations of EPR Parameters

NMR and EPR experiments detect the energy level splitting arising from the interaction of magnetic moments with an external magnetic field $\mathbf{B_0}$. Both approaches are based on the same fundamental principles of physics namely the precession of electron and nuclear spin dipole moments in a magnetic field and can thus be formulated in very similar ways.

The full quantum mechanical formulation of electron spin resonance is given in the textbook by Harriman [22] and its reproduction is beyond the scope of this chapter.

The **g**-tensor is defined as a second order property with derivatives to the external magnetic field $\mathbf{B_0}$ and the electron spin operator **S**.

$$g = g_e \mathbf{1} + \frac{1}{\mu_B}\left(\frac{\partial^2 E}{\partial \mathbf{B} \partial \mathbf{S}}\right)_{\mathbf{B},\mathbf{S}=0} \tag{8}$$

The contributions to deviations from the free electron value g_e are due to relativistic mass correction (RMC), gauge correction (GC) and a cross term between orbital Zeeman (OZ) and the spin–orbit coupling (SOC).

The spin–orbit coupling is the dominating term to the **g**-tensor and arises from the interaction between electron spin and orbital angular momentum. Various ways of approaching the **g**-tensor calculation using different approximations to the spin–orbit coupling are summarized in [23, 24], and in [25].

The procedure to include relativistic effects and to find the eigenfunctions and eigenvalues of the four-component Dirac equation in an all-electron basis is given in Chap. 5, Sect. 2.1. Methods that are based on an elimination of the small component are, for example, the regular approximations, the Breit-Pauli Hamiltonian or unitary transformations to decouple the electronic and positronic states such as Douglas–Kroll–Hess theory [25]. The Douglas–Kroll–Hess (DKH) method is a unique analytical expansion technique for the Dirac Hamiltonian in order to eliminate the small-component. Other elimination approaches are based on the Foldy–Wouthuysen transformation (see Chap. 5, Sect. 2.1). These lead to the Breit-Pauli Hamiltonian which can be used to perturbatively consider spin–orbit effects or the zero-order regular approximation (ZORA) to treat spin–orbit coupling self-consistently with an efficient relativistic correction for the region close to the nucleus. The exact two-component (X2C) relativistic Hamiltonian [26, 27] is based on an exact decoupling of the large and small components of the Dirac Hamiltonian in its matrix representation.

Computationally, it is possible to calculate the energy level splittings in a magnetic field and derive equations for spectroscopic parameters of the Spin Hamiltonian from electron spin resonance techniques using wavefunction-based approaches or Density Functional Theory (DFT).

The calculation of EPR parameters from first principles has extensively been presented and reviewed in [28–32]. State-of-the-art implementations of different approaches to calculate EPR Spin-Hamiltonian parameters are to be found in ADF [33, 34], ORCA [35, 36], Gaussian [37], Dalton [38] or MagRespect [39] programs which are the most common ones. An overview of different implementations in the codes is given in [28] and [40]. The various computational approaches to calculate EPR parameters have been thoroughly reviewed in [23, 28, 30, 32, 41] and are not repeated here. They range from first [42–44] to more recent ab initio computations [45–47], and from early work on DFT [48, 49] to more recent implementations [50–56], multireference spin–orbit configuration (MRSOCI) [57], CASPT2 [58], MRCI [59], coupled-cluster theory [60] and DMRG [61].

Despite all efforts, the calculation of **g**-tensors for transition metal complexes is still more challenging than for organic systems since the spin–orbit coupling term is dominating. It is therefore clear that an appropriate treatment of spin–orbit coupling is a pre-requisite for accurate calculations of $\Delta\mathbf{g}$. An accuracy of 30% for the calculated 'g-shift' is already considered to be satisfying [53]. Spin–orbit coupling has to be included either into the ground state calculation self-consistently or as a perturbation to the non-relativistic solution. A non-relativistic ground state DFT calculation with a perturbative approach to introduce spin–orbit coupling into the Breit-Pauli Hamiltonian was first formulated by [48, 62, 63]. One approach to consider SOC self-consistently is the zero-order approximation (ZORA; see Chap. 5) which also enables the calculations of **g**-tensors [49] and **A**-tensors [64]. ZORA was shown to be a good approximation to the fully relativistic treatment, in particular for properties that involve valence orbitals. To a certain extent, the ZORA and the DKH method to lowest order yield comparable results for many molecular properties. An efficient approximation to the full spin–orbit-coupling operator is the spin–orbit mean-field (SOMF) approximation developed by Hess et al. [65] and extended and implemented recently [66].

An accurate molecular geometry is an absolute necessity for a reliable relation between molecule structure and calculated magnetic resonance parameters. Density Functional Theory (DFT) is one computational method for the optimizations of large systems with 100–200 atoms and usually gives reliable structural parameters for transition metal systems. A correct description of the electron density in the system is also required for the calculation of spectroscopic parameters and other properties that rely on the electron density. It is thus always necessary to employ an all-electron and sufficiently large basis set with scalar relativistic effects included for the calculation of magnetic resonance parameters of transition metal systems.

In literature, quantum chemically calculated EPR **g**-tensors very often report only the principal values g_1, g_2, g_3 or an average g_{iso} value. There is, however, much more information in the principal axes system x, y, z which reveals spatially directed magnetic interactions and defines the magnetic axes of the system. In the presence of molecular symmetry, they may coincide with the symmetry axes. In biological systems, in the absence of molecular symmetry, the experimental assignment of principal axes and anisotropic magnetic interactions is more difficult and computational means are used to resolve ambiguities (see below).

4 Structural Information from the Anisotropy of Magnetic Interactions

4.1 *EPR Studies on Single Crystals*

In the laboratory axes system, the **g**-tensor can be obtained by measuring the electron Zeeman splitting as a function of the angle of the applied field and the crystallographic axes of a single crystal.

The full **g**-tensor can be measured from the angular variations of $\mathbf{g}^2$ in three mutually perpendicular planes of a single crystal

$$\mathbf{g}^2 = \left(g_{xx}^2 l_{xx}^2 + g_{yy}^2 l_{yy}^2 + g_{zz}^2 l_{zz}^2 \right) \tag{9}$$

where l_{xx}, l_{yy}, l_{zz} are the direction cosines between $\mathbf{B_0}$ and the three principal axes (see Fig. 4).

In the special case of an arbitrary crystal orientation and several paramagnetic sites per unit cell with symmetry relations between them, analysis of EPR spectra requires knowledge of the three axes systems

i. **L** the laboratory axes systems (1, 2, 3)
ii. **C** the crystal axes systems with crystallographic axes (a, b, c)
iii. **I** the intrinsic **g**-tensor principal axes system (x, y, z)

and the transformation matrices between them (see Fig. 4).

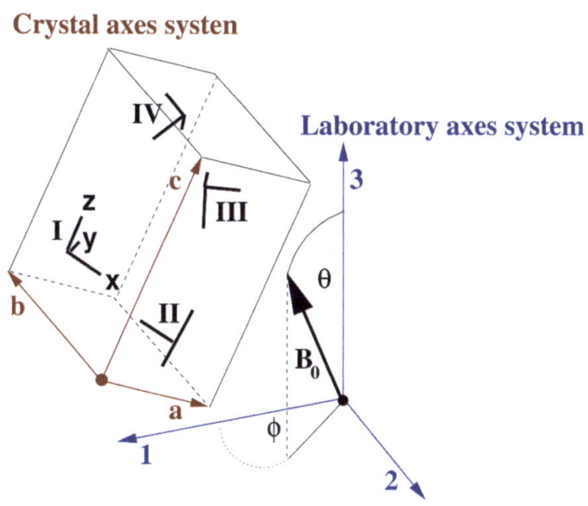

Fig. 4 Arbitrary orientation of a single crystal with four sites I–IV per unit cell in an external magnetic field $\mathbf{B_0}$. Definition of laboratory, crystal axes and the **g**-tensor principal axes systems

4.2 Model Complexes

Structural and functional model complexes are valuable systems to gain insight into the structure and electronic and spectroscopic properties of these small systems. Very often, model complexes are available in larger amount and higher purity compared to enzymatic systems. Thus, structural data (from e.g. X-ray crystallography) and magnetic resonance data are more facile to obtain and may be analyzed in more detail. The spectroscopic and computational results obtained for the biomimetic or bioinspired model may serve as a benchmark or comparison with an enzymatic system. In model compounds, magnetic and structural symmetry axes are very often coincident since both ligand field and magnetic field splittings are determined by ligand strength and coordination geometry.

4.2.1 Transition Metal Single Crystal EPR

In bis(malenonitriledithiolato)nickelate(III) ($Ni(mnt)_2^-$) the nickel atom possesses a first coordination sphere similar to that of the [NiFe]-hydrogenase enzyme. Nickel is in the +3 oxidation state as in the Ni–A, Ni–B, and Ni–C states in the enzyme. The central atom is bound to four sulfur atoms in a square-planar fashion in the model complex (with D_{2h} symmetry) (Fig. 5). Model complex and enzyme display low-spin $S = ½$ EPR spectra. Structural and functional nickel model systems have been reviewed in [67–69].

4.2.2 The Rhombic g-Tensor

The EPR spectrum of Ni(III) bis(malenonitrile)dithiolate was investigated in single crystals [70]. The magnetic axes were found to be coincidental with the symmetry axes of the complex: the z-axis is perpendicular to the molecule plane and the y-axis bisects the C=C bond of the ligand. The rhombic $S = ½$ EPR spectrum displays **g**-tensor principal values of 2.14 > 2.04 > 1.998. From extended Hückel calculations, the authors correctly predicted a $3d^7$ orbital occupancy with the $3d_{yz}$ orbital to be singly occupied in the ground state and an extensive interactions with the π-orbitals of the ligand. This was also investigated by DFT calculations [71].

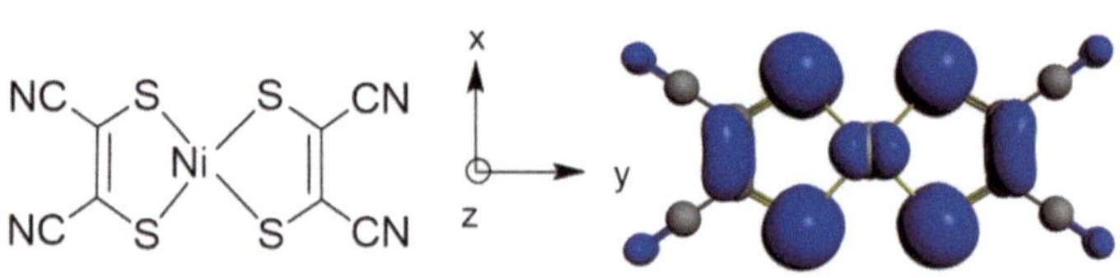

Fig. 5 Structure of a square-planar Ni(III) complex with an $S = ½$ electronic ground state: bis(malenonitriledithiolato)nickelate(III). The molecular axes and the **g**-tensor axes systems are coincident. The calculated spin density distribution shows a large degree of delocalization of the electron spin from the central transition metal to the ligand

Table 1 Comparison of calculated **g**-tensor principal values with experiment for $Ni(mnt)_2^-$

				Deviation from experiment
	g_x	g_y	g_z	$\Delta g_{i(exp\text{-}calc)}$ in ppt
Experimental[a]	2.14	2.04	1.99	
Calculated				
SO self-consistently				
BP86	2.103	2.037	1.985	37, 3, 5
B3LYP	2.124	2.067	1.997	16, −27, −4
B3LYP*	2.120	2.057	1.994	20, −17, −4
B3LYP**	2.114	2.049	1.991	26, −9, −1
PBE0	2.143	2.081	2.003	−3, −41, −13
SO perturbative				
BP86	2.106	2.039	1.987	34, 1, 3
B3LYP	2.126	2.070	1.996	14, −30, −6
B3LYP*	2.127	2.061	2.003	13, −21, −13
B3LYP**	2.118	2.052	1.995	22, −12, −5
PBE0	2.160	2.087	2.002	−20, −47, −12
WF multireference[b]				
CAS(13, 14)	2.583	2.067	1.983	−443, −27, 7
NEVPT2(13, 14)	2.487	2.076	1.991	−347, −36, −1

[a]From [72]
[b]From [73]

Table 1 compares calculated **g**-tensor principal values using different functionals. The **g**-tensor principal values from ZORA calculations with self-consistent inclusion of SOC in general reproduce well the rhombicity of the EPR spectrum. BP86 slightly underestimated g_x but performs very well for the g_y- and g_z-components. B3LYP performs only slightly better for g_x but overestimates g_y. By reducing the amount of Hartree-Fock exchange from 20 to 15 and 10%, a continuous decrease of all **g**-values can be seen. PBE0 only performs well for the x-component but significantly overestimates g_y and g_z. It has to be noted that the hybrid functionals do not outperform the GGA BP86 functional but a systematic underestimation of the **g**-values has to be noticed which cannot be corrected by using larger basis sets [71]. The absolute deviation between calculation and experiment increases with the deviation from g_e while the relative error remains about constant. In general, differences between different functionals are found to be pronounced for transition metal systems but still smaller than the systematic deviation from experiment [23] and hybrid functional do only slightly improve the accuracy of calculated **g**-values [51, 53].

In order to assess the dependence of the calculated **g**-tensor principal values on the treatment of spin–orbit coupling, a perturbation approach [48] to include spin–orbit coupling was also employed using identical structures, basis sets, and electron densities. The calculated **g**-tensor principal values are almost identical which shows that the spin–orbit treatment is not the most critical issue here, at least for this system with a rhombic **g**-tensor. The use of the X2C Hamiltonian, on the other hand, gives calculated **g**-tensor principal values of 2.148, 2.046, 1.974 and is in best agreement with experiment.

The challenges of multireference perturbation theory for calculating **g**-tensors of first row transition metal complexes were also investigated recently [73]. The scalar-relativistic complete active space self-consistent field (CASSCF) and N-electron valence perturbation theory (NEVPT2) methods were evaluated for a series of $S = ½$ transition metal complexes of which $Ni(mnt)_2^-$ was part of. CASSCF calculations based on active spaces that contain only metal-based orbitals significantly overestimate the **g**-values and are not able to give reliable results. The inclusion of dynamic correlation by means of the NEVPT2 method improves the results and a lowering in the g-shift is obtained. Still, CASSCF leads to an over description of the metal-ligand bond ionicity and, hence, the spin–orbit coupling matrix elements are too large. The results show that wavefunction theory has significantly progressed in computing the properties of large, open-shell transition-metal complexes. Some methodological work, however, is still required in order to be able to obtain accurate results.

Molybdenum is part of the active sites of a large number of enzymes such as sulfite oxidase (SO) and xanthine oxidase (XO) and usually shuttling between $Mo^{IV}/Mo^{V}/Mo^{VI}$ oxidation states. The Mo^{V} model compound $(N_2S_2)MoOCl$ shows similarities with certain catalytic intermediates and species of the sulfite and xanthine oxidases. The lack of a crystalline sample with sufficient size and purity has obstructed the determination of the full **g**- and **A**-tensors in the protein. Single crystal EPR spectra for the molybdenum(V) model compound *cis,trans*-$(L\text{-}N_2S_2)Mo^{V}OCl$ ($L\text{-}N_2S_2$=N,N′-dimethyl-N, N′-bis(mercaptophenyl)ethylenediamine) [74] revealed the orientation of the **g**-tensor principal axes in the coordination complex. The INDO/S-CI calculated principal values of the **g**-tensor of $g_1 = 1.985$, $g_2 = 1.963$, and $g_3 = 1.960$ compare reasonably well with the experimental frozen solution values of $g_1 = 2.004$, $g_2 = 1.963$, and $g_3 = 1.946$ (see Table 2). SOMF BP86/SV(P), TZP calculations gave principal **g**-values of 2.021, 1.964, 1.950, which are in good agreement with the experimental data and also the calculated **g**-tensor orientations were fully consistent with the experimental findings. The g_1 axis was found to be oriented along the Mo–O bond, with an angle of 30.8° for INDO/S and 14° for BP86 compared to ~10° found experimentally. The g_2 and g_3 axes are essentially oriented in the xy-plane of the complex. The g_2 axis forms an angle of 14° (INDO/S) and 18° (BP86/ZORA) with the Mo–S2 bond, compared to ~37° found experimentally. The g_3 axis is rotated by 14° (INDO/S) and 3° (BP86/ZORA) from the Mo–Cl bond, respectively, compared to ~38° in experiment (see Table 2).

Following an earlier benchmark study of small and medium Mo^{V} complexes [75], Fritscher et al. [76] performed computational studies of **g**- and molybdenum HFC tensors for series of larger Mo^{V} complexes. The best agreement with experimental data for both hyperfine and **g**-tensors was obtained with hybrid functionals containing approximately 30–40% Hartree–Fock exchange. Computed orientations of

Table 2 Comparison of experimental and calculated **g**-tensor principal values and axes orientation of *cis,trans*-(L–N_2S_2)Mo^VOCl. Reprinted figure with permission from [74]. Copyright (2005) American Chemical Society

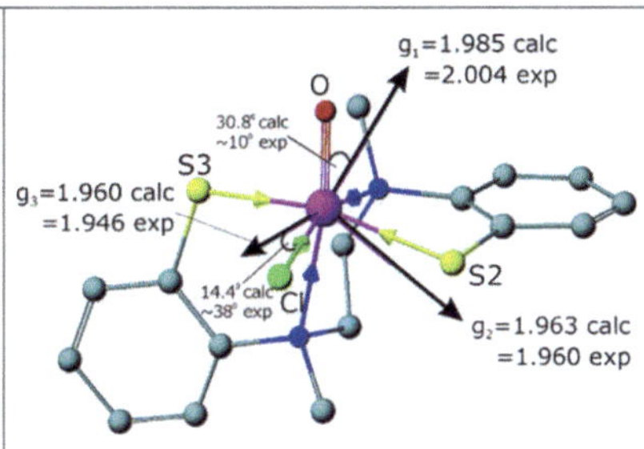

g-tensor principal values	g_1	g_2	g_3
Experimental	2.004	1.960	1.946
INDO/S-CI	1.985	1.963	1.960
BP86/ZORA SOMF	2.021	1.964	1.950
BPW91-40HF [75]	2.015	1.953	1.939
g-tensor principal axes orientation	<(Mo–O) bond	<(Mo–S) bond	<(Mo–Cl) bond
Experimental	10	38	38
INDO/S-CI	31	14	14
BP86/ZORA SOMF	14	18	3
BPW91-40HF [75]	10	10	7

g- and hyperfine tensors relative to each other and to the molecular framework for MoXLCl2 complexes provided good agreement between theory and single-crystal electron paramagnetic resonance experiments, where available. In these cases, the calculated molybdenum hyperfine tensor orientations are influenced only slightly by spin–orbit coupling effects. The *cis,trans*-(L-N_2S_2)Mo^VOCl complex of [74] was also re-investigated using unrestricted Kohn–Sham DFT together with hybrid functionals with neglect of scalar relativistic effects on HFC and **g**-tensors, as well as higher-order SO effects on **g**-tensors. The BP86 results (2.026, 1.967, 1.952) are almost identical to those of the previous work [74]. Admixture of 40% Hartree-Fock exchange to give the BPW91-40HF functional gave **g**-tensor principal values of 2.015, 1.953 and 1.939 which are slightly closer to the experimental results compared to INDO/S or BP86. The tensor components g_2 and g_3 become too small at even 20% of exact exchange whereas the shift of g_1 is still too large at 40% exchange.

4.2.3 Transition Metal Hyperfine Interactions

The ^{61}Ni hyperfine tensor in $Ni(mnt)_2^-$ was found to be collinear with the **g**-tensor and coincident with the molecular symmetry axes. The accuracy and challenges in the calculation of EPR hyperfine interactions is exemplified here for this $Ni(mnt)_2^-$ model complex. Ligand ^{33}S, ^{13}C and ^{14}N hyperfine interactions a_{iso} and $\mathbf{A_{dip}}$ can be calculated to within 1–2 MHz [71] from experimental EPR, HYSCORE and ESEEM investigations [72] using a standard non-relativistic approach. Spin–orbit coupling does not play a significant role here. In fact, quantum chemical calculations helped to resolve ambiguities in the sign of the experimental ^{33}S hyperfine tensor.

For the anisotropic ^{61}Ni hyperfine interaction, however, a first order calculation of the hyperfine interaction was clearly not sufficient. On the one hand, the isotropic hyperfine coupling could be well-reproduced using scalar-relativistic ZORA calculations, even without considering spin–orbit coupling effects (a_{iso} +18 MHz experimental; +17 MHz calculated). However, the experimental anisotropic ^{61}Ni hyperfine interaction [70] of A_{dip} (+27, −9, −18) MHz is not well reproduced in scalar relativistic calculations with (+53, −25, −28) MHz but overestimated by a factor of 2. Here, the consideration of spin–orbit coupling reduces the dipolar part to (+25, −7, −18) MHz and reproduces nicely the experimental ^{61}Ni hyperfine tensor. The hybrid functional B3LYP does not perform superior to BP86 with calculated ^{61}Ni hyperfine couplings of a_{iso} = +20 MHz and $\mathbf{A_{dip}}$ (+29, −14, −15) MHz. This example demonstrates the necessity for an appropriate treatment of second-order spin–orbit coupling contributions to the hyperfine interactions for transition metal ions (it makes up a factor of 2 here). In particular, the covalency of the Ni–S bond and the subsequent delocalization of the unpaired spin density are a challenge to describe correctly using DFT methods. BP86 gives spin densities of $\rho(Ni) = 0.27$ and $\rho(S) = 0.16$; B3LYP $\rho(Ni) = 0.28$ and $\rho(S) = 0.17$. The GGA results are closer to experimental results of $\rho(Ni) = 0.25$ and $\rho(S) = 0.13$ [70, 72] and show that also GGA functionals are able to correctly describe the degree of metal-ligand covalency.

$^{55}Mo^V$ hyperfine interactions were calculated for a series of six small [75] and ten larger [76] model complexes. For the large *cis,trans*-(L-N_2S_2)Mo^VOCl (L-N_2S_2=N, N′-dimethyl-N, N′-bis(mercaptophenyl)ethylenediamine) complex, the calculated isotropic Fermi-contact a_{iso} hyperfine interaction ranges from 56 MHz for BP86 to 95 MHz for BPW91-40HF. Consideration of the pseudo-contact correction term A_{PC} due to spin–orbit corrections to the hyperfine tensor typically amounts to 14–17% of a_{iso} and adds an extra 11 and 15 MHz, respectively. The sum of the two contributions to the isotropic ^{55}Mo hyperfine interaction gives 67 MHz for BP86 and 110 MHz for BPW91-40HF which compares well with the experimental value of 107 MHz. The consideration of second-order spin–orbit contributions to the metal hyperfine interaction leads to only a minor change in relative **g**-tensor and the hyperfine tensor orientation angles by 12, 7 and 13° for each of the principal axis.

4.3 *Transition Metal Containing Enzymes*

4.3.1 The g-Tensor Orientation in a Protein Single Crystal

The nickel atom of [NiFe]-hydrogenase active site gives rise to rhombic EPR spectra from an $S = ½$ spin system and reveal no hyperfine interaction (the hyperfine interaction is smaller than the EPR linewidth). Hydrogenase nickel EPR spectra were known in the literature for a very long time but the discovery of a second transition metal in the active site in the first protein crystal structure was not expected. The second metal was later identified by ^{57}Fe Mössbauer experiments to be a non-redox active iron atom. The 'Ni–C' EPR signal refers to a catalytic intermediate in the heterolytic splitting of H_2. Ni–C is light-sensitive and converted to the light-induced Ni–L state at low temperature (below 100 K). Upon illumination, the EPR spectrum converts from that of a Ni–C ($g_i = 2.20, 2.14, 2.01$) to that of a Ni–L species (2.30, 2.12, 2.05). Recently, Ni–L was discussed to be also involved in the catalytic mechanism.

The Ni–C and Ni–L states of the [NiFe]-hydrogenase from *D. vulgaris* Miyazaki F were generated in situ in protein single crystals by incubation with H_2. An analysis of the orientation-dependent EPR spectra yielded the full **g**-tensors and their orientations in the crystal axes system for both Ni–C and Ni–L forms simultaneously (see Fig. 6).

Fitting of the eight EPR transitions (see Fig. 6) from 2 species in the crystal with four molecules per unit cell each, the assignment of the magnetic **g**-tensor principal axes to the enzyme active site structure is not unambiguous.

The geometry of the catalytic center and the **g**-tensor orientations are closely related since the interaction of the unpaired electron spin at the nickel is mediated by the coordinating ligands and the crystal field thus determines the orientation of the magnetic axes. In the oxidized forms, the coordination of the Ni atom in the active site can be described as distorted octahedron with one empty ligand position. In absence of a protein structure of the reduced enzyme at that time, an active site geometry close to the oxidized forms was assumed. The shift of the smallest **g**-tensor component from $g_3 \approx 2.01$ to $g_3 \approx 2.05$ is indicative of a change in redox state of the nickel atom. Various candidates for models of the Ni–C and Ni–L forms were investigated by computations. The calculations reproduce well the upshift of g_3 to 2.05 upon photoreduction from Ni(III) to Ni(I) when dissociation of the bridging hydride is assumed. This detailed insight is not available from protein crystallography. The calculated **g**-tensor principal values from ZORA DFT calculations are in good agreement with experiment (see Table 3) and the deviation from experiment is in the typical range as was also found to the $Ni(mnt)_2^-$ model complex (see above).

DFT calculations of the orientation of the **g**-tensor principal axes in the molecular structure of the active site helped to assign the paramagnetic sites in the crystal (see Table 3). The calculated **g**-tensor orientation is in excellent agreement with only one possible experimental assignment and enabled the analysis and interpretation of the electronic structure of the Ni–C and Ni–L states simultaneously. It was suggested on the basis of computational work that Ni–L originates from the Ni(III)-μH^--

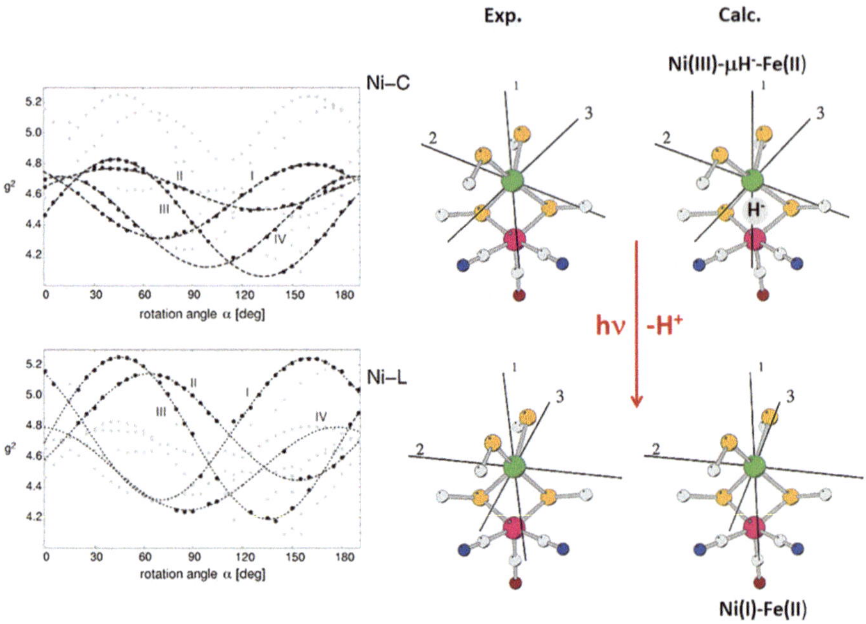

Fig. 6 Angular dependence of EPR spectra of the reduced and illuminated [NiFe]-cluster in a protein single crystal of the [NiFe]-hydrogenase from *D. vulgaris* Miyazaki F. Simultaneous fit of EPR transitions containing four sites per unit cell for Ni–C and Ni–L species (left). Orientation of the experimental and calculated **g**-tensor principal axes of Ni–C and Ni–L forms in the molecular structure of the active site (right). Adapted with permission from [77]. Copyright 2003 American Chemical Society

Table 3 Experimental and calculated **g**-tensor principal values and orientation of the principal axes in the Ni–C and Ni–L states of the [NiFe]-hydrogenase [77]

Ni–C	g_x	g_y	g_z
g-tensor principal values			
Experimental	2.20	2.14	2.01
Calculated	2.20	2.10	2.00
Tensor orientation/°	Experimental	Calculated	
$<g_x$, (Ni–SCys81)	12	15	
$<g_y$, (Ni–SCys84)	14	12	
$<g_z$, (Ni–SCys549)	5	4	
Ni–L			
g-tensor principal values			
Experimental	2.30	2.12	2.05
Calculated	2.26	2.10	2.05
Tensor orientation/°	Experimental	Calculated	
$<g_x$, (Ni–SCys81)	17	22	
$<g_y$, (Ni–SCys84)	19	13	
$<g_z$, (Ni-SCys549)	3	8	

Fe(II) Ni–C precursor state by photoreduction and dissociation of a proton to give a Ni(I)–Fe(II) state with a vacant bridging position. This formally corresponds to a two-electron reduction of the d^7 Ni(III) to a d^9 Ni(I) center.

In Ni–C, the g_z-axis orientation is identical to that of the oxidized Ni(III) states Ni–A and Ni–B. The calculated **g**-tensor axes orientation shows that the magnetic g_x and g_y-axes are interchanged in Ni–C compared to the oxidized Ni–B and Ni–A states [78] and related by a 90° rotation about the g_z-axis. This information is only available from calculations here and would not easily be resolved by analysis of the single crystal EPR spectra. The calculated orientation of the magnetic axes agrees to within 2–3° with the experimental ones (see Table 3). The g_z axis of Ni–C is almost perfectly oriented along the Ni–S(Cys549) bond (with an angle of 5° experimentally, and 4° in the calculations). The magnetic x-axis is roughly along the Ni–S(Cys81) bond (with an angle of 12° vs. 15°) and the orientation of g_y displays angles of 14° and 12° from the Ni–S(Cys84) bond direction.

For Ni–L, there are more assumptions going into the experimental assignment of the **g**-tensor principal axes orientation. Since the rotational behavior of the EPR transitions is similar to that of the enzyme in the Ni–C state (see Fig. 6), one possibility is the absence of a significant magnetic re-orientation upon photoreduction although the **g**-values do change. This would suggest a similar magnetic axes orientation to that of the Ni–C state. The orientation of the g_z-axis is also retained in Ni–L (within 2°), whereas those of g_x and g_y are rotated in plane by 10°. This assignment is supported by the DFT calculated **g**-tensor orientation to within 5–6° (see Table 3). Later, the ground state of Ni–L was interpreted to result from a re-hybridization of the nickel d-orbitals. In Ni–L, the nickel d_{x2} orbital points into the direction of the unoccupied bridging position and forms a metal–metal bond with the iron d_{x2} orbital. As opposed to the Ni–C state, the singly occupied molecular orbital in Ni–L is not a d_{z2} orbital but a d_{z2-y2} orbital, which results in significantly altered magnetic properties [79].

4.3.2 Proton Hyperfine Coupling Tensors in Protein Single Crystals

In an $S = ½$ and $I = ½$ case, the two ENDOR transition frequencies are

$$\nu_{\pm}^2 = \nu_N^2 + \frac{1}{4h^2}(\mathbf{I}\cdot\mathbf{A}\cdot\mathbf{A}\cdot\mathbf{I}) \mp \frac{\nu_N}{h}(\mathbf{I}\cdot\mathbf{A}\cdot\mathbf{I}) \tag{10}$$

with the upper sign referring to $M_s = +½$ and the lower signs to the $M_s = -½$ transitions. In a low-symmetry system, the **g**-tensor principal axes system that described the g-anisotropy is not necessarily coincident with the axes system that defines the hyperfine anisotropy.

The analysis of protein single crystal ENDOR is done in analogy to that of an orientation-dependent **g**-tensor study and requires an a priori EPR analysis of the single crystal. For the oxidized states Ni–A and Ni–B of the [NiFe]-hydrogenase enzyme, the orientation of the magnetic axes was determined from protein single crystal EPR investigations and a definite assignment of the respective paramagnetic site in the unit cell states from angle-dependent EPR spectra was feasible [78, 80].

In the high field approximation, the orientation-dependent hyperfine splitting is observed in the principal axes system *x*, *y*, *z* of the hyperfine tensor **A**

$$|\mathbf{A_{obs}}| = \left(A_{xx}^2 l_x^2 + A_{yy}^2 l_y^2 + A_{zz}^2 l_z^2\right)^{1/2} \tag{11}$$

in which A_{ii} are the principal values of the hyperfine tensor and l_i are the direction cosines of the hyperfine tensor with respect to the crystallographic axes. The fit and analysis of ENDOR transitions in a protein single crystal in an arbitrary orientation can be done in analogy to the rotation angle-dependent EPR spectra (see above). According to the EPR analysis, the protein crystal has a 7:3 ratio of Ni–B to Ni–A. Thus, the Ni–B ENDOR signals are more intense and their analysis will only be discussed here.

Two largely isotropic and thus almost angle-independent ^1H hyperfine interactions can be identified (**A1** and **A2** in Fig. 7). The isotropic hyperfine interactions of **A1** and **A2** indicate a significant amount of spin density at or in close vicinity to two protons. The small dipolar contribution is indicative of a short distance to the nickel or another atom with a fraction of spin. Quantum chemical calculations show that only one of the four coordination cysteine amino acids carries significant spin density (the bridging Cys533 residue; see Fig. 7).

The calculated unpaired spin density distribution assigns the ^1H hyperfine couplings **A1** and **A2** to originate from one of the bridging cysteines Cys_{533} (see Fig. 7). The **g**-tensor of the formal Ni(III) Ni–B state was shown to have its g_z-axis with a g-value close to g_e oriented along the Ni–S(Cys533) bond and the overlap of nickel and sulfur orbitals to contain at least a fraction of unpaired electron [78]. The isotropic hyperfine interactions of β–CH_2 protons follow a dihedral angle dependence $a_{iso} \sim \rho_S \cdot \cos^2(\theta + \varphi)$ with θ the dihedral angle and φ the periodicity of π/2. Since the isotropic hyperfine interaction of the β–CH_2 protons are very similar in magnitude here, one can even make a statement about the conformation of the β–CH_2 protons of the bridging cysteine. It can be concluded that the dihedral angle θ for both protons is almost identical.

Since the bridging cysteine Cys533 also carries a significant degree of the unpaired electron spin, the ^1H nuclei couple simultaneously to the p-orbital spin of the sulfur atom with ρ_S and the unpaired electron spin at the nickel atom with ρ_{Ni}. This is beyond the point-dipole approximation and hard to interpret from ENDOR spectra alone. Best agreement between spectral analysis and experiment were obtained for spin densities ρ at Ni 0.5 and sulfur 0.3 and the remaining 0.2 to be delocalized over the complete cluster. This is in excellent agreement with calculated atomic spin populations of 0.52 and 0.34 [82]. The calculated isotropic and dipolar hyperfine interactions for the β–CH_2 protons are in excellent agreement with experiment (see Table 4) and enable the assignment of **A1** and **A2**.

The third ^1H hyperfine interaction **A3** in Fig. 7 is smaller in magnitude and displays a larger anisotropic angular dependence. Its principal values can be reproduced by

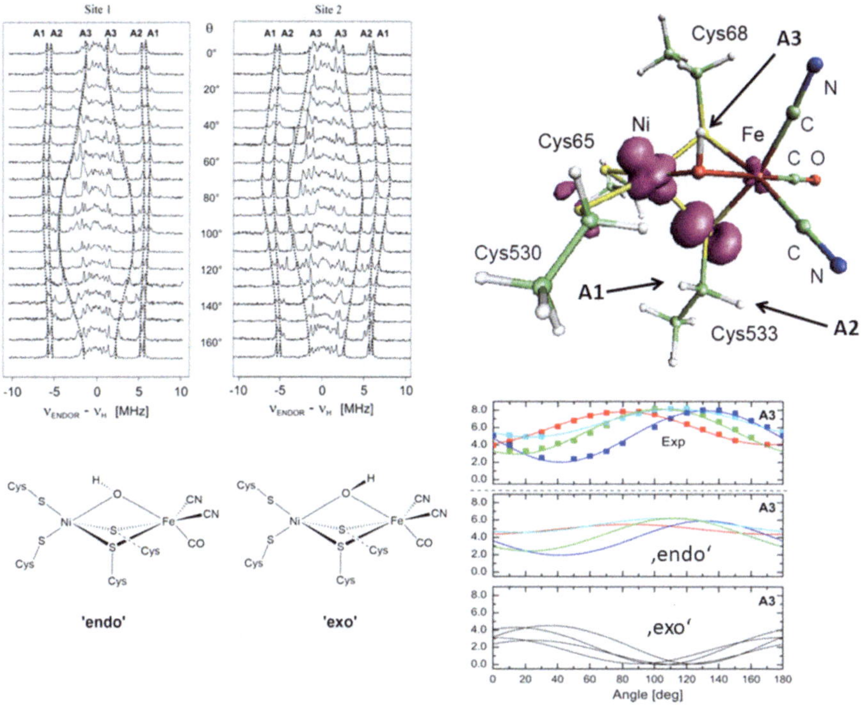

Fig. 7 ^{1}H-ENDOR spectra of the Ni–B state in protein single crystals of the [NiFe]-hydrogenase from *D. vulgaris* Miyazaki F. Tracing, analysis and structural assignment. Adapted from [81] with permission of the publisher. Copyright 2005, SBIC

Table 4 Comparison of experimental ^{1}H hyperfine interactions from ENDOR crystallography of protein single crystals and calculated hyperfine tensors in MHz of Ni–B in [NiFe]-hydrogenase [81]

a_{iso} (A_{aniso})	Experimental	Calculated	
		ZORA	SOMFI
$^1H_{\beta 1}$–Cys549	13.0 (−2.2, −1.7, 3.9)	14.1 (−1.7, −0.9, 2.6)	9.4 (−1.5, −0.9, 2.4)
$^1H_{\beta 2}$–Cys549	11.2 (−1.1, −1.1, 2.2)	12.9 (−2.3, −1.1, 3.4)	8.1 (−2.3, −1.4, 3.7)
μ–OH^1	−3.9 (−4.3, −3.1, 7.5)	Endo −1.5 (−4.4, −3.0, 7.4)	−0.8 (−5.8, −4.2, 9.9)
		Exo +1.6 (−3.8, −2.5, 6.3)	+4.6 (−5.0, −4.1, 9.1)

calculations when a μ-hydroxide (OH^-) in the bridging position between the nickel and iron atoms is assumed (see Table 4). QM calculations allow to probe different binding scenarios and then to elucidate the conformation of the hydroxide. The ^1H-ENDOR transition rotational dependence can only be explained when the proton of the hydroxide is in the 'endo' position (see Fig. 7). In the 'exo' position, the proton would be in a different orientation with respect to the d_{z2} orbital of the Ni, give rise to different ^1H hyperfine coupling signs, different hyperfine tensor principal axes orientations and different rotation angle-dependent ENDOR spectra. Here, the calculations do reproduce not only the ^1H hyperfine couplings in magnitude but also resolve experimental ambiguities in terms of the directionality of magnetic interactions and the orientation of the protonated bridging ligand. Table 4 summarizes the comparison of experimental and calculated ^1H hyperfine coupling constants. BP86 calculation using ZORA perform equally well as those with the B3LYP hybrid functional and the mean-field SOMFI operator. As a matter of fact, the isotropic hyperfine interactions from ZORA are in even better agreement with experiment, which might be due to the use of Slater-type basis sets and their correct description of the electron cusp close to the nucleus.

4.3.3 Manganese Zero-Field Splitting Parameters in Protein Single Crystals

The enolase enzyme catalyzes the reversible dehydration of D-2-phosphoglycerate to phosphopyruvate in the glycolytic pathway. The native enzyme uses Mg^{2+} in two binding sites I and II for the reaction but Mn^{2+}/Zn^{2+} substitutions also display catalytic activity in vivo. The enolase-PhAH inhibitor complex (PhAH = phosphonoacetohydroxamate) from yeast was investigated in manganese/zink-substituted protein single crystals using high-frequency W-band EPR and ^1H-ENDOR experiments [83]. The crystal structure of $P2_1$ symmetry was refined to a resolution of 1.54 Å. The rotational angle-dependent EPR spectra were recorded in two different protein crystal orientations parallel and perpendicular to a morphological crystal axis [83]. The ^1H-ENDOR results resolved the location of two water molecule ligands and showed their involvement in a hydrogen bonding network connecting the metal ion and the inhibitor. The Mn^{2+} ($S = 5/2$) zero-field splitting (ZFS) parameters D and D/E were determined experimentally and their assignment was based on DFT calculations which included spin-dependent two-electron second order contributions [84]. The D value is a very sensitive measure for any deviation from cubic symmetry and accurate structural models are required. The BP86/CP(PPP), TZVP, SV(P) calculations on a partially constrained cluster model with 81 atoms (see Fig. 8) and the spin–orbit mean field approximation significantly overestimated the experimental ZFS parameters by 50–70% and were then scaled by a factor of 2/3. The calculated high rhombicity E/D for site I was in agreement with its low coordination symmetry in the crystal structure. While the sign of D could not be determined experimentally, the DFT calculations predicted a negative zero-field splitting D and a larger absolute value of |D|. The calculated orientations of the principal axes of the ZFS for Mn^{2+}

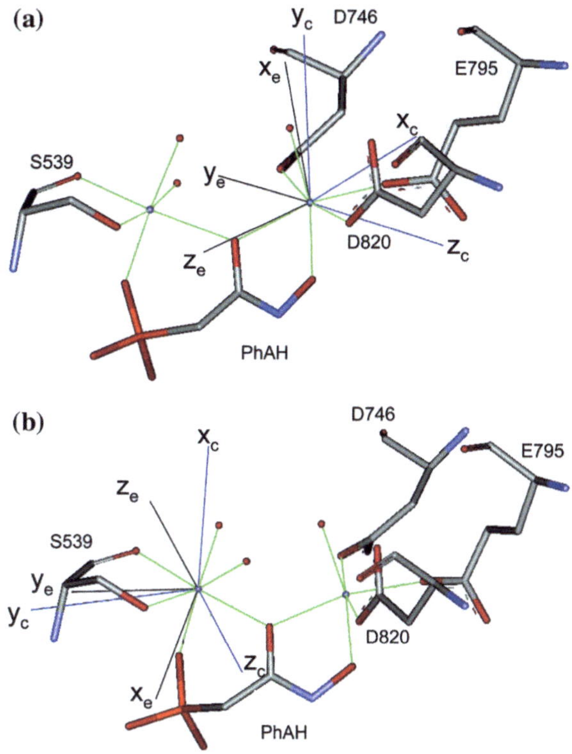

Fig. 8 The orientation of the ZFS principal axes of the Mn-substituted enolase from yeast. Experimental (x_e, y_e, z_e; black lines) and DFT-calculated (x_c, y_c, z_c; blue lines) principal axes of the ZFS tensors of **a** Mn_{A2} with respect to site I in molecule B, dimer I and **b** Mn_{B1} with respect to site II. Reprinted with permission from [83]. Copyright 2007 American Chemical Society

showed that none of the magnetic axes coincided with any of the Mn–ligand bonds which was also found experimentally (see Fig. 8). For site I, the deviations between ZFS splitting principal axes and Mn–ligand bonds were smallest and the calculated and experimental ZFS orientations were similar when using the not-optimized crystal structure in DFT single-point calculations. The deviation from experiment was more pronounced for partially optimized large cluster models with structural constraints.

4.3.4 Protein EPR Parameters from QM/MM Calculations

The function of heme proteins is fine-tuned by the choice of axial ligands, the overall structure and assembly of the binding pocket. Nitric oxide (NO) binding to myoglobin (MbNO) was investigated by performing reactions in the crystal which gave a number of different Fe–NO binding geometries. Advanced electron paramagnetic resonance methods allow a selective investigation of the active site only and can pro-

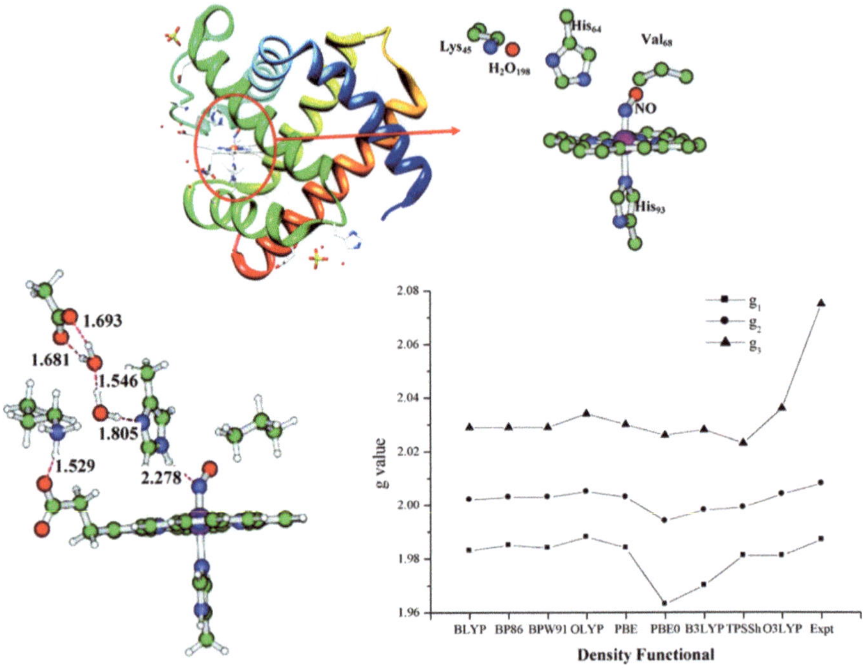

Fig. 9 Protein structure of NO-bound myoglobin and details of the heme binding site (top). Bottom: large structural model ('model 2'; left) which yields calculated EPR parameters in best agreement with experiment. Effect of choice of functionals for calculating **g**-tensor principal values (right). Adjusted and reprinted with permission from [85]. Copyright (2012) American Chemical Society

vide detailed information about slight structural changes during a reaction. MbNO displays two different EPR spectra: one with a rhombic **g**-tensor (the 'R-form') and one with an axial **g**-tensor (the 'A-form'); the structural classification and interconversion of those were not fully resolved. A large gas phase cluster model with 112 atoms including a significant part of the second coordination shell was necessary to replicate QM/MM structural parameters and amino acid protonation states (see Fig. 9). A large variety of density functional calculations was used to calculate EPR parameters from QM/MM structures and assign experimentally measured **g**-tensor principal values and hyperfine coupling constants [85]. The calculated EPR parameters were compared to experimental results for the **g**-tensor and the isotropic hyperfine coupling constants of ^{14}N of the axial NO, ^{14}N of His94 and ^{1}H of His64 [86]. Of all nine evaluated functionals, the hybrid functionals PBE0 and B3LYP significantly underestimated the **g**-values. By comparing calculated **g**-tensor principal values with experiment, several candidate structures were ruled out. The calculated **g**-tensor principal values, however, were only in qualitative agreement with experiment (2.075, 2.008, 1.987; see Fig. 9) but helped to structurally interpret the 'R-form'.

Only for one structural model ('model 2'; see Fig. 9), the calculated nitrogen and proton hyperfine coupling constants were in quantitative agreement with the experimental data. This statement, however, only holds for the GGA functionals. The B3LYP and PBE0 hybrid functionals severely underestimated the isotropic coupling a_{iso} of $^{14}N_{NO}$ and $^{14}N_{His93}$ by ~10 MHz. The best values from BP86 (34, 17, 13 MHz) and PBE (36, 18, 13 MHz) calculations for $^{14}N_{NO}$, $^{14}N_{His93}$ and $^{1}H_{His93}$ agree very well with experimental values of 38, 18 and 9 MHz, respectively [86], and allow an assignment of the rhombic 'R-form' EPR spectral properties to an active site conformation resembling 'model 2'.

5 Conclusion

The advancement of spectroscopic techniques to investigate larger and more complex chemical and biological systems affords detailed insight into the electronic structure and spatial arrangement of ligands and the protein environment. Electron spin resonance spectroscopy, due to the Zeeman splitting in an external magnetic field, yields the electronic **g**-tensor. Its principal values provide initial insight into the type of paramagnetic species (organic radical, spin and oxidation state of a transition metal) but the determination of the **g**-tensor principal axes and their orientations in the molecular structure reveal additional information about their directionality. Very often, the computation of EPR parameters solely focusses on reproducing experimental g-values or an isotropic g_{iso}. Despite recent progress in the development of computational approaches to include electron correlation and spin–orbit coupling as accurately as possible, it is not possible to generate computational results that perfectly match experiment. A systematic deviation between experimental and calculated **g**-tensor principal values is still apparent but its magnitude and whether calculations are over- or underestimating experiment cannot be stated in general. Additional confidence into computational results can be gained when also the orientation of the magnetic axes is compared with experiment. Current computational methods are able to reproduce well the orientation of magnetic principal axes for **g**-tensors, hyperfine tensors and ZFS tensors in a molecular geometry.

Additional interactions between the electron spin and the nuclear spin are a local probe of the distribution of the unpaired electron spin. The isotropic hyperfine coupling interaction yields information about the (non-uniform) spin density distribution; the scalar a_{iso} is a direct probe of the unpaired spin density at the nucleus and of conformational states of ligands and amino acids close to the spin center. The directional hyperfine interactions also probe the orientation and distance of nearby interacting partners, for example protons. This information is complementary to X-ray crystallography. In large systems with multiple paramagnetic sites or large spatial extension, the analysis and interpretation of electron-nuclear interactions is difficult and often not unambiguous. Current computational methods are able to give detailed insight into the unpaired electron spin distribution and produce hyperfine couplings (isotropic and dipolar) in very good agreement with experiment and thus assist the analysis and interpretation of magnetic resonance spectra from frozen solution or crystalline samples.

Carefully and accurately performed quantum chemical calculations of EPR parameters must not only try to reproduce **g**- and hyperfine tensor principal values but also consider information about their absolute and relative orientations. Despite recent progress in treating spin–orbit coupling and electron correlation appropriately, more basic research and development of methods for transition metal containing systems is required to provide reliable answers for complex systems from chemistry and biology.

Acknowledgements Support by the Max Planck Society for the Advancement of Science is acknowledged. Part of this work was also financed by the EU COST Action CM1305 'ECOSTBio' and the EU-program ERDF (European Regional Development Fund) of the German Federal State Saxony-Anhalt within the Research Center of Dynamic Systems (CDS).

References

1. Abragam A, Bleaney B (1970) Electron paramagnetic resonance of transition ions. Oxford University Press, Oxford
2. Mabbs FE, Collinson D (1992) Electron paramagnetic resonance of d transition metal compounds. Elsevier, Amsterdam
3. Boer JL, Mulrooney SB, Hausinger RP (2014) Nickel-dependent metalloenzymes. Arch Biochem Biophys 0:142–152
4. Walsh CT, Orme-Johnson WH (1987) Nickel enzymes. Biochemistry 26:4901–4906
5. Lubitz W, Ogata H, Rudiger O, Reijerse E (2014) Hydrogenases. Chem Rev 114:4081–4148
6. Lubitz W, Reijerse E, van Gastel M (2007) NiFe and FeFe hydrogenases studied by advanced magnetic resonance techniques. Chem Rev 107:4331–4365
7. Siegbahn PEM, Tye JW, Hall MB (2007) Computational studies of NiFe and FeFe hydrogenases. Chem Rev 107:4414–4435
8. Stein M, Lubitz W (2002) Quantum chemical calculations of NiFe hydrogenase. Curr Opin Chem Biol 6:243–249
9. Hille R (2002) Molybdenum and tungsten in biology. Trends Biochem Sci 27:360–367
10. Igarashi RY, Seefeldt LC (2003) Nitrogen fixation: the mechanism of the Mo-dependent nitrogenase. Crit Rev Biochem Mol Biol 38:351–384
11. Metz S, Thiel W (2011) Theoretical studies on the reactivity of molybdenum enzymes. Coord Chem Rev 255:1085–1103
12. Richards RL (1996) Reactions of small molecules at transition metal sites: studies relevant to nitrogenase, an organometallic enzyme. Coord Chem Rev 154:83–97
13. Kisker C, Schindelin H, Rees DC (1997) Molybdenum-cofactor-containing enzymes: structure and mechanism. Annu Rev Biochem 66:233–267
14. Hille R (1996) The mononuclear molybdenum enzymes. Chem Rev 96:2757–2816
15. Zelko IN, Mariani TJ, Folz RJ (2002) Superoxide dismutase multigene family: a comparison of the CuZn–SOD (SOD1), Mn–SOD (SOD2), and EC-SOD (SOD3) gene structures, evolution, and expression. Free Radic Biol Med 33:337–349
16. Nordberg J, Arner ESJ (2001) Reactive oxygen species, antioxidants, and the mammalian thioredoxin system. Free Radic Biol Med 31:1287–1312
17. Dismukes GC, Brimblecombe R, Felton GAN, Pryadun RS, Sheats JE, Spiccia L, Swiegers GF (2009) Development of bioinspired Mn4O4-Cubane water oxidation catalysts: lessons from photosynthesis. Acc Chem Res 42:1935–1943
18. Barber J (2006) Photosystem II: an enzyme of global significance. Biochem Soc Trans 34:619–631

19. Siegbahn PEM (2009) Structures and energetics for O_2 formation in photosystem II. Acc Chem Res 42:1871–1880
20. Griffith JS (1960) Some investigations in the theory of open-shell ions.1 the spin-hamiltonian. Mol Physics 3:79–89
21. Abragam A, Pryce MHL (1951) Theory of the nuclear hyperfine structure of paramagnetic resonance spectra in crystals. Proc R Soc London Ser Math Phys Sci 205:135–153
22. Harriman JE (1978) Theoretical foundations of electron spin resonance. Academic Press, New York
23. Patchkovskii S, Schreckenbach G (2004) Calculation of EPR g-tensors with density functional theory. In: Kaupp M, Bühl M, Malkin VG (eds) Calculation of NMR and EPR parameters
24. Autschbach J (2010) Relativistic effects on magnetic resonance parameters and other properties of inorganic molecules and metal complexes. In: Barysz M, Ishikawa Y (eds) Relativistic methods for chemists. Springer, Netherlands, Amsterdam, pp 521–598
25. Reiher M, Wolf A (2015) Relativistic quantum chemistry. Wiley, Weinheim
26. Dyall KG (1997) Interfacing relativistic and nonrelativistic methods. I. Normalized elimination of the small component in the modified dirac equation. J Chem Phys 106:9618–9626
27. Dyall KG (1994) An exact separation of the spin-free and spin-dependent terms of the dirac—coulomb–breit hamiltonian. J Chem Phys 100:2118–2127
28. Kaupp M, Bühl M, Malkin VG (2004) Calculation of NMR and EPR parameters. Wiley, Weinheim
29. Autschbach J (2010) Relativistic effects on magnetic resonance parameters and other properties of inorganic molecules and metal complexes, pp 521–598
30. Neese F (2017) Quantum chemistry and EPR parameters. In: Harris RK, Wasylishen RL (eds) eMagRes
31. Neese F (2009) Prediction of molecular properties and molecular spectroscopy with density functional theory: from fundamental theory to exchange-coupling. Coord Chem Rev 253:526–563
32. Neese F, Solomon EI (2002) Interpretation and calculation of spin-hamiltonian parameters in transition metal complexes. In: Miller JS, Drillon M (eds) Magnetism: molecules to materials IV. Wiley, Weinheim, pp 345–466
33. te Velde G, Bickelhaupt FM, Baerends EJ, Fonseca Guerra C, van Gisbergen SJA, Snijders JG, Ziegler T (2001) Chemistry with ADF. J Comput Chem 22:931–967
34. Baerends EJ, Ziegler T, Atkins AJ, Autschbach J, Bashford D, Baseggio O, Brces, A, Bickelhaupt FM, Bo C, Boerritger PM, Cavallo L, Daul C, Chong DP, Chulhai DV, Deng L, Dickson RM, Dieterich JM, Ellis DE, van Faassen M, Ghysels A, Giammona A, van Gisbergen SJA, Goez A, Gtz AW, Gusarov S, Harris FE, van den Hoek P, Hu Z, Jacob CR, Jacobsen H, Jensen L, Joubert L, Kaminski JW, van Kessel G, Knig C, Kootstra F, Kovalenko A, Krykunov M, van Lenthe E, McCormack DA, Michalak A, Mitoraj M, Morton SM, Neugebauer J, Nicu VP, Noodleman L, Osinga VP, Patchkovskii S, Pavanello M, Peeples CA, Philipsen PHT, Post D, Pye CC, Ramanantoanina H, Ramos P, Ravenek W, Rodrguez JI, Ros P, Rger R, Schipper PRT, Schlns D, van Schoot H, Schreckenbach G, Seldenthuis JS, Seth M, Snijders JG, Sol ADF (2017) SCM, theoretical chemistry. Vrije Universiteit, Amsterdam, The Netherlands. https://www.scm.com
35. Neese F (2011) The ORCA program system. Wiley Interdiscip Rev Comput Mol Sci 2:73–78
36. Neese F (2017) Software update: the ORCA program system, version 4.0. Wiley Interdiscip Rev Comput Mol Sci 8: e1327
37. Frisch MJ, Trucks GW, Schlegel HB, Scuseria GE, Robb MA, Cheeseman JR, Scalmani G, Barone V, Petersson GA, Nakatsuji H, Li X, Caricato M, Marenich AV, Bloino J, Janesko BG, Gomperts R, Mennucci B, Hratchian HP, Ortiz JV, Izmaylov AF, Sonnenberg JL, Williams, Ding F, Lipparini F, Egidi F, Goings J, Peng B, Petrone A, Henderson T., Ranasinghe D, Zakrzewski VG, Gao J, Rega N, Zheng G, Liang W, Hada M, Ehara M, Toyota K, Fukuda R, Hasegawa J, Ishida M, Nakajima T, Honda Y, Kitao O, Nakai H, Vreven T, Throssell K, Montgomery Jr, JA, Peralta JE, Ogliaro F, Bearpark MJ, Heyd JJ, Brothers EN, Kudin KN, Staroverov VN, Keith TA, Kobayashi R, Normand J, Raghavachari K, Rendell AP, Burant JC,

Iyengar SS, Tomasi J, Cossi M, Millam JM, Klene M, Adamo C, Cammi R, Ochterski JW, Martin RL, Morokuma K, Farkas O, Foresman JB, Fox DJ (2016) Gaussian, Inc., Wallingford CT, US

38. Aidas K, Angeli C, Bak KL, Bakken V, Bast R, Boman L, Christiansen O, Cimiraglia R, Coriani S, Dahle P, Dalskov EK, Ekström U, Enevoldsen T, Eriksen JJ, Ettenhuber P, Fernández B, Ferrighi L, Fliegl H, Frediani L, Hald K, Halkier A, Hättig C, Heiberg H, Helgaker T, Hennum AC, Hettema H, Hjertenæs E, Høst S, Høyvik I-M, Iozzi MF, Jansík B, Jensen HJA, Jonsson D, Jørgensen P, Kauczor J, Kirpekar S, Kjærgaard T, Klopper W, Knecht S, Kobayashi R, Koch H, Kongsted J, Krapp A, Kristensen K, Ligabue A, Lutnæs OB, Melo JI, Mikkelsen KV, Myhre RH, Neiss C, Nielsen CB, Norman P, Olsen J, Olsen JMH, Osted A, Packer MJ, Pawlowski F, Pedersen TB, Provasi PF, Reine S, Rinkevicius Z, Ruden TA, Ruud K, Rybkin VV, Sałek P, Samson CCM, de Merás AS, Saue T, Sauer SPA, Schimmelpfennig B, Sneskov K, Steindal AH, Sylvester-Hvid KO, Taylor PR, Teale AM, Tellgren EI, Tew DP, Thorvaldsen AJ, Thøgersen L, Vahtras O, Watson MA, Wilson DJD, Ziolkowski M, Ågren H (2013) The dalton quantum chemistry program system. Wiley Interdiscip Rev Comput Mol Sci 4:269–284
39. Malkin VG, Malkina OL, Reviakine R, Arbouznikov AV, Kaupp M, Schimmelpfennig B, Malkin I, Helgaker T, Ruud K (2010) Respect
40. Van Yperen-De Deyne A, Pauwels E, Van Speybroeck V, Waroquier M (2012) Accurate spin–orbit and spin-other-orbit contributions to the g-tensor for transition metal containing systems. Phys Chem Chem Phys 14:10690–10704
41. Neese F (2003) Quantum chemical calculations of spectroscopic properties of metalloproteins and model compounds: EPR and Moessbauer properties. Curr Opin Chem Biol 7:125–135
42. Lushington GH, Bundgen P, Grein F (1995) Ab-initio study of molecular g-tensors. Int J Q Chem 55:377–392
43. Lushington GH, Grein F (1996) Complete to second-order Ab initio level calculations of electronic g-tensors. Theor. Chim. Acta 93:259–267
44. Lushington GH (2000) Small closed-form CI expansions for electronic g-tensor calculations. J Phys Chem A 104:2969–2974
45. Jayatilaka D (1998) Electron spin resonance g tensors from general Hartree-Fock calculations. J Chem Phys 108:7587–7594
46. Vahtras O, Minaev B, Agren H (1997) Ab initio calculations of electronic g-factors by means of multi-configuration response theory. Chem Phys Lett 281:186–192
47. Rinkevicius Z, Telyatnyk L, Salek P, Vahtras O, Agren H (2003) Restricted density-functional linear response theory calculations of electronic g-tensors. J Chem Phys 119:10489–10496
48. Schreckenbach G, Ziegler T (1997) Calculation of the g-tensor of electron paramagnetic resonance spectroscopy using gauge-including atomic orbitals and density functional theory. J Phys Chem A 101:3388–3399
49. van Lenthe E, Wormer PES, van der Avoird A (1997) Density functional calculations of molecular g-tensors in the zero-order regular approximation for relativistic effects. J Chem Phys 107:2488–2498
50. Malkina OL, Vaara J, Schimmelpfennig B, Munzarová M, Malkin VG, Kaupp M (2000) Density functional calculations of electronic g-tensors using spin—orbit pseudo-potentials and mean-field all-electron spin—orbit operators. J Am Chem Soc 122:9206–9218
51. Kaupp M, Reviakine R, Malkina OL, Arbuznikov A, Schimmelpfennig B, Malkin VG (2002) Calculation of electronic g-tensors for transition metal complexes using hybrid density functionals and atomic mean field spin–orbit operators. J Comput Chem 23:794–803
52. Cherry PJ, Komorovsky S, Malkin VG, Malkina OL (2017) Calculations of the EPR g-tensor using unrestricted two- and four-component relativistic approaches within the HF and DFT frameworks. Mol Phys 115:75–89
53. Neese F (2001) Prediction of electron paramagnetic resonance g values using coupled perturbed Hartree-Fock and Kohn-Sham theory. J Chem Phys 115:11080–11096
54. Neese F (2001) Configuration interaction calculation of electronic g tensors in transition metal complexes. Int J Q Chem 83:104–114

55. Patchkovskii S, Ziegler T (2001) Calculation of the EPR g-tensors of high-spin radicals with density functional theory. J Phys Chem A 105:5490–5497
56. Neyman KM, Ganyushin DI, Matveev AV, Nasluzov VA (2002) Calculation of electronic g-tensors using a relativistic density functional Douglas-Kroll method. J Phys Chem A 106:5022–5030
57. Tatchen J, Kleinschmidt M, Marian CM (2009) Calculating electron paramagnetic resonance g-matrices for triplet state molecules from multi-reference spin–orbit configuration interaction wave functions. J Chem Phys 130:154106
58. Vancoillie S, Malmqvist P-Å, Pierloot K (2007) Calculation of EPR g tensors for transition-metal complexes based on multi-configurational perturbation Theory (CASPT2). Chem Phys Chem 8:1803–1815
59. Neese F (2007) Analytic derivative calculation of electronic g-tensors based on multi-reference configuration interaction wave functions. Mol Phys 105:2507–2514
60. Gauss J, Kállay M, Neese F (2009) Calculation of electronic g-tensors using coupled cluster theory. J Phys Chem A 113:11541–11549
61. Sayfutyarova ER, Chan GKL (2018) Electron paramagnetic resonance g-tensors from state interaction spin–orbit coupling density matrix renormalization group. J Chem Phys 148
62. Schreckenbach G, Ziegler T (1998) Density functional calculations of NMR chemical shifts and ESR g-tensors. Theor Chem Acc 99:71–82
63. Patchkovskii S, Ziegler T (1999) Prediction of electron paramagnetic resonance g-tensors of transition metal complexes using density functional theory: first applications to some axial d(1)MEX(4) systems. J Chem Phys 111:5730–5740
64. van Lenthe E, van der Avoird A, Wormer PES (1998) Density functional calculations of molecular hyperfine interactions in the zero order regular approximation for relativistic effects. J Chem Phys 108:4783–4796
65. Hess BA, Marian CM, Wahlgren U, Gropen O (1996) A mean-field spin–orbit method applicable to correlated wavefunctions. Chem Phys Lett 251:365–371
66. Neese F (2005) Efficient and accurate approximations to the molecular spin–orbit coupling operator and their use in molecular g-tensor calculations. J Chem Phys 122
67. Rauchfuss TB (2004) Synthesis of transition metal dithiolenes. In: Stiefel EI (ed) Progress in inorganic chemistry: synthesis, properties, and applications, pp 1–54
68. Harrop TC, Mascharak PK (2006) Model complexes of Ni-containing enzymes. Wiley, KGaA, pp 309–329
69. Schilter D, Camara JM, Huynh MT, Hammes-Schiffer S, Rauchfuss TB (2016) Hydrogenase enzymes and their synthetic models: the role of metal hydrides. Chem Rev 116:8693–8749
70. Maki AH, Davison A, Edelstein N, Holm RH (1964) Electron paramagnetic resonance studies of electronic structures of Bis(maleonitriledithiolato) copper(II)-nickel(III)-cobalt(II)-rhodium(II)-complexes. J Am Chem Soc 86:4580–4458X
71. Stein M, van Lenthe E, Baerends EJ, Lubitz W (2001) g- and a-tensor calculations in the zero-order approximation for relativistic effects of Ni complexes Ni(mnt)(2)(-) and Ni(CO)(3)H as model complexes for the active center of NiFe -hydrogenase. J Phys Chem A 105:416–425
72. Huyett JE, Choudhury SB, Eichhorn DM, Bryngelson PA, Maroney MJ, Hoffman BM (1998) Pulsed ENDOR and ESEEM Study of [Bis(maleonitriledithiolato)nickel]: an investigation into the ligand electronic structure. Inorg Chem 37:1361–1367
73. Singh SK, Atanasov M, Neese F (2018) Challenges in multireference perturbation theory for the calculations of the g-tensor of first-row transition-metal complexes. J Chem Theory Comput 14:4662–4677
74. Cosper MM, Neese F, Astashkin AV, Carducci MD, Raitsimring AM, Enemark JH (2005) Determination of the g-tensors and their orientations for cis, trans-(L-N2S2)MoVOX (X = Cl, SCH2Ph) by single-crystal epr spectroscopy and molecular orbital calculations. Inorg Chem 44:1290–1301
75. Fritscher J, Hrobárik P, Kaupp M (2007) Computational studies of electron paramagnetic resonance parameters for paramagnetic molybdenum complexes. 1. Method validation on small and medium-sized systems. J Phys Chem B 111:4616–4629

76. Fritscher J, Hrobárik P, Kaupp M (2007) Computational studies of epr parameters for paramagnetic molybdenum complexes. II. larger MoV systems relevant to molybdenum enzymes. Inorg Chem 46:8146–8161
77. Foerster S, Stein M, Brecht M, Ogata H, Higuchi Y, Lubitz W (2003) Single crystal EPR studies of the reduced active site of [NiFe] hydrogenase from Desulfovibrio vulgaris Miyazaki F. J Am Chem Soc 125:83–93
78. Trofanchuk O, Stein M, Gessner C, Lendzian F, Higuchi Y, Lubitz W (2000) Single crystal EPR studies of the oxidized active site of [NiFe] hydrogenase from Desulfovibrio vulgaris Miyazaki F. J Biol Inorg Chem 5:36–44
79. Kampa M, Pandelia M-E, Lubitz W, van Gastel M, Neese F (2013) A metal-metal bond in the light-induced state of [NiFe] hydrogenases with relevance to hydrogen evolution. J Am Chem Soc 135:3915–3925
80. Gessner C, Trofanchuk O, Kawagoe K, Higuchi Y, Yasuoka N, Lubitz W (1996) Single crystal EPR study of the Ni center of NiFe hydrogenase. Chem Phys Lett 256:518–524
81. van Gastel M, Stein M, Brecht M, Schröder O, Lendzian F, Bittl R, Ogata H, Higuchi Y, Lubitz W (2006) A single-crystal ENDOR and density functional theory study of the oxidized states of the [NiFe] hydrogenase from Desulfovibrio vulgaris Miyazaki F. J Biol Inorg Chem 11:41–51
82. Stein M, Lubitz W (2001) DFT calculations of the electronic structure of the paramagnetic states Ni–A, Ni–B and Ni–C of [NiFe] hydrogenase. Phys Chem Chem Phys 3:2668–2675
83. Carmieli R, Larsen TM, Reed GH, Zein S, Neese F, Goldfarb D (2007) The catalytic Mn2 + sites in the enolase-inhibitor complex: crystallography, single-crystal EPR, and DFT calculations. J Am Chem Soc 129:4240–4252
84. Neese F (2006) Importance of direct spin-spin coupling and spin-flip excitations for the zero-field splittings of transition metal complexes: a case study. J Am Chem Soc 128:10213–10222
85. Sundararajan M, Neese F (2012) Detailed QM/MM study of the electron paramagnetic resonance parameters of nitrosyl myoglobin. J Chem Theory Comput 8:563–574
86. Radoul M, Sundararajan M, Potapov A, Riplinger C, Neese F, Goldfarb D (2010) Revisiting the nitrosyl complex of myoglobin by high-field pulse EPR spectroscopy and quantum mechanical calculations. Phys Chem Chem Phys 12:7276–7289

Non-covalent Interactions in Selected Transition Metal Complexes

Filip Sagan and Mariusz P. Mitoraj

Abstract Chemical bonding in transition metal complexes is typically described by Dewar–Chatt–Duncanson model which separates donation (ligand → metal) and back-donation (metal → ligand) charge transfer processes—these are with no doubt crucial factors which determine a number of properties of metal complexes. This contribution highlights the importance of various non-covalent interactions including untypical homopolar dihydrogen contacts C–H•••H–C in metal complexes. The selected systems are: (1) Zn(II) species containing NTA (nitrotriacetic acid), NTPA (nitrotri-3-propanoic), BPy (2,2′-bipyridyl) ligands, (2) cis-NiL_2–hexane (L–thiourea-based ligand) complex, and (3) hydrogen storage materials $LiNMe_2BH_3$ and $KNMe_2BH_3$. It is shown consistently by various methods and bonding descriptors including for example the charge and energy decomposition scheme ETS-NOCV, Interacting Quantum Atoms (IQA), Reduced Density Gradient (NCI), Quantum Theory of Atoms in Molecules (QTAIM) and NMR spin-spin ^1J(C–H) coupling constants, that London dispersion dominated C–H•••H–C interactions and other more typical hydrogen bonds (e.g. C–H•••N, C–H•••O) driven mostly by electrostatics, are crucial for determination of the structures and stability of the selected metal complexes. Although London dispersion forces are the fundamental factor (~70% of the overall stabilization) contributing to C–H•••H–C interactions, the charge delocalization (outflow of electrons from the σ(C–H) bonds engaged in C–H•••H–C and the accumulation in the interatomic H•••H region) as well as electrostatic terms are also non-negligible (~30%). Remarkably, hydride–hydride interactions B–H•••H–B in $LiNMe_2BH_3$ are found to be repulsive due to dominant destabilizing electrostatic contribution as opposed to stabilizing C–H•••H–C.

F. Sagan · M. P. Mitoraj (✉)
Department of Theoretical Chemistry, Faculty of Chemistry, Jagiellonian University, Gronostajowa 2, 30-387 Kraków, Poland
e-mail: mitoraj@chemia.uj.edu.pl

F. Sagan
e-mail: filip.sagan@doctoral.uj.edu.pl

E. Broclawik et al. (eds.), *Transition Metals in Coordination Environments*, Challenges and Advances in Computational Chemistry and Physics 29,
https://doi.org/10.1007/978-3-030-11714-6_3

1 Introduction

Chemical bonding in transition metal complexes is usually described in terms of Dewar–Chatt–Duncanson model which accounts for donation (ligand → metal) and back-donation (metal → ligand) charge transfer processes [1, 2]. They are crucial for the molecular stability as well as for the determination of various catalytic and spectroscopic properties [1–5]. Very recently, however, more and more attention is paid to London dispersion forces and different types of non-covalent interactions which can, additionally to typical donor/acceptor phenomena, influence the chemistry of transition metal complexes [6, 7]. One shall emphasize that the DFT method together with the recent breakthrough developments of semi-empirical dispersion corrections by Grimme [8, 9] allows for identification and better understanding a number of very important physical phenomena in real materials [6, 7].

Transition metal complexes often contain sterically demanding ligands which traditionally are associated with the source of repulsion—however, the recent topical review by Schreiner and Wagner has suggested the necessity for "...reconsidering steric effects" because in many cases sterically demanding hydrophobic groups, often leading to formation of untypical homopolar C–H•••H–C non-covalent interactions, are truly London dispersion donors which can easily overcompensate Pauli repulsion [6]. Furthermore, Liptrot and Power nicely reviewed the importance of London dispersion forces in organometallic and inorganic complexes [7]. One could also reference that bulky ligands might control directly catalytic activity [10].

Despite significant progress in the identification of untypical homopolar X–H•••H–X non-covalent interactions in various systems including transition metal complexes, one must admit that this subject is still the matter of some discussion in the literature particularly as far as intramolecular X–H•••H–X contacts are taken into account. Here, one can cite the following debates on the stability of biphenyl (planar *versus* bent), [11–16] 2-butene isomers [17–19] or the nature of inter-molecular homopolar B–H•••H–B (hydride-hydride) and other similar contacts in hydrogen storage systems [20–27]. Apart from the above non-covalent interactions and well established now polar dihydrogen X–$H^{\delta+}$•••$^{-\delta}$H–Y (X $\neq$ Y) bonds (named also as proton–hydride contacts) [28–32], there are other non-conventional and quite unintuitive weak interactions, which have been discovered in recent time, such as anion•••π interactions [33–35] or various types of σ/π-hole bonds [36–40]. All these types of weak interactions as well as typical hydrogen bonds, π•••π stacking and others are now crucial forces for various branches of chemistry including transition metal complexes [41–45].

This chapter covers selected contributions which allowed to determine how various non-covalent interactions including controversial homopolar dihydrogen contacts X–H•••H–X and other types of chemical bonds affect the stability and other properties of selected transition metal complexes. In order to shed light on the bonding situations, various methods/descriptors will be applied including the quantum theory of atoms in molecules (QTAIM) [46], interacting quantum atoms (IQA) energy decomposition scheme [47], non-covalent Interactions (NCI) method [48], and (ETS-NOCV) energy decomposition scheme [49].

2 Methods

2.1 ETS-NOCV Charge and Energy Decomposition Scheme

The ETS-NOCV consists of the ETS energy decomposition scheme and NOCV method which allows for partitioning orbital charge delocalization term into chemically meaningful constituents (σ, π, δ, etc.) [49].

In ETS scheme, the total bonding energy (ΔE_{total}) is partitioned into the following contributions:

$$\Delta E_{\text{total}} = \Delta E_{\text{int}} + \Delta E_{\text{dist}} = \Delta E_{\text{elstat}} + \Delta E_{\text{Pauli}} + \Delta E_{\text{orb}} + \Delta E_{\text{disp}} + \Delta E_{\text{dist}}$$

where ΔE_{elstat} covers electrostatic interactions between fragments, ΔE_{Pauli} is responsible for the repulsion between electrons carrying the same spin, ΔE_{orb} reflects stabilizing component due to electron density reorganization upon bond formation, whereas positive ΔE_{dist} describes changes in fragments geometries due to the formation of a bond. Dispersion term ΔE_{disp} is accounted through the semi-empirical Grimme's D3 correction [8, 9].

Natural orbitals for chemical valence (NOCV) denoted as Ψ_i are eigenvectors diagonalizing the deformation density matrix $\Delta P = P - P_0$ (P—molecule's density matrix, P_0—promolecular density matrix):

$$\Delta P C_i = v_i C_i; \qquad \Psi_i = \sum_j^N C_{ij}\lambda_j$$

C_i is a vector of coefficients expanding NOCVs in the basis of fragment orbitals. Pairs (Ψ_{-i}, Ψ_i). NOCVs decompose the differential density ($\Delta\rho$) into chemically meaningful contributions ($\Delta\rho_{\text{k}}$):

$$\Delta\rho = \sum_{k=1}^{N/2} v_k\left[-\psi_{-k}^2 + \psi_k^2\right] = \sum_{k=1}^{N/2} \Delta\rho_k$$

Combining ETS and NOCV, it is possible to determine the energy corresponding to each eigenvalue pair and deformation density channel:

$$\Delta E_{\mathrm{orb}} = \sum_{i=1}^{N/2} v_i\left[-F_{-i,-i}^{TS} + F_{i,i}^{TS}\right] = \sum_{i=1}^{N/2} \Delta E_i^{\mathrm{orb}}$$

where $F_{i,i}^{TS}$ are Kohn–Sham matrix elements for the transition state, as formulated by ETS methodology [49].

2.2 Non-covalent Index (NCI)

Reduced density gradient (s) plots against electronic density (ρ):

$$s = \frac{1}{2(3\pi^2)^{1/3}} \frac{|\nabla\rho|}{\rho^{4/3}}$$

have been shown to be a useful tool to identify the presence of both inter- and intramolecular non-covalent interactions. On said plots, a characteristic spike at the low values on s and ρ indicates the existence of interaction. Sign of the eigenvalues (λi) of the Hessian ($\nabla^2\rho = \lambda 1 + \lambda 2 + \lambda 3$), precisely the sign of $\lambda 2$, indicates whether the interaction is bonding ($\lambda 2 < 0$) or not ($\lambda 2 > 0$). Also, plots of the contour of s colored by the sign of $\lambda 2$ are very informative since they show electronic exchange channels [48].

2.3 Quantum Theory of Atoms in Molecules (QTAIM)

In the QTAIM theory, molecular electron density ρ(r) is divided into atomic basins based on the zero-flux surface criterion [46]. Interacting atoms, each possessing own basin and electronic density maximum in the position of nucleus, are connected by the atomic interaction line (AIL)—a line of local maximum density, called also as a bond path. Due to a diagonalization of a Hessian matrix, one can obtain critical points of $\rho(r)$ [e.g., maximum of $\rho(r)$] including very often used a bond critical point (BCP)—the presence of BCP between atoms is attributed to the existence of bonding interactions. It allows further to create a molecular graph which shows which atoms are bonded to each other. Furthermore, values of electron density (and its laplacian) at BCPs are often discussed in terms of a bond strength.

2.4 Interacting Quantum Atoms (IQA) Energy Decomposition Scheme

IQA method [47] divides the total electronic energy E into atomic (E_{self}^{A}) and diatomic ($E_{\mathrm{int}}^{\mathrm{AB}}$) contributions:

$$E = \sum_{A} E_{\mathrm{self}}^{A} + \frac{1}{2} \sum_{A} \sum_{B \neq A} E_{\mathrm{int}}^{AB}$$

Diatomic part covers all interactions between particles of different atoms A and B: nucleus–nucleus ($V_{\mathrm{nn}}^{\mathrm{AB}}$), nucleus–electron ($V_{\mathrm{ne}}^{\mathrm{AB}}$) electron–nucleus ($V_{\mathrm{en}}^{\mathrm{AB}}$), and electron–electron ($V_{\mathrm{ee}}^{\mathrm{AB}}$):

$$E_{\mathrm{int}}^{\mathrm{AB}} = V_{\mathrm{nn}}^{\mathrm{AB}} + V_{\mathrm{en}}^{\mathrm{AB}} V_{\mathrm{ne}}^{\mathrm{AB}} V_{\mathrm{ee}}^{\mathrm{AB}} = V_{\mathrm{nn}}^{\mathrm{AB}} + V_{\mathrm{en}}^{\mathrm{AB}} + V_{\mathrm{ne}}^{\mathrm{AB}} + V_{\mathrm{eeC}}^{\mathrm{AB}} + V_{\mathrm{eeX}}^{\mathrm{AB}}$$

where $V_{\mathrm{eeC}}^{\mathrm{AB}}$ and $V_{\mathrm{eeX}}^{\mathrm{AB}}$ are Coulomb and exchange contributions of $V_{\mathrm{ee}}^{\mathrm{AB}}$.

2.5 Computational Details

Zinc complexes with NTA and NTPA were optimized at the B3LYP/6-311++G(d,p) level of theory in the Gaussian03.D01 software, with the solvent effects (water environment) taken into account by means of the CPCM model with UAKS cavities [50]. Minima were confirmed by frequency calculations. Topological AIM calculations were performed in the AIMALL package at the same level of theory (B3LYP/6-311++G(d,p)) [50]. ETS-NOCV analyses were performed in the Amsterdam density functional (ADF) package with BP86 functional in DZP basis set for all atoms except metal, where TZP basis was utilized [50].

Other zinc complexes, with the bipyridyl as ligands, were optimized by means of X3LYP/ATZP as implemented in ADF2010 package [51]. COSMO model was used to model a solvent in calculations. Results were confronted with the B97-D/ATZP and MP2/6-311++G(d,p) calculations to ensure their validity. Wavefunctions for QTAIM, IQA (later performed in AIMALL), and NCI (NCIPLOT) analyses were obtained with Gaussian09.B with X3LYP/6-311++G(d,p) [51].

Complexes of nickel were calculated in the geometry taken from the crystal cif files' data [52]. They were also reoptimized to confirm they are correct minima in B97-D3, B3LYP-D3, and M06. ETS-NOCV calculations were performed at BLYP-D3/TZP level of theory [52].

Main group metal complexes were also calculated in the geometry taken from the crystal structures [23]. They were also reoptimized to confirm they are correct minima in PBE-D3, BP86-D3, M06-2X, wB97XD, and MP2. ETS-NOCV calculations were performed at BLYP-D3/TZP level of theory [23].

3 Results and Discussion

It is known experimentally that Zn(II) ion prefers binding with five-member chelate rings (e.g., nitrotriacetic acid, NTA) rather than with six-member units (e.g., nitrotri-3-propanoic acid, NTPA)—the experimentally determined formation constants in water are $\log K = 5.3$ for ZnNTPA complex and $\log K = 10.45$ for ZnNTA, Fig. 1 [53, 54, 50]. Such difference in the stability is intuitively explained by the increased steric crowding (due to C–H•••H–C contacts between the adjacent C–H bonds) in the case of ZnNTPA [53, 54]. In order to shed some light on the origin of different stability between ZnNTPA and ZnNTA, we have performed an in-depth study of bonding situations in both complexes by the charge and energy decomposition method ETS-NOCV as well as by the QTAIM approach [50].

At first stage, the lowest energy conformations have been found (Fig. 1) followed by the computational determination of the formation constants—the higher stability of ZnNTA versus ZnNTPA has been reproduced as indicated by the computed $\log K =$ 5.3 (ZnNTA) versus 3.83 (ZnNTPA) [50]. Then, the complexes have been subjected to in-depth bonding analyses.

According to our QTAIM-based results, non-covalent interactions are only found in the case of ZnNTPA, Table 1. Namely, the classical intramolecular hydrogen bonds are formed between the ligand's C–H bonds and water species: CH35–O3H and CH24–O3H as indicated by the presence of the corresponding bond critical points, Table 1, Fig. 1. The most interestingly, the QTAIM revealed also bond critical points corresponding to non-classical homopolar dihydrogen interactions of the type CH31–H32C and CH27–H28C, Table 1, Fig. 1. It is very interesting result taking into account that lower stability of ZnNTPA versus ZnNTA is intuitively attributed to steric C–H•••H–C clashes [53, 54].

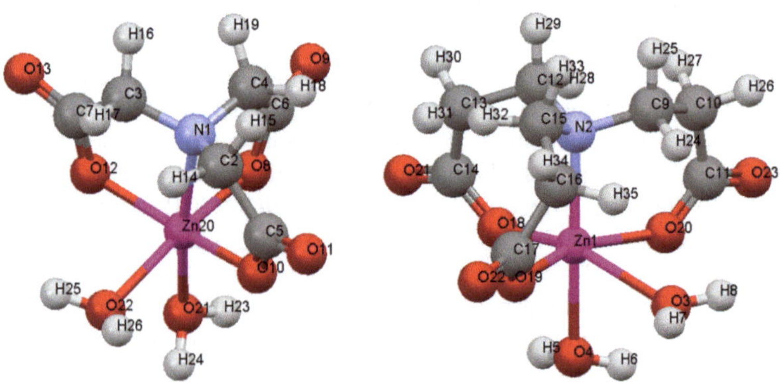

ZnNTA (logK=10.4) **ZnNTPA (logK=5.3)**

Fig. 1 Lowest energy conformers of ZnNTA and ZnNTPA complexes from B3LYP/6-311++G(d,p) in solvent (CPCM/UAKS). Reprinted with permission from [50]. Copyright (2011) American Chemical Society

Table 1 QTAIM results showing bond critical points (between selected atoms), their densities $\rho(r)$, and Laplacian $\nabla^2(r)$ for the lowest energy conformer of the ZnNTPA

Intramolecular bonds			
Atoms	ρ(r) au	∇^2(r) au	BL Å
CH35–O3H	0.0064	0.0198	2.769
CH24–O3H	0.0046	0.0150	2.952
CH31–H32C	0.0137	0.0482	1.962
CH27–H28C	0.0132	0.0499	2.062

*For labels, see Fig. 1

We have further confirmed due to the ETS-NOCV charge and energy decomposition-based study that there are indeed deformation density channels corresponding to homopolar C–H•••H–C interactions (H31•••H32, H27•••H28, H24•••H35) in ZnNTPA complex, Fig. 2. Clearly, an outflow of electron density from the occupied σ(C–H) bonds and the accumulation in the interatomic H•••H region is seen upon fragmentation of ZnNTPA into $CH_2CH_2COO^-$ arm and the rest of the molecule, black line in Fig. 2—such fragmentation allows to extract C–H•••H–C charge delocalizations between the adjacent carboxylic moieties within NTPA. They correspond to the overall stabilization by ca. $\Delta E_{orb} = -4.13$ kcal/mol, Fig. 2.

In order to extract ETS-NOCV-based information on typical dative bonds Zn–N and Zn–O, the following fragmentation patterns are applied L| $Zn(H_2O)_2$ and H_2O| $Zn(H_2O)L$ (where L = NTA, NTPA), respectively. It is found that both the ETS-NOCV-based and QTAIM-based results demonstrate stronger binding of NTPA versus NTA, Table 2. Namely, the calculated interaction energies are $\Delta E_{int} = -743.1$ kcal/mol for ZnNTPA versus $\Delta E_{int} = -732.8$ kcal/mol for ZnNTA. The electron densities of Zn–N BCPs follow the same relation (0.058 a.u. vs. 0.061 a.u.) [50]. It nicely correlates with the calculated Zn–N bonds, lengths which are longer for ZnNTA, by ca. 0.03Å. Furthermore, the vertical water molecule (labeled in Fig. 1 as O4H5H6 in ZnNTA and O21H23H24 in ZnNTPA) is also less efficiently bonded to Zn(II) ion in the case of ZnNTA, by $\Delta E_{int} = 2.04$ kcal/mol, Fig. 3. Furthermore, both types of dative bonds Zn–N and Zn–O are clearly mostly ionic (the dominance of the electrostatic terms ΔE_{elstat}) and the charge delocalization covalent-type channel is significantly less important, Fig. 3 and Table 2. All these results together with the already identified non-covalent interactions apparently would suggest higher stability of ZnNTPA than ZnNTA.

However, the situation changes dramatically when considering an energy penalty/distortion ΔE_{dist} which is required to change the optimal geometries of NTA/NTPA ($\Delta E_{dist\text{-}NTA/NTPA}$) and Zn-fragments ($\Delta E_{dist\text{-}Zn(H_2O)_2}$) to those adopted in the complexes, Table 2. It could be added, that, although significant geometry reorganization of NTPA versus NTA is quite expected (81.7 kcal/mol vs. 38.8 kcal/mol, respectively), please note, that the energy cost required to change the geometry of Zn-fragments is also significant [38.7 kcal/mol (ZnNTPA) vs. 23.4 kcal/mol (ZnNTA)], Table 2. They both contribute to the summarized distortion term ΔE_{dist}

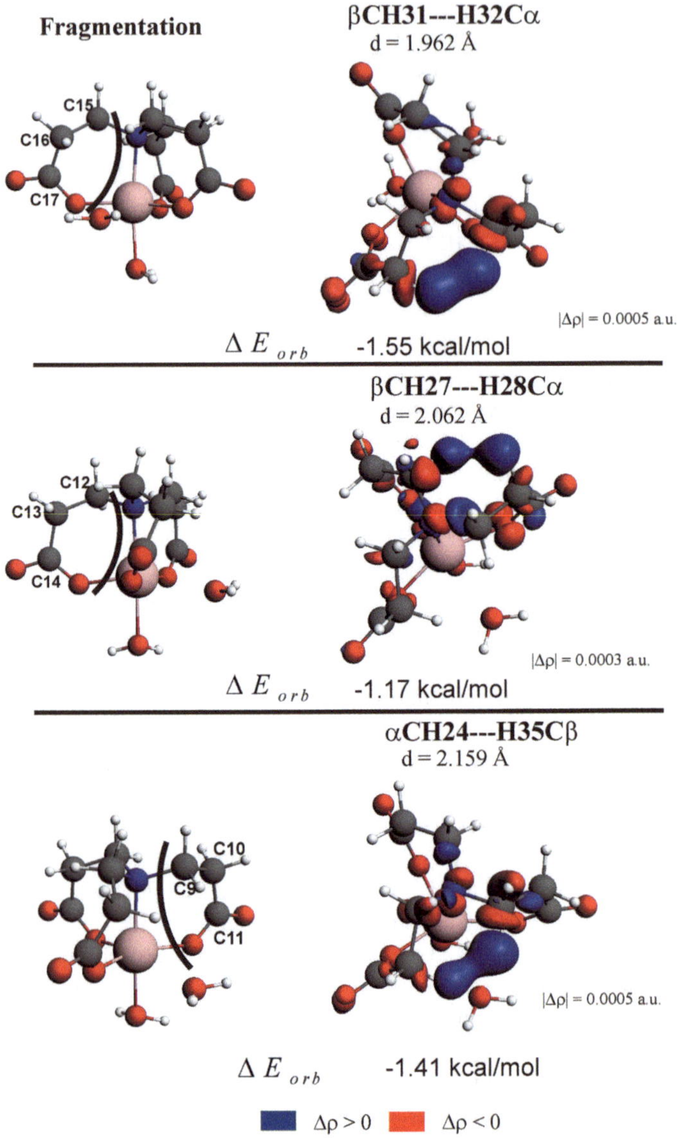

Fig. 2 Selected deformation density contributions together with the corresponding energies characterizing intramolecular interactions. Black circle lines in the structure of ZnNTPA indicate the fragmentation used in the bonding analysis. For clarity of the fragmentation, the carbon atoms of NTPA were labeled according to numbering in Fig. 1. Reprinted with permission from [50]. Copyright (2011) American Chemical Society

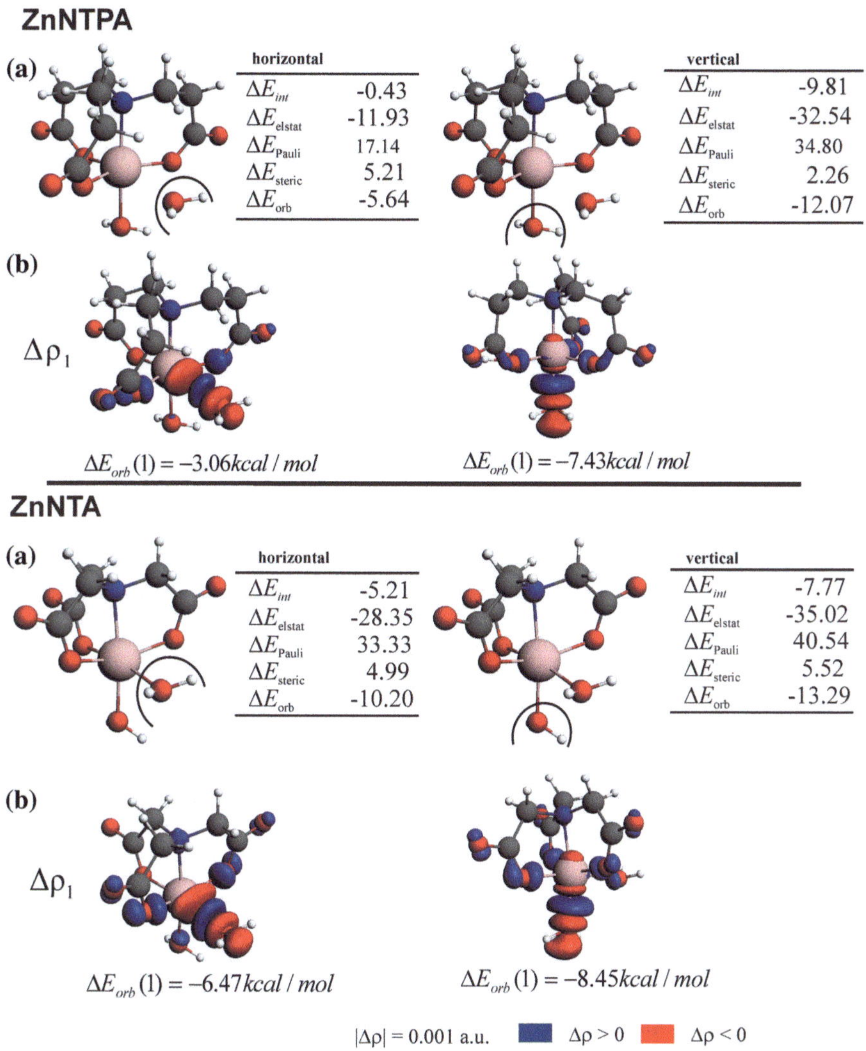

Fig. 3 ETS energy decomposition results describing the interaction of "vertical" (right) and "horizontal" (left) water molecules with the rest of the complex in ZnNTPA and ZnNTA complexes (panel **a**). In addition, the leading deformation densities ($\Delta\rho_1$) together with the corresponding energies ($\Delta E_{orb}[1]$) are presented based on ETS-NOCV method (panel **b**). Reprinted with permission from [50]. Copyright (2011) American Chemical Society

Table 2 ETS energy decomposition results (in kcal/mol) describing the interaction between NTA/NTPA ligands and the $Zn(H_2O)_2$ fragment in ZnNTA/ZnNTPA complexes. Interaction in the gas phase and the solvent is presented. Reprinted with permission from [50]. Copyright (2011) American Chemical Society

ETS results[a, b]	ZnNTA	ZnNTPA
ΔE_{orb}	−183.5	−210.4
ΔE_{Pauli}	140.9	166.8
ΔE_{elstat}	−690.2	−699.5
ΔE_{int}	−732.8	−743.1
$\Delta E_{dist\text{-}Zn(H_2O)_2}$	23.4	38.7
$\Delta E_{dist\text{-}NTA/NTPA}$	38.8	81.7
ΔE_{dist}	62.2	120.4
ΔE_{total}	−670.6	−622.7
ΔE_{int} (solvent)	−148.7	−172.1
$\Delta E_{dist\text{-}Zn(H_2O)_2}$ (solvent)	21.1	37.3
$\Delta E_{dist\text{-}NTA/NTPA}$ (solvent)	20.5	31.8
ΔE_{dist} (solvent)	41.6	69.1
ΔE_{total} (solvent)[c]	−107.1	−103.0

[a]kcal/mol

[b]$\Delta E_{total} = \Delta E_{dist} + \Delta E_{int} = \Delta E_{dist\text{-}Zn(H_2O)_2} + \Delta E_{dist\text{-}NTA/NTPA} + \Delta E_{elstat} + \Delta E_{Pauli} + \Delta E_{orb}$

[c]$\Delta E_{total}(\text{solvent}) = \Delta E_{int}(\text{solvent}) + \Delta E_{dist}(\text{solvent})$

= 120.4 kcal/mol for ZnNTPA and 62.2 kcal/mol for ZnNTA, Table 2. It leads to the overall bonding energy ΔE_{total} in favor of ZnNTA; ΔE_{total} = –670.6 kcal/mol for ZnNTA versus ΔE_{total} = –622.7 kcal/mol for ZnNTPA (gas phase), Table 2. An inclusion of solvent effects alleviates the energy differences obtained in the gas phase, but still the overall bonding energy is more negative by ca. 4.1 kcal/mol in favor of ZnNTA versus ZnNTPA. These results point at crucial role of the ligand's strain energy as well as stronger binding of the horizontal water molecules in explaining larger stability of ZnNTA versus ZnNTPA, despite the formation of non-covalent interactions (C–H•••O and C–H•••H–C) in the latter case. The latter weak interactions are clearly unable to overcome the large destabilization from the distortion term and the Pauli repulsion contribution which are discovered in ZnNTPA.

2,2′-bipyridyl ligand (abbreviated as BPy) is known to form a number of chelate complexes with transition metals—in order to do so, these ligands can adopt cis-configuration in which very close CH•••HC contacts (~2 Å) between 3,3′-hydrogen atoms are enforced [53–55]. The existence of purportedly repulsive CH•••HC contacts in BPy is often applied to rationalize trends in stability constants [55]. The forthcoming paragraphs provides in-depth analyses of weak non-covalent interactions and dative bonds in the complexes $[Zn(BPy)(H_2O)_4]^{2+}$, $[Zn(BPy)_2(H_2O)_2]^{2+}$ and $[Zn(BPy)_3]^{2+}$ (for simplicity, the abbreviations are applied ZnL, ZnL_2, and ZnL_3) [51], Fig. 4 (top). For the first time, four totally different bonding descriptors will be applied—[1] the quantum theory of atoms in molecules (QTAIM) [2, 46], the interacting quantum atoms (IQA) energy decomposition scheme [3, 47], the non-covalent

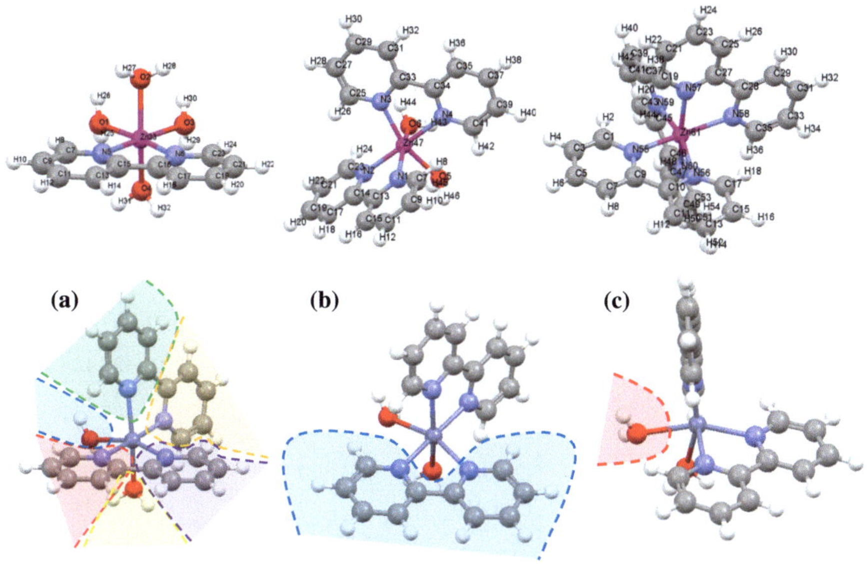

Fig. 4 (Top) Ball and stick representations of ZnL, ZnL_2 and ZnL_3 complexes. (Bottom) ZnL_2 complex together with the fragmentation patterns applied in ETS-NOCV analyses: **a** (7)-pyr, **b** (2)-bpy, and **c** (2)-OH_2 (the numbers denote a number of fragments). Reprinted with permission from [51]. Copyright (2014) American Chemical Society

Interactions (NCI) [48] method, and [4] (ETS-NOCV) [49] energy decomposition scheme.

It is determined, consistently through QTAIM, IQA, NCI, and ETS-NOCV, that the strength of dative bonds Zn–N systematically decreases from ZnL through ZnL_2 to ZnL_3 [51]. The same trend is valid for Zn–O connections [51]. The selected ETS-NOCV-based data in Table 3 clearly demonstrates a drop in Zn–N stabilization from $\Delta E_{orb} = -26.00$ kcal/mol (for ZnL), through $\Delta E_{orb} = -22.46$ kcal/mol (for ZnL_2) up to $\Delta E_{orb} = -18.70$ kcal/mol (for ZnL_3). Zn–O connections appeared to be weaker than Zn–N and similarly, their strength decrease from $\Delta E_{orb} = -11.33$ kcal/mol (for ZnL), to $\Delta E_{orb} = -8.57$ kcal/mol (for ZnL_2), Table 3. The same trend is valid when the overall Zn–N and Zn–O interaction energies are considered, what nicely correlates with the computed elongation of these distances when going from ZnL (e.g. Zn–N: 2.107Å) to ZnL_3 (Zn–N:2.218Å) [51].

Apparently, taking solely dative bonds into consideration would suggest the smallest stability constant for the most crowded ZnL_3 as compared with ZnL—the exactly opposite relation is valid experimentally where the stability decreases in the order $ZnL_3 > ZnL_2 > ZnL$ [55, 56]. The most important and striking findings, discovered consistently from ETS-NOCV, IQA, NCI, and QTAIM methods, are increasing a number of typical CH•••O, CH•••N, and unintuitive CH•••HC non-covalent interactions when going from ZnL to ZnL_3 [51]. Namely, the example ETS-NOCV results in Table 4 demonstrate the charge delocalization channels corresponding to effi-

Table 3 Orbital interaction energies for all Zn–L coordination bonds in Zn^{II} complexes with 2,2′-bipyridyl (L). Reprinted with permission from [51]. Copyright (2014) American Chemical Society

Complex	Atoms	Fragmentation scheme	NOCVs	ΔE_{orb}^{k*a}
ZnL	Zn–(N5,N6)	(2)-Bpy	1,2	−26.00
	Zn–O1	(2)-OH_2	1	−11.78
	Zn–O2	(2)-OH_2	1	−10.88
	Zn–O3	(2)-OH_2	1	−11.78
	Zn–O4	(2)-OH_2	1	−10.88
			Average	*−11.33*
ZnL_2	Zn–(N1,N2)	(2)-Bpy	1,2	−22.42
	Zn–(N3,N4)	(2)-Bpy	1,2	−22.50
			Average	*−22.46*
	Zn–O5	(2)-OH_2	1	−8.62
	Zn–O6	(2)-OH_2	1	−8.53
			Average	*−8.57*
ZnL_3	Zn–(N55,N56)	(2)-Bpy	1,2	−18.71
	Zn–(N57,N58)	(2)-Bpy	1,2	−18.71
	Zn–(N59,N60)	(2)-Bpy	1,2	−18.69
			Average	*−18.70*

[a]Describes a single Zn–L bond. In kcal/mol. For a fragmentation scheme, see Fig. 4

cient stabilization in ZnL_3 from CH•••HC contacts (CH8•••12HC, CH26•••30HC, CH44•••48HC), by ca. ΔE_{orb} = –7.36 kcal/mol, Table 4. Similar interactions are found in ZnL and ZnL_2, and they amount to ΔE_{orb} = –5.52 kcal/mol, ΔE_{orb} = –4.27 kcal/mol, respectively, Table 4. Similar trend in valid for intramolecular CH•••O interactions is shown in Table 4.

It is crucial to highlight that the real space-based IQA energy decomposition scheme consistently revealed exactly the same trends and also identified various non-covalent interactions including the stabilizing CH•••HC, Table 5. Furthermore, the strength of a single CH•••HC bond appeared to increase from ΔE_{int} = –2.5 kcal/mol (ZnL) up to –2.88 kcal/mol for ZnL_3, Table 5. The same trend, but with more efficient overall stabilization which ranges from –13 kcal/mol up to –16 kcal/mol depending on the system), is true for CH•••O and CH•••N, Table 5. It has been further determined that the quantum mechanical exchange-correlation contribution (XC) makes the overall CH•••HC interactions negative (stabilizing), Table 5. This component often correlates well with the orbital interaction term from the ETS-NOCV analysis [14]. The latter allowed to observe that formation of CH•••HC contacts leads to the outflow of electrons from the σ(C–H) bonds engaged in CH•••HC and the accumulation in the interatomic H•••H region. It has been further confirmed by the calculated NMR spin–spin [1] J(C–H) coupling constants which decrease from 177.06 Hz

Table 4 Averaged orbital interaction energies for CH•••O and CH•••HC intramolecular interactions in Zn^{II} complexes with 2,2′-bipyridyl (L). Reprinted with permission from [51]. Copyright (2014) American Chemical Society

Complex	Atoms	Fragmentation scheme	NOCVs[b]	ΔE_{orb}^{k*a}
ZnL	H8–O1, H24–O3	(2)-Bpy (N5,N6)	12	−0.91
			14,16	−0.57
			Sum	*−1.48*
	H14–H18	(7)-pyr	21,23	−4.27
ZnL_2	H42–O5	(2)-Bpy (N1,N2)	15	−1.11
			17,18	−0.57
			Sum	*−1.68*
	H24–O6	(2)-Bpy (N3,N4)	15	−1.13
			17,18	−0.55
			Sum	*−1.68*
	H16–H18, H36–H32	(7)-pyr	23,25,26,27	−5.52
ZnL_3	H8–H12, H26–H30, H44–H48	(7)-pyr	26,27,28,29,30	−7.36

[a]In kcal/mol
[b]A number which lists a given NOCV pair (printed in the output file)

(ZnL) to 173.87 Hz (ZnL_3) in support of an increase in the local CH•••HC stabilization from ZnL to ZnL_3 found from QTAIM, IQA, and ETS-NOCV [51]. These results shed novel light on factors which might determine the relative stability of ZnL_n complexes in the context of orthodox steric-based interpretation of CH•••HC contacts—namely, it is possible that an increase in a number of non-covalent interactions CH•••O, CH•••N, and CH•••HC (and their strength) together with the same trend in π-bonding and electrostatic contributions might overcompensate a decrease in the strength of Zn–N/Zn–O dative bonds resulting accordingly in an enhanced stability of ZnL_3 with respect to both ZnL_2 and ZnL [51].

In the previous paragraphs, we have identified various intramolecular non-covalent interactions in the Zn(II)-based complexes. In the forthcoming sections, inter-molecular CH•••HC and other types of bonds will be discussed in the newly synthesized quasi-tetrahedral nickel complex abbreviated as cis-NiL_2–hexane (L–thiourea-based ligand), Fig. 5 (top) [52]. It is formed by the reaction of N-thiophosphorylated thioureas containing iso-propyl (iPr) units with Ni(II) salt in basic condition crystallized from hexane solvent [52]. It has been also demonstrated that the crystals constituted from fully planar trans-NiL_2 units (without hexane molecules) can be formed provided that other more polar solvents are applied during crystallization [52]. ETS-NOCV calculations allowed to determine the stabilizing character of CH•••HC formed between hexane and the iPr units of cis-NiL_2—the overall CH•••HC interaction energies ΔE_{int} appeared to vary between –2 kcal/mol and

Table 5 Decomposition of two-bodied interaction energies within the IQA framework for all relevant bonds in Zn^{II} complexes with 2,2′-bipyridyl. Reprinted with permission from [51]. Copyright (2014) American Chemical Society

Complex	Atoms	d(A–B) Å	V_{ne}^{AB} *au*	V_{en}^{AB} *au*	V_{nn}^{AB} *au*	V_{XC}^{AB} kcal · mol^{-1}	E_{Int}^{AB} kcal · mol^{-1}
ZnL	CH•••HC	2.06	−0.2472	−0.2472	0.2569	−2.48	−2.5
	CH•••O	2.6	−1.7008	−1.7008	1.6285	−2.64	−14.5
	C–C	1.497	−11.4042	−11.4042	12.7218	−193.43	−97.74
	Zn–N5	2.145	−56.0668	−56.0668	51.8045	−40.14	−409.79
	Zn–N6	2.145	−56.0670	−56.0670	51.8046	−40.14	−409.79
	Zn–Ol	2.157	−61.8194	−61.8194	58.8847	−28.04	−326.47
	N5–N6	2.673	−11.8316	−11.8316	9.7005	−7.03	286.67
ZnL_2	CH•••HC	2.05	−0.2488	−0.2488	0.2581	−2.51	−2.74
	CH•••O	2.502	−1.7656	−1.7656	1.6917	−3.26	−15.73
	CH•••N	2.879	−1.3988	−1.3988	1.2866	−1.64	−11.56
	C–C	1.497	−11.4030	−11.4030	12.7264	−193.41	*−97.15*
	Zn–Nl	2.183	−55.1408	−55.1408	50.9086	−36.73	−390.14
	Zn–N2	2.182	−55.1524	−55.1524	50.9243	−36.76	−391.88
	Zn–06	2.236	−59.7008	−59.7008	56.7947	−22.97	−303.48
	N1–N2	2.676	−11.8153	−11.8153	9.6902	−7.15	285.1
ZnL_3	CH•••HC	2.073	−0.2463	−0.2463	0.2553	−2.36	−2.88
	CH•••N	2.746	−1.4669	−1.4669	0.2553	−2.26	−13.97
	C–C	1.496	−11.4046	−11.4046	0.5106	−193.41	−96.62
	Zn–N2	2.229	−54.0400	−54.0400	49.8607	−32.93	−371.44
	Zn–N23	2.228	−54.0550	−54.0550	49.8744	−32.89	−371.68
	N2–N23	2.690	−11.7953	−11.7953	9.6755	−7.24	284.11

–10 kcal/mol (depending on XC functional containing the Grimme D3 correction) [52]. The main contributor (~73%) is the dispersion term followed by the similarly important (~13.5%) electrostatic and charge delocalization terms, Fig. 5 [52]. As far the latter contribution is concerned, the charge outflow from the σ(C–H) bonds engaged in CH•••HC is clearly visible (Fig. 5b) together with the accumulation in the interatomic H•••H region. It is necessary to point out that our conclusions herein on relative weighs of various contributions to inter-molecular CH•••HC are in accord with the recent topical literature findings [6, 7, 10, 25, 57–64]. For example, recent findings by the groups of Echeverría and Shaik [57, 59] allowed to highlight that, although London dispersion forces are crucial for CH•••HC interactions, other bonding components including charge delocalization term are also important.

The comparison between the dimeric model of [6]-graphanes (six CH_2 units) bonded through CH•••HC with the larger one [65]-graphanes leads to the amplification of dispersion from ~5 kcal/mol up to ~90 kcal/mol, and at the same time, the two-way charge transfers σ(C–H) → σ*(C–H) cover ~15% of the overall stabiliza-

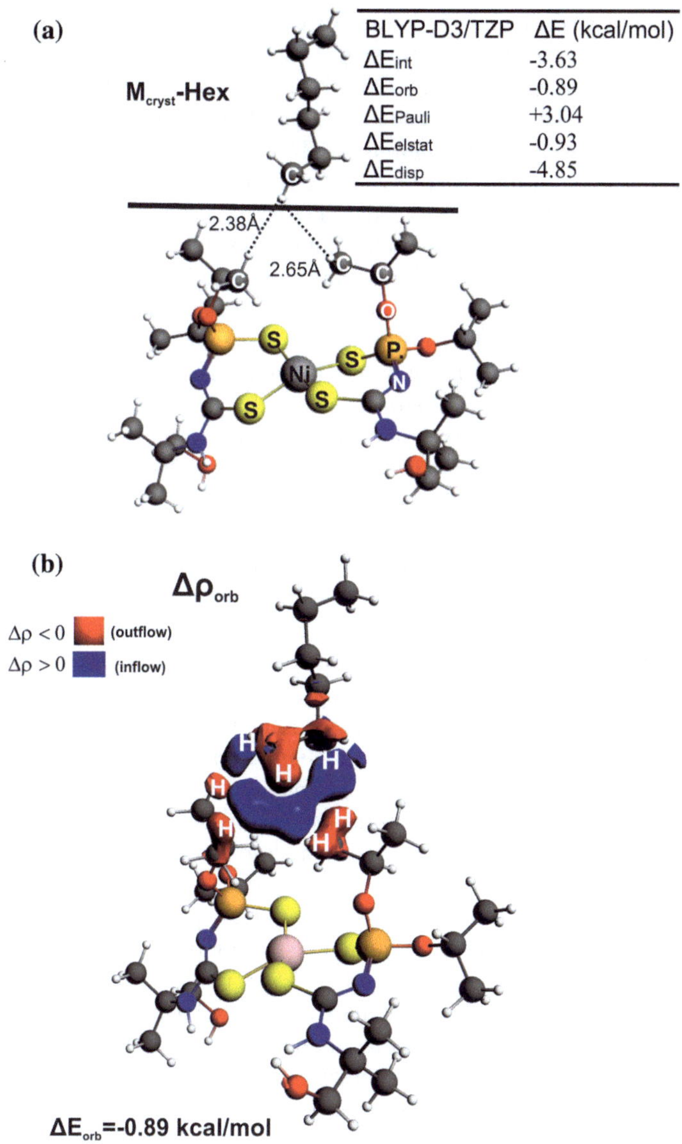

Fig. 5 Structure of Ni(II) complex cis-NiL_2 containing dihydrogen bonding CH•••HC formed between hexane and iPr units [52]. The synthon (M_{cryst}-Hex) extracted from the crystal structure [52] is depicted. Black line indicates the fragmentation applied in ETS-NOCV analyses. The contour value is 0.001a.u

tion which can reach the regime of typical covalently bonded species [59]. Echeverria has studied a series of homopolar dihydrogen interactions and also concluded on the importance of dispersion forces [24, 26, 27]. Furthermore, we have confirmed the stability of the M_{cryst}-Hex adduct bonded through CH•••HC by ab initio Born–Oppenheimer molecular dynamics simulations (DFT/BLYP-D3/TZP)—hexane was found to dynamically glide in the proximity of iPr ligands, and no spontaneous drifting apart was observed [52]. Furthermore, the calculations have shown that the sterically crowded cis-isomer exhibiting quasi-tetrahedral geometry is more stable than the corresponding planar trans-conformation at both energy and free energy levels, Fig. 6a. It is very striking result in the light of intuitively expectable steric repulsion between closely located iPr units in the case of cis-NiL_2, Fig. 6a. Substitution of bulky iPr units by smaller Me groups leads not only to the planar conformation of cis-NiL_2, but also the trans-isomer is now thermodynamically preferred [52], Fig. 6c. Therefore, the tetrahedral geometry of cis-NiL_2 originates from the existence of numerous cooperative intramolecular non-covalent interactions: CH•••HC, CH•••S CH•••Ni, Fig. 6b. In turn, it is related to the bulky iPr units that are close to each other in the cis-NiL_2. It is very beautiful example showing how two quite bulky groups being close to each other, classically considered as the source of steric repulsion, lead not only to overall stabilization, but also to the determination of the complex geometry. These data are perfectly in accord with recent topical findings on the importance of CH•••HC interactions and London dispersion forces in various branches of chemistry [6, 7, 10, 25, 52, 57–64]. It must be emphasized that dispersion contribution has been also recently recognized as a crucial factor (in addition to well-established charge transfer term) for agostic interactions (C–H•••metal) due to elegant and accurate energy decomposition DLPNO-CCSD(T) implemented in the Orca program [64].

Ammonia borane, named also as borazane, is considered nowadays as one of the most promising hydrogen storage materials predominantly due to significant hydrogen content (19.6%) as well as high melting point (104 °C). The latter property is attributed in the literature to the existence of polar (proton–hydride) dihydrogen bonds $N–H^{\delta+}$•••$^{-\delta}H–B$ between AB monomers. These types of interactions are crucial for hydrogen storage materials [28, 29, 31]. Very recently, McGrady and coworkers have published a series of high-quality papers which demonstrate thepreparation of various hydrogen storage materials including $LiN(CH_3)_2BH_3$ and $KN(CH_3)_2BH_3$ in which untypical hydride–hydride interactions $B–H^{\delta-}$•••$^{-\delta}H–B$ are observed from the QTAIM results [22, 66, 67]. These are very interesting suggestions since hydrogen atoms involved in such homopolar contacts $B–H^{\delta-}$•••$^{-\delta}H–B$ carry negative partial charges, what intuitively shall lead to overall repulsion due to destabilizing electrostatic contribution. In order to shed some light on the role of $B–H^{\delta-}$•••$^{-\delta}H–B$ and other types of chemical bonds in $LiN(CH_3)_2BH_3$ and $KN(CH_3)_2BH_3$, we have performed a comprehensive in-depth study of bonding situation based on ETS-NOCV, IQA, NCI methods, and molecular electrostatic potentials [23].

ETS-NOCV method allowed to determine that the major inter-molecular bonding in $LiN(CH_3)_2BH_3$ stems from B–H•••Li contacts—it is dominated by the electrostatics which covers 55% (ΔE_{elstat} = –35.56 kcal/mol) of the total stabilization, followed by the orbital interaction (ΔE_{orb} = –18.17 kcal/mol) and dispersion (17%, ΔE_{disp}

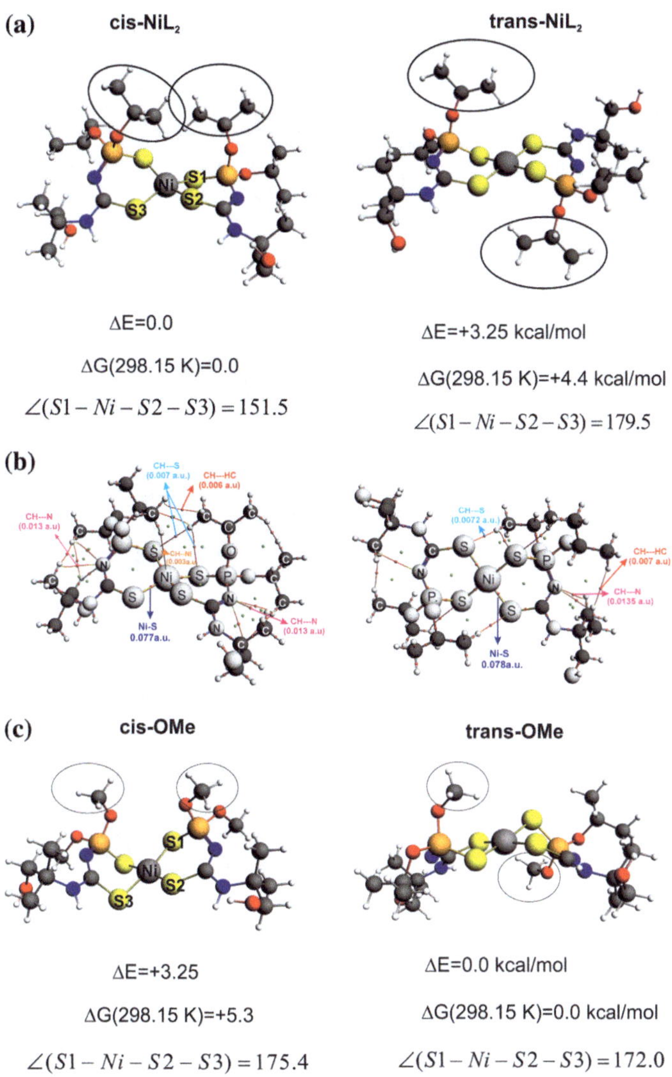

Fig. 6 Relative energies (and free energies) of cis-NiL_2 versus trans-NiL_2 together with dihedral angles which measure degree of planarity (part **a**). In part **b**, the QTAIM molecular graphs are depicted demonstrating formation of various non-covalent interactions. Part **c** depicts the relative energies (and free energies) of the models where the bulky OiPr units are replaced by OMe

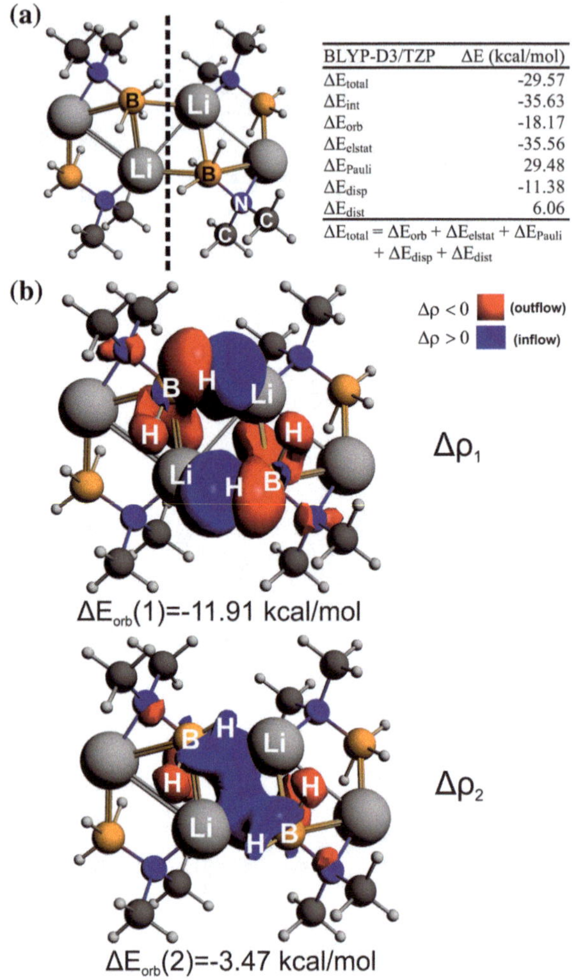

BLYP-D3/TZP	ΔE (kcal/mol)
ΔE_{total}	-29.57
ΔE_{int}	-35.63
ΔE_{orb}	-18.17
ΔE_{elstat}	-35.56
ΔE_{Pauli}	29.48
ΔE_{disp}	-11.38
ΔE_{dist}	6.06

$\Delta E_{total} = \Delta E_{orb} + \Delta E_{elstat} + \Delta E_{Pauli} + \Delta E_{disp} + \Delta E_{dist}$

Fig. 7 Optimized tetrameric cluster model of $LiN(CH_3)_2BH_3$ with fragmentation pattern and ETS energy decomposition results (part **a**) along with the two most important deformation density contributions describing B–H•••Li interactions. Adopted from [23]

= −11.38 kcal/mol) terms, Fig. 7. Electrostatics-based bonding is in line with the molecular electrostatic potential of the monomer—the borane moiety is negatively charged, while lithium cation is electrophilic, Fig. 8a. The bottom of Fig. 8 clearly identifies the ancillary intra-molecular B–H•••Li charge transfer in the monomers constituting the crystal.

The two NOCV channels which mostly (~85%) contribute to ΔE_{orb} describe the outflow of electron density from σ(B–H) orbitals and inflow into the proximity of Li^+ ions, what leads additionally to the charge accumulation into the BH•••HB bay region, $\Delta\rho_2$, Fig. 7. It is fully consistent with the QTAIM result of McGrady et al. [66] where bond critical points corresponding to BH•••HB are discovered. However, we have determined that removing the two monomers not engaged in B–H•••H–B interactions results in the formation of a system with the positive overall interaction energy

Fig. 8 Monomer of $LiN(CH_3)_2BH_3$ with its Molecular Electrostatic Potential in a.u. units (part **a**) and two most important deformation density channels depicting B–N bond and BH_3•••Li interacion (part **b**). Adopted from [23]

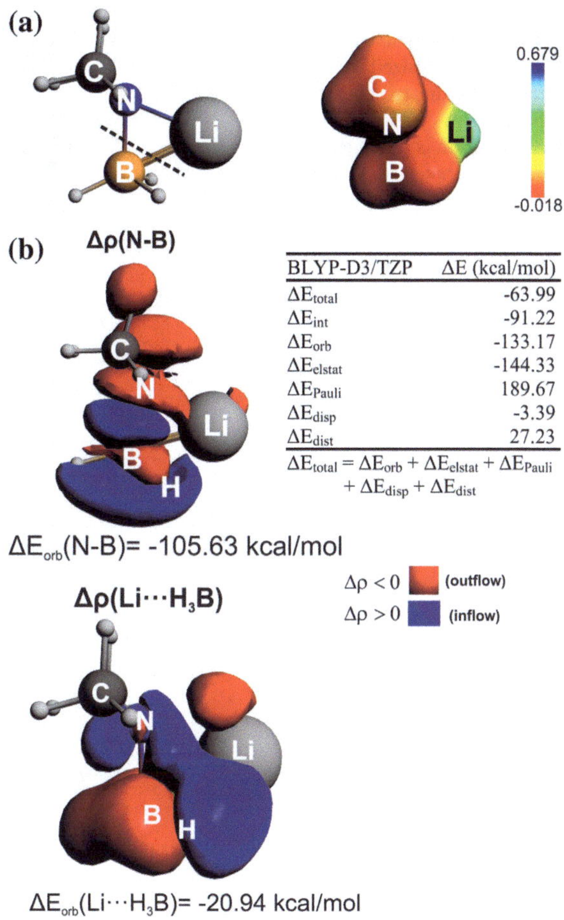

BLYP-D3/TZP	ΔE (kcal/mol)
ΔE_{total}	-63.99
ΔE_{int}	-91.22
ΔE_{orb}	-133.17
ΔE_{elstat}	-144.33
ΔE_{Pauli}	189.67
ΔE_{disp}	-3.39
ΔE_{dist}	27.23

$\Delta E_{total} = \Delta E_{orb} + \Delta E_{elstat} + \Delta E_{Pauli} + \Delta E_{disp} + \Delta E_{dist}$

ΔE_{total} = +4.9 kcal/mol, Fig. 9. Although some stabilization from the orbital overlapping exists (ΔE_{orb} = –1.86 kcal/mol), as well as from dispersion effects (ΔE_{disp} = –2.30 kcal/mol), in line with the QTAIM data [66], significant electrostatic and Pauli repulsion (ΔE_{elstat} = +4.90 kcal/mol, ΔE_{Pauli} = +4.16 kcal/mol), overcompensate the stabilizing effect leading to positive (destabilizing) ΔE_{total} = +4.9 kcal/mol, Fig. 9. The same conclusion is reached by us when considering the point charges (which mimic the Li ions) [23]. Moreover, no stabile minimum featuring solely B–H•••H–B interactions have been found upon geometry optimization. To this end, these ETS-NOCV-based data points at rather destabilizing nature of B–H•••H–B interactions in this system, contrary to the analogous C–H•••H–C contacts which are found to be significantly stabilizing in $LiN(CH_3)_2BH_3$ and $KN(CH_3)_2BH_3$, Figs. 10, 11 [23]. It is determined herein that the overall C–H•••H–C interaction energy in $LiN(CH_3)_2BH_3$ is ΔE_{int} = –4.34 kcal/mol and ΔE_{int} = –17.45 kcal/mol for $KN(CH_3)_2BH_3$ (Figs. 10, 11), which is quite comparable to typical hydrogen bonds [e.g., ΔE_{int} for water dimer

Fig. 9 Dimer of $LiN(CH_3)_2BH_3$ (part **b**) from the optimized tetramer as indicated in part **a**. ETS energy decomposition is presented in part **b**. Adopted from [23]

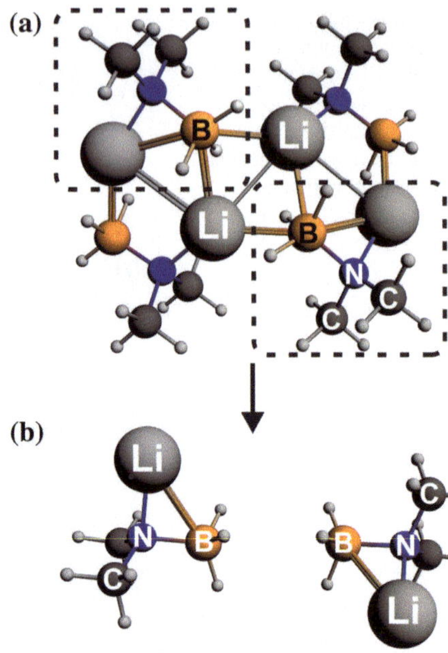

BLYP-D3/TZP	ΔE (kcal/mol)
ΔE_{total}	4.90
ΔE_{orb}	-1.86
ΔE_{elstat}	4.90
ΔE_{Pauli}	4.16
ΔE_{disp}	-2.30

$$\Delta E_{total} = \Delta E_{orb} + \Delta E_{elstat} + \Delta E_{Pauli} + \Delta E_{disp}$$

is ~ –5 kcal/mol, for adenine–thymine base pair is ~ –12 kcal/mol, and for FH•••HLi is ~ –13 kcal/mol] [44].

Further studies of the tetrameric model by means of IQA method were carried out in order to further characterize Li•••HB CH•••HC and BH•••HB interactions, Table 6. Expectedly, the strongest bond appeared to be Li•••HB with E_{int}(Li•••H–B) = –98.8 kcal/mol. In line with the ETS-NOCV results, homopolar CH•••HC is weakly stabilizing, whereas the similar B–H•••H–B interactions are indeed destabilizing as indicated by the large positive value of E_{int}(BH•••HB) =+49.4 kcal/mol, Table 6. It is due to electrostatic electron–electron repulsion term, Table 6, which is in line with the ETS-NOCV-based results (Fig. 9). These outcomes [23] pointing at destabilizing nature of B–H•••H–B in $LiNMe_2BH_3$ are in accord with the numerous experimental papers [20, 21, 68, 69, 65]. The excellent review by McGrady et al. [22] on possible role of stabilizing hydride–hydride interactions in hydrogen storage materials has been recently published. Therefore, definitely more works (from both theoretical

(a)

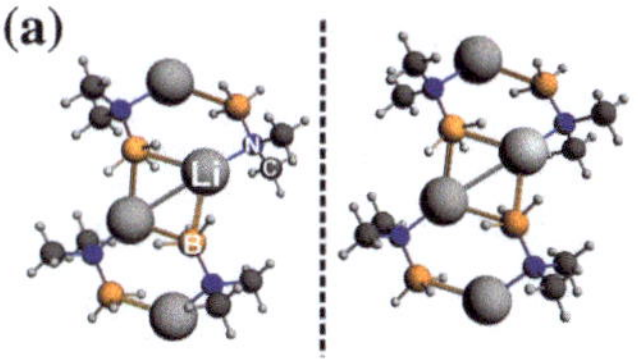

BLYP-D3/TZP	ΔE (kcal/mol)
ΔE_{total}	-4.34
ΔE_{orb}	-0.94
ΔE_{elstat}	-1.68
ΔE_{Pauli}	2.91
ΔE_{disp}	-4.63

$\Delta E_{total} = \Delta E_{orb} + \Delta E_{elstat} + \Delta E_{Pauli} + \Delta E_{disp}$

(b)

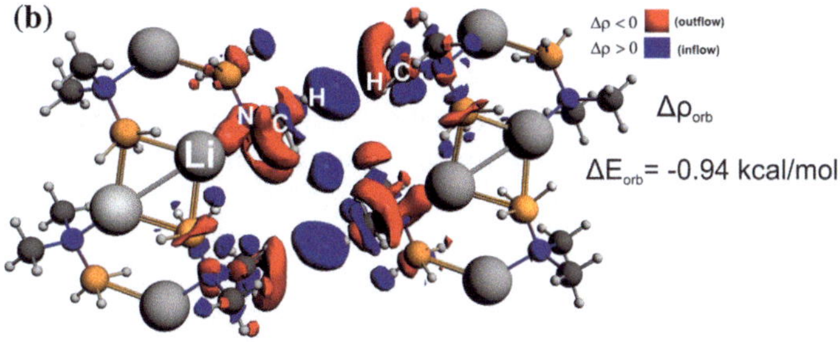

Fig. 10 Cluster model containing eight monomers of $LiN(CH_3)_2BH_3$ along with energy decomposition results describing CH•••HC interactions between the two selected fragments (marked by black line), part **a**. In part **b**, the overall deformation density Δ_{orb} is depicted together with the corresponding stabilization ΔE_{orb}. Adopted from [23]. The contour value is 0.001a.u

(a)

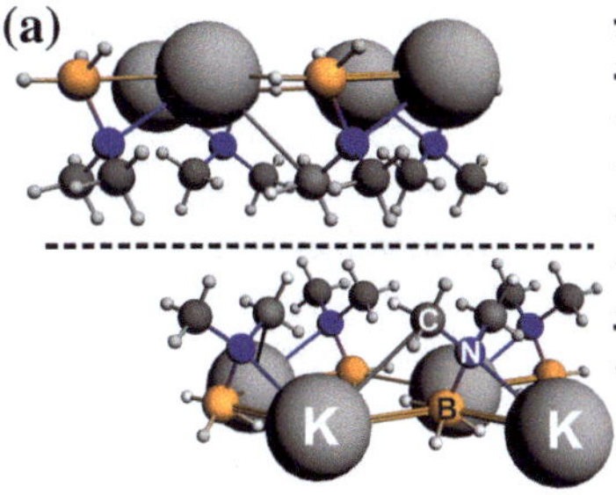

BLYP-D3/TZP	ΔE (kcal/mol)
ΔE_{total}	-17.45
ΔE_{orb}	-3.50
ΔE_{elstat}	-4.68
ΔE_{Pauli}	9.51
ΔE_{disp}	-18.78

$\Delta E_{total} = \Delta E_{orb} + \Delta E_{elstat} + \Delta E_{Pauli} + \Delta E_{disp}$

(b)

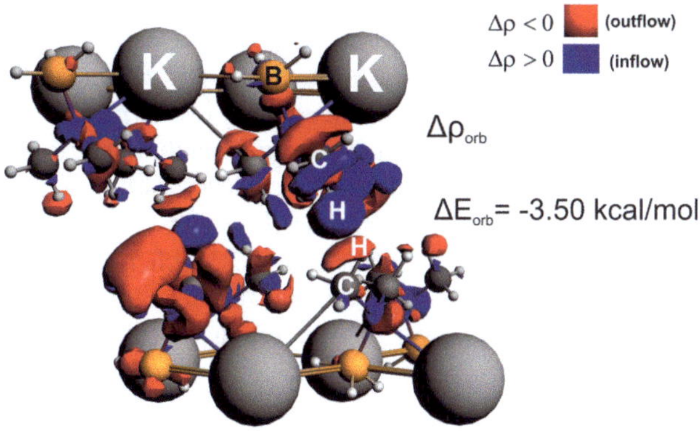

Fig. 11 Cluster model containing eight monomers of $KN(CH_3)_2BH_3$ along with energy decomposition results describing CH•••HC interactions between the two selected fragments (marked by black line), part **a**. In part **b**, the overall deformation density Δ_{orb} is depicted together with the corresponding stabilization ΔE_{orb}. Adopted from [23]. The contour value is 0.001a.u

Table 6 IQA energy decomposition results (in kcal/mol) describing the two atomic interactions in $LiNMe_2BH_3$. Adopted from [23]

IQA(X•••Y)	V_{ne}^{AB}	V_{en}^{AB}	V_{nn}^{AB}	V_{ee}^{AB}	V_{eeC}^{AB}	V_{eeX}^{AB}	E_{int}^{AB}
Li•••H–B	−805.9	−338.4	480.1	565.4	568.2	−2.8	−98.8
CH•••HC	−140.6	−138.2	130.8	147.2	148.6	−1.4	−0.80
BH•••HB	−198.9	−198.6	119.7	327.2	330.0	−2.7	+49.4

and experimental laboratories) are needed to identify and fully unveil the nature of different types of X–H•••H–X contacts in various systems.

4 Conclusions

It has been demonstrated in this chapter that non-covalent interactions including untypical homopolar C–H•••H–C, despite being relatively weak (as compared to typical dative or covalent bonds), might play very important role in transition metal systems. Zn(II) complexes with various ligands NTA, NTPA, and BPy have been stabilized not only by typical electrostatically dominated dative–covalent bonds (e.g. Zn–N, Zn–O), but additionally through a number of typical hydrogen bonds CH•••O, CH•••N and predominantly unintuitive CH•••HC interactions—the latter have been shown to be stabilizing as opposed to traditional steric repulsion-based interpretation [23, 50, 51, 52]. We have determined that dispersion dominated CH•••HC can be as strong as typical hydrogen bonds [23, 50, 51, 52]. Although London dispersion forces are the prevailing factor, the charge delocalization contribution (outflow of electrons from the σ(C–H) bonds engaged in CH•••HC and the accumulation in the interatomic H•••H region) and electrostatic term are also non-negligible [23, 50, 51, 52]. Interestingly, similar to CH•••HC, hydride–hydride interactions BH•••HB in $LiNMe_2BH_3$ are found to be repulsive [23]. We have further proven that the two bulky alkyl groups being close to each other in Ni(II) complex, classically considered as the source of steric repulsion, lead not only to overall stabilization (due to the formation of multitude non-covalent interactions including CH•••HC), but also to determination of the complex geometry [52]. These results perfectly fit very recent topical findings which highlight the crucial role of non-covalent interactions including London dispersion forces in various branches of chemistry including transition metal complexes [6, 7]. Although significant progress has been made recently in terms of identification of non-covalent interactions in real materials [6, 7], there are still many known systems where the importance of London dispersion forces has not been yet recognized as nicely emphasized by Liptrot and Power [7].

Acknowledgements DFT calculations were partially performed using the PL-Grid Infrastructure and resources provided by the ACC Cyfronet AGH (Cracow, Poland). M. P. M. acknowledges the financial support of the Polish National Science Center within the Sonata Bis Project 2017/26/E/ST4/00104.

References

1. (a) Dewar, MJS (1951) Bull Soc Chim 18:C71–C79 (b) Chatt J, Duncanson JA (1953) J Chem Soc 3:2939–2943 (c) Frenking G, Shaik S (eds) (2014) The chemical bond. Fundamental aspects of chemical bonding (part 1), chemical bonding across the periodic table (part 2), Wiley-VCH, Weinheim
2. Zhao L, von Hopffgarten M, Andrada DM, Frenking G (2018) WIREs Comput Mol Sci 8:e1345. https://doi.org/10.1002/wcms.1345
3. Brocławik E, Załucka J, Kozyra P, Mitoraj MP, Datka J (2011) Catal Today 169(1):45–51
4. Broclawik E, Załucka J, Kozyra P, Mitoraj MP, Datka J (2010) J Phys Chem C 114(21):9808–9816
5. Rejmak P, Mitoraj MP, Broclawik E (2010) Phys Chem Chem Phys 12(10):2321–2330
6. Wagner JP, Schreiner PR (2015) Angew Chem Int Ed 54:14
7. Liptrot DJ, Power PP (2017) Nat Rev Chem 1:0004
8. Grimme S, Antony J, Ehrlich S, Krieg H (2010) J Chem Phys 132:154104
9. Grimme S, Ehrlich S, Goerigk L (2011) J Comput Chem 32:1456
10. Wolters LP, Koekkoek R, Bickelhaupt FM (2015) ACS Catal 5:5766
11. Matta CF, Hernández-Trujillo J, Tang TH, Bader RFW (2003) Chem Eur J 9:1940
12. Poater J, Solà M, Bickelhaupt FM (2006) Chem Eur J 12:2889
13. Bader RFW (2006) Chem Eur J 12:2896
14. Pendás AM, Francisco E, Blanco MA, Gatti C (2007) Chem Eur J 13:9362
15. Eskandari K, Alsenoy CV (2014) J Comput Chem 35:1883
16. Cukrowski I (2015) Comput Theor Chem 1066:62
17. Weinhold F, Schleyer PR, McKee WC (2014) J Comput Chem 35:1499
18. Matta CF, Sadjadi SA, Braden DA, Frenking G (2016) J Comput Chem 37:143
19. Cukrowski I, Sagan F, Mitoraj MP (2016) J Comput Chem 37:2783
20. Ravindran P, Vajeeston P, Vidya R, Kjekshus A, Fjellvåg H (2002) Phys Rev Lett 89:106403
21. Schouwink P, Hagemann H, Embs JP, Anna VD, Černy R (2015) J Phys: Condens Matter 27:265403
22. Wolstenholme DJ, Dobson JL, McGrady GS (2015) Dalton Trans 44:9718 and references therein
23. Sagan F, Filas R, Mitoraj MP (2016) Crystals 6:28
24. Echeverría J, Aullón G, Alvarez S (2017) Dalton Trans 46:2844
25. Yourdkhani S, Jabłoński M, Echeverría J (2017) Phys Chem Chem Phys 19:28044
26. Echeverría J (2017) Cryst Growth Des 17:2097
27. Echeverría J, Aullón G, Alvarez S (2017) Int J Quantum Chem 117:e25432
28. Custelcean R, Jackson JE (2001) Chem Rev 101:1963
29. Mitoraj MP (2011) J Phys Chem A 115:14708
30. Grabowski SJ, Ruipérez F (2016) Phys Chem Chem Phys 18:12810
31. Sagan F, Piękoś Ł, Andrzejak M, Mitoraj MP (2015) Chem Eur J 21:15299
32. Belkova NV, Epstein LM, Filippov OA, Shubina ES (2016) Chem Rev 116:8545
33. Gamez P, Mooibroek TJ, Teat SJ, Reedijk J (2007) Acc Chem Res 40:435
34. Frontera A, Gamez P, Mascal M, Mooibroek TJ, Reedijk J (2011) Angew Chem Int Ed. 50:9564
35. Safin DA, Pialat A, Leitch AA, Tumanov NA, Korobkov I, Filinchuk Y, Brusso JL, Murugesu M (2015) Chem Commun 51:9547

36. Politzer P, Murray JS, Clark T (2010) Phys Chem Chem Phys 12:7748
37. Scheiner S (2013) Acc Chem Res 46:280
38. Bauzá A, Mooibroek T, Frontera A (2016) Chem Rec 16:473
39. Grabowski SJ, Sokalski WA (2017) Phys Chem Chem Phys 18:1569
40. Gleiter R, Haberhauer G, Werz DB, Rominger F, Bleiholder C (2018) Chem Rev 118:2010
41. Grabowski SJ (2006) Hydrogen bonding—new insights. Springer, Dordrecht
42. Scheiner S (1997) Hydrogen bonding a theoretical perspective. Oxford University Press Inc., New York
43. Grabowski SJ (2016) Crystals 6:59
44. Grabowski SJ (2011) Chem Rev 111:2597 and references therein
45. Petrović P, Djukic JP, Hansen A, Bannwarth C, Grimme S (2016) Non-covalent stabilization in transition metal coordination and organometallic complexes, Editors(s): Abel M. MaharramovKamran T. MahmudovMaximilian N. KopylovichArmando J. L. Pombeiro, Wiley, Print ISBN:9781119109891
46. Bader RFW (1990) Atoms in molecules: a quantum theory. Oxford University Press, Oxford
47. Blanco MA, Pendás AM, Francisco E (2005) J Chem Theory Comput 1:1096
48. Johnson ER, Keinan S, Mori-Sánchez P, Contreras-García J, Cohen AJ, Yang W (2010) J Am Chem Soc 132:6498
49. Mitoraj M, Michalak A, Ziegler T (2009) J Chem Theory Comput 5:962
50. Cukrowski I, Govender KK, Mitoraj MP, Srebro M (2011) J Phys Chem A 115:12746
51. Cukrowski I, de Lange JH, Mitoraj MP (2014) J Phys Chem A 118:623
52. Safin DA, Babashkina MG, Robeyns K, Mitoraj MP, Kubisiak P, Garcia Y (2015) Chem Eur J 21:16679
53. Hancock RD, de Sousa AS, Walton GB, Reibenspies JH (2007) Inorg Chem 46:4749
54. Hambley TW (1986) J Chem Soc Dalton Trans 565
55. Hancock RD, Martell AE (1989) Chem Rev 89:1875
56. Smith RM, Martell AE (eds) (2004) NIST standard reference database 46. NIST critically selected stability constants of metal complexes database; Version 8.0; US Department of Commerce, National Institute of Standards and Technology: Gaithersburg, MD
57. Echeverría J, Aullón G, Danovich D, Shaik S, Alvarez S (2011) Nat Chem 3:323
58. Krapp A, Frenking G, Uggerud E (2008) Chem Eur J 14:4028
59. Wang C, Mo Y, Wagner JP, Schreiner PR, Jemmis ED, Danovich D, Shaik S (2015) J Chem Theory Comput 11:1621
60. Jahiruddin S, Mandal N, Datta A (2018) Chem Phys Chem 19:67
61. Dutta B, Pratik SM, Jana S, Sinha C, Datta A, Hedayetullah Mir M (2018) ChemistrySelect 3:4289
62. Mandal N, Pratik SM, Datta A (2017) J Phys Chem B 121:825
63. de Almeida LR, Carvalho Jr PS, Napolitano HB, Oliveira SS, Camargo AJ, Figueredo AS, de Aquino GLB, Carvalho-Silva VH (2017) Cryst Growth Des 17(10):5145
64. (a) Rösel S, Quanz H, Logemann C, Becker J, Mossou E, Canadillas-Delgado L, Caldeweyher E, Grimme S, Schreiner PR (2017) J Am Chem Soc 139(22):7428 (b) Schneider WB, Bistoni G, Sparta M, Saitow M, Riplinger C, Auer AA, Neese FJ (2016) Chem Theory Comput 12:4778–4792 (c) Lu Q, Neese F, Bistoni G (2018) Angew Chem Int Ed 57:4760–4764
65. Černý R, Ravnsbæk DB, Schouwink P, Filinchuk Y, Penin N, Teyssier J, Smrčok L, Jensen TR (2012) J Phys Chem C 116:1563
66. Wolstenholme DJ, Flogeras J, Che FN, Decken A, McGrady GS (2013) J Am Chem Soc 135:2439

67. Wolstenholme DJ, Traboulsee KT, Hua Y, Calhoun LA, McGrady GS (2012) Chem Commun 48:2597
68. Černý R, Kim KC, Penin N, D'Anna V, Hagemann H, Sholl DS (2010) J Phys Chem C 114:19127
69. Ravnsbæk D, Filinchuk Y, Cerenius Y, Jakobsen HJ, Besenbacher F, Skibsted J, Jensen TR (2009) Angew Chem Int Ed 48:6659

Applications of the Density Matrix Renormalization Group to Exchange-Coupled Transition Metal Systems

Vera Krewald and Dimitrios A. Pantazis

Abstract Transition metal complexes containing magnetically interacting open-shell ions are important for diverse areas of molecular science. The reliable prediction and computational analysis of their electronic structure and magnetic properties, either in qualitative or quantitative terms, remain a central challenge for theoretical chemistry. The use of multireference methods is in principle the ideal approach to the inherently multireference problem of exchange coupling in oligonuclear transition metal complexes; however, the applicability of such methods has been severely restricted due to their computational cost. In recent years, the introduction of the density matrix renormalization group (DMRG) to quantum chemistry has enabled the multireference treatment of chemical problems with previously unattainable numbers of active electrons and orbitals. This development also paved the way for the first-principles multireference treatment of magnetic properties in the case of exchange-coupled transition metal systems. Here, the first detailed applications of DMRG-based methods to exchange-coupled systems are reviewed and the lessons learned so far regarding the applicability, apparent limitations, and future promise of this approach are discussed.

1 Introduction

Systems with multiple interacting open-shell transition metal ions are encountered in areas of science as diverse as active sites of metalloenzymes and synthetic molecular complexes or solid-state inorganic systems. The defining feature of these systems

V. Krewald
Fachbereich Chemie, Technische Universität Darmstadt, Alarich-Weiss-Straße 4, 64287 Darmstadt, Germany
e-mail: krewald@chemie.tu-darmstadt.de

D. A. Pantazis (✉)
Max-Planck-Institut Für Kohlenforschung, Kaiser-Wilhelm-Platz 1, 45470 Mülheim an der Ruhr, Germany
e-mail: dimitrios.pantazis@kofo.mpg.de

E. Broclawik et al. (eds.), *Transition Metals in Coordination Environments*, Challenges and Advances in Computational Chemistry and Physics 29,
https://doi.org/10.1007/978-3-030-11714-6_4

is that the interaction between the spin of unpaired electrons, often called magnetic or exchange coupling, gives rise to unique magnetic and spectroscopic properties that could not arise from the isolated transition metal centers. The magnitude and the nature of the coupling, for example, ferromagnetic or antiferromagnetic, have profound impact on the properties and reactivity of these systems. One of the great challenges for quantum chemistry is to understand these interactions in the context of electronic structure theory, connect the fundamental description with the phenomenological models often employed in the analysis of experiments, and finally predict the relevant parameters that describe the properties of such exchange-coupled systems with high accuracy and reliability.

The spin states associated with this situation (magnetic levels) typically arise from a single electronic configuration, but can be formally described only with linear combinations of multiple determinants. In contrast to spin states that correspond to distinct configurations of d electrons, such as, a low-spin and high-spin configuration of a transition metal ion, the magnetic levels of an exchange-coupled system span a narrow energy range of a few tens or hundreds of wavenumbers [1]. The demands imposed on quantum chemical calculations that target magnetically coupled states are therefore of the order of a wavenumber, and hence much higher than the usual definitions of "chemical accuracy" related to the prediction of common thermodynamic properties (1 cm^{-1} = 0.00286 kcal/mol = 0.01196 kJ/mol, or 1 kJ/mol = 0.239 kcal/mol = 83.593 cm^{-1}). The computational method of choice must therefore be able to predict the energies of all spin states of the magnetically coupled system equally well and converge them to the same accuracy.

Several quantum chemical approaches have been proposed to achieve qualitative and quantitative descriptions of magnetic coupling in molecular complexes with open-shell transition metal ions [2–6]. Density functional theory (DFT) based on the broken-symmetry approach [7–10] has been used for a wide range of systems over several decades with varying levels of success. However, the problem of exchange coupling is inherently a multireference problem that should be formally treated with multireference methods. These have also a long history in the field of exchange-coupled transition metal systems, but their applicability has been severely limited to small dinuclear systems with very few unpaired electrons, for example, Cu(II) dimers [2, 11]. This is due to the steeply increasing cost of multireference calculations for problems with more than a few electrons in a few orbitals. The key challenge of how to enable treatment of large active spaces, for example, in complete active space self-consistent field (CASSCF) calculations, is of direct relevance for the treatment of exchange-coupled transition metal systems, where the presence of more than two metal ions, of many unpaired electrons, or the necessity to include electrons and orbitals of bridging ligands in the active space quickly renders such calculations entirely impossible.

The focus of this chapter is on a method that was introduced relatively recently to the theoretical chemistry community, the density matrix renormalization group (DMRG) [12, 13]. From the point of view of applied quantum chemistry, DMRG can be considered as a method that enables the use of large active spaces in multireference calculations. It has already been employed in configuration interaction calculations, DMRG-CI, as well as in calculations involving orbital optimization, DMRG-SCF, to a range of chemical questions (for example [14–17]). DMRG-SCF has been reported for dinuclear and even tetranuclear complexes with open-shell transition metals [18–21], although these studies have focused on specific electronic states of the systems of interest without addressing explicitly the problem of exchange coupling. It is important to stress this point because the methodological and technical requirements for the application of DMRG-based approaches are neither obvious nor necessarily transferrable from other wavefunction-based approaches. At the time of this writing, very few studies have used DMRG to predict the relative energies of spin states that arise from magnetic coupling in transition metal clusters. Our aim is to review two of these very first case studies [22, 23] in order to understand the technical and methodological challenges encountered in applications of DMRG to problems of magnetic coupling, as well as to highlight the emerging opportunities that DMRG brings for the computational treatment of these systems. The point of view adopted here is of application-oriented quantum chemistry; the reader interested in the theoretical foundations of the methods and in current theoretical developments is directed to existing excellent reviews [24–27].

2 Theoretical Treatment of Exchange Coupling

The phenomenological Heisenberg–Dirac–van Vleck (HDvV) Hamiltonian is typically used to model the energy spacing between the magnetic levels in terms of pairwise exchange coupling constants and additional parameters. For two centers with spins $\mathbf{S}_\mathrm{A}$ and $\mathbf{S}_\mathrm{B}$, the simplest form of the HDvV Hamiltonian can be written as:

$$\hat{\boldsymbol{H}}_\mathrm{HDvV} = -2J\mathbf{S}_\mathrm{A}\mathbf{S}_\mathrm{B}$$

This is often the leading or the only term considered and the exchange coupling constant J determines the nature of the fictitious magnetic interaction, ferromagnetic for positive, and antiferromagnetic for negative values. In this case, the energies of adjacent energy levels with total coupled spin $S = S_\mathrm{A} + S_\mathrm{B}$, $S_\mathrm{A} + S_\mathrm{B} - 1$, ..., $|S_\mathrm{A} - S_\mathrm{B}|$ conform to the Landé interval rule:

$$E(S) - E(S-1) = -2JS$$

Additional terms are used in order to model deviations from isotropic behavior. These include, for example, the biquadratic term $j(\mathbf{S}_\mathrm{A} \cdot \mathbf{S}_\mathrm{B})^2$, double exchange $\pm B(S + 1/2)$ in the case of some mixed-valence systems, zero-field splitting terms for total $S \geq 1$, etc. The interested reader is referred to the landmark book of Bencini and Gatteschi for in-depth discussions [28]. Experimental data on the lowest energy levels, such as those derived from magnetic susceptibility measurements, electron paramagnetic resonance spectroscopy, or polarized neutron diffraction, are fitted with such a phenomenological Hamiltonian as appropriate to the chemical system at hand, yielding numerical values for the terms introduced above. In order to make the fitting problem tractable and reasonably defined, simplifying assumptions are often made regarding the relative magnitudes of particular coupling constants and the magnetic topology of a compound. Importantly, for a sufficiently complex system the fitting cannot be unique, not even if the Hamiltonian is restricted to a single term [29]. Instead of treating the quantities that appear in the HDvV Hamiltonian as merely numerical parameters to be fitted, quantum chemistry attempts to assign physical meaning to these parameters by connecting them with fundamental aspects of the electronic structure, thus enabling both interpretation and prediction by first principles.

The magnetic coupling problem is inherently a multireference problem: Even if the ground state of an exchange-coupled system is described by a single electronic configuration, that is, a unique distribution of electrons among a set of metal-based orbitals, the resulting spin states are multideterminantal. Nevertheless, the use of approximate treatments based on single-determinant methods has a long tradition in computational studies of exchange-coupled transition metal systems. With the exception of approaches that allow local spins to be non-collinear, single-reference treatments are mostly restricted to broken-symmetry DFT (BS-DFT). A Kohn–Sham determinant can formally represent only the magnetically coupled state with maximum total spin multiplicity (e.g., for a dinuclear complex with local spins $\mathbf{S}_A$ and $\mathbf{S}_B$, $S_{max} = S_A + S_B$), referred to as the high-spin (HS) solution. For all other rungs of the spin ladder with $S < S_{max}$, more than one determinant is required. The broken-symmetry (BS) formalism was introduced to circumvent this problem [6–8, 30, 31]. Here, an unrestricted determinant is constructed with an M_S value equal to that of the antiferromagnetically coupled state ($S_{min} = |S_A - S_B|$). In the BS determinant, the singly occupied orbitals of opposite spin ("magnetic orbitals") are allowed to localize at the spin centers while retaining overlap "tails" [7, 32, 33]. The BS determinant is not a spin eigenfunction, and hence, it has no defined spin quantum number S; it can be seen as a weighted mixture [34] of all spin states that contain magnetic sublevels with the same magnetic quantum number M_S.

A central question is how to interpret the energy of the BS solution. Several mapping procedures have been proposed and they all use the energy difference between the HS and BS determinants, relying on assumptions regarding a valid form of a phenomenological Hamiltonian, focusing chiefly on isotropic bilinear exchange [7, 10, 35, 36]. A popular expression for two-spin systems was proposed by Yamaguchi, who used the total spin angular momentum expectation values of the HS and BS determinants to provide a consistent description for weakly to strongly coupled systems [10, 36]:

$$J = -\frac{E_{HS} - E_{BS}}{\langle S^2 \rangle_{HS} - \langle S^2 \rangle_{BS}}$$

Having obtained a value for the exchange coupling constant J, the spin-state ordering and the relative energies of all rungs of the spin ladder are deduced through the HDvV Hamiltonian. The generalization to oligonuclear systems with N spin centers is straightforward but of rapidly increasing complexity as one needs to determine the values of up to $N(N-1)/2$ pairwise exchange coupling constants J_{ij}. These are accessible through the computation of up to 2^{N-1} distinct broken-symmetry determinants. For more than three non-symmetry-related spin centers, the number of possible BS determinants can exceed the number of pairwise exchange coupling constants. This leads to an overdetermined system of equations, which can be solved via singular value decomposition [37] to obtain a set of exchange coupling constants J_{ij} that is unique in the least-squares sense [29, 38]. It is noted that a generalized spin projection method has also been introduced for oligonuclear systems [39].

Despite the extensive use of BS-DFT [40–52], the approach has significant limitations. In terms of energetics, the application of the method suffers by the pronounced sensitivity on the density functional and relies on empirical benchmarking against experimental data [4, 5, 53–55]. Although the charge density of the system described by the broken-symmetry determinant is often reliable, the spin density of any state other than the high-spin solution is qualitatively incorrect [4, 56]. The intermediate spin states are not accessible at all by the broken-symmetry formalism; only their energies relative to the HS or BS energy can be predicted, and this only indirectly [4]. This necessitates the use of approximate spin projection methods for predicting spin-dependent properties. Moreover, the interpretation of magnetic coupling based on BS determinants is often limited to qualitative analysis or visualization of magnetic orbitals via the corresponding orbital transformation of Amos and Hall [33, 57] which is not obviously extendable beyond dinuclear species [47].

Multireference wavefunction-based calculations present a distinct quantum chemical alternative, because they offer an opposite approach to the problem. Instead of trying to approximate the HDvV solution space based on a much more limited and approximate number of single-determinant solutions, one can work directly with the (approximate) solutions of the Schrödinger equation. In this case, no assumptions are required regarding the form and the terms of a phenomenological HDvV Hamiltonian, and hence the problem can be approached from the opposite direction than that represented by BS-DFT. CI (if only the coefficients of distinct configuration state functions are optimized) and CASSCF [58, 59] (if the orbitals are also optimized) are examples of multireference methods by which all individual spin states of the magnetically coupled system can be accessed directly.

To describe a magnetically coupled system at the very least, the magnetic orbitals have to be considered in the construction of a minimal active space. A common expansion of the active space in systems with first-row transition metal ions is to include unoccupied d orbitals. The so-called double shell, $3d'$ or $4d$ orbitals are important for an adequate description of radial electron correlation [60, 61]. Considering the Anderson model of superexchange, by which bridging ligands mediate the transfer of spin between the individual spin sites, it is obvious that to describe magnetically coupled systems larger active spaces are needed than in cases that are dominated by the local properties of an individual transition metal ion. A logical extension of the minimal valence space is thus to include orbitals of the bridging ligands [2], which turns on various types of charge-transfer excitations that contribute to charge and spin polarization effects and adjust the weight of neutral and ionic determinants to better describe the low-energy region of the spin ladder for the exchange-coupled system.

The total size of the active space is commonly abbreviated with the ($N_{\text{electrons}}$, N_{orbitals}) notation, e.g., (12, 10) denotes an active space with 12 electrons in 10 orbitals. Given that the upper limit for an active space size that can be practically treated with CASSCF is around 16–18 orbitals, the method may be inapplicable even for relatively simple dinuclear exchange-coupled systems. The selection of orbitals that should enter the active space in addition to the magnetic orbitals, the details of orbital preparation and optimization, the number of states targeted, and other technical choices are crucial factors for the design and ultimately for the success of a computational study. Still, a CASSCF treatment does not afford quantitative predictions, and may even fail qualitatively, because despite the formally correct multideterminantal description of the states, dynamic electron correlation is absent. Some of this may be recovered by applying second-order perturbation theory to the CASSCF wavefunction (complete active space second-order perturbation theory, CASPT2 [62, 63], or N-electron valence second-order perturbation theory, NEVPT2 [64, 65]). In contrast to these perturbational methods, difference-dedicated configuration interaction (DDCI) is a variational approach, in which particular classes of CT-excitations are included explicitly in the wavefunction [2, 11, 66–71]. DDCI was suggested to have considerably better performance and robustness for exchange-coupled systems over CASPT2 [72], but its applicability remains severely restricted to minimalistic problems because of its high computational cost.

Although by no means the only issue that has to be addressed, increasing the size of the active space appears as the major obstacle to applications of multiconfigurational SCF methods in exchange-coupled transition metal systems. One way of dealing with this problem has been to use partitioning or truncation schemes, as represented for example by the restricted active space (RAS) [73, 74], the generalized active space (GAS) [75], and the split GAS [76, 77] approaches. Alternatively, the active space limitations are attacked through novel algorithmic approaches, such as the stochastic full configuration interaction quantum Monte Carlo (FCIQMC) [78, 79] technique, and the DMRG approach that is the subject of this chapter.

3 The Density Matrix Renormalization Group Approach

The DMRG algorithm, its implementation, and the extraction of (chemical) observables have been discussed in many papers and reviews [12, 13, 25, 26, 80–88]. Its importance and relevance in particular to inorganic complexes lies in enabling the description of large active spaces in CASCI and CASSCF calculations. Here, we present a qualitative description of the fundamental concepts and highlight practical considerations for the application to open-shell transition metal complexes. DMRG was first used in physics to describe spin-spin correlations in a more refined way than can be achieved with a mean-field approach [12]. The DMRG algorithm was originally devised for chain-like systems or lattices, and thus the first applications aimed at predicting magnetically coupled systems involved linear chains of open-shell metals [89–91]. However, in these studies the electronic states are not constructed as many-electron wavefunctions as is the case in quantum chemical calculations. As such, DMRG is not a fundamentally new method for describing molecules, but rather a new algorithm that is effectively used as a CI-solver.

The algorithm differs from other CI approaches in that it stores the wavefunction in a different numerical representation than that typically encountered in CASSCF calculations. The striking feature of DMRG is that in principle the full CI solution can be approximated to an accuracy typically required for chemically relevant systems with a computational cost that is normally lower scaling than other multireference methods. While the computational cost of conventional multireference approaches is exponential, and DMRG can approximate the correct solution with polynomial cost [26] for chemically relevant systems.

The DMRG algorithm benefits from proper choice of orbital shape and ordering to achieve a desired level of accuracy while minimizing the computational cost in solving the CI problem. A key aspect is that only a few orbitals are treated exactly during each substep of the iterative procedure, and the other orbitals are either part of the so-called active subsystem or the complementary subsystem. Thus, the orbitals sequentially become part of what is known as the "exactly represented subsystem," a set of neighboring spatial orbitals. Once each orbital has been treated exactly, i.e., after a series of microiterations, a macroiteration or *sweep* is completed. The wavefunction is represented in the product space of all orbitals, restricted to the desired total number of electrons and spin state. The number of basis states in the active and complementary subsystems is denoted by M. For two spatial orbitals in the exactly represented subsystem, the number of states is 16, as both orbitals can be in any of four occupations (doubly occupied, singly occupied spin up, singly occupied spin down, and unoccupied). The algorithm involves a step known as *blocking*, where the active subsystem is enlarged by the adjacent orbital from the exactly represented system, leading to a dimension of $4M$ for the increased active subsystem. To obtain a system size of M again, the following step is to transform the system to a new many-particle basis and thus reduce its size. This step, the transformation from a $M \times 4M$ matrix to an $M \times M$ matrix, is called *renormalization* and involves the diagonalization of the reduced density matrix. The choice of which elements to discard is based on the weights of the corresponding eigenstates, and the effect is measured as the so-called *discarded weight*. With the renormalized system, the next microiteration can start, in which the active system is enlarged by one, the exactly represented system loses one old member and gains one new member, and the complementary subsystem is diminished by one. Once the algorithm has reached the final pair of orbitals in the exactly represented subsystem, one sweep is completed. Usually several sweeps in alternating directions are performed to improve the accuracy of the DMRG representation by optimizing the representation of the complementary subsystem. The number of sweeps at a given discarded weight can be adjusted. Most DMRG implementations have specific *sweep schedules* in which the number of renormalized basis states and sweeps is adjusted until the user-defined energy thresholds are reached.

As noted above, the DMRG algorithm in quantum chemistry packages is used as a CI-solver, and thus the result of a DMRG-SCF calculation, where DMRG-CI and orbital optimization steps alternate until convergence is achieved, should in principle be identical to that of a CASSCF calculation. The key difference between a CASSCF and a DMRG calculation lies in an additional parameter that needs to be carefully monitored and adjusted by the user in any DMRG calculation: the number of renormalized block states M, closely connected to the discarded weight and thus the accuracy of the calculation. Because a DMRG calculation does not simply converge to a preset energy criterion as a DFT or CASSCF calculation normally would, the user has to run several DMRG calculations with increasing values of M until the energy has converged to the required accuracy. The ideal value of M depends on the nature of the chemical species under investigation, on the size and character of the active space, as well as on the number of roots requested, their spin states and the type of state averaging required.

Depending on the chemical property that is targeted and the nature of the computed energies, one may be able to apply extrapolation techniques, where the full CI energy is linearly extrapolated from several calculations with increasing M [25]. It has to be noted that extrapolation from DMRG wavefunctions obtained at small M can be problematic due to the "noise" introduced deliberately in the algorithm's initial phases [25]. The applicability of extrapolation techniques for magnetically coupled systems will be discussed in more detail in the context of the case studies presented in this chapter.

The choice of orbitals to be included in the active space, the origin of these orbitals, the localization or not, and their initial ordering are crucial decisions for a DMRG-SCF calculation and influence the convergence of the DMRG algorithm [80]. For the choice of orbitals, similar arguments can be followed as in CASSCF calculations [92]. Reiher and coworkers noted that natural orbitals from a CASSCF calculation may be better suited as starting orbitals than orbitals derived from a preceding Hartree–Fock calculation [26]. More powerful approaches rely on automated selection procedures [84, 93, 94]. For example, Stein and Reiher proposed an algorithm that employs orbital entanglement or orbital entropy measures derived from a low-accuracy, large-CAS calculation and selects the most highly entangled orbitals for the active space of the production-level calculation [84]. The initial ordering of orbitals is an important technical aspect. In general, orbitals that are more entangled should be placed closely together, but unlike in chain-like systems the optimal way of doing this is not obvious for complex non-linear molecules. Current implementations of DMRG software in quantum chemistry usually optimize and update the order of orbitals through automated reordering procedures [25, 80, 95]. Similarly, localized orbitals can improve the performance of DMRG as they help to reduce the entanglement of the system, which implies that the number of renormalized basis states to reach a certain accuracy will be lower.

DMRG enables CASCI and CASSCF calculations with active spaces containing tens of orbitals; however, only a small part of dynamic electron correlation can be recovered by extending the active space. Therefore, additional treatments must still be applied to the DMRG-SCF wavefunction, such as second-order perturbation theory [96–99]. An example of the effect of perturbative treatment of a DMRG-SCF wavefunction in the case of a magnetically coupled system will be discussed in one of the case studies below.

DMRG has already seen a number of applications in transition metal chemistry [15, 16, 19, 20, 100–107] and its ability to handle large active spaces in exchange-coupled transition metal systems has been showcased in two papers that deal with tetramanganese cluster complexes (Fig. 1). The first one studied a minimal model of the oxygen-evolving complex of Photosystem II, a tetramanganese–calcium cluster with five oxo-bridges embedded in a protein environment composed mainly of carboxylate ligands [19]. An active space of 44 electrons in 35 orbitals was constructed from all Mn 3*d* orbitals and the oxygen 2*p* orbitals of all five bridges. A single root with the experimentally known spin multiplicity was calculated with this (44, 35) active space using DMRG-SCF and its energy was converged to 0.16 kJ mol^{-1}. Furthermore, the quantum entanglement of the cluster [83] was analyzed [19]. In another example, Paul et al. [21] studied a synthetic tetramanganese–calcium complex [108] that is considered a structural mimic [109] of the oxygen-evolving complex in Photosystem II [110], albeit lacking one flexible oxo-bridge in the center of the inorganic core [111, 112]. DMRG-SCF with a (37, 32) active space containing all Mn 3*d* and O 2*p* orbitals was used to distinguish between two isomeric forms of the complex. A single root was calculated for each isomer and the energies were converged to 10^{-4} kcal mol^{-1}. It is also worth mentioning a DMRG-SCF study by Sharma et al. [20] of biologically ubiquitous [113] iron–sulfur systems, specifically Fe_2S_2 dimers and Fe_4S_4 clusters. DMRG allowed the use of a (32, 30) active space for the dimers, i.e., including all Fe 3*d*, 4*s*, 4*d* and S 3*p* orbitals, and energies of individual spin states were converged to 0.1 kcal mol^{-1} (35 cm^{-1}). For the Fe_4S_4 cluster, a Fe 3*d* and S 3*p* (54, 36) active space could be used for specific roots in DMRG-CI calculations. These three studies either did not attempt or did not conclusively address the problem of magnetic coupling in the tetranuclear systems, but the impressive feat of performing multireference calculations on systems of this size nevertheless demonstrates the impressive new possibilities offered by DMRG. At the time of this writing, only two detailed studies of the performance of DMRG-SCF for the exchange coupling problem per se exist in the literature, both on exchange-coupled transition metal dimers. In the remainder of this chapter, we present and discuss the content and insights gained from these studies.

4 Case Studies: Magnetic Coupling in Dinuclear Complexes

Studies of exchange coupling in transition metal complexes using DMRG-based multireference approaches are still rare. Consequently, the optimal ways of constructing and handling the large active spaces enabled by the DMRG approach, as

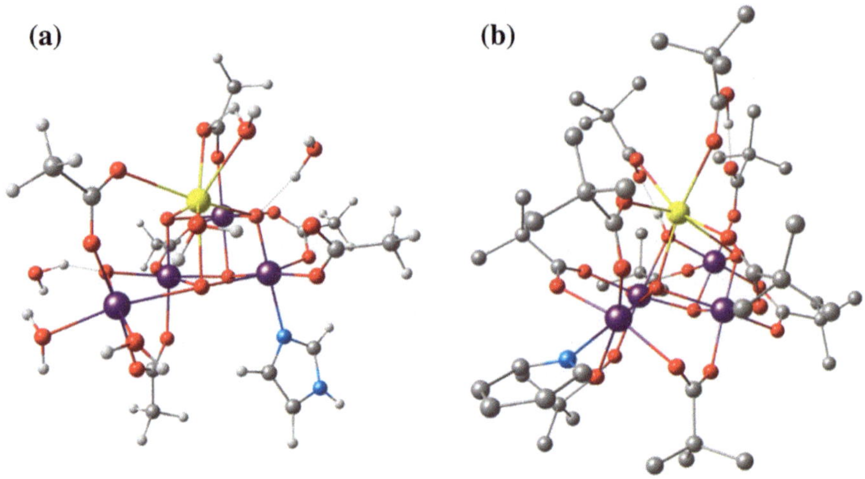

Fig. 1 Examples of oligonuclear exchange-coupled transition metal systems for which CASSCF calculations with large active spaces including all metal d and bridging ligand orbitals became tractable through the use of DMRG: **a** 79-atom simplified model of the tetranuclear Mn_4O_5Ca cluster in the oxygen-evolving complex of photosystem II studied by Kurashige et al. [19] (Mn: purple; Ca: yellow; O: red; N: blue; C: gray; H: white). **b** 182-atom synthetic analogue of the OEC with a Mn_4O_4Ca core [108] studied by Paul et al. [21] (right, hydrogen atoms omitted for clarity). Single-root DMRG-SCF calculations for these two systems were reported with (44, 35) and (37, 32) active spaces, respectively

well as the technical parameters that define the best usage of the method remain under investigation. In the following, we will discuss two landmark case studies on exchange-coupled dinuclear transition metal complexes that have contributed toward clarifying these points.

4.1 Fe_2 and Cr_2 Mono-μ-Oxo Complexes

For two mono-μ-oxo-bridged dinuclear complexes exhibiting antiferromagnetic coupling, $[Fe_2OCl_6]^{2-}$ and $[Cr_2O(NH_3)_{10}]^{4+}$ (Fig. 2), Harris et al. studied the effects of basis set choice, number of renormalized basis states M, and active space composition on the predicted exchange coupling constant J [22].

In the iron complex, the single oxo–bridge can engage in σ- and π-bonding with the Fe(III) ions. The authors chose a considerably bent geometry in which the Fe–μ–O bond lengths are 1.761 Å, and the Fe–O–Fe angle is 144.6°. It should be noted that this might have not been an optimal choice of either reference system or geometry because an earlier paper by Lledós et al. had shown that the bent form is the result of weak intermolecular interactions in the crystal, suggesting that only linear Fe–O–Fe conformations are found in solution [114]. Experimentally, the magnetic susceptibil-

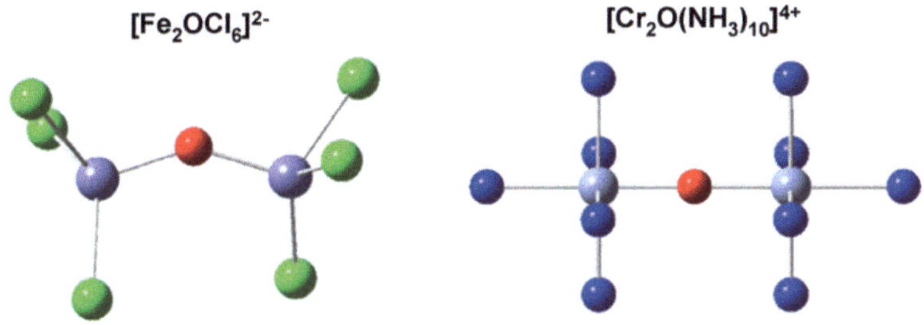

Fig. 2 Two mono-μ-oxo-bridged complexes studied by Harris et al. The hydrogen atoms of the NH_3 ligands have been omitted for clarity. Reprinted from [22] with the permission of AIP publishing

ity was very difficult to fit due to the simultaneous presence of linear and bent forms in the powdered sample. Based on B3LYP or BHLYP-derived absolute magnitudes and relative differences between the linear and bent forms, Lledós et al. suggested fitted exchange coupling constants of -117 or -119 cm^{-1} for the bent and -133 or -130 cm^{-1} for the linear form. The magnetic coupling constants computed with BHLYP were in fact significantly smaller ($J_{lin} = -84$ cm^{-1}, $J_{bent} = -73$ cm^{-1}) than the B3LYP ones ($J_{lin} = -145$ cm^{-1}, $J_{bent} = -161$ cm^{-1}) [114]. The coupling strength used as the "experimental" reference value by Harris et al. was -117 cm^{-1} [22].

The Fe(III) ions have locally high-spin d^5 configurations. The Heisenberg spin ladder produced by the coupling of the two local $S_A = S_B = 5/2$ spins thus consists of the six spin states $S = 0, 1, 2, 3, 4$, and 5. The minimal active space is composed of 10 electrons in 10*d* orbitals, (10, 10). Including the occupied μ-O bridge O(2*p*), orbitals result in a (16, 13) full-valence active space. Both of these spaces can be treated at the CASSCF level. Assuming a regular Landé spacing, that is, an energy difference of $2J$ between the $S = 0$ and the $S = 1$ states, the predicted magnetic coupling constant was -39.7 cm^{-1} for the minimal active space and -58.6 cm^{-1} for the full-valence active space. Both fall short of the reference value of -117 cm^{-1}. Without further active space expansion, the experimental value can be approached using the (16, 13) active space with multireference configuration interaction calculations including the Davidson correction (MRCI+Q), which yields a value of -115.3 cm^{-1} for the exchange coupling constant.

Expansion of the active space with unoccupied metal and ligand orbitals leads to active space sizes that can only be described with the DMRG approach. Upon inclusion of the ten 4*d* orbitals to the metal-only (10, 10) active space, leading to a CAS(10, 20), the antiferromagnetic exchange coupling is strengthened to -49.0 cm^{-1}. Inclusion of only the μ-O 3*p* orbitals on top of the full-valence active space, i.e., (16, 16), leads to a coupling constant of -57.7 cm^{-1} and is hence insufficient for a quantitative agreement with experiment. However, inclusion of both metal and bridge virtual orbitals to the full-valence active space, resulting in a (16, 26) active space, was shown to yield a projected magnetic coupling constant of -117.4 cm^{-1}, in quantitative agreement with the experimental value.

The above observations can be rationalized in terms of charge-transfer configurations when virtual orbitals are included in the active space. By adding bridge 3*p* orbitals, metal-to-ligand charge transfer (MLCT) configurations are considered. Analogously, the inclusion of a metal "double shells" adds ligand-to-metal charge transfer (LMCT) configurations and introduces radial dynamic correlation. In a traditional CASSCF calculation, the individual configurations and their relative contributions can be analyzed and quantified in a relatively straightforward manner; however, the weights of contributing configurations [88, 115] were not reported from the DMRG calculations. Nevertheless, as an indirect or summative effect of all CT configurations, the orbital contour plots revealed the contribution of LMCT states as a more pronounced delocalization on the bridging ligands.

The effect of basis set choice on the exchange coupling constant was studied in some detail. The calculations employed relativistic atomic natural orbital basis sets (ANO-RCC) with a series of contractions. For both the CASSCF(16, 13) and DMRG-SCF(16, 26) calculations, a larger basis set led to weaker exchange coupling. For the CASSCF approach, the results were converged with a quintuple-ζ basis set for iron and oxo-bridge, and a quadruple-ζ basis set for the peripheral chloride ligands. For the DMRG approach, the convergence behavior is less clear, as there is an additional strong dependency on the M value: larger basis sets require a larger M, but as this creates higher memory demands the calculations are not always feasible. The basis set convergence is very similar for the CASSCF and DMRG approaches as long as M is sufficiently large, i.e., up to a quadruple-ζ basis for all elements. It should be noted that although no chloride orbitals enter the active space, increasing the basis set size from double-ζ to quadruple-ζ was reported to change the predicted exchange coupling constants by ca. 5 cm^{-1}. This may be related to the π-bonding between chloride and the iron ions. Taking both accuracy and efficiency into account, Harris et al. [22] opted for a triple-ζ basis set for iron and oxygen, and a double-ζ basis set for the peripheral chloride ligands.

The number of renormalized basis states, M, must be sufficiently large to ensure that the energy converges to the correct value for a given choice of active space. Furthermore, larger active spaces require larger values of M to converge properly, implying that across a series of calculations with varying active space sizes for the same system, different numbers of renormalized basis states will be needed. Because the energy of a system converges to the exact value for increasing values of M, i.e., decreasing discarded weights, a linear extrapolation can be used to find the exact energy based on several calculations with different numbers of renormalized basis states.

The convergence with M is different for different spin states. This is apparent, for example, in the case of the iron dimer (Fig. 3) [22]. The extrapolation of the $S = 0$ state has a steeper slope than the extrapolation of the $S = 1$ state, both based on two energies calculated with $M = 512$ and $M = 1000$. It is also noted that the corresponding discarded weights differ for calculations of different spin states with the same M value.

When aiming to predict accurate energies of different spin states of an exchange-coupled system to subsequently extract coupling constants measured in cm^{-1} units

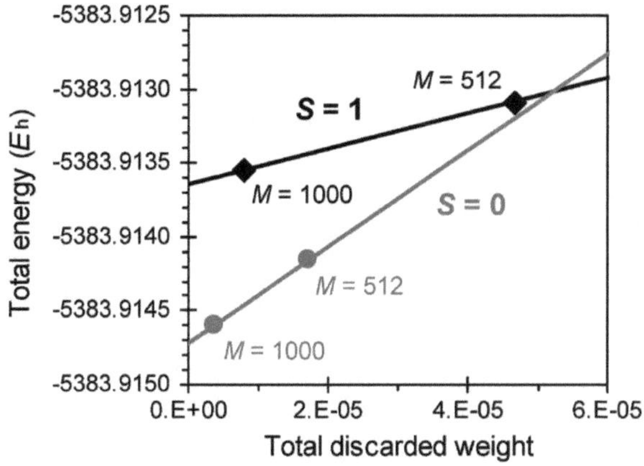

Fig. 3 Extrapolation of DMRG(12, 26) energies to infinite M for the singlet and triplet states of $[Fe_2OCl_6]^{2-}$. Reprinted from [22] with the permission of AIP publishing

of energy, one needs to take into account that a 1 cm^{-1} energy difference corresponds to 4.556×10^{-6} Eh. The question of which M is sufficient in the case of the μ-oxo-bridged dimer was studied by comparing the CASSCF(16, 13) energies with the DMRG-SCF energies calculated with $M = 64, 128, 256$. A micro-Hartree energy difference is achieved between $M = 128$ and $M = 256$, and the latter energy is within 0.2 μEh of the CASSCF energy. Of course, the CASSCF energy difference itself is insufficient to extract quantitatively correct exchange coupling constants, but it can serve as a valid reference point. For larger active spaces without CASSCF reference energies, Harris et al. increased M until the change in energy was less than 1 μEh, resulting in $M = 1000$ for (16, 16), whereas for the (10, 20) active space $M = 256$ was considered sufficient [22]. For the (16, 26) active space, the energy change for $M = 512$ and $M = 1000$ was ca. 500 μEh, but the authors opted to extrapolate these two values to deduce the exchange coupling constant. This resulted in the value that gives the best agreement with the experimental reference of -117 cm^{-1}.

Given that different spin states converge differently with respect to M, it seems strongly advisable to test the convergence behavior of all spin states of a magnetically coupled system instead of only two. As will be discussed in greater detail for the second case study on a manganese dimer, low M values can lead to non-Landé spin-state patterns, meaning that without having calculated the full spin ladder, it cannot be known whether the $S = 0$ and $S = 1$ states are actually correctly computed with respect to the other states, or indeed if they actually are the lowest energy states predicted by the method for the given choice of active space and M.

The second example in the study of Harris et al. was the antiferromagnetically coupled chromium dimer $[Cr_2O(NH_3)_{10}]^{4+}$, with a linear Cr–O–Cr angle and Cr–O bond lengths of 1.821 Å [22]. The ammonia ligands were placed at unoptimized average crystallographic distances of 2.12 Å from the chromium ions, and the hydrogen atoms were optimized. The chromium ions are in their +III oxidation states, with a d^3 electronic configuration ($S_A = S_B = 3/2$), leading to a spectrum of four coupled spin states, $S = 0$, 1, 2, and 3. The experimentally determined exchange coupling constant is $-225\ \text{cm}^{-1}$.

Similar to the iron dimer, a triple-ζ basis set for chromium and oxygen, and a double-ζ basis set for the peripheral ammonia ligands was chosen. The minimal active space is a (6, 6) active space containing only the magnetic orbitals. CASSCF with this active space produces an exchange coupling constant of $-52.4\ \text{cm}^{-1}$. The expansion of the active space to include occupied bridge orbitals as well as virtual orbitals was described in this study as much more challenging. Upon inclusion of the occupied 2*p* as well as the virtual 3*p* oxo-bridge orbitals, the exchange coupling strength increases to $-60\ \text{cm}^{-1}$ with CASSCF. The authors attempted to include the empty Cr *d* orbitals in the active space, but only a (12, 13) CAS could be converged, which contained an orbital delocalized over the entire core. It had $d(z^2)$ character on both metal centers, and symmetrical nodal planes with a lobe of s-character on the oxo-bridge. Although still not containing all metal *d* orbitals, the exchange coupling constant increases to $-92.8\ \text{cm}^{-1}$. The authors attributed this improvement to a more balanced description of the Cr–O σ bonds, achieved by symmetrical mixing of the orbitals dominated by O $p(z)$ and Cr $d(z^2)$ character. From a pure MO theory point of view, one would expect three orbitals in total to be of importance for the σ bonds: one dominated by the O $p(z)$ atomic orbital (no nodal plane), and two dominated by Cr $d(z^2)$ character (one and two nodal planes).

Using an active space of (12, 25), containing the magnetic orbitals and their double shells as well as the oxo-bridge 2*p*, 3*p*, and 3*d* orbitals, a stronger antiferromagnetic coupling can be achieved. With $M = 512$ the exchange coupling constant J is $-166.9\ \text{cm}^{-1}$, but increasing M to 1000 results in an exchange coupling constant that is significantly weaker, $J = -137.9\ \text{cm}^{-1}$. Extrapolation of the individual state energies to zero discarded weight results in an exchange coupling constant of $-123.6\ \text{cm}^{-1}$, ca. $100\ \text{cm}^{-1}$ lower than the experimental value. In terms of possible charge-transfer excitations with this active space, the MLCT excitations would be expected to be adequately represented given that all magnetic orbitals and relevant virtual ligand orbitals are included in the active space. In contrast, LMCT excitations must be more limited, given that the virtual chromium 3*d* orbitals and their double shells are not taken into account, and therefore dynamic correlation is recovered incompletely compared to other systems.

An active space that contains a larger number of chromium 3*d* and 4*d* orbitals is the (12, 32) active space. All 3*d* orbitals, the $4d(xy)$, $4d(xz)$, $4d(yz)$, and one $4d(z^2)$ orbital are included. Additionally, the 2*p*, 3*s*, 3*p*, 3*d*, and 4*p* orbitals of the central oxo-bridge are taken into account. With this active space, an exchange coupling constant of $-165.9\ \text{cm}^{-1}$ is achieved, with $M = 1000$ and a very small basis set (Cr: double-ζ, O: triple-ζ, N, H: single-ζ). Extrapolation was not possible in this case due to convergence problems of the $M = 512$ calculation. It was also shown that the basis set effect in this system is considerable. For the (12, 25) active space, a $11.8\ \text{cm}^{-1}$ difference results between the double-ζ and triple-ζ basis sets on Cr and O. With a larger basis set, it might be hoped that an improved exchange coupling constant could be achieved; however, the available results strongly indicate that this active

Fig. 4 Deviation from the Landé interval rule for the $[Fe_2OCl_6]^{2-}$ complex, calculated with DMRG(16, 26) and $M = 1000$. The computed data are fit with a biquadratic term in the Hamiltonian. Reprinted from [22] with the permission of AIP publishing

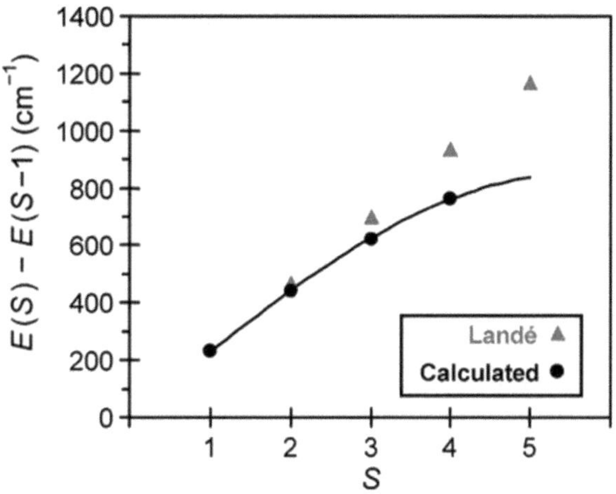

space is inadequate to approximate the experimental value of $-225\ \mathrm{cm}^{-1}$, regardless of basis set size and M convergence.

In all of the above results and the related discussion, the exchange coupling constant was calculated exclusively from the energy difference of the $S = 0$ and $S = 1$ states. As described in the introduction, deviations from the Landé interval rule can be indicators of biquadratic exchange [116, 117]. For the iron dimer, Harris et al. reported significant deviations from the expected splitting, whereby for the CASSCF(10, 10) calculations the exchange coupling constant calculated from the $S = 1$ and $S = 0$ energy difference is $-41.2\ \mathrm{cm}^{-1}$, whereas from the $S = 4$ and $S = 5$ energy difference J is calculated as $-27.9\ \mathrm{cm}^{-1}$. Similarly, for the largest active space treated at DMRG level, (16, 26), the range of magnetic coupling constants predicted for different adjacent energy levels ranges from -95.4 to $-116.8\ \mathrm{cm}^{-1}$, although the real span might be slightly larger considering that, curiously, the calculation for the ferromagnetic (and hence single determinantal) $S = 5$ state could not be completed in this study (Fig. 4). The deviation was fit with a biquadratic term; however, no experimental data are available to verify whether this is necessary or physically valid. For the chromium dimer, similar trends are found: the exchange coupling constants for the smallest active space range from -45.5 to $-52.4\ \mathrm{cm}^{-1}$ depending on the spin-state interval they are derived from, and for the DMRG-SCF(12, 25) active space the magnetic coupling constants range between -115.4 and $-137.9\ \mathrm{cm}^{-1}$.

Importantly, this latter finding is at odds with a subsequent paper by Spivak et al. [118], who studied precisely the same system with an approach that combined state-averaged CASSCF orbitals with partially contracted N-electron valence second-order perturbation theory (NEVPT2) calculations. In the study by Spivak et al. [118], it was reported that the exchange coupling constants derived from the different pairs of spin states (singlet–triplet, triplet–quintet, and quintet–septet) are all very similar and that no significant deviations from the Landé pattern were observed, in stark contrast to the DMRG results of Harris et al. The crucial difference between the two studies appears to be the method of orbital optimization: Harris et al. reported results from state-specific calculations, whereas Spivak et al. used state-averaged orbitals, i.e., a common set of orbitals for all spin states arising from the magnetic coupling of the two ions. Among other methodological points, the study of the Mn dimer by Roemelt et al. [23] to be discussed in the following investigated this issue explicitly, strongly suggesting that the use of state-averaged orbital optimization is necessary to avoid spurious and often unphysical deviations from the regular spacing of magnetic levels.

4.2 Mn_2 Bis-μ-Oxo/μ-Acetato Complex

Another example of a detailed investigation of DMRG-based approaches comes from Roemelt et al. [23], who presented an in-depth study of a mixed-valence bis-μ-oxo/μ-acetato-bridged Mn(III/IV) dimer (Fig. 5). This complex was synthesized and characterized by Bossek et al. [119], and features one 1, 4, 7-trimethyl-1, 4, 7-triazacyclononane, and two additional acetates as terminal ligands. Owing to the combination of mixed-valence and asymmetric ligation, the Mn ions adopt distinct coordination environments. The Mn(III) site features a strong axial pseudo-Jahn–Teller elongation, a hallmark of occupation of the σ-antibonding orbital of $d(z^2)$ parentage, that leads to an approximately square–pyramidal coordination geometry. This type of system is of particular interest in inorganic and bioinorganic chemistry because the high-valent nature of the Mn ions, the chemical nature of the ligands and the bridging topology are of direct relevance to manganese systems encountered widely in molecular magnetism and bioinorganic catalysis [108, 120, 121]. A prominent example in the latter case is the oxo/carboxylato-bridged Mn_4CaO_5 cluster of the oxygen-evolving complex of photosystem II, the site of water oxidation in biological photosynthesis [112, 122].

The two ions have local high-spin configurations, d^4 for Mn(III) and d^3 for Mn(IV), with corresponding local spins $S_A = 2$ and $S_B = 3/2$. These couple to produce a ladder of four spin states with total spin $S = 7/2$, 5/2, 3/2, and 1/2. The complex exhibits antiferromagnetic coupling, therefore the spin doublet is the ground state. Magnetic susceptibility measurements led to a fitted value for the exchange coupling constant of $J = -90.0$ cm^{-1}. Assuming the ideal case of an isotropic bilinear term in the Heisenberg Hamiltonian, the energy splittings for the spin states correspond to the Landé pattern, i.e., the $S = 3/2$ state is $3J$ higher than the $S = 1/2$ ground state, the $S = 5/2$ state is $5J$ higher than the $S = 3/2$ state, and the ferromagnetic $S = 7/2$ state is $7J$ higher than the $S = 5/2$ state, yielding a total span of $15J$ for the spin ladder.

The minimal active orbital space consists of the metal 3*d* orbitals, i.e., an active space of 7 electrons in 10 orbitals, (7, 10). Regardless of the origin of the starting orbitals or of the method used in producing localized input orbitals (e.g., Pipek–Mezey or Foster–Boys), a simple CASCI treatment was reported to lead always to strong ferromagnetic coupling ($S = 7/2$ ground state) with an exchange coupling constant J of almost +180 cm^{-1}, in profound qualitative disagreement

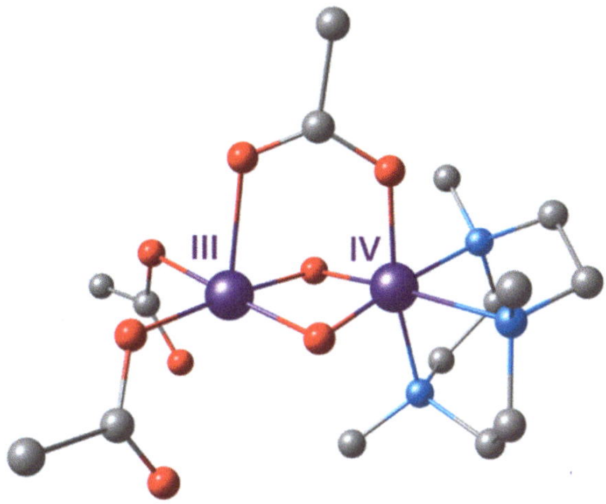

Fig. 5 Structure of the bis-μ-oxo/μ-acetato-bridged manganese dimer studied by Roemelt et al. Hydrogen atoms are omitted for clarity. Reprinted with permission from [23]. Copyright 2018 American chemical society

with experiment. This indicates that orbital optimization is essential for a physically meaningful treatment of the problem. CASSCF(7, 10) calculations indeed change the picture drastically, leading to an $S = 1/2$ ground state. However, at this point a very important observation was made with respect to the method of orbital optimization.

Specifically, when the orbitals of each spin state were optimized individually (state-specific orbital optimization), the relative energies of the four spin states did not follow a regular pattern: the $S = 1/2$ ground state was followed by the ferromagnetic $S = 7/2$ at 12 cm^{-1}, then the $S = 3/2$ state at 16 cm^{-1} and finally the $S = 5/2$ at 28 cm^{-1}. This order of states was confirmed to be the converged result of state-specific CASSCF(7, 10) calculations irrespective of various technical and methodological details. Hence, even though the spin-doublet state turns out to be the lowest in energy, the description of the electronic structure is fundamentally deficient and the results are of no use in the discussion of magnetic properties.

The alternative to state-specific orbital optimization is the state-averaged approach, where a common set of orbitals is obtained as the result of the CASSCF procedure, assigning equal weights to the four states that are optimized simultaneously. Note that this state-averaged approach does not refer to averaging over multiple roots of the same spin multiplicity, but averaging over the lowest root of all the different spin multiplicities that are relevant to the spin-coupling problem, in the present case the lowest root of the $S = 1/2$, $S = 3/2$, $S = 5/2$, and $S = 7/2$ states simultaneously. These state-averaged CASSCF(7, 10) calculations correctly predict antiferromagnetic coupling with a regular Landé progression and spacing of spin states. However, the computed antiferromagnetic coupling at this level was extremely weak ($J = -1.6$ cm^{-1}) and hence the spin ladder was predicted to be highly compressed, spanning merely 24 cm^{-1}.

In comparison to the state-specific CASSCF(7, 10) results, the qualitative success of the state-averaged CASSCF(7, 10) calculations was directly attributable to the use of a common set of orbitals for all states. On the other hand, the quantitative failure

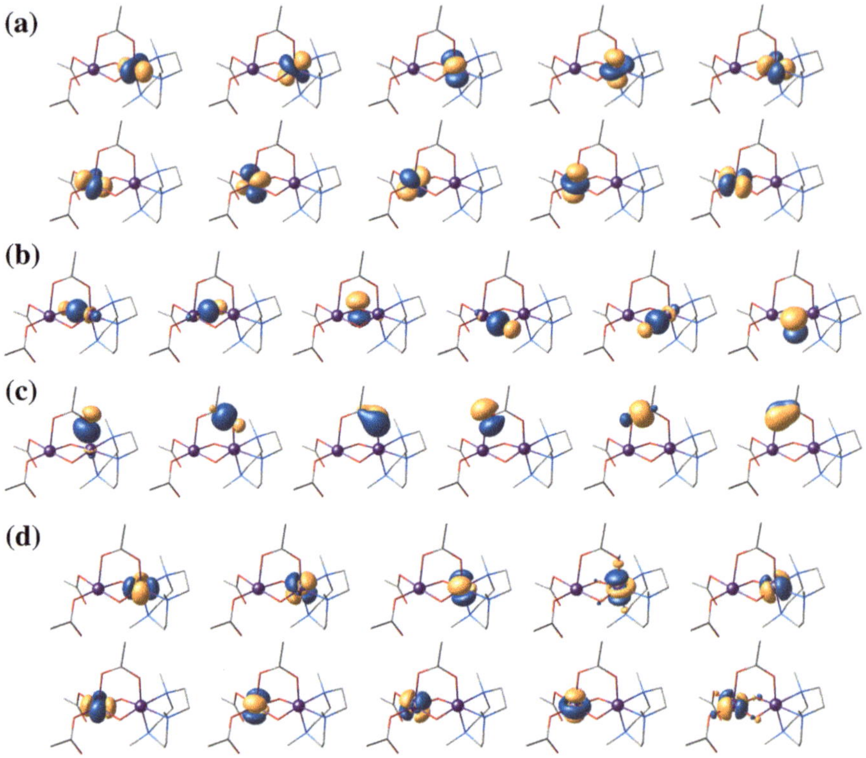

Fig. 6 Localized orbitals employed in the construction of the various active spaces described in the study of Roemelt et al. **a** Mn 3*d* orbitals; **b** O 2*p* orbitals; **c** OAc 2*p* orbitals; and **d** Mn 4*d* orbitals. Reprinted with permission from [23]. Copyright 2018 American chemical society

to approximate the experimental magnitude of the antiferromagnetic coupling was attributed to the exclusion of bridging orbitals from the active space. Subsequent calculations extended the active space to include orbitals of the oxo-bridges, of the acetato bridge, as well as "double-shell" 4*d* orbitals of the Mn ions and 3*p* orbitals of the oxo-bridges (Fig. 6), relying on DMRG to enable multireference calculations with exceedingly large active spaces.

Inclusion of the valence 2*p* orbitals of the oxo-bridges leads to a (19, 16) active space. DMRG-CI calculations without reoptimizing the CASSCF (7, 10) metal-based orbitals or the newly introduced localized orbitals of the oxo-bridges led to a considerable increase in the magnitude of the antiferromagnetic coupling, from less than -2 to $-29\ \text{cm}^{-1}$. Subsequent orbital optimization with state-averaged DMRG-SCF calculations at increasing M values (see below) eventually yielded a converged value for J of almost $-59\ \text{cm}^{-1}$, i.e., approximately two-thirds of the experimental exchange coupling constant. Extension of the active space by inclusion of acetato orbitals, leading to a (31, 22) active space, did not afford any further improvement. In terms of physical insight into the specific system, the above results demonstrate that antiferromagnetic interaction in the Mn dimer arises exclusively through superexchange via the oxo-bridges and that the acetato ligand plays no role in mediating the spin coupling.

A crucial methodological point that was encountered again, even with the more extended active spaces, concerns the use of state-specific versus state-averaged orbital optimization procedures. In contrast to the state-specific calculations with the metal-only (7, 10) active space, which did not produce a qualitatively correct order of states, spate-specific DMRG-SCF(19, 16) calculations correctly produce the ordering of spin states from $S = 1/2$ as the lowest to $S = 7/2$ as the highest. Nevertheless, the spin-state energy spacings deviate strongly from the Landé pattern that is approximated very closely by state-averaged calculations. State-specific calculations produce a strong compression of the spin ladder at progressively higher spin levels, which is particularly exaggerated at low values of M. A crucial observation was that even with fully individually converged absolute energy values (at $M = 2000$) state-specific calculations underestimate the stability of the intermediate $S = 3/2$ and $S = 5/2$ states and overestimate the stability of the high-spin $S = 7/2$ state. The result is that at the level of accuracy required for description of magnetic levels, no meaningful exchange coupling constant can be extracted because the J values computed from energy differences between adjacent levels range from -75 cm^{-1} for the energy difference between the two lowest spin states to -44 cm^{-1} for the energy difference between the two highest spin states.

Importantly, by simply using the energy difference between the two lowest states from state-specific calculations, a deceptively "good" value for J would result, and this effect would be exaggerated at low M values (<1000). These results clearly demonstrate that the method of orbital optimization and the careful examination of convergence with M for all states of the spin ladder are essential for successful applications of DMRG-based approaches to exchange coupling problems.

A second methodological point concerns the convergence with M of the exchange coupling values obtained by state-averaged DMRG-SCF(19, 16) calculations. Table 1 reproduces some of the results from the Roemelt et al. study [23], showing the evolution of the energy levels with the number of renormalized states. The smallest value reported was $M = 250$ because smaller values either led to numerically scattered results or failed to converge. The $M = 250$ results are not physically meaningful as they show no reasonable relation between the spin levels, strongly underestimating the stability of the intermediate $S = 3/2$ and $S = 5/2$ states. $M = 500$ is an improvement but must be considered similarly unusable because of the large differences of the J value obtained for different pairs of states. For $M \geq 1000$, the average J value is converged, but further small improvements are observed up to the highest tested $M = 3000$ with respect to the energy of individual spin states, particularly the $S = 3/2$ state.

In an attempt to further improve the numerical result for the exchange coupling constant, virtual orbitals of the Mn ions (the "double shell" of 4d orbitals) were included in the active space. DMRG-CI(19, 26) calculations demonstrated a small increase in the antiferromagnetic interaction ($J = -65$ cm^{-1}). This improvement was however negated by the further expansion of the active space to include the 3p

Table 1 Energy differences (cm^{-1}) between spin states of the mixed-valence Mn dimer reported by Roemelt et al. [23]. The results are from state-averaged DMRG-SCF calculations with a (19, 16) active space composed of Mn 3*d* and O 2*p* orbitals for different numbers of renormalized states *M*. Pairwise exchange coupling constants and the average *J* value (cm^{-1}) are reported for each *M*. Adapted with permission from [23]. Copyright 2018 American chemical society

S	$M = 250$	$M = 500$	$M = 1000$	$M = 2000$	$M = 3000$
7/2	775.3	868.2	878.7	879.9	879.9
5/2	694.8	486.3	476.3	475.9	475.9
3/2	490.1	215.0	182.6	180.6	180.4
1/2	0.0	0.0	0.0	0.0	0.0
$J_{(7/2–5/2)}$	−11.5	−54.6	−57.5	−57.7	−57.7
$J_{(5/2–3/2)}$	−40.9	−54.3	−58.7	−59.1	−59.1
$J_{(3/2–1/2)}$	−163.4	−71.7	−60.9	−60.2	−60.1
J	−71.9	−60.2	−59.0	−59.0	−59.0

orbitals of the oxo-bridges, as DMRG-CI(19, 32) calculations yielded a value of $J = -60$ cm^{-1}. These results suggest that at the CASCI level the virtual orbitals play a secondary role and their inclusion is not sufficient to overcome the principle limitations of the approach, namely the incomplete account of dynamic correlation.

Full state-averaged orbital optimization could not be completed with the (19, 32) active space; however, DMRG-SCF calculations could be successfully driven to completion with the (19, 26) active space. The latter were reported to be extremely challenging in terms of convergence, but even when converged, the results revealed a new complication. This was the highly increased sensitivity to the *M* value that resulted in strong deviations from the normal inter-level energy spacing even at the highest applicable $M = 1500$. Clearly, what constitutes a sufficient value for *M* is entirely dependent on the nature and composition of the active space. As an example, although $M = 1000$ was perfectly adequate in DMRG-SCF calculations with the (19, 16) active space, when the 4*d* metal orbitals were included in the active space the pairwise values for the exchange coupling constants at $M = 1000$ were $J_{(7/2–5/2)} = -39$ cm^{-1}, $J_{(5/2–3/2)} = -73$ cm^{-1}, and $J_{(3/2–1/2)} = -142$ cm^{-1}, precluding any interpretation. At $M = 1500$, these values improved to $J_{(7/2–5/2)} = -61$ cm^{-1}, $J_{(5/2–3/2)} = -75$ cm^{-1}, and $J_{(3/2–1/2)} = -97$ cm^{-1}, which is still far from convergence. Thus, it was concluded that values of *M* significantly higher than what was technically possible at that point would be necessary to produce converged results. On the other hand, it was noticed that the incoherent results were mostly due to the relative energies of the two intermediate spin states and that the total span of the ladder, i.e., the energy difference between $S = 1/2$ and $S = 7/2$, was less sensitive to the increasing *M*. This observation allowed the estimation that the full effect of the Mn 4*d* orbitals would be a ca. 10 cm^{-1} enhancement of the antiferromagnetic coupling.

The study of the manganese dimer concluded that the enormously increased cost and effort of obtaining reasonably converged state-averaged DMRG-SCF results for active spaces considerably larger than the full-valence (19, 16) space of Mn 3*d* and O 2*p* orbitals, particularly for active spaces that include virtual orbitals, is neither justified by the limited numerical improvements nor expected to eventually produce quantitatively satisfying results. Instead it was suggested that similarly to the standard use of dynamic correlation methods such as CASPT2 and NEVPT2 on top of a CASSCF reference, the full-valence DMRG-SCF wavefunction could be used in subsequent DMRG-NEVPT2 calculations.

Indeed, DMRG-NEVPT2 calculations with the (19, 16) active space produced an average J value of $-85\ \mathrm{cm}^{-1}$ (converged at $M' = 1500$), in very good agreement with the experimental value. Compared to the DMRG-SCF calculations with the same active space, a larger number of retained states were required for satisfactory convergence of the NEVPT2 calculations, because of the requirement to calculate reduced density matrices for more than two active electrons. Importantly, the variations of J values obtained from different spin-state pairs by DMRG-NEVPT2 was less than $1\ \mathrm{cm}^{-1}$, confirming the prediction of Heisenberg behavior by the state-averaged DMRG-SCF calculations.

5 General Remarks

5.1 Active Space Composition

A common choice in the case studies mentioned above is that the smallest useful active space in practice contains the metal *d* orbitals as well as the valence orbitals of the bridging ligands. Although the results obtained from DMRG-SCF calculations with a metal-only active space are typically not numerically or qualitatively useful in themselves, this does not mean that an active space composed of only metal-based orbitals is a meaningless choice in principle. In certain types of application, this can indeed form a well-defined starting point for certain computational approaches, such as the difference-dedicated configuration interaction approach (DDCI) that attempts to introduce a posteriori all the important corrections which are by definition absent from the small reference wavefunction. However, these approaches lack generality because of their extremely restricted field of application given their enormous cost. The point of using DMRG is precisely that the severe restrictions on the size of the active space can be lifted at the reference level, which not simply extends the applicability of multireference methods to any exchange-coupled transition metal system but, importantly, allows explicit inclusion of a large part of the required physics directly into the reference wavefunction. Therefore, we consider the inclusion of all metal and ligand valence orbitals to be the natural minimal choice in DMRG-based studies of exchange-coupled systems.

On the other hand, the question of whether virtual orbitals of the metal ions and the bridging ligands are required in the active space, for example, the double shell of 4*d* orbitals in the case of 3*d* metal ions, does not appear to have a straightforward and universal answer based on the existing studies. It is clear that these orbitals have an effect, which can be attributed in part to further recovery of dynamic correlation or incorporation of specific excitation classes, but the magnitude of this effect appears to be system-dependent. Therefore, this type of active space extension should be evaluated in combination with the estimated gains and in relation to the relative cost. This point became obvious with the MRCI+Q results in the study of Harris et al. [22] and was further discussed in the study of Roemelt et al. where it was judged that the steep increase in cost outweighs the gains, particularly if one considers that the "missing" part of the exchange coupling interaction is unlikely to be sufficiently recovered by the double-shell extension of the active space and can be more conveniently accounted for with the DMRG-NEVPT2 approach [97].

However, this is likely not a general result. The conclusion may be in part due to the fact that the d shell of the Mn ions is less than half-filled. In cases where the double-shell effect is strong, it is expected that it has to be treated as part of the static correlation [123] and included directly in the orbital optimization. In addition, CASPT2/NEVPT2 energies in general react more sensitively to the double-shell effect than CASSCF. Although no general guideline with respect to the treatment of virtual orbitals can be given at this stage, we expect that the increasing use of automated or semi-automated active space selection procedures based on entanglement measures [84] will enable a more efficient and systematic approach to determining active space composition for DMRG-SCF calculations on exchange-coupled systems.

5.2 *Orbital Optimization, State Selection and Convergence*

Besides the definition of the active space, a decision that crucially determines the nature of results and conclusions concerns the states targeted and the method employed for orbital optimization, given that the form of the active orbitals has a crucial effect on the prediction of magnetic properties [124]. Of the case studies discussed above, only the work of Roemelt et al. [23] contrasted explicitly the results of state-specific versus state-averaged orbital optimization. The study of the Mn dimer vividly demonstrated that state-specific CASSCF calculations with a metal-only active space lead to erratic results, while state-specific DMRG-SCF calculations with a full-valence active space, even though not producing qualitatively unreasonable values, still lead to large and experimentally incompatible deviations from the Landé spacing of spin states. It was concluded that state-averaging is the preferred approach because it minimizes these artificial deviations from the Landé pattern. This means that the orbital optimization procedure should produce simultaneously orbitals equally good for *all* states of the spin ladder that describe the magnetic interaction in the exchange-coupled system.

As a corollary, extraction of a hypothetical exchange coupling constant J based on only two computed states using state-specific calculations is unreliable because the assumption of a Landé pattern may not hold at all. With the energies of only two states, whether they are assumed to be the two lowest energy ones or the states with highest and lowest spin multiplicity for a given spin-coupling situation, one could still remain blind to potentially fundamental deficiencies in the description of the electronic structure of the system. State-averaged orbital optimization can obviously be more expensive than a state-specific approach, but as a counterweight it usually has the advantage of more efficient convergence of the CASSCF procedure, associated with the treatment of all states arising from a given electronic configuration. In this respect, it should be realized that both the inability to adequately approximate an experimental exchange coupling constant and the encounter of convergence problems for specific magnetic states may not be due to active space selection but due to the choice of the orbital optimization procedure. This may relate, for example, to the observation of Harris et al. that certain spin states of the chromium dimer could not be converged (in state-specific calculations) for specific choices of M and active space [22].

State-averaged orbital optimization does not automatically eliminate Landé deviations; there is still a strong dependence of the relative energies of the spin states on the number of retained states M, as shown in Table 1. Therefore, state-averaged orbital optimization should be combined with careful examination of the convergence with M of pairwise energy differences between the different spin states in order to determine the point where converged values for the exchange coupling problem can be obtained. It is important to note that at small values of M, the results on the Mn dimer can be considered numerically unstable [23], but results obtained with such small M values have been used in extrapolating spin-state energies in the study of Harris et al. [22]. It will be interesting to see the effect of state-averaged orbital optimization in the case of the Fe and Cr dimers. It is clear in both studies that regardless of the method employed for orbital optimization, different spin states converge at different rates with increasing M. The energies of individually optimized spin states can in principle be extrapolated to infinite M (see Fig. 3), whereas it is not clear how this can be performed in the case of state-averaged results, especially when the discarded weight for all states becomes negligible at high enough M values [23]. Further studies of exchange-coupled systems will be required to better evaluate the various methodological parameters relating to spin state selection, orbital optimization, and convergence of relative energetics with M. An important question to clarify for large-active-space DMRG-SCF calculations is whether it will be possible to avoid or relax the requirement for state-averaged orbital optimization over the complete span of the spin ladder, because this seriously limits the nuclearity of the complexes that can be successfully treated.

5.3 Deviations from Heisenberg Behavior

From the preceding discussion, it can be concluded that if a state-averaged approach is required for the qualitatively correct description of magnetic levels, then the use of state-specific DMRG-SCF calculations to predict *deviations* from Heisenberg behavior is unjustified. Such deviations can of course be perfectly physical: the isotropic bilinear Heisenberg Hamiltonian is but an idealized approximation and further terms can be invoked to account for specific observed deviations in the magnetic behavior of a system, e.g., non-Hund states [2, 125]. As such, the reproduction of a Landé pattern should not be considered as a desirable "target" for quantum chemical treatments of exchange-coupled dimers. On the other hand, the analysis of Landé deviations for the systems discussed above suggests that such deviations result from specific methodological choices, and at least in certain cases they can be viewed as artifacts of either a small active space, state-specific orbital optimization, small M, or a combination of the above, with state-averaged orbital optimization being the most important factor in avoiding such artifacts.

The computed energetics in the study of Fe and Cr dimers suggested a progressive compression of the spin ladder at higher S states (see Fig. 4) [22], which Harris et al. considered to be a genuine and physically meaningful demonstration of non-Heisenberg behavior. Consequently, they fitted the computed energies with an additional biquadratic term in the Hamiltonian. However, de Graaf and coworkers did not observe this type of deviation when they revisited one of the dimers of the Harris et al. study [118]. The critical difference in this subsequent study was the use of state-averaged orbitals, which apparently eliminated the artificial compression of the spin ladder. The direct comparison between state-specific and state-averaged orbitals by Roemelt et al. for the Mn dimer quantified explicitly the significant differences between state-specific and state-averaged results, concluding that only the latter afford a valid description of the spin ladder. One might argue that the use of state-averaged orbitals could introduce a bias toward isotropic behavior. However, this would be incorrect for two reasons: first, because the magnetic levels do normally arise from a single principle electronic configuration, and second, because the use of state-averaged orbitals does not impose an idealized isotropic spacing of the magnetic levels anyway, as demonstrated clearly from the results on the Mn dimer.

The question, therefore, is when should computed deviations be considered physically meaningful? One safe conclusion so far is that state-specific calculations are inappropriate for this problem. This has wider implications for any study that attempts to employ large-active-space DMRG calculations for the analysis of exchange-coupled systems, but even more so for studies that aspire to directly predict such deviations. An example of the latter is the investigation of double exchange in a series of iron–sulfur systems by Sharma et al. [20], who employed DMRG to study a series of iron–sulfur systems that can be considered models of the Fe/S cofactors in biological electron transfer. Among the complexes investigated was the $[Fe_2S_2(SCH_3)_4]^{2-/3-}$ pair. In the oxidized form both ions are high-spin Fe(III), with local spins $S_A = S_B = 5/2$, which lead to the same set of total spin states as for the iron dimer discussed above, i.e. $S = 0, 1, 2, 3, 4$, and 5. The reduced dimer formally contains one high-spin Fe(III) ion ($S_A = 5/2$) and one Fe(II) ($S_B = 2$). This is a particularly challenging case because the additional electron can delocalize between the two iron sites. In the classic picture, this leads to the splitting of the individual Heisenberg spin levels S by a "double-exchange" term $\pm B(S + 1/2)$. Sharma et al. investigated the low-energy spectrum for this complex, albeit without performing spin-state averaging of orbitals. They noticed that the low-lying states could not be adequately described by the classical double-exchange Hamiltonian because their number was greater than what is accounted for by the classical phenomenological model as a result of multi-orbital double-exchange processes [20]. It remains to be seen how the choice of orbital optimization affects the conclusions in such electronic situations, electronically more complex than the exchange-coupled dimers discussed by Harris et al. and Roemelt et al., but potentially also more sensitive to methodological choices. At least in the more straightforward examples of the dimers discussed as case studies herein, where the spin manifold is sufficiently separated from excited electronic configurations, the quality of the results is adversely affected by the use of spin-state-specific energies.

5.4 Analysis of Exchange Coupling

The investigation by Roemelt et al. into the effect of different bridging ligands in the case of the Mn dimer by including subsets of bridge-localized orbitals in the active space established that the oxo-bridges mediate exchange coupling but that the acetato bridge plays only a structural and not a magnetic role [23, 45]. This application serves to demonstrate an important use of DMRG in exchange-coupled systems, namely that by selective inclusion of localized orbital subspaces in the multireference treatment one can systematically evaluate the contribution of distinct magnetic pathways. Thus, even if the absolute value of the computed exchange coupling constants is not in quantitative agreement with experiment, this approach enables the mapping of the "magnetic topology" of a complex. Analysis of broken-symmetry determinants along the lines of the Amos–Hall corresponding orbital transformation have long been used for qualitative analysis of superexchange pathways in DFT studies of exchange-coupled systems [33, 45], but is not directly applicable to systems with more than two spin sites [47]. By contrast, the DMRG-driven analysis based on selective active space inclusion of valence orbitals of specific groups is general and can be applied in any chemical context. Evidently, this approach is not unique to DMRG approaches. For example, Domingo et al. [126] have used standard CASPT2 calculations with localized orbitals to investigate the influence of different parts of the molecule on the overall magnetism of a series of dinuclear molecules. Still, DMRG enables this treatment to be applied to much larger systems than previously possible in terms of size, nuclearity, magnetic topology, type, and number of bridging ligands. This type of treatment should nevertheless not be viewed as "quantitative," because of the missing contributions from dynamic correlation and possible cross-interactions.

A potentially more powerful way of analyzing exchange coupling interactions would be to use orbital entanglement measures, again in combination with localized orbital subspaces. To our knowledge, the capabilities of such an approach for investigating and defining the magnetic topology of exchange-coupled transition metal complexes are still unexplored.

6 Summary and Perspectives

The new capabilities offered by the DMRG algorithm in terms of handling large active spaces in multiconfigurational SCF calculations of exchange-coupled transition metal systems have already led to pioneering applications in dinuclear complexes. The studies discussed in this chapter demonstrate that the active space limitations of traditional CASSCF approaches can largely be lifted. This promises that a multireference description is in principle achievable for any transition metal dimer regardless of the nature and oxidation state or electron configuration of the metal ions, with active spaces that at the very least are "valence-complete" in terms of including orbitals of all bridging ligands that could mediate superexchange. However, the existing in-depth studies on dinuclear complexes also highlight a number of issues that need to be taken into account and some possible problems that need to be addressed in future applications.

First of all, it is clear that in dealing with magnetic coupling orbital optimization is essential and CI approaches do not lead to useful results. At the same time, it appears that state-averaged calculations over all different spin states of the spin ladder are required to obtain correct relative energies. This can be a major obstacle in extending DMRG-SCF to systems of higher nuclearity, not necessarily because the active space would become exceedingly large, but because state-averaged orbital optimization might not be feasible over hundreds or thousands of roots encompassing all possible spin multiplicities. On the other hand, state-specific orbital optimizations, and perhaps state-averaged but spin-specific orbital optimizations, run the risk of introducing large errors in the relative energetics that can seriously undermine the quality of the results. It is not clear at this point how this conundrum can be answered, but it will certainly be an important target of future studies. Finally, it should be recognized that DMRG-SCF calculations reported to date have not demonstrably converged in a systematically defined manner to the experimental exchange coupling constants for all dinuclear complexes investigated. It is expected that, in general, quantitative predictions will still require treatment of dynamic electron correlation to achieve high accuracy. The use of DMRG-NEVPT2 has already been demonstrated in the case of a manganese dimer but other methods need to be explored and evaluated [127–132]. DMRG-based multiconfigurational approaches offer undoubtedly a new basis not simply for obtaining numerically useful results for exchange-coupled systems but for analyzing their electronic structure, investigating their magnetic topology, and enabling an improved understanding of their magnetic properties. The case studies discussed in this chapter simply break the ground for what is expected to be a challenging, exciting, and richly rewarding new field of quantum chemistry.

References

1. Kahn O (1993) Molecular magnetism. Wiley, New York
2. Malrieu JP, Caballol R, Calzado CJ, de Graaf C, Guihéry N (2014) Chem Rev 114:429–492
3. Moreira IDPR, Illas F (2006) Phys Chem Chem Phys 8:1645–1659
4. Neese F (2009) Coord Chem Rev 253:526–563
5. Illas F, Moreira IDPR, de Graaf C, Barone V (2000) Theor Chem Acc 104:265–272
6. Caballol R, Castell O, Illas F, Moreira IDPR, Malrieu JP (1997) J Phys Chem A 101:7860–7866
7. Noodleman L (1981) J Chem Phys 74:5737–5743
8. Noodleman L, Davidson ER (1986) Chem Phys 109:131–143
9. Yamaguchi K, Tsunekawa T, Toyoda Y, Fueno T (1988) Chem Phys Lett 143:371–376
10. Yamanaka S, Kawakami T, Nagao H, Yamaguchi K (1994) Chem Phys Lett 231:25–33
11. De Loth P, Cassoux P, Daudey JP, Malrieu JP (1981) J Am Chem Soc 103:4007–4016
12. White SR, Martin RL (1999) J Chem Phys 110:4127–4130
13. Schollwöck U (2011) Ann Phys 326:96–192
14. Moritz G, Wolf A, Reiher M (2005) J Chem Phys 123:184105
15. Marti KH, Ondík IM, Moritz G, Reiher M (2008) J Chem Phys 128:014104
16. Freitag L, Knecht S, Keller SF, Delcey MG, Aquilante F, Pedersen TB, Lindh R, Reiher M, Gonzalez L (2015) Phys Chem Chem Phys 17:14383–14392
17. Hu W, Chan GKL (2015) J Chem Theory Comput 11:3000–3009
18. Kurashige Y, Yanai T (2009) J Chem Phys 130:234114
19. Kurashige Y, Chan GKL, Yanai T (2013) Nat Chem 5:660–666
20. Sharma S, Sivalingam K, Neese F, Chan GKL (2014) Nat Chem 6:927–933
21. Paul S, Cox N, Pantazis DA (2017) Inorg Chem 56:3875–3888
22. Harris TV, Kurashige Y, Yanai T, Morokuma K (2014) J Chem Phys 140:054303
23. Roemelt M, Krewald V, Pantazis DA (2018) J Chem Theory Comput 14:166–179
24. Chan GKL, Head-Gordon M (2002) J Chem Phys 116:4462–4476
25. Olivares-Amaya R, Hu W, Nakatani N, Sharma S, Yang J, Chan GKL (2015) J Chem Phys 142:034102
26. Keller SF, Reiher M (2014) Chimia 68:200–203
27. Knecht S, Hedegård ED, Keller S, Kovyrshin A, Ma Y, Muolo A, Stein CJ, Reiher M (2016) Chimia 70:244–251
28. Bencini A, Gatteschi D (1990) EPR of exchange coupled systems. Springer Verlag, Berlin
29. Pantazis DA, Orio M, Petrenko T, Zein S, Bill E, Lubitz W, Messinger J, Neese F (2009) Chem Eur J 15:5108–5123
30. Ginsberg AP (1980) J Am Chem Soc 102:111–117
31. Mouesca JM, Noodleman L, Case DA, Lamotte B (1995) Inorg Chem 34:4347–4359
32. Noodleman L, Post D, Baerends EJ (1982) Chem Phys 64:159–166
33. Neese F (2004) J Phys Chem Solids 65:781–785
34. Dai D, Whangbo M-H (2003) J Chem Phys 118:29–39
35. Ruiz E, Cano J, Alvarez S, Alemany P (1999) J Comput Chem 20:1391–1400
36. Yamaguchi K, Takahara Y, Fueno T (1986) Ab-initio molecular orbital studies of structure and reactivity of transition metal-oxo compounds. In: Smith VH Jr, Scheafer HF III, Morokuma K (eds) Applied quantum chemistry. D. Reidel, Boston, pp 155–184
37. Golub GH, Reinsch C (1970) Numer Math 14:403–420

38. Pantazis DA, Orio M, Petrenko T, Zein S, Lubitz W, Messinger J, Neese F (2009) Phys Chem Chem Phys 11:6788–6798
39. Shoji M, Koizumi K, Kitagawa Y, Kawakami T, Yamanaka S, Okumura M, Yamaguchi K (2006) Chem Phys Lett 432:343–347
40. Desplanches C, Ruiz E, Alvarez S (2003) Eur J Inorg Chem 2003:1756–1760
41. Hubner O, Fink K, Klopper W (2007) Phys Chem Chem Phys 9:1911–1920
42. Coulaud E, Malrieu J-P, Guihéry N, Ferré N (2013) J Chem Theory Comput 9:3429–3436
43. Gutierrez L, Alzuet G, Real JA, Cano J, Borrás J, Castiñeiras A (2000) Inorg Chem 39:3608–3614
44. Rodríguez-Fortea A, Alemany P, Alvarez S, Ruiz E (2001) Chem Eur J 7:627–637
45. Pantazis DA, Krewald V, Orio M, Neese F (2010) Dalton Trans 39:4959–4967
46. Coulaud E, Guihéry N, Malrieu J-P, Hagebaum-Reignier D, Siri D, Ferré N (2012) J Chem Phys 137:114106
47. Krewald V, Neese F, Pantazis DA (2013) J Am Chem Soc 135:5726–5739
48. Orio M, Pantazis DA, Petrenko T, Neese F (2009) Inorg Chem 48:7251–7260
49. Ciofini I, Daul CA (2003) Coord Chem Rev 238–239:187–209
50. Bencini A, Totti F (2009) J Chem Theory Comput 5:144–154
51. Ruiz E (2004) Struct Bond 113:91–102
52. Comba P, Hausberg S, Martin B (2009) J Phys Chem A 113:6751–6755
53. Illas F, Moreira IDPR, Bofill JM, Filatov M (2004) Phys Rev B 70:132414
54. Illas F, Moreira IDPR, Bofill JM, Filatov M (2006) Theor Chem Acc 116:587–597
55. Zein S, Poor Kalhor M, Chibotaru LF, Chermette H (2009) J Chem Phys 131:224316
56. Neese F (2006) J Biol Inorg Chem 11:702–711
57. Amos AT, Hall GG (1961) Proc R Soc Lond A 263:483–493
58. Malmqvist P-Å, Roos BO (1989) Chem Phys Lett 155:189–194
59. Roos BO, Taylor PR, Siegbahn PEM (1980) Chem Phys 48:157–173
60. Andersson K, Roos BO (1992) Chem Phys Lett 191:507–514
61. Pierloot K, Persson BJ, Roos BO (1995) J Phys Chem 99:3465–3472
62. Andersson K, Malmqvist PA, Roos BO, Sadlej AJ, Wolinski K (1990) J Phys Chem 94:5483–5488
63. Finley J, Malmqvist P-Å, Roos BO, Serrano-Andrés L (1998) Chem Phys Lett 288:299–306
64. Angeli C, Cimiraglia R, Evangelisti S, Leininger T, Malrieu JP (2001) J Chem Phys 114:10252–10264
65. Angeli C, Cimiraglia R, Malrieu J-P (2002) J Chem Phys 117:9138–9153
66. Miralles J, Daudey JP, Caballol R (1992) Chem Phys Lett 198:555–562
67. Miralles J, Castell O, Caballol R, Malrieu JP (1993) Chem Phys 172:33–43
68. Castell O, Caballol R (1999) Inorg Chem 38:668–673
69. Calzado CJ, Cabrero J, Malrieu JP, Caballol R (2002) J Chem Phys 116:2728–2747
70. Calzado CJ, Cabrero J, Malrieu JP, Caballol R (2002) J Chem Phys 116:3985–4000
71. Calzado CJ, Angeli C, Taratiel D, Caballol R, Malrieu JP (2009) J Chem Phys 131:044327
72. Queralt N, Taratiel D, de Graaf C, Caballol R, Cimiraglia R, Angeli C (2008) J Comput Chem 29:994–1003
73. Olsen J, Roos BO, Jørgensen P, Jensen HJA (1988) J Chem Phys 89:2185–2192
74. Malmqvist PA, Rendell A, Roos BO (1990) J Phys Chem 94:5477–5482
75. Ma D, Li Manni G, Gagliardi L (2011) J Chem Phys 135:044128
76. Li Manni G, Aquilante F, Gagliardi L (2011) J Chem Phys 134:034114
77. Li Manni G, Ma D, Aquilante F, Olsen J, Gagliardi L (2013) J Chem Theory Comput 9:3375–3384
78. Thomas RE, Sun Q, Alavi A, Booth GH (2015) J Chem Theory Comput 11:5316–5325
79. Li Manni G, Smart SD, Alavi A (2016) J Chem Theory Comput 12:1245–1258
80. Moritz G, Hess BA, Reiher M (2005) J Chem Phys 122:024107
81. Zgid D, Nooijen M (2008) J Chem Phys 128:014107
82. Chan GKL, Sharma S (2011) Annu Rev Phys Chem 62:465–481
83. Boguslawski K, Tecmer P, Legeza Ö, Reiher M (2012) J Phys Chem Lett 3:3129–3135

84. Stein CJ, Reiher M (2016) J Chem Theory Comput 12:1760–1771
85. Ma Y, Knecht S, Keller S, Reiher M (2017) J Chem Theory Comput 13:2533–2549
86. Baiardi A, Stein CJ, Barone V, Reiher M (2017) J Chem Theory Comput 13:3764–3777
87. Battaglia S, Keller S, Knecht S (2018) J Chem Theory Comput 14:2353–2369
88. Moritz G, Reiher M (2007) J Chem Phys 126:244109
89. Sobczak P, Barasiński A, Drzewiński A, Kamieniarz G, Kłak J, Bieńko A, Mroziński J (2009) Polyhedron 28:1838–1841
90. Barasiński A, Sobczak P, Drzewiński A, Kamieniarz G, Bieńko A, Mroziński J, Gatteschi D (2010) Polyhedron 29:1485–1491
91. Schnack J, Ummethum J (2013) Polyhedron 66:28–33
92. Keller S, Boguslawski K, Janowski T, Reiher M, Pulay P (2015) J Chem Phys 142:244104
93. Sayfutyarova ER, Sun Q, Chan GK, Knizia G (2017) J Chem Theory Comput 13:4063–4078
94. Bao JJ, Dong SS, Gagliardi L, Truhlar DG (2018) J Chem Theory Comput 14:2017–2025
95. Barcza G, Legeza Ö, Marti KH, Reiher M (2011) Phys Rev A 83
96. Saitow M, Kurashige Y, Yanai T (2015) J Chem Theory Comput 11:5120–5131
97. Guo S, Watson MA, Hu W, Sun Q, Chan GKL (2016) J Chem Theory Comput 12:1583–1591
98. Roemelt M, Guo S, Chan GKL (2016) J Chem Phys 144:204113
99. Yanai T, Saitow M, Xiong XG, Chalupský J, Kurashige Y, Guo S, Sharma S (2017) J Chem Theory Comput 13:4829–4840
100. Boguslawski K, Tecmer P, Barcza G, Legeza Ö, Reiher M (2013) J Chem Theory Comput 9:2959–2973
101. Wouters S, Bogaerts T, Van Der Voort P, Van Speybroeck V, Van Neck D (2014) J Chem Phys 140:241103
102. Kurashige Y, Saitow M, Chalupský J, Yanai T (2014) Phys Chem Chem Phys 16:11988–11999
103. Chalupský J, Rokob TA, Kurashige Y, Yanai T, Solomon EI, Rulíšek L, Srnec M (2014) J Am Chem Soc 136:15977–15991
104. Ma Y, Knecht S, Reiher M (2017) Chem Phys Chem 18:384–393
105. Phung QM, Domingo A, Pierloot K (2018) Chem Eur J 24:5183–5190
106. Phung QM, Pierloot K (2018) Phys Chem Chem Phys 20:17009–17019
107. Dong G, Phung QM, Hallaert SD, Pierloot K, Ryde U (2017) Phys Chem Chem Phys 19:10590–10601
108. Zhang C, Chen C, Dong H, Shen JR, Dau H, Zhao J (2015) Science 348:690–693
109. Paul S, Neese F, Pantazis DA (2017) Green Chem 19:2309–2325
110. Umena Y, Kawakami K, Shen J-R, Kamiya N (2011) Nature 473:55–60
111. Pantazis DA, Ames W, Cox N, Lubitz W, Neese F (2012) Angew Chem Int Ed 51:9935–9940
112. Krewald V, Retegan M, Pantazis DA (2016) Top Curr Chem 371:23–48
113. Beinert H, Holm RH, Münck E (1997) Science 277:653–659
114. Lledós A, Moreno-Mañas M, Sodupe M, Vallribera A, Mata I, Martínez B, Molins E (2003) Eur J Inorg Chem 2003:4187–4194
115. Boguslawski K, Marti KH, Reiher M (2011) J Chem Phys 134:224101
116. Bencini A, Totti F (2008) Inorg Chim Acta 361:4153–4156
117. Bastardis R, Guihéry N, de Graaf C (2008) J Chem Phys 129:104102
118. Spivak M, Angeli C, Calzado CJ, de Graaf C (2014) J Comput Chem 35:1665–1671
119. Bossek U, Weyhermueller T, Wieghardt K, Nuber B, Weiss J (1990) J Am Chem Soc 112:6387–6388
120. Kanady JS, Tsui EY, Day MW, Agapie T (2011) Science 333:733–736
121. Mukherjee S, Stull JA, Yano J, Stamatatos TC, Pringouri K, Stich TA, Abboud KA, Britt RD, Yachandra VK, Christou G (2012) Proc Natl Acad Sci USA 109:2257–2262
122. Krewald V, Retegan M, Neese F, Lubitz W, Pantazis DA, Cox N (2016) Inorg Chem 55:488–501
123. Stein CJ, von Burg V, Reiher M (2016) J Chem Theory Comput 12:3764–3773
124. Angeli C, Calzado CJ (2012) J Chem Phys 137:034104
125. Bjornsson R, Lima FA, Spatzal T, Weyhermüller T, Glatzel P, Bill E, Einsle O, Neese F, DeBeer S (2014) Chem Sci 5:3096–3103

126. Domingo A, Specklin D, Rosa V, Mameri S, Robert V, Welter R (2014) Eur J Inorg Chem 2014:2552–2560
127. Kurashige Y, Yanai T (2011) J Chem Phys 135:094104
128. Saitow M, Kurashige Y, Yanai T (2013) J Chem Phys 139:044118
129. Sharma S, Chan GKL (2014) J Chem Phys 141:111101
130. Yanai T, Kurashige Y, Mizukami W, Chalupský J, Lan TN, Saitow M (2015) Int J Quantum Chem 115:283–299
131. Veis L, Antalík A, Brabec J, Neese F, Legeza Ö, Pittner J (2016) J Phys Chem Lett 7:4072–4078
132. Phung QM, Wouters S, Pierloot K (2016) J Chem Theory Comput 12:4352–4361

New Strategies in Modeling Electronic Structures and Properties with Applications to Actinides

Aleksandra Leszczyk, Paweł Tecmer and Katharina Boguslawski

Abstract This chapter discusses contemporary quantum chemical methods and provides general insights into modern electronic structure theory with a focus on heavy-element-containing compounds. We first give a short overview of relativistic Hamiltonians that are frequently applied to account for relativistic effects. Then, we scrutinize various quantum chemistry methods that approximate the N-electron wave function. In this respect, we will review the most popular single- and multi-reference approaches that have been developed to model the multi-reference nature of heavy element compounds and their ground- and excited-state electronic structures. Specifically, we introduce various flavors of post-Hartree–Fock methods and optimization schemes like the complete active space self-consistent field method, the configuration interaction approach, the Fock-space coupled cluster model, the pair-coupled cluster doubles ansatz, also known as the antisymmetric product of 1 reference orbital geminal, and the density matrix renormalization group algorithm. Furthermore, we will illustrate how concepts of quantum information theory provide us with a qualitative understanding of complex electronic structures using the picture of interacting orbitals. While modern quantum chemistry facilitates a quantitative description of atoms and molecules as well as their properties, concepts of quantum information theory offer new strategies for a qualitative interpretation that can shed new light onto the chemistry of complex molecular compounds.

A. Leszczyk · P. Tecmer · K. Boguslawski (✉)
Faculty of Physics, Astronomy and Informatics, Institute of Physics,
Nicolaus Copernicus University in Torun, Grudziadzka 5, 87-100 Toruń, Poland
e-mail: k.boguslawski@fizyka.umk.pl

A. Leszczyk
e-mail: all@fizyka.umk.pl

P. Tecmer
e-mail: ptecmer@fizyka.umk.pl

K. Boguslawski
Faculty of Chemistry, Nicolaus Copernicus University in Torun, Gagarina 7,
87-100 Toruń, Poland

E. Broclawik et al. (eds.), *Transition Metals in Coordination Environments*,
Challenges and Advances in Computational Chemistry and Physics 29,
https://doi.org/10.1007/978-3-030-11714-6_5

1 Introduction

One of the main goals of quantum chemistry is to understand the physicochemical properties of atoms, molecules, and materials using the first principles. This knowledge can be further used to interpret and explain existing experimental data or to design new compounds with much sought-after properties. However, the molecules under investigation usually contain numerous interacting electrons, which leads to a complex computational problem with a large number of degrees of freedom. The interplay between relativistic effects, the correlated motion of electrons, and the basis set quality is the main difficulty that limits the possibility to express the electronic wave function in exact form. Various quantum chemical methods have been successfully applied to molecular systems where these effects play a minor role. However, molecules containing heavy elements like actinides or other d- and f-block elements still pose a challenge to quantum chemistry as both correlation and relativistic effects have a dominant contribution to their electronic structure.

In this chapter, we review conventional and unconventional quantum chemical theories that are applicable to heavy-element chemistry like actinide-containing compounds. Our discussion starts with presenting the properties of actinides as an example of complex many-electron systems. Then, we briefly summarize some popular approaches that account for relativistic effects, followed by electronic structure methods that optimize (approximate) electronic wave functions for ground and excited states. Furthermore, we outline how information from the electronic wave function can be extracted to obtain a qualitative interpretation of electronic structures. Specifically, our analysis covers concepts of quantum information theory. Finally, we present some challenging examples of computational actinide chemistry that highlight the difficulty in describing the electronic structure of actinide-containing compounds.

2 A Brief Overview of Actinides and Their Complex Electronic Structure

Heavy elements with atomic numbers ranging from 89 to 103 form a distinct group in the periodic table known as actinides. This series includes actinium, the early actinides (thorium, protactinium, uranium, and neptunium), the middle actinides (plutonium, americium, curium, berkelium, and californium), and the late actinides (einsteinium, fermium, mendelevium, nobelium, and lawrencium). All elements are radioactive metals and almost all of them are characterized by short lifetimes. Only some isotopes of thorium and uranium elements have long lifetimes and thus can be found in nature. Thorium, uranium, neptunium, plutonium, americium, and curium have important applications in the nuclear industry, whereas thorium and uranium are also exploited in catalysis.

In actinide elements, the $5f$ electrons strongly interact with the remaining valence electrons as well as with each other. This interaction implicates an electronic structure composed of many quasi-degenerate electronic configurations. Examples are the $5f^n7s^2$ or $5f^{n-1}6d7s^2$ series of electronic configurations, where n represents the number of electrons in the $5f$ shell and is defined as $Z - 88$, with Z being the atomic number. These numerous energetically close-lying electronic configurations are, however, of different character across the actinide series, which causes irregularities in the electronic ground and excited state energies of actinide elements.

Similar to transition metals, the highest (formal) oxidation state of the early actinides equals the total number of electrons that can be removed from the valence shell, that is, from the $6d$ and $5f$ atomic orbitals. Furthermore, the early actinides resemble transition metals also in terms of orbitals and valence properties. The main reason for the close resemblance of actinides and transition metals is that the actinide $6d$ orbitals do participate in chemical bonding with other elements [102, 137]. Recently, Wilson et al. [164] observed the energetic crossing of the $5f$ and $6d$ atomic states for protactinium, making the protactinium atom a potential crossing point of valence properties that are characteristic for either transition metals or actinides. Using quantum chemistry methods, the authors provided numerical evidence that both the $5f$ and $6d$ orbitals participate in the chemistry of Pa and that the participation of the $5f$ orbitals increases for the middle actinides.

Unfortunately, experimental manipulations with actinide species are very limited, primarily because most actinide atoms are unstable, feature a large number of various oxidation states, or are radiotoxic. Despite these technical difficulties, experimental actinide chemistry remains an active field of research that mainly focuses on molecular synthesis of compounds containing thorium and uranium as well as spectroscopic studies of such compounds [5, 34, 42, 52, 53]. Due to these difficulties, theoretical modeling of actinide-containing compounds can complement experimental studies and provide the much sought-after insights into the physico-chemical properties of actinide complexes and clusters, their reaction mechanisms, and thermochemistry. However, theoretical modeling of actinide chemistry is challenging for present-day quantum chemistry as our theoretical model has to account for (i) relativistic effects and (ii) the correlated motion of electrons.

Due to the large atomic number present in actinide atoms, relativistic effects considerably affect the electronic structure of actinide-containing compounds and may change the character of the principle configuration compared to calculations where relativistic effects are ignored. For instance, the relativistic mass correction to the core electrons causes the contraction of their corresponding orbital radii, while the valence orbitals are expanded leading to elongated chemical bonds [6]. Furthermore, spin-orbit interactions, which are comparable in magnitude to the electron-electron repulsion energy, reduce the degeneracies of states with non-zero angular momentum [86]. To appropriately model the correlated motion of the electrons, our electronic structure method has to include all degenerate or quasi-degenerate, low-lying electronic configurations resulting from the energetic proximity of the actinide valence 5*f*, 6*d*, and 7*s* orbitals. Such calculations are usually rather expensive. Hence, various approximations have been introduced in quantum chemistry that allow us to efficiently treat (quasi-)degeneracies.

3 Electronic Structure Methods in Quantum Chemistry

In the standard formulation of quantum chemistry, the quantum state of atoms and molecules consisting of N electrons and M nuclei is described by the total wave function $\Psi(\boldsymbol{x}, \boldsymbol{R})$, which depends on the spatial and spin coordinates $\boldsymbol{x} \equiv \{\boldsymbol{x}_1, \boldsymbol{x}_2, ..., \boldsymbol{x}_N\}$ of all electrons as well as on the spatial coordinates of all nuclei $\boldsymbol{R} \equiv \{\boldsymbol{R}_1, \boldsymbol{R}_2, ..., \boldsymbol{R}_M\}$. In quantum chemistry, we are usually interested in the electronic part of the wave function at a given molecular geometry, for instance the equilibrium structure. Within the so-called Born-Oppenheimer approximation, the total wave function is then written as a product of a nuclear part and an electronic part. In particular, the electronic part of the total wave function $\Psi_{\text{el}}(\boldsymbol{x}; \boldsymbol{R})$ depends on all electronic coordinates, while the positions of the nuclei remain fixed and enter the wave function as parameters. In non-relativistic quantum chemistry, the electronic wave function is obtained by solving the time-independent, electronic Schrödinger equation

$$H_{\text{el}}\Psi_{\text{el}}(\boldsymbol{x}; \boldsymbol{R}) = E_{\text{el}}\Psi_{\text{el}}(\boldsymbol{x}; \boldsymbol{R}), \tag{1}$$

where H_{el} denotes the Hamiltonian of the system, whose eigenvalues E_{el} are the electronic energies. Typically, the non-relativistic electronic Hamiltonian H_{el} of a molecular system containing N electrons and M nuclei is given in Hartree atomic units ($\hbar = m_e = 4\pi\varepsilon_0 = 1$) and reads

$$H_{\text{el}} = -\sum_{i=1}^{N}\frac{1}{2}\nabla_i^2 - \sum_{i=1}^{N}\sum_{J=1}^{M}\frac{Z_J}{r_{iJ}} + \sum_{i=1}^{N}\sum_{j>i}^{N}\frac{1}{r_{ij}}, \tag{2}$$

with $r_{ij} = |\boldsymbol{r}_i - \boldsymbol{r}_j|$ being the distance between any two particles (electrons or nuclei) and Z_J indicating the charge of nuclei J. In the above equation, the first term is the kinetic energy of the electrons, the second term describes the electron–nucleus attraction (also referred to as the external potential), while the last term corresponds to the potential energy of the repulsion between electrons. Usually, the nucleus–nucleus repulsion term $\sum_{I<J}^{M}\frac{Z_I Z_J}{R_{IJ}}$ is included in the electronic Hamiltonian and manifests itself as a constant shift in the electronic energy.

When the speed of the electrons becomes comparable to the speed of light, relativistic effects have to be included into the equation, which has to be invariant under Lorentz transformation. In the framework of relativistic quantum chemistry, any free particle with spin of 1/2 is described by the time-independent Dirac equation [35] (again in atomic units)

$$H_{\mathrm{fp}}\psi(\boldsymbol{x}) = \left(c\sum_{n=1}^{3}\boldsymbol{\alpha}_n p_n + \boldsymbol{\beta}c^2\right)\psi(\boldsymbol{x}) = E\psi(\boldsymbol{x}), \tag{3}$$

where $\boldsymbol{\alpha}_n$ and $\boldsymbol{\beta}$ are Dirac matrices, c is the speed of light, and the wave function $\psi(\boldsymbol{x})$ is a four-component (spinor) vector. Specifically, $\boldsymbol{\alpha}_n$ are written in terms of the Pauli matrices $\boldsymbol{\sigma}_n$ and $\boldsymbol{\beta}$ contains the 2×2 identity matrix,

$$\boldsymbol{\alpha}_n = \begin{pmatrix} 0 & \boldsymbol{\sigma}_n \\ \boldsymbol{\sigma}_n & 0 \end{pmatrix}, \quad \boldsymbol{\beta} = \begin{pmatrix} \boldsymbol{I}_2 & 0 \\ 0 & -\boldsymbol{I}_2 \end{pmatrix}. \tag{4}$$

For atoms and molecules, the relativistic Hamiltonian can be written as a sum of one- and two-electron operators, similar to non-relativistic theory. The one-electron part is the sum of the one-electron Dirac Hamiltonian H_D for all electrons in the quantum system. Specifically for the hydrogen atom (as for all one-electron systems) the Dirac Hamiltonian H_D can be written in closed form and reads

$$H_D = \boldsymbol{\beta}c^2 + c\boldsymbol{\alpha} \cdot \boldsymbol{p} + V, \tag{5}$$

where V is the Coulomb potential (electron-nuclear interaction). Although the Dirac equation is rigorous only for one-electron systems, it provides a starting point for further routines to introduce relativistic effects for molecular systems.

3.1 Introducing Relativistic Effects

In actinide chemistry, the most important relativistic effects include the so-called scalar relativistic effects and spin-orbit coupling. Specifically, scalar relativistic effects are responsible for the contraction of s and p orbitals and the expansion of d and f orbitals compared with the non-relativistic Schrödinger equation. Spin-orbit coupling originates from interactions between the magnetic field produced by the orbital motion of a charged particle and its spin. Both scalar relativistic and spin-orbit effects are important in actinide compounds, while other higher-order effects are typically neglected [151].

The most rigorous procedure to include relativistic effects is to find the eigenfunctions and eigenvalues of the four-component Dirac equation in an all-electron basis. This many-particle equation is built on a top of the Dirac equation for a single fermion. Specifically, the many-electron relativistic Hamiltonian combines the one-electron Dirac operators from (5), the electron–electron repulsion term as given in (2), and the Breit operator [23],

$$g_{ij}^{\mathrm{Breit}} = -\frac{c\boldsymbol{\alpha}_i \cdot c\boldsymbol{\alpha}_j}{2c^2 r_{ij}} - \frac{(c\boldsymbol{\alpha}_i \mathbf{r}_{ij}) \cdot (c\boldsymbol{\alpha}_j \mathbf{r}_{ij})}{2c^2 r_{ij}^3}, \tag{6}$$

(or the simplified Gaunt operator $g_{ij}^{\text{Gaunt}} = -\frac{c\alpha_i \cdot c\alpha_j}{c^2 r_{ij}}$ [48]) that mimics the retardation of the potentials due to the finite speed of light. Although the corresponding equation is Lorentz-invariant only approximately, it describes relativistic effects most accurately. The drawback of the so-called Dirac–Coulomb–Breit Hamiltonian is the large computational cost, which makes this approach computationally infeasible for molecules with a large number of electrons. In practical applications the four-component Dirac–Coulomb Hamiltonian is used at the SCF level and the correlated calculations are performed within the so-called "no-pair" approximation, where projection operators remove any Slater determinant containing negative-energy orbitals from the Dirac–Coulomb Hamiltonian [129]. In this approach both one- and two-electron contributions to spin–orbit coupling are accounted for. It is possible to further reduce the computational cost and approximate the "full" spin–orbit operator using either atomic or molecular mean field theories [133].

Computationally less expensive methods work within a two-component framework, where the small component of the Dirac equation is eliminated. However, this decoupling is not straightforward for many-electron systems and a number of routines have been developed during the past decades to transform the four-component form of the many-particle Dirac equation into an equation with at most two components [9, 153]. One popular approach includes the so-called regular approximations. The four-component state vector is divided into a large-component spinor $\psi^L(\boldsymbol{r})$ and a small-component spinor $\psi^S(\boldsymbol{r})$ [26, 156]. The atomic balance relation between these two spinors,

$$\psi^S(\boldsymbol{r}) = \left(1 + \frac{E - V}{2c^2}\right)^{-1} \frac{\boldsymbol{\sigma} \cdot \boldsymbol{p}}{2c} \psi^L(\boldsymbol{r}), \tag{7}$$

allows us to eliminate the small component from the Dirac equation and solve the Dirac equation for the large component only, which represents a two-component equation. The most simple flavour of the regular approximation is the zeroth order regular approximation (ZORA), where the ZORA Hamiltonian for the large component reads

$$H^{\text{ZORA}} = \frac{1}{2}(\boldsymbol{\sigma} \cdot \boldsymbol{p}) \left(1 - \frac{V}{2c^2}\right)^{-1} (\boldsymbol{\sigma} \cdot \boldsymbol{p}) + V. \tag{8}$$

The above (truncated) Hamiltonian includes parts of the Darwin term and all spin-orbit interactions arising from the nuclei. However, the ZORA Hamiltonian is not gauge invariant. This deficiency can be fixed by appropriate scaling procedures or inclusion of higher order approximations [9, 158].

A different family of approaches aims at decoupling the electronic and positronic solutions of the Dirac Hamiltonian using a unitary transformation U, which makes the Dirac Hamiltonian H_D block-diagonal with respect to the large (h_+) and small component (h_-),

$$\bar{H}_D = U^\dagger H_D U = \begin{pmatrix} h_+ & 0 \\ 0 & h_- \end{pmatrix}. \tag{9}$$

The resulting blocks in the transformed Hamiltonian $\bar{H}_D$ are two-component Hamiltonians and act on electronic and positronic states only. The exact form of the unitary transformation U is, however, only known for the free-particle Dirac equation and is called the Foldy–Wouthuysen transformation [40]. An approximate decoupling scheme for the many-electron Dirac equation in quantum chemistry was proposed by Hess. The so-called Douglas–Kroll–Hess (DKH) method [38, 61, 123] is based on the Foldy–Wouthuysen transformation [40] and represents an order-by-order expansion (in the external potential V), where the electronic and positronic components of the Dirac equation are separated iteratively. The DKH transformed Hamiltonian of $(n+1)$-th order has the general form

$$H_{n+1} = U_n^\dagger U_{n-1}^\dagger \dots U_2^\dagger U_1^\dagger H_1 U_1 U_2 \dots U_{n-1} U_n, \tag{10}$$

where H_1 is the free-particle Foldy–Wouthuysen (fpFW) transformed Dirac Hamiltonian $H_1 = U_{\mathrm{fpFW}}{}^\dagger H_D U_{\mathrm{fpFW}}$. Thus, different orders of approximations are obtained by applying subsequent unitary transformations to the relativistic Dirac Hamiltonian [123, 124, 157, 165]. Specifically, the second-order Douglas–Kroll–Hess (DKH2) Hamiltonian is most commonly used in quantum chemistry as it provides satisfactory results for conventional chemical problems. In DKH2, only one unitary transformation U_1 has to be applied. We should note that the explicit form of the unitary transformation U does not affect lower order DKH Hamiltonians and hence the operators U_i can be represented in different ways, using, for instance, a power series expansion of an (anti-Hermitian) operator.

The exact two-component (X2C) relativistic Hamiltonian is based on exact decoupling of the large and small components of the Dirac Hamiltonian in its matrix representation. Specifically, the X2C method exploits the non-symmetric Algebraic Riccati Equation (nARE) [75, 76], a quadratic matrix equation. The nARE approach was used for the Dirac Hamiltonian for the first time by Ilias and Saue [65] and introduced as the X2C method. Most importantly, the eigenvalues of the X2C Hamiltonian are identical to the positive energy branch of the four-component Dirac Hamiltonian.

One should stress that in the majority of quantum chemical applications, these two-component Hamiltonians account only for scalar relativistic effects and thus only have a one-component form. Due to this one-component nature, such Hamiltonians can be easily interfaced with standard quantum chemistry codes. Spin-orbit coupling effects can be included a posteriori using the spin-orbit configuration-interaction approach, where the relativistic Hamiltonian is diagonalized in the spin-free basis [97, 154]. To further decrease the computational cost, the spin-orbit integrals are often calculated within the atomic mean-field intergrals (AMFI) approach [62, 98, 130].

The computationally most efficient way of including relativistic effects in the Schrödinger equation is to introduce scalar relativistic effects using relativistic effective core potentials (RECP) [37]. Such a crude approximation is usually sufficiently accurate for chemistry as the influence of the core electrons on the valence shell (that is the shell containing electrons of relevance in chemical processes) is rather indirect and can be accurately modelled using parametrized effective pseudopotentials in conjunction with scalar relativistic interactions [37]. Besides being computationally inexpensive and fast to compute, RECPs provide reliable results if spin-orbit coupling is negligible. Spin–orbit corrections can be added a posteriori on top of RECP [37, 168].

3.2 Solving the Electronic Problem

Since the Schrödinger or Dirac equation cannot be solved exactly for many-electron systems, many approximate methods have been introduced to quantum chemistry that aim at solving the electronic problem as accurately as possible. The simplest—and probably the most important—model is the molecular orbital approximation, where each electron occupies exactly one orbital. The total electronic wave function is then constructed as an antisymmetric product of these spin orbitals $\chi_i(\boldsymbol{x}_j)$ that depend on the spatial coordinates $\boldsymbol{r}_j$ and spin coordinate σ_j of one electron. The antisymmetric product of spin orbitals is called a Slater determinant (or electronic configuration),

$$\Psi_{\text{el}}(\boldsymbol{x}_1, \boldsymbol{x}_2, \ldots, \boldsymbol{x}_N) = \frac{1}{\sqrt{N!}} \begin{vmatrix} \chi_1(\boldsymbol{x}_1) & \chi_2(\boldsymbol{x}_1) & \ldots & \chi_N(\boldsymbol{x}_1) \\ \chi_1(\boldsymbol{x}_2) & \chi_2(\boldsymbol{x}_2) & \ldots & \chi_N(\boldsymbol{x}_2) \\ \vdots & \vdots & \ddots & \vdots \\ \chi_1(\boldsymbol{x}_N) & \chi_2(\boldsymbol{x}_N) & \ldots & \chi_N(\boldsymbol{x}_N) \end{vmatrix}, \tag{11}$$

where $\chi_i(\boldsymbol{x}_j)$ is the ith spin orbital populated by the jth electron and N is the total number of electrons. In quantum chemistry, the Hartree–Fock method optimizes a single Slater determinant and represents a common starting point for more elaborated approaches. Using the notation of second quantization [60], a Slater determinant can be written in a very compact form,

$$\Psi_{\text{el}} = \prod_i a_i^\dagger |\rangle, \tag{12}$$

where $a_i^\dagger$ is the fermionic creation operator, which creates an electron in spin orbital i, and $|\rangle$ is the vacuum state. For simplicity, we have dropped the dependence of Ψ_{el} on the electronic coordinates. Note that a Slater determinant contains only occupied orbitals. If the number of one-electron functions (that is, spin orbitals) is greater than the total number of electrons in the system, it is possible to construct more than one Slater determinant. If the electronic wave function is expanded as a sum of all possible Slater determinants Φ_k that can be constructed by distributing N electrons in K orbitals,

$$\Psi_{\text{el}}^{\text{FCI}} = \sum_k c_k \Phi_k = \sum_k c_k \left(\prod_{i_k} a_{i_k}^\dagger |\rangle \right), \tag{13}$$

we obtain the so-called Full Configuration Interaction expansion (for a given finite basis with K orbitals), where c_k are some expansion coefficients.

The energy difference between the FCI solution and the electronic energy corresponding to a single Slater determinant (SD),

$$E_{\text{el}}^{\text{corr}} = E_{\text{el}}^{\text{FCI}} - E_{\text{el}}^{\text{SD}}, \tag{14}$$

is defined as the correlation energy and originates from the correlated motions of the electrons that cannot be described within Hartree–Fock theory (except of exchange correlation). Thus, in order to account for correlation effects, we have to include more than one Slater determinant in the wave function expansion. Although FCI allows us to solve the Schrödinger (or Dirac) equation exactly (within a given finite orbital basis), it is computationally feasible only for the smallest systems, containing up to, say, 20 electrons. Since actinide atoms and actinide-containing molecules usually contain much more than 20 electrons, the FCI ansatz cannot be applied in computational actinide chemistry. Furthermore, since electron correlation effects are crucial for a reliable description of chemical properties and chemical reactions involving actinide compounds, we have to find suitable wave function models that allow us to approximate the FCI wave function as accurate as possible by reducing the number of degrees of freedom in the optimization problem. This can be done by either restricting the number of Slater determinants by truncating the FCI expansion or by using more efficient parameterizations of the CI expansion coefficients c_k (or any combinations of those two strategies).

3.2.1 Accounting for Electron Correlation Effects in the Ground-State Electronic Wave Function

In quantum chemistry, we usually distinguish between single- and multi-reference approaches. The former employ some reference configuration Φ_0 to construct a truncated CI expansion. Multi-reference methods do not refer to a single Slater determinant but employ a set of selected determinants that are chosen due to some criterion. Both single- and multi-reference methods are commonly applied in computational actinide chemistry to model ground- and excited-states properties. In the following, we will briefly discuss some conventional and unconventional electronic structure methods that have been used to study heavy-element-containing compounds.

3.2.2 Truncated Configuration Interaction

One single-reference approach, where the FCI wave function is systematically truncated, represents truncated configuration interaction (CI). In truncated CI, only those Slater determinants are included in the wave function expansion that differ by one, two, three, etc. orbitals with respect to the reference determinant Φ_0. The electronic wave function is then a linear expansion containing the reference determinant and all singly, doubly, triply, etc. substituted configurations. Most commonly, the FCI expansion is truncated to include only single and double excitations leading to the CI Singles Doubles (CISD) wave function,

$$\Psi_{\mathrm{el}}^{\mathrm{CISD}} = \Phi_0 + \sum_{i}^{\mathrm{occ}} \sum_{a}^{\mathrm{virt}} c_i^a a_a^\dagger a_i \Phi_0 + \sum_{i<j}^{\mathrm{occ}} \sum_{a<b}^{\mathrm{virt}} c_{ij}^{ab} a_a^\dagger a_b^\dagger a_j a_i \Phi_0. \tag{15}$$

In the above equation, we have used the conventional notation of quantum chemistry, where indices $i, j, \ldots$ indicate occupied (spin) orbitals, while $a, b, \ldots$ run over all virtual (spin) orbitals of the reference determinant Φ_0. a_i is the fermionic annihilation operator and depopulates the i-th orbital. One drawback of CISD (or any truncated CI method) is its lack of size-extensivity and size-consistency. The size-consistency error can be reduced using, for instance, the Davidson correction [87].

3.2.3 Single-Reference Coupled Cluster Theory

A different single-reference method that is frequently applied in actinide chemistry is coupled cluster (CC) theory. In the CC method, the electronic wave function is written using an exponential ansatz,

$$\Psi_{\mathrm{el}}^{\mathrm{CC}} = e^T \Phi_0, \tag{16}$$

where T is the so-called cluster operator and can be expressed as a sum of excitation operators $T = T_1 + T_2 + T_3 + \cdots$. As in truncated CI, the excitation operators excite one, two, three, etc. electrons from occupied orbitals to virtual orbitals,

$$T_1 = \sum_i \sum_a t_i^a a_a^\dagger a_i, \quad T_2 = \frac{1}{(2!)^2} \sum_{ij} \sum_{ab} t_{ij}^{ab} a_a^\dagger a_b^\dagger a_j a_i,$$
$$T_3 = \frac{1}{(3!)^2} \sum_{ijk} \sum_{abc} t_{ijk}^{abc} a_a^\dagger a_b^\dagger a_c^\dagger a_k a_j a_i, \tag{17}$$

and so on, where $t_i^a, t_{ij}^{ab}, \ldots$ are the CC singles, doubles, etc. amplitudes. In conventional electronic structure calculations, the full cluster operator is approximated and restricted to include only some lower-order excitation operators. Specifically, in the CC Singles and Doubles (CCSD) approach, we have $T = T_1 + T_2$. In truncated CC methods, the wave function expansion still contains all Slater determinants of the FCI expansion, yet the expansion coefficients c_k are approximated by only a subset of cluster amplitudes. These conventional single-reference methods typically break down when orbitals become (quasi-)degenerate and hence cannot be unambiguously separated into an occupied and virtual space. In such strongly-correlated cases, we can switch to a multi-reference description of electronic structures. There exist, however, extensions (or simplifications) of conventional single-reference methods that allow us to model strongly-correlated quantum states within a single-reference framework. Examples are spin-flip CC [80] and pair-CCD [136].

3.2.4 Multi-reference Complete Active Space Self-consistent Field Theory

The complete active space self-consistent field (CASSCF) method is a variant of the multi-configurational SCF (MCSCF) approach. The model wave function,

$$\Psi_{\text{el}}^{\text{CASSCF}} = e^{-\kappa} \sum_i c_i \Phi_i, \tag{18}$$

has a similar form as the FCI wave function being a linear expansion in terms of Slater determinants (or configuration state functions) Φ_i with expansion coefficients c_i. The operator $e^{-\kappa}$ performs unitary transformation of the spin orbitals, where κ is the generator of orbital rotations,

$$\kappa = \sum_{p>q} \kappa_{pq} (a_p^\dagger a_q - a_q^\dagger a_p). \tag{19}$$

Thus, in contrast to FCI, the orbital basis is optimized self-consistently within CASSCF. The ground-state wave function is obtained by minimizing the electronic energy with respect to all variational parameters,

$$E_{\text{el}}^{\text{CASSCF}} = \min_{\kappa, c} \frac{\left\langle \Psi_{\text{el}}^{\text{CASSCF}}(\kappa, \mathbf{c}) \right| H \left| \Psi_{\text{el}}^{\text{CASSCF}}(\kappa, \mathbf{c}) \right\rangle}{\left\langle \Psi_{\text{el}}^{\text{CASSCF}}(\kappa, \mathbf{c}) \middle| \Psi_{\text{el}}^{\text{CASSCF}}(\kappa, \mathbf{c}) \right\rangle}, \tag{20}$$

So far, we have made no assumptions about the configurational space of CASSCF and the above equations are valid for any MCSCF wave function. Since we have to optimize both the expansion coefficients c_i and the spin orbitals, MCSCF-type methods are computationally expensive. To reduce the computational cost, the configurational space is heavily truncated. Specifically in CASSCF, the molecular orbitals are divided into three subsets: (1) doubly-occupied inactive (frozen, core) orbitals, (2) active orbitals, and (3) unoccupied external (virtual) orbitals. In each electronic configuration (Slater determinant), the inactive orbitals are always doubly occupied, while the external orbitals remain unoccupied. Only the orbital occupations of the active orbitals are allowed to differ in each Slater determinant. Furthermore, all possible ways of distributing the active electrons in the active space orbitals are permitted in the CASSCF wave function, which makes the active space complete in terms of the CI expansion. Thus, CASSCF represents a FCI expansion in the active space orbitals.

In general, the active space should compromise all chemically important orbitals for a given molecular system. For small molecules, an energetic criterion can be used to select the active space orbitals. In actinide chemistry, conventional selection procedures might be ineffective as actinide complexes feature many quasi-degenerate orbitals and it remains ambiguous which metal and ligand orbitals have to be included in the active space. Novel approaches based on quantum information theory allow us to identify the chemically most important orbitals and can be applied to develop a black-box like selection procedure of active spaces for MCSCF-type calculations. Such an approach will be discussed in Sect. 3.2.11. The CASSCF method removes the problem of (quasi-)degeneracies and allows us to model static correlation effects. However, it does not account for dynamical correlation effects attributed to electron excitations beyond the active space orbitals. CASSCF, thus, provides a spin-adapted zero-order wave function, where dynamical and core-valence electron correlation can be added using various a posteriori corrections such as complete-active-space second-order perturbation theory (CASPT2) [3, 4] or multi-reference configuration-interaction (MRCI) [115].

3.2.5 The Density Matrix Renormalization Group

The density matrix renormalization group (DMRG) [25, 77, 90, 91, 99, 131, 142, 166, 170] algorithm represents a computationally efficient variant of MCSCF theory where the evaluation of the electronic energy scales only polynomially with system size. Due to its low computational scaling, the DMRG protocol allows us to approach the FCI limit of an N-particle Hilbert space constructed from L orbitals for large molecules, where FCI calculations are computationally unfeasible. In contrast to conventional ab initio methods, DMRG optimizes a special type of quantum states, so-called matrix product states (MPS), that allow us to efficiently reparameterize the electronic wave function using a significantly smaller number of variational parameters. In the MPS representation, the CI wave function expansion (13) is rewritten in terms of a product of matrices that replaces the CI expansion coefficients,

$$\Psi_{\mathrm{el}}^{\mathrm{DMRG}} = \sum_{k_1,k_2,\ldots,k_L} \boldsymbol{A}_1^{k_1} \boldsymbol{A}_2^{k_2} \ldots \boldsymbol{A}_L^{k_L} |k_1, k_2, \ldots, k_L\rangle, \tag{21}$$

where L is the number of spatial orbitals in some active space, $\{\boldsymbol{A}_1^{k_1}, \boldsymbol{A}_2^{k_2}, \ldots, \boldsymbol{A}_L^{k_L}\}$ is a set of matrices that are optimized by the algorithm, and $\{k_1, k_2, \ldots, k_L\}$ are the occupations of the orbitals (either unoccupied, singly occupied, or doubly occupied) written in terms of the occupation number representation, where each occupation number vector $|k_1, k_2, \ldots, k_L\rangle$ represents a Slater determinant.

The DMRG wave function and its many-particle basis is optimized in a sweeping procedure. One sweep contains $(L - q - 2)$ microiterations, where q is the number of exactly-represented orbitals (either 1 or 2). To perform the sweeping algorithm, the orbitals have to be aligned on a one-dimensional lattice. Thus, the DMRG algorithm is best suited to describe one-dimensional problems. There exist different approaches to order the orbitals along a one-dimensional lattice. Specifically, concepts of quantum information theory allow us to select an optimal orbital ordering in a black-box-like fashion [90]. One microiteration includes three distinct steps: (1) blocking, (2) diagonalization, and (3) decimation. First, the L-orbital space is partitioned into three subspaces: a system block, an environment block, and q exactly represented orbitals in between represented by 4 basis functions in the case of spatial orbitals. In the blocking step, the system and, if $q = 2$, the environment are enlarged by one of the exactly represented orbitals. The many-particle basis states are defined as a tensor product of basis states of the subsystem block and the neighbouring exactly-represented orbital. In the second step, the Hamiltonian of the *superblock* (enlarged system+enlarged environment block) is constructed and diagonalized. Usually, we are only interested in the ground-state wave function and hence only one root of the superblock Hamiltonian needs to be computed. In the third and last step, the dimensionality of the enlarged system and enlarged environment blocks is reduced to prevent the many-particle basis from growing exponentially (due to the blocking step). The number of basis functions is reduced to a limit indicated by a parameter m. In DMRG, this so-called renormalization step is performed in a specific way. From the superblock wave function Ψ^{SB}, we calculate the many-particle reduced density matrix of the enlarged active system block $\rho^{\mathrm{s}} = \mathrm{Tr}_{m_{\mathrm{e}}} \left|\Psi^{\mathrm{SB}}\right\rangle\left\langle\Psi^{\mathrm{SB}}\right|$, where m_{e} indicate states defined on the (enlarged) environment block. This reduced density matrix ρ^{s} is then diagonalized and the eigenvectors corresponding to the m largest eigenvalues form the new many-particle basis of the enlarged system block. In the final renormalization step, all matrix representations of operators are transformed into this new basis. Specifically, the computed transformation matrices correspond to the $\boldsymbol{A}$ matrices of the MPS ansatz. Thus, each microiteration step optimizes exactly one MPS matrix $\boldsymbol{A}$ and we have to sweep through the lattice to obtain an (approximate) full representation of the MPS. After the decimation step, the algorithm starts again with the blocking procedure, where the new system block is enlarged by one orbital, while the new environment is reduced by one orbital. We should note that the choice of m is crucial to find a balance between accuracy and computational cost. There is, however, no straightforward formula that indicates the best value of m and several calculations with different values of m are required to converge the wave function with satisfactory accuracy.

Typically, an MPS is represented in its canonical form containing so-called left- and right-normalized matrices. The DMRG algorithm optimizes a mixed-canonical MPS that is composed of both left- and right-normalized matrices,

$$\Psi_{\mathrm{el}}^{\mathrm{MPS}} = \sum_{k_1,k_2,\ldots,k_L} \boldsymbol{A}_1^{k_1} \boldsymbol{A}_2^{k_2} \ldots \boldsymbol{A}_{l-1}^{k_{l-1}} \boldsymbol{\Psi}^{k_l k_{l+1}} \boldsymbol{A}_{l+2}^{k_{l+2}} \ldots \boldsymbol{A}_L^{k_L} \left|k_1, k_2, \ldots, k_L\right\rangle. \tag{22}$$

In the above equation, the left-normalized matrices are defined for orbitals $k_1, \ldots,$ k_{l-1}, while the right-normalized matrices are obtained for orbitals $k_{l+2}, \ldots, k_L$. The matrix $\boldsymbol{\Psi}^{k_l k_{l+1}}$ is obtained during the diagonalization step of the superblock Hamiltonian. Since one matrix of the MPS representation contains at most m^2 elements, the total number of variational parameters is at most $4Lm^2$ if we work in a spatial orbital basis with 4 possible occupations. Thus, the high-dimensional CI coefficient tensor, which scales binomially with system size, has been replaced by a product of lower-dimensional tensors.

The DMRG algorithm is a powerful tool to approximate FCI wave functions in a given active space that are computationally not accessible for conventional quantum mechanical methods like CASSCF. Most importantly, it is suitable for strongly-correlated systems and hence allows us to accurately model heavy-element compounds, like transition metal- or actinide-containing molecules. Although originally formulated to tackle one-dimensional problems, DMRG has been successfully applied to describe strong correlation in general 3-dimensional systems. The missing dynamical correlation effects can be added a posteriori using the same corrections as developed for traditional MCSCF methods. Examples are second order complete active space perturbation theory (DMRG-CASPT2), [82] the multi-reference configuration interaction approach with internal contraction of DMRG (DMRG-icMRCI), [128] canonical transformation theory (CT), [106] or perturbation theory formulated in terms of matrix product states [132].

3.2.6 Geminal-Based Approaches

All electronic structure methods discussed above use one-electron functions (orbitals) to construct the Slater determinants that span the N-particle Hilbert space. A conceptually different approach to account for electron correlation effects is to use two-electron functions as fundamental building blocks of the electronic wave function [21, 73, 96]. In second quantization, a (singlet) two-electron operator $\psi_i^\dagger$, also called geminal, can be written as a linear combinations of electron-pair creators,

$$\psi_i^\dagger = \sum_{q=1}^{M} C_q^i a_q^\dagger a_{\bar{q}}^\dagger, \tag{23}$$

where $a_q^\dagger$ ($a_{\bar{q}}^\dagger$) are the fermionic creation operators for α (β) electrons and M is the number of one-electron functions used to construct geminal i. In the above equation, C_q^i are the geminal coefficients that link the geminal creation operator with the underlying one-particle basis (represented by $a_p^\dagger$). Thus, geminals are quasi-particles and the corresponding geminal creation operators are electron pair creators. The geminal-based electronic wave function is a product of the geminal creation operators acting on the vacuum state,

$$\Psi_{\text{el}} = \prod_i^P \psi_i^\dagger \,|\rangle, \tag{24}$$

with $P = N/2$ being the number of electron pairs. If the geminal creation operators have the general form of (23), we obtain the antisymmetric product of interacting geminals (APIG) wave function [96]. Although APIG includes correlations between orbital pairs, it is computationally intractable for larger systems. Substituting (23) in the electronic wave function (24), we can rewrite the APIG wave function using a linear expansion of Slater determinants,

$$\Psi_{\mathrm{el}}^{\mathrm{APIG}} = \sum_{\{m_i=0,1|P\}} |\boldsymbol{C}(\boldsymbol{m})|^{+}(a_1^{\dagger}a_{\bar{1}}^{\dagger})^{m_1}(a_2^{\dagger}a_{\bar{2}}^{\dagger})^{m_2}\dots(a_K^{\dagger}a_{\bar{K}}^{\dagger})^{m_M}\,|\rangle, \tag{25}$$

where $\boldsymbol{C}(\boldsymbol{m})$ is the geminal coefficient matrix, $|\boldsymbol{A}|^{+}$ indicates the permanent of matrix $\boldsymbol{A}$, and

$$P = \sum_{k=1}^{M} m_k \quad \text{with} \quad P < M. \tag{26}$$

Specifically, $\boldsymbol{C}(\boldsymbol{m})$ is a $P \times P$ matrix and contains only those columns for which $m_k = 1$. In order to evaluate the coefficients in front of the wave function expansion of (25), we have to evaluate the permanent of $\boldsymbol{C}(\boldsymbol{m})$ (similar to the determinant with all $-$ signs replaced by $+$ signs). Since the evaluation of the permanent of a matrix is #P-hard and the Slater determinant expansion of (25) includes a factorial number of determinants, the APIG model is computationally expensive. To make geminal-based models applicable to larger systems, we have to introduce constraints that allow us to evaluate the permanent efficiently. One simplified geminal-based wave function is the antisymmetric product of strongly orthogonal geminals (APSG) [84, 85, 107, 108, 119], where the geminal creation operators create two-electron states that are orthogonal to each other. Specifically, the sum of (23) is restricted to run over mutually exclusive subspaces M_i of orbitals,

$$\psi_i^{\dagger} = \sum_{q=1}^{M_i} C_q^i a_q^{\dagger i} a_{\bar{q}}^{\dagger i} \quad \text{with} \quad \sum_q C_q^i C_q^k = \delta_{ik}.$$

The partitioning of M into disjoint subspaces M_i is equivalent to associating subsets of orbitals to specific geminals, that is, each orbital may belong to only one geminal. Although the strong orthogonality constraint allows us to efficiently optimize the wave function using the variational principle, we miss electron correlation effects between the orbital subsets (as they are disjoint).

A promising geminal-based model that has been successfully applied to actinide chemistry is the antisymmetric product of 1-reference orbital geminal (AP1roG) [19–21, 46, 149]. In AP1roG, the strong orthogonality constraint is relaxed and inter-geminal correlations are introduced in the geminal ansatz,

$$\psi_i^{\dagger} = a_i^{\dagger}a_{\bar{i}}^{\dagger} + \sum_{a}^{\mathrm{virt}} c_i^a a_a^{\dagger} a_{\bar{a}}^{\dagger}, \tag{27}$$

where the sum runs over all virtual orbitals with respect to some reference determinant (like the HF determinant). The second term of the above equations assigns (virtual) orbitals to all geminals and accounts for inter-geminal correlations. The main feature of the AP1roG wave function is that the first term of (27) selects some reference determinant, that is, one specific orbital is occupied by an α and β electron in one specific geminal. The corresponding geminal coefficient matrix has a special form,

$$\boldsymbol{C}_{\text{AP1roG}} = \begin{pmatrix} 1 & \cdots & 0 & 0 & c_{1;P+1} & c_{1;P+2} & \cdots & c_{1;K} \\ 0 & 1 & \cdots & 0 & c_{2;P+1} & c_{2;P+2} & \cdots & c_{2;K} \\ \vdots & \vdots & \ddots & \vdots & \vdots & \vdots & \ddots & \vdots \\ 0 & \cdots & 0 & 1 & c_{P;P+1} & c_{P;P+2} & \cdots & c_{P;K} \end{pmatrix}, \tag{28}$$

where each row is one geminal and the left block contains the $P \times P$ identity matrix due to the first term in (27).

In addition to the above mentioned geminal models, different geminal-based wave functions have been introduced in quantum chemistry, like generalized-valence-bond perfect-pairing (GVB-PP) [49, 57, 64] and the particle-number projected Hartree–Fock–Bogoliubov model [30]. However, none of these geminal-based models have been applied to actinide chemistry and hence will not be discussed in this chapter. In the following, we will have a closer look at the AP1roG method, its optimization schemes, and possible extensions, as it has been proven to properly describe the static correlation in certain (heavy-element containing) molecules such as UO_2^{2+} and ThO_2 [152].

Although the structure of the AP1roG coefficient matrix allows us to efficiently evaluate the permanent $|\boldsymbol{C}|^+$, we still have to deal with a factorial number of Slater determinants when optimizing the electronic wave function. To obtain a computationally efficient optimization method, we can rewrite the AP1roG wave function using an exponential ansatz,

$$\begin{aligned} \Psi_{\text{el}}^{\text{AP1roG}} &= \prod_i \psi_i^\dagger \, |\rangle \\ &= \prod_i \left(a_i^\dagger a_{\bar{i}}^\dagger + \sum_a^{\text{virt}} c_i^a a_a^\dagger a_{\bar{a}}^\dagger \right) |\rangle \\ &= \prod_i \left(1 + \sum_a^{\text{virt}} c_i^a a_a^\dagger a_{\bar{a}}^\dagger a_{\bar{i}} a_i \right) a_i^\dagger a_{\bar{i}}^\dagger \, |\rangle \\ &= e^{\sum_i^{\text{occ}} \sum_a^{\text{virt}} c_i^a a_a^\dagger a_{\bar{a}}^\dagger a_{\bar{i}} a_i} \Phi_0, \end{aligned} \tag{29}$$

where $\Phi_0 = \prod_i a_i^\dagger a_{\bar{i}}^\dagger \, |\rangle$. Thus, the AP1roG method optimizes a coupled cluster-type wave function where the cluster operator T is restricted to electron pair excitations T_{p} [136],

$$T_{\text{p}} = \sum_i^{\text{occ}} \sum_a^{\text{virt}} c_i^a a_a^\dagger a_{\bar{a}}^\dagger a_{\bar{i}} a_i. \tag{30}$$

Since T_p excites an electron pair, it can be considered as a simplified T_2 operator and hence AP1roG is a simplified version of the CCD method. Due to its exponential ansatz, the AP1roG wave function is also known as the pair coupled cluster doubles (pCCD) wave function,

$$\Psi_\mathrm{el}^\mathrm{AP1roG} = \Psi_\mathrm{el}^\mathrm{pCCD} = e^{T_\mathrm{p}}\Phi_0. \tag{31}$$

Furthermore, the geminal coefficients $\{c_i^a\}$ of AP1roG are equivalent to the pCCD amplitudes and we can use the optimization techniques of single-reference coupled cluster theory to solve for $\{c_i^a\}$. Specifically, the electron-pair amplitudes are optimized using the projected Schrödinger equation, where the projection manifold is restricted to electron-pair excited determinants $\Phi_{i\bar{i}}^{a\bar{a}} = a_a^\dagger a_{\bar{a}}^\dagger a_{\bar{i}} a_i \Phi_0$, [73]

$$\left\langle \Phi_{i\bar{i}}^{a\bar{a}} \right| H \left| \Psi_\mathrm{el}^\mathrm{AP1roG} \right\rangle = E \left\langle \Phi_{i\bar{i}}^{a\bar{a}} \right| \left| \Psi_\mathrm{el}^\mathrm{AP1roG} \right\rangle = E c_i^a, \tag{32}$$

or using the similarity transformed Hamiltonian of coupled cluster theory, [136]

$$\left\langle \Phi_{i\bar{i}}^{a\bar{a}} \right| e^{-T_\mathrm{p}} H e^{T_\mathrm{p}} \left| \Phi_0 \right\rangle = 0. \tag{33}$$

Both optimization procedures are equivalent and (32) and (33) yield similar working equations.

Although the AP1roG wave function does not contain determinants with unpaired electrons or so-called "broken" electron pairs, reducing the CCD model to electron-pair terms not only decreases the computational cost, but also provides a better description of strongly-correlated systems. The AP1roG method is thus suitable for systems with quasi-degenerate electronic states, transition-state structures, and bond-breaking processes. The composition of the ansatz has been validated by Bytautas et al. [24, 73], who have shown that the correct description of strong correlation effects depends mainly on determinants with a small number of unpaired electrons.

In contrast to CCD, the AP1roG method is sensitive to rotations among the occupied orbitals and among the virtual orbitals as the pairing schemes are not equivalent. Thus, electronic energies and properties depend on the choice of the molecular orbital basis and two different molecular orbital sets can yield different results even though the reference determinant remains unaffected (note that the HF determinant is invariant under rotations of the occupied or virtual orbital space, respectively). In order to resolve the problem of non-size-consistency, the molecular orbital basis and hence the pairing scheme need to be optimized. The optimization of the orbital-pairing scheme allows us to obtain accurate results that almost reproduce doubly-occupied self consistent field (DOSCF) results [96]. Computational studies suggest that a variational orbital optimization protocol provides the most robust and reliable orbital optimization procedure in comparison to other investigated non-variational methods [20, 21]. The optimal set of orbitals is obtained by minimizing the AP1roG energy functional subject to the constraint that the AP1roG coefficient equations (32) are satisfied. The energy Lagrangian, thus, reads

$$\mathscr{L} = \langle \Phi_0 | e^{\kappa} H e^{-\kappa} | \Psi_{\text{el}}^{\text{AP1roG}} \rangle + \sum_{i,a} \lambda_i^a \left(\langle \Phi_{i\bar{i}}^{a\bar{a}} | e^{\kappa} H e^{-\kappa} | \Psi_{\text{el}}^{\text{AP1roG}} \rangle - E c_i^a \right), \quad (34)$$

where κ is again the generator of orbital rotations as defined in (36) and $\{\lambda_i^a\}$ are the Lagrange multipliers. The Lagrange multipliers are obtained from equations that are analogous to the Λ-equations of coupled cluster theory, where we require the partial derivative of the Lagrangian with respect to the geminal coefficients $\{c_i^a\}$ to equal zero, $\frac{\partial \mathscr{L}}{\partial c_i^a}|_{\kappa=0} = 0$. The geminal coefficients are obtained by making $\mathscr{L}$ stationary with respect to the Lagrange multipliers $\{\lambda_i^a\}$, $\frac{\partial \mathscr{L}}{\partial \lambda_i^a}|_{\kappa=0} = 0$ [20]. The orbital gradient is the partial derivative of $\mathscr{L}$ with respect to the orbital rotation coefficients $\{\kappa_{pq}\}$ evaluated for the current set of orbitals ($\kappa = 0$),

$$\begin{aligned} \left.\frac{\partial \mathscr{L}}{\partial \kappa_{pq}}\right|_{\kappa=0} = g_{pq}|_{\kappa=0} = &\langle \Phi_0 + \sum_{i,a} \lambda_i^a \Phi_{i\bar{i}}^{a\bar{a}} | [(a_p^\dagger a_q - a_q^\dagger a_p), H] | \Psi_{\text{el}}^{\text{AP1roG}} \rangle \\ &- \sum_{i,a} \langle \Phi_0 | [(a_p^\dagger a_q - a_q^\dagger a_p), H] | \Psi_{\text{el}}^{\text{AP1roG}} \rangle \sum_{i,a} \lambda_i^a c_i^a. \end{aligned} \quad (35)$$

After the orbital gradient and (approximate) orbital Hessian $\boldsymbol{A}$ are evaluated, the matrix representation of κ can be determined from

$$\boldsymbol{\kappa} = -\boldsymbol{A}\boldsymbol{g} \quad (36)$$

and the orbital basis can be transformed using the unitary transformation matrix $e^{-\kappa}$. For reasons of computational efficiency, the orbital Hessian is typically approximated by its diagonal, $\boldsymbol{A}_{pq,pq} = \left.\frac{\partial g_{pq}}{\partial \kappa_{pq}}\right|_{\kappa=0}$.

Although AP1roG captures a significant amount of the strong electron correlation energy and represents a very promising reference wave function in actinide chemistry, it misses a large fraction of the dynamic (weak) correlation energy. Dynamic electron correlation effects on top of the geminal wave function can be included in the wave function ansatz a posteriori using, for instance, perturbation theory [70, 95, 120, 127, 141], extended random phase approximation [27, 47, 109, 114], density functional theory (DFT) [44, 45, 47], and coupled cluster theory [136] or its linearized version [13, 47, 88]. Numerical studies indicate that the perturbation theory corrections with an AP1roG reference function do not provide reliable electronic structures and properties for actinide-containing compounds. To reliably model actinide chemistry, we can use various coupled cluster corrections on top of the AP1roG wave function.

One possible way to extend AP1roG is to apply an AP1roG-tailored CC formalism. In AP1roG-tailored CC theory, the electron-pair amplitudes of the CC singles, doubles, triples, etc. equations are substituted by the AP1roG geminal coefficients and not optimized, that is kept frozen, during the optimization procedure. Note that the tailored CC amplitude equations are similar to the conventional CC working equations, except that some selected amplitudes (the tailored amplitudes) are not varied. Thus, a tailored CC calculation represents only a minor modification in any CC code.

In the case of AP1roG-tailored CC, the corresponding CC corrections are referred to as frozen-pair (fp) CCD, fpCCSD, fpCCSDT, etc. In the fpCCD and fpCCSD methods, the single and non-pair double amplitudes can provide a balanced description of electron correlation effects when both CCD and CCSD fail in describing strongly-correlated systems.

A different CC corrections on top of AP1roG employs a linearized coupled cluster (LCC) ansatz and represents a simplification of any frozen-pair CC approach. In the LCC correction, we approximate the exponential coupled cluster ansatz with an AP1roG reference as

$$\Psi_{\text{el}}^{\text{AP1roG-LCC}} = e^{T}\Psi_{\text{el}}^{\text{AP1roG}} \tag{37}$$

$$\approx (1+T)\Psi_{\text{el}}^{\text{AP1roG}}, \tag{38}$$

where T is a general cluster operator. The Schrödinger equation for this wave function ansatz reads

$$\begin{aligned} H\left|\Psi_{\text{el}}^{\text{AP1roG-LCC}}\right\rangle &= E\left|\Psi_{\text{el}}^{\text{AP1roG-LCC}}\right\rangle \\ e^{-T}He^{T}\left|\Psi_{\text{el}}^{\text{AP1roG}}\right\rangle &= E\left|\Psi_{\text{el}}^{\text{AP1roG}}\right\rangle, \end{aligned} \tag{39}$$

where we have used (37) and multiplied from the left by e^{-T}. In the LCC correction, the left-hand-side of (39) is approximated to contain only linear terms in the Baker–Campbell–Hausdorff expansion,

$$(H + [H, T])\left|\Psi_{\text{el}}^{\text{AP1roG}}\right\rangle = E\left|\Psi_{\text{el}}^{\text{AP1roG}}\right\rangle. \tag{40}$$

If we now substitute the exponential form of the AP1roG wave function (31) in the above equation, we can bring the AP1roG-LCC Schrödinger equation into the familiar form

$$(H + [H, T] + [[H, T], T_{\text{p}}])\left|\Phi_0\right\rangle = E\left|\Phi_0\right\rangle \tag{41}$$

of single-reference CC theory. Furthermore, in AP1roG-LCC, the cluster operator is restricted to contain electron excitations (singles, broken-pair doubles, etc.) beyond electron-pair excitations. For instance, in the case of double excitations, we must have $T = T_2 - T_{\text{p}}$, which results in the AP1roG-LCCD method. Note that (41) is the Schrödinger equation for the non-pair amplitudes as the electron-pair amplitudes have been already optimized within AP1roG. Although being simplifications of conventional CC methods, the linearized and frozen-pair CC corrections feature a similar computational scaling as their single-reference counter parts.

3.2.7 Kohn-Sham Density Functional Theory

Density functional theory (DFT) is the most popular electronic structure method due to its rather low computational cost and conceptual simplicity. In DFT, the molecular system and its properties are determined by its electron density $\rho(\mathbf{r})$ instead of the electronic wave function. Specifically, Hohenberg and Kohn [63] proved that the non-degenerate ground-state wave function is uniquely determined by the electron density that corresponds to some external potential $v_{\text{ext}}(\mathbf{r})$.

The most common implementation of this method is within the Kohn-Sham formalism (KS) [78]. Specifically, in KS-DFT, an artificial reference system of non-interacting electrons is introduced that yields exactly the same electron density as the fully interacting system. Furthermore, the electronic energy is a functional of the density and is decomposed into different contributions,

$$E[\rho] = T_s[\rho] + V_{\text{ext}}[\rho] + J[\rho] + E_{\text{xc}}[\rho], \tag{42}$$

where $T_s[\rho]$ is the kinetic energy of the non-interacting system, $V_{\text{ext}}[\rho]$ is the potential energy due to some external potential, $J[\rho]$ is the classical Coulomb interaction, and $E_{xc}[\rho]$ is the so-called exchange–correlation functional and accounts for all non-classical contributions to the electron–electron interaction as well as a correction term for the kinetic energy that corresponds to the difference in kinetic energy of the fully-interacting and non-interacting system. However, the exact form of $E_{\text{xc}}[\rho]$ in (42) is unknown and approximations thereof have to be used. Due to their approximate nature, some density functional approximations (DFA) are appropriate for only certain types of molecules or particular properties [29, 31, 81, 145]. One major drawback of DFAs is the so-called self-interaction error attributed to the interactions between an electron and its own electric field [111]. This error is an artifact of the approximate nature of the DFT exchange–correlation functional. Paradoxically, the self-interaction error may be partly balanced by other deficiencies in the energy functional yielding electronic energies and molecular properties that agree well with experimental results due to cancellation of errors.

In KS-DFT, we have to solve a set of one-particle equations (the KS equations),

$$\Big(-\frac{1}{2}\nabla^2 + v(\mathbf{r}_i)\Big)\chi_i(\mathbf{r}_i) = \varepsilon_i\chi_i(\mathbf{r}_i), \tag{43}$$

which optimize the KS orbitals $\chi_i(\mathbf{r}_i)$. In the above equation, $v(\mathbf{r}_i)$ is an effective potential and determined as the variation of the energy functional $E[\rho]$ with respect to the electron density. After the KS equations are solved, the electron density can be expressed in terms of the optimized KS orbitals,

$$\rho(\mathbf{r}) = \sum_i^N |\chi_i(\mathbf{r})|^2. \tag{44}$$

Due to its low computational cost, KS-DFT has been extensively used in actinide chemistry, including its time dependent extensions (TD-DFT) to model electronically excited states [55, 67, 68, 83, 145, 167]. Specifically, molecular structures can be accurately calculated using generalized gradient approximation (GGA) functionals, such as BP86, [10, 110] while electronic energies (thermochemistry) and excitation energies are best determined using so-called hybrid exchange–correlation functionals, like PBE0 [1, 112, 113], B3LYP [10, 89], or CAM-B3LYP [169] that provide good agreement with experimental data or high-level wave-function-based methods [51, 79, 117, 146, 160].

3.2.8 Targeting Excited States with Wave-Function-Based Approaches

Since electronic spectra are used to identify the oxidation states and ligand effects in actinide species, reliable theoretical predictions of excitation energies of actinide compounds are highly important. In CI-type methods, such as MCSCF or DMRG, the electronic excitation energies are usually obtained by calculating higher roots of the eigenvalue problem. However, in order to compute excited states in coupled cluster theory, we have to define a new ansatz. The most popular approaches applied to actinides are the equation-of-motion (EOM) and linear response (LR) coupled cluster formulations [11, 12, 122, 148]. In this chapter, we will focus on a different approach that allows us to include strong correlation effects in excited states: the Fock-space coupled cluster (FSCC) approach.

3.2.9 Strongly-Correlated Excited States with Fock-Space Coupled Cluster Theory

The FSCC method belongs to the group of state-universal multi-reference coupled cluster theories and operates in the Fock space. The key idea behind the FSCC approach is to find an effective Hamiltonian in a low-dimensional model P space, with eigenvalues that reliably approximate the desired eigenvalues of the real (physical) Hamiltonian. In the FSCC method, the P space (also called the model space) contains all active valence orbitals directly involved in the electronic excitations, while the Q space (also called auxiliary or complementary space) includes all remaining orbitals. Thus, only a few eigenvalues out of the whole spectrum are calculated, reducing the expensive step of diagonalizing the Hamiltonian matrix. In many practical applications, however, the P and Q spaces are not well separated from each other, which might result in intruder state problems. They usually manifest as convergence difficulties for large P spaces, which are particularly desired for modeling electronic structure of actinides. Such divergencies might occur for a specific molecule, a given molecular geometry, or basis set. To remedy this problem, the intermediate Hamiltonian (IH) formulation of the FSCC method has been introduced, which imposes a buffer space between the desired and undesired states. Thus, the P space is further divided into a main P_m space and an intermediate P_i space. The intermediate space serves as a buffer between the P_m and Q spaces, for which various numerical procedures have been developed [101].

A characteristic feature of the FSCC approach is the partitioning into sectors (k, l) depending on the number of electrons removed from or attached to the reference state. Within the hole-particle formalism, the cluster operator T_n is expressed as

$$T_n = T_n^{(0,0)} + T_n^{(0,1)} + T_n^{(1,0)} + T_n^{(1,1)} + \cdots + T_n^{(k,l)}, \tag{45}$$

where the $T_n^{(0,0)}$ represents the ground state (zero holes and zero particles). In the above equation, $T_n^{(0,1)}$ corresponds to the system with one additional electron (zero holes and one particle), $T_n^{(1,0)}$ reduces the number of electrons by one (one hole and zero particles), and $T_n^{(1,1)}$ is a single excitation (one hole and one particle). The Hamiltonian is decomposed in the same way as the cluster operator T_n yielding electronic energies for the individual sectors, e.g., the ground-state energy for sector (0, 0), electron affinities for sector (0, 1), ionization potentials for sector (1, 0), and excitation energies for sector (1, 1). Electronic spectra can also be obtained as a double electron attachment, that is, from sector (0, 2) of the Fock space [105, 144, 147]. Higher order sectors have also been explored, but they are not commonly used. FSCC calculations require a reference determinant that dominates in the wave function expansion. Non-degenerate closed-shell states or high-spin open-shell states are usually the right choice for the reference determinant.

The advantage of the FSCC method is the size-extensiveness of ground-state energies and size-intensivity of excitation energies. The method allows us to obtain several electronic excited states of molecules with a common Fermi vacuum in a single run. Finally, the FSCC approach includes correlation effects of core and valence electrons, while its relativistic version is appropriate for actinide-containing molecules [66, 122, 144, 145, 151].

3.2.10 Embedding Wave Function Theory in Density Functional Theory

Reliable modeling of electronic spectra of actinide species with wave function-based methods is rather expensive and therefore usually limited to small model compounds. One way to overcome this problem is to combine wave function theory (WFT) with density functional theory within the so-called WFT-in-DFT approach. Within the WFT-in-DFT framework, the whole quantum system is partitioned into a system part and into an environment part. The system is represented by the WFT-based method, while the environment is modeled by the (usually less accurate, but significantly less expensive) DFT approach. The combination of these two methods will allow us to reliably account for static and dynamic electron correlation effects in large molecular systems, yet including environmental effects at the DFT level. Particularly for actinides, the embedding approach allows us to account for the chemical environment, such as ligand and crystal effects, in a cost effective way [50, 51, 147]. In the simplest WFT-in-DFT embedding scheme, the DFT embedding is accounted for as a static external potential and the orthogonality between the system and environment is neglected. Such an embedding potential includes the electrostatic potentials of the nuclei and the electron density of the environment, as well as contributions originating from the non-additive part of the exchange–correlation energy and from the non-additive part of the kinetic energy,

$$\begin{aligned} v^{\mathrm{emb}}[\rho_{\mathrm{WFT}}, \rho_{\mathrm{DFT}}] = & \; v^{\mathrm{nuc}}_{\mathrm{DFT}}(\mathbf{r}) \\ & + \int d\mathbf{r}' \frac{\rho_{\mathrm{DFT}}(\mathbf{r}')}{|\mathbf{r}-\mathbf{r}'|} + \frac{\delta E^{\mathrm{nadd}}_{\mathrm{xc}}[\rho_{\mathrm{WFT}}, \rho_{\mathrm{DFT}}]}{\delta \rho_{\mathrm{WFT}}} \\ & + \frac{\delta T^{\mathrm{nadd}}_{\mathrm{s}}[\rho_{\mathrm{WFT}}, \rho_{\mathrm{DFT}}]}{\delta \rho_{\mathrm{WFT}}}. \end{aligned} \tag{46}$$

This one-body term is then coupled with a given WFT model [59]. Such a WFT-in-DFT model can also be used to calculate excitation energies. However, the corresponding excitation spectra should be treated with care as excited states in the system might require coupling to the environment.

3.2.11 Interpretation of Electronic Wave Functions

Within molecular orbital theory, the electronic wave function is constructed from one-electron functions. This formalism provides a convenient description of molecular systems, where the electrons occupy specific orbitals and hence are localized in certain spatial regions of molecules. The contribution of individual orbitals to electronic structures and properties can be assessed using, for instance, concepts of quantum information theory (QIT) [7, 8, 14, 15, 17, 92, 125]. Specifically, QIT provides us with tools that allow us to interpret electronic wave function using the popular picture of interacting orbitals.

If a (pure) quantum state whose wave function is given by (13) cannot be written as product of states of its components (here, orbitals), $\Psi_{\mathrm{el}} \neq \psi_1 \otimes \psi_2 \otimes \cdots \otimes \psi_N$, we say that the quantum state is entangled. Thus, a single determinant wave function does not and cannot describe an entangled state. To simplify our discussion, we will focus on a bipartite system AB, that is, a quantum state that is composed of two parts. Note, however, that our analysis can be extended to quantum states that are composed of more than two subsystems $A, B, C, \ldots$. For a bipartite system AB, the wave function Ψ_{el}^{AB} of an entangled quantum state can only be written as a series of tensor products of basis states defined on the individual subsystems, $\Psi_{\mathrm{el}}^{AB} = \sum_{pq} c_{pq} \Psi_{\mathrm{el},p}^{A} \otimes \Psi_{\mathrm{el},q}^{B}$. Furthermore, while the quantum state of the composite system is well-defined, the states of its components cannot be determined unambiguously, that is, the subsystems A and B are correlated and cannot be treated independently. Quantum entanglement is an important feature in correlated systems such as actinide complexes and provides new perspectives on traditional quantum-chemical tools to interpret electronic structures.

A quantitative measure of the entanglement between any two subsystems is described by the von Neumann entropy and reads

$$S_{A|B} = -\mathrm{Tr}(\rho_A \ln \rho_A), \tag{47}$$

where ρ_A is the reduced density matrix (RDM) for subsystem A given by

$$\rho_A = \mathrm{Tr}_B \left|\Psi^{AB}\right\rangle\left\langle\Psi^{AB}\right| \tag{48}$$

for any pure state. Thus, ρ_A is obtained by tracing out all degrees of freedom from subsystem B and *vice versa*. Since the von Neumann entropy corresponds to the Shannon entropy in information theory, it quantifies how much information about subsystem A is encoded in subsystem B and *vice versa*. The entanglement entropy (47) is determined by the eigenvalue spectrum of the RDMs.

In this chapter, we aim at quantifying the interactions between orbitals. Thus, our subsystems should be composed of the molecular orbitals that are used to construct the Slater determinants in our wave function expansion. For that purpose, let us rewrite the FCI wave function (13) in occupation number form (dropping the superscript)

$$\Psi_{\mathrm{el}} = \sum_{k_1,k_2,\ldots,k_L} c_{k_1,k_2,\ldots,k_L} \left|k_1, k_2, \ldots, k_L\right\rangle, \tag{49}$$

where $c_{k_1,k_2,\ldots,k_L}$ are the expansion coefficients for each determinant $\left|k_1, k_2, \ldots, k_L\right\rangle$ and the sum runs over all occupation number vectors in the corresponding Hilbert space. Furthermore, we will consider only two different partitionings of our orbital space: (1) one subsystem contains exactly one orbital, while the other subsystem (here called environment) contains the remaining $L-1$ orbitals and (2) one subsystem contains exactly two orbitals, while the environment is constructed from the other $L-2$ orbitals. More general partitioning schemes have been investigated in the literature [143], however, focusing on one- and two-orbital entanglement measures will be sufficient to provide first insights into electronic structures of molecular systems. For the first case, we explicitly write the wave function of (49) in its bipartite form

$$\Psi_{\mathrm{el}}^{i,e} = \sum_{k_1,k_2,\ldots,k_L} \tilde{c}_{k_1,k_2,\ldots,k_L} \left|k_i\right\rangle \otimes \left|e\right\rangle, \tag{50}$$

where $\left|e\right\rangle = \left|k_1, k_2, \ldots, k_{i-1}, k_{i+1}, \ldots, k_L\right\rangle$ is a many-electron state vector containing environment orbitals and $\tilde{c}_{k_1,k_2,\ldots,k_L}$ are the expansion coefficients that may differ from $c_{k_1,k_2,\ldots,k_L}$ by a phase factor. This N-electron state vector is then used to construct the reduced density matrix for orbital i, the so-called one-orbital RDM, according to (48) with elements

$$\rho_{i,i'} = \sum_e \left\langle e\right| \left\langle k_i\right| \left|\Psi_{\mathrm{el}}^{i,e}\right\rangle\left\langle\Psi_{\mathrm{el}}^{i,e}\right| \left|k_{i'}\right\rangle \left|e\right\rangle, \tag{51}$$

where we sum over all many-electron states composed of the environment orbitals. The index i denotes all possible spin-occupations of a spatial orbital i and includes empty orbital (−), doubly occupied orbitals (↑↓), orbitals with spin-up electron (↑) and orbitals with spin-down electron (↓). Thus, the one-orbital RDM is a 4×4 matrix and is used to calculate the entanglement entropy of a single orbital, the so-called single-orbital entropy, given by

$$s_i = -\sum_{\alpha=1}^{4} \omega_{\alpha,i} \ln \omega_{\alpha,i}, \tag{52}$$

where $\omega_{\alpha,i}$ are the eigenvalues of the ith orbital RDM and the sum runs over all four possible occupations of a spatial orbitals. The single-orbital entropy reaches a maximum value of ln(4).

The entanglement entropy between an orbital pair ij and the remaining orbitals is obtained in a similar way. For our second case, the environment states are defined as $|e\rangle = |k_1, k_2, \ldots, k_{i-1}, k_{i+1}, \ldots, k_{j-1}, k_{j+1}, \ldots, k_L\rangle$, while the quantum state for this orbital partitioning reads

$$\Psi_{\text{el}}^{i,j,e} = \sum_{k_1,k_2,\ldots,k_L} \tilde{c}_{k_1,k_2,\ldots,k_L} |k_i, k_j\rangle \otimes |e\rangle. \tag{53}$$

The matrix elements of the two-orbital RDM are determined in a similar way and have the elements

$$\rho_{(i,j),(i',j')} = \sum_e \langle e| \langle k_i, k_j| |\Psi_{\text{el}}^{i,j,e}\rangle \langle \Psi_{\text{el}}^{i,j,e}| |k_{i'}, k_{j'}\rangle |e\rangle. \tag{54}$$

The indices i and j encode all possible occupations of orbitals i and j in the two-orbital Fock space that is spanned by 16 states (for spatial orbitals): $(-\,-)$, $(\uparrow\,-)$, $(\downarrow\,-)$, $(-\,\uparrow)$, $\ldots$, $(\uparrow\downarrow\uparrow\downarrow)$. Thus, the two-orbital RDM can be expressed as a 16×16 matrix and determines the two-orbital entropy $s_{i,j}$ [18]. Specifically, the two-orbital entropy quantifies the entanglement between the environment orbitals and a particular orbital pair ij and is given by

$$s_{i,j} = -\sum_{\alpha=1}^{16} \omega_{\alpha,i,j} \ln \omega_{\alpha,i,j} \tag{55}$$

where $\omega_{\alpha,i,j}$ are the eigenvalues of the two-orbital RDM. Given the one- and two-orbital RDMs, we can calculate the so-called mutual information between any orbital pair ij,

$$I_{i|j} = s_i + s_j - s_{i,j}. \tag{56}$$

Most importantly, the mutual information is a measure of correlation and describes both classical and quantum correlations. $I_{i|j}$ takes values in the range of [0, ln 16], where 0 is obtained for uncorrelated wave functions such as a single Slater (or the HF) determinant. We should note that evaluating the one- and two-orbital RDMs using the general (51) and (54) might be cumbersome due to the phase factors that have to be accounted for in (50) and (53). For practical calculations, the one- and two-orbital RDMs can be expressed in terms of conventional N-particle reduced density matrices [15, 16]. Specifically, $\rho_{i,i'}$ requires only the 1- and 2-particle RDMs, while $\rho_{(i,j),(i',j')}$ requires in addition some elements of the 3- and 4-particle RDMs. In

Table 1 Different types of electron correlation effects and the corresponding values of the single-orbital entropy and the orbital-pair mutual information [15, 16]

Type of correlation	s_i	$I_{i\vert j}$
Nondynamic	>0.5	$\approx 10^{-1}$
Static	$0.1 - 0.5$	$\approx 10^{-2}$
Dynamic	<0.1	$\approx 10^{-3}$
Weak (dispersion, etc.)	≈ 0	$\leq 10^{-4}$

conventional electronic structure methods, the N-particle RDMs are either already available or can be easily determined. Thus, the evaluation of the single- and two-orbital entropy as well as the mutual information does not pose a computational difficulty.

The single-orbital entropy and orbital-pair mutual information are particularly useful to classify electron correlation effects into different contributions. Large values of the entropic measures appear in molecules where strong (static and nondynamic) correlation effects dominate. Dynamic (weak) correlation is characterized by smaller values for both s_i and $I_{i|j}$, while the single-orbital entropy is close to zero for dispersion interactions. As there is no unique definition of the different contributions to electron correlation effects, a distinction between them is rather arbitrary. Boguslawski et al. [15, 16] proposed to dissect electron correlation effects according to the values of the single-orbital entropy and the orbital-pair mutual information which are given in Table 1.

s_i and $I_{i|j}$ can provide many additional insights in electronic structure theory calculations. Examples are elucidating chemical bonding, [18, 103] monitoring bond-formation processes, [39], identifying transition states, [104] and defining stable active orbital spaces in MCSCF-type calculations [15, 16, 22, 74]. The last point may be particularly important in actinide chemistry as the number of chemically relevant orbitals is difficult to predict a priori. A particularly straightforward selection protocol to define stable and reliable active spaces in correlation calculations was proposed recently [15–17] and applied to plutonium [22] and neptunium [86] compounds that have not been investigated using MCSCF-type approaches on a routine basis. Similar approaches have been proposed for transition metals [135]. The proposed selection procedure exploits the orbital-pair mutual information as sole selection criterion. Since the orbital-pair mutual information measures orbital-pair correlations, the corresponding active orbital spaces should provide a balance description of electron correlation effects even for unknown compounds. The protocol to obtain correlation-based active spaces includes the following steps:

1. Perform a *large* active space calculation with a quantum chemical method of your choice, like the DMRG algorithm, and determine the orbital-pair correlations. Note that the wave function for this large active space calculations does not have to be fully converged. For instance, in DMRG calculations, already 4–6 sweeps are sufficient to calculate the orbital-pair mutual information with sufficient accuracy
2. Choose a cutoff threshold for the orbital-pair mutual information, for instance $I_{i|j} > 10^{-2}$. Such a threshold allows us to describe static/nondynamic electron correlation

3. Select those orbitals where $I_{i|j}$ is above the threshold
4. Optimize the wave function for this correlation-based active space
5. Determine the corresponding orbital-pair mutual information
6. Compare the values of $I_{i|j}$ to those of the reference calculation in step 1. If the small active space calculation can accurately reproduce the orbital-pair correlation profile, the small active space can be considered as one *optimal* choice, otherwise the acceptance threshold for $I_{i|j}$ has to be further reduced (for instance, to 10^{-3}) and steps 3–5 have to be repeated until convergence.

Although, such correlation-based active orbital spaces represent a step towards true black-box MCSCF-type calculations, the selection criteria might have to be extended so that they allow us to consider all orbitals important for bond-breaking processes at all points of the dissociation pathway as the magnitude of orbital-pair correlations might change along the reaction coordinate. However, technical limitations, like stability of active space calculations, cannot be excluded and will restrict all automatic active space selection protocols.

4 Challenging Examples in Computational Actinide Chemistry

In the following, we briefly review some challenging case studies where computational chemistry allowed us to explain some peculiar or unexpected properties of actinide-containing compounds.

4.1 Symmetric Dissociation of UO_2^{2+}

The uranyl cation is a small building block of plenty uranium-containing compounds [34, 51, 58]. This molecule is characterized by a linear geometry and a singlet ground-state electronic structure. The energetically close lying 5*f*, 6*d*, and 7*s* orbitals are crucial to describe the strongly-correlated valence electrons. In addition, the uranium 6p orbitals are "pushed from below" by oxygen 2p electrons and thus complicate the electronic structure as 6p orbitals are easily polarizable and mix with 5f orbitals [32, 33, 69, 100, 116, 117, 121, 122, 151, 152]. While the bonding mechanism in UO_2^{2+} is well described by single-reference CC theory for molecular structures close to the equilibrium, conventional quantum chemistry methods, like CCD, CCSD, CCSD(T), and DFT, usually fail for elongated U–O bonds. Furthermore, the CASSCF method does not allow us to define stable and consistent active spaces along the whole dissociation pathway. Specifically for UO_2^{2+}, a minimal active space (CAS(12,12)SCF) around the equilibrium should contain all σ-, σ^*-, π-, and π^*-orbitals as shown in Fig. 1a. For stretched U–O bonds, however, the ϕ_u and δ_u orbitals become partially occupied and should be included in the active space (see

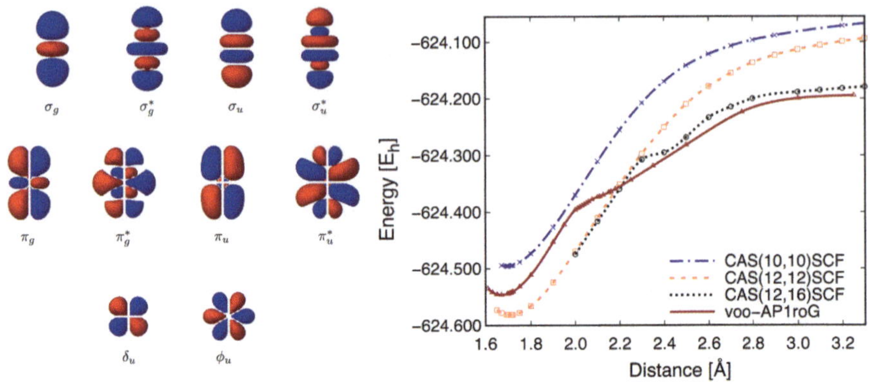

(a) Valence molecular orbitals (b) Potential energy surfaces

Fig. 1 Valence molecular orbitals and dissociation curves for the symmetric stretching of UO_2^{2+}. Subfigure (**b**) has been reproduced from [152] with permission from the PCCP Owner Societies

Fig. 1). Since these orbitals are unoccupied around the equilibrium, they cannot be included in the active space due to convergence difficulties in CASSCF calculations. Thus, CASSCF either predicts the wrong dissociation limit (minimal active space) or does not provide a smooth potential energy surface (CAS(12,16)SCF). Including ϕ_u and δ_u orbitals into the active space results in a qualitative change in the shape of the PES featuring a shoulder around 2 Å. In contrast to CASSCF calculations, AP1roG allows us to include all orbitals in the active space and provides a smooth dissociation curve. In addition, the corresponding potential energy surface features a similar shoulder as predicted by CAS(12,16)SCF. Thus, AP1roG can capture (static/nondynamic) electron correlation effects along the dissociation pathway without imposing active spaces.

4.2 Excitations of NUN

The NUN complex is the isoelectronic analogue of UO_2^{2+} and has been formed in noble gas matrices and as a free molecule [163]. This compound is particularly interesting because of its possible applications in the nuclear industry. In its equilibrium geometry, the ground-state of NUN is closed-shell, similar to the isoelectronic UO_2^{2+}. The U–N triple bonds (1.73–1.76 Å) are slightly longer than the U–O distance in UO_2^{2+} (1.70–1.72 Å). In the spin-free formalism, the ground-state wave function is dominated by a single determinant (with a weight of about 0.9 for the principal determinant) with small contributions from doubly excited determinants. The energies of the upper bonding molecular orbitals are distributed equidistantly [163]. Furthermore, the δ and ϕ virtual orbitals are equally important for excited states as they lie close in energy [145]. The spin–orbit electronic spectrum of NUN using

Table 2 15 lowest-lying IH-FSCCSD vertical excitation energies of the NUN molecule ($r_{U-N} = 1.739$ Å). Excitation energies are given in eV [151]

Ω	Character (from DC)	DC	DC(G)	X2C/AMF	X2C/MMF	X2C(G)/MMF
2_g	52% $\sigma_{1/2u}\phi_{5/2u}$ + 26% $\pi_{1/2u}\phi_{5/2u}$	0.956	0.923	0.936	0.957	0.927
3_g	50% $\sigma_{1/2u}\phi_{5/2u}$ + 24% $\pi_{1/2u}\phi_{5/2u}$	1.103	1.068	1.083	1.103	1.072
1_g	45% $\sigma_{1/2u}\delta_{3/2u}$ + 20% $\pi_{1/2u}\delta_{3/2u}$	1.134	1.094	1.106	1.134	1.098
2_g	30% $\sigma_{1/2u}\delta_{3/2u}$ + 15% $\pi_{1/2u}\delta_{3/2u}$	1.398	1.355	1.374	1.398	1.358
4_g	49% $\sigma_{1/2u}\phi_{7/2u}$ + 23% $\pi_{1/2u}\phi_{7/2u}$ + 16% $\sigma_{1/2u}\phi'_{7/2u}$	1.699	1.645	1.680	1.698	1.646
3_g	43% $\sigma_{1/2u}\delta_{5/2u}$ + 20% $\pi_{1/2u}\delta_{5/2u}$	1.757	1.704	1.739	1.757	1.705
3_g	38% $\sigma_{1/2u}\phi_{7/2u}$ + 22% $\pi_{1/2u}\phi_{7/2u}$	2.076	2.028	2.059	2.076	2.029
2_g	33% $\sigma_{1/2u}\delta_{5/2u}$ + 17% $\pi_{1/2u}\delta_{5/2u}$	2.519	2.476	2.502	2.519	2.478
1_u	40% $\sigma_{1/2u}\delta_{3/2g}$ + 24% $\pi_{1/2u}\delta_{3/2g}$	2.669	2.696	2.680	2.669	2.696
0_u^+	41% $\sigma_{1/2u}\sigma_{1/2g}$ + 30% $\pi_{1/2u}\sigma_{1/2g}$	2.709	2.757	2.740	2.709	2.755
1_u	41% $\pi_{3/2u}\sigma'_{1/2g}$ + 29% $\pi'_{3/2u}\sigma''_{1/2g}$	2.711	2.759	2.743	2.711	2.757
1_g	80% $\pi_{3/2u}\phi_{5/2u}$	2.711	2.675	2.690	2.711	2.679
2_u	34% $\sigma_{1/2u}\delta_{3/2g}$ + 24% $\pi_{1/2u}\delta_{3/2g}$	2.749	2.767	2.758	2.749	2.768
4_g	84% $\pi_{3/2u}\phi_{5/2u}$	2.844	2.808	2.823	2.844	2.811
2_u	73% $\sigma_{1/2g}\phi_{5/2u}$	2.895	2.857	2.875	2.895	2.860

different relativistic Hamiltonians is presented in Table 2 [151]. DC denotes the standard Dirac–Coulomb Hamiltonian, DC(G) is the DC Hamiltonian augmented with the Gaunt operator at the SCF level, X2C/AMF and MMF correspond to the X2C Hamiltonian with the atomic and molecular mean field approximations to spin–orbit coupling, and X2C(G) is again the X2C Hamiltonian augmented with the Gaunt operator at the SCF level. Based on the data presented in Table 2, we can conclude that NUN possesses significant multi-reference character and a rather complex electronic spectrum [151]. Furthermore, including the Gaunt operator in the Hamiltonian has only negligible effect on the electronic spectra of NUN, while the X2C Hamiltonian represents a computationally cheaper alternative to the full DC Hamiltonian. Specifically, the spin–orbit electronic spectrum calculated within X2C/MMF is almost identical to the DC spectrum.

4.3 CUO Diluted in Noble Gas Matrices

As noble gases are known to be inert, they constitute an ideal environment to investigate properties of single molecules. Diluted in a noble gas matrix, the studied substance should not interact with the environment and thus its electronic structure should remain unaffected. A peculiar observation has been made for the CUO molecule and was reported by Andrews et al. [93, 94, 171]. Experiments revealed a blue shift in the asymmetric U–O and U–C vibrational spectra when the composition of the noble gas matrix—a mixture of neon and argon—was changed. To explain the observed shifts, experimentalist anticipated that the noble gas environment may not be inert and may interact with the CUO unit differently depending on its composition [93, 94].

To understand the specific interaction between CUO and noble gas matrices, various quantum chemistry methods have been applied both to the bare CUO unit and to small model systems including noble gas atoms. A new perspective has been provided by ab initio calculations using scalar-relativistic high-level wave-function based methods with spin-orbit corrections added a posteriori [150]. A computational study covering spin-orbit (SO)-CASSCF and SO-DMRG calculations indicate that the interaction of CUO with Ne atoms does not change the order of states and the ground state remains a $^1\Sigma^+$ state (Fig. 2). However, the energy gap between the two lowest-lying states becomes negligible in $CUOAr_4$ and the molecule requires a multi-reference treatment. Reference [150] was the first work dissecting electron correlation effects using quantum information theory measures in $CUONe_4$ and $CUOAr_4$ molecules. The interactions between CUO and Ar valence electrons has been confirmed supporting experimental and theoretical anticipations. Furthermore, numerical results suggest that a (thermal) spin crossover may occur.

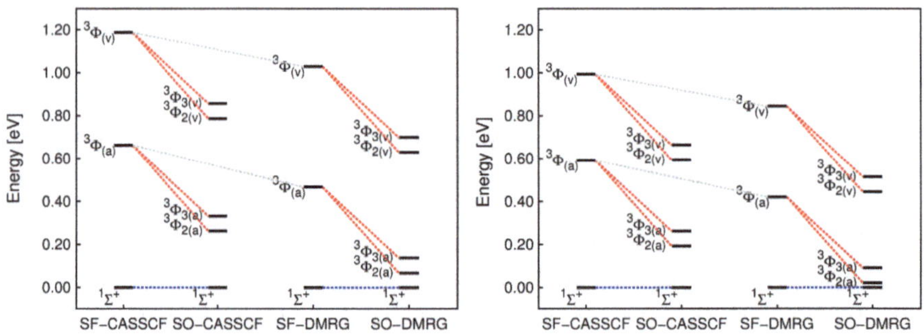

(a) $CUONe_4$ (b) $CUOAr_4$

Fig. 2 Spin state energy splittings of CUO diluted in different noble gas matrices. All calculations have been performed using the spin-free (SF) DKH Hamiltonian of 10th order, while a spin-orbit (SO) correction was added a posteriori. $^3\Phi_{(v)}$: vertically excited $^3\Phi$ state. $^3\Phi_{(a)}$: adiabatically excited $^3\Phi$ state. Reproduced from [150] with permission from the PCCP Owner Societies

4.4 Cation–cation Attraction in $[NpO_2]_2^{2+}$

The attractive interaction between two cations is a characteristic feature of heavy-element-containing molecules and has been observed for the first time in uranyl perchlorate solution and aqueous chlorate media [138–140]. This so-called cation–cation interactions (CCI) are exploited in the synthesis of new crystalline structures [2, 5, 36, 41, 71, 72, 134, 159, 161, 162]. Most importantly, CCIs pose a technical difficulty when reprocessing spent nuclear fuel.

Another example for this peculiar interaction are neptunyls. Specifically, neptunyl CCIs feature an end-on or side-on interaction producing the T-shaped or diamond-shaped dimers. The CCIs structures are stable primarily because of the bonding interaction between the oxygen and the neptunium atoms of two neighbouring complexes, where the effective charge of the oxygen atoms is negative [155] in contrast to the effective positive charge localised on the actinide atoms [28]. The stability of CCIs is strongly influenced by the Np–O bond distance, [54] which changes in different environments [43, 56, 118, 126]. In order to describe the ground- and excited-state properties of such CCIs, the theoretical model needs to account for environmental effects originating from the solvent. Figure 3a shows two different explicit solvation models for the T-shaped (i and ii) and diamond-shaped (iii and iv) clusters. Including both explicit and implicit solvation models in quantum chemistry calculations allows us to reproduce the experimentally measured Np–Np distance within (spin-free) DFT calculations. However, in order to accurately model the ground state (and

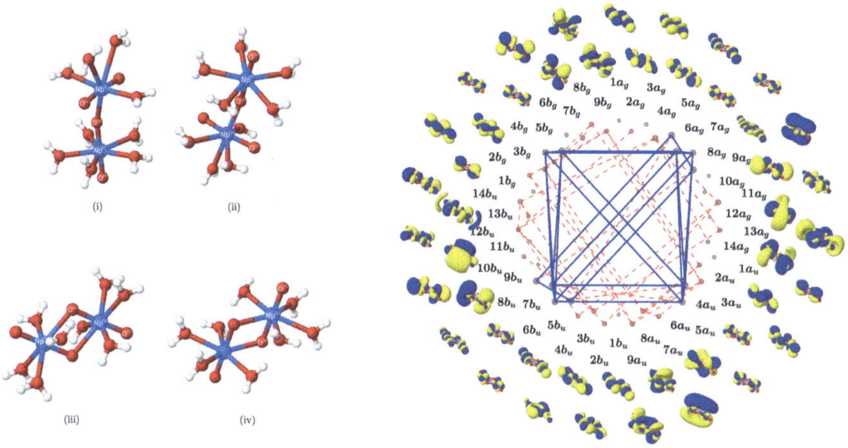

(a) Molecular geometries of neptunyl CCIs (b) Orbital-pair correlations in the diamond-shaped CCI

Fig. 3 **a** Molecular geometries of neptunyl CCIs including explicit water molecules and **b** orbital-pair correlations of the diamond-shaped (bottom) $[NpO_2]_2^{2+}$. The values of the single-orbital entropy are coded by the size of the dots corresponding to each orbital. The strongest correlated orbitals are connected by blue lines ($I_{i|j} > 10^{-1}$), followed by orbital-pair correlations marked by red lines ($10^{-1} > I_{i|j} > 10^{-2}$)

also excited state) electronic structure, we have to account for both relativistic and correlation effects on an equal footing. This poses a particular challenge for conventional electronic structure methods primarily because we have to deal with two heavy-element centers.

The strong multi-reference nature of the neptunyl CCIs becomes evident when we perform an orbital correlation analysis. The orbital-pair mutual information for the ground-state of the diamond-shaped neptunyl CCI $[NpO_2]_2^{2+}$ is shown in Fig. 3b. The correlation between orbital pairs is indicated by lines, while its strength is color-coded: strong correlations are shown in blue, medium-sized correlations in red, etc. Specifically for the diamond-shaped $[NpO_2]_2^{2+}$, the σ_g- and σ_g^*-type orbitals are as important as δ_u- and ϕ_u-type orbitals. Note that π-type orbitals are only moderately correlated with each other. The orbital-correlation analysis, thus, suggests that a balanced active space for neptunyl-containing CCIs that allows us to describe both nondynamic and static correlation (or moderately and strongly correlated orbitals) should contain approximately 30 orbitals (δ_u-, ϕ_u-, bonding and antibonding combinations of σ_g-, σ_u-, π_u-, and π_g-type orbitals of each monomer). However, such large active spaces are difficult to handle with conventional multi-reference methods. In a first approximation, we can consider active spaces where only the strongest orbital-pair correlations are accounted for, while the remaining correlations are treated a posteriori using, for instance, perturbation theory. Such a study has been presented recently in [86] and highlights the interplay of relativistic and correlations effects in neptunyl CCIs.

5 Summary

This chapter reviewed different quantum chemistry approaches applicable to actinide chemistry. Actinide-containing compounds are particularly challenging as both relativistic effects and correlation effects have to be accounted for on an equal footing. Specifically, we have focused on the most common relativistic Hamiltonians and wave-function-based methods that allow us to reliably model actinide chemistry. Particularly interesting are novel and unconventional methods, like the DMRG algorithm or geminal-based approaches as they allow us to include a large number of orbitals in active space calculations. This feature is desirable especially for multi-centered actinide-containing compounds.

Our numerical examples show the strengths and pitfalls of present-day quantum chemistry methods when the systems under study contain one or more heavy-elements. While DFT can accurately provide molecular geometries, wave-function-based methods have to be applied to describe electronic structures of ground- and excited-states. Furthermore, conventional methods like CASSCF fail in describing potential energy surfaces for stretched actinide–ligand bonds. Such calculations can only be accomplished using modern and unconventional methods like DMRG or AP1roG.

Acknowledgements A. Ł. and K. B. acknowledge financial support from the National Science Centre, Poland (SONATA BIS 5 Grant No. 2015/18/E/ST4/00584). K. B. gratefully acknowledges funding from a Marie-Skłodowska-Curie Individual Fellowship project no. 702635–PCCDX and a scholarship for outstanding young scientists from the Ministry of Science and Higher Education. P. T. thanks the POLONEZ fellowship program of the National Science Center, Poland, No. 2015/19/P/ST4/02480. This project had received funding from the European Union's Horizon 2020 research and innovation programme under the Marie Skłodowska–Curie grant agreement No. 665778.

References

1. Adamo C, Barone V (1999) Toward reliable density functional methods without adjustable parameters: the PBE0 model. Phys Rev Lett 110:6158–6170
2. Almond PM, Skanthakumar S, Soderholm L, Burns PC (2007) Cation-cation interactions and antiferromagnetism in $Na[Np(V)O_2(OH)_2]$: synthesis, structure, and magnetic properties. Chem Mater 19:280–285
3. Andersson K, Malmqvist PÅ, Roos BO, Sadlej AJ, Woliński K (1990) Second-order perturbation theory with a CASSCF reference function. J Phys Chem 94:5483–5488
4. Andersson K, Malmqvist PÅ, Roos BO (1992) Second-order perturbation theory with a complete active space self-consistent field reference function. J Chem Phys 96:1218–1226
5. Arnold PL, Love JB, Patel D (2009) Pentavalent uranyl complexes. Coord Chem Rev 253:1973–1978
6. Autschbach J, Siekierski S, Seth M, Schwerdtfeger P, Schwarz WHE (2002) Dependence of relativistic effects on electronic configuration in the neutral atoms of d- and f-block elements. J Comput Chem 23:804–813
7. Barcza G, Legeza Ö, Marti KH, Reiher M (2011) Quantum-information analysis of electronic states of different molecular structures. Phys Rev A 83:012508
8. Barcza G, Noack R, Sólyom J, Legeza Ö (2014) Entanglement patterns and generalized correlation functions in quantum many-body systems. Phys Rev B 92:125140
9. Barysz M (2003) Two-component methods. In: Kaldor U, Wilson S (eds) Theoretical chemistry and physics of heavy and superheavy elements. Springer, pp 349–397. Chapter 9
10. Becke AD (1988) Density-functional exchange-energy approximation with correct asymptotic behavior. Phys Rev A 38:3098–4000
11. Boguslawski K (2016) Targeting excited states in all-trans polyenes with electron-pair states. J Chem Phys 145:234105
12. Boguslawski K (2017) Erratum: "targeting excited states in all-trans polyenes with electron-pair states". J Chem Phys 147(13):139901
13. Boguslawski K, Ayers PW (2015) Linearized coupled cluster correction on the antisymmetric product of 1-reference orbital geminals. J Chem Theory Comput 11:5252–5261
14. Boguslawski K, Reiher M (2014) Chemical bonding in open-shell transition-metal complexes. In: Frenking G, Shaik S (eds) The chemical bond: Chemical bonding across the periodic table, 1st edn. Wiley-VCH Verlag GmbH & Co KGaA, Weinheim, chap 8, pp 219–252
15. Boguslawski K, Tecmer P (2015) Orbital entanglement in quantum chemistry. Int J Quantum Chem 115:1289–1295
16. Boguslawski K, Tecmer P (2017) Erratum: orbital entanglement in quantum chemistry. Int J Quantum Chem 117:e25455
17. Boguslawski K, Tecmer P, Legeza Ö, Reiher M (2012) Entanglement measures for single- and multireference correlation effects. J Phys Chem Lett 3:3129–3135
18. Boguslawski K, Tecmer P, Barcza G, Legeza Ö, Reiher M (2013) Orbital entanglement in bond-formation processes. J Chem Theory Comput 9:2959–2973

19. Boguslawski K, Tecmer P, Ayers PW, Bultinck P, de Baerdemacker S, van Neck D (2014a) Efficient description of strongly correlated electrons with mean-field cost. Phys Rev B 89:201106
20. Boguslawski K, Tecmer P, Limacher PA, Johnson PA, Ayers PW, Bultinck P, de Baerdemacker S, van Neck D (2014b) Nonvariational orbital optimization techniques for the AP1roG wave function. J Chem Theory Comput 10:4873–4882
21. Boguslawski K, Tecmer P, Limacher PA, Johnson PA, Ayers PW, Bultinck P, de Baerdemacker S, van Neck D (2014c) Projected seniority-two orbital optimization of the antisymmetric product of one-reference orbital geminal. J Chem Phys 140:214114
22. Boguslawski K, Réal F, Tecmer P, Duperrouzel C, Gomes ASP, Legeza Ö, Ayers PW, Vallet V (2017) On the multi-reference nature of plutonium oxides: PuO_2^{2+}, PuO_2, PuO_3, and $PuO_2(OH)_2$. Phys Chem Chem Phys 19:4317–4329
23. Breit G (1929) The effect of retardation on the interaction of two electrons. Phys Rev 34:553–573
24. Bytautas L, Henderson TM, Jiménez-Hoyos CA, Ellis JK, Scuseria GE (2011) Seniority and orbital symmetry as tools for establishing a full configuration interaction hierarchy. J Chem Phys 135:044119
25. Chan GKL, Sharma S (2011) The density matrix renormalization group in quantum chemistry. Annu Rev Phys Chem 62:465–481
26. Chang C, Pelissier M, Durand P (1986) Regular two-component Pauli-like effective hamiltonians in Dirac theory. Phys Scr 34:394–404
27. Chatterjee K, Pastorczak E, Jawulski K, Pernal K (2016) A minimalistic approach to static and dynamic electron correlations: amending generalized valence bond method with extended random phase approximation correlation correction. J Chem Phys 113:2960–2963
28. Choppin GR, Rao LF (1984) Complexation of pentavalent and hexavalent actinides by fluoride. Radiochim Acta 37:143–146
29. Cohen AJ, Mori-Sánchez P, Yang W (2012) Challenges for density functional theory. Chem Rev 112:289–320
30. Coleman AJ (1965) Structure of fermion density matrices. II. Antisymmetrized geminal powers. J Math Phys 6:1425–1431
31. Cremer D (2001) Density functional theory: coverage of dynamic and non-dynamic electron correlation effects. Mol Phys 99:1899–1940
32. de Jong WA, Visscher L, Nieuwpoort WC (1999) On the bonding and the electric field gradient of the uranyl ion. J Mol Struct (Theochem) 458:41–52
33. Denning RG (1992) Electronic structure and bonding in actinyl ions, vol 79. Springer, Berlin, Heidelberg, pp 313–336
34. Denning RG (2007) Electronic structure and bonding in actinyl ions and their analogs. J Phys Chem A 111:4125–4143
35. Dirac PAM (1928) The quantum theory of the electron. Proc Roy Soc Lond A 117:610–624
36. Diwu J, Wang S, Albrecht-Schmitt TE (2012) Periodic trends in hexanuclear actinide clusters. Inorg Chem 51:4088–4093
37. Dolg M, Cao X (2012) Relativistic pseudopotentials: their development and scope of applications. Chem Rev 112:403–480
38. Douglas N, Kroll NM (1974) Quantum electrodynamical corrections to fine-structure of helium. Ann Phys 82:89–155
39. Duperrouzel C, Tecmer P, Boguslawski K, Barcza G, Legeza Ö, Ayers PW (2015) A quantum informational approach for dissecting chemical reactions. Chem Phys Lett 621:160–164
40. Foldy LL, Wouthuysen SA (1950) On the Dirac theory of spin 1/2 particles and its non-relativistic limit. Phys Rev 78:29–36
41. Forbes TZ, Burns PC, Soderholm L, Skanthakumar S (2006) Crystal structures and magnetic properties of $NaK_3(NpO_2)_4(SO_4)_4(H_2O)_2$ and $NaNpO_2SO_4H_2O$: cation-cation interactions in a neptunyl sulfate framework. Chem Mater 18:1643–1649
42. Fortier S, Hayton TW (2010) Oxo ligand functionalization in the uranyl ion (UO_2^{2+}). Coord Chem Rev 254:197

43. Gainar I, Sykes KW (1964) The spectra and stability of some neptunium complex ions in water and methanol. J Chem Soc 9:4452–4459
44. Garza AJ, Bulik IW, Henderson TM, Scuseria GE (2015a) Range separated hybrids of pair coupled cluster doubles and density functionals. Phys Chem Chem Phys 17:22412–22422
45. Garza AJ, Bulik IW, Henderson TM, Scuseria GE (2015b) Synergy between pair coupled cluster doubles and pair density functional theory. J Chem Phys 142:044109
46. Garza AJ, Sousa Alencar AG, Scuseria GE (2015c) Actinide chemistry using singlet-paired coupled cluster and its combinations with density functionals. J Chem Phys 143:244106
47. Garza AJ, Bulik IW, Alencar AGS, Sun J, Perdew JP, Scuseria GE (2016) Combinations of coupled cluster, density functionals, and the random phase approximation for describing static and dynamic correlation, and van der Waals interactions. Mol Phys 114:997–1018
48. Gaunt JA (1929) The triplets of helium. Proc Roy Soc Lond A 122:513–532
49. Goddard WA, Dunning TH Jr, Hunt WJ, Hay PJ (1973) Generalized valence bond description of bonding in low-lying states of molecules. Acc Chem Res 6:368–376
50. Gomes ASP, Jacob CR, Visscher L (2008) Calculation of local excitations in large systems by embedding wave-function theory in density-functional theory. Phys Chem Chem Phys 10:5353–5362
51. Gomes ASP, Jacob CR, Réal F, Visscher L, Vallet V (2013) Towards systematically improvable models for actinides in condensed phase: the electronic spectrum of uranyl in $Cs_2UO_2Cl_4$ as a test case. Phys Chem Chem Phys 15:15153–15162
52. Goncharov V, Han J, Kaledin LA, Heaven MC (2005) Ionization energy measurements and electronic spectra for ThO. J Chem Phys 122:204311
53. Goncharov V, Kaledin LA, Heaven MC (2006) Probing the electronic structure of UO^+ with high-resolution photoelectron spectroscopy. J Phys Chem 125:133202
54. Guillaume B, Hobart DE, Bourges JY (1981) "Cation-cation" complexes of pentavalent actinides—II Spectrophotometric study of complexes of americium(V) with uranyl(VI) and neptunyl(VI) ions in aqueous perchlorate solution. J Inorg Nucl Chem 43:3295–3299
55. Häller LJL, Kaltsoyannis N, Sarsfield MJ, May I, Cornet SM, Redmond MP, Helliwell M (2007) A structural and theoretical investigation of equatorial cis and trans uranyl phosphinimine and uranyl phosphine oxide complexes $UO_2Cl_2(Cy_3PNH)_2$ and $UO_2Cl_2(Cy_3PO)_2$. Inorg Chem 46:4868–4875
56. Halperin J, Oliver JH (1983) Sulfate complexation of neptunium(V) in aqueous solution. Radiochim Acta 33:29–33
57. Hay PJ, Hunt WJ, Goddard WA (1972) Generalized valence bond wavefunctions for the low lying states of methylene. Chem Phys Lett 13:30–35
58. Hayton TW (2013) Uranium chemistry: an actinide milestone. Nat Chem 5:451–452
59. Hégely B, Nagy PR, Ferenczy GG, Kállay M (2016) Exact density functional and wave function embedding schemes based on orbital localization. J Chem Phys 145(6):064107
60. Helgaker T, Jørgensen P, Olsen J (2000) Molecular electronic-structure theory. Wiley, New York
61. Hess BA (1986) Relativistic electronic-structure calculations employing a two-component no-pair formalism with external-field projection operators. Phys Rev A 33:3742–3748
62. Hess BA, Marian CM, Wahlgren U, Gropen O (1996) A mean-field spin-orbit method applicable to correlated wavefunctions. Chem Phys Lett 251:365–371
63. Hohenberg P, Kohn W (1964) Inhomogeneous electron gas. Phys Rev 136:B864–B871
64. Hurley AC, Lennard-Jones J, Pople JA (1953) The molecular orbital theory of chemical valency XIV. Paired electrons in the presence of two unlike attracting centres. Proc R Soc Lond Ser A 220:446–455
65. Iliaš M, Saue T (2007) An infinite-order two-component relativistic Hamiltonian by a simple one-step transformation. J Chem Phys 126:064102
66. Infante I, Gomes ASP, Visscher L (2006) On the performance of the intermediate Hamiltonian Fock-space coupled-cluster method on linear triatomic molecules: the electronic spectra of NpO_2^+, NpO_2^{2+}, and PuO_2^{2+}. J Chem Phys 125:074301

67. Ingram KIM, Kaltsoyannis N, Gaunt AJ, Neu MP (2007) Covalency in the f-element-chalcogen bond computational studies of $[M(N(EPH_2)_2)_3]$ (M = La, U, Pu; E = O, S, Se, Te). J Alloys Compd 444–445:369–375
68. Ingram KIM, Tassell MJ, Gaunt AJ, Kaltsoyannis N (2008) Covalency in the f element-chalcogen bond. Computational studies of $M[N(EPR_2)_2]_3$ (M = La, Ce, Pr, Pm, Eu, U, Np, Pu, Am, Cm; E = O, S, Se, Te; R = H, (i)Pr, Ph). Inorg Chem 47:7824–7833
69. Jackson VE, Craciun R, Dixon DA, Peterson K, de Jong WB (2008) Prediction of Vibrational Frequencies of UO_2^{2+} at the CCSD(T) Level. J Phys Chem A 112:4095–4099
70. Jeszenszki P, Nagy PR, Zoboki T, Szabados A, Surján PR (2014) Perspectives of APSG-based multireference perturbation theories. Int J Quantum Chem 114:1048–1052
71. Jin GB, Skanthakumar S, Soderholm L (2011a) Cation-cation interactions: crystal structures of neptunyl(V) selenate hydrates, $(NpO_2)_2(SeO_4)(H_2O)_n$ (n = 1, 2, and 4). Inorg Chem 50:5203–5214
72. Jin GB, Skanthakumar S, Soderholm L (2011b) Two new neptunyl(V) selenites: a novel cation-cation interaction framework in $(NpO_2)_3(OH)(SeO_3)(H_2O)_2{\cdot}H_2O$ and a uranophane-type sheet in $Na(NpO_2)(SeO_3)(H_2O)$. Inorg Chem 50:6297–6303
73. Johnson PA, Ayers PW, Limacher PA, de Baerdemacker S, van Neck D, Bultinck P (2013) A size-consistent approach to strongly correlated systems using a generalized antisymmetrized product of nonorthogonal geminals. J Chem Theor Comput 1003:101–113
74. Keller S, Boguslawski K, Janowski T, Reiher M, Pulay P (2015) Selection of active spaces for multiconfigurational wavefunctions. J Chem Phys 142:244104
75. Kędziera D (2006) Solving of the infinite-order two-component method equations. In: Recent progess in computational sciences and engineering, VSP BV-C/O BRILL ACAD PUBL, Leiden, The Netherlands, Lecture series on computer and computational sciences, vol 7A–B, pp 252–255
76. Kędziera D, Barysz M (2007) Non-iterative approach to the infinite-order two-component (IOTC) relativistic theory and the non-symmetric algebraic Riccati equation. Chem Phys Lett 446:176–181
77. Knecht S, Keller S, Autschbach J, Reiher M (2016) A nonorthogonal state-interaction approach for matrix product state wave functions. J Chem Theory Comput 12:5881–5894
78. Kohn W, Sham LJ (1965) Self-consistent equations including exchange and correlation effects. Phys Rev 140:A1133–A1138
79. Kovács A, Konings RJM, Raab J, Gagliardi L (2008) A theoretical study of AmO_n and CmO_n (n = 1, 2). Phys Chem Chem Phys 10:1114–1117
80. Krylov AI (2006) Spin-flip equation-of-motion coupled-cluster electronic structure method for a description of excited states, bond breaking, diradicals, and triradicals. Acc Chem Res 39:83–91
81. Kümmel S, Kronik L (2008) Orbital-dependent density functionals: theory and applications. Rev Mod Phys 80:3–60
82. Kurashige Y, Yanai T (2011) Second-order perturbation theory with density matrix renormalization group self-consistent field reference function: theory and application to the study of chromium dimer. J Chem Phys 135:094104
83. Kutepov AL (2007) The effect of exact calculation of exchange interaction upon calculated electronic structure of actinides. J Alloys Compd 444–445:174–176
84. Kutzelnigg W (1964) Direct determination of natural orbitals and natural expansion coefficients of many-electron wavefunctions. I. Natural orbitals in the geminal product spproximation. J Chem Phys 40:3640–3647
85. Kutzelnigg W (1965) On the validity of the electron pair approximation for the beryllium ground state. Theor Chim Acta 3:241–253
86. Łachmanska A, Tecmer P, Legeza Ö, Boguslawski K (2018) Elucidating cation–cation interactions in neptunyl dications using multireference ab initio theory. Phys Chem Chem Phys 21:744–759
87. Langhoff SR, Davidson ER (1974) Configuration interaction calculations on the nitrogen molecule. Int J Quantum Chem 8:61–72

88. Lawler KV, Beran GJO, Head-Gordon M (2008) Symmetry breaking in benzene and larger aromatic molecules within generalized valence bond coupled cluster methods. J Chem Phys 128:024107
89. Lee C, Yang W, Parr RG (1988) Development of the Colle-Salvetti correlation-energy formula into a functional of the electron density. Phys Rev B 37:785–789
90. Legeza Ö, Sólyom J (2003) Optimizing the density-matrix renormalization group method using quantum information entropy. Phys Rev B 68:195116
91. Legeza Ö, Noack R, Sólyomand J, Tincani L (2005) Applications of quantum information in the density-matrix renormalization group. In: Fehske H, Schneider R, Weiße A (eds) Computational many-particle physics. Springer, Berlin/Heidelberg, chap 24, pp 653–664
92. Legeza Ö, Barcza G, Noack RM, Sólyom J (2013) Entanglement topology of strongly correlated systems. Korrelationstage MPIPKS, Dresden
93. Li J, Bursten BE, Liang B, Andrews L (2002) Noble gas-actinide compounds: complexation of the CUO molecule by Ar, Kr, and Xe atoms in noble gas matrices. Science 295:2242–2245
94. Liang B, Andrews L, Li J, Bursten BE (2002) Noble gas-actinide compounds: evidence for the formation of distinct $CUO(Ar)_{4-n}(Xe)_n$ and $CUO(Ar)_{4-n}(Xe)_n$ (n = 1, 2, 3, 4) complexes. J Am Chem Soc 124:9016–9017
95. Limacher P, Ayers P, Johnson P, de Baerdemacker S, van Neck D, Bultinck P (2014) Simple and inexpensive perturbative correction schemes for antisymmetric products of nonorthogonal geminals. Phys Chem Chem Phys 16:5061–5065
96. Limacher PA, Ayers PW, Johnson PA, de Baerdemacker S, van Neck D, Bultinck P (2013) A new mean-field method suitable for strongly correlated electrons: computationally facile antisymmetric products of nonorthogonal geminals. J Chem Theor Comput 9:1394–1401
97. Malmqvist PÅ, Roos BO, Schimmelpfennig B (2002) The restricted active space (RAS) state interaction approach with spin-orbit coupling. Chem Phys Lett 357:230–240
98. Marian CM, Wahlgren U (1996) A new mean-field and ECP-based spin-orbit method. Applications to Pt and PtH. Chem Phys Lett 251:357–364
99. Marti KH, Reiher M (2010) The density matrix renormalization group algorithm in quantum chemistry. Z Phys Chem 224:583–599
100. Matsika S, Zhang Z, Brozell SR, Bladeau JP, Pitzer RM (2001) Electronic structure and spectra of actinyl ions. J Phys Chem A 105:3825–3828
101. Meissner L, Musiał M (2010) Intermediate Hamiltonian formulations of the Fock-space coupled-cluster method: details, comparisons, examples. In: Recent progress in coupled cluster methods. Springer, p 395
102. Mikheev NB, Kulyukhin SA, Melikhov IV (2007) Lanthanides and actinides among other groups of elements of the periodic table. Radioch 49:449–463
103. Mottet M, Tecmer P, Boguslawski K, Legeza Ö, Reiher M (2014) Quantum entanglement in carbon-carbon, carbon-phosphorus, and silicon-silicon bonds. Phys Chem Chem Phys 16:8872–8880
104. Murg V, Verstraete F, Schneider R, Nagy PR, Legeza O (2015) Tree tensor network state with variable tensor order: an efficient multireference method for strongly correlated systems. J Chem Theory Comput 11:1027–1036
105. Musiał M, Bartlett RJ (2011) Multi-reference fock space coupled-cluster method in the intermediate hamiltonian formulation for potential energy surfaces. J Chem Phys 135(4):044121
106. Neuscamman E, Yanai T, Chan GKL (2010) A review of canonical transformation theory. Int Rev Phys Chem 29:231–271
107. Parks JM, Parr RG (1958) Theory of separated electron pairs. J Chem Phys 28:335–345
108. Parr RG, Ellison FO, Lykos PG (1956) Generalized antisymmetrized product wave functions for atoms and molecules. J Chem Phys 24:1106–1106
109. Pastorczak E, Pernal K (2015) ERPA-APSG: a computationally efficient geminal-based method for accurate description of chemical systems. Phys Chem Chem Phys 17:8622–8626
110. Perdew JP (1986) Density-functional approximation for the correlation energy of the inhomogeneous electron gas. Phys Rev B 33:8822–8824

111. Perdew JP, Zunger A (1981) Self-interaction correction to density-functional approximations for many-electron systems. Phys Rev B 23:5048
112. Perdew JP, Burke K, Ernzerhof M (1996) Generalized gradient approximation made simple. Phys Rev Lett 77:3865–3868
113. Perdew JP, Burke K, Ernzerhof M (1997) Generalized gradient approximation made simple. Phys Rev Lett 78:1396
114. Pernal K (2014) Intergeminal correction to the antisymmetrized product of strongly orthogonal geminals derived from the extended random phase approximation. J Chem Theory Comput 10:4332–4341
115. Peyerimhoff SD, Buenker RJ (1969) Study of the geometry and spectra of the allylic systems by ab initio methods. J Chem Phys 51:2528–2537
116. Pierloot K, van Besien E (2005) Electronic spectrum of UO_2^{2+} and $[UO_2Cl_4]^{2-}$. J Phys Chem 123:204309
117. Pierloot K, van Besien E, van Lenthe E, Baerends EJ (2007) Electronic structure and spectrum of UO_2^{+2} 2 and $UO_2Cl_4^{2-}$ calculated with time-dependent density functional theory. J Chem Phys 126:194311
118. Rao PRV, Gudi NM, Bagawde SV, Patil SK (1979) The complexing of neptunium(V) by some inorganic ligands. J Inorg Nucl Chem 41:235–239
119. Rassolov VA (2002) A geminal model chemistry. J Chem Phys 117:5978–5987
120. Rassolov VA, Xu F, Garashchuk S (2004) Geminal model chemistry II. Perturbative corrections. J Chem Phys 120:10385–10394
121. Réal F, Vallet V, Marian C, Wahlgren U (2007) On the bonding and the electric field gradient of the uranyl ion. J Phys Chem 127:214302
122. Réal F, Gomes ASP, Visscher L, Vallet V, Eliav EJ (2009) Benchmarking electronic structure calculations on the bare UO_2^{2+} ion: how different are single and multireference electron correlation methods? J Phys Chem A 113:12504–12511
123. Reiher M (2006) Douglas-Kroll-Hess Theory: a relativistic electrons-only theory for chemistry. Theor Chem Acc 116:241–252
124. Reiher M (2012) Relativistic Douglas-Kroll-Hess theory. WIREs Comput Mol Sci 2:139–149
125. Rissler J, Noack RM, White SR (2006) Measuring orbital interaction using quantum information theory. Chem Phys 323:519–531
126. Roesch F, Dittrich S, Buklanov GV, Milanov M, Khalkin VA, Dreyer R (1990) Electromigration of carrier-free radionuclides. 12. Reactions of neptunium-239(V) with acetate and citrate ligands in neutral solutions. Radiochim Acta 49:29–34
127. Rosta E, Surján PR (2002) Two-body zeroth order hamiltonians in multireference perturbation theory: the APSG reference state. J Chem Phys 116:878–889
128. Saitow M, Kurashige Y, Yanai T (2013) Multireference configuration intaraction theory using cumulant reconstruction with internal contraction of density matrix renormalization group wave function. J Chem Phys 139:044118
129. Saue T (2012) Relativistic hamiltonians for chemistry: a primer. Chem Phys Chem 3:3077–3094
130. Schimmelpfennig B, Maron L, Wahlgren U, Teichteil C, Fagerli H, Gropen O (1998) On the combination of ECP-based CI calculations with all-electron spin-orbit mean-field integrals. Chem Phys Lett 286:267–271
131. Schollwöck U (2005) The density-matrix renormalization group. Rev Mod Phys 77:259–315
132. Sharma S, Chan G (2014) Communication: a flexible multi-reference perturbation theory by minimizing the Hylleraas functional with matrix product states. J Chem Phys 141:111101
133. Sikkema J, Visscher L, Saue T, Iliaš M, (2009) The molecular mean-field approach for correlated relativistic calculations. J Chem Phys 131:124116
134. Skanthakumar S, Antonio MR, Soderholm L (2008) A comparison of neptunyl(V) and neptunyl(VI) solution coordination: the stability of cation-cation interactions. Inorg Chem 47:4591–4595
135. Stein CJ, Reiher M (2016) Automated selection of active orbital spaces. J Chem Theory Comput 12:1760–1771

136. Stein T, Henderson TM, Scuseria GE (2014) Seniority-based coupled cluster theory. J Chem Phys 140:214113
137. Straka M, Hrobarik P, Kaupp M (2005) Understanding structure and bonding in early actinide $6d^0 5f^0$ MX_6^q (M = Th-Np; X = H, F) complexes in comparison with their transition metal $5d^0$ analogues. J Am Chem Soc 127:2591–2599
138. Sullivan JC (1962) Complex-ion formation between cations. Spectra and identification of a Np(V)-Cr(III) complex. J Am Chem Soc 84:4256–4259
139. Sullivan JC, Hindman JC, Zielen AJ (1960) Kinetics of the reduction of neptunium(VI) by uranium(IV). J Am Chem Soc 82:5288–5292
140. Sullivan JC, Hindman JC, Zielen AJ (1961) Specific interaction between Np(V) and U(VI) in aqueous perchloric acid media. J Am Chem Soc 83:3373–3378
141. Surján PR, Jeszenszki P, Szabados Á (2015) Role of triplet states in geminal-based perturbation theory. Mol Phys 113:2960–2963
142. Szalay S, Pfeffer M, Murg V, Barcza G, Verstraete F, Schneider R, Legeza Ö (2015) Tensor product methods and entanglement optimization for ab initio quantum chemistry. Int J Quantum Chem 115:1342–1391
143. Szalay S, Barcza G, Szilvási T, Veis L, Legeza Ö (2017) The correlation theory of the chemical bond. Sci Rep 7:2237
144. Tecmer P, González-Espinoza CE (2018) Electron correlation effects of the ThO and ThS molecules in the spinor basis. A relativistic coupled cluster study of ground and excited states properties. Phys Chem Chem Phys 20:23424–23432
145. Tecmer P, Gomes ASP, Ekström U, Visscher L (2011) Electronic spectroscopy of UO_2^{2+}, NUO^+ and NUN: an evaluation of time-dependent density functional theory for actinides. Phys Chem Chem Phys 13:6249–6259
146. Tecmer P, Bast R, Ruud K, Visscher L (2012a) Charge-transfer excitations in uranyl tetrachloride ($[UO_2Cl_4]^{2-}$): How reliable are electronic spectra from relativistic time-dependent density functional theory? J Phys Chem A 116:7397–7404
147. Tecmer P, van Lingen H, Gomes ASP, Visscher L (2012b) The electronic spectrum of CUONg4 (Ng = Ne, Ar, Kr, Xe): new insights in the interaction of the CUO molecule with noble gas matrices. J Chem Phys 137:084308
148. Tecmer P, Govind N, Kowalski K, de Jong WA, Visscher L (2013) Reliable modeling of the electronic spectra of realistic uranium complexes. J Chem Phys 139:034301
149. Tecmer P, Boguslawski K, Johnson PA, Limacher PA, Chan M, Verstraelen T, Ayers PW (2014a) Assessing the accuracy of new geminal-based approaches. J Phys Chem A 118:9058–9068
150. Tecmer P, Boguslawski K, Legeza Örs, Reiher M (2014b) Unravelling the quantum-entanglement effect of noble gas coordination on the spin ground state of CUO. Phys Chem Chem Phys 16:719–727
151. Tecmer P, Gomes ASP, Knecht S, Visscher L (2014c) Communication: Relativistic fock-space coupled cluster study of small building blocks of larger uranium complexes. J Chem Phys 141:041107
152. Tecmer P, Boguslawski K, Ayers PW (2015) Singlet ground state actinide chemistry with geminals. Phys Chem Chem Phys 17:14427–14436
153. Tecmer P, Boguslawski K, Kędziera D (2017) Relativistic methods in computational quantum chemistry. In: Leszczyński J (ed) Handbook of computational chemistry, vol 2. Springer. Netherlands, Dordrecht, pp 885–926
154. Vallet V, Maron L, Teichteil C, Flament JP (2000) A two-step uncontracted determinantal effective Hamiltonian-based SO-CI method. J Chem Phys 113:1391–1402
155. Vallet V, Privalov T, Wahlgren U, Grenthe I (2004) The mechanism of water exchange in $AmO_2(H_2O)_5^{(2+)}$ and in the isoelectronic $UO_2(H_2O)_5^{(+)}$ and $NpO_2(H_2O)_5^{(2+)}$ complexes as studied by quantum chemical methods. J Am Chem Soc 126:7766–7767
156. van Lenthe E, Baerends EJ, Snijders JG (1994) Relativistic total energy using regular approximations. J Chem Phys 101:9783

157. van Wüllen C (2002) Relation between different variants of the generalized Douglas-Kroll transformation through sixth order. J Chem Phys 120:7307–7313
158. Visscher L (2017) An introduction to relativistic quantum chemistry. In: Reine S, Saue T (eds) European summerschool in quantum chemistry 2017—book III, 10th edn. ESQC committee—2017, chap 7, pp 605–646
159. Vlaisavljevich B, Miró P, Ma D, Sigmon GE, Burns PC, Cramer CJ, Gagliardi L (2013) Synthesis and characterization of the first 2D neptunyl structure stabilized by side-on cation-cation interactions. Chem Eur J 19:2937–2941
160. Wang D, van Gunsteren WF, Chai Z (2012a) Recent advances in computational actinoid chemistry. Chem Soc Rev 41:5836–5865
161. Wang S, Alekseev EV, Depmeier W, Albrecht-Schmitt TE (2011) Surprising coordination for plutonium in the first plutonium(III) borate. Inorg Chem 50:4692–4694
162. Wang S, Diwu J, Alekseev EV, Jouffret LJ, Depmeier W, Albrecht-Schmitt TE (2012b) Cation-cation interactions between neptunyl(VI) units. Inorg Chem 51:7016–7018
163. Wei F, Wu G, Schwarz WH, Li J (2011) Geometries, electronic structures, and excited states of UN_2, NUO^+, and UO_2^{2+}: A combined CCSD(T), RAS/CASPT2 and TDDFT study. Theor Chem Acc 129:467–481
164. Wilson RE, Sio SD, Vallet V (2018) Protactinium and the intersection of actinide and transition metal chemistry. Nat Chem 9:622
165. Wolf A, Reiher M, Hess BA (2002) The generalized Douglas-Kroll transformation. J Chem Phys 117:9215–9226
166. Wouters S, van Neck D (2014) The density matrix renormalization group for ab initio quantum chemistry. Eur Phys J D 68:272
167. Wåhlin P, Danilo C, Vallet V, Réal F, Flament JP, Wahlgren U (2008) An investigation of the accuracy of different DFT functionals on the water exchange reaction in hydrated uranyl(VI) in the ground state and the first excited state. J Chem Theory Comput 4:569–577
168. Yabushita S, Zhang Z, Pitzer RM (1999) Spin-orbit configuration interaction using the graphical unitary group approach and relativistic core potential and spin-orbit operators. J Phys Chem A 103:5791–5800
169. Yanai T, Tew DP, Handy NC (2004) A new hybrid exchange-correlation functional using the Coulomb-attenuating method (CAM-B3LYP). Chem Phys Lett 393:51–57
170. Yanai T, Kurashige Y, Mizukami W, Chalupský J, Lan TN, Saitow M (2015) Density matrix renormalization group for ab initio calculations and associated dynamic correlation methods: a review of theory and applications. Int J Quantum Chem 115:283–299
171. Zhou MF, Andrews L, Li J, Bursten BE (1999) Reaction of laser-ablated uranium atoms with CO: infrared spectra of the CUO, CUO^-, OUCCO, $(\eta^2\text{-}C_2)UO_2$, and $U(CO)_x$ (x = 1–6) molecules in solid neon. J Am Chem Soc 121:9712–9721

Computational Versus Experimental Spectroscopy for Transition Metals

Maja Gruden, Wesley R. Browne, Marcel Swart and Carole Duboc

Abstract Knowledge of the electronic structure of transition-metal complexes is increasingly being obtained through joint efforts by theory and experiments. Here, we describe a variety of examples where spectroscopy is being used to determine, e.g., the oxidation state, spin state, or coordination environment around redox-active metal ions such as iron, manganese, or nickel. Both enzymatic and biomimetic systems are included, from the literature and from our own laboratories. It is shown that the combined efforts of wet and dry laboratories lead to a more profound understanding, and allows for systematic exploration of coordinate chemistry around the central metal atom.

1 Introduction

The past decades have seen enormous advances in understanding how Nature uses metals in the daily functioning of human life on earth. Processes such as respiration to transform oxygen and sugar into carbon dioxide and water, and its corresponding counterpart photosynthesis to transform carbon dioxide back into oxygen, would not

M. Gruden
Faculty of Chemistry, University of Belgrade, Studentski trg 12–16, 11000 Belgrade, Serbia
e-mail: gmaja@chem.bg.ac.rs

W. R. Browne
Faculty of Science and Engineering, Stratingh Institute for Chemistry, Nijenborgh 4, 9747AG Groningen, The Netherlands
e-mail: w.r.browne@rug.nl

M. Swart (✉)
ICREA, Pg. Lluís Companys 23, 08010 Barcelona, Spain
e-mail: marcel.swart@gmail.com

IQCC, Universitat de Girona, Campus Montilivi, 17003 Girona, Spain

C. Duboc
Departement de Chimie Moleculaire, University Grenoble Alpes, CNRS, DCM, Grenoble, France
e-mail: carole.duboc@univ-grenoble-alpes.fr

E. Broclawik et al. (eds.), *Transition Metals in Coordination Environments*, Challenges and Advances in Computational Chemistry and Physics 29,
https://doi.org/10.1007/978-3-030-11714-6_6

be viable on the scale needed for human survival in the absence of transition metals. The chameleon-like capability of transition metals to switch between different oxidation and spin states enables these chemical transformations to be performed at ambient temperatures, efficiently, and often selectively. The topic of spin states has only recently come to the forefront [1] leading to a collaborative approach within the formation of a European network of researchers to study spin states in a variety of chemical systems (enzymes, spin-crossover complexes, biomimetic complexes). In all cases, a joint effort of both wet laboratories (experiments [2, 3]) and dry laboratories (theory) was needed to delve deeper into the origins of rate enhancements, spectroscopic properties, structural changes, etc. The scientific production resulting from this European network is too large to be highlighted in a few sentence (a representative summary can be found in a special issue in *Chemistry-a European Journal* [4]). Instead, within this chapter, we will give some representative examples of how theory and experiment go hand in hand in order to reach a deep understanding of these transition-metal complexes, how their spectroscopic features arise, and how can we make adjustments to further improve them. For obvious reasons, these examples are taken mainly from our own laboratories, although we also include some from the literature.

2 Structure of the FeMoco Cofactor of Nitrogenase in Comparison with the Oxygen-Evolving Complex of Photosystem II

The story of the structure of the active species in the nitrogenase enzyme is one of brave researchers, and intense discussions, which finally was solved more than a decade later because of a joint study by DeBeer (spectroscopy) and Neese (theory). But, first things first. The nitrogenase enzyme catalyzes the transformation of the inert nitrogen molecule into ammonia:

$$N_2 + 6 \cdot H^+ + 6 \cdot e^- \rightarrow 2 \cdot NH_3 \quad (1)$$

This transformation is not an easy task and involves a large number of steps in the catalytic cycle (>12). Interestingly enough, in the active site of the enzyme, an iron-sulfur complex was found, but unlike the typical iron-sulfur clusters ($[FeS_4]^{2-}$, $[Fe_2S_2]^{2+/+}$, $[Fe_3S_4]^{+/0}$, $[Fe_4S_4]^{2+/+}$) [5–7], two more exotic ones were present: the FeMoco cofactor $[Fe_7Mo_1S_8(SCH_3)]^q$ and the P-cluster $[Fe_8S_7(SCH_3)_6]^q$. The FeMoco cofactor is proposed to be involved in the transformation (1), while the P-cluster is probably involved in the electron-transfer process.

The elucidation of the crystal structure in 1992 [8] (2.7 Å resolution) showed an empty space within the FeMoco cofactor (see Fig. 1, left). Within the 1.16 Å resolution of 2002 [9], an interstitial atom was observed (Fig. 1, middle), and proposed to be nitrogen. This was subsequently discarded by Yang and coworkers [12] based

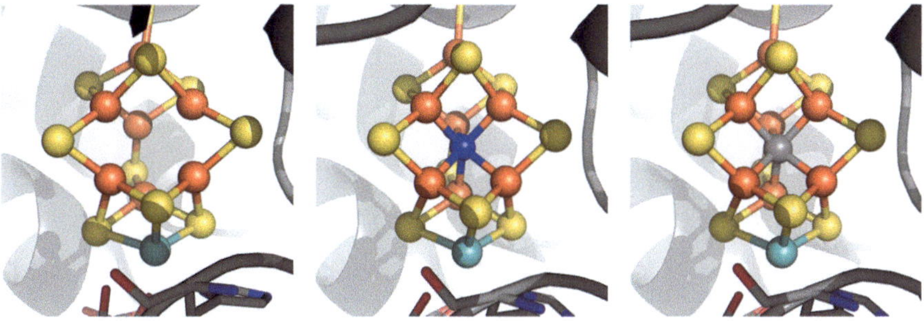

Fig. 1 Crystal structures for FeMoco cluster in nitrogenase enzyme as obtained in 1992 [8] (left), 2002 [9] (middle), and 2011 [10, 11] (right)

on ENDOR and ESEEM spectroscopy, who demonstrated that the interstitial atom could not be nitrogen.

It was not until 2011 when Lancaster and coworkers [13] used X-ray emission spectroscopy that for the first time clear evidence was found that the elusive interstitial atom is instead carbon. This finding was corroborated by Spatzal and coworkers [11] using a higher-resolution X-ray structure together with ^{13}C-labeled proteins (Fig. 1, right). The key piece of evidence by Lancaster and coworkers was the side-by-side comparison of X-ray emission spectroscopy, measured versus the computed ones for three situations with as interstitial atom either carbon, nitrogen or oxygen (see Fig. 2c).

In follow-up studies, the same authors showed that during the catalytic cycle the molybdenum changes oxidation state from the anticipated +IV state into an unusual +III state [14]. Apart from the molybdenum cofactor, also an enzyme with molybdenum replaced by vanadium was obtained [15], as such getting one step closer to the realization of a CoVFeFe cluster. This replacement of Mo by V changed dramatically the redox potential and reactivity. The reaction mechanism taking place in the nitrogenase is therefore difficult to predict, as was shown as well by recent computational chemistry studies [16–20].

The presence of a second metal atom (Mo or V) within the coordination sphere of the presumably active species (FeMoco), which might be acting as a Lewis acid, was shown to be vital as well for the oxygen-evolving complex (OEC) of Photosystem II involved in the respiration process. Early on [21], it was observed that within the active site of the enzyme a manganese-oxide (Mn_4O_4) cluster was present, together with a calcium ion next to it. The role of the calcium and the oxidation states of the manganese ions during the cycle have been studied extensively. It was found that calcium can only be replaced by strontium, the replacement by any other metal led to a nonfunctional enzyme. Furthermore, there were two scenarios proposed for the oxidation states of manganese during the catalytic cycle, either the **high** or the **low** route (see Fig. 3). A clear indication of the oxidation states of manganese ions during the catalytic cycle was obtained through again a combination of theory and spectroscopy from the Max-Planck Institute in Mülheim [22]. With extensive calculations

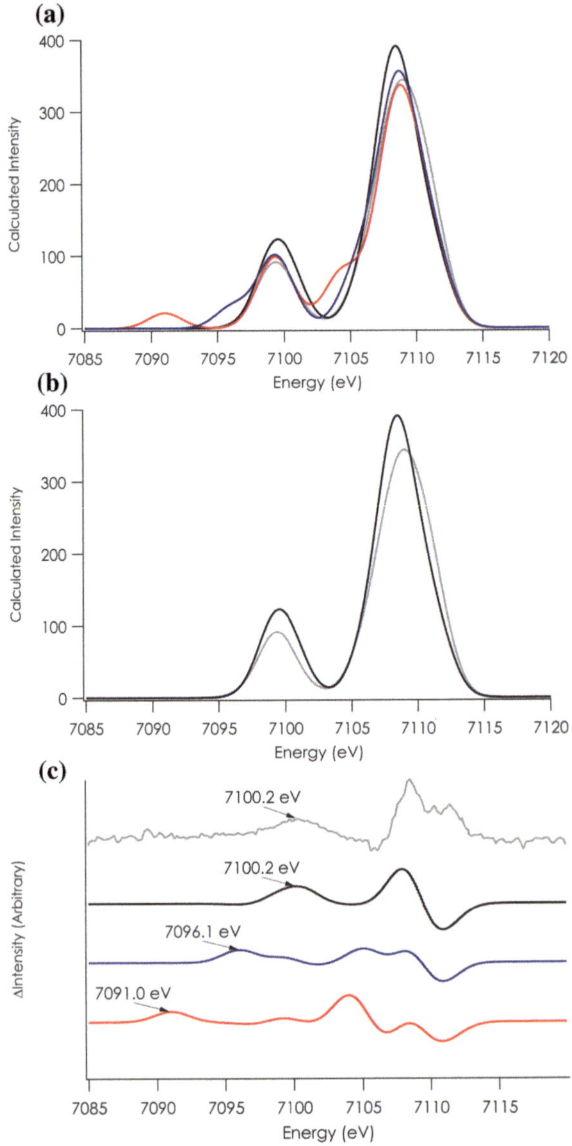

Fig. 2 **a**, **b** Calculated XES spectra for FeMoco with carbon (black), oxygen (red), nitrogen (blue) and of P-clusters (gray); **c** measured (top, gray) versus computed XES spectra for nitrogenase enzyme with as interstitial atom carbon (black), nitrogen (blue), and oxygen (red). From Lancaster et al. [13]; Reprinted with permission from AAAS

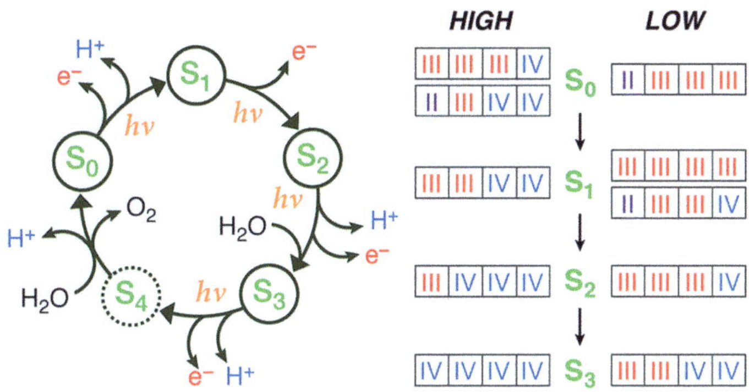

Fig. 3 Schematic catalytic cycle of the oxygen-evolving complex enzyme, indicating the manganese oxidation states within the **high** and **low** scenarios (from Krewald et al. [22]; Reprinted with permission)

on the possible spin states, geometries, and corresponding effects on the computed spectroscopy, the authors made a convincing case for the prevalence of the **high**-valent scenario. Furthermore, they were able to show that the Mn_4O_4 cluster along the reaction path was opening up, allowing thereby the bonding of to-be-transformed water molecules to the manganese. A decisive point in the argumentation for the most valid mechanistic proposals was the combination of theory and spectroscopy, which allows for fingerprinting the active species in the reaction.

3 Lewis-Acid Capped Iron-Oxygen and Copper-Nitrogen Species

The previous section already introduced the presence of metal ions in the vicinity of a high-valent metal species within enzymes. These examples inspired Fukuzumi and Nam to explore if similar effects might be obtained with biomimetic iron complexes. Indeed, after adding scandium triflate ([Sc(OTf)$_3$]) to a solution containing a stable Fe^{IV}-oxo complex [23] ([Fe^{IV}(O)(TMC)]$^{2+}$, TMC = 1,4,8,11-tetra–methyl-1,4,8,11-tetraazacyclotetradecane], they obtained [24] crystals that could be used to determine their three-dimensional structure. In this structure, the Fe–O moiety was capped from above by the scandium triflate (see Fig. 4, left). Moreover, it was accompanied by three remarkable features: (i) the axial solvent molecule, coordinated to Fe^{IV} in the starting material [23], was no longer present; (ii) the scandium picked up a fourth triflate and an additional axial group (either water or hydroxyl, vide infra); (iii) the methyl groups of the TMC ligand, pointing downward (or *anti* to the oxo) in the starting material, suddenly pointed upward (*syn* to oxo) in the scandium-capped complex. Initially, the oxidation state of iron within the scandium-capped complex

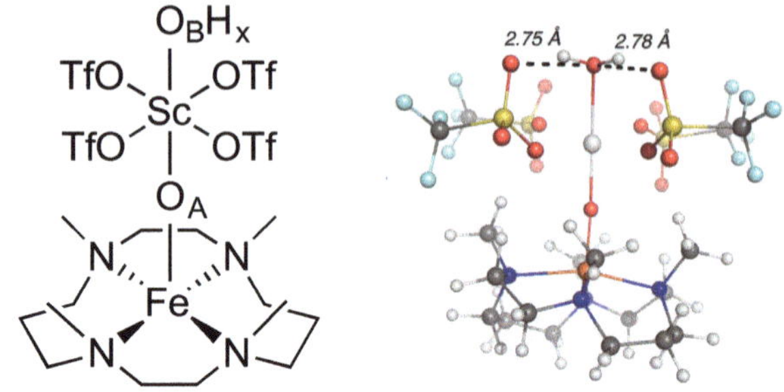

Fig. 4 Structure of Sc^{3+}-capped iron–oxygen complex

was proposed to remain +IV, however, based on the rather long Fe–O distance (1.76 Å), Que and coworkers proposed the oxidation state to be +III instead [25].

One of us took up the challenge by studying the complex with computational chemistry [26]. Already directly after its publication, an initial study was performed in order to determine the spin ground state for the complex; in the original paper, no proposal was made for this. Even though the computational setup had proven to be reliable (and in particular for spin states and geometries of iron complexes) [27], the results obtained were disappointing: the iron-oxygen and the axial scandium-oxygen distances were totally different. After concerns were raised about the oxidation state of iron within the scandium-capped complex [25], a fresh attempt was started through a systematic approach. First of all, a set was constructed with all $Fe^{III/IV}$ complexes with oxo, peroxo, hydroperoxo, or hydroxo ligands, whose structure had been determined by X-ray or EXAFS, and whose electronic structure was proven with Mössbauer spectroscopy. A computational setup was made where the geometries were obtained with PBE-D/TZ2P [28, 29], spin ground states were determined by the spin-state consistent SSB-D functional [30], and Mössbauer parameters were obtained by Noodleman's parameterization with the OPBE functional [31, 32]. The computed iron-oxygen distances showed to be very close (0.01–0.02 Å difference) [26] to the experimentally observed ones (except for two complexes, which were subsequently shown to be resulting from a wrong assignment; revisiting the original references [33, 34] brought the apparent larger deviations down [35] to the 0.01–0.02 Å range). The spin ground states were correctly predicted, as was to be expected, and the Mössbauer isomer shift was in almost cases accurate to within 0.05–0.10 mm s^{-1}; the quadrupole splitting showed some larger differences in some cases, but for most cases was within 0.2–0.3 mm s^{-1} accuracy. Based on these encouraging results, the next step was to tackle the scandium-capped complex.

Given that there were no counter ions or other molecules present in the crystal structure, the scandium-capped iron-oxygen complex must be charge neutral overall. The oxidation state of iron is therefore directly connected to the nature of the axial ligand to scandium (OH_x): with a +IV oxidation state on iron, this must be a hydroxyl (as originally proposed by Nam and Fukuzumi). Instead, if the oxidation state of iron is +III, the axial ligand to scandium must be a water molecule. The spin state of the

Table 1 Metal–ligand distances[a] (Å) for different systems

	$[Sc^{3+}(OTf)_4(X)]^{2-/-}$			Fe^{IV}–O–Sc^{3+}(OH)			Fe^{III}–O–$Sc^{3+}(OH_2)$		
	X = OH	X = OH_2	**Exp.**	LS	IS	HS	LS	IS	**HS**
Fe–N_{av}	–	–	**2.175**	2.073	2.076	2.125	2.070	2.108	**2.204**
Fe–O_A	–	–	**1.754**	1.627	1.626	1.692	1.741	1.779	**1.765**
O_A–Sc	–	–	**1.934**	2.281	2.289	2.190	1.879	1.892	**1.914**
Sc–O_B	1.895	2.096	**2.188**	1.898	1.898	1.913	2.289	2.284	**2.265**

[a]Computed values obtained at PBE-D/TZ2P, including ZORA scalar relativistic corrections and COSMO solvent effects (both self-consistently)

complex was also unknown; therefore, for both the Fe^{IV} and Fe^{III} complexes, all three spin states were explored.

A clear indication of the oxidation state of iron in the scandium-capped complex was already obtained by focusing on the scandium moiety alone. The two possible options are $[Sc^{3+}(OTf)_4(OH)]^{2-}$ and $[Sc^{3+}(OTf)_4(OH_2)]^{-}$ (see Table 1). The X-ray structure showed a Sc–O_B distance of 2.19 Å, which is incompatible with the distance obtained for X = OH (1.90 Å); instead, it is fully consistent with the axial ligand being a water molecule (2.10 Å).

For the full scandium-capped iron-oxygen complex $[Fe^q(O)(Sc^{3+}(OTf)_4OH_x]$, for both Fe^{III} and Fe^{IV}, the spin ground state was observed to be high spin ($S = 5/2$ for Fe^{III}, $S = 2$ for Fe^{IV}), with the other spin states higher in energy by 4–35 kcal mol^{-1} (SSB-D/TZ2P). As was observed for the isolated scandium moiety, for the full complex the computed scandium-oxygen distances are totally different from those observed in the X-ray structure: the Sc–O_B (2.19 Å) and Sc–O_A (1.91 Å) distances for Fe^{IV} are reversed compared to the X-ray structure (Sc–O_B 1.93 Å; Sc–O_A 2.19 Å). Instead, the Fe^{III} distances match very well (Sc–O_B 1.91 Å; Sc–O_A 2.27 Å). The same is true for the Fe–O_A distances: 1.75 Å (X-ray), 1.69 (Fe^{IV}), 1.76 (Fe^{III}). Finally, out of the six computational structures, there is only one that confirms the long equatorial Fe–N distances (2.18 Å X-ray, 2.20 Å HS–Fe^{III}). Based on this overwhelming evidence of spin ground states and structures, it was suggested that the oxidation state of iron in the scandium-capped complex should be revised to be HS–Fe^{III} ($S = 5/2$).

Mössbauer spectroscopy is the method of choice for determining the oxidation and spin state of iron complexes, however, up until then the scandium-capped complex had proven to be too fragile and sensitive for it. Therefore, based on the structure of the computed HS–Fe^{III} ($S = 5/2$) complex, a prediction was made for its isomer shift (0.39 mm s^{-1}) and quadrupole splitting (−0.99 mm s^{-1}) to serve as a reference point for future experimental studies. A major breakthrough was achieved when Que, Münck, and coworkers reported early in 2015 that they had been able to obtain a closely related complex in sufficient quantity to perform EPR, Mössbauer, and other spectroscopies. It was shown that the replacement of the axial water by acetonitrile did not have an effect on the structure or spectroscopy of the iron part. Moreover,

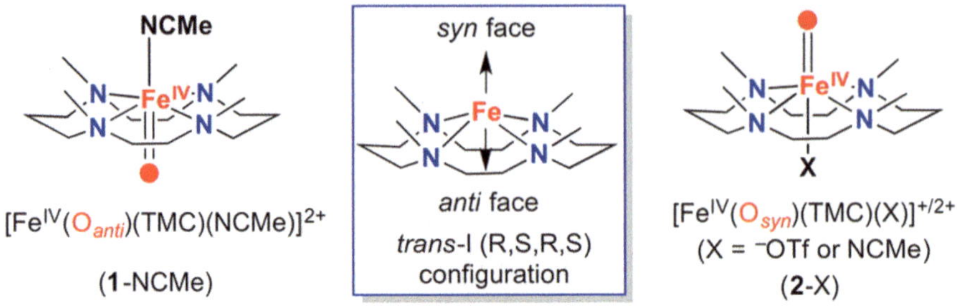

Fig. 5 Oxo-binding modes in $[Fe^{IV}(O)(TMC)(X)]^q$; reprinted with permission from Prakash et al. [37]. Copyright (2015) American Chemical Society

the Mössbauer spectroscopy showed isomer shift (0.36 mm s^{-1}) and quadrupole splitting (−1.02 mm s^{-1}) values that were on top of the predicted values by theory.

Que and coworkers then explored in more detail the possibility of a change in the oxidation state of iron by bubbling dioxygen into a mixture of $[Fe^{II}(TMC)(OTf)_2]^0$ and $[Cr^{II}(OTf)_2]^0$, which resulted in the formation of a $[(TMC)Fe^{III}–O–Cr^{III}(OTF)_4]^0$ species [36]. The same species was observed by adding $[Cr^{II}(OTf)_2]^0$ to $[Fe^{IV}(O)(TMC)(NCCH_3)]^{2+}$. No reaction was observed in the absence of chromium-triflate, nor was the $[Fe^{II}(TMC)(OTf)_2]^0$ oxidized by dioxygen alone. This indicated clearly that dioxygen reacted with chromium, and the corresponding Cr^{III}-superoxide species was then trapped by iron. It helped in understanding how the bimetallic complex was formed through an inner-sphere electron transfer, but there remained one mystery to be solved: how the methyl groups went from being positioned at the *anti* face in the Fe^{IV}=O species to the *syn* face in Fe^{III}–O–Sc/Cr^{III} (see Fig. 5).

In the first study [37], Que and coworkers explored the reaction of $[Fe^{II}(TMC)(OTf)_2]^0$ with a different iodosoarene, and observed the formation of a species that showed spectroscopic characteristics that differed from both the original $[Fe^{IV}(O)(TMC)]^{2+}$ species [23] and its inverted isomer [38]. In particular, the NMR spectra were considerably different, which might explain small peaks observed in the original spectra.

Now that both faces of the TMC ligand had become accessible for binding of the oxo and other groups, in a follow-up study [39] Que and coworkers reported the formation of $[(TMC)Fe^{III}–O_{syn}–Cr^{III}(OTf)_4(NCCH_3)]$, complementing the O_{anti} isomer reported previously. The corresponding structures were obtained from DFT calculations, corroborating the coordination sphere around iron. Furthermore, they were able to prepare crystal structures for diferric oxygen species (both $Fe^{III}–O_{anti}–Fe^{III}$ and $Fe^{III}–O_{syn}–Fe^{III}$), giving additional insights into how the ligand topology can affect the coordination chemistry of the central iron, and what effect this has on spectroscopy. In all of these cases, the interplay between theory and experiment was shown to be beneficial for obtaining a profound understanding of the electronic structure of these transition-metal complexes.

4 Transient Species: The Case of Nickel(IV) Tris-μ-Oxido

The identification of transient intermediates, i.e., species with short lifetimes, relies heavily on time-resolved spectroscopy, especially where such species are not amenable to trapping in crystalline form. In contrast to photochemically generated intermediates (such as those of photosystem II, in which a substantial population of an intermediate is generated by the actinic pulse), in thermal reactions the steady-state population depends on the reaction kinetics, and is controllable only to a limited extent by temperature and solvent.

In thermal reactions, the elucidation of molecular structures from spectroscopy is essential and relies primarily on precedent: we associate spectral features observed with those of species that are known, and we use analogy to guide us. However, what if the structure has no real precedent, or is suspected to be similar to a hypothesized but never observed structure? In such a situation, we are faced with a challenge to recognize the new species, and we turn to experience and bring to bear as many appropriate spectroscopic techniques as we can to gather the evidence needed to unmask the structure and character of the intermediate. For these cases, theory provides us with a powerful tool through its ability to test our hypothesized structures both for thermodynamic feasibility and for prediction of spectroscopic properties. In this section, we aim to show, however, that rather than being a separate aspect of such a study, theory is integral to the process. The example taken, a recently identified nickel intermediate [40], is of course anecdotal and our aim is not to summarize the findings but rather to focus on the approach taken and the order in which experiments were actually done.

The study began with the observation of a pronounced but short-lived (seconds) visible absorption band (Fig. 6) when near-stoichiometric NaOCl(aq) was added to methanol solutions of the known complex $[(Ni^{II})_2(Cl)_3(tmtacn)_2]^{2+}$ (where tmtacn is 1,4,7-trimethyl-1,4,7-triazacyclononane). The species was longer lived in acetonitrile but the scatter caused by precipitated NaCl in it, more or less canceled the benefit in regard to spectroscopic features.

Our anecdote begins at this point, with the immediate curiosity as to what species had been observed, i.e., what is the molecular structure of the visible chromophore. The obvious choice of techniques, at least to us, was to use ESI-MS and Raman spectroscopy. ESI-MS relies on the species being charged and sufficiently stable to remain intact during generation of the electrospray, while the use of Raman spectroscopy relies on signal enhancement (resonance enhancement) that can occur when the wavelength of excitation is coincident with the visible absorption band.

ESI-MS was useful to show that the dinuclear structure of $[(Ni^{II})_2(Cl)_3(tmtacn)_2]^{2+}$ is dynamic, i.e., we should not view the complex in solution as being kinetically inert to exchange of the chlorido ligands. That is, the ESI-MS spectra obtained just after adding NaOCl (the active component of bleach) indicated a new doubly charged ion with $[Ni_2(O)_3(tmtacn)_2]^{2+}$ as chemical formula. It showed the expected isotope sensitivity when NaOCl equilibrated (which occurs essentially instantaneously upon mixing) in $H_2^{18}O$. Of course, ESI-MS can say little

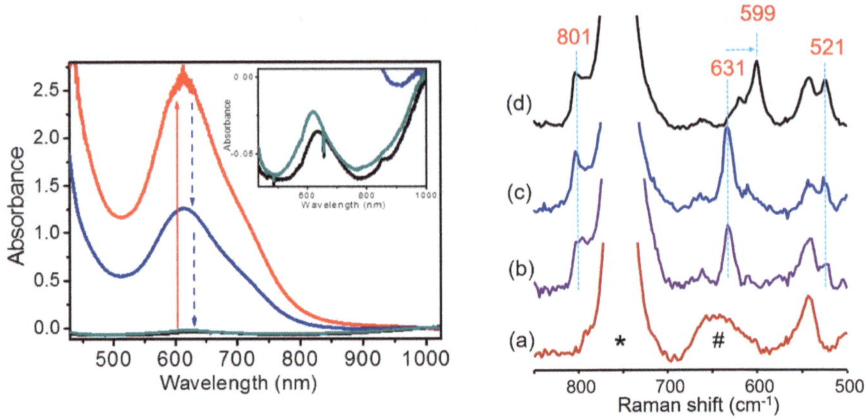

Fig. 6 UV–vis absorption spectrum (left) and resonance Raman spectrum (right, at 532 nm) of nickel chlorido complex (3.5 mM) in acetonitrile (**a**) and with **b** 4.5 equiv. of NaOBr, **c** 4.5 equiv. of $Na^{16}OCl$ and **d** 4.5 equiv. of $Na^{18}OCl$. *Solvent band. # Raman band from quartz; reprinted with permission from Padamati et al. [40]. Copyright (2017) American Chemical Society

about formal oxidation states or the nature of the bonding in such a species. Raman spectroscopy at 532 nm (in resonance with the visible absorption band) showed a number of low-frequency bands that were considerably enhanced in intensity (furthermore, they were completely absent in Raman spectra recorded off-resonance, e.g., at 785 nm). At this point, one would normally consult tables of data of related species as one does now for all kinds of iron, copper, cobalt, and chromium oxygen species; however, such tables do not yet exist for Ni–O species. Hence, assignment of the bands must be made by ^{18}O labeling and analogy to structures with other metals.

The chemical formula from ESI-MS provided a basis for proposing structures and to begin the cycle of setting up experiments to test these proposed structures and subsequent elimination that is central to any mechanistic study. Three potential candidates were considered initially on the basis of analogy to known structures and the initial complex $[(Ni^{II})_2(\mu\text{-}Cl)_3(tmtacn)_2]^{2+}$. A key point that we had to take into account was that the Ni^{III} oxidation state would already be considered a high-valent species, and proposing a Ni^{IV} species was, to put it mildly, stretching the borders of what was reasonable.

Nevertheless, the $[(Ni^{IV})_2(\mu\text{-}O)_3(tmtacn)_2]^{2+}$ complex (Fig. 7) is isostructural with the parent complex $[(Ni^{II})_2(\mu\text{-}Cl)_3(tmtacn)_2]^{2+}$ and, especially, the well-known highly stable complexes $[(Mn^{IV})_2(\mu\text{-}O)_3(tmtacn)_2]^{2+}$ and $[(Ru^{IV})_2(\mu\text{-}O)_3(tmtacn)_2]^{2+}$ described structurally by Wieghardt and coworkers [41, 42] already in the 1980s. Alternatively, a potentially more reasonable structure was the oxo–peroxo complex $[(Ni^{III})_2(\mu\text{-}O)(\mu\text{-}O_2)(tmtacn)_2]^{2+}$ (**3a** in Fig. 7) which also has a precedent in an isostructural manganese complex $[(Mn^{IV})_2(O)(O_2)(tmtacn)_2]^{2+}$. The formation of the O–O bond with NaOCl is then simply the reverse of the reaction of

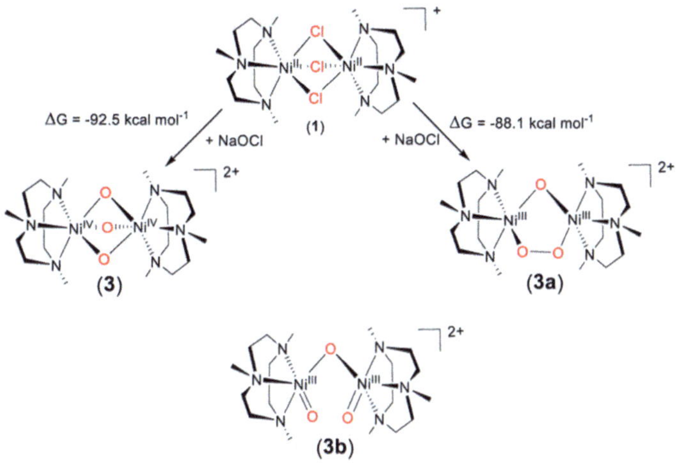

Fig. 7 Proposed oxido species; reprinted with permission from Padamati et al. [40]. Copyright (2017) American Chemical Society

H_2O_2 with NaCl. Finally, a third more fanciful possibility was also included, which is the mono-μ-oxo-bis-terminal-oxo $[(tmtacn)(Ni^{IV})(O)–(\mu\text{-}O)–(Ni^{IV})(O)(tmtacn)]^{2+}$ (**3b** in Fig. 7).

It is at this point, the proposition of structures that theory began to play a role. Initially, the focus was on the relative energies of both the complexes and their many spin states, for which a spin-state consistent density functional (S12g) was used. The manifold of possible spin states, ranging from ferromagnetically coupled high spin ($S = 4$) to anti-ferromagnetically (AFM) coupled open-shell ($S = 0$) or closed-shell ($S = 0$), was studied. Different ground states were observed for the three isomers: a closed-shell state for **3**, an AFM-coupled (locally doublet, d^7) open-shell singlet for **3a**, and an AFM-coupled (locally triplet, d^6) for **3b**. For each of these isomers, the other spin states were higher in energy by at least 10 kcal mol^{-1}, and the difference in stability between the isomers was also substantial with **3** being clearly the most favored, with **3a** (+7.5 kcal mol^{-1}) and **3b** (+46.1 kcal mol^{-1}) much less stable.

Although these computational data corroborated the expected tris-μ-oxo isomer as being the most stable isomer, additional evidence was obtained by comparing the computed vibrational spectra with the experimental ones. Raman data (Fig. 6, right) showed a peak at 631 cm^{-1} after the addition of NaOCl (or NaOBr), which shifted to 599 cm^{-1} upon ^{18}O labeling. Furthermore, two more enhanced bands were observed at 521 and 801 cm^{-1}, which were hardly affected by the ^{18}O labeling. The computed vibrational spectra for the three isomers showed additional proof that the only isomer consistent with the experimental data was **3**. Both the peak position (638 cm^{-1}) and isotope shift for ^{18}O (to 609 cm^{-1}) were in excellent agreement with experiment; neither **3a** nor **3b** showed Ni–O bands at this position, but were instead found at 892 cm^{-1}/663 cm^{-1} (**3a**), and 742 cm^{-1}/700 cm^{-1} and 689 cm^{-1} (**3b**). Furthermore, the peaks at 521 and 801 cm^{-1} were presented for **3**, but absent for **3a/3b**. Also mixed labeling studies were performed (varying the amount of ^{16}O and ^{18}O, in 3:0, 2:1, 1:2, 0:3 ratios). These experiments showed further confirmation of the correctness of our mechanistic study, with four bands shown for the four isotopologues that matched the computational data.

Since we were dealing with high-valent (formal) oxidation states, ranging from +II for the starting material, to +III/+ IV in the high-valent species, we also explored the use of X-ray absorption spectroscopy (XAS) . For iron, it was shown that a change in oxidation state is accompanied by a shift of the near-edge XAS position (XANES) [43]. Instead for nickel, there exists an ambiguous situation where, e.g., Pirovano and coworkers showed [44, 45] that XANES seemed to be unable to distinguish between Ni^{II} and Ni^{III}. Nevertheless, in order to be consistent, and answer one of the questions from a reviewer, we asked our colleague and XAS expert Moniek Tromp to perform the experiments. As anticipated, Ni K edge XANES spectra taken of starting material and just after adding NaOCl show a nondescript edge, i.e., no significant pre-edge features. This indicated the presence of a near-octahedral, six-coordinate, geometry around Ni for all species present; this is consistent with complexes **1**, **3** and **3a**, but does not allow assignment to any one of them.

The final piece in the puzzle was made by EPR spectroscopy, which would allow to (dis)prove the presence of paramagnetic mononuclear Ni^{III} complexes. The absence of EPR (X-band) signals at 77 K at any time suggested that such mononuclear species were not present at any time.

The interplay between theory and experiment for the characterization of the short-lived $[(Ni^{IV})_2(\mu\text{-}O)_3(\text{tmtacn})_2]^{2+}$ species was therefore essential for arriving at the final result. It all started with the observation of the pronounced absorption band, for which possible structures were determined by theory. The spin ground states of the three isomers then led to a variety of spectroscopic measurements to (dis)prove the proposed candidates, including isotope labeling experiments in the wet and dry laboratory. In the end, a consistent picture emerged that showed evidence for the existence of a high-valent Ni^{IV} species, in reaction conditions that appear so simple that it was surprising to see that such a transition-metal complex had not been discovered earlier.

5 Magnetic Anisotropy in Transition-Metal Complexes

The final part of this chapter is dedicated to the investigation of magnetic anisotropy using various spectroscopic techniques for its experimental determination in combination with quantum chemistry for its theoretical understanding. Such an approach could lead to the definition of magnetostructural correlation essential for the rational design of complexes with targeted magnetic properties. Nearly two decades ago, researchers discovered that certain molecules exhibit slow relaxation of the magnetization after removal of an applied dc magnetic field [46]. Such species can thus behave at low temperature as classical magnets but of molecular origin, and have come to be known as single-molecule magnets (SMMs). Slow relaxation present in

SMMs is related to the existence of two states separated by an energy barrier (U_{eff}) with $U_{\text{eff}} = |D| \cdot S^2$ (S being the spin ground state of the complex and D describing the magnetic anisotropy). SMMs have been proposed as potential candidates for several technological applications that require highly controlled thin films and patterns (high-density information storage and quantum computing) [47, 48].

However, most of the existing molecular systems display extremely low transition temperatures to act as SMMs and then can not be used for applications [49, 50]. Recently, part of the origin of this problem has been unveiled as it has been shown that the energy barrier does not depend on the total spin of the molecule as it was previously believed, but on the magnetic anisotropy [51, 52]. Thus, the control of the energy barrier is reduced to understand the so-called zero-field-splitting (ZFS) parameters that determine the magnetic anisotropy of an isolated transition-metal complex. The ZFS parameter tensor is defined by the D and E parameters, which represent its axial and rhombic contributions, respectively. Several experimental techniques have been developed to determine D including indirect methods such as magnetometry, variable temperature, variable-field magnetic circular dichroism, and Mössbauer spectroscopy, or direct methods such as inelastic neutron scattering and frequency domain magnetic resonance spectroscopy [53]. However, the most appropriate technique to accurately define the ZFS parameters remains electronic paramagnetic resonance (EPR) spectroscopy. A precise determination of D generally necessitates high-field limit conditions, i.e., when the energy provided by the EPR spectrometer is much larger than the magnitude of D. Therefore, for systems with $|D| > 0.2\ \text{cm}^{-1}$, high-field EPR (HF-EPR) is required. In parallel, important theoretical efforts have been carried out to develop suitable methodologies to predict the ZFS parameters and to rationalize the experimental observations. For such investigations, both contributions to D, i.e., the spin-orbit coupling (SOC) and the electron–electron spin–spin interaction (SS) have been accurately calculated [54]. Multiconfigurational wavefunction approaches [55], including ab initio frameworks, generally give satisfactory predictions of D. Several multi-determinantal methods, especially like the complete active space self-consistent field (CASSCF) and N-electron valence second-order perturbation theory of second-order (NEVPT2) approaches lead to calculated D values in good agreement with the experimental data, but are computationally much more expensive than density functional theory (DFT) [56], significantly hampering their ability to apply them to medium and large transition-metal complexes. Two DFT approaches rooted in very different physical foundations can be in principle used. The first one is based on the coupled perturbed SOC method (CP-DFT) and closely related to the work of Perderson and Khanna [57, 58]. It should be pointed out that DFT perturbational approaches like, e.g., Pederson–Khanna, cannot be used when orbital degeneracy is present. The second method is the ligand-field DFT (LF-DFT) approach by Daul et al. [59], based on a multideterminant description of the transition metal's multiplet fine structure that is able to tackle many difficult problems including orbital degeneracy.

Since the most famous SMM is a Mn-based cluster, $[Mn_{12}O_{12}(CH_3CO_2)_{16}(H_2O)_4]$, composed by eight Mn^{III} and four Mn^{IV} ions [60], special interests have been devoted to these specific ions. However, while the

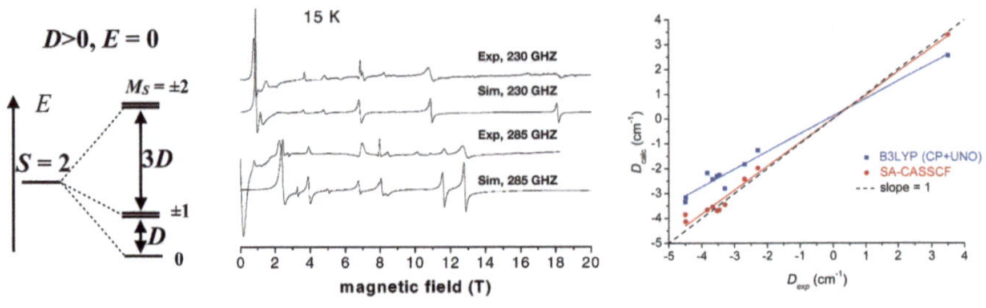

Fig. 8 (left) Schematic representation of the splitting of the Zeeman levels for a $S = 2$ spin state with a positive D value, (middle) experimental and simulated powder HF-EPR spectra recorded on a the mononuclear Mn(III) complex, (left) correlation between the calculated and experimental D values from DFT and ab initio approaches [61]

magnetic anisotropy of Mn^{III} has been extensively explored through experimental determination and quantum chemical prediction of the ZFS parameters of series of complexes [61, 62], this is not the case for Mn^{IV} [63, 64]. Mn^{III} is generally high spin with an integer spin, $S = 2$, characterized by moderate magnetic anisotropy with $|D|$ lying in the small range from 2.5 to 4.5 cm^{-1} [62]. This is thus the perfect case to apply EPR for the precise determination of the ZFS parameters by using multifrequency (95–500 GHz) HF-EPR spectroscopy (Fig. 8) [65, 66]. Interestingly, the sign of D, which should be negative for a system that displays SMM properties, can be determined by analyzing low-temperature EPR spectra. By studying series of Mn^{III} complexes, it has been confirmed experimentally that the sign of D is consistent with the nature of the JT distortion, i.e., axial elongation (compression) for $D < 0$ ($D > 0$). Regarding the trend in the $|D|$-value, it has been shown that (i) low-coordinate complexes (4 and 5) generate smaller D with respect to six-coordinate complexes, (ii) oxygen-based ligands leads to larger D than N-based ligands, and (iii) unexpectedly large SOC contribution of ligands tends to decrease the magnetic anisotropy, i.e., $|D_{Br}| < |D_{Cl}| < |D_F|$ [61]. In this case, while CASSCF ab initio leads to the accurate prediction of D, the cheaper CP-DFT calculations corroborate experimental trends, and both approaches give the origin of magnetic anisotropy (Fig. 8). Especially, it has been revealed that the spin–spin contribution to D should not be neglected as well as contributions of the first excited states [54].

Recently, the first systematic investigation of the electronic structure of a series of octahedral Mn^{IV} complexes has been published [63, 64]. Since Mn^{IV} is a Kramers $S = 3/2$ ion with moderate D-magnitudes (between 0.25 and 2.29 cm^{-1}), EPRs can be observed in spectra recorded at X- and Q-band frequencies. However, because these data are not recorded in high-field limit conditions, precise determination of D cannot be achieved based on a single EPR frequency and HF-EPR spectroscopy remains more reliable for such study (Fig. 9) [63, 67]. The accurate determination of magnetic anisotropy in this series of complexes has thus been performed through high-field EPR experiments and LF- and CP-DFT calculations have been carried out to obtain insight into its origin (Fig. 9) [63]. The major conclusions are: (i) the

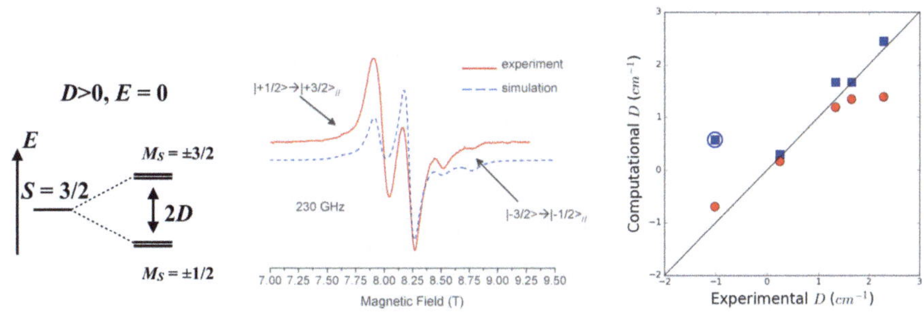

Fig. 9 (left) Schematic representation of the splitting of the Zeeman levels for a $S = 3/2$ spin state with a positive D value, (middle) experimental and simulated powder HF-EPR spectra recorded on a mononuclear Mn^{IV} complex [67]; (left) comparison between LF-DFT (blue squares) and CP-DFT (red circles) calculated D values and experimental data [63]

magnitude of D in Mn^{IV} ion can be of the same order of magnitude as that in the non-Kramers Mn^{III} ion contrary to a common belief, (ii) LF-DFT reveals a good ability to correctly predict $|D|$-values and the orientations of the principal magnetic axis, (iii) the analysis of the different contributions to D shows that it is mainly controlled by transitions within the t_{2g}-like manifold, and (iv) Mn^{IV} naturally tends toward having an easy magnetization plane ($D > 0$). As for Mn^{III}, CP-DFT calculations lead to a systematic underestimation of D with a low impact on the quality of the prediction from the choice of the functional. Besides, the LF-DFT shows that the doublet-excited states related to d–d transitions mainly contribute to D.

In the cases of Mn^{III} and Mn^{IV}, we have highlighted the fact that the combination of experiments and theory allows us to define and to rationalize magnetostructural correlations, because such systems are easy to synthesize in series. This is not the case for several types of systems such as those with intermediate ground spin state that require an appropriate coordination sphere to be stabilized. Interestingly, two series of intermediate spin-state complexes have been recently described with the same ligands, i.e., mononuclear isostructural $S = 1$ Co^{III} and $S = 3/2$ Fe^{III} complexes of formula [MLX] (with L, a N2S2 ligand and X = Cl^-, Br^- and I^-), in a square pyramidal environment (Fig. 10) [68, 69]. This investigation starts with the Co^{III}-based series. Analysis of the magnetic susceptibility data as a function of temperature shows that the magnetic anisotropy of these complexes is sensitive to the nature of the halide, but in an unexpected way: the largest $|D|$-value is found for the Cl^- derivate ($D = 35$ cm^{-1}), while Cl^- displays the smallest SOC constant in the series. This result shows that factors other than the SOC of the heavier ligand notably contribute to the ZFS and that all contributions need to be understood.

This is related to the fact that the magnetic anisotropy results from a balance between the metal ion and halide SOC contributions, for which it is difficult to identify the precise respective weight [70]. Therefore, depending on the metal ion and the halide, the SOC contribution of the heavier ligand and the metal ion can compensate each other, leading to an inverse tendency with respect to the SOC of

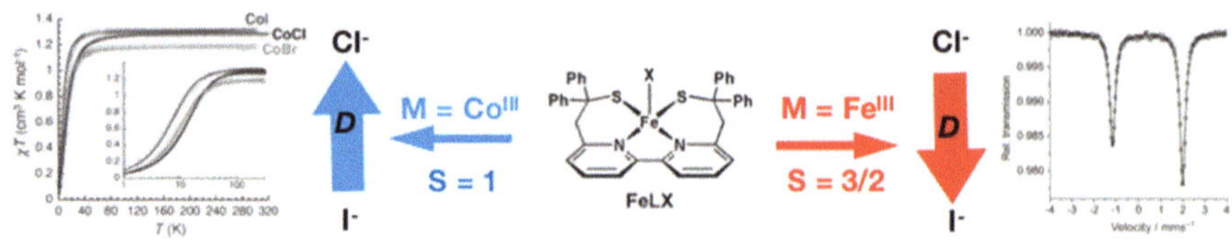

Fig. 10 Schematic representation of [MLX] complex (middle), (left) temperature dependence of the χT product for the series of [CoLX] complexes [68] and (right) experimental and simulated powder Mössbauer spectra of [FeLCl] [69]

the halide, as found in this series of Co^{III} complexes. This is also reflected in the spin-state energy spectrum: the first excited spin state is $S = 2$ for the Cl^- and Br^- derivates, while it is $S = 0$ for the I^- one. Furthermore, it was possible to define a magnetostructural correlation rationalized by theory. The complex displaying the more distorted square pyramidal geometry exhibits the largest magnetic anisotropy.

We expected the same trend and rationalization in the Fe^{III} series as in the case of Co^{III}. Susceptibility measurements, powder *cw* X- and Q-band EPR spectra, and zero-field powder Mössbauer spectra showed that all complexes display distinct magnetic anisotropy but with an inverse trend: the I^- derivate displaying the largest magnetic anisotropy ($D = 11.5\ cm^{-1}$) and the Cl^- one ($D = 3.7\ cm^{-1}$) the smallest. Mössbauer data and CP-DFT calculations evidenced that the electronic structure of these complexes, and hence magnitude and sign of D, is mainly determined by the metal–ligand covalency and coordination mode. In the series from Cl^- to I^-, covalency increases, i.e., more covalent character is observed for heavier halides. In the simple ligand-field model, increasing of covalency leads to decreasing electron—electron repulsion, which can be seen from decreasing of Racah's parameters [56]. In particular, for the iodide complex, there is a dramatic reduction of the parameter C, leading to the stabilization of the low-spin state. Following only this, one may argue that this increase of covalency would cause diminishing importance of the metal SOC, and therefore lowering D, contrary to what is observed [69]. However, at the same time, more covalent character of the metal–halide bond implies that the importance of the spin-orbit contribution of the heavier halides cannot be neglected. Large covalent character of metal–ligand bonds makes CASSCF (with second-order perturbation correction, i.e., CASPT2 or NEVPT2) not appropriate to study this system as a very large active space, including ligand orbitals, is required even to get the correct ground spin state. LF-DFT faced problems because the SOC contribution of heavier halides such as Br^- and I^- is not included in the present model. CP-DFT, on the other hand, is well suited for analysis of the ZFS in these covalent systems, and a good agreement with experimental values has been obtained. Overall ZFS is a consequence of a delicate balance of different effects, i.e., increased covalency, importance of the SOC of the metal ion and heavier halides and coupling of excited states with different multiplicities to the ground state.

The main conclusion of all these studies is that the origin of the magnetic anisotropy, and even the trend in values depend strongly on the spin system and analogies cannot be drawn a priori with systems that are yet to be investigated.

The dream of any coordination chemist is the rational design of transition-metal complexes with desired properties. Studies by Ruamps et al. [71] have shown that controlling the geometry of a transition-metal complex is the way to chemically control their magnetic properties. This study shows that spin–orbit coupling can be used to tune the magnetic anisotropy if symmetry considerations allow it. Knowing the fact that both coordination of the ligands, and the partial quenching of the SOC due to Jahn–Teller (JT) distortions reduce |*D*| from its ideal, the free ion value led us to develop a simple model to define the main chemical factors that control the magnetic anisotropy [72].

From that perspective, five coordinated Ni^{II} trigonal-bipyramidal complexes fulfill all the above-mentioned requirements. We focused first on a case corresponding to a D_{3h} point group, where the nickel ion is coordinated to three equivalent ligands in the plane and two axial ligands, which differ from them. Since in Ni^{II}, the 3d orbitals are populated with eight electrons in D_{3h} symmetry it can be either high ($S = 1$), or low ($S = 0$) spin, depending on the splitting between the d_{z2} orbital from the degenerate d_{x2-y2} and d_{xy} ones. Since low-spin complexes are diamagnetic, we focused only on high-spin complexes that appear with ligands like Cl^- or F^-, with $^3E'$ as the ground state. As is well known, orbitally degenerate systems are subject to the JT effect, which lowers the energy by reducing the symmetry to C_{2v} and hence lifts the degeneracy of the e' orbitals. In the case of Ni^{II} trigonal–bipyramidal complexes [72], symmetry considerations indicate that as the degenerate orbitals (d_{x2-y2} and d_{xy}) expand over the equatorial plane, the most important distortion is described by the angle α controlling the distortion of the equilateral triangle formed by equatorial ligands into an obtuse or acute one. The second important motion is given by β, the bending of the axial ligands. It is important to note that the distortions described by α and β are strongly coupled. The JT effect pushes the axial ligands to relax into the plane, favoring β to be smaller than 180, which in turn pushes the equatorial ligands toward smaller α values. Consequently, by choosing bulky axial ligands, which cannot fit in between the equatorial ligands, the magnetic anisotropy is expected to be enhanced by quenching the JT distortions (Fig. 11).

On predicting the |*D*|-values in theoretical models, we found that the maximum value for magnetic anisotropy is always obtained for the high-symmetry configuration but that JT distortions quickly reduce this value. In particular, we revealed that complexes with small ligands (like F^- or NH_3) or ligands that break the symmetry (like OH_2), are prone to large distortion, quenching the value of |*D*|. On the other hand, complexes with large ligands are much less distorted leading to large magnetic anisotropy, even in the low symmetry state.

We searched for an appropriate candidate in the Cambridge Structural Database [73] with a structure that met our requirements. We found the experimentally well-characterized complex $[NiCl_3(Hdabco)_2]^+$ (dabco is 1,4-diazabicyclo[2.2.2]-octane), with (dabco) as axial ligands and Cl^- as equatorial ligands (Fig. 11) [74, 75]. Our calculations both at LF-DFT and MRCI levels revealed that this complex

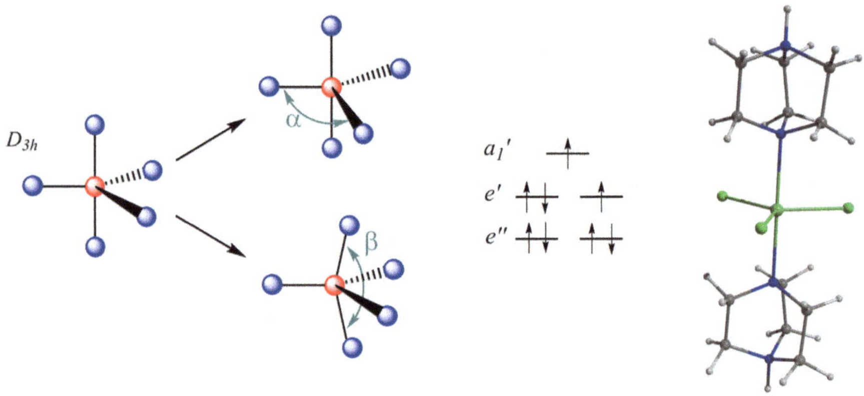

Fig. 11 Scheme representing the Jahn–Teller distortions for a NiX_5 TBP complex (left); Orbital splitting and electronic configurations associated with a D_{3h} d^8 HS configuration (middle); structure of $[NiCl_3(Hdabco)_2]^+$ (right)

displays a remarkably large magnetic anisotropy, |*D*| about 524 cm^{-1}, when compared to other Ni^{II}-based systems. This giant magnetic anisotropy arises from the bulkiness of the dabco ligand that favors a complete destruction of the JT distortions. At that time, we did not have any experimental proof that Ni(II) complexes can act as single-ion SMMs; so, it was only stated "Even though no Ni^{II} single-ion SMM has been found, the properties of this complex make it susceptible to sustain a magnetization over time under the influence of a small magnetic field. If experimentally confirmed, this complex would be the first of its kind" [72]. Just a year after this theoretical work, the magnetic properties of this complex were investigated by an experimental group [76], revealing the largest magnetic anisotropy never observed for a mononuclear Ni^{II} complex, thus fully validating our theoretical findings. Later, our approach also motivated another experimental group to synthesize and characterize a series of complexes displaying similar structural properties with bulky constrained pentadentate Schiff-based ligands [77].

6 Concluding Remarks

The examples in this chapter have one thing in common: through a joint effort by theory and experiment a deeper understanding is obtained of the electronic structure in these fascinating molecules. The spectroscopic features can nowadays be obtained both in the wet laboratory and in the dry laboratory, thereby reinforcing the structural effect of the coordination environment on the redox-active metal, and how these determine—or alternatively, are determined by—the spin and oxidation state. Computational chemistry has reached the point where meaningful results can be obtained about the actual transition-metal complex (instead of small model systems), which together with vibrational and paramagnetic spectroscopy bring forward new insights into how this behavior can be further tuned and extended. Indeed, transition-metal chemistry is promising to remain an interesting playground for understanding the key processes in life.

7 Further Reading

Spin states in biochemistry and inorganic chemistry: Influence on structure and reactivity, M. Swart, M. Costas (Eds.); Wiley UK, 2015

Physical methods in bioinorganic chemistry, L. Que Jr. (Ed.); University Science Books, 2000

Practical approaches to biological inorganic chemistry, R. Crichton, R. Louro (Eds.); Elsevier, 2013

Acknowledgements The COST association action CM1305 ECOSTBio (STSM grant 34080), the European Research Council (ERC 279549, WRB), the labex arcane (ANR-11-LABX-003), MINECO (CTQ2014-59212-P, CTQ2015-70851-ERC, CTQ2017-87392-P, MS), GenCat (2014SGR1202, MS), FEDER (UNGI10-4E-801, MS), and the Serbian Ministry of Science (Grant No. 172035) are acknowledged for financial support.

References

1. Swart M, Costas M (eds) (2015) Spin states in biochemistry and inorganic chemistry: influence on structure and reactivity. Wiley, Oxford. https://doi.org/10.1002/9781118898277
2. Que Jr L (2000) Physical methods in bioinorganic chemistry. University Science Books
3. Crichton R, Louro R (2013) Practical approaches to biological inorganic chemistry. Elsevier
4. Ray K, Duboc C (2018) ECOSTBio: explicit control over spin states in technology and biochemistry. Chem Eur J 24:5003–5005. https://doi.org/10.1002/chem.201801041
5. Bergeler M, Stiebritz MT, Reiher M (2013) Structure–property relationships of Fe_4S_4 clusters. ChemPlusChem 78:1082–1098
6. Carvalho ATP, Swart M (2014) Electronic structure investigation and parameterization of biologically relevant iron-sulfur clusters. J Chem Inf Model 54:613–620
7. Sharma S, Sivalingam K, Neese F, Chan GK-L (2014) Low-energy spectrum of iron–sulfur clusters directly from many-particle quantum mechanics. Nat Chem 6:927–933. https://doi.org/10.1038/nchem.2041
8. Kim J, Rees DC (1992) Structural models for the metal centers in the nitrogenase molybdenum-iron protein. Science 257:1677–1682. https://doi.org/10.1126/science.1529354
9. Einsle O, Tezcan FA, Andrade SLA, Schmid B, Yoshida M, Howard JB, Rees DC (2002) Nitrogenase MoFe-protein at 1.16 Å resolution: a central ligand in the FeMo-cofactor. Science 297:1696–1700. https://doi.org/10.1126/science.1073877
10. Ramaswamy S (2011) One atom makes all the difference. Science 334:914–915. https://doi.org/10.1126/science.1215283
11. Spatzal T et al (2011) Evidence for interstitial carbon in nitrogenase FeMo cofactor. Science 334:940–940. https://doi.org/10.1126/science.1214025

12. Yang T-C, Maeser NK, Laryukhin M, Lee H-I, Dean DR, Seefeldt LC, Hoffman BM (2005) The interstitial atom of the nitrogenase FeMo-cofactor: ENDOR and ESEEM evidence that it is not a nitrogen. J Am Chem Soc 127:12804–12805. https://doi.org/10.1021/ja0552489
13. Lancaster KM et al (2011) X-Ray emission spectroscopy evidences a central carbon in the nitrogenase iron-molybdenum cofactor. Science 334:974–977. https://doi.org/10.1126/science.1206445
14. Bjornsson R et al (2014) Identification of a spin-coupled Mo(III) in the nitrogenase iron—molybdenum cofactor. Chem Sci 5:3096–3103. https://doi.org/10.1039/c4sc00337c
15. Sippel D et al (2018) A bound reaction intermediate sheds light on the mechanism of nitrogenase. Science 359:1484–1489. https://doi.org/10.1126/science.aar2765
16. Cao L, Ryde U (2018) Influence of the protein and DFT method on the broken-symmetry and spin states in nitrogenase. Int J Quantum Chem 118:e25627. https://doi.org/10.1002/qua.25627
17. Cao L, Caldararu O, Ryde U (2017) Protonation states of homocitrate and nearby residues in nitrogenase studied by computational methods and quantum refinement. J Phys Chem B 121:8242–8262. https://doi.org/10.1021/acs.jpcb.7b02714
18. Siegbahn PEM (2018) A major structural change of the homocitrate ligand of probable importance for the nitrogenase mechanism. Inorg Chem 57:1090–1095. https://doi.org/10.1021/acs.inorgchem.7b02493
19. Siegbahn PEM (2016) Model calculations suggest that the central carbon in the femo-cofactor of nitrogenase becomes protonated in the process of nitrogen fixation. J Am Chem Soc 138:10485–10495. https://doi.org/10.1021/jacs.6b03846
20. Siegbahn PEM (2018) Is there computational support for an unprotonated carbon in the E4 state of nitrogenase? J Comput Chem 39:743–747. https://doi.org/10.1002/jcc.25145
21. Dau H, Iuzzolino L, Dittmer J (2001) The tetra-manganese complex of photosystem II during its redox cycle—X-ray absorption results and mechanistic implications. BBA-Bioenergetics 1503:24–39. https://doi.org/10.1016/s0005-2728(00)00230-9
22. Krewald V et al (2015) Metal oxidation states in biological water splitting. Chem Sci 6:1676–1695
23. Rohde JU et al (2003) Crystallographic and spectroscopic characterization of a nonheme Fe(IV)-O complex. Science 299:1037–1039. https://doi.org/10.1126/science.299.5609.1037
24. Fukuzumi S, Morimoto Y, Kotani H, Naumov P, Lee Y-M, Nam W (2010) Crystal structure of a metal ion-bound oxoiron(IV) complex and implications for biological electron transfer. Nat Chem 2:756–759. https://doi.org/10.1038/nchem.731
25. McDonald AR, Que Jr L (2013) High-valent nonheme iron-oxo complexes: synthesis, structure, and spectroscopy. Coord Chem Rev 257:414–428. https://doi.org/10.1016/j.ccr.2012.08.002
26. Swart M (2013) A change in oxidation state of iron: scandium is not innocent. Chem Commun 49:6650–6652. https://doi.org/10.1039/C3CC42200C
27. Swart M (2008) Accurate spin-state energies for iron complexes. J Chem Theory Comp 4:2057–2066
28. Perdew JP, Burke K, Ernzerhof M (1996) Generalized gradient approximations made simple. Phys Rev Lett 77:3865–3868
29. Grimme S (2006) Semiempirical GGA-type density functional constructed with a long-range dispersion correction. J Comput Chem 27:1787–1799
30. Swart M, Solà M, Bickelhaupt FM (2009) A new all-round DFT functional based on spin states and S_N2 barriers. J Chem Phys 131:094103
31. Han W-G, Liu T, Lovell T, Noodleman L (2006) DFT calculations of ^{57}Fe Mössbauer isomer shifts and quadrupole splittings for iron complexes in polar dielectric media: applications to methane monooxygenase and ribonucleotide reductase. J Comput Chem 27:1292–1306
32. Güell M, Solà M, Swart M (2010) Spin-state splittings of iron(II) complexes with trispyrazolyl ligands. Polyhedron 29:84–93. https://doi.org/10.1016/j.poly.2009.06.006
33. MacBeth CE et al (2004) Utilization of hydrogen bonds to stabilize M-O(H) units: synthesis and properties of monomeric iron and manganese complexes with terminal oxo and hydroxo ligands. J Am Chem Soc 126:2556–2567. https://doi.org/10.1021/ja0305151

34. Cho J et al (2011) Structure and reactivity of a mononuclear non-haem iron(III)–peroxo complex. Nature 478:502–505
35. Swart M, Gruden M (2016) Spinning around in transition-metal chemistry. Acc Chem Res 49:2690–2697. https://doi.org/10.1021/acs.accounts.6b00271
36. Zhou A, Kleespies ST, Van Heuvelen KM, Que Jr L (2015) Characterization of a heterobimetallic nonheme Fe(III)–O–Cr(III) species formed by O_2 activation. Chem Commun 51:14326–14329. https://doi.org/10.1039/c5cc05931c
37. Prakash J, Rohde GT, Meier KK, Münck E, Que Jr L (2015) Upside down! Crystallographic and spectroscopic characterization of an $[Fe^{IV}(O_{syn})(TMC)]^{2+}$ complex. Inorg Chem 54:11055–11057. https://doi.org/10.1021/acs.inorgchem.5b02011
38. Ray K, England J, Fiedler AT, Martinho M, Münck E, Que Jr L (2008) An inverted and more oxidizing isomer of $[Fe(IV)(O)(tmc)(NCCH3)]^{2+}$. Angew Chem Int Ed 47:8068–8071. https://doi.org/10.1002/anie.200802219
39. Zhou A et al (2017) The two faces of tetramethylcyclam in iron chemistry: distinct Fe−O−M complexes derived from $[Fe^{IV}(O_{anti/syn})(TMC)]^{2+}$ isomers. Inorg Chem 56:518–527. https://doi.org/10.1021/acs.inorgchem.6b02417
40. Padamati SK et al (2017) Transient formation and reactivity of a high-valent nickel(IV) oxido complex. J Am Chem Soc 139:8718–8724. https://doi.org/10.1021/jacs.7b04158
41. Wieghardt K et al (1988) Synthesis, crystal structures, reactivity, and magnetochemistry of a series of binuclear complexes of manganese(II), -(III), and -(IV) of biological relevance. The crystal structure of $[L'Mn^{IV}(m\text{-}O)_3Mn^{IV}L'](PF_6)_2{\cdot}H_2O$ containing an unprecedented short Mn···Mn distance of 2.296 Å. J Am Chem Soc 110:7398–7411. https://doi.org/10.1021/ja00230a021
42. Neubold P, Della Vedova BSPC, Wieghardt K, Nuber B, Weiss J (1990) Novel cofacial bioctahedral complexes of ruthenium: syntheses and properties of the mixed-valence species [LRu2.5(.mu.-X)3Ru2.5L]2 + (X = Cl, Br, I, OH). Crystal structures of [LRu2.5(.mu.-OH)3Ru2.5L](PF6)2.cntdot.H2O and [LRuIV(.mu.-O)3RuIVL](PF6)2.cntdot.H2O (L = 1,4,7-trimethyl-1,4,7-triazacyclononane). Inorg Chem 29:3353–3363. https://doi.org/10.1021/ic00343a020
43. Rohde J-U et al. (2004) Structural insights into nonheme alkylperoxoiron(III) and oxoiron(IV) intermediates by X-ray absorption spectroscopy. J Am Chem Soc 126:16750–16761. https://doi.org/10.1021/ja047667w
44. Pirovano P, Farquhar E, Swart M, Fitzpatrick AJ, Morgan GG, McDonald AR (2015) Characterization and reactivity of a terminal nickel(III)-oxygen adduct. Chem Eur J 21:3785–3790. https://doi.org/10.1002/chem.201406485
45. Pirovano P, Farquhar ER, Swart M, McDonald AR (2016) Tuning the reactivity of terminal nickel(III)–oxygen adducts for C–H bond activation. J Am Chem Soc 138:14362–14370. https://doi.org/10.1021/jacs.6b08406
46. Gatteschi D, Sessoli R, Villain J (2006) Molecular nanomagnets. Oxford University Press, New-York
47. Mannini M et al (2009) Magnetic memory of a single-molecule quantum magnet wired to a gold surface. Nat Mater 8:194–197. https://doi.org/10.1038/nmat2374
48. Bogani L, Wernsdorfer W (2008) Molecular spintronics using single-molecule magnets. Nat Mater 7:179–186. https://doi.org/10.1038/nmat2133
49. Craig GA, Murrie M (2015) 3d single-ion magnets. Chem Soc Rev 44:2135–2142. https://doi.org/10.1039/c4cs00439f
50. Frost JM, Harriman KLM, Murugesu M (2016) The rise of 3-d single-ion magnets in molecular magnetism: towards materials from molecules? Chem Sci 7:2470–2491. https://doi.org/10.1039/c5sc03224e
51. Waldmann O (2007) A criterion for the anisotropy barrier in single-molecule magnets. Inorg Chem 46:10035–10037. https://doi.org/10.1021/ic701365t
52. Neese F, Pantazis DA (2011) What is not required to make a single molecule magnet. Faraday Discuss 148:229–238. https://doi.org/10.1039/C005256F

53. Duboc C, Gennari M (2015) Experimental techniques for determining spin states. In: Spin states in biochemistry and inorganic chemistry. Wiley, pp 59–83. https://doi.org/10.1002/9781118898277.ch4
54. Neese F (2006) Importance of direct spin-spin coupling and spin-flip excitations for the zero-field splitting of transition metal complexes: a case study. J Am Chem Soc 128:10213–10222. https://doi.org/10.1021/ja061798a
55. Sousa C, de Graaf C (2015) Ab initio wavefunction approaches to spin states. In: Spin states in biochemistry and inorganic chemistry. Wiley, pp 35–57. https://doi.org/10.1002/9781118898277.ch3
56. Daul CA, Zlatar M, Gruden-Pavlovic M, Swart M (2015) Application of density functional and density functional based ligand field theory to spin states. In: Spin states in biochemistry and inorganic chemistry. Wiley, pp 7–34. https://doi.org/10.1002/9781118898277.ch2
57. Pederson MR, Khanna SN (1999) Magnetic anisotropy barrier for spin tunneling in $Mn_{12}O_{12}$ molecules. Phys Rev B 60:9566–9572
58. Neese F, Solomon EI (1998) Calculation of zero-field splittings, g-values, and the relativistic nephelauxetic effect in transition metal complexes. Application to high-spin ferric complexes. Inorg Chem 37:6568–6582
59. Atanasov M, Daul CA, Rauzy C (2003) New insights into the effects of covalency on the ligand field parameters: a DFT study. Chem Phys Lett 367:737–746. https://doi.org/10.1016/s0009-2614(02)01762-1
60. Sessoli R, Gatteschi D, Caneschi A, Novak MA (1993) Magnetic bistability in a metal-ion cluster. Nature 365:141–143
61. Duboc C, Ganyushin D, Sivalingam K, Collomb M-N, Neese F (2010) Systematic theoretical study of the zero-field splitting in coordination complexes of Mn(III). Density functional theory versus multireference wave function approaches. J Phys Chem A 114:10750–10758. https://doi.org/10.1021/jp107823s
62. Duboc C (2016) Determination and prediction of the magnetic anisotropy of Mn ions. Chem Soc Rev 45:5834–5847. https://doi.org/10.1039/C5CS00898K
63. Zlatar M et al (2016) Origin of the zero-field splitting in mononuclear octahedral Mn^{IV} complexes: a combined experimental and theoretical investigation. Inorg Chem 55:1192–1201. https://doi.org/10.1021/acs.inorgchem.5b02368
64. Leto DF, Massie AA, Colmer HE, Jackson TA (2016) X-Band electron paramagnetic resonance comparison of mononuclear Mn^{IV}-oxo and Mn^{IV}-hydroxo complexes and quantum chemical investigation of Mn^{IV} zero-field splitting. Inorg Chem 55:3272–3282. https://doi.org/10.1021/acs.inorgchem.5b02309
65. Barra AL, Gatteschi D, Sessoli R, Abbati GL, Cornia A, Fabretti AC, Uytterhoeven MG (1997) Electronic structure of manganese(III) compounds from high-frequency EPR spectra. Angew Chem Int Ed 36:2329–2331
66. Goldberg DP, Telser J, Krzystek J, Montalban AG, Brunel L-C, Barrett AGM, Hoffman BM (1997) EPR spectra from "EPR-silent" species: high-field EPR spectroscopy of manganese(III) porphyrins. J Am Chem Soc 119:8722–8723. https://doi.org/10.1021/ja971169o
67. Duboc C, Collomb MN (2009) Multifrequency high-field EPR investigation of a mononuclear manganese(IV) complex. Chem Commun 2715–2717. https://doi.org/10.1039/b901185d
68. Brazzolotto D et al (2016) An experimental and theoretical investigation on pentacoordinated cobalt(III) complexes with an intermediate $S = 1$ spin state: how halide ligands affect their magnetic anisotropy. Chem Eur J 22:925–933. https://doi.org/10.1002/chem.201502997
69. Wang LK et al (2018) Experimental and theoretical identification of the origin of magnetic anisotropy in intermediate spin iron(III) complexes. Chem Eur J 24:5091–5094. https://doi.org/10.1002/chem.201705989
70. Duboc C, Phoeung T, Zein S, Pécaut J, Collomb MN, Neese F (2007) Origin of the zero-field splitting in mononuclear octahedral dihalide Mn-II complexes: an investigation by multifrequency high-field electron paramagnetic resonance and density functional theory. Inorg Chem 46:4905–4916. https://doi.org/10.1021/ic062384l

71. Ruamps R et al (2013) Giant Ising-type magnetic anisotropy in trigonal bipyramidal Ni(ii) complexes: experiment and theory. J Am Chem Soc 135:3017–3026. https://doi.org/10.1021/ja308146e
72. Gruden-Pavlovic M, Peric M, Zlatar M, Garcia-Fernandez P (2014) Theoretical study of the magnetic anisotropy and magnetic tunnelling in mononuclear Ni(II) complexes with potential molecular magnet behavior. Chem Sci 5:1453–1462. https://doi.org/10.1039/c3sc52984c
73. Groom CR, Bruno IJ, Lightfoot MP, Ward SC (2016) The Cambridge structural database. Acta Cryst B 72:171–179. https://doi.org/10.1107/s2052520616003954
74. Rozell WJ, Wood JS (1977) Crystal and molecular-structures of 5-coordinate complexes $[Ni(LCH_3)_2Cl_3]ClO_4$ and $[Cu(LCH_3)_2Cl_3]ClO_4$, where L^+CH_3 = N-methyl-1,4-diazabicyclo[2.2.2]octonium ion. Inorg Chem 16:1827–1833. https://doi.org/10.1021/ic50174a001
75. Pritchard RG, Ali M, Munim A, Uddin A (2006) Effedcts d-orbital occupancy on the geometry of the trigonal-bipyramidal complexes (MCl3)-Cl-II(Hdabco)(dabco) (n), where M is Mn, Co, Ni or Cu and dabco is 1,4-diazabicyclo 2.2.2-octane. Acta Crystallogr Sect C Crystal Struct Commun 62:M507–M509. https://doi.org/10.1107/s0108270106037504
76. Marriott KER, Bhaskaran L, Wilson C, Medarde M, Ochsenbein ST, Hill S, Murrie M (2015) Pushing the limits of magnetic anisotropy in trigonal bipyramidal Ni(ii). Chem Sci 6:6823–6828. https://doi.org/10.1039/c5sc02854j
77. Nemec I, Herchel R, Svoboda I, Boca R, Travnicek Z (2015) Large and negative magnetic anisotropy in pentacoordinate mononuclear Ni(II) Schiff base complexes. Dalton Trans 44:9551–9560. https://doi.org/10.1039/c5dt00600g

Multiconfigurational Approach to X-ray Spectroscopy of Transition Metal Complexes

Marcus Lundberg and Mickaël G. Delcey

Abstract Close correlation between theoretical modeling and experimental spectroscopy allows for identification of the electronic and geometric structure of a system through its spectral fingerprint. This is can be used to verify mechanistic proposals and is a valuable complement to calculations of reaction mechanisms using the total energy as the main criterion. For transition metal systems, X-ray spectroscopy offers a unique probe because the core-excitation energies are element specific, which makes it possible to focus on the catalytic metal. The core hole is atom-centered and sensitive to the local changes in the electronic structure, making it useful for redox active catalysts. The possibility to do time-resolved experiments also allows for rapid detection of metastable intermediates. Reliable fingerprinting requires a theoretical model that is accurate enough to distinguish between different species and multiconfigurational wavefunction approaches have recently been extended to model a number of X-ray processes of transition metal complexes. Compared to ground-state calculations, modeling of X-ray spectra is complicated by the presence of the core hole, which typically leads to multiple open shells and large effects of spin–orbit coupling. This chapter describes how these effects can be accounted for with a multiconfigurational approach and outline the basic principles and performance. It is also shown how a detailed analysis of experimental spectra can be used to extract additional information about the electronic structure.

M. Lundberg (✉) · M. G. Delcey (✉)
Department of Chemistry - Ångström Laboratory, Uppsala University, 751 21 Uppsala, Sweden
e-mail: marcus.lundberg@kemi.uu.se

M. G. Delcey
e-mail: mickael.delcey@kemi.uu.se

E. Broclawik et al. (eds.), *Transition Metals in Coordination Environments*, Challenges and Advances in Computational Chemistry and Physics 29,
https://doi.org/10.1007/978-3-030-11714-6_7

1 X-ray Spectroscopy for Transition Metals

First-row transition metals are key components of many catalytic systems. Insights into their mechanisms can help in improving their efficiency and stability. Theoretical chemistry is frequently used to predict mechanisms of transition metal-catalyzed reactions. This is typically done by using the relative energy to distinguish between different geometric and electronic structures. To reliably identify a given species requires that the deviations from experiment are smaller than the energy differences between the alternative species. However, this can be very difficult to achieve for systems with several states close in energy, as is often the case for transition metal complexes. Mechanistic predictions also rely on the total charge of the system, which can be difficult to assign a priori. Despite the successful efforts to improve calculations of relative energies, modeling results often require validation by evaluating spectroscopic signatures of key intermediates.

In this chapter, we outline how X-ray spectroscopy, in combination with theoretical modeling, can be used to identify and characterize the electronic structure of transition metal systems. In X-ray spectroscopy, a high-energy photon interacts with the sample and when the photon energy matches the energy required to excite a core electron, the absorption intensity gets an edge-like increase. For first-row transition metals, the most commonly studied core excitations are from the L shell (mainly 2*p*) and are called L-edges, and the K shell (1*s*), called K edges, see Fig. 1 [20]. X-ray spectroscopy has a number of advantageous properties compared to other experimental techniques. The energy required to excite core electrons is element specific, which makes it possible to selectively study the catalytic metal in a complex system. Relevant examples are solar fuel systems that catalyze the formation of chemical fuels from solar energy, with plant photosynthesis being the most well-known system. Here intense transitions in the chromophores designed to maximize light absorption obscure many spectral probes of the catalyst itself [96]. As seen in Fig. 2, the core hole excitation is very localized and X-ray spectroscopy thus selectively probes charge and spin density on the metal [20, 49]. This makes it a widely used tool to extract oxidation and spin state of catalytic metals. X-ray spectra can be obtained

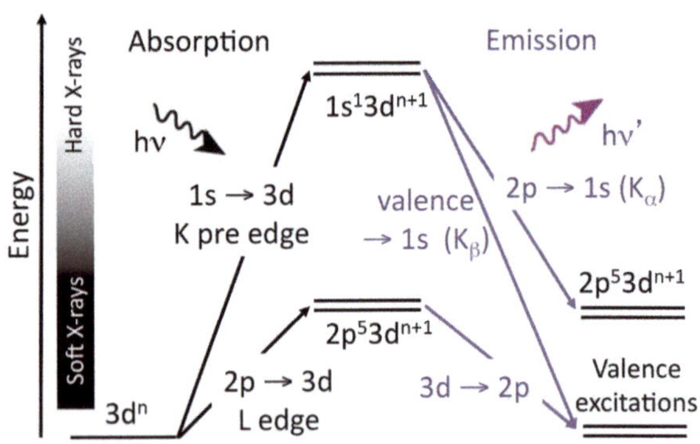

Fig. 1 Selected X-ray processes directly involving the metal 3*d* orbitals in both hard (high-energy) and soft (low-energy) X-ray regions, including X-ray absorption, X-ray emission and resonant inelastic X-ray scattering. Relative energies of different states are not to scale

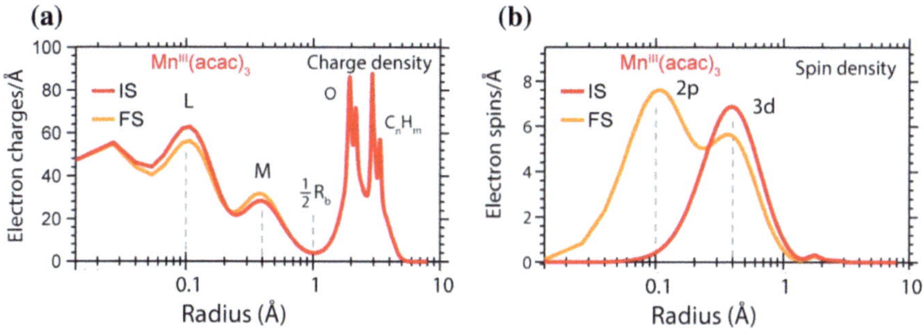

Fig. 2 Radial extension of the 2*p* core hole in manganese. **a** Radial charge densities (RCD) from restricted active-space (RAS) calculations of $Mn^{III}(acac)_3$ in the initial state (IS) and averaged over selected final states (FS). **b** Radial spin densities (RSD) from RAS of $Mn^{III}(acac)_3$ in the initial state and averaged over selected final states. The distance scale is logarithmic to enhance visibility at shorter distances. Reproduced from [49] with permission from the Royal Society of Chemistry

with femtosecond resolution to study transient intermediates in ultrafast chemical reactions [93].

The most direct technique to probe the 3*d* orbitals involved in metal–ligand binding and redox reactivity is through metal L-edge ($2p \rightarrow 3d$ excitations) X-ray absorption spectroscopy (XAS), see Fig. 1. L-edge X-ray photoelectron spectroscopy (XPS), where the 2*p* core electron is excited into the continuum, also gives information about the valence electrons through their interactions with the core hole. Although edge energies are element specific, soft X-ray photons in the metal L-edge energy range (hundreds of eV) also have high probability of photoexciting 1*s* electrons of lighter elements like carbon, nitrogen, and oxygen. This background absorption leads to challenges in extracting the metal signal as well as reduction of the metal site by excess photoelectrons [50, 89]. Metals in enzymes and solution systems are, therefore, often probed using hard X-rays (thousands of eV) in the metal K pre-edge ($1s \rightarrow 3d$ excitations). With the development of new intense X-ray sources, both synchrotron and X-ray free-electron lasers (XFELs), it has also become possible to perform the X-ray equivalent of resonance Raman, often called resonant inelastic X-ray scattering (RIXS). With RIXS, it is possible to reach final states corresponding to both core and valence excitations, see Fig. 1.

These various X-ray spectroscopy techniques allow a wealth of information to be gathered from first-row transition metals. However, the resulting spectra are often complicated to interpret. This is especially true for final states with 2*p* holes, as there are strong interactions between the valence electrons and the core hole, as well as a strong spin–orbit coupling in the 2*p* shell. Theoretical models are, thus, necessary to correlate experimental data and electronic structure. Together with the development of new experimental capabilities, there has been an intense effort to develop theoretical models that include all relevant interactions. This chapter will describe the multiconfigurational wavefunction approach based on the complete active-space (CAS) paradigm [76]. After a short introduction to different X-ray modeling approaches, we will give a step-by-step explanation of the multiconfigurational approach for transition metal systems. This will be followed by examples of how the combination of theory and experiment can give new insights into transition metal chemistry, in areas ranging from femtosecond spectroscopy to biological cofactors [39, 74]. The current limits of the multiconfigurational approach are systems with two transition metals and the final section describes potential improvements to handle more complex systems.

2 Theoretical Simulations of X-ray Spectra

Modeling X-ray spectroscopy includes the same challenges involved in accurate descriptions of valence states, but the presence of a core hole introduces further complications. Taking the metal L-edge XAS spectra of an open-shell system as an example, the final states are affected by both strong $3d$-$3d$ and $2p$-$3d$ correlation, as well as $2p$ and $3d$ spin–orbit coupling (SOC). This complicates the mapping of electronic structure to spectral shape.

On the other hand, due to the local nature of the core hole, atomic models can provide a very efficient description of many X-ray spectroscopies. A standard modeling approach based on this idea is the charge-transfer multiplet (CTM) method [31, 87]. In this model, initial and final states are calculated from an atomic full configuration interaction (CI) including $2p$ and $3d$ orbitals, i.e., taking into account all possible $3d$ electron configurations. Ligands are described by an empirical ligand-field splitting and to model more covalent interactions, ligand-to-metal charge-transfer (LMCT) or metal-to-ligand charge-transfer (MLCT) configurations can be added in the CI. This method is not only conceptually simple but also computationally inexpensive and has historically been one of the dominant ways to theoretically reproduce X-ray spectra of transition metals and to interpret them in intuitive terms. However, the CTM method includes parameters that are fitted to the experimental spectrum, which makes it less suitable for predictive purposes. Additionally, the number of model parameters increases with decreasing symmetry, and the semiempirical CTM approach thus works best for complexes with a high degree of symmetry.

If the aim is to predict the spectral fingerprint of a molecule and distinguish between different electronic structure alternative, ab initio approaches are preferable as they are independent from semiempirical parameters. A detailed review of ab initio methods for X-ray spectroscopy simulation can be found in [67]; for our purpose here, we will only provide a short overview. Formally, any quantum chemistry method able to describe valence excitations can be extended to the X-ray regime. However, most methods typically generate excited states in energy ordering, so the main change needed in the formalism is a way to target the proper energy range without first having to compute all valence states. Various ideas have been proposed and implemented such as core–valence separation [14, 84], efficient energy-specific eigenvalue solvers [54], and the complex polarization propagator (CPP) [22]. Those formalisms are general and have been applied to many standard quantum chemistry methods such as density functional theory (DFT) [84], algebraic diagrammatic construction (ADC) [38], and coupled cluster [17]. Yet, as is the case for ground state chemistry and UV–visible spectroscopy, transition metal complexes are particularly challenging, due to their open-shell electronic structures and strong correlation, and some of the methods mentioned above cannot be trusted in this context.

Thanks to its low cost and reasonable accuracy, time-dependent density functional theory (TD-DFT) is widely used. DFT can also be used in conjunction with wavefunction methods such as in the DFT/restricted open-shell CI singles (ROCIS), where a CI with single excitations is built and DFT correlation is added using three system-independent parameters [75]. These DFT-based methods typically fail if the ground states have significant multi-reference character, which is frequent for first-row transition metals. In addition, there is often a strong functional dependence, reducing the predictive power of the method as different functionals can lead to different conclusions. In particular, self-interaction error has a strong effect on the calculated spectrum, the low spatial overlap between core and valence orbitals creating what can effectively be considered to be a charge-transfer excitation, which is an important weakness of some DFT functionals and leads to a critical dependence on the amount of Hartree–Fock exchange [65].

An electronic structure approach that is well suited for transition metal systems is the multiconfigurational (MC) self-consistent field (SCF) method, among which the complete active-space (CASSCF) version is the most widely used [76]. Multiconfigurational methods can be adapted to X-ray processes by including also core electrons in the excitation space [1, 2, 40]. As the number of excitations from the core orbitals can typically be restricted to one, it becomes convenient to use a restricted active space (RAS) wavefunction [62]. This approach has become a leading method to simulate X-ray spectra of smaller transition metal complexes [10, 42, 43, 71, 72]. It can, with minor adaptations, be applied to L-edge XAS and RIXS dominated by electric dipole transitions between bound states [6, 71, 92]. By including second-order terms in the wave vector expansion, electric dipole forbidden transitions in metal K pre-edge XAS and RIXS can be described [32, 33]. With these developments, multiconfigurational methods have now been used to describe all X-ray processes shown in Fig. 1. With recent extensions to continuum excitations, it is also possible to calculate XPS [29, 43]. The combination of an ab initio philosophy with good accuracy provides a powerful predictive tool for the analysis of X-ray spectra. In the following section, we will discuss the basic principles of the multiconfigurational approach, and how to design calculations to get accurate and reliable results.

3 Multiconfigurational Approach to X-ray Processes

The multiconfigurational active-space methods are based on the division of the orbital space into a small set of so-called active orbitals and a larger set of inactive orbitals [79]. Within the active orbitals, electron correlation is treated accurately with CI.

Both CI coefficients and orbital shapes are optimized. Correlation outside of the active space can be treated with a low-level method, typically second-order perturbation theory (PT2). Basic equations for the active-space methods can be found in Chap. 5 of this book. Although multiconfigurational methods are essentially ab initio, their accuracy and computational cost can be tuned through the choice of a number of simulation parameters, with the most critical choice being the choice of orbitals in the active space. This flexibility, together with the relatively high-computational cost, necessitates an understanding of the effect of model choices on the cost and accuracy of the calculation. To demonstrate the methodology and the impact of the different parameters, we will in this section extensively use examples from L-edge XAS modeling of ferric ($3d^5$) reference systems with well-known electronic structures, namely high-spin $[FeCl_6]^{3-}$ (ferric chloride) and low-spin $[Fe(CN)_6]^{3-}$ (ferricyanide) [23, 71, 73].

3.1 System Selection

Before starting the modeling, as with any theoretical chemistry calculation, the first step is the choice of the system. This choice is constrained by the cost of the calculation. The cost of multiconfigurational methods depends on both the total system size and the size of the active space. The active space will be discussed in more details in the next paragraph, but the severe scaling with respect to the number of active orbitals typically restricts calculations to a single transition metal center, as opposed to DFT where large clusters [60], and even extended systems can be described [13]. On the other hand, the scaling with system size is less drastic, especially if PT2 calculations can be avoided. For example, X-ray calculations of the heme iron systems with more than forty heavy atoms have been performed using RAS [34, 74]. Additionally, due to the locality of X-ray spectroscopy, the convergence of the spectrum with the system size is expected to be rapid, allowing more crude models than would otherwise be recommended. Because of the prohibitive cost of geometry optimization with correlated multiconfigurational methods, starting geometries are often taken from experiments or from another level of theory, typically DFT. It is worth noting that in some cases, especially for very covalent metal–ligand bonds with strong multiconfigurational effects, the starting geometry can be of insufficient quality for accurate spectrum calculations. In such cases, reoptimization of a few geometrical parameters can be performed with multiconfigurational perturbation theory [85]. Finally, environment effects can be included in the same way as for calculations of valence states, such as the polarizable continuum model to describe solvent effects [18, 50].

3.2 Active-Space Selection

The second step in the design of the calculation is the choice of the active space, the trademark of multiconfigurational methods. The general rules for any application are to include in the active space any orbital participating in strong correlation. For electronic excitations, any orbital whose occupation is expected to change significantly should also be included. In practice, this selection requires both expertise and experience. Aiming at making multiconfigurational calculations more black box, there have been developments toward automated active-space selection where the selection criteria are meant to optimize the description of strong correlation [83]. However, such criteria cannot be directly applied to spectrum calculations as the orbitals important to describe the photoexcitation process are not necessarily the same that contribute most to correlation. In practice, the final choice of active space is still driven by chemical intuition and experience.

For transition metals, there is a wealth of experience on the choice of active space to describe strong correlation and intuitive rules have been compiled [69]. For X-ray spectroscopy, where the range of final states span over several eV, the target accuracy is typically lower than most other applications and the active space can be reduced somewhat. One can often satisfactorily restrict the selection to the metal 3*d* orbitals and any ligand orbitals forming strong covalent bonds with the metal. Local symmetry, either strict or approximate, can significantly help reducing the number of ligand orbitals included. To this space, one should add orbitals that are excited to or from in the X-ray process, which typically are virtual orbitals with some metal content and the core orbitals. When using the RAS formalism, the core orbitals are conveniently put in the ras1 space allowing a single excitation as most processes include a single core hole. The orbitals involved in metal–ligand bonding are typically put in ras2 to allow all possible configurations, see Fig. 3a.

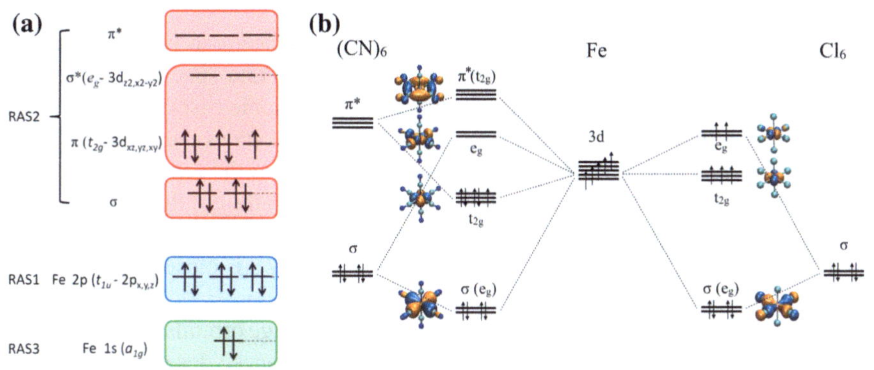

Fig. 3 **a** Active space for RAS calculation of 1*s*2*p* RIXS of iron hexacyanides. Reproduced from [32] with permission from the American Chemical Society. **b** Schematic orbital diagram of $[Fe(CN)_6]^{3-}$ and $[FeCl_6]^{3-}$. Only selected ligand orbitals are shown

The orbital diagrams of ferricyanide and ferric chloride are shown in Fig. 3b. An active space for valence calculations of $FeCl_6$ would include the five $3d$ orbitals and the two e_g ligand orbitals combinations forming the σ bond with the metal. For accurate energy calculations, it is recommended to include an additional set of metal d-type orbitals, the double shell, for a total of twelve active orbitals [69]. However, for X-ray calculations, neglecting the double shell gives small differences in the spectrum [71]. As the complex is relatively ionic, the ligand-dominated σ orbitals have limited metal d character, and one could imagine limiting the active space to the five $3d$ orbitals. Yet, with this choice of orbitals, the subsequent second-order perturbation theory calculation fails to converge (see Sect. 3.5), a typical sign of an ill-balanced active space. This example thus shows that the final choice of active space is usually a dialogue between the user and the program.

For $Fe(CN)_6$, the situation is similar except that this time, the ligands have empty π^* orbitals able to form a bond with the metal t_{2g} orbitals through π-backdonation. This stabilizes the t_{2g} orbitals and thus increases the gap with the e_g, which favors the low-spin configuration. As the π^* orbitals are empty and have some $3d$ character, they can be populated by X-ray absorption and are needed in the active space, consisting thus of ten orbitals.

In some cases, the X-ray process involves two different core levels. One example is $1s2p$ RIXS, where $1s \rightarrow 3d$ absorption is followed by monitoring the strongest emission channel, $2p \rightarrow 1s$, see Fig. 1. Modeling of this process requires independent control of the core-hole occupations. In a RAS calculation, this can be achieved by placing the two sets of core orbitals in different ras spaces, typically $2p$ in ras1 and $1s$ in ras3 and thus, both $1s$ and $2p$ core-hole states can be computed with the same active space, see Fig. 3a [32].

3.3 Generating Core-Hole States

To obtain a spectrum, a large number of excited states needs to be computed. As mentioned in the previous section, the target is to generate specific core-excited states, without computing all possible valence states. In some cases, such as L-edge spectra of centrosymmetric complexes, the $2p$ core-hole states can be in different symmetries (ungerade) than the valence states (gerade) and the separation is trivial. For the general case, a simple technique is to remove from the configuration interaction all configurations with fully occupied core orbitals, the so-called core–valence separation (CVS) [14]. For active-space methods, the CVS is closely related to the generalized active-space method [33, 59].

Orbital optimization is typically done using state-average orbitals. This avoids the separate optimization of every state, while ensuring a balanced description of all the states. The main drawback is that the results depend strongly on the number of states, which needs to be taken into account, in particular when comparing calculations with different number of states. During orbital optimization, the algorithm may lower the state average energy by replacing the core orbital by an occupied orbital of higher energy, for example 3p instead of $2p$. Unless this is prevented using e.g., a restricted step algorithm [40], the core orbitals have to be frozen during the orbital optimization. A downside to this is that this prevents the expected contraction of those orbitals upon core excitation, but the main effect is expected to be a global shift of the final states to higher energy. Often the relative edge position is more important than the absolute one, and the frozen-core approximation has been used to predict oxidation state shifts with errors of 0.3 eV when using the same active space and number of states [33, 49].

3.4 Simulating Light-Matter Interaction

The last step to obtain a spectrum is to compute transition intensities at different wavelengths. For bound states, the intensity can be calculated from the matrix element of the operator representing the light-matter interaction between the initial and final states. Most often, the plane wave representing the light is approximated by an electric dipole. If core-hole states have been optimized separately from the valence states, transition intensities can still be calculated correctly by taking advantage of a biorthogonalization scheme [61]. From the computed intensities, the final spectrum is generated using a Lorentzian lifetime broadening convoluted with a Gaussian broadening to account for the experimental resolution. If only absorption intensities are considered, this corresponds to a spectrum collected in transmission mode. However, the transmission spectra and those obtained by measuring the photoelectron current, the total electron yield, are similar. RIXS spectra can be calculated from the transition intensities for both absorption and emission processes according to the Kramers–Heisenberg formula [25]. L-edge XAS spectra of metal complexes in solution are in many cases collected by measuring the fluorescence from the core-excited states, the partial fluorescence yield (PFY) mode. The PFY-XAS spectra can be calculated from the RIXS cross sections by integrating the relevant emission channels for each incident energy [27, 28, 47].

Results obtained for L-edge XAS of ferric chloride and ferricyanide are shown in Fig. 4 [36, 73, 91]. The spectra are divided into two separate regions, L_3 and L_2, split by the $2p$ spin–orbit coupling, as will be discussed in detail below. RAS calculations can be used to correlate spectra and electronic structure. In ferric chloride, the ligand field is weak and different configurations mix strongly, making it difficult to assign transitions to specific $3d$ orbitals. One exception is the high-energy peak in the L_3 edge that has been identified as a $2p \rightarrow 3d$ transition, combined with a $\sigma \rightarrow 3d$ LMCT [71, 91]. In ferricyanide the ligand field is strong, and a molecular orbital picture becomes more relevant. The sharp first peak corresponds to a $2p$ electron filling the t_{2g} hole and the second peak is an excitation to e_g. The third peak is a signature of π-backbonding and is typically labeled after the π^* molecular orbital, see Fig. 3, but is in reality a strong mix of e_g and π^* contributions [36, 71].

The 1s $\rightarrow$ 3d transitions of K pre-edge are dipole forbidden in centrosymmetric complexes and still relatively weak in many other systems. In the former case,

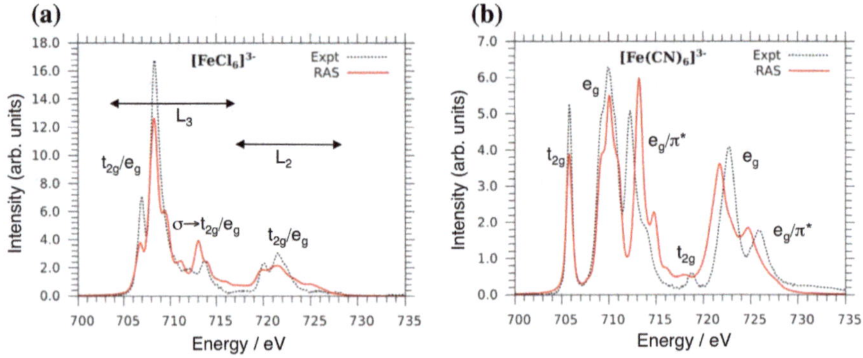

Fig. 4 Comparison between theory and experiment for the iron L-edge X-ray absorption spectra of **a** $[FeCl_6]^{3-}$ and **b** $[Fe(CN)_6]^{3-}$. Adapted from [73] with permission from Wiley

they only gain intensity from what is typically referred to as electric quadrupole transitions. To model these transitions requires a second-order expansion of the electromagnetic wavevector [8], or use of the exact form [55, 56]. Both alternatives have also been used to calculate iron K pre-edges using RAS wavefunctions [33, 81, 82].

While multiconfigurational methods are most often used to describe excitations to bound states, they can also be extended to describe ejection of electrons into the continuum, such as for photoelectron spectroscopy. XPS provides a wealth of information on the electronic structure and can for example be used to study specific solute–solvent interactions of metal complexes in solutions [88]. From a modeling perspective, the key is often to compute the Dyson orbitals, which corresponds to the overlap between the initial N-electron wavefunction and the final $N-1$-electron wavefunction of the ionized molecule. To compute the intensities, one common method is the so-called sudden approximation (SA) that neglects the dependency on the kinetic energy of the outgoing photon and simply estimates the intensity as the norm of the Dyson orbital. However, this approximation is not valid for low-energy photoelectrons, and instead, a more sophisticated approach is to explicitly model the free electron and compute the transition intensity with the Dyson orbital with the help of numerical integration. An efficient implementation of the Dyson orbitals using a biorthonormal basis has recently been implemented in the CAS/RAS framework [29, 30].

XPS spectra for $[Fe(H_2O)_6]^{2+}$ have been calculated using a minimal active space including only the 3d orbitals in the valence as shown in Fig. 5a. There is good agreement between RAS modeling and experiment, with much improved results for the full formalism over the sudden approximation, see Fig. 5b. The Dyson orbitals can additionally be analyzed to understand the relation between the experimental features and the electronic structure, see Fig. 5c.

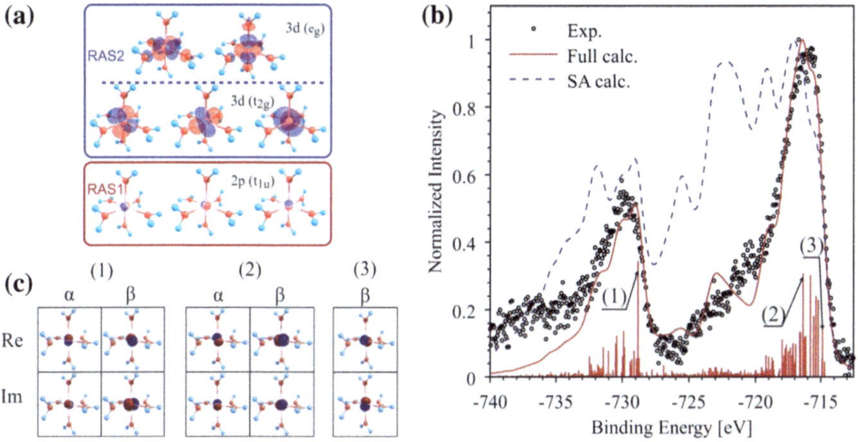

Fig. 5 **a** Orbitals in the active space of $[Fe(H_2O)_6]^{2+}$. **b** Experimental (2M $FeCl_2$ aqueous solution) and calculated ($[Fe(H_2O)_6]^{2+}$ cluster) XPS for incoming photon energy of 925 eV. Full calculation corresponds to numerical integration of XPS matrix element, SA means sudden approximation. **c** Real and imaginary parts of α and β spin contributions to the Dyson orbitals for selected transitions. Adapted from [29], with the permission of AIP Publishing

3.5 *Number of States, Correlation Level and Basis Set*

The simulated spectra shown in Fig. 4 are sensitive to several different modeling parameters, among them the number of states, the level of electron correlation, and the choice of basis set. Even if the computation of all valence states can be avoided, as discussed above, the number of excited states needed to describe an X-ray spectrum can be very large. In transition metal complexes, the density of states tends to be very high, and an X-ray absorption spectrum typically spans 10 eV or more, and often several hundred states are required. As an example, the ferricyanide spectrum displays a strong peak that is associated with the ligand-dominated π^* orbitals, see Fig. 3. However, the goal of the orbital optimization is to minimize the energy, not to reproduce an X-ray spectrum. Unless enough states are included to excite to these orbitals, the optimization prefers to include $4d$-type orbitals that correlate well with the t_{2g} $3d$ (double-shell effect), but are not particularly relevant to the spectrum. As shown in Fig. 6, at least 320 states were needed to reach the π-backbonding orbitals and reproduce the corresponding peak in the spectrum. Even more states were required to fully converge the spectrum [73]. The large number of states constitutes one of the major limitations of the method, as the cost of the calculation increases at least linearly with this parameter. Additionally, it is difficult to estimate in advance how many states are required. A simple convergence test is to increase the number until the spectrum features remain approximately fixed.

While calculations at the multiconfigurational self-consistent field level often give qualitatively correct descriptions, higher numerical accuracy can be reached

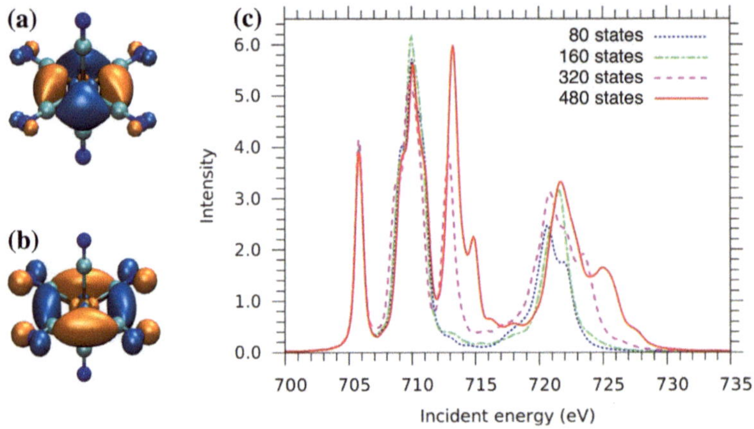

Fig. 6 Effects of number of states on RAS modeling. **a** High-lying t_{2g} orbitals of 4*d* character included with fewer than 80 states. **b** π^* orbital included in the active space with 80 states or more. **c** L-edge XAS spectra of $[Fe(CN)_6]^{3-}$ calculated using RASPT2/ANO-RCC-VTZP with different number of states per spin multiplicity. Adapted from [73] with permission from Wiley

by including dynamical correlation through second-order perturbation theory, either using the second-order complete active-space perturbation theory (CASPT2) [3], its counterpart for a restricted active space (RASPT2) [64], or the N-electron valence perturbation theory (NEVPT2) [4]. The latter two have both been used for calculations of X-ray spectra [16, 42, 73].

The effect of adding a perturbation correction can be seen in Fig. 7. For ferric chloride, focusing on the L_3 edge, the spectrum remains relatively similar, and only the minor $\sigma \rightarrow 3d$ LMCT peak sees a significant shift in position. On the other hand, the effect on ferricyanide is more significant. The t_{2g}-e_g splitting decreases by 1 eV, giving almost perfect agreement with experiment, while the overestimation of the position of the π^* peak drops from 4 to 1 eV. The large effect of the perturbative step

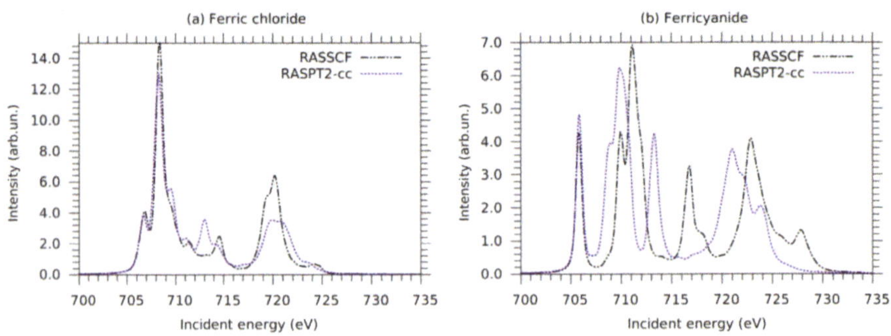

Fig. 7 L-edge XAS spectra of **a** $[FeCl_6]^{3-}$ and **b** $[Fe(CN)_6]^{3-}$ calculated at the RASSCF and RASPT2 levels. Reproduced from [73] with permission from Wiley

on the π backbonding peak actually indicates a missing orbital in the active space, specifically the ligand π which strongly correlates with the π^*. In ferrocyanide ($[Fe(CN)_6]^{4-}$), the error for the corresponding π^* peak drops from 2.0 to 0.6 eV with PT2, while the error in the relative energy of different states in the e_g peak decrease from 0.6 to 0.1 eV [32].

A well-known problem with CASPT2/RASPT2 is the presence of possible intruder states. To reduce this problem, an imaginary shift of 0.3–0.5 hartree can be applied [24]. It is still important to check that the reference weights, i.e., the weight of the RASSCF state in the total correlated wavefunction, are consistent for all core-hole states. Despite being a perturbative approach, the RASPT2 equations are solved iteratively and low reference weights often translate into convergence difficulties and/or inaccurate results. Low weights despite a small imaginary shift most commonly stem from improper active-space selection.

The formulation of CASPT2/RASPT2 also includes one empirical parameter called the ionization potential electron affinity (IPEA) shift, which was introduced to fix a systematic error when dealing with open-shell configurations [26]. There is no consensus on the optimal value of this shift and recent studies suggest that it strongly depends on system, active space and basis set [97]. The IPEA should not be used as an empirical parameter to improve the match with the experimental spectrum, and a large effect of changing the IPEA value indicates that the active space may be too limited, as can be seen for the previously discussed π^* peak in ferricyanide [73]. NEVPT2 does not include any such shift in its formulation and is also less sensitive to intruder states. However, the correlation contribution to the spin-state energetics of some transition metal complexes showed larger deviations for NEVPT2 compared to CASPT2 [70].

Second-order perturbation theory is a correlated method, and thus in theory displays a slow basis set convergence. However, for X-ray calculations no major changes in the spectrum have been observed going beyond a triple-zeta basis set, and in many cases good results are obtained already at the double-zeta level [73]. On the other hand, standard contracted basis sets do not provide much flexibility for core electrons to contract upon excitation or for core correlation. This leads to errors in absolute edge positions of around 3–4 eV for L-edges and up to 18–20 eV for K edges when using a triple-zeta basis set, slightly depending on the active space and the number of states [33, 49, 73]. The errors in the absolute L-edge position can be improved to 0.75 eV with the use of an uncontracted basis set [16], but this is very expensive and only applicable to small systems. In the frozen-core approximation, the quality of the core basis set is less important. However, relative energies between complexes with similar ligand environments and active-space selections can still be reproduced within 0.3 eV [33, 49].

When it comes to spectral shape, RASPT2 calculations typically predict all major peaks in the L_3 edge with at most 30% error in intensity. The largest error in relative energy, around 1 eV, is seen for cases with incomplete active spaces, as in the ferricyanide L-edge XAS π^* peak. That energetic error might seem large, but should be compared to the 30 eV range of the full spectrum. The accuracy gives, in most cases, sufficient predictive power to identify the charge, spin, or electronic structure of a chemical species [34].

3.6 *Relativistic Effects*

Even for first-row transition metals, because of the direct involvement of core orbitals, relativistic effects are very significant. Scalar relativistic effects affect the energy and radial extent of the core orbitals and thus have a significant effect of the spectrum, though mostly as a global shift. In our methodology, scalar effects are included using a second-order Douglas–Kroll–Hess (DKH) Hamiltonian [21, 35], coupled with a basis set designed specifically to be used in conjunction with DKH, namely the ANO-RCC basis [77, 78].

When dealing with 2*p* core holes, the description of spin–orbit coupling is also essential because the 2*p* orbitals split into 2 spin–orbit levels, $P_{1/2}$ and $P_{3/2}$ separated by around 10 eV (depending on the metal). A computationally efficient way to include spin–orbit coupling in active-space calculations is to compute core-hole states with spin multiplicities $S = 0, \pm 1$ relative to the ground state, and diagonalizing an approximate spin–orbit hamiltonian in the basis of those states [63]. This is equivalent to using Russell-Saunders (LS) coupling. It is only an approximation of the correct four-component solution, but it is significantly simpler and sufficiently accurate for most purposes. A full four-component multiconfigurational code has been applied to X-ray spectroscopy, but only to systems with a small number of active orbitals and without dynamical correlation [7].

The L-edge spectrum from a model low-spin d^5 system in O_h symmetry offers a clear and extensive demonstration of the effect of spin–orbit coupling, see Fig. 8a.

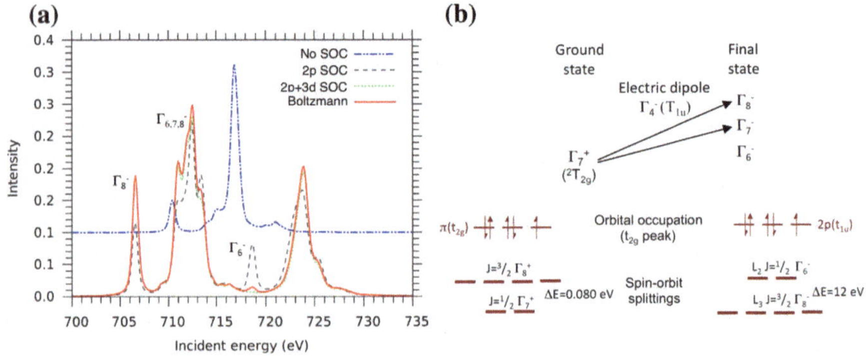

Fig. 8 **a** RAS L-edge XAS spectra of the Fe^{3+} ion with different treatments of 2*p* and 3*d* SOC. Boltzmann referes to a Boltzmann distribution of different SOC ground states. **b** Selection rules for electric dipole transitions using Bethe notation for double groups. Adapted from [71] with the permission of AIP Publishing

As expected, without any SOC, there is a single edge. Inclusion of $2p$ SOC not only splits the spectrum into L_3 ($J_{2p} = \frac{3}{2}$) and L_2 ($J_{2p} = \frac{1}{2}$) edges but also leads to major changes in spectral shape because of the mixing of states with different multiplicity. This shows that the spin–orbit effect cannot always be modeled by simply duplicating the spin-free spectrum and shifting the two edges away from each other. It is also important to note that to form the correct spin–orbit states requires all three $2p$ orbitals in the active space.

The $3d$ SOC constant is much weaker, 0.05 eV for iron, than the $2p$ one (8 eV). However, taking $3d$ SOC into account leads to visible changes in the calculated spectrum, especially in the intensities of the two $2p \rightarrow t_{2g}$ peaks, see Fig. 8. This can be explained by the selection rules, see Fig. 8b. Ignoring Jahn–Teller distortions, which have only minor effects on the energy levels [71], there is triple orbital degeneracy in the ground state. This degeneracy is lifted by spin–orbit coupling, and the lowest spin–orbit states have different selection rules compared to the low-lying excited states and thus generate different spectra.

3.7 *Simulating X-ray Processes with Molcas*

The X-ray calculations described in this chapter have almost exclusively been performed using the Molcas program [5], which is a leading program for multiconfigurational quantum chemistry. The same capabilities are also available in the open-source distribution OpenMolcas. To facilitate future calculations, Fig. 9 shows the different steps of a RAS X-ray calculation in OpenMolcas. The program is composed of several modules, each performing a specific task with their own input and communicating together through files.

The active space is defined in the input to the `RASSCF` program. As shown in Fig. 3, it is common to place core orbitals in ras1 allowing for at most one excitation. To avoid calculating a large number of valence excited states with filled core orbitals, core–valence separation can be invoked using the *hexs* keyword. To avoid that the

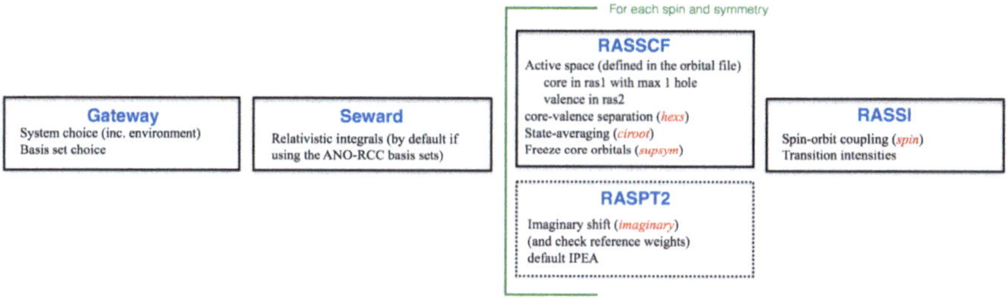

Fig. 9 Calculation template for X-ray simulations with the RAS method in OpenMolcas. The name of the boxes are the names of the OpenMolcas module corresponding to the specific parameters. The red italic text indicates specific keywords

hole rotates out of the target core orbitals, these can be frozen during the orbital optimization using *supsym*. In the `RASPT2` program, both imaginary and IPEA shifts can be specified, with default values being 0.1 and 0.25 eV, respectively. Spin–orbit coupled states are formed in a RAS state interaction (SI) algorithm using the *spin* keyword in the `RASSI` program. `RASSI` also calculates transition matrix elements between all computed states. The default is the electric dipole approximation but a complete second-order expansion as well as the exact form of the wavevector are also implemented [81, 82]. With most practical aspects of the modeling covered, the next sections will describe applications of X-ray spectroscopy in different fields of chemistry.

4 Electronic Structure from X-ray Spectra

In this section, we will describe how the combination of X-ray experiments and multiconfigurational modeling can be used to extract very detailed information about electronic structure. The first examples show how modeling can be used to characterize the total spin and oxidation state of a complex. The next examples focus on the structure of individual orbitals involved in metal–ligand binding, often through studies of charge-transfer and ligand-field transitions. The examples include both ground-state electronic structures and time-resolved studies of transient reaction intermediates at the femtosecond timescale. The final examples describe the electronic structure at even finer detail by looking into splittings between states with the same formal orbital occupation, in the X-ray field usually called multiplet splittings. These examples illustrate how multiconfigurational methods give a correct description of these different states, and that this can be used to extract detailed orbital information.

4.1 Spin and Oxidation State

Reaction mechanisms of redox reactions can at the most basic level be described in terms of changes of the spin and oxidation states of the metal. X-ray spectroscopy is ideal to observe these effects because the core hole is a very local probe of the metal site as shown in Fig. 2. Multiconfigurational calculations accurately predict oxidation state dependent spectral changes, as has been shown for ferrous ($3d^6$) and ferric ($3d^5$) complexes [10, 32], as well as a series of photocatalytically relevant Mn systems [11]. The predictive power of the simulations makes it possible to identify problems in the experimental data, which is important as many samples easily photodamage in intense X-ray beams [50].

In addition to fingerprinting reaction intermediates, calculations can also be used to understand how oxidation state is reflected in the L-edge XAS spectrum because both redox and core-excitation processes can be treated at an equal level. One of the

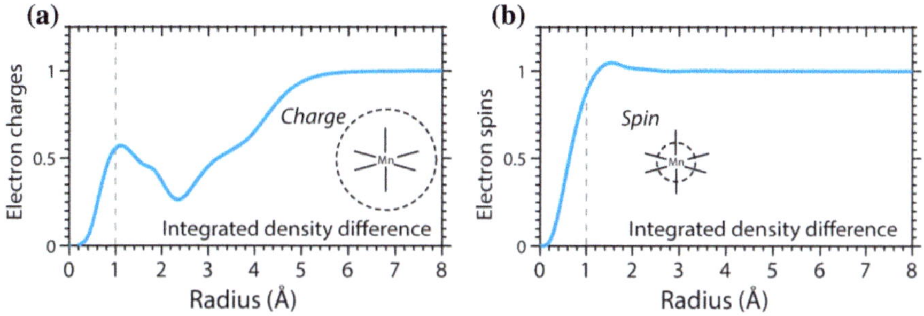

Fig. 10 Changes in spin and charge density upon reduction of $Mn^{III}(acac)_3$. **a** Integrated radial charge density (RCD) difference (RCD of $[Mn^{II}(acac)_3]^{1-}$ minus RCD of $Mn^{III}(acac)_3$. **b** Integrated radial spin density (RSD) difference (RSD of $[Mn^{II}(acac)_3]^{1-}$ minus RSD of $Mn^{III}(acac)_3$. The dashed vertical lines indicate half the Mn-O bond length R_b. Adapted from [49] with permission from the Royal Society of Chemistry

main signatures of increasing metal oxidation state in L-edge spectroscopy is a shift of the absorption edge to higher energy, together with significant changes in spectral shape [19]. These changes can be used to identify species in ultrafast reactions, even for systems as complex as the four-manganese oxygen-evolving complex in photosystem II [46]. Oxidation state should also be possible to identify from the total absorption cross sections, which should be roughly proportional to the number of holes in the 3*d* orbitals [47, 48].

The clear effects of formal oxidation state on X-ray spectra are somewhat intriguing because quantum chemistry calculations show that the charge density of the transition metal does not strongly correlate with its formal oxidation state. Instead the spin density provides a more reliable signature [9, 41]. This can be illustrated for the reduction of the well-known model complex $Mn^{III}(acac)_3$ [49]. Figure 10 shows the calculated changes in charge and spin density upon addition of an electron, while keeping the geometry constant. Charge density is delocalized over the whole molecule due to Coulomb repulsion, even for this relatively ionic complex. In contrast, changes in spin density are localized to the metal atom due to favorable exchange interactions.

In reality, reduction of $Mn^{III}(acac)_3$ leads to geometry changes from octahedral to tetrahedral coordination $Mn^{II}(acac)_2$. The experimental and RAS calculated PFY-XAS spectra of $Mn^{II}(acac)_2$ and $Mn^{III}(acac)_3$ are shown in Fig. 11a, b. The overall agreement is good, with the exception of the position and intensity of the L_2 edge, partially due to problems to describe fluorescence in this edge. Experimentally, the maximum of the L_3 edge is shifted to higher energies by 2.0 eV upon oxidation and the simulations reproduce this shift with only a minor error of 0.3 eV.

Interestingly, the spectral shape can be partially explained by looking at the position of the different spin-state contributions, see Fig. 11c, d. As a result, one can expect that the spectral shape is strongly affected by exchange interactions. As the spin density is strongly localized on the metal, the spectral shape remains atomic

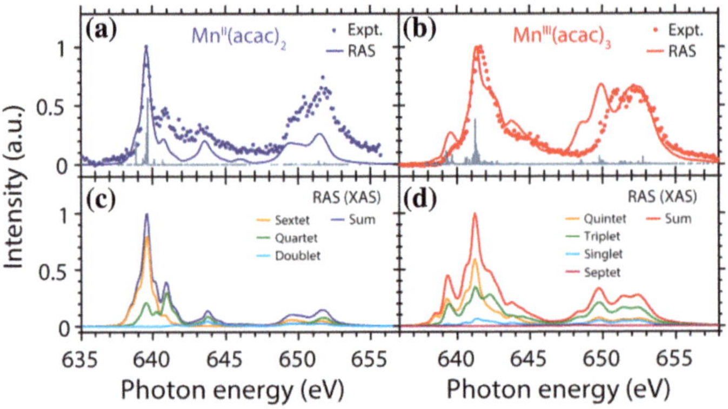

Fig. 11 Experimental and RAS modeling of Mn(acac) complexes. **a, b** Calculated RAS and experimental PFY-XAS spectra of **a** $Mn^{II}(acac)_2$ and **b** $Mn^{III}(acac)_3$. **c, d** Calculated absorption spectra (XAS not PFY) decomposed into the relative contributions of the (spin) multiplicities in the final states for **c** $Mn^{II}(acac)_2$ and **d** $Mn^{III}(acac)_3$. Reproduced from [49] with permission from the Royal Society of Chemistry

like and provides a signature for the number of spins on the metal. However, the nature of the shift requires a more in-depth analysis, and it was shown that it is due to changes in Coulomb interactions, which depend on the charge density. By studying the charge density changes during core excitation, it is proposed that the core excitation increases the electron affinity in the final state, which leads to lower excitation energies for Mn(II) compared to Mn(III) [49].

The results show that multiconfigurational calculations are able to reproduce spectral changes due to changes in oxidation state, which is important for reliable fingerprinting of reaction intermediates. At the same time, the simulations give additional insights into the molecular origin of these changes and how they are linked to charge and spin density.

4.2 Molecular Orbitals in Metal–Ligand Binding

X-ray spectroscopy does not only give information about spin and oxidation states but also provides detailed information about metal–ligand interactions. This sensitivity has been used to extract ground-state electronic structure by fitting parameters in the parameterized CTM model to the spectrum [36, 91]. For nonparameterized methods like RAS, calculating spectra does not give any new information compared to accurate ground-state calculations. Instead, RAS offers the possibility to rationalize how different spectral features are connected to the electronic structure. However, the strong interactions with the 2*p* hole in the final states lead to complicated electronic structures that, although they can be correctly described in a multiconfigurational model, are difficult to interpret.

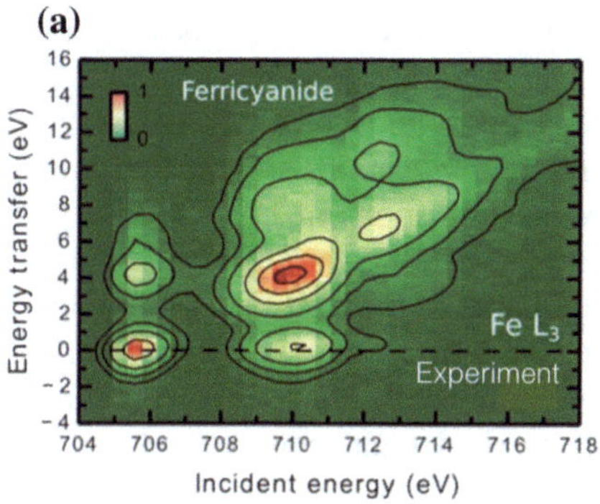

Fig. 12 RIXS maps of ferricyanide at the Fe L_3 edge from **a** experiment and **b** RAS simulations. Adapted from [53] with permission from the American Chemical Society

To get direct access to valence states, it is possible to use L-edge RIXS, see Fig. 1. This is a two-photon process, where absorption of an incident photon (hν) leads to emission of a scattered photon of a different wavelength (hν'). By varying the energy difference between incident and emitted photons, i.e., the energy transfer, different valence states can be accessed. The two-dimensional RIXS spectra provide more information than the one-dimensional XAS spectra and have been used to identify reaction intermediates in ultrafast chemical reactions [39, 51, 52, 66, 90, 93]. To aid in fingerprinting, theoretical models can be used to assign spectral features and extract electronic structure information. The first RIXS applications of RAS targeted ligand-field excitations of metal ions in water [6, 42, 92]. Later studies have focused on highly covalent complexes like $Fe(CO)_5$ and $Fe(CN)_6$ [23, 53, 86, 93]. In this and the following subsection, we will show how RIXS modeling can be used to study molecular orbital interactions in both ground and short-lived excited states of iron hexacyanides.

Ground-state L-edge RIXS spectra have been analyzed for both ferro- and ferricyanide [23, 53], but here only results for the ferric complex will be discussed. Experimental and RAS simulated L-edge RIXS spectra of ferricyanide are shown in Fig. 12. In the two-dimensional spectra, the incident energy axis is the same as in the L-edge absorption. Although the RIXS spectra have been collected over the full incident energy range, only the L_3 edge is shown to better highlight spectral features. As in the L-edge XAS, the resonances along the incident energy axis can be conveniently labeled t_{2g}, e_g, and π^*.

The additional information in the RIXS experiment comes from the energy transfer axis, which corresponds to the valence excitation energies. Starting with the t_{2g} resonance, the peak at 0 eV corresponds to elastic transitions where an electron from the newly closed t_{2g} shell fills the $2p$ hole. The second peak at 4 eV corresponds to emission from a filled orbital. With the help of RAS calculations, this orbital was identified as the ligand-dominated σ orbital shown in Fig. 3. The resonance can thus be assigned as a $\sigma \rightarrow t_{2g}$ LMCT transition.

Proceeding along the incident energy direction, the next resonance is the e_g peak, for which several features along the energy transfer axis can be resolved. After the elastic peak, there is a broad and intense resonance around 4 eV that corresponds to $t_{2g} \rightarrow e_g$ transitions, see Fig. 12. These ligand-field transitions are the most intense in the RIXS spectrum, in contrast to the weak transitions in UV/Vis absorption spectroscopy. This is due to the differences in selection rules. The single-photon $g \rightarrow g$ transition is parity forbidden and only gain intensity through vibronic coupling, while the two-photon $g \rightarrow u \rightarrow g$ transition is parity allowed. In addition, the strong spin–orbit coupling in the intermediate state breaks the spin selection rules, and calculations indicate that both transitions to singlet and triplet final states have appreciable magnitude [53].

Notice that although the $\sigma \rightarrow t_{2g}$ and $t_{2g} \rightarrow e_g$ transitions have similar final-state energies, these resonances are clearly separated along the incident energy in the RIXS map. RIXS thus includes more information than a single-photon absorption, partly due to the enhancement of ligand-field transitions, but also facilitates the assignment of these resonances to different molecular orbital transitions. With the help of electronic structure calculations, the RIXS plane can be used to map out the entire set of valence orbitals.

4.3 Transient Intermediates from Charge-Transfer Excitations

Full understanding of catalytic reactions requires knowledge of intermediates along the reaction pathway. The development of intense XFELs with time resolution in the femtosecond range has opened up new ways to study short-lived intermediates. A prominent example is how the combination of femtosecond RIXS with RAS modeling has given detailed insight into the spin and ligand-exchange dynamics of photoexcited $Fe(CO)_5$ [51, 52, 93]. In general, valence excited states of iron complexes have attracted considerable scientific interest, as charge separation in these states can be used in light-harvesting applications [57]. Again, iron hexacyanide serves as a suitable model system to understand how information about electronic, spin and structural dynamics can be extracted from the combination of modeling and experiment [39, 66].

In the experiment, ferricyanide absorbs a photon from the UV/Vis probe, which leads to an LMCT excitation that fills the t_{2g} shell and at the same time creates a hole on the ligand. The time evolution of this excited state is then followed using femtosecond RIXS [39]. Figure 13a shows the difference spectrum of the LMCT state compared to the ground state of ferricyanide (shown in Fig. 12a). A clear fingerprint of the LMCT state is the complete loss of the t_{2g} peak in the RIXS spectrum, because the hole in that orbital is filled in the valence excitation.

RAS calculations have been used to predict spectra of potential species along the reaction pathway and offer fingerprints for the dynamics [66]. They can also explain

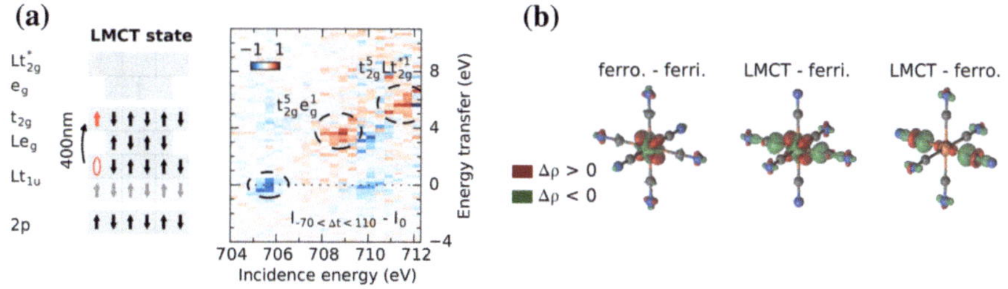

Fig. 13 **a** Valence electronic occupation of the LMCT state and difference map in the range of 70–110 fs. **b** Charge density differences (CDDs) of the LMCT state and ferrocyanide taken with respect to ferricyanide. To isolate ligand-hole effects, the CDD of the LMCT state with respect to ferrocyanide is additionally shown. All differences are calculated at the CASPT2 level at the optimized ferricyanide geometry. Adapted from [39] with permission from the American Chemical Society

how changes in the RIXS spectrum relates to changes in metal–ligand interactions of the ferrocyanide LMCT state compared to the ground states of ferro- and ferricyanide. Although the electronic structure of the excited state can be directly obtained from calculations, the comparison to experiment can verify the predicted changes in metal–ligand interactions. RAS calculations of the charge density difference between the LMCT state and the ferrocyanide ground state, which both have the same nominal $t_{2g}^6 e_g^0$ configuration, show an increase in charge density on the iron along the metal–ligand bond axis, see Fig. 13b. This indicates a net increase in σ-donation in the LMCT state. At the same time, π-backdonation remains largely constant, which gives overall stronger metal–ligand binding in the LMCT state compared to ferrocyanide and a reduced Fe-C bond length [66]. The predicted changes in electronic structure are consistent with a shift in the onset of the edge to lower energies, as well as an increase in the ligand-field strength [39]. This example demonstrates how time-resolved RIXS can give detailed insight into the properties of short-lived excited states in metal complexes, and how calculations can rationalize the relation between spectra and metal–ligand orbital interactions.

4.4 *Multiconfigurational Description of Multiplet Splittings*

After showing how X-ray modeling can be used to get insights into molecular orbital interactions, the next level of detail is to look at the different electronic states that arise from a given electron configuration. These states are split by differences in spin and spatial orientation of the electrons, here referred to as multiplet splittings. If these states can be resolved, this gives the most detailed information about the electronic structure of a metal complex. These concepts will be illustrated by first looking at iron K pre-edge XAS, with focus on ferricyanide [33]. This is followed by

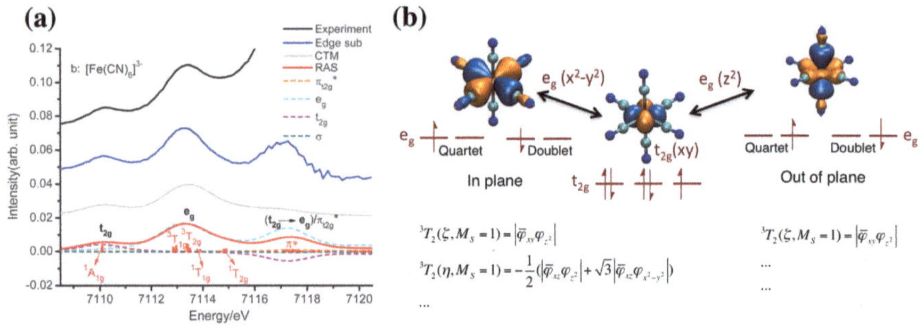

Fig. 14 **a** Iron K pre-edge XAS spectra of ferricyanide. Experimental spectra before and after subtraction of the rising edge are shown in black and blue. Theoretical simulations using CTM and RAS are shown in gray and red. Dashed lines shows the changes in orbital occupation number during the pre-edge process scaled by the intensity of the transition. **b** Orbital interactions in the $t_{2g}^5 e_g^1$ configuration leading to $^{1,3}(T_{1g}, T_{2g})$ different states. **c** Selected wavefunctions of the $M_s = 1$ triplet component, without considering spin-orbit coupling. O_h symmetry has been used for labeling of the orbitals. Reproduced from [33] with permission from the Royal Society of Chemistry

an illustration of how the $2p$-$3d$ multiplet interactions in $1s2p$ RIXS directly relates to the strength of σ-bonding in ferrocyanide [32].

Iron K edge XAS corresponds to excitations from the $1s$ orbital. It is commonly used for metal complexes in solution because hard (high-energy) X-rays are only weakly absorbed by the environment. The lowest resonances are typically assigned to $1s \rightarrow 3d$ transitions, see Fig. 1. For centrosymmetric complexes, these transitions are electric dipole forbidden, and for most systems they appear as a weak pre-edge before the rising edge dominated by electric dipole allowed $1s \rightarrow 4p$ transitions. The K pre-edge spectrum of ferricyanide is shown in Fig. 14a. After subtracting the rising edge, three resonances can be identified. These resonances can, as was done for the L-edge XAS spectrum, be labeled t_{2g}, e_g, and a mixed e_g/π^* peak, see Fig. 14a.

The t_{2g} transition results in a closed valence shell, so there is only one final state in this region. The second resonance consists of $1s \rightarrow e_g$ transitions, and the relative position of t_{2g} and e_g resonances reflects the ligand-field strength. A closer analysis shows that resonance is composed of a large number of transitions to different states of the $t_{2g}^5 e_g^1$ configuration, see Fig. 14a. The important $1s$ core hole states are all doublets, like the ground state. However, the relative spin orientations of t_{2g} hole and the e_g electrons can give both singlet and triplet valence states. These are split by differences in exchange interactions. States are further split by the differences in the relative orientation of the e_g electron and the t_{2g} hole. The T_{1g} states represent the energetically more favorable situation where hole and electron are in the same plane, while in the T_{2g} states they are in different planes, see Fig. 14b. A correct description of the properties of these states requires a multiconfigurational approach. It is well known that open-shell singlet states cannot be described by a single determinant. However, some of the wavefunctions of the valence triplet states also require two or more determinants, see Fig. 14c.

The multiplet splittings are directly related to the structure of the molecular orbitals. The t_{2g}-e_g interactions, and thus the multiplet splittings, are largest if both orbitals are localized on the metal, i.e., if they are ionic. The size of the splitting is thus related to the extent of orbital delocalization in the molecule. In practice, the short lifetime of the 1s hole gives rise to large lifetime broadenings which can make it difficult to accurately determine the energy of all the states. This limitation can be overcome with the use of RIXS.

4.5 Metal–Ligand Covalency from Multiplet Splittings

RIXS can achieve higher resolution than XAS because the lifetime broadening in the energy transfer direction is determined by the lifetime of the final state after emission. L-edge RIXS can under the right experimental conditions resolve different mutiplet states in the valence region [80], but this requires better resolution than in the study discussed above [53]. Instead, multiplet splittings will be illustrated using 1s2p RIXS where the final state has a 2p core hole, see Fig. 1 [45, 58]. The same approach has already been used to study how the metal ligands modulate electron transfer in cytochrome c, a key component in cell respiration [44].

1s2p RIXS spectra of ferro- and ferricyanide are shown in Fig. 15 [32, 58, 68]. All spectra have two separate regions, stretching roughly diagonally across the plane. The region at lower energy transfer corresponds to states in the L_3 edge of the XAS spectrum, while the upper region corresponds to the L_2 edge. The calculated RAS spectra do not include the intense transitions in the rising edge, but reproduce the structure of the pre-edge. The incident energy resonances are the same as in K pre-edge XAS. In ferrocyanide, there are two pre-edge resonances, $1s \rightarrow e_g$ and $1s \rightarrow e_g/\pi^*$, with the latter being hidden under the rising edge in the experimental spectrum [33]. Ferricyanide also has a low-energy $1s \rightarrow t_{2g}$ resonance, and a broad e_g peak split by multiplet interactions as shown in Fig. 14a.

It is most instructive to look at the e_g resonance in ferrocyanide. Along the incident energy axis, it does not contain much information because it corresponds to a single state. More information can be obtained from the energy transfer direction. The $2p \rightarrow 1s$ emission from the intermediate state lead to $2p^5t_{2g}^6e_g^1$ final states, nominally the same as in L-edge XAS. An L-edge-like spectrum is obtained by taking a vertical cut along constant incident energy (CIE) through the maximum of the e_g resonance, see Fig. 15. With only a single incident energy resonance, it could be expected that the e_g part of the CIE cut and the L-edge XAS should look similar. Instead, the width of the e_g resonance increases from the 0.8 eV in the L-edge spectrum to 1.5 eV in the CIE spectrum, see Fig. 16a. As the experimental broadenings are similar in the two experiments, the explanation is instead the differences in selection rules [58].

The $2p^5t_{2g}^6e_g^1$ configuration gives T_{1u} and T_{2u} states. The single-photon electric dipole transitions in XAS only reach T_{1u} states from the $^1A_{1g}$ ground state, while the two-photon RIXS process reaches both T_{1u} and T_{2u} final states, see Fig. 16b.

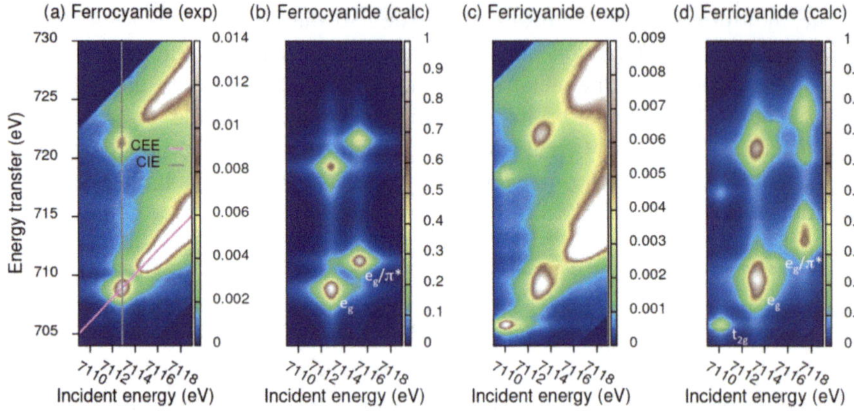

Fig. 15 Experimental and RASPT2 calculated $1s2p$ RIXS planes of iron hexacyanides [32, 58]. Cuts along constant incident energy (CIE) and constant emission energy (CEE) are shown with gray and pink lines, respectively. The RAS spectra only describe pre-edge absorption and the rising edges are not included. Reproduced from [32] with permission from the American Chemical Society

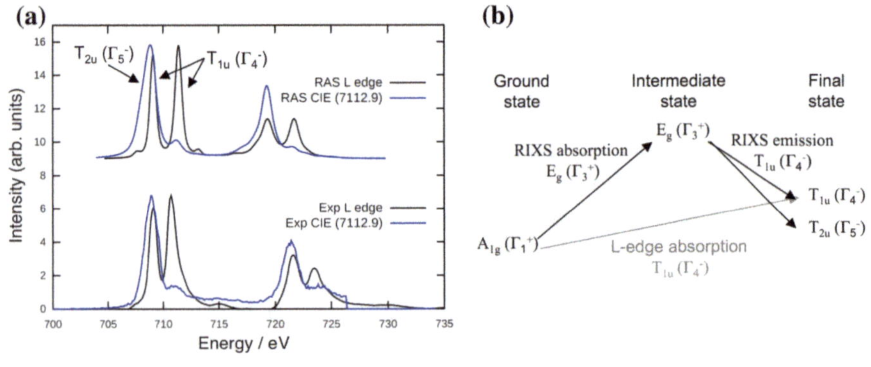

Fig. 16 **a** Iron L-edge XAS and CIE cut through the e_g pre-edge peak for ferrocyanide from RAS modeling (top) and experiment (bottom). **b** Relevant selection rules in O_h symmetry for $1s2p$ RIXS and L-edge XAS from the A_{1g} ground state in low-spin ferrous complexes. Adapted from [32] with permission from the American Chemical Society

This effect is captured in RAS simulations, and the molecular orbital representation can be used to visualize the differences between these states. In analogy to the different $t_{2g}^5 e_g^1$ valence states in Fig. 14, the T_{2u} states are lower in energy because of more favorable in-plane interactions between the $2p$ hole and the e_g electron. As the $2p$ core hole probe is completely localized on the metal, the strength of the interaction measures the amount of metal character in the e_g orbital, with more metal character corresponding to lower metal–ligand covalency. This has been shown experimentally by comparisons between ferrocyanide and ferrous tacn (tacn =1,4,7 triazacyclononane). The latter ligand is a weaker σ donor, leading to less covalent bonds and more localized 3*d* orbitals, which is seen in a significantly larger width of the e_g resonance of ferrous tacn [58]. Notice that the individual states are actually not resolved in the experiment. It is instead the differences in selection rules that makes it possible to identify the different energy regions for T_{1u} and T_{2u} states.

5 Extensions to Metal Dimers and Complex Systems

All systems previously discussed in this chapter have been relatively small and included no more than one transition metal atom. Many catalytic systems include two or more metal atoms, but multiconfigurational calculations of X-ray processes for such systems are challenging. Including two instead of one metal basically doubles the number of core and valence orbitals and leads to large active spaces. This in turn leads to a very large number of states within the energy range covered by the X-ray spectra [74]. Here two different approaches to RAS modeling of metal dimers are presented. First, a heme dimer with intermolecular coupling between metal atoms is discussed, followed by a μ-oxo bridged metal dimer with covalent coupling [74, 81].

5.1 *Intermolecular Coupling*

Heme systems play important roles in many biological systems including oxygen transport and catalysis. In many spectral probes, the intense transitions in the porphyrin obscure information about the electronic structure of the iron. This limitation can be overcome with a suitable X-ray probe, and iron L-edge XAS has been successfully used to probe the electronic structure of the Fe–O_2 bond [95]. Another interesting characteristic is that hemes are prone to complexation in solvent. This gives rise to π-π interactions between the porphyrins, as well as resonant coupling of close-lying electronic states of the monomers. These interactions should be detectable in the X-ray signature [74].

RAS simulations have been made of hemin dimers that form in water solvent, see Fig. 17. To avoid treating the full supermolecule, the relevant valence and core-hole states of each monomer are calculated first. The configurations with energies close to X-ray resonances are then extracted. XAS and RIXS correspond to one and two-particle excitations correspondingly, and the full set of states necessary to model these processes can at a first approximation be modeled using a configuration interaction model including singles and doubles (CISD) [74]. After reduction of the size of the interaction matrix by ignoring some contributions, diagonalization gives the states of the full dimer from which X-ray intensities can be calculated.

The simulated spectra of three different dimer orientations are shown as CIE cuts through the L-edge RIXS planes, see Fig. 17. Some resonances show distinct changes, including the elastic peak at 0 eV energy transfer. The magnitude of these

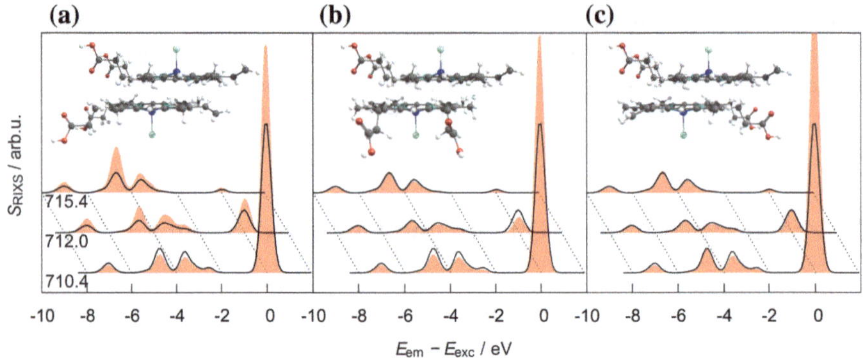

Fig. 17 CIE cuts of the simulated L-edge RIXS spectra of the [Heme B-Cl]^0 dimer (red filled curves) with different orientations of the COOH groups **a** 0, **b** 90, and **c** 180 for three incident energy resonances energies. The monomer spectra are shown as black lines for comparison. Reproduced from [74] and made available under a Creative Commons 4.0 license

changes depends on the molecular orientation, with larger effects in transitions that involve orbitals oriented out of the plane of the porphyrin. These calculations show the sensitivity of the RIXS probe to heme dimerization, but a direct comparison to experiment probably requires more extensive sampling of different orientations [74].

5.2 Intramolecular Coupling

Metal complexes with strong covalent coupling between metals are important in many catalytic systems. For these systems, it is more difficult to separate the active spaces of the two metal centers, which puts severe limitations on the modeling. This is illustrated for the iron K pre-edge spectra of the (hedta)Fe$^{III}\mu$–OFe^{III}(hedta) (hedta = N-hydroxyethyl-ethylenediamine-triacetic acid) metal dimer, see Fig. 18a [81]. The K pre-edge is sensitive to both geometric and electronic structure [94]. In iron dimers, deviations from centrosymmetry caused by the metal–metal interactions lead to electric dipole contributions in addition to what is usually referred to as electric quadrupole transitions.

The RAS spectrum was calculated with 13 valence orbitals in the active space, three Fe(3d)–O(p) bonding orbitals, seven metal-3d-dominated orbitals and three antibonding iron–oxygen orbitals, see Fig. 18b. The ground state has antiferromagnetic coupling between two high-spin 3d^5 centers, giving an open-shell singlet. However, due to the challenges to calculate the very large number of singlet states, simulations were instead made using the ferromagnetically coupled undectet, which lies 0.1 eV above the ground state. This leads to a significant reduction in the number of possible states and enables the calculation of the full K pre-edge spectrum.

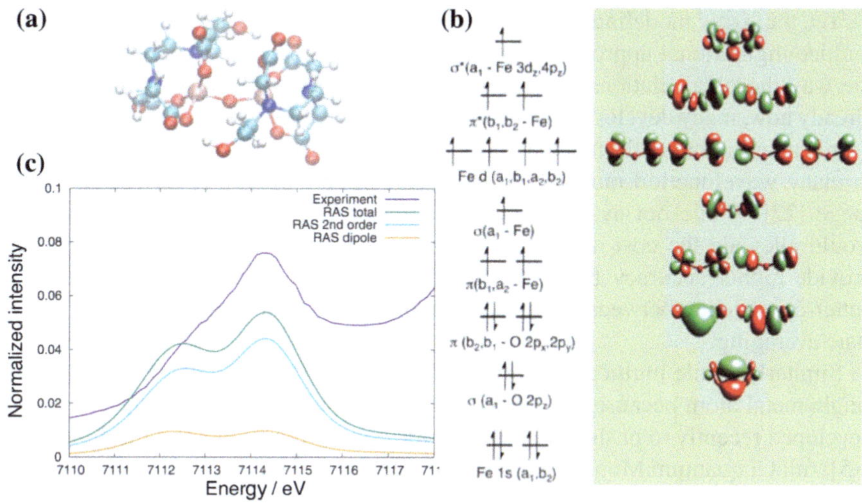

Fig. 18 **a** Structure of [(hedta)FeOFe(hedta)]. RAS active space for Fe_2O. Experimental and RASSCF/ANO-RCC-VTZP simulated K pre-edge spectra of [(hedta)FeOFe(hedta)] [81, 94]. Adapted from [81] with permission from Taylor and Francis

The experimental K pre-edge of the hedta dimer has two discernible features with an energy splitting around 1.7 eV, see Fig. 18 [94]. The RASSCF spectrum also shows two distinct pre-edge features, with a more intense peak at higher energy. According to the simulations, there are non-negligible contributions from electric dipole contributions, but the largest intensity still comes from quadrupole contributions. The energy splitting is overestimated by 0.4 eV and the low-energy peak appears more intense in the simulated spectrum. These deviations could possibly decrease with use of PT2 corrections, but this was not tested due to the high-computational cost. The challenges in modeling X-ray spectra of covalently linked metal clusters illustrate the need for further development of the multiconfigurational approach.

6 Conclusions and Outlook

By its position at the intersection of theory and experiment, the field of ab initio X-ray simulations combines the strengths of both. Theory provides insight into the chemical process while the experiment can be used to verify the theoretical findings. This is particularly relevant for transition metal catalysts, where accurate theoretical predictions are often difficult. In recent years, multiconfigurational calculations have become a reference for accurate X-ray simulations for transition metal complexes. Thanks to the inherent flexibility of the method, and helped by constant developments, most X-ray spectroscopies can now be simulated and many interesting applications have already been performed, showcasing the strong promises of this field.

Yet, the X-ray modeling field is still evolving rapidly. This is certainly true for the multiconfigurational approach, with new method developments constantly shaping the way these calculations are performed. This process is likely to continue and already now, many developments in related fields offer great promises to lift some of the main limitations of the method. As an example, the CPP approach has been applied to many wavefunction models to efficiently compute the spectrum at any energy range [22]. While not available yet, an efficient CPP-CAS/RAS implementation would alleviate the cost associated with the high density of states. It could also provide higher accuracy by, for example, allowing core relaxation and ensuring better consistency between different calculations by removing artifacts caused by state averaging.

Similarly, while multiconfigurational simulations have mainly been limited to a single metal atom because of active-space restrictions, many techniques have been developed recently to push this limit, e.g., the density matrix renormalization group [15], full CI quantum Monte Carlo [12], or Heat-Bath CI [37]. While those methods still have not been used to compute X-ray spectra, and some technical difficulties are still left to be overcome, the potential to calculate X-ray spectra of some of the fascinating natural and synthetic multi-metallic complexes at the multiconfigurational level is certainly very appealing. Those developments, and others yet unforeseen, will shape the future of the field and push the limits of what can be done, hopefully matching the significant advances in the experimental techniques. This can only improve the already strong complementarity between theory and experiment and deepen our insights into the captivating world of transition metal catalysis.

Acknowledgements We acknowledge financial support from the foundation Olle Engkvist Byggmastare and the Knut and Alice Wallenberg Foundation (Grant No. KAW-2013.0020). We thank Meiyuan Guo and Michael Odelius for useful discussions.

References

1. Ågren H, Jensen HJA (1987) An efficient method for the calculation of generalized overlap amplitudes for core photoelectron shake-up spectra. Chem Phys Lett 137(5):431–436
2. Ågren H, Flores-Riveros A, Jensen HJA (1989) An efficient method for calculating molecular radiative intensities in the vuv and soft x-ray wavelength regions. Phys Scr 40(6):745
3. Andersson K, Malmqvist PÅ, Roos BO, Sadlej AJ, Wolinski K (1990) Second-order perturbation theory with a casscf reference function. J Phys Chem 94(14):5483–5488
4. Angeli C, Cimiraglia R, Evangelisti S, Leininger T, Malrieu JP (2001) Introduction of n-electron valence states for multireference perturbation theory. J Chem Phys 114(23):10252–10264. https://doi.org/10.1063/1.1361246
5. Aquilante F, Autschbach J, Carlson RK, Chibotaru LF, Delcey MG, De Vico L, Fdez Galván I, Ferre N, Frutos LM, Gagliardi L et al (2016) Molcas 8: new capabilities for multiconfigurational quantum chemical calculations across the periodic table. J Comput Chem 37(5):506–541
6. Atak K, Bokarev SI, Gotz M, Golnak R, Lange KM, Engel N, Dantz M, Suljoti E, Kühn O, Aziz EF (2013) Nature of the chemical bond of aqueous Fe2+ probed by soft X-ray spectroscopies and ab initio calculations. J Phys Chem B 117(41):12613–12618

7. Bagus PS, Nelin CJ, Ilton ES, Sassi MJ, Rosso KM (2017) Analysis of X-ray adsorption edges: L2,3 edge of $FeCl_4^-$. J Chem Phys 147(22):224306. https://doi.org/10.1063/1.5006223
8. Bernadotte S, Atkins AJ, Jacob CR (2012) Origin-independent calculation of quadrupole intensities in X-ray spectroscopy. J Chem Phys 137(20):204106
9. Blomberg MR, Siegbahn PE (1997) A comparative study of high-spin manganese and iron complexes. Theor Chem Acc 97(1–4):72–80
10. Bokarev SI, Dantz M, Suljoti E, Kühn O, Aziz EF (2013) State-dependent electron delocalization dynamics at the solute-solvent interface: soft-x-ray absorption spectroscopy and ab initio calculations. Phys Rev Lett 111(8):083002–083007
11. Bokarev SI, Khan M, Abdel-Latif MK, Xiao J, Hilal R, Aziz SG, Aziz EF, Kühn O (2015) Unraveling the electronic structure of photocatalytic manganese complexes by L-edge X-ray spectroscopy. J Phys Chem C 119(33):19192–19200
12. Booth GH, Thom AJW, Alavi A (2009) Fermion monte carlo without fixed nodes: a game of life, death, and annihilation in slater determinant space. J Chem Phys 131(5):054106. https://doi.org/10.1063/1.3193710
13. Bunău O, Joly Y (2012) Full potential x-ray absorption calculations using time dependent density functional theory. J Phys: Condens Matter 24(21):215502. http://stacks.iop.org/0953-8984/24/i=21/a=215502
14. Cederbaum LS, Domcke W, Schirmer J (1980) Many-body theory of core holes. Phys Rev A 22:206–222. https://doi.org/10.1103/PhysRevA.22.206
15. Chan GKL, Sharma S (2011) The density matrix renormalization group in quantum chemistry. Annu Rev Phys Chem 62(1):465–481. https://doi.org/10.1146/annurev-physchem-032210-103338 (pMID: 21219144)
16. Chantzis A, Kowalska JK, Maganas D, DeBeer S, Neese F (2018) Ab initio wave function-based determination of element specific shifts for the efficient calculation of x-ray absorption spectra of main group elements and first row transition metals. J Chem Theory Comput 14(7):3686–3702. https://doi.org/10.1021/acs.jctc.8b00249 (pMID: 29894196)
17. Coriani S, Christiansen O, Fransson T, Norman P (2012) Coupled-cluster response theory for near-edge x-ray-absorption fine structure of atoms and molecules. Phys Rev A 85:022507. https://link.aps.org/doi/10.1103/PhysRevA.85.022507
18. Cossi M, Barone V (2000) Solvent effect on vertical electronic transitions by the polarizable continuum model. J Chem Phys 112(5):2427–2435
19. Cramer S, DeGroot F, Ma Y, Chen C, Sette F, Kipke C, Eichhorn D, Chan M, Armstrong W (1991) Ligand field strengths and oxidation states from manganese L-edge spectroscopy. J Am Chem Soc 113(21):7937–7940
20. De Groot F (2001) High-resolution x-ray emission and x-ray absorption spectroscopy. Chem Rev 101(6):1779–1808
21. Douglas M, Kroll NM (1974) Quantum electrodynamical corrections to the fine structure of helium. Ann Phys 82(1):89–155
22. Ekström U, Norman P, Carravetta V, Ågren H (2006) Polarization propagator for x-ray spectra. Phys Rev Lett 97:143001. http://link.aps.org/doi/10.1103/PhysRevLett.97.143001
23. Engel N, Bokarev SI, Suljoti E, Garcia-Diez R, Lange KM, Atak K, Golnak R, Kothe A, Dantz M, Kühn O, Aziz EF (2014) Chemical bonding in aqueous ferrocyanide: experimental and theoretical X-ray spectroscopic study. J Phys Chem B 118(6):1555–1563
24. Forsberg N, Malmqvist PÅ (1997) Multiconfiguration perturbation theory with imaginary level shift. Chem Phys Lett 274(1):196–204
25. Gel'mukhanov F, Ågren H (1999) Resonant x-ray raman scattering. Phys Rep 312(3–6):87–330
26. Ghigo G, Roos BO, Malmqvist PÅ (2004) A modified definition of the zeroth-order Hamiltonian in multiconfigurational perturbation theory (CASPT2). Chem Phys Lett 396(1):142–149
27. Golnak R, Bokarev SI, Seidel R, Xiao J, Grell G, Atak K, Unger I, Thürmer S, Aziz SG, Kühn O et al (2016a) Joint analysis of radiative and non-radiative electronic relaxation upon X-ray irradiation of transition metal aqueous solutions. Sci Rep 6:24659
28. Golnak R, Xiao J, Atak K, Unger I, Seidel R, Winter B, Aziz EF (2016b) Undistorted X-ray absorption spectroscopy using s-core-orbital emissions. J Phys Chem A 120(18):2808–2814

29. Grell G, Bokarev SI, Winter B, Seidel R, Aziz EF, Aziz SG, Kühn O (2015) Multi-reference approach to the calculation of photoelectron spectra including spin-orbit coupling. J Chem Phys 143(7):074104
30. Grell G, Bokarev SI, Winter B, Seidel R, Aziz EF, Aziz SG, Kühn O (2016) Erratum: multi-reference approach to the calculation of photoelectron spectra including spin-orbit coupling. J Chem Phys 143:074104 (2015). J Chem Phys 145(8):089901
31. de Groot F (2005) Multiplet effects in X-ray spectroscopy. Coord Chem Rev 249(1):31–63
32. Guo M, Källman E, Sørensen LK, Delcey MG, Pinjari RV, Lundberg M (2016a) Molecular orbital simulations of metal 1s2p resonant inelastic X-ray scattering. J Phys Chem A 120(29):5848–5855
33. Guo M, Sørensen LK, Delcey MG, Pinjari RV, Lundberg M (2016b) Simulations of iron K pre-edge X-ray absorption spectra using the restricted active space method. Phys Chem Chem Phys 18(4):3250–3259
34. Guo M, Källman E, Pinjari RV, Couto RC, Sørensen LK, Lindh R, Pierloot K, Lundberg M (2019) Fingerprinting electronic structure of heme iron by ab initio modeling of metal L-edge X-ray absorption spectra. J Chem Theory Comput 15(1):477–489. https://doi.org/10.1021/acs.jctc.8b00658
35. Hess BA (1986) Relativistic electronic-structure calculations employing a two-component no-pair formalism with external-field projection operators. Phys Rev A 33(6):3742
36. Hocking RK, Wasinger EC, de Groot FM, Hodgson KO, Hedman B, Solomon EI (2006) Fe L-edge XAS studies of $K_4[Fe(CN)_6]$ and K3$[Fe(CN)_6]$: a direct probe of back-bonding. J Am Chem Soc 128(32):10442–10451
37. Holmes AA, Tubman NM, Umrigar CJ (2016) Heat-bath configuration interaction: an efficient selected configuration interaction algorithm inspired by heat-bath sampling. J Chem Theory Comput 12(8):3674–3680. https://doi.org/10.1021/acs.jctc.6b00407 (pMID: 27428771)
38. Jan W, Michael W, Andreas D (2014) Calculating core-level excitations and x-ray absorption spectra of medium-sized closed-shell molecules with the algebraic-diagrammatic construction scheme for the polarization propagator. J Comput Chem 35(26):1900–1915. https://doi.org/10.1002/jcc.23703, https://onlinelibrary.wiley.com/doi/abs/10.1002/jcc.23703
39. Jay RM, Norell J, Eckert S, Hantschmann M, Beye M, Kennedy B, Quevedo W, Schlotter WF, Dakovski GL, Minitti MP, Hoffmann MC, Mitra A, Moeller SP, Nordlund D, Zhang W, Liang HW, Kunnus K, Kubiek K, Techert SA, Lundberg M, Wernet P, Gaffney K, Odelius M, Föhlisch A (2018) Disentangling transient charge density and metalligand covalency in photoexcited ferricyanide with femtosecond resonant inelastic soft X-ray scattering. J Phys Chem Lett 9(12):3538–3543. https://doi.org/10.1021/acs.jpclett.8b01429 (pMID: 29888918)
40. Jensen HJA, Jørgensen P, Ågren H (1987) Efficient optimization of large scale MCSCF wave functions with a restricted step algorithm. J Chem Phys 87(1):451–466
41. Johansson MP, Blomberg MR, Sundholm D, Wikström M (2002) Change in electron and spin density upon electron transfer to haem. Biochim Biophys Acta-Bioenerg 1553(3):183–187
42. Josefsson I, Kunnus K, Schreck S, Föhlisch A, de Groot F, Wernet P, Odelius M (2012) Ab initio calculations of X-ray spectra: atomic multiplet and molecular orbital effects in a multiconfigurational SCF approach to the L-Edge spectra of transition metal complexes. J Phys Chem Lett 3(23):3565–3570. https://doi.org/10.1021/jz301479j
43. Klooster R, Broer R, Filatov M (2012) Calculation of x-ray photoelectron spectra with the use of the normalized elimination of the small component method. Chem Phys 395:122–127
44. Kroll T, Hadt RG, Wilson SA, Lundberg M, Yan JJ, Weng TC, Sokaras D, Alonso-Mori R, Casa D, Upton MH, Hedman B, Hodgson KO, Solomon EI (2014) Resonant inelastic X-ray scattering on ferrous and ferric bis-imidazole porphyrin and cytochrome c: nature and role of the axial methionine-Fe bond. J Am Chem Soc 136(52):18087–18099
45. Kroll T, Lundberg M, Solomon EI (2016) X-ray absorption and RIXS on coordination complexes. In: Van Bokhoven JA, Lamberti C (eds) X-ray absorption and X-ray emission spectroscopy: theory and applications. Wiley, Chichester, pp 407–435
46. Kubin M, Kern J, Gul S, Kroll T, Chatterjee R, Lchel H, Fuller FD, Sierra RG, Quevedo W, Weniger C, Rehanek J, Firsov A, Laksmono H, Weninger C, Alonso-Mori R, Nordlund DL,

Lassalle-Kaiser B, Glownia JM, Krzywinski J, Moeller S, Turner JJ, Minitti MP, Dakovski GL, Koroidov S, Kawde A, Kanady JS, Tsui EY, Suseno S, Han Z, Hill E, Taguchi T, Borovik AS, Agapie T, Messinger J, Erko A, Föhlisch A, Bergmann U, Mitzner R, Yachandra VK, Yano J, Wernet P (2017) Soft x-ray absorption spectroscopy of metalloproteins and high-valent metal-complexes at room temperature using free-electron lasers. Struct Dyn 4(5):054307. https://doi.org/10.1063/1.4986627

47. Kubin M, Guo M, Ekimova M, Baker ML, Kroll T, Källman E, Kern J, Yachandra VK, Yano J, Nibbering ET, Lundberg M, Wernet P (2018) Direct determination of absolute absorption cross sections at the L-Edge of dilute Mn complexes in solution using a transmission flatjet. Inorg Chem 57(9):5449–5462
48. Kubin M, Guo M, Ekimova M, Källman EJ, Kern J, Yachandra VK, Yano J, Nibbering ET, Lundberg M, Wernet P (2018) Cr L-edge X-ray absorption spectroscopy of CrIII $(acac)_3$ in solution with measured and calculated absolute absorption cross sections. J Phys Chem B. https://doi.org/10.1021/acs.jpcb.8b04190
49. Kubin M, Guo M, Kroll T, Löchel H, Källman E, Baker ML, Mitzner R, Gul S, Kern J, Föhlisch A, Erko A, Bergmann U, Yachandra VK, Yano J, Lundberg M, Wernet P (2018) Probing the oxidation state of transition metal complexes: a case study on how charge and spin densities determine Mn L-Edge X-ray absorption energies. Chem Sci. https://doi.org/10.1039/C8SC00550H
50. Kubin M, Kern J, Guo M, Källman E, Mitzner R, Yachandra VK, Lundberg M, Yano J, Wernet P (2018) X-ray-induced sample damage at the Mn L-edge: a case study for soft X-ray spectroscopy of transition metal complexes in solution. Phys Chem Chem Phys 20:16817–16827
51. Kunnus K, Josefsson I, Rajkovic I, Schreck S, Quevedo W, Beye M, Grbel S, Scholz M, Nordlund D, Zhang W, Hartsock RW, Gaffney KJ, Schlotter WF, Turner JJ, Kennedy B, Hennies F, Techert S, Wernet P, Odelius M, Fhlisch A (2016) Anti-stokes resonant x-ray Raman scattering for atom specific and excited state selective dynamics. New J Phys 18(10):103011. http://stacks.iop.org/1367-2630/18/i=10/a=103011
52. Kunnus K, Josefsson I, Rajkovic I, Schreck S, Quevedo W, Beye M, Weniger C, Grbel S, Scholz M, Nordlund D, Zhang W, Hartsock RW, Gaffney KJ, Schlotter WF, Turner JJ, Kennedy B, Hennies F, de Groot FMF, Techert S, Odelius M, Wernet P, Fhlisch A (2016) Identification of the dominant photochemical pathways and mechanistic insights to the ultrafast ligand exchange of $Fe(CO)_5$ to $Fe(CO)_4EtOH$. Struct Dyn 3(4):043204. https://doi.org/10.1063/1.4941602
53. Kunnus K, Zhang W, Delcey MG, Pinjari RV, Miedema PS, Schreck S, Quevedo W, Schroeder H, Föhlisch A, Gaffney KJ, Lundberg M, Odelius M, Wernet P (2016) Viewing the valence electronic structure of ferric and ferrous hexacyanide in solution from the Fe and cyanide perspectives. J Phys Chem B 120(29):7182–7194. https://doi.org/10.1021/acs.jpcb.6b04751
54. Liang W, Fischer SA, Frisch MJ, Li X (2011) Energy-specific linear response TDHF/TDDFT for calculating high-energy excited states. J Chem Theory Comput 7(11):3540–3547. https://doi.org/10.1021/ct200485x
55. List NH, Kauczor J, Saue T, Jensen HJA, Norman P (2015) Beyond the electric-dipole approximation: a formulation and implementation of molecular response theory for the description of absorption of electromagnetic field radiation. J Chem Phys 142(24):244111
56. List NH, Saue T, Norman P (2017) Rotationally averaged linear absorption spectra beyond the electric-dipole approximation. Mol Phys 115(1–2):63–74
57. Liu Y, Persson P, Sundström V, Wärnmark K (2016) Fe N-heterocyclic carbene complexes as promising photosensitizers. Acc Chem Res 49(8):1477–1485
58. Lundberg M, Kroll T, DeBeer S, Bergmann U, Wilson SA, Glatzel P, Nordlund D, Hedman B, Hodgson KO, Solomon EI (2013) Metal-ligand covalency of iron complexes from high-resolution resonant inelastic X-ray scattering. J Am Chem Soc 135(45):17121–17134. https://doi.org/10.1021/ja408072q (pMID: 24131028)
59. Ma D, Li Manni G, Gagliardi L (2011) The generalized active space concept in multiconfigurational self-consistent field methods. J Chem Phys 135(4):044128. https://doi.org/10.1063/1.3611401

60. Maganas D, DeBeer S, Neese F (2014) Restricted open-shell configuration interaction cluster calculations of the L-Edge X-ray absorption study of TiO_2 and CaF_2 solids. Inorg Chem 53(13):6374–6385. https://doi.org/10.1021/ic500197v (pMID: 24871209)
61. Malmqvist PÅ (1986) Calculation of transition density matrices by nonunitary orbital transformations. Int J Quantum Chem 30(4):479–494. https://doi.org/10.1002/qua.560300404
62. Malmqvist PÅ, Rendell A, Roos BO (1990) The restricted active space self-consistent-field method, implemented with a split graph unitary group approach. J Phys Chem 94(14):5477–5482. https://doi.org/10.1021/j100377a011
63. Malmqvist PÅ, Roos BO, Schimmelpfennig B (2002) The restricted active space (RAS) state interaction approach with spin-orbit coupling. Chem Phys Lett 357(3):230–240
64. Malmqvist PÅ, Pierloot K, Shahi ARM, Cramer CJ, Gagliardi L (2008) The restricted active space followed by second-order perturbation theory method: theory and application to the study of CuO_2 and Cu_2O_2 systems. J Chem Phys 128(20):204109
65. Nakata A, Imamura Y, Otsuka T, Nakai H (2006) Time-dependent density functional theory calculations for core-excited states: assessment of standard exchange-correlation functionals and development of a novel hybrid functional. J Chem Phys 124(9):094105. https://doi.org/10.1063/1.2173987
66. Norell J, Jay RM, Hantschmann M, Eckert S, Guo M, Gaffney KJ, Wernet P, Lundberg M, Föhlisch A, Odelius M (2018) Fingerprints of electronic, spin and structural dynamics from resonant inelastic soft X-ray scattering in transient photo-chemical species. Phys Chem Chem Phys 20(10):7243–7253
67. Norman P, Dreuw A (2018) Simulating X-ray spectroscopies and calculating core-excited states of molecules. Chem Rev 118(15):7208–7248
68. Penfold TJ, Reinhard M, Rittmann-Frank MH, Tavernelli I, Rothlisberger U, Milne CJ, Glatzel P, Chergui M (2014) X-ray spectroscopic study of solvent effects on the ferrous and ferric hexacyanide anions. J Phys Chem A 118(40):9411–9418
69. Pierloot K (2003) The CASPT2 method in inorganic electronic spectroscopy: from ionic transition metal to covalent actinide complexes. Mol Phys 101(13):2083–2094
70. Pierloot K, Phung QM, Domingo A (2017) Spin state energetics in first-row transition metal complexes: contribution of (3s3p) correlation and its description by second-order perturbation theory. J Chem Theory Comput 13(2):537–553. https://doi.org/10.1021/acs.jctc.6b01005 (pMID: 28005368)
71. Pinjari RV, Delcey MG, Guo M, Odelius M, Lundberg M (2014) Restricted active space calculations of L-edge X-ray absorption spectra: From molecular orbitals to multiplet states. J Chem Phys 141(12):124116
72. Pinjari RV, Delcey MG, Guo M, Odelius M, Lundberg M (2015) Erratum: restricted active space calculations of L-edge X-ray absorption spectra: from molecular orbitals to multiplet states. J Chem Phys 141:124116 (2014)]. J Chem Phys 142(6):069901
73. Pinjari RV, Delcey MG, Guo M, Odelius M, Lundberg M (2016) Cost and sensitivity of restricted active-space calculations of metal L-edge X-ray absorption spectra. J Comput Chem 37(5):477–486
74. Preuße M, Bokarev SI, Aziz SG, Kühn O (2016) Towards an ab initio theory for metal L-edge soft X-ray spectroscopy of molecular aggregates. Struct Dynam 3(6):062601
75. Roemelt M, Maganas D, DeBeer S, Neese F (2013) A combined DFT and restricted open-shell configuration interaction method including spin-orbit coupling: application to transition metal L-edge X-ray absorption spectroscopy. J Chem Phys 138(20):204101
76. Roos BO (1980) The complete active space SCF method in a fock-matrix-based super-CI formulation. Int J Quantum Chem 18(S14):175–189. https://doi.org/10.1002/qua.560180822
77. Roos BO, Lindh R, Malmqvist PÅ, Veryazov V, Widmark PO (2004) Main group atoms and dimers studied with a new relativistic ANO basis set. J Phys Chem A 108(15):2851–2858
78. Roos BO, Lindh R, Malmqvist PÅ, Veryazov V, Widmark PO (2005) New relativistic ANO basis sets for transition metal atoms. J Phys Chem A 109(29):6575–6579
79. Roos BO, Lindh R, Malmqvist PÅ, Veryazov V, Widmark PO (2016) Multiconfigurational quantum chemistry. Wiley, New York

80. van Schooneveld MM, Juhin A, Campos-Cuerva C, Schmitt T, de Groot FM (2013) Origin of low energy d-d excitations observed on wet chemically prepared cobalt bearing nanoparticles by 2p3d resonant X-ray emission spectroscopy. J Phys Chem C 117(27):14398–14407
81. Sørensen LK, Guo M, Lindh R, Lundberg M (2017) Applications to metal K pre-edges of transition metal dimers illustrate the approximate origin independence for the intensities in the length representation. Mol Phys 115(1–2):174–189
82. Sørensen LK, Kieri E, Srivastava S, Lundberg M, Lindh R (2019) Implementation of the exact semiclassical light-matter interaction using the Gauss-Hermite quadrature: A simple alternative to the multipole expansion. Phys Rev A 99(013419):1–11. https://doi.org/10.1103/PhysRevA.99.013419
83. Stein CJ, Reiher M (2016) Automated selection of active orbital spaces. J Chem Theory Comput 12(4):1760–1771. https://doi.org/10.1021/acs.jctc.6b00156 (pMID: 26959891)
84. Stener M, Fronzoni G, de Simone M (2003) Time dependent density functional theory of core electrons excitations. Chem Phys Lett 373(1):115–123. https://doi.org/10.1016/S0009-2614(03)00543-8
85. Stenrup M, Lindh R, Fdez Galván I (2015) Constrained numerical gradients and composite gradients: practical tools for geometry optimization and potential energy surface navigation. J Comput Chem 36(22):1698–1708
86. Suljoti E, Garcia-Diez R, Bokarev SI, Lange KM, Schoch R, Dierker B, Dantz M, Yamamoto K, Engel N, Atak K, Kuhn O, Bauer M, Rubensson JE, Aziz EF (2013) Direct observation of molecular orbital mixing in a solvated organometallic complex. Angew Chem Int Ed 52(37):9841–9844. https://doi.org/10.1002/anie.201303310, https://onlinelibrary.wiley.com/doi/abs/10.1002/anie.201303310
87. Tanaka A, Jo T (1994) Resonant 3d, 3p and 3s photoemission in transition metal oxides predicted at 2p threshold. J Phys Soc JPN 63(7):2788–2807
88. Thürmer S, Seidel R, Eberhardt W, Bradforth SE, Winter B (2011) Ultrafast hybridization screening in Fe3+ aqueous solution. J Am Chem Soc 133(32):12528–12535
89. Van Schooneveld M, DeBeer S (2015) A close look at dose: toward l-edge xas spectral uniformity, dose quantification and prediction of metal ion photoreduction. J Electron Spectrosc Relat Phenom 198:31–56
90. Wang H, Bokarev SI, Aziz SG, Kühn O (2017) Ultrafast spin-state dynamics in transition-metal complexes triggered by soft-X-ray light. Phys Rev Lett 118(2):023001
91. Wasinger EC, De Groot FM, Hedman B, Hodgson KO, Solomon EI (2003) L-edge X-ray absorption spectroscopy of non-heme iron sites: experimental determination of differential orbital covalency. J Am Chem Soc 125(42):12894–12906
92. Wernet P, Kunnus K, Schreck S, Quevedo W, Kurian R, Techert S, de Groot FM, Odelius M, Föhlisch A (2012) Dissecting local atomic and intermolecular interactions of transition-metal ions in solution with selective X-ray spectroscopy. J Phys Chem Lett 3(23):3448–3453
93. Wernet P, Kunnus K, Josefsson I, Rajkovic I, Quevedo W, Beye M, Schreck S, Gruebel S, Scholz M, Nordlund D, Zhang W, Hartsock RW, Schlotter WF, Turner JJ, Kennedy B, Hennies F, de Groot FMF, Gaffney KJ, Techert S, Odelius M, Foehlisch A (2015) Orbital-specific mapping of the ligand exchange dynamics of $Fe(CO)_5$ in solution. Nature 520(7545):78–81
94. Westre TE, Kennepohl P, DeWitt JG, Hedman B, Hodgson KO, Solomon EI (1997) A multiplet analysis of Fe K-edge 1s 3d pre-edge features of iron complexes. J Am Chem Soc 119(27):6297–6314
95. Wilson SA, Kroll T, Decreau RA, Hocking RK, Lundberg M, Hedman B, Hodgson KO, Solomon EI (2013) Iron L-Edge X-ray absorption spectroscopy of oxy-picket fence porphyrin: experimental insight into $Fe-O_2$ bonding. J Am Chem Soc 135(3):1124–1136
96. Yano J, Yachandra V (2014) Mn_4Ca cluster in photosynthesis: where and how water is oxidized to dioxygen. Chem Rev 114(8):4175–4205
97. Zobel JP, Nogueira JJ, Gonzalez L (2017) The ipea dilemma in caspt2. Chem Sci 8:1482–1499. https://doi.org/10.1039/C6SC03759C

Assessing Electronically Excited States of Cobalamins via Absorption Spectroscopy and Time-Dependent Density Functional Theory

Megan J. Toda, Pawel M. Kozlowski and Tadeusz Andruniów

Abstract In the field of B_{12} chemistry, absorption spectroscopy, hand in hand with computational modeling, has played an important role in describing electronically excited states of vitamin B_{12} derivatives, also known as cobalamins. This chapter focuses on the current understanding of absorption properties of cobalamins from both spectroscopic and computational points of views. The main emphasis is on methylcobalamin (MeCbl), adenosylcobalamin (AdoCbl), and cyanocobalamin (CNCbl). In addition, we will discuss some other unique derivatives including antivitamins, non-alkyl cobalamins, as well as reduced and super-reduced forms. Due to the complexity and the size of these systems, computational analysis is almost exclusively represented by density functional theory (DFT) and time-dependent DFT (TD-DFT) methods. Proper DFT functional choice is paramount in predicting electronic transitions and simulating the full spectrum reliably. At this juncture in the field of B_{12} chemistry, it is indisputable that the BP86 functional is the proper choice for the assessment of the electronically excited states of cobalamins.

1 Introduction

Vitamin B_{12} and its family of derivatives, otherwise known as cobalamins (Fig. 1), are among the most complex organometallic naturally occurring compounds. Cobalamins contain over one hundred eighty atoms. There are certain features of these

M. J. Toda · P. M. Kozlowski
Department of Chemistry, University of Louisville, 2320 South Brook Street, Louisville, KY 40292, USA
e-mail: megan.toda@louisville.edu

P. M. Kozlowski
e-mail: pawel.kozlowski@louisville.edu

T. Andruniów (✉)
Advanced Materials Engineering and Modelling Group, Department of Chemistry, Wroclaw University of Science and Technology, Wyb. Wyspianskiego 27, 50-370 Wrocław, Poland
e-mail: tadeusz.andruniow@pwr.edu.pl

E. Broclawik et al. (eds.), *Transition Metals in Coordination Environments*, Challenges and Advances in Computational Chemistry and Physics 29,
https://doi.org/10.1007/978-3-030-11714-6_8

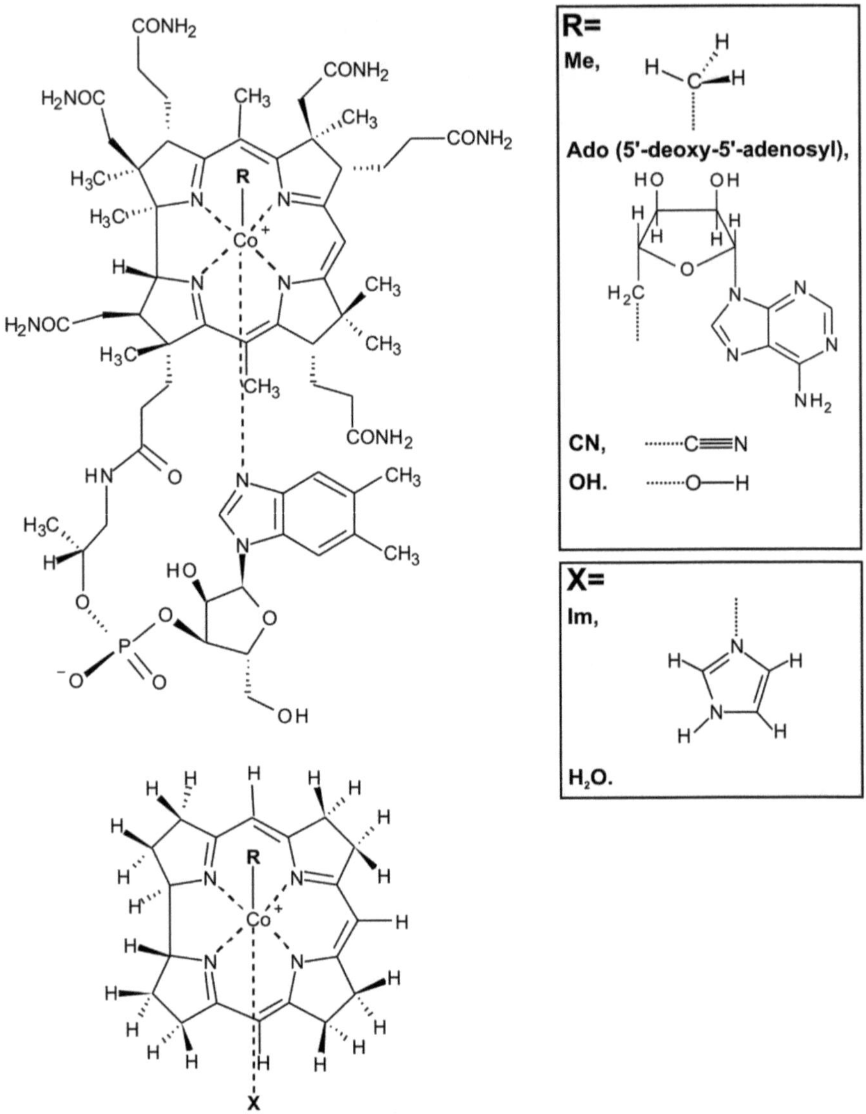

Fig. 1 Molecular structure of cobalamins (top) and truncated models (bottom) used in DFT and TD-DFT calculations. Reprinted with permission from [42]. Copyright 2016 Royal Society of Chemistry

compounds that remain constant across the entire family of derivatives. All B_{12} derivatives contain a corrin macrocycle with a central cobalt (Co) ion that is equatorially ligated to four nitrogens of the corrin ring. There are several side chains attached to the corrin ring that include methyl, acetamide, and propionamide groups. In the Co(III) form, the metal center is axially ligated to an upper and lower ligand. The lower ligand is a dimethylbenzimidazole (DBI) base in what is known as the base-on form. The upper axial ligand is variable and can be used to distinguish cobalamins from each other. For the case of vitamin B_{12} (CNCbl), also known as cyanocobalamin, the upper ligand is a cyano group (CN). There are several other upper axial ligands that are known to occur naturally as well as various synthetic derivatives. Herein, we will focus on a few of them, such as the free base corrin (FBC) and the biochemically relevant CNCbl, methylcobalamin (MeCbl), and adenosylcobalamin (AdoCbl). In addition, we will discuss some other unique derivatives including antivitamins B_{12}, non-alkyl cobalamins, as well as reduced and super-reduced forms. These will all be discussed in later sections of this chapter.

The field of B_{12} chemistry is diverse and has a long and rich history. The discovery of vitamin B_{12}, the elucidation of its role in metabolism, structural characterization, total synthesis, as well as the discovery of B_{12}-dependent enzymes occurred in distinct phases over more than 100 years. It was the subject of two separate Nobel Prizes. Vitamin B_{12} was discovered in 1926 by Georg R. Minot and William P. Murphy. In 1934, both scientists, as well as George Whipple, won a Nobel Prize for their work in the treatment of pernicious anemia. Vitamin B_{12} was not officially isolated until 1948. The vitamin was isolated by two groups simultaneously and was crystallized and characterized in the laboratory of Dorothy Hodgkin in 1954–1955. Her contributions to X-ray crystallography, through the determination of the structures of important biochemical substances, including vitamin B_{12}, were recognized when she was awarded the Nobel Prize in chemistry in 1964.

There have been a number of extensive reviews and a few books that summarize various aspects of B_{12}-related chemistry and biochemistry [4–9, 16, 25, 67, 69, 75, 77, 80, 89, 95]. However, a synopsis concerning the role of the methods of computational chemistry in advancing this field beyond the twentieth century has not yet been completed. This book chapter aims to shed light on the use of density functional theory (DFT) and time-dependent density functional theory (TD-DFT) [10, 11, 74] in understanding the electronically excited states of these compounds in the context of absorption (Abs) spectroscopy. Herein, we will follow an overview of B_{12} chemistry with a discussion of the importance of electronically excited states and early efforts to understand these states. For thorough discussions of the electronic spectra of B_{12}, we encourage the curious reader to refer to Giannotti [24] and Pratt's [68] contributions. We will also introduce DFT and TD-DFT and their application to electronic structure calculations. Important conclusions from benchmark studies of various cobalamins for both ground- and excited-state properties will be summarized. The bulk of this chapter will be devoted to discussing specific cobalamins and what is known about their Abs properties from both spectroscopic and computational point of views.

2 Electronic and Structural Properties of Cobalamins

As has been briefly outlined in the Introduction, cobalamins typically exist as low-spin octahedral Co(III) complexes with d^6 electronic configuration of Co. The splitting of Co *d* orbitals is more in line with axially perturbed square planar geometry, with local D_{4h} symmetry. The equatorial corrin ligand resembles the porphyrin macrocycle, while the upper and lower positions can be occupied by a variety of different ligands. Although cobalamins tend to exist in the Co(III) form, other coordination environments are important to consider. These other low oxidation states include the Co(II) and Co(I) species, which have roles in the context of enzymatic catalysis and in various photochemical reactions. The Co(II) form, where paramagnetic Co center has d^7 electronic configuration, is produced by one-electron reduction of Co(III)-based cobalamins and is a five-coordinate species. It can be also generated upon homolytic cleavage of the Co–C bond, thus forming cob(II)alamin. Cob(II)alamin is a common intermediate formed in enzymatic reactions catalyzed by coenzyme B_{12} (aka AdoCbl). In all (AdoCbl)-dependent enzymes, the presence of Co(II) can be detected via electron paramagnetic resonance (EPR) spectroscopy. Alternatively, Co(II)-based species can also be generated photochemically when the Co–C bond is cleaved with light. Regardless of environment, solution versus enzyme, homolytic cleavage of the Co–C bond will lead to a radical pair (RP) formation. One-electron reduction of a Co(II)-based cobalamin leads to a low-spin Co(I) compound, which has complex electronic structure involving mixture of Co(d^8) and Co(d^7)/corrin(π^*)^1 configurations. The four-coordinate Co(I) form is known as the super-reduced form of vitamin B_{12} and has also been deemed a super-nucleophile due to its high reactivity. Heterolytic cleavage of the Co–C bond in methylcobalamin (MeCbl)-dependent enzymes leads to a Co(I) species. The resulting cob(I)alamin plays key mechanistic roles in B_{12}-dependent methyltransferases such as methionine synthase (MetH).

It is worthy to further discuss the importance of the axial base, namely DBI. Co(III)-based cobalamins can either exist in the base-on or base-off forms. In solution, the attachment of the axial base can be modulated by pH. So, in highly acidic environments such as pH 2, the axial base will be detached when the DBI is protonated. The Co will remain Co(III) by becoming weakly coordinated to a water molecule, thus adopting the base-off form. Alternatively, the lower ligand may be water or can be a histidine (His) residue in certain enzymatic environments. In cases where the axial base becomes detached in an enzyme, a His residue will coordinate to the Co. This is known as the base-off/His-on form, and the Co will coordinate to the nitrogen of the histidine. In the base-on form, whether in solution or in an enzyme, the Co will coordinate to the nitrogen of the benzimidazole ring.

3 Importance of Abs, CD, and MCD Spectroscopy

It has been known for a long time that the features of Abs spectra of cobalamins arise from the oxidation state of the Co as well as the nature of the upper and lower axial ligands. The changes in the spectrum that accompany changes in oxidation and ligation have been intensively scrutinized. Again, we refer to Giannotti [24] and Pratt's [68] contributions for a more detailed discussion. Herein, for convenience of presentation, it is important to mention characteristic features of Abs spectra of several cobalamins, such as CNCbl, AdoCbl, cob(II)alamin, or cob(I)alamin as shown in Fig. 2 [83]. The features in the UV-visible spectra of Co corrinoids are usually designated by, α, β, γ, and δ. The α and β bands are typically referred to as the α/β region, and these are located in the visible region (ε ~8,000 – 10,000 M^{-1} cm^{-1}). The γ band is found in the UV (ε ~25,000 M^{-1} cm^{-1}) and can also be referred to as the Soret band. The δ band occurs in the region less than 300 nm. In addition, low-intensity bands around 400 nm are referred to as D and E as these are separate transitions from those designated with the Greek letters. As a side note, the naming system for the Abs spectra of Co corrinoids is based on the Abs nomenclature used for porphyrins.

Cobalamins' Abs spectra can be designated as normal or anomalous [20] and sometimes as typical or unique [85]. Normal Abs spectra are marked by an α band that is more intense than the β band. For anomalous spectra, this is reversed and the β band is more intense than the α band. However, the major difference between the normal and anomalous Abs spectra is with the γ band. In normal spectra, the γ band is quite obviously the Soret band and is the most intense feature. An intense γ band is not present in anomalous Abs spectra. Rather, transitions that would correspond to an intense γ band in a normal spectrum are distributed over a larger range of wavelengths in the UV region. Notably, the Abs spectra for CNCbl and aquocobalamin (H_2OCbl^+)

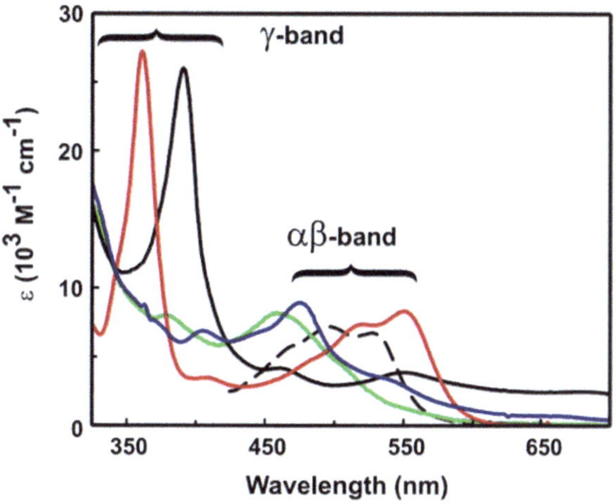

Fig. 2 Absorption spectra of cob(I)alamin (black), cob(II)alamin (blue), CNCbl (red), base-off AdoCbl (green), and cyanocobinamide (gray dashed) measured in water. . Reprinted with permission from [83]. Copyright 2006 American Chemical Society

are representative of the normal type, while MeCbl and AdoCbl Abs spectra can be classified as anomalous. The spectra of some lesser known derivatives, NO_2Cbl, SCNCbl, ClCbl, N_3Cbl, CNCbl, and (i-Pr)$_2$-phosphitoCbl, could be classified as typical and those of SO_3Cbl, SeCNCbl, MeCbl, AdoCbl and isoamylCbl as unique.

A switch of the axial group can result in a change in the spectra of the cobalamin. This can be further explained through the lens of electronic structure theory, but there are two different explanations of why there is a change in the spectra just by changing the axial group. One was proposed by Brunold and co-workers [85], while other by Marques and co-workers, [66] each proposing different reasons for the change.

Brunold and co-workers [85] have indicated that the unique spectra of cobalamins with an alkyl ligand in the upper axial position arises from a combination of two effects. The first is that there is a net increase in the number of donor molecular orbitals available for electronic transitions. The second is that there is a mixing between Co 3d and corrin π orbitals. The increase of intensity of the β band relative to the α band from typical to unique cobalamin Abs spectra is a result of the additional excited-state distortion (the corrin-based HOMO has Co–C σ-bonding character which is lost upon HOMO$\rightarrow$LUMO excitation distorting along both coordinates in the corresponding excited state).

According to Marques and co-workers, [66] there is actually no fundamental difference between the typical and unique spectra of cobalamins, but the reason for the difference is the relative position of the γ-region components of the cobalamin spectra. As the donor power of the axial ligand increases, the electronic transitions of cobalamins move to longer wavelengths. The γ-region moves apart in response to the increase in donation of electronic density from the axial ligand, which causes the difference. Also, the bond length increases as the donor power of the axial ligand increases.

The differences in Abs spectra as they relate to oxidation state of Co and the nature of the upper axial ligand can be easily understood upon inspection of Fig. 2. CNCbl exhibits the normal spectrum with a very intense γ band and a prominent α/β band. The Abs spectrum for the base-off AdoCbl is markedly different. It does not contain the Soret band, and the α/β region is blue shifted in comparison with the α/β band of CNCbl. It has been summarized that strong σ-donor ligands produce a red-shifted γ band, while weak σ-donor ligands produce γ bands that are blue shifted (Fig. 3) [91]. Clearly, the nature of the axial ligands contributes to the Abs spectra, providing evidence that $\pi\rightarrow\pi^*$ transitions from the corrin macrocycle are not the only contributor to spectral features. Further differences are evident when comparing the spectra of cob(III)alamin species like CNCbl and AdoCbl to other relevant oxidation states like Co(II) and Co(I). The cob(I)alamin spectrum is similar to that of CNCbl with respect to the intensity of the γ band, although this band is slightly red shifted compared to CNCbl. Although the Co(I) species does not contain axial ligands, the intensity of the γ band can be attributed to the lone pair of electrons acting as a sort of strong ligand which shifts the γ band further to the red [91]. The major difference between cob(I)alamin and CNCbl is with the intensity of the α/β band. The α/β band is much less pronounced for cob(I)alamin. The blue-shifted α/β

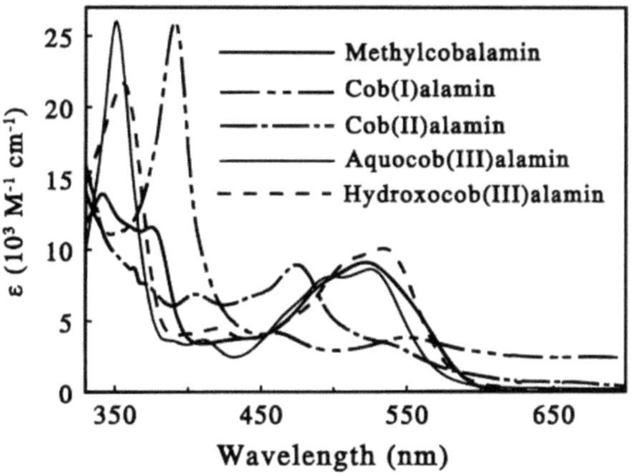

Fig. 3 Steady-state spectra of methylcobalamin, cob(I)alamin, cob(II)alamin, hydroxocob(III)alamin, and aquocob(III)alamin. Reprinted with permission from [91]. Copyright 1998 American Chemical Society

band of cob(II)alamin is nearly as intense as the α/β band for CNCbl. Additionally, there is a strong band around 312 nm in cob(II)alamin.

Circular dichroism (CD) and magnetic CD (MCD) can also be used to provide additional insights into the electronic structure of cobalamins. These are valuable tools to understand electronic structure, although the interpretation of their spectra as it pertains to cobalamins is not as straightforward as for Abs spectra. Generally, there is a correlation between the wavelengths of the Abs bands and the bands that arise in CD spectra of cobalamins [16]. CD and MCD spectra tend to show additional bands and other features, in particular in the lowest energy part.

In addition to the insights about Abs spectra described above, Brunold and co-workers also applied spectral decomposition of CD and MCD spectra to further understand the electronically excited states of several cobalamins including MeCbl, AdoCbl, Ado-cobinamide ($AdoCbi^+$), H_2OCbl^+, and CNCbl [85]. $AdoCbi^+$ is a cob(III)alamin with the axial base detached and a water molecule coordinated to the Co in its place. All of the experimental Abs, CD, and MCD states were systematically fit to the fewest number of Gaussian bands in order to resolve the major electronic transitions. There are a few important details from the CD and MCD spectroscopy to summarize here. For H_2OCbl^+, there is a weak feature observed in the CD spectrum that is lower in energy than the α band, but this feature does not appear to correlate to a band in the Abs spectrum. This observation led to the conclusion that for H_2OCbl^+, the corresponding transition would be magnetic dipole like and involve a Co 3d orbital. TD-DFT calculations suggest that this weak feature in the CD spectrum is a corrin $\pi \rightarrow$ Co $3d_{z^2}$ charge-transfer transition originating from the HOMO. This feature is not observed in the CD spectrum of CNCbl. The fitting of the Abs, CD, and MCD spectra of MeCbl indicate that two sets of four total bands contribute to the α/β region. The oppositely signed sets of bands in the MCD spectrum for MeCbl indicates that the two sets are overlapping vibrational progressions that are associated with two distinct electronic transitions. CD and MCD spectroscopy certainly can provide additional insights to electronically excited states of cobalamins, but Abs spectra is widely more understood to date.

4 Transient Absorption Spectroscopy

While UV-visible spectroscopy can be used to identify the oxidation state and the axial ligation associated with cobalamins, transient absorption spectroscopy (TAS) can be used to characterize the intermediate states involved in the photolysis mechanism. From several of these time-resolved studies, it is apparent that the photolysis mechanism related to the cleavage of the Co–C bond is dependent on the nature of the upper axial ligand and the solvation environment [83]. Generally speaking, non-alkyl cobalamins such as CNCbl are photostable. The important alkyl B_{12} cofactors, such as MeCbl and AdoCbl, are photolabile, and photolysis results in cob(II)alamin and the corresponding alkyl radical pairs (RP).

5 Early Attempts to Analyze and Assign Electronically Excited States

In the field of B_{12} chemistry, Abs spectroscopy has played an important role in understanding the electronically excited states of cobalamins. There are common features in Abs spectra across the family of cobalamin derivatives. These were mentioned briefly in Sect. 3. The α band in the region of 420–600 nm has been associated with π transitions related to the corrin ring. The β band at 550 nm has been typically referred to as vibrational fine structure. The combined α/β band has been attributed to the forbidden $\pi_7 \rightarrow \pi_8^*$ transitions based on early studies by Eckert and Kuhn [17]. The intense γ band transitions between 350–420 nm correspond to π transitions that are symmetry allowed. The antisymmetric forbidden transitions are designated as C and D. The δ bands (300–330 nm) are of lower intensity than the Soret band and have been considered as γ vibrational bands. In B_{12} Volume 1: Chemistry, C. Giannotti [24] indicated five transition types that would be present in cobalamins including:

1. d-d transitions that are specific to the Co ion
2. Internal transitions corresponding to the equatorial corrin ligand
3. Internal transitions of the axial ligand
4. Charge-transfer transitions from the corrin ring to Co or vice versa
5. Charge-transfer transitions from an axial ligand to Co or vice versa.

As spectroscopic techniques and interpretation from simulations have progressed, the traditional assignment of peaks in Abs spectra of cobalamins has been further refined. It was once generally accepted that Abs spectra of cobalamins can be interpreted based on the electronic transitions associated entirely with the corrin ring.

This conclusion was based on studies of a FBC, i.e., a cobalamin containing no Co, which will be discussed more thoroughly later.

As cobalamins have become more intensely scrutinized experimentally and computationally, a number of inconsistencies have been noticed. Consider MeCbl as an example of cobalamin with an anomalous spectrum, with the α band of lesser intensity than the β band. Traditionally, the α/β band has been viewed as arising from $\pi \rightarrow \pi^*$ excitations associated with vibrational progression of the C=C stretching of the corrin ring. In other words, the α/β band has been interpreted as a single $S_0 \rightarrow S_1$ electronic transition. TAS has indicated that the S_1 state in MeCbl results from metal-to-ligand charge-transfer (MLCT) transitions. Wave function-based ab initio methods confer the MLCT description emphasizing importance of Co d orbitals [39]. As shall be discussed later, hybrid functionals in the TD-DFT framework agree with the earlier interpretation of an S_1 state arising from $\pi \rightarrow \pi^*$ transitions, while gradient-corrected pure functionals give a description consistent with the MLCT assignment pointing out the importance of the Co d orbitals. This is just one example of how recent developments in computations have led to a more complete understanding of cobalamin spectroscopy, and this will be further noted in Sect. 7 where Abs spectra will be discussed for several cobalamins.

Beyond the theoretical interpretations, there are several experimental considerations that have led to a deeper understanding of cobalamin Abs spectra. Various parameters can affect the position of Abs bands including temperature, solvation environment, axial ligation, and modifications to the corrin ring, and these can be summarized as follows. A reduction in temperature leads to a sharpening of all the bands. Cooling also leads to an increase in the intensity of the α band in comparison to the β band. Increases in temperature lead to the displacement of the DBI base. There is a correlation between the solvent polarity parameter and the absorption maxima of the α/β band. More simply put, the intensity of the α/β band is affected by the solvation environment. Solvent also plays an important role in terms of time-resolved TAS. Particularly, cage dynamics in photolytic processes are solvent-dependent, even for cobalamins [94]. It has been shown that cage escape for RP occurs over various time scales based on solvent fluidity and radical size [87]. The size of the upper axial ligand does not only have a determinate effect on cage escape time scale, but the nature of this ligand also effects the static Abs spectrum. Inherently, there will be changes in properties of cobalamins based on the σ-donor strength of the upper axial ligand. The σ-donor strength is the amount of negative charge donated to the Co ion through the Co-R_{ax} bond. Differences in σ-donor strength affect Abs spectra. Shifts of the α band to longer wavelengths should be proportional to the electron-donating ability of the axial ligand. In addition, the wavelength of the γ band shifts with electronegativity of the upper axial ligand [16]. The γ band will shift to shorter wavelengths with more electronegative ligands like CN^-, whereas less electronegative ligands like CH_3 will shift the γ band to longer wavelengths. This can be essentially simplified as follows: The greater the σ-donor strength of the axial ligand, the longer the wavelength of the γ band. Furthermore, Abs spectra are less altered when changes are made to the side chains of the corrin macrocycle when compared to the changes noticed in axial ligation. However, more noticeable changes are apparent when substituents of the conjugated ring are changed [16].

6 Importance of Electronically Excited States: Relevance of DFT and TD-DFT in Electronic Structure Calculations

The ground- and excited-state properties of cobalamins can be represented well by using DFT and TD-DFT methods, respectively. Although wave function-based methods are more accurate, they are simply far too expensive to apply to systems as large as cobalamins. Alternatively, DFT is a method not based on the wave function but is based on inhomogeneous electron gas, typically referred to as electron density. DFT efficiently scales with system size and is applicable to large systems. TD-DFT can be viewed as an extension of DFT but is used to study properties related to time-dependent potentials. Analogous to the Hohenberg–Kohn theorem in DFT is the Runge–Gross theorem in TD-DFT [74]. In TD-DFT, the many-body time-dependent Schrodinger equation is replaced with a set of time-dependent single-particle equations [10]. TD-DFT has become a widely used tool to study the electronically excited states of complex systems, including cobalamins. Vertical excitation energies and transition dipole moments at a particular geometry can be calculated using TD-DFT. Excitation energies tend to agree with experiment within 0.3 eV, but typically calculated excitation energies require a shift to the red to yield much better agreement with experiment [65]. With that being said, the proper description of electronically excited states within the TD-DFT framework is dependent on functional choice. A common practice is to rely on benchmark studies to determine the most appropriate functional to use for a particular system. There have been several studies that sought to determine the proper functionals to use for both ground- and excited-state properties of cobalamins, and these will be discussed in detail in Sects. 7 and 8.

7 Co–C Bond Strength: Key to Theoretical Benchmarks

There are several challenges to overcome when studying cobalamins computationally, including the system size and the presence of a transition metal. The former can be resolved by using a truncated cofactor in simulations (Fig. 1). This typically involves replacing all side chains with hydrogens, replacing the lower axial DBI base with a much simpler imidazole (Im) ligand, while maintaining the structural integrity of the upper axial ligand. These truncated models have less than 70 atoms and provide a good agreement with relevant known structural details from high-quality crystal structures. Due to system size, DFT has become the method of choice for studying cobalamins. The ground-state prediction by DFT of various cobalamins is well understood. This is exemplified by the results of a theoretical investigation of the

Table 1 Experimental values of Co–C BDEs [kcal/mol] in selected cobalamins

Cobalamin	BDE	Method	References
MeCbl	37±3	Thermolysis	[58, 59]
MeCbl	36±4	Calorimetry	[30]
AdoCbl	31.5±1.3	Thermolysis	[19, 27]
AdoCbl	30.9±4.1	Calorimetry	[55]

inverse and normal trans influence in alkylcobalamins [46]. The geometries of 28 corrin-containing models were optimized, and it was concluded that the inverse trans influence is not a general characteristic of B_{12} compounds as was supposed based on experimental results. The BP86/6-31G(d) level of theory yielded bond lengths that were comparable to experimental data. It was also shown that inverse trans influence is observed when the upper axial ligand is enlarged or has an increased electron-donating ability.

As far as determining the proper functional to use within the DFT framework to study cobalamins, one must rely on benchmark calculations. With access to over three hundred density functionals, it is not automatically apparent which functional is appropriate to study a particular system, including cobalamins. Thus, a number of thorough benchmark studies have been reported where different functionals have been applied to estimate dissociation of the Co–C bond in cobalamins [34, 37, 41, 45, 76]. Reproducing this energy (Table 1) accurately is a key standard for evaluating whether a particular theoretical method or density functional, in the case of DFT, is suitable to analyze these complex bioinorganic systems. In order to draw meaningful conclusions in studies where understanding various aspects of catalysis or photochemistry of cobalamins are the target, only functionals that can determine the strength of the Co–C bond accurately should be considered. Toward this, several benchmark studies have been completed for the bond dissociation energy (BDE) of MeCbl and AdoCbl [34, 37, 41, 76]. The major conclusion of these studies is that hybrid DFT functionals significantly underestimate the experimentally determined BDE, while pure GGA functionals like BP86 with dispersion correction (D3) provide BDE with reliable agreement with experiment.

A benchmark analysis of the Co–C_{Me} BDE in MeCbl was performed in order to determine the most appropriate functional to use within the DFT framework for studying ground-state properties [41]. In this study, coupled cluster (CC), DFT, complete active space self-consistent field (CASSCF), and CASSCF-based second-order perturbation theory (CASPT2) were used to predict BDE in the Im-[Co^{III}-corrin]-Me^+ model system, a representative of MeCbl. There are several important conclusions that are of note. The benchmark ab initio potential energy curve (PEC) for the Co–C_{Me} dissociation was determined using the canonical CC approach with singles and doubles (CR-CC(2,3)/CCSD) (Fig. 4). Upon introduction of zero-point energy (ZPE) and basis set superposition error corrections (BSSE), BDEs were produced in excellent agreement with experiment. For instance, with the 6-311++G**

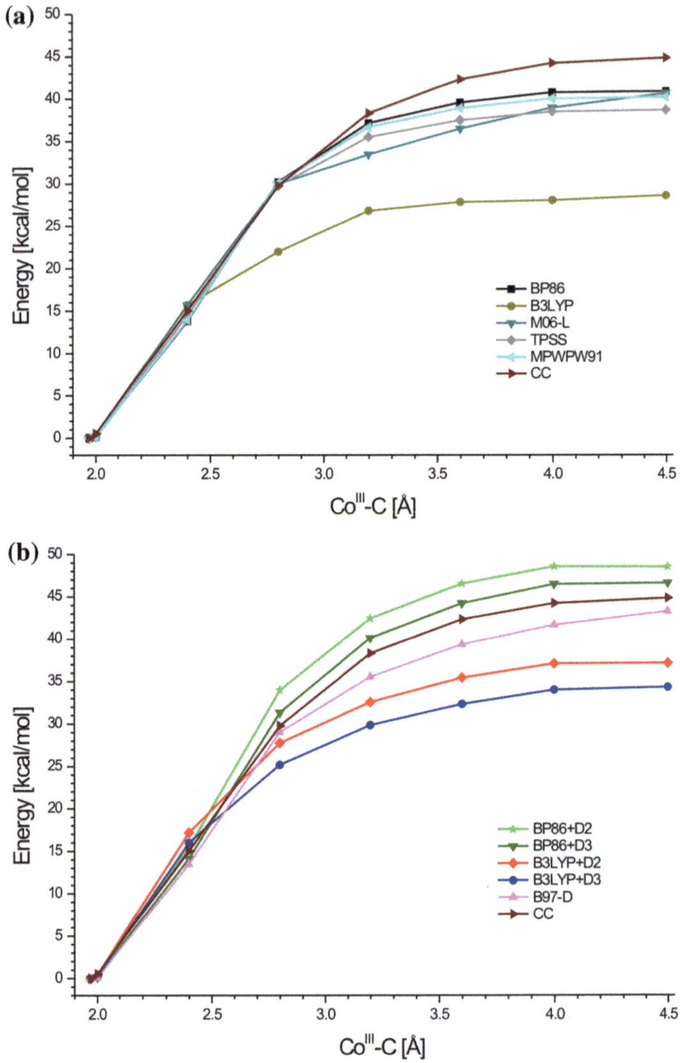

Fig. 4 Comparison of Co–C bond dissociation curves computed using selected DFT functionals without (**a**) and with (**b**) a dispersion correction with a reference coupled cluster (CR-CC(2,3)/CCSD method denoted as CC) curve for MeCbl model (Im-[Co^{III}-corrin]-Me^+). 6-31G(d) basis set was used in all calculations. Reprinted with permission from [41]. Copyright 2012 American Chemical Society

basis set, BDE was calculated to be 38 kcal/mol [41]. The experimental range for Co–C_{Me} BDE for MeCbl falls within the range of 32–40 kcal/mol, including error (Table 1) [30, 58, 59]. It was found that among all of the hybrid functionals tested, including the popular B3LYP, the strength of the Co–C_{Me} bond is always underestimated (Fig. 5). The underestimation correlates with the percentage of Hartree–Fock (HF) exchange that is included in calculations. It has been suggested that the neglect of dispersion interactions at shorter Co–C_{Me} bond lengths is a contributing factor to the poor performance of the hybrids. Alternatively, pure functionals give much better agreement with experiment, and addition of dispersion corrections tends to bring the predicted BDEs to even closer agreement with experiment. The best-performing DFT functionals, with the 6-311++G** basis set, were B97-D and BP86 corrected for dispersion as these predicted Co–C_{Me} BDE to be 35 and 40 kcal/mol, respectively. A further conclusion of this study is that high-level calculations such as CASSCF and CASPT2 seem to be far too expensive for any realistic study of cobalamins because appropriate active space would be prohibitively large.

Previously, the Co–C bond strength was studied in the context of determining why B3LYP, the density functional of choice for an enormous majority of computational studies, performed so poorly in determining BDE in MeCbl [34]. It was suggested that this error comes from the HF exchange and the LYP functional. This study also indicated that BP86 is an appropriate functional to use for the analysis of cobalamins, especially over MP2 and B3LYP, citing several reasons. BP86 is much less expensive than MP2. There is a smaller basis set dependence for DFT methods than for ab initio methods and finally BP86 gives better agreement with experimentally determined structural details than B3LYP and MP2. It was further suggested that the B3LYP functional is a problematic choice in general for determining homolytic metal-carbon BDEs in tetrapyrroles and other highly conjugated systems.

Co–C BDE was also studied in AdoCbl [36, 45]. These benchmark studies yielded similar conclusions to those where MeCbl was the cofactor under study. Namely, it was found that BP86 is an appropriate functional to predict BDE in AdoCbl. B3LYP significantly underestimates the strength of the Co–C_{Ado} bond. Beyond the poor performance of B3LYP in predicting Co–C_{Ado} BDE, this functional does not properly reproduce axial bond distances from crystal structures in cobalamins in the base-on form. It was also shown that B3LYP underestimates the energies of the bonding orbitals and overestimates the energies of the antibonding orbitals.

8 Benchmarks for Electronically Excited States

In order to use calculations to aid in the interpretation of experimental spectroscopic data, benchmarks for electronically excited states within the TD-DFT framework are critical. Results from benchmarks can be relied upon to choose the proper functional that will ensure meaningful conclusions. These have been performed for both CNCbl [38] and MeCbl [28, 39, 41]. In particular for CNCbl, which has been referred to as the paradigm system for the evaluation of excited states of cobalamins, a benchmark analysis targeting the manifold of low-lying excited states was performed [38]. TD-DFT calculations were compared with high-level ab initio calculations and experimental results in order to analyze the nature of the low-lying excited states. The performance of three density functionals, B3LYP, BP86, and LC-BLYP, represent-

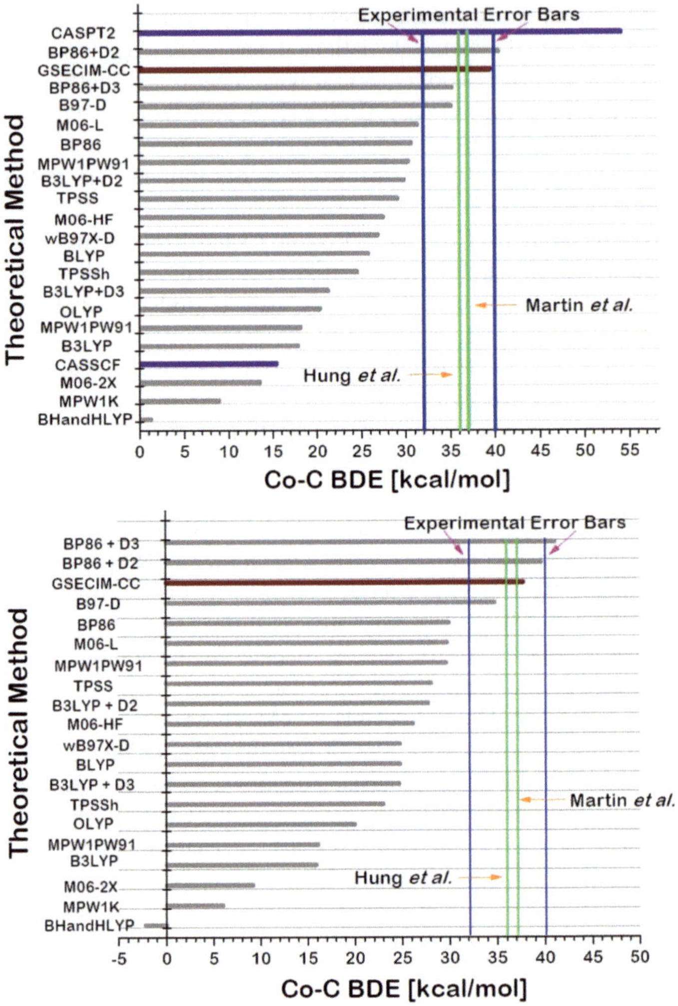

Fig. 5 Performance of various theoretical methods in reproduction of experimental value of Co–C BDE of MeCbl. In theoretical studies, Im-[Co^{III}-corrin]-Me^+ model of MeCbl was utilized. Calculations were performed using 6-31G* (top panel) and 6-311++G** (bottom panel) basis sets. Reprinted with permission from [41]. Copyright 2012 American Chemical Society

ing hybrids, GGAs, and range-separated functionals, respectively, was investigated (Fig. 6) [38]. Both BP86 and LC-BLYP, with range-separation parameter $\mu \sim 0$, yielded results that were consistent with experimental results and high-level ab initio calculations. The electronic properties of CNCbl were most appropriately described when employing the BP86 functional. The α/β region of the Abs spectrum was once interpreted as a vibrational progression of one intense $\pi \rightarrow \pi^*$ electronic transition. This is not necessarily supported by CD and MCD spectra. Rather this region of

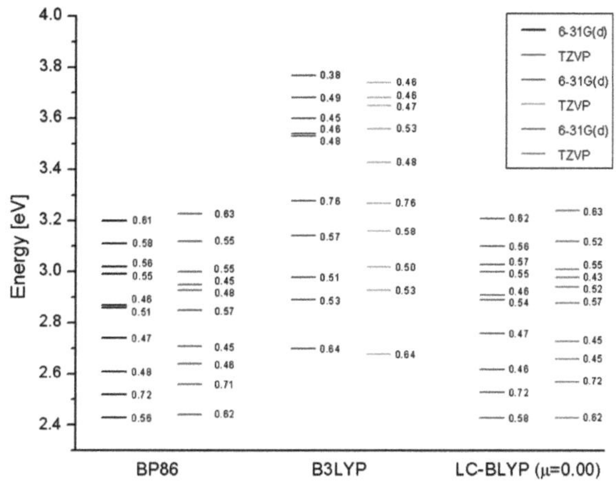

Fig. 6 TD-DFT-based ten lowest electronic transitions computed using three different functionals and various basis sets. For LC-BLYP functional, a corresponding Λ diagnostic parameter is provided (as numbers). Reprinted with permission from [38]. Copyright 2011 American Chemical Society

the spectrum can be dominated by both d/$\pi \rightarrow \pi^*$ and $\pi \rightarrow \pi^*$ transitions. The S_4 transition, which is primarily a $\pi \rightarrow$d excitation, may also contribute the α/β band. The BP86 functional can reliably predict these excitations.

Additional insights can be gained from studies associated with cob(I)alamin. Electronically excited states of this compound can be assessed through various types of spectroscopy including Abs, CD, and MCD spectroscopy techniques [49]. Each of these types of spectra can be readily simulated using TD-DFT. For the test case of cob(I)alamin, Abs, CD, and MCD spectra were simulated using a GGA functional, BP86, and a hybrid functional, CAM-B3LYP (Fig. 7) [40]. Extended second-order multiconfigurational quasi-degenerate perturbation theory (XMCQDPT2), a high-level ab initio method, was used as a reference to explore the nature of the low-lying electronic transitions. It is apparent from this study, [40] involving both a truncated model and calculations with the full structure, that BP86 performs better than the hybrid functional in describing the excitations associated with Co d and corrin π localized transitions. Another important implication from this work was that the lowest energy band was associated with metal-to-ligand charge-transfer (MLCT) excitations, a finding that is distinct from the widely presumed assignment of a single $\pi \rightarrow \pi^*$ transition followed by a vibrational progression for this state. This will be discussed further in Sect. 7.

TD-DFT calculations involving various functionals were also benchmarked against XMCQDPT2 and equation-of-motion coupled-cluster singles and doubles (EOM-CCSD) calculations with MeCbl as a test case [39]. These were compared with the low-lying excited states with the aim of determining the proper density functional to describe the S_1 state as MLCT, a finding that has been experimentally confirmed via TAS and resonance Raman spectroscopy [12, 26, 82, 91].

The choice of functional is paramount in providing the proper description of the character of the S_1 state. The hybrid functionals tested (B3LYP, MPW1PW91, and TPSSh) provide a description of the S_1 state representative of a pure $\pi \rightarrow \pi^*$ tran-

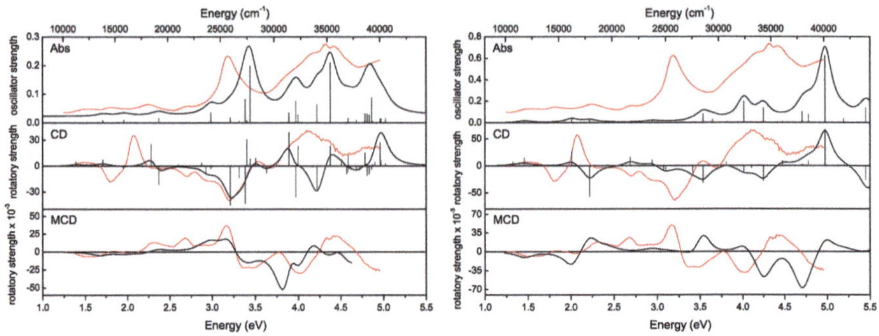

Fig. 7 Abs, CD, and MCD spectra of cob(I)alamin model computed with TD-DFT/BP86/6-31G(d) (left) and TD-DFT/CAM-B3LYP/6-31G(d) (right). Experimental spectra (in red) taken from [49]. Reprinted with permission from [40]. Copyright 2013 American Chemical Society

sition localized on the corrin. Alternatively, the GGA (BP86, BLYP, MPWPW91), the meta-GGA (TPSS), and range-separated LC-BLYP describe the S_1 consistently with experiment as MLCT. In summary, it would seem that, at the time this chapter was written, DFT and TD-DFT employing the BP86 functional is the most economical and reliable way to study cobalamins in both the ground and excited states, respectively.

9 Absorption Features Across Specific Systems: Theory and Experiment

9.1 Free Base Corrin

The foundation for the understanding of cobalamin Abs spectra is the 1965 work of Toohey [88] who studied a B_{12} derivative that contained no Co (FBC). Toohey isolated a corrinoid compound from the bacterium Chromatium. He found that the isolated compound has a pH-dependent Abs spectrum while having similarities with other B_{12} vitamins including a sharp, intense band in the upper UV region and two relatively weak bands close together in the visible region. Maxima in the visible region were very similar to maxima for aquocobamides. On the other hand, the maximum in the upper UV region is considerably shifted to lower wavelengths in comparison to the maxima shown by other B_{12} vitamins. Alkali conditions cause deprotonation of the corrin ring and changes in the electronic spectrum. Toohey suggested that the general configuration of Abs bands is not dependent on the Co but appears to be associated with the conjugate double-bond system of the corrinoid ring structure. This is similar to porphyrin Abs spectra.

This problem was revisited 38 years later from a different perspective where both protonated and deprotonated metal-free corrins were analyzed using the TD-

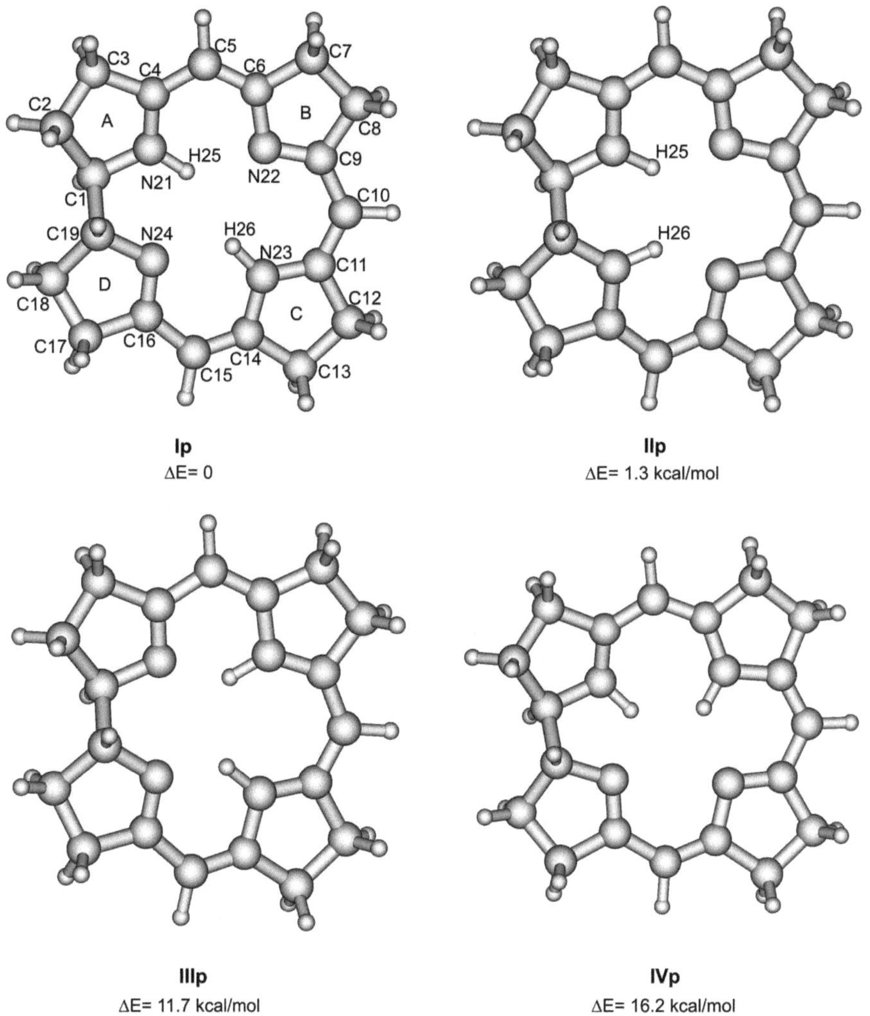

Fig. 8 Molecular structures of protonated free base corrins obtained from DFT/B3LYP calculations. Reprinted with permission from [32]. Copyright 2003 American Chemical Society

DFT method [32]. The electronic spectrum of both forms of the metal-free corrins was simulated using TD-DFT B3LYP and was compared to the Abs spectrum of dicyanocobinamide and the synthetic 1,2,2,7,7,12,12-heptamethylcorrin, which is very similar to Toohey's [88] isolated compound. For the calculations in this study, structures with all possible configurations of two protons among the four nitrogen atoms of the corrin macrocycle were considered, leading to four protonated structures under analysis. These were designated based on the protonated nitrogens as follows, Ip for N23 and N21, IIp for N21 and N24, IIIp for N22 and N23, and IVp for N21

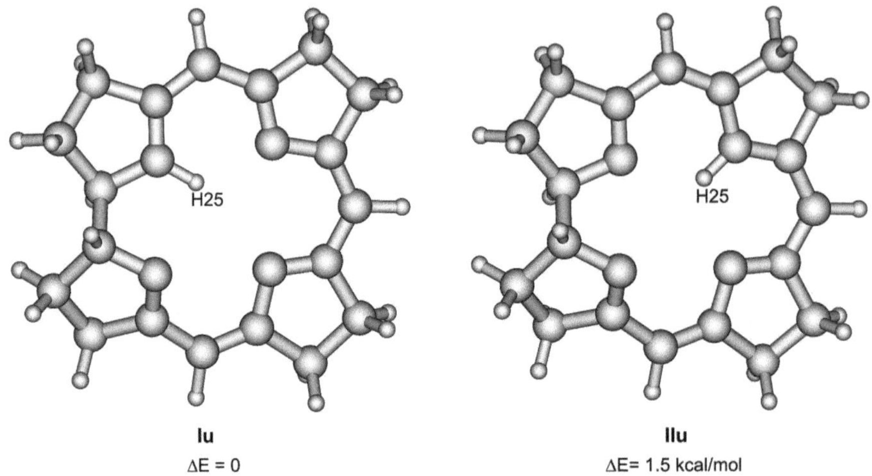

Fig. 9 Molecular structures of unprotonated free base corrins obtained from DFT/B3LYP calculations. Reprinted with permission from [32]. Copyright 2003 American Chemical Society

and N22 (Fig. 8) [32]. For the unprotonated structures with the neutral corrin, two possible forms were considered. These were designated Iu for N21 and IIu for N22 protonation (Fig. 9) [32]. At neutral pH, both the protonated and unprotonated forms may exist in equilibrium with each other.

For the electronic spectra of the protonated forms the spin-allowed, singlet transitions were considered for Ip and IIp [32]. Both forms were expected to contribute to the electronic spectra as these were close in energy. Alternatively, the energies for IIIp and IVp were much higher and were not considered as major contributors to the electronic spectra. Thus, the simulated electronic spectrum for the protonated FBC was based on structure Ip and IIp. The calculated transition at 453 nm from IIp and 428 nm from Ip is the result of a $\pi_7 \rightarrow \pi_8{}^*$ excitation and correlates to the α/β band of experimental spectra of both dicyanocobinamide and 1,2,2,7,7,12,12-heptamethylcorrin. The calculated transitions at 327 nm from Ip and 362.5 nm from IIp were considered to be related to the experimental D, E, and γ bands of dicyanocobinamide. It was further suggested that these arise from asymmetric and antisymmetric combinations of the $\pi_6 \rightarrow \pi_8{}^*$ transition for Ip and the $\pi_7 \rightarrow \pi_9{}^*$ transition for IIP. The δ band of the dicyanocobinamide also appeared to be accounted for in the simulated spectrum via the band at 299 nm for IIP. The δ band was considered to arise from symmetric combinations of $\pi_7 \rightarrow \pi_9{}^*$ and $\pi_6 \rightarrow \pi_8{}^*$ excitations. Overall, the simulated spectrum of the protonated corrin is in good agreement with the experimental Abs spectra of the synthetic corrin 1,2,2,7,7,12,12-heptamethylcorrin and the dicyanocobinamide (see Fig. 10) [32]. It would appear that there are four key orbitals responsible for the major bands including π_6, π_7, π_8^*, and π_9^*, and this is in line with semi-empirical calculations [14, 15].

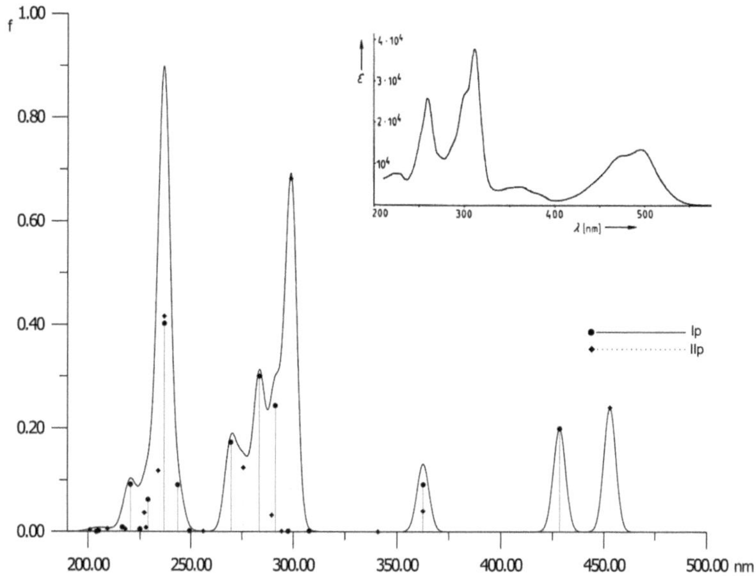

Fig. 10 Electronic spectrum of Ip and IIp calculated using TD-DFT/B3LYP method. Experimental spectrum [48] shown in the inset. Reprinted with permission from [32]. Copyright 2003 American Chemical Society

In addition, it was noted that the bands located at lower wavelengths are the result of mostly $\pi \rightarrow \pi^*$ excitations. It was also observed that n$\rightarrow\pi^*$ transitions are present throughout the simulated spectrum of the protonated corrin, albeit with a low oscillator strength.

For the neutral FBC, also referred to as the unprotonated form, there are two possible forms that were considered (Iu and IIu; see Fig. 9) [32]. The energy difference between Iu and IIu is small, so transitions from both forms were considered contributors to the Abs spectrum. The simulated Abs spectrum is quite similar to the simulated spectrum for the protonated forms. The difference is in the shifting of transitions to shorter wavelengths. The most significant difference with respect to the protonated compounds simulated spectrum is the more pronounced contribution of n$\rightarrow\pi^*$ excitations. This is not surprising as the unprotonated corrin has one lone electron pair localized on the nitrogen atoms, while the protonated form has two lone pairs. As a result, the unprotonated forms have three n orbitals, whereas the protonated forms only have two.

To summarize, for both the protonated and neutral forms, the main calculated transitions between 450 and 290 nm are from excitations between π_6, π_7 and π_8^*, π_9^* orbitals with contributions from other electronic transitions as well. Transitions at higher energies are mainly $\pi \rightarrow \pi^*$ excitations. The n$\rightarrow\pi^*$ transitions play a more prominent role in the unprotonated Abs spectrum due to one more lone electron pair on the nitrogen atoms. This theoretical study [32] has confirmed that the main Abs bands of the free base corrin are primarily the result of $\pi \rightarrow \pi^*$ transitions, but these alone are not sufficient to describe the full spectrum.

9.2 Cyanocobalamin

CNCbl is by far the most well-studied cobalamin to date. It has been called the paradigm system for the evaluation of electronically excited states of B_{12} derivatives [92]. There is a wealth of both experimental and theoretical data for this compound since the elucidation of its crystal structure, but the electronically excited states of CNCbl continue to be the topic of current research as advanced spectroscopic techniques such as X-ray absorption near-edge structure (XANES) [61] progress into the field of B_{12} chemistry. Several theoretical studies have sought to understand the electronic Abs spectrum.

Recall from Sect. 3 that CNCbl exhibits a typical Abs spectrum. Brunold and co-workers [85] suggested the following for the assignment of the major peaks for CNCbl. The α/β region of typical type is the result of corrin-based HOMO→LUMO transition that is polarized along the C5$\cdots$C15 vector. The γ band can be ascribed to a corrin-based $\pi \rightarrow \pi^*$ transition. A corrin-based $\pi \rightarrow \pi^*$ transition is responsible for the most intense band in the D/E region, and this is polarized along the Co$\cdots$C10 vector. There is a notable comparison to discuss. For the Abs spectra of CNCbl and H_2OCbl^+, it is important to consider that CN is a much stronger σ-donor than H_2O. As a result, the Co $3d_{z^2}$ for CNCbl is energetically higher than H_2OCbl^+. Additionally, there is an increased contribution of the unoccupied Co $3d_{z^2}$ to the corrin-based HOMO in CNCbl than H_2OCbl^+. It would seem that the α/β and γ bands in typical cob(III)alamin Abs spectra shift to lower energy with increasing σ-donor strength of the upper axial ligand.

The electronically excited states of CNCbl were investigated using TD-DFT with the B3LYP functional [1]. It was found that electronic excitations were systematically overestimated, but the application of a scaling procedure can bring better agreement with experimental results. Two models were used in this study, denoted CN-[Co^{III}-corrin]-CN^+, with CN as the lower axial base and Im-[Co^{III}-corrin]-CN^+ with Im as the lower axial base (their simulated absorption spectra can be seen in Fig. 11), analogous to dicyanocobinamide (DCC) and CNCbl, respectively. In these models, all side chains and the nucleotide loop were replaced by hydrogens. In order to provide good agreement with experiment, the transition energies for both models were systematically shifted to the red. It would seem that the need to apply a scaling procedure is that the B3LYP functional insufficiently treats electron correlation and that gas-phase calculations were compared with solution data. In agreement with semi-empirical methods [14, 15, 43, 63], the α band was assigned as mainly HOMO→LUMO. It was observed that the α band shifts to the red when the lower axial ligand is CN instead of Im. In order to describe the β band properly the Co *d* orbitals must be included. For both models the β band was ascribed to the $S_0 \rightarrow S_6$

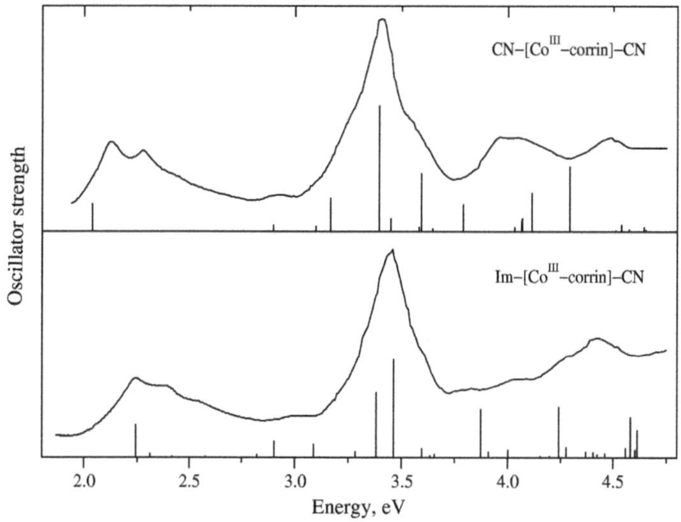

Fig. 11 Comparison of simulated (TD-DFT/B3LYP) and experimental absorption spectra of dicyanocobinamide and cyanocobalamin. In theoretical work, Im-[Co^{III}-corrin]-CN$^+$ and CN-[Co^{III}-corrin]-CN$^+$ models of cyanocobalamin and dicyanocobinamide, respectively, were utilized. Reprinted with permission from [1]. Copyright 2001 American Institute of Physics

transition. This transition involves a combination of π orbitals of the corrin ring and the d_{yz} Co orbital but not axial ligand orbitals.

In a TD-DFT study of simulated Abs, CD, and MCD spectra for CNCbl, [84] it was found that the BP86 GGA-type functional performed much better than CAM-B3LYP in describing electronically excited states. The Abs, CD, and MCD spectra simulated with BP86 agreed very well with experiment and further agreement could be achieved by including solvation with a water polarized continuum model (PCM). A key finding of this study was related to the assignment of the low-energy α/β band. Using the BP86 functional, it was found that the α/β band results from multiple electronic transitions. It was previously thought that this band arises from a single electronic excitation followed by a vibrational progression [85]. However, this was based on the use of hybrid functionals that do not reproduce experimental CD and MCD spectra well. Based on the simulated CD spectrum for CNCbl the D band was assigned to a d/$\pi \rightarrow \pi^*$ excitation. It is apparent from the simulated Abs spectrum that the γ band arises from two main transitions that are of $\pi \rightarrow \pi^*_{DBI}$ and $\pi \rightarrow \pi^*$ character.

The most relevant statement in terms of photophysics is that CNCbl is photostable. This means that upon excitation with light the axial ligands will not dissociate. Instead excitation with light ends in internal conversion (IC) to the ground state (S_0). The photochemistry and photophysics of CNCbl have been investigated with TAS after 400 and 520 nm excitation [92]. Under this single photon excitation, the Co–C bond will not cleave. To further understand the photostability of CNCbl, the low-lying singlet and triplet excited states have been further investigated via TD-DFT [52, 53]. This investigation included thorough analysis of potential energy curves (PECs) that were constructed as a function of axial bond lengths for S_0, singlet, and triplet states as well as, potential energy surfaces (PESs) that were constructed as a function of both axial bond lengths for the S_0, S_1, and T_1. There are some important conclusions to highlight. The excited states computed along the Co–C bond length show a repulsive triplet state with $^3(\sigma_{Co\text{–}C}\rightarrow\sigma^*_{Co\text{–}C})$ character that does not become dissociative. Two energy minima were located on the S_1 PES. The first has $\pi\pi^*$/MLCT/σ-bond–ligand CT (SBLCT) character and the second can be characterized as ligand-to-metal CT/ligand-to-σ-bond CT (LMCT/LSBCT).

The photophysics of CNCbl, based on experiment and simulations, can be summarized as follows. The initial excitation of CNCbl leads to a population of higher $\pi\rightarrow\pi^*$ states. From this Franck–Condon point, there is a decay to a $\pi\rightarrow d$ intermediate LMCT state that corresponds to only modest elongation of the axial bonds. This short-lived intermediate further relaxes to a LMCT minimum characterized as $\pi\rightarrow\sigma^*$ (d_{z^2}). As the axial bond lengths continue to elongate, IC occurs as the S_0 crosses the S_1. As a result of this crossing, photodissociation does not occur as axial bonds elongate and thus CNCbl is designated as photostable. From an experimental perspective, the decay to the S_0 is dependent on both temperature and solvent. Less polar solvents increase the excited-state lifetimes.

Recently, a combined theoretical and experimental study involved the analysis of CNCbl with XANES to further characterize the photochemistry of CNCbl [61]. Evidence of very good agreement between experiment and theory in terms of nature of geometrical changes in electronically excited states was exhibited. This experiment confirmed that the formation of a Franck–Condon excited state is the initial photoinduced event. There is an IC to the $\pi^1(\sigma^* d_{z^2})^1$ intermediate. This intermediate is marked by elongated axial bonds. The major conclusion from this work is that, upon photoexcitation of CNCbl, the largest structural changes that occur are related to axial bond lengths. As a result, PESs constructed as a function of axial bond lengths are an important tool to use in order to understand the mechanisms associated with photodissociation of other cobalamins.

9.3 Methylcobalamin

Like CNCbl, the electronically excited states of MeCbl have been thoroughly investigated using both experimental and computational methods. MeCbl exhibits a unique Abs spectrum. Brunold and co-workers [85] proposed that two factors are primarily responsible for the features in unique spectra, (1) a net increase in the number of donor MOs available for electronic transitions and (2) extensive mixing of the Co 3d and corrin π orbitals. The α/β band in MeCbl and AdoCbl Abs spectra is a HOMO→LUMO transition polarized along the C5···C15 vector. Specifically, for MeCbl, the corrin-based HOMO has significant σ-bonding character. The more intense β band, in comparison to the typical spectra, is the result of excited-state distortion that lengthens the Co–C bond. The second intense feature in the α/β region of MeCbl arises from a corrin-based $\pi \rightarrow \pi^*$ transition and a Co $3d_{xz} \rightarrow$ corrin π^* transition. The γ region in unique Abs spectra results from significant mixing between occupied Co 3d and corrin π orbitals. As a result of this mixing, there is distribution of the relatively intense transitions across a larger range of the UV region and a single intense γ band is not observed. To summarize, there is a high degree of σ-donation from the alkyl ligands (Me or Ado, for example) to the Co and this destabilizes the Co 3d orbitals. The contribution of the Co $3d_{z^2}$ orbital to the HOMO increases with increasing σ-donor strength of the upper axial ligand [85].

Later, another TD-DFT analysis shed further light on the assignment of the bands in MeCbl Abs, CD, and MCD spectra [84]. The BP86 functional was used in this study, and it is now well established that for cobalamins this functional provides much better agreement with experiment than others such as hybrids [38, 45]. For MeCbl, the α band is less intense than the β band. Like in CNCbl, the α/β band of MeCbl does not appear to be based on a vibrational progression. Rather, BP86 calculations indicate that the α/β band arises from several electronic transitions with the first two characterized as $\pi/d \rightarrow \pi^*$ and $d/\pi \rightarrow \pi^*$ for the S_1 and S_2, respectively. This is consistent with results from TAS that indicate that the S_1 state is MLCT [82]. Based on the BP86 simulated Abs spectrum, it would also seem that the S_2 state gives rise to the α band. The β band assignment can be determined upon inspection of both Abs and CD spectra. The S_3 electronic excitation appears to be responsible for the β band, and the transition has been characterized as $\pi/d \rightarrow \pi^*$. The γ band in MeCbl, which is much broader than in the typical Abs spectra, was ascribed to a manifold of transitions in the BP86 simulated Abs spectrum where the occupied molecular orbitals are a mixture of d and π character. This is in stark contrast to Brunold's work [85] where this region was described as solely $\pi \rightarrow \pi^*$ transitions.

Returning to the nature of the lowest energy part of the spectrum of MeCbl, the α/β band, there have been two divergent interpretations for the assignment of this region. Historically, this region was considered to be dominated by $\pi \rightarrow \pi^*$ excitations [24, 68]. Another more recent study indicated that there was a second electronic transition involved [85]. TAS measurements indicate that these low-energy excitations are MLCT. Consequently, a study based on several theoretical methods was conducted to resolve this discrepancy [39]. The technical details were described in Sect. 8. The major conclusion from this work was that, a pure GGA like BP86, a meta-GGA, or a LC-BLYP functional should be employed in order to properly describe the S_1 state in MeCbl as having MLCT character. Hybrid functionals interpret the S_1 state as having $\pi \rightarrow \pi^*$ character, which is now widely understood as a misinterpretation. The reasons for these varying interpretations between BP86 and B3LYP are mainly due to the differences associated with how these functionals describe bonding. B3LYP tends to describe corrin $\pi \rightarrow \pi^*$ excitations well, while BP86 describes $d/\pi \rightarrow \pi^*/d$ excitations (MLCT) properly.

In addition to work related to the S_1 state and its comparison with TAS data, [39] the Abs spectrum for MeCbl as well as its base-off derivative (MeCbi-H_2O) was simulated [2]. Two functionals, BP86 and B3LYP, were used in this study, and as has been seen in other cases, these functionals offer different insights for the electronic excitations. Starting with MeCbl, the α/β band has been ascribed to two transitions located at 520 and 496 nm based on BP86 calculations. The calculated transition at 520 nm has mixed character (d→π^* and $\pi\rightarrow\pi^*$) and is the HOMO→LUMO excitation. The calculated transition at 496 nm also has mixed character that includes d orbitals. Based on results using the BP86 functional, the γ band for MeCbl arises from three electronic transitions that have d/$\pi\rightarrow\pi^*/\sigma^*$ character. The α/β region appears to arise from transitions that are predominately $\pi\rightarrow\pi^*$ when B3LYP is employed. Using B3LYP, the γ band experimentally observed at 316 nm arises from the transitions that were calculated at 293 and 279 nm with d→π^* and d→σ^* character. For MeCbi-H_2O, which represents the base-off configuration that persists in low pH environments, differences can also be noted in the calculated electronic transitions with the two different functionals. Like the base-on case, transitions with d character have very low oscillator strengths when B3LYP is employed. However, B3LYP is reliable in reproducing the blue-shift of the lowest-intensity band that is observed experimentally for MeCbi-H_2O in relation to MeCbl. BP86 predicts the transitions with d orbital character with high oscillator strengths for calculated transitions at 457, 461, and 405 nm. The calculated transition at 315 nm corresponds to the experimentally observed band at 315 nm and is of mixed character including π/d→π^*/σ^*. For the experimental band at 264 nm, the BP86 results suggest this arises from primarily $\sigma/\pi\rightarrow\sigma^*/\pi^*$ excitations as well as n→π^* which involves corrin nitrogen lone pairs. Clearly, regardless of the model (base-on/base-off MeCbl), B3LYP and BP86 provide unique insights into the nature of the electronic transitions responsible for the Abs spectra.

There are some additional conclusions to note based on the comparison between the TD-DFT/B3LYP and TD-DFT/BP86 results [2]. First, both functionals can reproduce Abs spectra in good accordance with experiment. B3LYP reproduced the intensity of the bands in a better fashion than BP86, although it suggested these transitions were basically $\pi\rightarrow\pi^*$. While BP86 predicts Co–C BDE much better than B3LYP, it is less effective than B3LYP for describing the corrin-based electronically excited states. These differences are related to the bonding descriptions provided by the two functionals. BP86 describes the Co–C bond with proper energy for its covalent nature, while the energy produced by B3LYP is far too low. All of this indicates that BP86 is the better functional choice for studies involving the cleavage of the Co–C bond. Also, BP86 is better for describing MLCT electronic transitions that involve the Co *d* orbitals while B3LYP does a reliable job for corrin $\pi\rightarrow\pi^*$ transitions. Taking all of this into account, this study suggests that two different functionals are required to provide a proper description of electronically excited states. Practically speaking, determining a reliable functional for studies of complex systems like cobalamins comes down to the question that is being asked. When photolytic cleavage of the Co–C bond is the target, BP86 is certainly the more reliable choice. However, B3LYP is a good option for reproducing peak intensities well.

Even with the use of the proper functional, it still may be necessary to apply a scaling procedure to obtain agreement between experiment and theory. An important implication that arose from TD-DFT analysis of MeCbl Abs spectra was the introduction of a dependable scaling procedure [2]. A previous scaling technique was designed under the premise that the B3LYP functional requires a uniform shift to the red, while BP86 requires a uniform shift to the blue to simulate Abs spectra with good alignment with experiment [85]. However, this method is problematic because it operated under the assumption that the energy separation between Abs maximum bands is correlated between theory and experiment and this is not the case. The technique suggested by Kozlowski and co-workers [2] resolves this issue and involves two parameters, one for the low-energy excitations and another for the high-energy excitations. The scaling procedure is marked by the equation:

$$E^{i}_{\text{scaled}} = \zeta E^{i}_{TD-DFT} + E_{\text{shift}} \tag{1}$$

where E^{i}_{TD-DFT} is the TD-DFT computed ith electronic excitation. ζ and E_{shift} are empirical parameters that will adjust the simulated spectrum. The values of ζ and E_{shift} are dependent upon the electronic transitions that are selected. The empirical parameters are further denoted as:

$$\zeta = (E^{\gamma}_{\text{exp}} - E^{\alpha}_{\text{exp}})/(E^{\gamma}_{TD-DFT} - E^{\alpha}_{TD-DFT}) \tag{2}$$

and

$$E_{\text{shift}} = (E^{\gamma}_{\text{exp}} + E^{\alpha}_{\text{exp}})/2 - \zeta(E^{\gamma}_{TD-DFT} + E^{\alpha}_{TD-DFT})/2 \tag{3}$$

This scaling procedure can be applied to other cobalamin systems. In fact, it was applied in a TD-DFT analysis of AdoCbl which will be discussed in Sect. 9.4.

In addition to these Abs studies, MeCbl has been investigated using TAS. An important conclusion to note is that the photolysis of MeCbl is wavelength dependent [82]. Upon excitation at 400 nm, a partitioning between prompt bond homolysis and a metastable cob(II)alamin photoproduct is observed. With excitation at 520 nm, no prompt bond homolysis is observed. The reason for the wavelength dependence for the photolysis of MeCbl is still not well understood from a theoretical point of view.

The electronically excited states of ethylcobalamin (EtCbl) were also investigated in the comparison with MeCbl [51]. The EtCbl model was derived by replacing the methyl group in MeCbl with ethyl. Overall, the simulated Abs spectrum for EtCbl and MeCbl are quite similar (Fig. 12). The difference in these species is in the mechanism of photolysis associated with TAS. For both EtCbl and MeCbl, after excitation to the α/β band, internal conversion (IC) will result in the population to the S_1 state where the photolytic fate is either photolysis or IC to the ground state. The barrier for photolysis is lower for EtCbl, so photolysis is more prevalent. The reverse is true for MeCbl where only partial photolysis is observed.

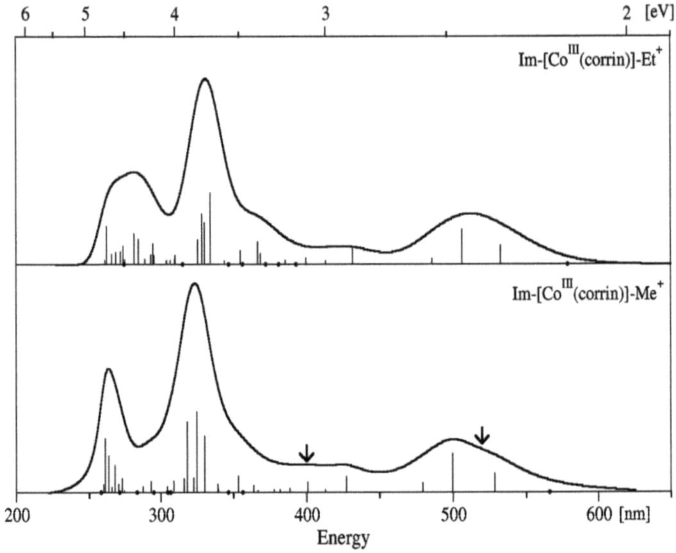

Fig. 12 Simulated electronic absorption spectrum of EtCbl (top) and MeCbl (bottom) models employing TD-DFT/BP86/6-31G(d) method. Arrows indicate laser excitation energies at 520 nm and 400 nm, respectively. Reprinted with permission from [51]. Copyright 2009 American Chemical Society

9.4 Adenosylcobalamin

AdoCbl has been the subject of several spectroscopic and theoretical studies with the target of elucidating the electronically excited states. This cofactor is quite complex compared to CNCbl and MeCbl due to its large upper axial ligand. Researchers have taken several approaches to deal with this large 5-deoxyadenosyl (Ado) upper axial ligand in theoretical studies of the electronic structure of AdoCbl. In one approach, the Ado ligand was replaced with the much simpler methyl group with the idea that methyl could be representative of Ado [85]. This approach will certainly simplify TD-DFT calculations. Unfortunately, it is not an appropriate approach as AdoCbl and MeCbl are quite different structurally and electronically, especially in terms of charge transfer. In addition, the photochemistry of AdoCbl and MeCbl are very different [12, 78]. The two derivatives cannot be interchanged without consequence in theoretical studies.

An important thing to consider in all studies of the electronically excited states of AdoCbl is the occurrence of long-range charge transfer (LRCT). This is probably the most significant challenge in studies of AdoCbl [42]. There is the propensity for a manifold of charge-transfer transitions between the Ado and corrin ligands in AdoCbl. Accordingly, MeCbl is not a suitable model structure for AdoCbl theoretical studies.

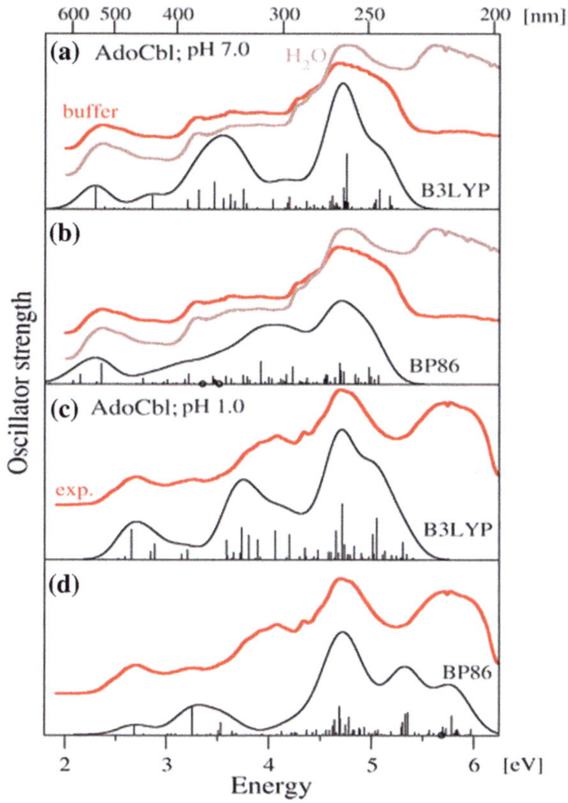

Fig. 13 Comparison of experimental (red and brown) and calculated TD-DFT (black) absorption spectra of AdoCbl in the neutral (top two panels) and acidic pH condition (bottom two panels). Im-[Co^{III}-corrin]-Ado^+ and H_2O-[Co^{III}-corrin]-Ado^+ structural models represent AdoCbl complexes in pH = 7 and pH = 1, respectively. BP86/6-31G(d) and B3LYP/6-31G(d) functionals were employed. Spectra were computed using a set of scaling parameters as in Table 1 of [3]. Reprinted with permission from [3]. Copyright 2009 American Institute of Physics

The electronically excited states of AdoCbl were investigated in another TD-DFT-based study using a model that contained the full Ado ligand [3]. This model can be denoted, Im-[Co^{III}-corrin]-Ado^+ with all side chains simplified by replacement with hydrogen and the axial base replaced with an Im. B3LYP and BP86 were used, similarly to the TD-DFT analysis of MeCbl [2]. Simulated Abs spectra (Fig. 13) were adjusted using the scaling procedure described in the above section. Here, it is important to recall that both MeCbl and AdoCbl exhibit unique Abs spectra. Specifically, for AdoCbl, the α/β band is located at 590–450 nm, the D/E band above 300 nm, the γ band between 400–300 nm, and the δ band below 300 nm. At very low pH (pH = 1), the α/β band blue shifts. Not surprisingly, B3LYP and BP86 result in divergent assignments of the key peaks in the AdoCbl Abs spectrum. Considered judiciously, results from these two functionals can be seen as complimentary, in this case. For brevity of presentation, only key conclusions will be highlighted concerning these differences.

The B3LYP functional assigns the lowest energy transitions of the α/β region to be of $\pi/d_{z^2}\rightarrow\pi^*$, $\pi/d\rightarrow\sigma^*$, $\pi/d\rightarrow d$ character. A LRCT transition was observed at 420 nm in the B3LYP simulated spectrum and can be characterized as $\pi_{Ado}\rightarrow\pi^*_{corrin}$. The

D/E band is π/d$\to\pi^*$ type. The broad γ band clearly results from multiple transitions, and those with significant oscillator strengths are characterized as mostly π/d$\to\pi^*$, π/d$\to$d and π/d$\to\sigma^*$. The δ band arises from $\pi\to\pi^*$ transitions. For BP86, the α/β region arises from d/$\pi\to\pi^*$ and d$\to\pi^*$ transitions. The D/E band is a mixture of several transitions including π/d$\to$d, π/d$\to$d/π^*, and π/d$\to\sigma^*$ type. The γ band results from several transition types and the ones with the most significant oscillator strengths of the simulated spectrum being π/d$\to\sigma^*$ and d$\to\pi^*$/d at 385 nm, π/d$\to\pi^*$ at 339 and 327 nm, $\sigma\to\pi^*$ at 310 nm, and finally d$\to\pi^*$ at 307 nm. The δ band arises from transitions that are $\sigma\to\pi^*$/d character.

The electronic spectrum for a model of the base-off form of AdoCbl (AdoCbi-H_2O) was also simulated in this study (Fig. 13) [3]. This model denoted H_2O-[Co^{III}-corrin]-Ado^+ is representative of the low pH situation where the lower axial DBI group is replaced by water. The α/β region is shifted to higher energies in the AdoCbi-H_2O spectrum. B3LYP properly predicted this shifting of bands to shorter wavelengths in accordance with experiment. The B3LYP assignments for AdoCbi-H_2O are as follows. The α/β band arises from three transitions that are primarily π/d$\to\pi^*$. The D/E band results from two d/$\pi\to\pi^*$-type transitions. As in the AdoCbl base-on case, the γ region is the result of several transitions which are mainly π/d$\to\pi^*$, but contributions from π/d$\to\sigma^*$ and $\sigma\to\pi^*$ persist as well. The δ band arises from several transitions of varying oscillator strengths. Those with the largest oscillator strengths are $\pi\to\pi^*$ transitions of the adenosine moiety and of π/d$\to\pi^*$/d character. The electronic Abs spectrum was also computed with BP86; the following details are of note. As was the case for the Im-[Co^{III}-corrin]-Ado^+ model, the first transition is LRCT when employing BP86. Again there are some differences in the assignments of key regions when comparing the assignments of the key features of AdoCbi-H_2O simulated spectra in terms of B3LYP and BP86 methods. The α/β region arises from d/$\pi\to\pi^*$ and π/d$\to\pi^*$-type transitions. The D/E band is a mixture of two transitions that can be characterized as d/$\pi\to\pi^*$ and d$\to\sigma^*$. These transitions are more complex than at first glance. The D/E transitions involve excitation from free electron pairs from the adenine group to the σ^*(Co-Ado) orbital or the π^* orbital of the corrin. The γ band results from several transition types and the ones with the most significant oscillator strengths being π/d$\to\pi^*$ and π/d$\to\sigma^*$ character. There are also contributions from Ado(n)$\to\sigma^*$ and Ado(n)$\to$d excitations. The δ band arises from transitions that are $\sigma\to\pi^*$/d character. The δ part arises from transitions of several types including $\sigma\to\sigma^*$, $\sigma\to\pi^*$, d/$\pi\to$Ado(π^*), $\pi\to\pi^*$ of the adenosine, Ado(n)$\to\pi^*$, and finally $\sigma\to\pi^*$ types. Clearly, the interpretation of Abs spectrum of AdoCbl and AdoCbi-H_2O is quite involved and requires a thorough analysis of the electronic structure and relevant molecular orbitals.

A major conclusion of this work [3] is that B3LYP and BP86 can be used in a complimentary manner to provide a more complete picture of the electronically excited states. B3LYP is reliable in simulating the spectral pattern of the experimental Abs spectra for AdoCbl even though it performs poorly in replicating accurate axial bond lengths. Because the two functionals indicate a different orbital ordering and HOMO/LUMO character, the lowest energy transitions are predicted to be of different types. For B3LYP, the lowest energy transitions are of π/d$\to\pi^*$, π/d$\to\sigma^*$, and

π/d$\rightarrow$d character. For BP86, these transitions are d/$\pi\rightarrow\pi^*$ and d$\rightarrow\pi^*$ character. BP86 indicates more transitions in the α/β region than does B3LYP. LRCT at low energies occurs more noticeably when BP86 is employed, but the use of a solvent model eliminates these. Overall, both functionals pick up on the shift of the lowest intense transition to higher energies that occurs when the Im is exchanged for water in low pH. Again, it is apparent the assessment of the electronically excited states of cobalamins is not completely straightforward but requires careful interpretation of TD-DFT results.

In TD-DFT studies of the photodynamics of AdoCbl, which can be experimentally probed through TAS, a simplified model of AdoCbl named ribosylcobalamin (RibCbl) was used [50]. A major concern of theoretical studies of the photocleavage of the Co–C bond in AdoCbl is predicting the BDE properly. As even truncated models of AdoCbl are computationally demanding, it was important to consider if a RibCbl model could provide a reliable representation of AdoCbl. Comparisons between RibCbl simulated Abs spectrum and the experimental AdoCbl Abs spectrum offered a good opportunity to investigate the relevance of this simple structural model to excited-state descriptions. Later studies of the photodissociation of AdoCbl would involve the full Ado ligand [22]. The simulated Abs spectra of RibCbl calculated with TD-DFT/BP86 in both the gas phase and in the water PCM are pertinent to discuss in this chapter [50]. As was pointed out in the study, the simulated Abs spectrum for RibCbl cannot be directly compared to the experimental Abs spectrum of AdoCbl, but important correlations can be noted. First, the simulated Abs spectrum of AdoCbl discussed above and the RibCbl spectrum reasonably agree. So, it would seem that the RibCbl model can provide information related to the experimental spectroscopic features of AdoCbl. For the gas-phase simulated spectrum, the assignments are as follows. The α/β bands primarily involved d/$\pi\rightarrow\pi^*$ transitions with the LUMO being a pure corrin π^* orbital. The D/E and γ bands resulted from several transitions with contributions from the ribose group playing a significant role. The energy of the γ band is the same in gas and in water. The α/β band for the RibCbl simulated spectrum in water is narrower compared to the gas-phase simulation.

9.5 *Antivitamins B_{12}*

Antivitamins B_{12} are cobalamin derivatives that counteract the physiological effects of their cobalamin analogues. They have received attention in recent years due to their potential applications in the medical field [90, 96]. Two antivitamins B_{12}, alkynylcobalamin phenylethynylcobalamin (PhEtyCbl) and arylcobalamin 4-ethylphenylcobalamin (EtPhCbl), have been developed [72, 73]. These strongly resemble the structures of the biochemically relevant derivatives while being metabolically inert themselves. These maintain all the structural features of cobalamins with the only difference in the nature of the upper axial ligand. Like the other well-studied vitamin B_{12} derivatives, the Abs spectra of EtPhCbl and PhEtyCbl have been collected and the photodynamics have been assessed with TAS [60].

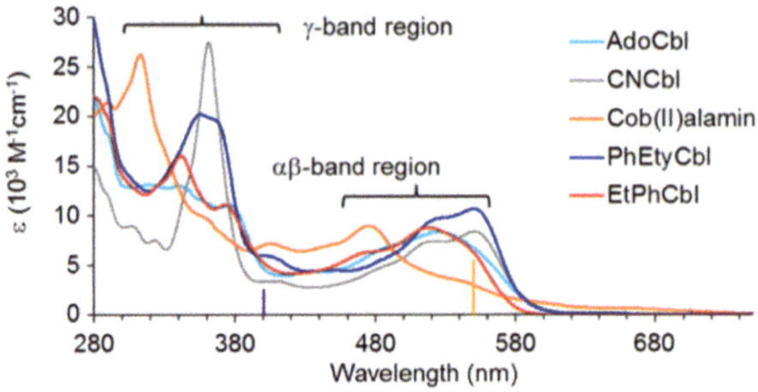

Fig. 14 Absorption spectra of three cobalamins and two antivitamins of B_{12} (PhEtyCbl and EtPhCbl). The excitation wavelengths are indicated by vertical lines at 400 nm and 550 nm. Reprinted with permission from [60]. Copyright 2016 American Chemical Society

The Abs spectra for these antivitamins B_{12} exhibit similarities to the other derivatives discussed (see Fig. 14). For PhEtyCbl, the experimental Abs spectrum is quite similar to CNCbl. Both have a nearly identically structured α/β band. PhEtyCbl also contains a prominent γ band, but it is not as narrow or as intense as the γ band of a typical spectrum. The Abs spectrum of EtPhCbl exhibits a α/β region much like AdoCbl. Rather uniquely, the γ region of for EtPhCbl is marked by two separate peaks that are less intense than the γ band of CNCbl or PhEtyCbl. The γ band of the antivitamins B_{12} is a somewhat combination of the style of typical and unique γ bands.

TAS measurements also have provided further insight. For PhEtyCbl, TAS shows a weak broad Abs that extends into the near infrared region and a blue-shifted α/β band. Despite no evidence of photodissociation of either axial ligands, the TAS observations are consistent with those of CNCbl. Upon excitation, it would seem that the axial bonds elongate and the LMCT state is populated via the Co d_{z^2} orbitals. PhEtyCbl is photostable, and upon excitation, IC to the ground state is complete. Conversely, EtPhCbl is a conditional antivitamin B_{12}, and it is slightly photolabile. TAS measurements indicate that the α/β region peaks are <450 nm. This is consistent to a ligand field region with a very elongated or dissociated axial ligand. In this case, the lower axial ligand can dissociate. The strong γ band indicates that the Co–C bond to the upper axial group is elongated. Sension and co-workers [60] have proposed that the photolysis mechanism for EtPhCbl is very similar to MeCbl. Theoretical studies are underway to assign the peaks in the antivitamins B_{12} Abs spectra and to elucidate the nature of the electronically excited states involved in the photodynamic processes [54].

9.6 *Non-alkyl Cobalamins*

Aquacobalamin (H_2OCbl^+), where water is the upper axial ligand, exhibits a typical Abs spectrum much like CNCbl. Accordingly, there are three bands that are important to characterize (Figs. 3 and 15). These include the α/β region, the D/E bands, and the γ band [85]. Pratt's chapter [68] indicated that the α/β region is the result of a sim-

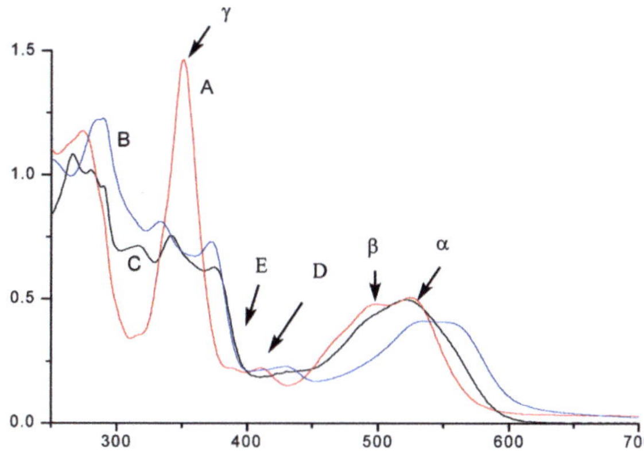

Fig. 15 UV-visible spectra of cobalamins: **a** H_2O-Cbl, **b** GS-Cbl, **c** MeCbl at pH = 5. Reprinted with permission from [18]. Copyright 2012 American Chemical Society

ple HOMO→LUMO transition of $\pi \rightarrow \pi^*$. The Brunold and co-workers study [85] confirmed that a single electronic transition gives rise to the α/β region. Here, it was also suggested that the D/E region likely arises from at least four different transitions with two being magnetic dipole in character, while the other are distinctly electronic transitions. However, the most intense transition in the D/E region is corrin-based $\pi \rightarrow \pi^*$. The γ band is a single intense band, in the style of a typical Abs spectrum. There are three different one-electron excitations that contribute to this band that can be characterized as corrin-based $\pi \rightarrow \pi^*$. Overall, H_2OCbl^+ and CNCbl yield very similar Abs spectra; however, one difference can be identified when comparing the CD spectra of the two. There is a low-energy feature in the CD spectrum of H_2OCbl^+ that is not exhibited in the CD spectrum of CNCbl. As σ-donor strength increases, α/β and γ regions tend to shift to lower energy in typical Abs spectra and this is confirmed upon a comparison of CNCbl and H_2OCbl^+ spectra.

Hydroxocobalamin (HOCbl) is another cob(III)alamin and contains a hydroxyl group as the upper axial ligand. The Abs spectrum of HOCbl is much like a typical Abs spectrum such as CNCbl (Fig. 3). It nearly resembles the Abs spectrum of H_2OCbl^+ with a few notable differences. The γ band for HOCbl is less intense than for H_2OCbl^+. The reverse is true for the α/β band and in HOCbl, this band is more intense than for H_2OCbl^+. The photophysics of HOCbl has been probed using TAS, especially with its potential use in photogeneration of radicals in mind [93]. The S_1 spectrum resembles the Abs spectrum of HOCbl (see Fig. 4 in [93]). PESs of the S_1 state based on axial bond elongation provide further insights into this excited state. The S_1 PES minimum is predominately of $p_{OH}/d \rightarrow \pi^*$ type, but as the Co-OH bond elongates, a crossing of states occurs. At this crossing, there is a switch to $p_{OH}/d \rightarrow \sigma^*$ character. So, at long bond distances the S_1 surface is mostly $p_{OH} \rightarrow \sigma^*$ with lesser contributions from Co *d* orbitals [93]. While PESs can be used to understand TAS data, additional theoretical work is necessary to understand and assign the features of the Abs spectrum of HOCbl.

Cobalamins containing a Co–S bond to the upper axial ligand have also received attention. In particular, the Abs spectrum and electronic structure of glutathionylcobalamin (GSCbl) have been investigated [18]. The Abs spectrum for GSCbl (Fig. 15) is more in line with unique spectra of MeCbl or AdoCbl. In particular, the γ region of MeCbl and GSCbl are quite similar. Generally speaking, for cobalamins, the shift of the γ band is due to the σ-donor strength of the upper axial ligand [67]. For GSCbl, the γ band is shifted to longer wavelengths and the α/β region is shifted to the red compared to MeCbl. Also, the upper axial ligand for GSCbl is responsible for the split into two bands in the γ region.

Not surprisingly, there have been two different theoretical approaches to elucidate the nature of the electronic transitions associated with the Abs spectrum. Brunold and co-workers analyzed GSCbl with Abs, CD, and MCD spectroscopy as well as with TD-DFT [13]. Interestingly, they chose to model the ground state of GSCbl using a combined quantum mechanical/molecular mechanics (QM/MM) approach. Upon ground-state optimization, they truncated the model to be much more cost effective for TD-DFT. The upper axial ligand was simplified to be SCH_3. The simulated Abs spectrum was produced with the B3LYP functional, and comparisons were drawn between the Abs spectra of MeCbl and GSCbl.

In a more thorough TD-DFT study, both B3LYP and BP86 were employed to simulate Abs spectra of GSCbl; only this time, the upper axial ligand was not truncated [18]. As in the case for the other cobalamins discussed, the two functionals predict different characters for the relevant transitions that give rise to the spectral features. To summarize, Birke and co-workers [18] conclude that BP86 is the proper functional to use to understand the electronically excited states of GSCbl. Overall, BP86 simulated a more accurate Abs spectrum than B3LYP. BP86 indicated more participation of *d* orbitals than B3LYP. An interesting conclusion from this study was that the experimentally observed peaks at 612 and 529 nm were suggested to arise from vibronic transitions.

9.7 Reduced Cobalamins

The Abs spectra of the reduced cobalamins, cob(II)alamin and cob(I)alamin, have also been studied from an experimental [49, 64, 86] and theoretical [21, 33, 40, 44, 49, 64, 86] point of view. The colors of each compound are indicative of their oxidation state. CNCbl is red, cob(II)alamin is orange, and cob(I)alamin is green.

One-electron reduction of cob(III)alamin leads to cob(II)alamin. This species yields a unique Abs spectrum. In order to understand the experimental Abs spectrum more fully, two different truncated models representative of the Co(II) compound were used (Fig. 16) [21]. One was four-coordinate and the other five-coordinate. The accuracy of the TD-DFT calculations was compared to XMCQDPT2 results. Both methods yielded good agreement with each other and further validated the idea that TD-DFT is a less expensive and a reliable choice for studying the electronically excited states of the cobalamins. The α/β region is not as pronounced as in the other cobalamins but is narrower and is found between 450–500 nm. This region results from a mixture of transitions that included *d* orbitals. A sharp γ peak is not present. A

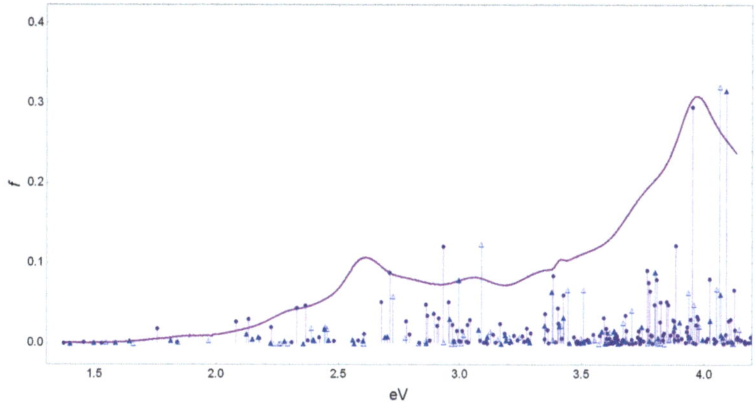

Fig. 16 Comparison of experimental (solid line) and TD-DFT/BP86-derived spectra of cob(II)alamin. Full structural model treated at BP86/SDD (circles), two truncated models (filled and empty triangles). Reprinted with permission from [21]. Copyright 2016 Royal Society of Chemistry

Fig. 17 Absorption spectrum of cob(I)alamin. Reprinted with permission from [49]. Copyright 2006 American Chemical Society

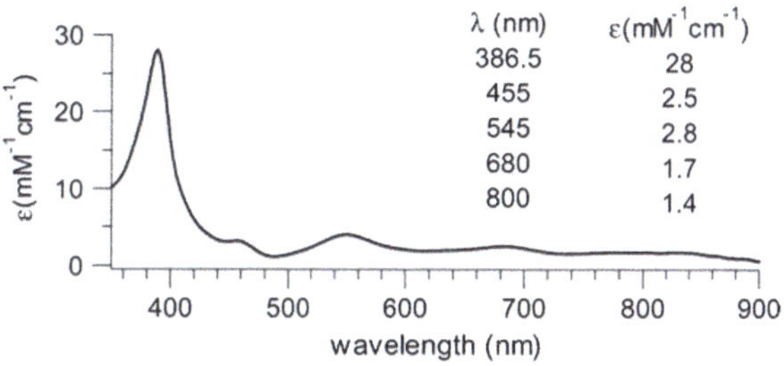

common theme in this study was that regardless of the basis set, method, or structural model, the D_1 state was characterized by a transition to the d_{z^2} orbital.

For cob(I)alamin, also known as the super-reduced B_{12} cofactor, the Abs spectrum is quite distinct from the other cob(III)alamins discussed earlier. Its electronic spectrum is marked by a single sharp peak located at 385 nm that would be analogous to the γ band of CNCbl or H_2OCbl^+ (Fig. 17). There are at least four bands of very weak-intensity throughout the rest of the spectrum. Cob(I)alamin has been the subject of several theoretical studies. For instance, early studies used B3LYP to simulate spectra [31]. Later, CASPT2 calculations were used to characterize the nature of the most intense peaks in the Abs spectrum [33]. Brunold and co-workers also studied the two-electron reduced forms in terms of Abs, CD, and MCD spectra experimentally and with TD-DFT using the PBE functional [49]. Considered together, these studies painted different pictures of the electronic structure and electronically excited states. Both the B3LYP and PBE results indicated that the low-lying excited states are primarily MLCT in character. However, the PBE results suggested a closed-shell $Co(d^8)$ configuration, while previous studies indicated that cob(I)alamin should be described as open shell in character. Kornobis et al. [40] set out to reach a consensus on the character of transitions giving rise to the cob(I)alamin Abs spectra using TD-DFT. The Abs spectra were simulated with CAM-B3LYP and BP86 and were compared to the experimental spectrum. BP86 performed better than the hybrid in exhibiting good agreement with experimentally determined Abs, CD, and MCD spectra. It was concluded that the lowest energy band was MLCT in characterand not a single $\pi \rightarrow \pi^*$ transition followed by a vibrational progression as was previously thought [49]. In addition, the idea that the α/β region arises from a vibrational progression was not supported in this study rather both functionals indicated that multiple electronic transitions are involved in this region.

10 Summary and Future Directions

In sum, Abs spectroscopy is an important tool to probe the electronically excited states of cobalamins. It is evident that Abs spectra can be used as a diagnostic tool to distinguish cobalamins from each other based on oxidation state and the nature of the upper axial ligand. While experimental Abs spectroscopy is of critical importance to understand the light sensitivity of B_{12} derivatives, this alone is not enough to understand the nature of the transitions that give rise to the distinctive spectral features. TD-DFT is an appropriate computational tool to elucidate the nature of the excitations responsible for a particular B_{12} Abs spectrum. However, proper functional choice is paramount in predicting these transitions and simulating the full spectrum reliably. At this juncture in the field of B_{12} chemistry, it is indisputable that the BP86 functional are the proper functional for the assessment of the electronically excited states of the cobalamins. The specific stories for each of the cobalamins discussed in this chapter certainly corroborate the use of BP86. This GGA-type functional is especially dependable in accounting for the contributions from Co. Now that details of the electronically excited states of the cobalamins are widely understood through the lens of Abs spectroscopy, it is time to look ahead to see how this tool can be harnessed to solve practical problems in the field of B_{12} chemistry.

Abs spectroscopy could be an important tool to understand light-induced RP formation. This has applications across various disciplines, but recent attention has been given to the generation of OH radicals, drug delivery, antivitamins B_{12}, the transcription regulator CarH, and mimicking enzymatic catalysis. It has been shown that hydroxyl radicals can be generated in situ from the aerobic photolysis of hydroxocobalamin (HOCbl), and this can be used to study DNA structure and binding [35]. Cobalamins can also be synthetically altered to be used as scaffolds for the targeted release of therapeutics via light [80, 81]. Recently, two synthetic antivitamins B_{12}, namely EtPhCbl and PhEtyCbl, have been studied and represent an interesting group of derivatives that can counteract the physiological effects of the biologically relevant forms of B_{12} by inhibition [60]. EtPhCbl has been used to induce B_{12}-deficiency in laboratory animals in studies of pernicious anemia [62]. It is expected that these compounds can be used to investigate various aspects of B_{12} pathophysiology that remain poorly understood. EtPhCbl and PhEtyCbl also represent an important step

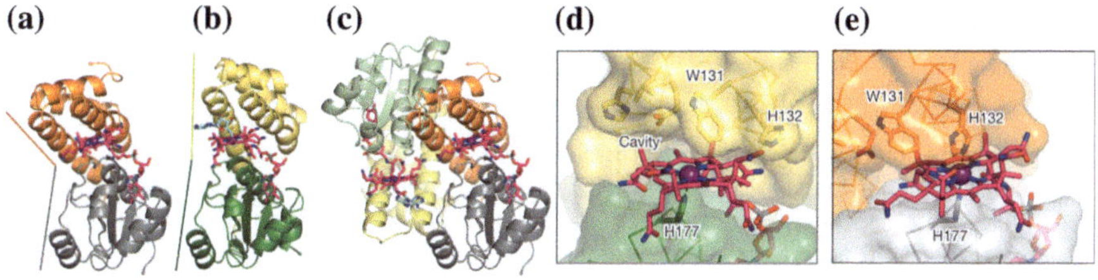

Fig. 18 Light-induced conformational changes in CarH. **a** Structure of light-exposed CarH. **b** Structure of CarH protomer in the dark state. **c** Tetramer disassembly after light-induced helix bundle movement. **d** Departure of 5′-dAdo group after light exposure. **e** Helix bundle movement brings His132 to the cobalt. In all figures, cobalamin's carbons are in pink, cobalt in purple, and 5′-dAdo group in cyan, respectively. Reprinted with permission from [47]. Copyright 2015 Springer Nature

toward the design of cobalamin-based agents for medicinal purposes as both can bind to all of the B_{12} related enzymes in humans [90, 96].

The photochemistry of B_{12} goes beyond synthetic applications as nature has already harnessed the light sensitivity of B_{12} as exemplified by the activity of the transcription regulator CarH [47]. There is still much to be understood, but the main challenge for theory would be to explain how the dark state converts to the light state (Fig. 18). The most unusual aspect is an ionic rather than radical mechanism. There is also a debate in the field of B_{12} chemistry as to whether or not B_{12} photochemistry can be used to mimic or to elucidate details of native enzymatic catalysis. For instance, the enzymes ethanolamine ammonia-lyase (EAL) and glutamate mutase (GLM) represent two AdoCbl-dependent enzymes that have been studied with advanced spectroscopic techniques in order to assess the photolytic properties of AdoCbl bound to enzymes with the hope that photochemical analyses will shed light on the native catalytic reactions [57, 70, 71, 79]. Recently, the photolytic properties of EAL and MetH have been investigated computationally [23, 56].

Finally, it should be pointed out that the full understanding of photophysics of cobalamins would be greatly enhanced through investigations of the crossings between ground/excited and/or excited/excited electronic states (conical intersections). Unfortunately, these features of the PES require multiconfigurational wave function treatment, for instance, the CASSCF wave function, because TD-DFT methodology leads to artifacts. As was shown for a small retinal model, the deficiencies of TD-DFT can be reduced by employing spin-restricted ensemble DFT (REKS), especially its state-averaged (SA-REKS) and state-interaction SA-REKS (SI-SA-REKS) variants [29]. It remains to be seen how this method can be adapted and how it performs in cobalamins.

References

1. Andruniów T, Kozlowski PM (2001) Theoretical analysis of electronic absorption spectra of vitamin B_{12} models. J Chem Phys 115:7522–7533
2. Andruniów T, Jaworska M, Lodowski P, Zgierski MZ, Dreos R, Randaccio L, Kozlowski PM (2008) Time-dependent density functional theory study of cobalt corrinoids: electronically excited states of methylcobalamin. J Chem Phys 115(085):101
3. Andruniów T, Jaworska M, Lodowski P, Zgierski MZ, Dreos R, Randaccio L, Kozlowski PM (2009) Time-dependent density functional theory study of cobalt corrinoids: electronically excited states of coenzyme B_{12}. J Chem Phys 115(105):105
4. Banerjee R (1997) The yin-yang of cobalamin biochemistry. Chem Biol 4:175–186
5. Banerjee R, Gherasim C, Padovani D (2009) The tinker, tailor, soldier in intracellular B_{12} trafficking. Curr Opin Chem Biol 13:484–491
6. Bonnett R (1963) The chemistry of the vitamin B_{12} group. Chem Rev 63:573–605
7. Brearley AE, Gott H, Hill HAO, O'Riordan M, Pratt JM, Williams RJP (1971) The chemistry of vitamin B_{12}. Part XIV. Reaction of vitamin B_{12s} with nitrobenzene and its reduction products. J Chem Soc A 12:612–614
8. Bridwell-Rabb J, Drennan CL (2017) Vitamin B_{12} in the spotlight again. Curr Opin Chem Biol 37:63–70
9. Brown KL (2005) Chemistry and enzymology of vitamin B_{12}. Chem Rev 105:2075–2150
10. Burke K, Werschnik J, Gross EKU (2005) Time-dependent density functional theory: past, present, and future. J Chem Phys 123(062):206
11. Casida ME (1996) Recent developments and application of modern density functional theory. Elsevier, Amsterdam
12. Cole AG, Yoder LM, Shiang JJ, Anderson NA, Walker LA, Holl B, M M, Sension RJ, (2002) Time-resolved spectroscopic studies of B_{12} coenzymes: a comparison of the primary photolysis mechanism in methyl-, ethyl-, n-propyl-, and 5'-deoxyadenosylcobalamin. J Am Chem Soc 124:434–441
13. Conrad KS, Brunold TC (2011) Spectroscopic and computational studies of glutathionylcobalamin: nature of Co-S bonding and comparison to Co-C bonding in coenzyme B_{12}. Inorg Chem 54:8755–8766
14. Day P (1967a) The electronic structure and spectrum of vitamin B_{12}. Coord Chem Rev 2:99–108
15. Day P (1967b) A theory of the optical properties of vitamin B_{12} and its derivatives. Theor Chim Acta 7:328–341
16. Dolphin D (1982) B_{12} Volume 1: chemistry. Wiley, New York
17. Eckert R, Kuhn H (1960) Richtunger der übersgangmomente der absorptionsbanden von polyenen, cyaninen und vitamin b_{12} aus dichroismus und fluoreszenzpolarisation. Z Electrochem 64:356–364
18. Eisenberg AS, Likhtina IVZ, S V, Birke RL, (2012) Electronic spectroscopy and computational studies of glutathionylcob(III)alamin. J Phys Chem A 116:6851–6869
19. Finke RG (1998) Coenzyme B_{12}-based chemical precedent for Co-C bond homolysis and other key elementary step. In: Kräutler B, Arigoni D, Golding BT (eds) Vitamin B_{12} and B_{12}-proteins. Wiley-VCH, Weinheim, pp 383–402
20. Firth RA, Hill HAO, Pratt JM, Williams RJP, Jackson WR (1967) The circular dichroism and absorption spectra of some vitamin B_{12} derivatives. Biochemistry 6:2178–2189
21. Garabato BD, Kumar N, Lodowski P, Jaworska M, Kozlowski PM (2016) Electronically excited states of cob(II)alamin: Insights from CASSCF/XMCQDPT2 and TD-DFT calculations. Phys Chem Chem Phys 18:4513–4526
22. Garabato BD, Lodowski P, Jaworska M, Kozlowski PM (2016) Mechanism of Co-C photodissociation in adenosylcobalamin. Phys Chem Chem Phys 18:19,070–19,082
23. Ghosh AP, Mamun AA, Lodowski P, Jaworska M, Kozlowski PM (2018) Mechanism of the photo-induced activation of Co-C bond in methylcobalamin dependent methionine synthase. J Photochem Photobiol B: Biol 18:306–317

24. Giannotti C (1982) Electronic spectra of B_{12} and related systems. In: Dolphin D (ed) B_{12} volume: chemistry. Wiley, New York, pp 393–430
25. Giedyk M, Goliszewska K, Gryko D (2015) Vitamin B_{12} catalysed reactions. Chem Soc Rev 44:3391–3404
26. Harris DA, Stickrath AB, Carroll EC, Sension RJ (2007) Influence of environment on the electronic structure of cob(III)alamins: time-resolved absorption studies of the S_1 state spectrum and dynamics. J Am Chem Soc 129:7578–7585
27. Hay BP, Finke RG (1988) Thermolysis of the Co-C bond in adenosylcobalamin (coenzyme B_{12} - IV. Products, kinetics, and Co-C bond dissociation energy studies in ethylene glycol. Polyhedron 7:1469–1481
28. Hirao H (2011) Which DFT functional performs well in the calculation of methylcobalamin? Comparison of the BLYP and BP86 functionals and evaluation of the impact of empirical dispersion correction. J Phys Chem A 115:9308–9313
29. Huix-Rotllant M, Filatov M, Gozem S, Schapiro I, Olivucci M, Ferré N (2013) Assessment of density functional theory for describing the correlation effects on the ground and excited state potential energy surfaces of a retinal chromophore model. J Chem Theory Comput 9:3917–3922
30. Hung RR, Grabowski JJ (1999) Listening to reactive intermediates: application of photoacoustic calorimetry to vitamin B_{12} compounds. J Am Chem Soc 115:1359–1364
31. Jaworska M, Lodowski P (2003) Electronic spectrum of Co-corrin calculated with the TDDFT method. Mol Struct (THEOCHEM) 631:209–223
32. Jaworska M, Kazibut G, Lodowski P (2003) Electronic spectrum of cobalt-free corrins calculated by TDDFT method. J Phys Chem A 107:1339–1347
33. Jensen KP (2005) Electronic structure of cob(I)alamin: the story of an unusual nucleophile. J Phys Chem B 109:10,505–10,512
34. Jensen KP, Ryde U (2003) Theoretical prediction of the Co-C bond strength in cobalamins. J Phys Chem A 107:7539–7545
35. Jones AR (2017) The photochemistry and photobiology of vitamin B_{12}. Photochem Photobiol Sci 16:820–834
36. Kepp KP (2014) Co-C dissociation of adenosylcobalamin (coenzyme B_{12}): role of dispersion, induction effects, solvent polarity, and relativistic and thermal corrections. J Phys Chem A 118:7104–7117
37. Kobylianskii IJ, Widner FJ, Kräutler B, Chen P (2013) Co-C bond energies in adenosylcobinamide and methylcobinamide in the gas phase and in silico. J Am Chem Soc 135:13,648–13,651
38. Kornobis K, Kumar N, Wong BM, Lodowski P, Jaworska M, Andruniów T, Ruud K, Kozlowski PM (2011) Electronically excited states of vitamin B_{12}: benchmark calculations including time-dependent density functional theory and correlated ab initio methods. J Phys Chem A 115:1280–1292
39. Kornobis K, Kumar N, Lodowski P, Jaworska M, Piecuch P, Lutz JJ, Wong BM, Kozlowski PM (2013) Electronic structure of the S_1 state in methylcobalamin: Insight from CASSCF/MC-XQDPT2, EOM-CCSD, and TD-DFT calculations. J Comp Chem 34:987–1004
40. Kornobis K, Ruud K, Kozlowski PM (2013) Cob(I)alamin: Insight into the nature of electronically excited states elucidated via quantum chemical computations and analysis of absorption, CD and MCD data. J Phys Chem A 117:863–876
41. Kozlowski PM, Kumar M, Piecuch P, Li W, Bauman NP, Hansen JA, Lodowski P, Jaworska M (2012) The cobalt-methyl bond dissociation in methylcobalamin: new benchmark analysis based on density functional theory and completely renormalized coupled-cluster calculations. J Chem Theory Comput 8:1870–1894
42. Kozlowski PM, Garabato BD, Lodowski P, Jaworska M (2016) Photolytic properties of cobalamins: a theoretical perspective. Dalton Trans 45:4457–4470
43. Kuhn H, Drexhage KH, Martin H (1965) The light absorption of vitamin B_{12}. Proc Roy Soc A 288(288):348–350

44. Kumar M, Kozlowski PM (2017) Electronic and structural properties of cob(I)alamin: ramifications for B_{12}-dependent processes. Coord Chem Rev 333:71–81
45. Kuta J, Patchkovskii S, Zgierski MZ, Kozlowski PM (2006) Performance of DFT in modeling electronic and structural properties of cobalamins. J Comput Chem 27:1429–1437
46. Kuta J, Wuerges J, Randaccio L, Kozlowski PM (2009) Axial bonding in alkylcobalamins: DFT analysis of the inverse versus normal trans influence. J Phys Chem A 113:11,604–11,612
47. Kutta RJ, Hardman SJ, Johannissen LO, Bellina B, Messiha HL, Ortiz-Guerrero JM, Elias-Arnanz M, Padmanabhan S, Barran P, Scrutton NS, Jones AR (2015) The photochemical mechanism of a B_{12}-dependent photoreceptor protein. Nat Commun 6:7907
48. Lewis NJ, Pfaltz A, Eschenmoser A (1983) Acid-catalyzed demetalation of nickel-hydrocorphin and cobalt-corrin complexes with 1,3-propoanedithiol. Angew Chem Int Ed 22:735–736
49. Liptak MD, Brunold TC (2006) Spectroscopic and computational studies of Co^{1+} cobalamin: spectral and electronic properties of the “superreduced” B_{12} cofactor. J Am Chem Soc 128:9144–9156
50. Liu H, Kornobis K, Lodowski P, Jaworska M, Kozlowski PM (2014) TD-DFT insight into photodissociation of the Co-C bond in coenzyme B_{12}. Front Chem 1:1–12
51. Lodowski P, Jaworska M, Andruniów T, Kumar M, Kozlowski PM (2009) Photodissociation of Co-C bond in methyl- and methylcobalamin: an insight from TD-DFT calculations. J Phys Chem B 113:6898–6909
52. Lodowski P, Jaworska M, Kornobis K, Andruniów T, Kozlowski PM (2011) Electronic and structural properties of low-lying excited states of vitamin B_{12}. J Phys Chem B 115:13,304–13,319
53. Lodowski P, Jaworska M, Andruniów T, Garabato BD, Kozlowski PM (2014) Mechanism of the S_1 excited state internal conversion in vitamin B_{12}. Phys Chem Chem Phys 16:18,675–18,679
54. Lodowski P, Ciura K, Toda MJ, Jaworska M, Kozlowski PM (2017) Photodissociation of ethylphenylcobalamin antivitamin B_{12}. Phys Chem Chem Phys 19:30,310–30,315
55. Luo LB, Li G, Chen HL, Fu SW, Zhang SY (1998) Laser-induced photoacoustic calorimetric determination of enthalpy and volume changes in photolysis of 5’-deoxyadenosylcobalamin and methylcobalamin. J Chem Soc Dalton Trans pp 2103–2107
56. Mamun AA, Toda MJ, Lodowski P, Jaworska M, Kozlowski PM (2018) Mechanism of light induced radical pair formation in coenzyme B_{12}-dependent ethanolamine ammonia-lyase. ACS Catalysis 8:7164–7178
57. Marsh EN, Melendez GD (2012) Adenosylcobalamin enzymes: theory and experiment begin to converge. Biochim Biophys Acta 1824:1154–1164
58. Martin BD, Finke RG (1990) Co-C homolysis and bond-dissociation energy studies of biological alkylcobalamins: methylcobalamin, including a $\geq 10^{15}$ Co-CH_3 homolysis rate enhancement at 25°c following one-electron reduction. J Am Chem Soc 112:2419–2420
59. Martin BD, Finke RG (1992) Methylcobalamins full-strength vs half-strenght cobalt-carbon sigma-bonds and bond-dissociation enthalpies $\geq 10^{15}$ Co-CH_3 homolysis rate enhancement following one-antibonding-electron reduction of methylcobalamin. J Am Chem Soc 114:585–592
60. Miller NA, Wiley TE, Spears KG, Ruetz M, Kieninger C, Krautler B, Sension RJ (2016) Toward the design of photoresponsive conditional antivitamins B_{12}: a transient absorption study of an arylcobalamin and an alkynylcobalamin. J Am Chem Soc 138:14,250–14,256
61. Miller NA, Deb A, Alonso-Mori R, Garabato BD, Glownia JM, Kiefer LM, Koralek J, Sikorski M, Spears KG, Wiley TE, Zhu D, Kozlowski PM, Kubarych KJ, Penner-Hahn JE, Sension RJ (2017) Polarized XANES monitors femtosecond structural evolution of photoexcited vitamin B_{12}. J Am Chem Soc 139:1894–1899
62. Mutti E, Ruetz M, Birn H, Krütler B, Nexo E (2013) 4-ethylphenyl-cobalamin impairs tissue uptake of vitamin B_{12} and causes vitamin B_{12} deficiency in mice. PLOS ONE 8(e75):312
63. Offenhartz PO, Offenhartz BH, Fung MM (1970) Theoretical analysis of corrin optical spectra. J Am Chem Soc 92:2966–2973

64. Park K, Brunold TC (2013) Combined spectroscopic and computational analysis of the vibrational properties of vitamin B_{12} in its Co^{3+}, Co^{2+}, and Co^{1+} oxidation states. J Phys Chem B 117:5397–5410
65. Peach MJ, Benfield P, Helgaker T, Tozer DJ (2008) Excitation energies in density functional theory: an evaluation and a diagnostic test. J Chem Phys 128(044):118
66. Perry CB, Marques HM (2005) Probing the cis and trans influrence in cobalamin chemistry by electronic spectroscopy. S Afr J Chem 58:9–15
67. Pratt JM (1972) Inorganic chemistry of vitamin B_{12}. Academic Press, London
68. Pratt JM (1999) Electronic structure and spectra of B_{12}: from trans effects to protein conformation I and II. In: Banerjee R (ed) Chemistry and biochemistry of B_{12}. Wiley, New York, pp 113–164
69. Ragsdale SW (2006) Metals and their scaffolds to promote difficult enzymatic reactions. Chem Rev 106:3317–3337
70. Robertson WD, Wang M, Warncke K (2011) Characterization of protein contributions to cobalt-carbon bond cleavage catalysis in adenosylcobalamin-dependent ethanolamine ammonia-lyase by using photolysis in the ternary complex. J Am Chem Soc 133:6968–77
71. Roman-Melendez GD, von Glehn P, Harvey JN, Mulholland AJ, Marsh ENL (2014) Role of active site residues in promoting cobalt-carbon bond homolysis in adenosylcobalamin-dependent mutases revealed through experiment and computation. Biochemistry 53:169–77
72. Ruetz M, Gherasim C, Gruber K, Fedosov S, Banerjee R, Kräutler B (2013) Access to organometallic arylcobaltcorrins through radical synthesis: 4-ethylphenylcobalamin, a potential "antivitamin B_{12}". Angew Chem Int Ed 52:2606–2610
73. Ruetz M, Salchner R, Wurst K, Fedosov S, Kräutler B (2013) Phenylethynylcobalamin: a lightstable and thermolysisresistant organometallic vitamin B_{12} derivative prepared by radical synthesis. Angew Chem Int Ed 52:11,406–11,409
74. Runge E, Gross EKU (1984) Density-functional theory for time-dependent systems. Phys Rev Lett 52:997–1000
75. Rury AS, Wiley TE, Sension RJ (2015) Energy cascades, excited state dynamics, and photochemistry in cob(III)alamins and ferric porphyrins. Acc Chem Res 48:860–867
76. Ryde U, Mata RA, Grimme S (2011) Does DFT-D estimate accurate energies for the binding of ligands to metal complexes? Datlon Trans 40:11,176–11,183
77. Sandala GM, Smith DM, Radom L (2010) Modeling the reactions catalyzed by coenzyme B_{12}-dependent enzymes. Acc Chem Res 43:642–651
78. Sension RJ, Harris DA, Cole AG (2005a) Time-resolved spectroscopic studies of B_{12} coenzymes: Comparison of the influence of solvent on the primary photolysis mechanism and geminate recombination of methyl-, ethyl-, n-propyl, and 5'-deoxyadenosylcobalamin. J Phys Chem B 109:21,954–21,962
79. Sension RJ, Harris DA, Stickrath A, Cole AG, Fox CC, Marsh EN (2005b) Time-resolved measurements of the photolysis and recombination of adenosylcobalamin bound to glutamate mutase. J Phys Chem B 109:18,146–18,152
80. Shell TA, Lawrence DS (2015) Vitamin B_{12}: a tunable, long wavelength, light-responsive platform for launching therapeutic agents. Acc Chem Res 48:2866–2874
81. Shell TA, Shell JR, Rodgers ZL, Lawrence DS (2014) Tunable visible and near-IR photoactivation of light-responsive compounds by using fluorophores as light-capturing antennas. Angew Chem Int Ed 53:875–878
82. Shiang JJ, Walker LA, Anderson NA, Cole AG, Sension RJ (1999) Time-resolved spectroscopic studies of B_{12} coenzymes: The photolysis of methylcobalamin is wavelength dependent. J Phys Chem B 103:10,532–10,539
83. Shiang JJ, Cole AG, Sension RJ, Hang K, Weng Y, Trommel JS, Marzilli LG, Lian T (2006) Ultrafast excited-state dynamics in vitamin B_{12} and related cob(III)alamins. J Am Chem Soc 128:801–808
84. Solheim H, Kornobis K, Ruud K, Kozlowski PM (2011) Electronically excited states of vitamin B_{12} and methylcobalamin: theoretical analysis of absorption, CD, and MCD data. J Phys Chem B 115:737–748

85. Stich TA, Brooks AJ, Buan NR, Brunold TC (2003) Spectroscopic and computational studies of Co^{3+}-corrinoids: spectral and electronic properties of the B_{12} cofactors and biologically relevant precursors. J Am Chem Soc 125:5897–5914
86. Stich TA, Buan NR, Brunold TC (2004) Spectroscopic and computational studies of Co^{2+}-corrinoids: spectral and electronic properties of the biologically relevant base-on and base-off forms of Co^{2+}cobalamin. J Am Chem Soc 126:9735–9749
87. Stickrath AB, Carroll EC, Dai X, Harris DA, Rury A, Smith B, Kc Tang, Wert J, Sension RJ (2009) Solvent-dependent cage dynamics of small nonpolar radicals: lessons from the photodissociation and geminate receombinations of alkylcobalamins. J Phys Chem A 113:8513–8522
88. Toohey JIA (1965) Vitamin B_{12} compound containing no cobalt. Proc Natl Acad Sci USA 54:934–942
89. Toraya T (2003) Radical catalysis in coenzyme B_{12}-dependent isomerization (eliminating) reactions. Chem Rev 103:2095–2128
90. Waibel R, Treichler H, Schaefer NG, van Staveren DR, Mundwiler S, Kunze S, Kuenzi M, Alberto R, Nuesch J, Knuth A, Moch H, Schibli R, Schubiger PA (2008) New derivatives of vitamin B_{12} show preferential targeting of tumors. Cancer Res 68:2904–2911
91. Walker LA, Jarrett JT, Anderson NA, Pullen SH, Matthews RG, Sension RJ (1998) Time-resolved spectroscopic studies of B_{12} coenzymes: the identification of a metastable cob(III)alamin photoproduct in the photolysis of methylcobalamin. J Am Chem Soc 120:3597–3603
92. Wiley TE, Arruda BC, Miller NA, Lenard M, Sension RJ (2015) Excited electronic states and internal conversion in cyanocobalamin. Chin Chem Lett 26:439–443
93. Wiley TE, Miller WR, Miller NA, Sension RJ, Lodowski P, Jaworska M, Kozlowski PM (2016) Photostability of hydroxocobalamin: ultrafast excited state dynamics and computational studies. J Phys Chem Lett 7:143–7
94. Yoder LM, Cole AG, Walker LA, Sension RJ (2001) Time-resolved spectroscopic studies of B_{12} coenzymes: influence of solvent on the photolysis of adenosylcobalamin. J Phys Chem B 105:12,180–12,188
95. Zelder F, Zhou K, Sonnay M (2013) Peptide B12: emerging trends at the interface of inorganic chemistry, chemical biology and medicine. Dalton Trans 42:854–862
96. Zelder F, Sonnay M, Prieto L (2015) Antivitamins for medicinal applications. Chem Bio Chem 16:1264–1278

Photodeactivation Channels of Transition Metal Complexes: A Computational Chemistry Perspective

Daniel Escudero

Abstract A detailed molecular-level understanding of the excited-state (ES) decay dynamics of transition metal complexes (TMCs) is vital to develop the next generation of light-active components in a wide variety of applications related to photochemistry, including optoelectronics, photocatalysis, dye-sensitized solar cells, artificial photosynthesis, photonics sensors and switches, and bioimaging. After photoexcitation, TMCs can undergo a plethora of interconnected relaxation processes, which compete to each other and are controlled by the subtle interplay of electronic and geometrical rearrangements that take place during the ES deactivation dynamics at different timescales. Intrinsic factors such as (i) the spin and character of the electronically ES involved in the process and (ii) the energetic alignment and effective couplings between these states do play a protagonist role in determining the preferred deactivation channels. Extrinsic factors, such as temperature, pressure, excitation wavelength, and environmental effects, can often strongly modify the outcome of the photochemical processes. As kinetic control is always at play, only the fastest processes among all possible deactivation channels are generally observed. Due to their high density of ES of various characters, TMCs usually display rich and chameleonic ES and photochemical properties. Computational chemistry is a powerful and unique tool to provide a microscopic and time-resolved description of these complex processes, and it often constitutes the fundamental ingredient for the interpretation of time-resolved absorption and emission spectroscopic measurements. This chapter provides first a general overview on this complex topic, followed by an overview of the state-of-the-art quantum chemical and reaction dynamics methods to study the photodeactivation dynamics of TMCs and finally illustrates the progress and challenges in this field with recent examples from the literature. Importantly, these examples cover the ultrafast ES decay regime but also the long-lived photodeactivation from thermally equilibrated ES.

D. Escudero (✉)
Department of Chemistry, KU Leuven, Celestijnenlaan 200F, 3001 Leuven, Belgium
e-mail: daniel.escudero@kuleuven.be

E. Broclawik et al. (eds.), *Transition Metals in Coordination Environments*, Challenges and Advances in Computational Chemistry and Physics 29,
https://doi.org/10.1007/978-3-030-11714-6_9

1 General Overview

In a molecular system, the chemistry in the electronically excited states (ES) tends to significantly differ from that of the ground state (GS). First, as the photon absorption conducts the molecular system to a high-energy level, the ESs are generally short-lived and significantly more reactive than the GS. The molecular system in their ESs will tend to dissipate this excess of energy in a radiative or radiationless manner. Second, the GS and the ES's molecular properties and thereto their chemical reactivity strongly differ. GS molecular properties can easily be analyzed in terms of basic concepts of bonding, anti-bonding, and/or non-bonding interactions, and furthermore, numerous (i) structure–property relationships and (ii) rules to predict chemical reactivity are available for the GS. In contrast, ES molecular properties and ES reactivity are hardly predictable. This complex behavior is even more dramatic in the case of transition metal complexes (TMCs), which compared with organic molecules, possess (i) a much higher density of ESs in the near-UV/visible spectral regions, (ii) easily populated triplet ESs due to larger spin–orbit couplings (SOCs), and (ii) a larger variety of ESs of different character [1, 2]. Among these types of ESs, I highlight: (i) metal-centered (MC) states, i.e., fully localized on the transition metal; (ii) ligand-centered (LC) states, i.e., localized on a single ligand; (iii) ligand-to-metal charge transfer (LMCT) states, where the ligand acts as a donor and the transition metal as an acceptor; (iv) metal-to-ligand charge transfer (MLCT) states; and (v) ligand-to-ligand charge transfer (LLCT) states [3, 4]. In some cases, the ES character is readily recognizable from a visual inspection of the involved orbitals. However, it is often found that there is a certain degree of admixture between different ES, leading to, e.g., mixed MLCT/LC ES. To alleviate these issues and avoid subjective assignments of the ES characters, several quantitative analysis tools have been developed in recent years [5–7].

Let us now recall the possible pathways of a given photochemical reaction. These processes are schematically presented in the Jablonski diagram depicted in Fig. 1. Such a textbook representation is chosen herein for several reasons: first because it is generally valid for small organic chromophores and second because it is presumably well known by most of the readers. Note that the photodeactivation pathways highlighted in Fig. 1 and their corresponding timescales might not always apply for TMCs. Upon photoexcitation, which is a very fast process taking place on a timescale of ca. 10^{-15} s, and thus the nuclei remain fixed during this process (i.e., Franck–Condon (FC) approximation), the manifold of singlet ESs (S_m) is populated vertically. According to the Kasha rule [8], which it is generally assumed to always hold true, radiationless deactivation would occur next through internal conversion (IC) and vibrational relaxation processes, which ultimately lead to the population of the lowest vibrational level of the S_1 state. The latter processes are generally very fast; i.e., the k_{IC} rate constant is usually higher than 10^{12} s^{-1}, so that other possible deactivation processes (such as $S_m \rightarrow T_n$ intersystem crossing (ISC) processes and/or anti-Kasha $S_m \rightarrow T_0$ radiative deactivation) are often not competitive to IC. This is the reason that Kasha rule holds true for many molecular systems. In this chapter, we will showcase several anti-Kasha scenarios for TMCs (see, e.g., Sects. 3.2 and 3.3). Next, starting from the thermally equilibrated S_1 state, fluorescence can then occur if the emission of a photon is competitive to other radiationless processes depopulating this state, such as $S_1 \rightarrow S_0$ IC and/or $S_1 \rightarrow T_n$ ISC processes (note that the k_f values typically range between 10^{11} and 10^8 s^{-1}). In the case that ISC is fast enough to compete with fluorescence and IC, triplet ESs will be populated with a certain yield. Importantly, for many TMCs, due to: (i) their high density of near-degenerate ES of different spin symmetry and (ii) the large size of the SOCs between these states; it often leads to an important grade of admixture between these pure states, and thus resulting in mixed spin–orbit states. This situation makes often meaningless not only (i) the classical nomenclature of "pure" (e.g., singlet, triplet, or any other multiplicity) ESs, (ii) but also the fundamental difference between an IC process and an ISC process. Moreover, in these systems, SOCs might not always be the main driving force for ISC, but instead it could be controlled by vibronic effects [9]. Let us assume now that the "pure" lowest triplet ES is finally predominantly populated. Once the well of T_1 is populated, long-lived phosphorescence with a rate constant k_r typically ranging between 10^6 and 10^{-2} s^{-1} will occur from this thermally equilibrated ES if it is fast enough to compete with $T_1 \rightarrow S_0$ nonradiative decay. An important feature of triplet states is that they consist of three different spin sublevels (T_x, T_y, and T_z), in accordance with the three different M_s quantum numbers (+1, 0, and −1), and that result in the three different components of their associated rates (see Fig. 1) [10]. To complete this plethora of interconnected relaxation processes, the manifold of singlet and triplet ES may additionally be depopulated through many other photophysical and/or photochemical processes. Among others, we highlight: (i) reverse ISC (rISC) processes leading to the repopulation of the S_1 state, and which are responsible of the thermally activated delayed fluorescence (TADF) phenomenon that has recently found tremendous interest for OLED applications [11], (ii) photochemical reactions leading to photoproducts, and (iii) intermolecular photoinduced energy and electron transfer processes occurring due to the interaction of the molecular system with its environment. Having in mind the above discussion, one can easily recognize how complex and intricate the ES dynamics is, being the final outcomes of the photochemical reaction controlled by a subtle interplay of electronic and geometrical rearrangements that take place during the ES deactivation dynamics [12]. While intrinsic factors of the molecular systems, such as (i) the spin and character of the ES involved in the process and (ii) the energetic alignment and effective couplings between these states, do play a protagonist role in determining the preferred deactivation channels, other factors of extrinsic nature, such as temperature, pressure, excitation wavelength, and environmental effects, can often strongly modify the outcome of the photochemical processes. To illustrate this complexity, computational works covering the early-time photophysics but also the long-lived ES decay regime are highlighted in this book chapter (see selected examples in Sect. 3).

Approaching the study of the ES dynamics of TMCs from an experimental viewpoint requires techniques that can provide sensitivity to the electronic states, their spin and the vibrational dynamics. Obtaining a simultaneous description of all these variables within a single technique is challenging, and thus, a combination

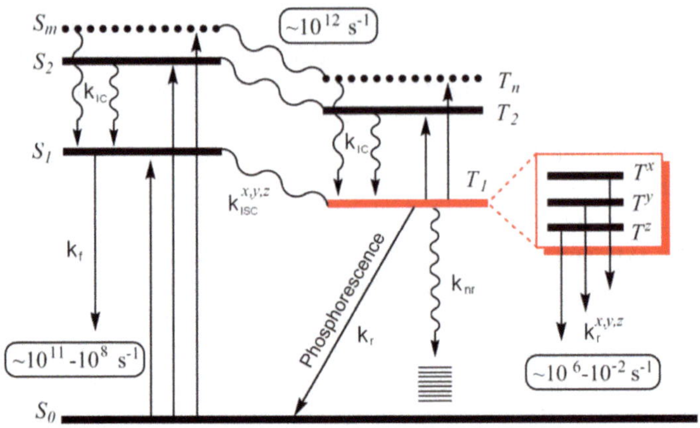

Fig. 1 Schematical Jablonski diagram highlighting the main ES decay channels. Reprinted with permission from [10]. Copyright 2018 American Chemical Society

of different experimental approaches is often needed. While steady-state fluorescence/phosphorescence spectroscopy and photoluminescence quantum yield determinations have become nowadays routine investigations, the recent advances during the last decades in techniques with high-spectral and time resolution enabled to provide a leap in our understanding of TMC photochemistry [13]. Among these techniques, optical pump-probe techniques, i.e., time-resolved emission (fluorescence or phosphorescence) and transient optical absorption spectroscopies [14], track the ES dynamics through changes in the spectral features over time. These techniques permitted for instance to attain quantitative determinations of ISC processes in TMCs. For instance, using these techniques, ISC was found to occur in an ultrafast manner in the case of, e.g., Ir(ppy)$_3$ (<100 fs) [15], Re(Cl)(CO)$_3$(bpy) (80 fs) [16], and $[Ru(bpy)_3]^{2+}$ (<30 fs) [17, 18], or in a slowed down fashion in the case of the binuclear $[Pt_2(P_2O_5H_2)_4]^{4-}$ complex (29 ns) [19]. Other important spectroscopic techniques for TMCs include (i) time-resolved vibrational spectroscopy, especially nonlinear Raman spectroscopies, such as femtosecond stimulated Raman scattering [20], which provided further insights into the nature of ISC of $[Ru(bpy)_3]^{2+}$ [21] and a Fe(II) spin crossover system [22] and (ii) modern time-resolved X-ray spectroscopies [23]. The use of X-ray spectroscopy is especially suited for TMCs, as this radiation can deliver not only electronic and spin state information but also structural information. The latter techniques enabled to unambiguously track the deactivation cascade in $[Fe(bpy)_3]^{2+}$ [24], which involves two-spin transitions that ultimately lead to the population of a non-emissive quintet MC state in less than 100 femtoseconds. In addition, it is also worth mentioning the progresses on multidimensional electronic and infrared spectroscopies [25], which provide information about the coupling between the electronic and vibrational degrees of freedom, respectively, and can alleviate problems in the case of overlapping absorption/emission bands. As a representative example, two-dimensional electronic spectroscopy (2D-ES) was used to track the ES decay dynamics of $[IrBr_6]^{2-}$ [26]. Finally, these techniques can often be used to analyze the effect of external stimulus on the photochemical properties, such as temperature, pressure, environmental effects (importantly aggregation effects), and excitation wavelength. For instance, the effect of temperature on the ISC timescales and yields in $[Pt_2(P_2O_5H_2)_4]^{4-}$ was deeply analyzed in [19]. In this study, temperature was found to play a very important role in controlling the ISC processes and thereto in determining the fluorescence/phosphorescence ratio of this binuclear complex at a given temperature. Other recent examples highlighting the effect of extrinsic factors on TMCs photochemistry are given in Sect. 3.3.

All the time-resolved techniques described above require very expensive and intricate experimental setups that prevent their use in a systematic manner. This is why theoretical and computational investigations have become increasingly popular in the field, as they offer an access to the ES properties and decay dynamics at a reasonable cost. In this regard, since the pioneer work on TMCs photochemistry of Daniel and coworkers in the 1990s [27, 28], the boost in the theoretical activity during recent years would have not been possible without the developments done on: (i) quantum chemical methods for the ES, (ii) algorithms to locate the optimal geometries of the relevant stationary points on the ES potential energy surfaces (PESs), (iii) ES decay rate theories, and (iv) ES reaction dynamics methods. In analogy with the experimental techniques, there is not a single state-of-the-art methodology able to capture all the inherent complexities of a photochemical reaction, so that often a combination of them is required. Thus, depending on the problematic to be solved one has to focus on attaining an accurate description of the ESs (e.g., modeling absorption and emission processes), or on obtaining accurate ES PESs (e.g., modeling photoreactivity), or conversely on choosing an adequate ES reaction dynamic method to model the early-time photophysics of TMCs. The absorption spectra of TMCs and the associated ES properties (i.e., ES dipoles, densities, and charges) at the FC point can now be routinely modeled [3], but still not with same accuracy as the ESs of organic systems (see Sect. 2.1). Conversely, accessing emission properties and vibronic effects remains more difficult and costly. Still, for some quantum chemical methods for the ES, such as time-dependent density functional theory (TD-DFT) [29], both analytic first (gradient) and second (Hessian) derivatives are already available, so that ES geometry optimizations and calculations of the ES vibrational levels are possible at a reduced computational cost. This has facilitated the calculation of a large panel of emissive (long-lived) properties in molecular systems containing up to ca. 100 atoms [30], including fluorescence spectra, k_f decay rates, and vibrationally resolved emission band shapes. Additionally, one can compute additional magnitudes, such as the adiabatic energy differences and the nonadiabatic couplings and SOCs between the electronic states of interest, which can be used in combination with ES decay rate formalisms (mainly Fermi-Golden rule-based) for the calculation of nonradiative decays rate, i.e., k_{IC} and k_{ISC}. Note that since Fermi-Golden expressions are based on perturbation theory, their application is strictly valid only when the size of the coupling is significantly smaller than the adiabatic energy gap between the final and initial states, so that these expressions should cautiously be used when dealing with close-lying ESs. The latter calculations, along with the calculation of the k_f and k_r decay rates, can ultimately provide access to the quantum yields of photoluminescence. Further details of these ES decay rate formalisms and recent examples of its application for TMCs will be presented in Sects. 2.2 and 3.1, respectively. Unfortunately, these investigations are yet scarce and can hardly be applied in a systematic manner. This is so because the calculation of particularly ES $\rightarrow$ GS nonradiative rates still pose significant problems, as (i) the actual mechanisms are often unknown without a prior exploration of the involved PES and (ii) more than one ES might be involved in these nonradiative decay processes. The possible radiationless deactivation pathways are schematically shown in Fig. 2. In the weak coupling limit (Fig. 2a), the PESs of the ES and GS are weakly displaced, so that both surfaces will only intersect at regions far away from the ES minimum. Therefore, the nonradiative decay is mainly driven by the overlap between the vibrational wavefunctions of the ES and the GS, and thus, it is relatively slow and it generally obeys the energy gap law [31]. This weak coupling limit is found for aromatic hydrocarbons, but also for strongly emissive molecular systems. In Fig. 2b, the strong coupling limit is depicted, which is characterized by largely displaced ES and GS minima. Due to these large displacements, the ES and GS PESs intersect at regions near to the ES minimum, leading to nonadiabatic photochemical events. These surface crossings are mediated by conical intersections (CoIns) [32] and provide very efficient funnels for the radiationless deactivation to the GS, which typically occur on a subpicosecond scale for these molecular systems [3]. The MC states of TMCs do belong to this strong coupling limit [33]. In a nutshell, if one aims to study the nonradiative processes of a given molecular system, one should first explore their ES and GS PESs and locate the relevant stationary points along the deactivation pathways to assess whether the strong or the weak limit applies. In addition, this situation may become even more complex due to the population of close-lying ES; see Fig. 2c. In Fig. 2c, ES_1 is weakly coupled to the GS and it may thus decay back to the GS in a radiative or nonradiative manner. Conversely, ES_2 is strongly coupled to the GS and it will thus be exclusively involved in nonradiative processes. This situation is very common for TMCs, being ES_1 an emissive state of fundamentally MLCT character and ES_2 a MC state that acts as a quencher of photoluminescence [34]. Depending on the relative energetic alignment between ES_1 and ES_2, an activation barrier to populate ES_2 might appear (see E_{act} in Fig. 2), leading to strongly temperature-dependent nonradiative decay rates. Cyclometalated Ir(III) complexes do usually belong to this latter scenario [35]. Importantly, these situations require from complex ES kinetic models, since the involved ESs are thermally equilibrated. In this regard, the development of realistic kinetic models is crucial to attain quantitative determinations of the photoluminescence quantum yields (PLQY) in these systems. Recent computational works combining the calculation of ES decay rate constants at the first principles level with kinetic modeling enabled the calculations of lifetimes and PLQY values of Ir(III) complexes [36, 37]. These strategies are further elaborated in Sect. 3.1.

The above static approach is best suited to assess the competition between radiative and nonradiative processes occurring at long timescales (i.e., from the nanosecond regime, where fluorescence typically occurs, up to the micro or the millisecond regime, and thus also covering phosphorescence), and indeed, this approach remains

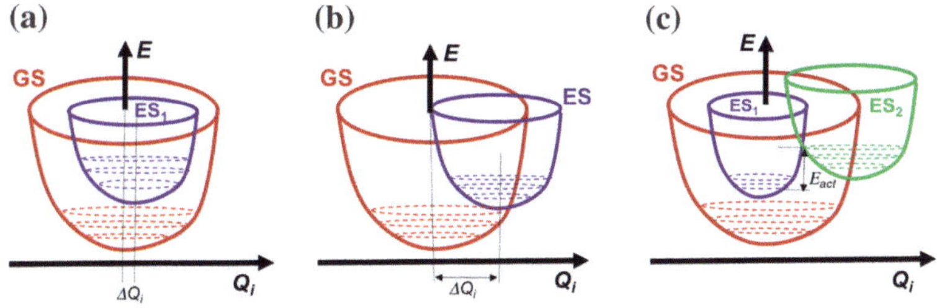

Fig. 2 The weak coupling limit **a** and at the strong coupling limit (**b**, **c**) A three-state model involving a strongly and weakly coupled ESs besides the GS

to date the only possible alternative, as the simulation of these long timescales poses significant challenges for other complementary approaches, such as ES reaction dynamics simulations, that can only be run up to the picosecond regime due to their increased computational cost. However, the use of a static approach brings two major drawbacks: (i) It does not provide real time resolution, and (ii) it is inherently limited by the calculation of energies and properties at fixed nuclear coordinates (typically GS and ES minima). In this regard, a much more detailed description of the fate of the ES population can be obtained with ES reaction dynamics methods [9, 38], which explicitly include time and temperature. Two kinds of reaction dynamics methods (see Sect. 2.3) are commonly used to track the ES dynamics: quantum wavepacket dynamics [9] and *on-the-fly* nonadiabatic molecular dynamics [38–40], and both methods have already provided important insights into the photodeactivation dynamics of TMCs. The fine interplay between the spin, electronic, and nuclear degrees of freedom can only be attained with the most expensive quantum wavepacket dynamics methods, since both nuclei and electrons are fully treated quantum mechanically. Conversely, *on-the-fly* nonadiabatic molecular dynamics offer the possibility to perform simulations on full-dimensional PES and to even include the atomistic descriptions of the environment. Running dynamical simulations beyond the picosecond regime are still challenging, and thus, these methods were mainly used to study the primary events occurring within the first few picoseconds after photoexcitation of TMCs. Despite these limitations, these techniques helped unraveling the subtle effects that drive the complicated ES dynamics of TMCs. In this context, one of the most relevant examples is the study of their ultrafast IC and ISC kinetics involving the high density of ES of various multiplicities and characters, which enabled to unambiguously unravel the ES decay mechanisms and highlighted the interplay between spin and vibronic interactions [9]. In these dynamics simulations, the evolution of the ES species can be followed in real time, so that rate constants can be directly attained from quantum dynamics simulations or statistically derived by averaging independent trajectories in the case of nonadiabatic molecular simulations. These methods are discussed in Sect. 2.2, and a recent application of *on-the-fly* nonadiabatic molecular dynamics is shown in Sect. 3.2.

One important aspect when modeling ES is that the impact of the environment is often larger in the ES than in the GS, and thus, the outcome of the photochemical processes can strongly vary from the gas phase to the condensed phase and/or in the presence of complex environments. This poses significant challenges from a modeling viewpoint, as the explicit quantum chemical consideration of the environment can become easily prohibitive for the ES. Thus, the most popular approaches to mimic the environment are (i) hybrid techniques such as quantum mechanical/molecular mechanics (QM/MM), which provide an atomistic description of the environment at a reasonable cost [41], and (ii) polarizable continuum models (such as the PCM approach [42]), where the solvent is modeled as a polarizable continuum rather than with its atomistic description. PCM models are commonly used to mimic solvation effects and provide the best compromise between accuracy and computational cost. Problems of the PCM approach may arise (i) in the presence of protic solvents, where specific solvent–solute hydrogen bonding interactions may arise and (ii) for excitations involving large density rearrangements. A first shell of explicit solvent molecules in combination with the bulk PCM effects may alleviate the first problem, as exemplarily shown for a Fe(II) complex [43], while the latter problem appears in linear response approaches and requires the use of advanced state-specific (SS) PCM models [44, 45]. The SS-PCM approaches are best suited to model emission processes. This is so because photon emission is often a "slow" process, involving not only the solute geometry relaxation in the ES but also the reorganization of the solvent molecules to adapt to the solute charge distribution in the ES (an example of the interplay between temperature and solvation effects is presented in Sect. 3.3). Alternatively, QM/MM methods can be used straightforwardly in systems where one can clearly define the partitioning between the photoactive part and the environment, such as protein embedding, modeling caging, and solvent effects, as well as in the case of molecular crystals. However, it is not always possible to split the level of accuracy of the different parts of the system, since for instance, in photosynthetic systems many photoactive units are spread within the protein, and therefore, the whole system would require to be described quantum mechanically. Other related problems may arise from the appearance of excimers and exciplexes due to aggregation in the condensed phase. These situations require the inclusion of several photoactive units in the QM part. An alternative to study large multichromophoric systems is to use exciton models [46]. Aiming at an accurate recovery of the environmental effects for the ES in a QM/MM framework, increased efforts in the community have been recently devoted to the development of polarizable QM/MM schemes [47], where both QM and MM parts polarize each other. Finally, as a prominent alternative for QM/MM, subsystem and embedding quantum chemical schemes have found increased interest in recent years [48].

In this chapter, in Sect. 2 an overview of the state-of-the-art theoretical/computational methods to unveil photodeactivation channels in TMCs is presented, while selected recent computational works are highlighted in Sect. 3. These examples were selected to illustrate the progress achieved in deciphering photodeactivation channels of TMCs and which importantly cover the ultrafast ES decay regime (Sect. 3.2) but also longer timescales including long-lived phosphorescence (Sects. 3.1 and 3.3). Finally, Sect. 4 is devoted for conclusions and outlook.

2 State-of-the-Art Theoretical and Computational Methods for the Study of the ES Deactivation Channels of TMCs

This section is split into three subsections, and it covers in the following order the state-of-the-art quantum chemical methods for the ES and computational techniques to study the ES PES (Sect. 2.1), ES decay rate theories (Sect. 2.2), and ES reaction dynamic methods (Sect. 2.3). The objective of this section is not to present an exhaustive description of these methods, as this topic has extensively been covered elsewhere [3, 9, 40], but to present a concise overview that focuses on the recent advances in theoretical and computational methods to describe the ES of TMCs.

2.1 Quantum Chemical Methods for the ES

As highlighted in Sect. 1, the accurate description of the ESs of TMCs poses significant challenges, as these systems concentrate multiple complexities inherent to their theoretical study (e.g., relativistic effects, near degeneracies, high density of ESs of various spins and characters). Due to their capability to recover large amounts of dynamic and non-dynamic electron correlation, multi-configurational wave function approaches, such as the complete-active-space second-order perturbation theory (CASPT2) [49] and the restricted-active-space PT2 (RASPT2) [50] methods, are capable to deal with all kinds of electronic structures, including the complex ESs of TMCs. In fact, the errors of these methods for vertical excitation energies are generally well below 0.3 eV for TMCs, and within 0.15 eV for most of the cases [51, 52]. However, these methods have not yet acquired a black-box status, since (i) the selection of a chemically sound active space is not always straightforward, and (ii) their computational cost still restricts their use for medium- and large-size TMCs, which require the use of very large active spaces. To alleviate the first problem, general rules and even automatic approaches have been developed to choose a chemically sound active space for TMCs [53–55]. The second problem has a less likely solution within these approaches. However, recent advances on the use of matrix renormalization group (DMRG) [56, 57] and quantum Monte Carlo [58] linked to CASPT2 permit the study in practical terms of large TMCs with 30–50 active orbitals [59, 60].

Methods based on DFT, such as constricted variational DFT (CV-DFT) [61], spin-flip and/or ΔSCF-DFT [62], and TD-DFT in its linear response version [29], appear as computationally cheaper alternatives to treat the ES of TMCs, but they are generally less accurate than multi-configurational methods. In opposition to organic molecules, for which abundant benchmark data, and systematic TD-DFT studies evaluating the performance of the different exchange-correlation (xc) functionals are common in the literature, systematic TD-DFT studies on TMCs are more scattered and less common, as the attainment of reliable benchmark data is more complicated in these systems. This makes difficult to develop general recommendations. In general, difficulties at describing on the same footing the ESs of different character have been described for TD-DFT [3]. In addition, TD-DFT results are often functional dependent, particularly in the description of long-range CT states and MLCT states. Among the different xc functionals, it is generally observed that hybrid functionals, and especially those bearing intermediate amounts of exact exchange (such as B3LYP and PBE0), outperform the rest of pure and range-separated functionals, which tend to under- and over-estimate, respectively, the excitation energies. This trend is overall observed for the singlet ESs of positively charged Ru(II), Pt(II), Os(II), Ir(III), Re(I), Au(I) TMCs (to mention some relevant examples) [3, 63–67]. In contrast, pure functionals might be more suited to treat the ESs of negatively charged TMCs (bearing, e.g., a Fe(II) and Cr central atom) [63, 68], but also to generally provide more accurate singlet–triplet gaps, as recently shown in the case of $[Ru(bpy)_3]^{2+}$ [69, 70]. Note that if one aims to study the ISC processes, attaining accurate estimations of singlet–triplet gaps might be more crucial than describing the UV-Vis is absorption bands with a higher accuracy (this is the reason that the pure PBE functional is recommended for $[Ru(bpy)_3]^{2+}$ in [69]). As the inclusion of exact exchange usually exacerbates the triplet instability problems, and thereto impairing the accurate calculation of singlet–triplet gaps, an alternative approach is the use of the Tamm–Dancoff approximation within TD-DFT, which avoids problems with spurious low-lying triplet ES. In a nutshell, future work should be devoted to evaluate statistical errors and extract further conclusions regarding the performance of TD-DFT for TMCs.

With the final goal of an effective recovery of correlation effects at a reduced computational cost, other alternative methods are those combining wave function theory with DFT. Among these methods, three different approaches are highlighted, namely (i) the multi-configuration pair-density DFT (MC-pDFT) method [71], which it may be considered as a multi-configurational analog of Kohn-Sham (KS)-DFT, (ii) the MC-srDFT approach [72] of Jensen and coworkers, and (iii) the DFT/MRCI approach [73]. The latter approach is a DFT-based method for electronically ES that makes use of KS orbitals in a multi-reference configuration interaction (MRCI) framework. Hence, non-dynamic (long-range) correlation effects are ensured at the MRCI level while KS-DFT captures the dynamic (short-range) electron correlation. While the performance of the MC-pDFT method should still be explored for describing the ES of TMCs, the DFT/MRCI method appears to be superior as compared with TD-DFT for these types of systems, especially in providing the correct ordering of their ES and in attaining a balanced description of all type of ES [62]. A redesigned spin-invariant DFT/MRCI Hamiltonian was recently developed [74].

SOCs importantly impact the photophysical and photochemical properties of TMCs. Indeed, without SOCs, the electronically ES of different spin angular moment would not couple, and thus, the ISC probability in classical terms would be zero [75].

The most rigorous description of SOC requires the use of the four-component Dirac equation, from which SOC naturally emerges. However, four-component calculations are hardly tractable even for the smaller molecular systems; and therefore, for practical applications, the spin–orbit operator is often added ad-hoc to nonrelativistic electronic structure theories. The most used approximated SOC operators arise from the formulation of the one- and two-electron SOC Breit–Pauli operator [76]. As in TMCs the one-electron term dominates, effective one-electron SOC operators are commonly used for (i) the calculation of SOC matrix elements (SOCMEs) between the involved states, and (ii) related spin–orbit properties. The calculation of SOCMEs is nowadays available with multiple electronic structure theories, often being the results not extremely sensitive to the chosen level of theory [75].

A final aspect to be mastered by the theoretical investigations is the calculation of the geometrical rearrangements occurring during the course of photodeactivation. The calculations of (i) stationary point geometry optimizations in the ES PES and (ii) ES vibrational levels are considerably more expensive than those for the GS [30]. In practical terms, these calculations require the implementation of analytical first and second derivatives within the electronic structure theories of choice. As those are available within TD-DFT, this latter method is often used for such purposes and has found many applications in TMCs photochemistry. However, due to its single reference character, special caution should be taken in the surroundings of surface crossings, such as CoIns. Alternatively, in these situations, the use of multi-configuration wave function approaches is best suited. Recent implementations of analytical first derivatives within some of these methods (such as for CASPT2 [77] and MC-pDFT [78]) should alleviate not only to perform nonadiabatic molecular dynamics simulations but also the optimization of CoIns with the latter methods. Among other critical points of relevance, minimum energy crossing points (MECP) between states of different multiplicity are highlighted. Efficient algorithms to locate MECPs are available since the late 1990s [79]. MECPs are not only relevant in two- and multistate reactivity but also in TMCs photochemistry (see, e.g., for $[Ru(bpy)_3]^{2+}$ [80] and for cyclometalated Ir(III) complexes [36, 37]). In the case that the MECP involves the lower-lying states of their given multiplicity, these calculations can be performed with DFT. The recent automated algorithms to search for MECP (and/or minimum energy CoIns) geometries developed by Maeda and coworkers [81, 82] should help to automatically map out the intricate PES of TMCs, as recently shown for a Re(I) complex [83].

2.2 ES Decay Rate Theories

As discussed in Sect. 1, ES decay rate formalisms are best suited to assess the competition between radiative and nonradiative deactivation channels. In the context of TMCs, these formalisms are often used to model the long-lived phosphorescence arising from T_1. The validity of these formalisms might be questioned for these systems if near-degeneracy situations between different low-lying ES apply, or in the case of modeling "pure" fluorescence, which is rarely found in TMCs. Nevertheless, and because of pedagogical purposes, let us start the discussion with the fluorescence decay rate (k_f), which is determined in an approximated way by the following Einstein relationship

$$k_f = f \Delta E^2 / 1.5 \tag{1}$$

where ΔE is the vertical emission energy (in cm^{-1}) and f is the oscillator strength (in dimensionless units). Note that for very bright singlet ES, f ~ 1.0, thus k_f values are typically between 10^8 and 10^{11} s^{-1} (i.e., occurring in the nanosecond regime). Both magnitudes are easily accessible from ES calculations at many different levels of theory, so that rough estimates of the fluorescence decay rates can be derived using (1). Alternatively, k_f values can also obtained by integration over the whole computed emission spectrum using, e.g., the thermal vibration correlation function (TVCF) formalism developed by Shuai and coworkers. [84] The emission spectra calculations require the calculations of the GS and ES vibrational levels and the concomitant evaluation of FC factors, Herzberg-Teller effects as well as Duschinsky rotation effects of the PESs. For attaining tractable calculations, the harmonic oscillator approximation is often considered within these calculations.

Phosphorescence decay rates (k_r) are also accessible through analogous Einstein-like expressions to those used for fluorescence. As mentioned in Sect. 1, the three spin sublevels of T_1 contribute to the global phosphorescence decay rate (see Fig. 1), and hence, k_r values are temperature-dependent, especially at cryogenic temperatures (see more details in Sect. 3.1). Conversely, at the high temperature limit, thermal equilibration between the spin sublevels is at play and only weighted phosphorescence is observed; thus

$$k_r = \frac{1}{3} \sum_{i=1}^{3} k_r^i \tag{2}$$

where i stands for the x, y, and z spin sublevels of T_1. The main challenge from a computational viewpoint is to compute accurate estimates for the f values of phosphorescence, which are mainly determined by the electric transition dipole moment of the radiative $T_1 \rightarrow$ GS transition. Using perturbation theory, the j-axis projection of the dipole transition moment between the GS and the ith spin sublevel of T_1 is given by

$$M_j^i = \sum_{n,m=0}^{\infty} \frac{\langle S_0 | \hat{\mu}_j | S_m \rangle \langle S_m | \hat{H}_{SO} | T_1^i \rangle}{E(S_m) - E(T_1)} + \frac{\langle S_0 | \hat{H}_{SO} | T_n \rangle \langle T_n | \hat{\mu}_j | T_1^i \rangle}{E(T_n) - E(S_0)}, i \in \{x, y, z\} \tag{3}$$

where phosphorescence activity gains dipole activity through the matrix elements of the electronic spin–orbit operator ($\hat{H}_{SO}$) over intermediate singlet (S_m, see the term on the left side of (3)) and triplet (T_n, see right-side term) ESs [10, 85]. Often the left-side term contributes more to the intensity borrowing. Equation (3) is a sum-over-states expression that can be solved using the quadratic response (QR) methodology [86], which replaces (3) by solutions of sets of linear equations, and which combined with TD-DFT calculations makes tractable the phosphorescence calculations in TMCs. Additionally, phosphorescence decay rates are accessible with other alternative approaches, namely self-consistent SOC-TD-DFT and perturbative SOC-TD-DFT (pSOC-TD-DFT). In general, perturbative schemes yield systematically lower rates than self-consistent approaches [87].

As stated in Sect. 1, the calculations of ES $\rightarrow$ GS nonradiative rates still pose significant challenges. Two main processes, i.e., k_{IC} and k_{ISC}, contribute to the global nonradiative rates in TMCs (we recall that IC and ISC processes might be indistinguishable in the case that strong spin–vibronic coupling applies). The expressions for k_{IC} and k_{ISC} require a more elaborated treatment as compared to their radiative counterparts. Second-order perturbation theory (i.e., Fermi-Golden rule like) is usually a good approximation for the calculations of the nonradiative rates within the local harmonic approximation, where distortions, displacements, as well as Duschinsky rotation effects of the PESs are taken into account. The effective couplings (nonadiabatic electronic coupling and SOCs) are often calculated at a particular nuclear geometry (typically the ES and/or GS minimum). Evaluation of k_{IC} and k_{ISC} can be done in the energy or in the time domain, the latter being preferred for its reduced associated computational cost. Among the time-domain approaches, I highlight the TVCF formalism of Shuai and coworkers [84, 88] and the time-dependent integration schemes developed by Marian and coworkers [89]. Importantly, vibronic contributions to the k_{ISC} values can also be computed [88]. The nonradiative rate calculations might be sensitive to the computed relative energies between the final and initial states and to the geometric rearrangements of the involved PES, and thus, these calculations may require a prior assessment of the chosen electronic structure methods to describe both the ES and GS properties. Additionally, further problems may arise from (i) the use of the harmonic approximation, which might be inappropriate for strongly distorted ESs and (ii) the interference with other close-lying ESs. Despite these limitations, these approaches have been successfully applied for many TMCs, importantly for Ir(III) complexes [37, 90] (see an example in Sect. 3.1), but also in $[Fe(bpy)_3]^{2+}$ [91]. In these examples, the computed rates are on the same order than the measured ones. Finally, these schemes can be extended to include thermal effects and high-order spin-vibronic interactions, the latter ones being more relevant in cases of negligible SOCMEs between the initial and final states.

2.3 ES Reaction Dynamics Methods

As detailed in Sect. 1, an alternative approach to ES decay rate theories to attain time-resolved information are ES reaction dynamic methods. These methods provide a solution to the time-dependent Schrödinger equation and do treat the nuclear motion over a PES in an explicit manner. Hence, they generally yield an accurate picture of the ES dynamics, and in principle, more accurate rates calculations can be derived from these simulations. Two suites of reaction dynamic methods are generally used to follow the fate of ES, namely quantum wavepacket dynamics and *on-the-fly* nonadiabatic molecular dynamics. For an exhaustive description of these methods, the readers are referred to exceptional reviews on the field [9, 40].

Let us start the discussion with quantum dynamics. The multi-configuration time-dependent Hartree (MCTDH) method [92] belongs to this class of methods and it is especially well suited to treat cases where strong vibronic and/or SOCs applies. This is because the equations of motion in the MCTDH method are purely derived from quantum mechanics. In addition, purely quantum effects like tunneling are inherently described. In MCTDH, one should build model PES before running the dynamic simulations, and thus, these methods require a prior evaluation of the important electronic and nuclear degrees of freedom. Although these simulations have not yet acquired a routine status, MCTDH simulations have already provided important insights into the photodeactivation dynamics of TMCs. Notably, a MCTDH study of Daniel and coworkers in the $[Re(bpy)(CO)_3(X)]$ (where X = Br, I) series [83, 93], highlighted that a spin-vibronic mechanism involving several intermediate singlet and triplet states of different characters is responsible for the counterintuitive heavy atom effect measured experimentally, namely a faster decay for the bromine substituted complex as compared to the iodine substituted one. The computed lifetimes for the ultrafast decay to T_1 agree well with the measured evidence. Regarding its future prospects, the biggest challenge for MCTDH is to calculate the multidimensional PES, as there is a limitation in the number of nuclear degrees of freedom (DOFs) to account for, and importantly, these DOFs are often approximated as harmonic potentials. These approximations might be insufficient to treat large TMCs and especially, to treat large geometrical deformations (e.g., ligand dissociation), which are not uncommon in TMCs photochemistry. Additionally, it is challenging to propagate these dynamic beyond the picosecond regime.

Conversely, nonadiabatic *on-the-fly* molecular dynamics offer the possibility to perform simulations on full-dimensional PES. This is so because these mixed classical-quantum approaches are based on trajectories (which can be easily parallelized, as each trajectory is independent of the ensemble), where the nuclei are propagated classically following Newton's equations of motion and the electronic properties (e.g., energies, gradients, SOCs, and nonadiabatic couplings) are computed *on-the-fly* at each time step and only when it is required, and hence reducing the computational efforts as compared to quantum dynamic approaches. Among such methods, fewest switches Tully's trajectory surface hopping (TSH) [94] is one of the most commonly used approaches. The initial formulation of the TSH algorithm has been recently expanded to treat both SOCs and nonadiabatic dynamical effects, such as in the SHARC suite of programs developed by González and coworkers [95, 96], and hence giving access to model ISC processes. In TSH approaches, the nonadiabatic population transfer occurring between the PES, and which is strictly forbidden within a classical dynamics framework, is conducted using an stochastical algorithm which is controlled by the nonadiabatic couplings and SOCs. The main drawbacks of these methods are that purely quantum effects such as quantum tunneling and/or "nuclear" coherences are not included, being especially the latter of relevance in the photophysics of TMCs. The limitations in the electronic part will depend on the inherent limitations of the electronic structure method of choice, and thus, they are system-dependent and they will need some pre-calibration. Using these dynamical approaches, the ES ultrafast decay of several TMCs has already been modeled, notably for $[Ru(bpy)_3]^{2+}$ (see Sect. 3.2) [69, 97], but also for Ir(III) complexes [98].

3 Photodeactivation Channels of TMCs: Selected Recent Computational Works

3.1 ES Decay Rate Theory and ES Kinetic Modeling: Calculation of the Temperature-Dependent Photoluminescence Lifetimes and Efficiencies of Cyclometalated Ir(III) Complexes

This recent work [37] is selected as a representative example of how the application of a static approach based on state-of-the-art quantum chemical calculations along with several ES decay rate theories and kinetic modeling enables the calculations of global photoluminescence lifetimes and efficiencies of cyclometalated Ir(III) complexes. Additionally, it is the perfect playground to show how extrinsic factors (temperature in this case) strongly influence the long timescale ES decay dynamics. Ir(III) complexes are used as dopant materials in state-of-the-art organic light-emitting diode (OLED) devices. Therefore, as lifetimes and radiative efficiencies are often optimized experimentally in a trial-and-error manner, the development of an accurate theoretical approach to attain predictions of these magnitudes is acknowledged by the OLEDs industry. In [37], a general approach to compute temperature-dependent lifetimes and PLQY of Ir(III) complexes was developed, and its validity was proved for a blue (**1**), a green (**2**), and a yellow/orange (**3**) emitter (see Fig. 3a). Additionally, the origins of the photoluminescence quenching mechanisms with increasing temperatures, which importantly reduce the efficiency of blue emitters at OLEDs standard working conditions, were disclosed. As described in Sect. 2.2, the calculation of phosphorescence rates (k_r in Fig. 3b) at the first principles level is currently attainable based on the Einstein spontaneous emission formula and relying on DFT and TD-DFT computed data (note that the B3LYP *xc* functional was used in this study and SOCs were treated as a perturbation). Two main factors lead to large k_r values: substantial ^{3}MLCT-S_0 SOCs and efficient intensity borrowing from the manifold of singlets to the triplets (see Sect. 2.2). Note that the predominant ^{3}MLCT character of the T_1 emissive state guarantees very efficient radiative processes, which are often of the order of 10^5 s^{-1} for Ir(III) complexes. Boltzmann statistics of the three spin sublevels is applied to calculate the overall radiative rate at a given temperature

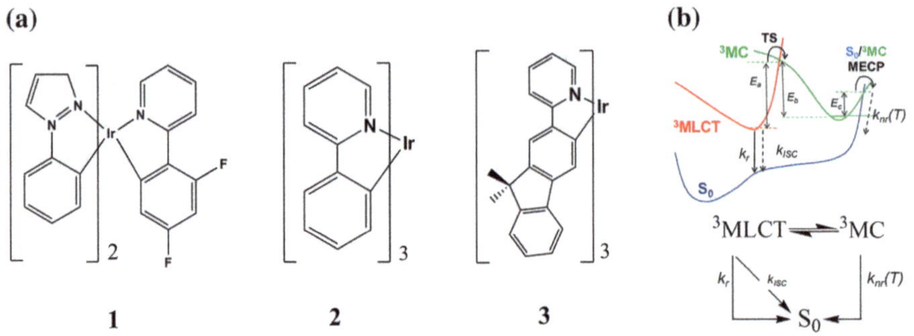

Fig. 3 **a** Chemical structures of complexes 1–3. **b** Competing deactivation channels and kinetic model used for the ES decay kinetics. Reprinted with permission from [37]. Copyright 2018 American chemical society

$$k_r = \frac{k_{r,I} + k_{r,II}\exp\left(-E_{II,I}/k_B T\right) + k_{r,III}\exp\left(-E_{III,I}/k_B T\right)}{1 + \exp\left(-E_{II,I}/k_B T\right) + \exp\left(-E_{III,I}/k_B T\right)} \tag{4}$$

where k_B is the Boltzmann constant, T is the given temperature, and $E_{II,I} = E_{\mathrm{vert},II} - E_{\mathrm{vert},I}$ and $E_{III,I} = E_{\mathrm{vert},III} - E_{\mathrm{vert},I}$ stand for the energetic differences between the triplet spin sublevels. Consequently, the k_r values are slightly temperature-dependent. This slight dependency does not explain why some blue phosphors are highly efficient at cryogenic temperatures but nearly non-emissive at room temperature. As experimentally and computationally corroborated [36, 37], this is due to the presence of close-lying ^{3}MC states, which are involved in thermally activated nonradiative channels (cf. Fig. 2c). These ^{3}MLCT and ^{3}MC are often thermally equilibrated at room temperature. The competing deactivation channels, their respective rate constants, and the associated kinetic model for the ES decay kinetics are shown in Fig. 3b. Note that singlet ESs are not included within this model, as the ISC processes occur in the unity of quantum yield for these complexes.

Having in mind this kinetic model, the temperature-dependent photoluminescence efficiencies (5) and global lifetimes (6) can be expressed as

$$\Phi_p(T) = \frac{k_r}{k_r + k_{\mathrm{ISC}} + k_{\mathrm{nr}}(T)}, \tag{5}$$

$$\tau(T) = \frac{1}{k_r + k_{\mathrm{ISC}} + k_{\mathrm{nr}}(T)}, \tag{6}$$

where k_{ISC} is the $^3\mathrm{MLCT} \rightarrow \mathrm{GS}$ nonradiative ISC decay rate, which is moderately temperature-dependent through vibration population, and $k_{\mathrm{nr}}(T)$, which corresponds to the strongly temperature-dependent rate associated with the population of the ^{3}MC state. Overall, emission from the ^{3}MLCT state (k_r) will be observed if this rate is fast enough to compete with k_{ISC} and with $k_{\mathrm{nr}}(T)$ nonradiative decays. The k_{ISC}

values were computed with the TVCF formalism (see Sect. 2.2) using B3LYP and TD-B3LYP data (in practical terms, one should perform the calculation of energy differences between the involved states and their effective SOC and nonadiabatic couplings along with the calculation of vibrational frequencies and normal modes for both the ^{3}MLCT and GS states). Three main factors lead to enhanced k_{ISC} values: (i) a small ^{3}MLCT-S_0 adiabatic energy difference, (ii) large ^{3}MLCT-S_0 SOCs, and (iii) large reorganization energies, which measure the extent of vibronic coupling between these two electronic states. For these complexes, the k_{ISC} values only slightly depend with temperature, and this is because the relevant vibrational modes promoting ISC (notably, C=C stretchings and in-plane deformations of the ligands) are of high frequency. The $k_{nr}(T)$ values were for the first time computed in [37]. This latter channel is characterized by the population of the ^{3}MC state through a transition state (TS, see Fig. 3b) and the irreversible recovery of the GS geometry through a MECP. The rate-limiting step is often the ^{3}MLCT → ^{3}MC transformation (see E_a in Fig. 3b). Under these circumstances, assuming a thermally equilibrated scenario and using the steady-state approximation the $k_{nr}(T)$ can be expressed in the following Arrhenius-like form

$$k_{nr}(T) = A_0 A \exp(-E_a/k_B T) \tag{7}$$

where A_0 is a temperature-dependent prefactor that accounts for the thermal equilibrium (it is controlled by the E_a, E_b, and E_c values; see Fig. 3b) and A is the pre-exponential factor for the ^{3}MLCT → ^{3}MC transformation. Therefore, to evaluate the $k_{nr}(T)$ rates, in a first step, the geometries of the relevant stationary points (minima, TS, MECP) along this pathway were optimized with UB3LYP, as this method succeeds in obtaining a continuous adiabatic description of the lowest triplet PES. Note that to refine the calculation of E_a, the dispersion-corrected double-hybrid PWPB95-D3 functional was chosen, which shows a better agreement with the experimental evidences. In a second step, canonical variational transition state theory (CVT) dynamics were performed based on the intrinsic reaction coordinate (IRC) calculations for the ^{3}MLCT → TS → ^{3}MC pathway, which give access to the A value and thereto to the $k_{nr}(T)$ rates at any given temperature. These calculations evidenced how strongly influences temperature the $k_{nr}(T)$ values, especially in the case of small E_a values. For instance, in the case of the blue complex (**1**), the $k_{nr}(T)$ at 77 K only amounts up to 4.43 × 10^{-12} s^{-1}, clearly not competitive neither with k_r nor with k_{ISC}. Conversely, the $k_{nr}(T)$ channel becomes the most relevant deactivation channel at 298 K (5.69 × 10^6 s^{-1}) and consequently leads to a quench of photoluminescence at this latter temperature. In Fig. 4, the experimental (a) and computed (b) temperature-dependent photoluminescence lifetimes using (6) are plotted, which further highlights the relevance of the $k_{nr}(T)$ channel. Thus, by switching off the contribution of the $k_{nr}(T)$ component to (6) (dashed line in Fig. 4b), the sigmoid-like temperature-dependence observed experimentally is lost.

Furthermore, the PLQY values can be calculated at any given temperature with (5). The computed PLQY values for **1–3** showed a reasonable agreement with the

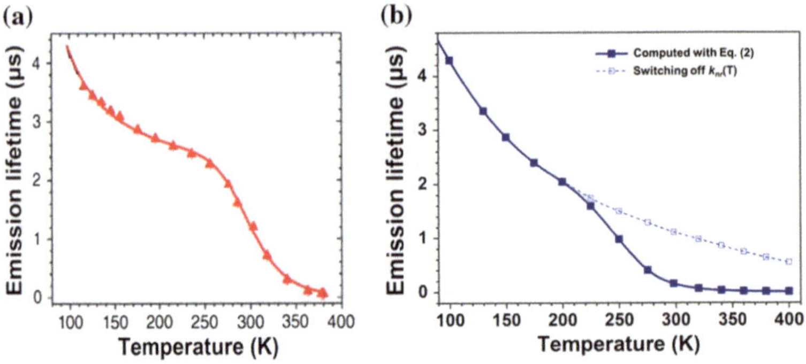

Fig. 4 Experimental **a** and computed **b** temperature-dependent global photoluminescence decay for 1. Reprinted with permission from [37]. Copyright 2018 American chemical society

experimental observations. Note that obtaining accurate PLQYs' estimations still remains difficult due to the inaccuracies on the computed DFT energies and because the energies are exponentially related with the decay rates. Nevertheless, since all possible kinetic scenarios are included, the validity of this general approach is valid for a wide chemical space, i.e., from deep-blue to yellow-to-red phosphors. Another important conclusion of this study is that a E_a value of ca. 4000 cm^{-1} is enough to impede the population of the $k_{nr}(T)$ channel at RT, which might help in the design of the new generation of efficient blue phosphors for PhOLEDs.

3.2 ES Dynamics: TSH Dynamics Including ISC in [Ru(bpy)$_3$]$^{2+}$

This selected work [69] underlines the in-deep understanding one can attain with state-of-the-art *on-the-fly* nonadiabatic molecular dynamics simulations for TMCs. As highlighted in Sect. 1, these methods permit to track the evolution of the ES species in real time and have provided an in-depth understanding of the ultrafast IC and ISC processes of TMCs, otherwise difficult to obtain due to the high density of ES of various multiplicities and characters for these molecular systems. In [69], González and coworkers performed TSH dynamics coupled to LR-TD-PBE calculations including explicit nonadiabatic couplings and SOCs, as implemented in the SHARC suite [95, 96], to model the ultrafast IC and ISC dynamics in $[Ru(bpy)_3]^{2+}$. Note that, as mentioned in Sect. 2.1, special caution should be put when using LR-TD-DFT methods to treat near-degenerate ES. Since these near-degeneracy regions are ubiquitous in $[Ru(bpy)_3]^{2+}$, it is fair alerting on the reliability of these results, although it is also fair noting that TD-DFT remains to date the only possibility in practical terms to perform TSH dynamics in TMCs. Experimentally, it was found that ISC processes occur in less than 30 fs with almost unity yield for this TMC. In

Fig. 5a, it is shown the TD-PBE UV-Vis is absorption spectrum generated from a Wigner distribution of ca. 1500 initial geometries. The gray area in Fig. 5a stands for the energetic excitation window used in the simulations, which mimics the experimental conditions. This energetic window was chosen to account for the possible variance of the incident wavelength and the inaccuracy in the computed excitation energies. The GGA functional PBE was chosen to correctly describe singlet–triplet energy gaps [70]. As shown in Fig. 5a, a high density of bright singlet ESs is found within this excitation energy range. Therefore, a large number of trajectories starting from different ESs are needed to mimic the experimental excitation conditions. 101 permissible random trajectories were run for 30 fs, within 15 singlets and 15 triplets (that means 60 states in total, as the triplet states split into their three spin sublevels). These trajectories were initially distributed in the absorbing states (S_6–S_{14}), following their ratio of computed relative intensities. The time evolution of all possible singlet and triplet ES normalized for all trajectories is displayed in panel A of Fig. 5b, which clearly shows that triplet ESs are rapidly populated from the starting singlet ES. This is even more clear in panel B of Fig. 5b, where the time evolution of the summed up triplet and singlet populations is shown. Thus, the triplet population amounts up to ca. 65% after 30 fs, and the global fitting of this kinetics provides a time constant of 26 ± 3 fs for ISC, in good agreement with the experimental evidences. Besides the quantitative agreement with the experimental ISC rates, these TSH simulations provided an in-depth understanding of the ISC processes taking place in $[Ru(bpy)_3]^{2+}$, as it is not a single ^{1}MLCT and ^{3}MLCT state that participates in these processes but instead a manifold of near-degenerate singlet and triplet ES that are responsible for the ultrafast population transfer between ES. In addition, S_1 and T_1 populations remain well below 3% after 30 fs, implying that ISC processes occur in an "horizontal" manner between high-lying singlet and triplet ES and thus breaking Kasha's rule. The authors further investigated whether it was only the confluence of two important factors: (i) the high density of ESs and (i) the large size of the SOCs between the involved ES (which amounted up to 350 cm^{-1}) the sole ingredients to promote rapid ISC in $[Ru(bpy)_3]^{2+}$, or contrarily, if other factors also played an important role. Toward this latter aim, further sets of TSH simulations were performed, and they are displayed in Panels C-E of Fig. 5b. In Panel C, the results of freezing the initial Wigner geometries during the dynamics are shown, while in Panels D and E, the results for the simulations starting both at the FC geometry, with (D) or without (E) allowing for geometrical relaxation are shown. The frozen dynamics (C and E) clearly result in a significant decrease of triplet population with respect to their non-frozen counterparts (e.g., compare 15% in C with 65% in B after 30 fs). This is because IC is not operative in the frozen dynamics, being the population transfer only driven by successive spin–orbit-mediated transitions. One can thus conclude that the ultrafast ISC processes in $[Ru(bpy)_3]^{2+}$ are not only due to the high density of ES and the large SOCs, but dynamical molecular relaxation leads to a substantial enhancement of the ISC rates.

In an effort to identify which are the internal motions that promote ISC in $[Ru(bpy)_3]^{2+}$, time-resolved normal mode (see Fig. 6a) and principal component analyses of the averaged trajectories were performed. These analyses permitted iden-

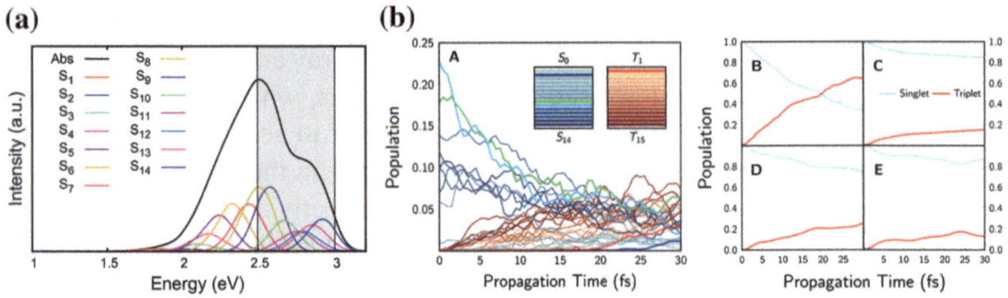

Fig. 5 **a** Calculated UV-vis absorption spectrum of $[Ru(bpy)_3]^{2+}$, the gray area being the excitation energy range of the launched trajectories. **b** Time-resolved (over 30 fs) normalized singlet (bluish) and triplet (brownish) populations. Panels B–E display normalized total singlet and triplet populations under different conditions (see text). Reprinted with permission from [69]. Copyright 2018 American chemical society

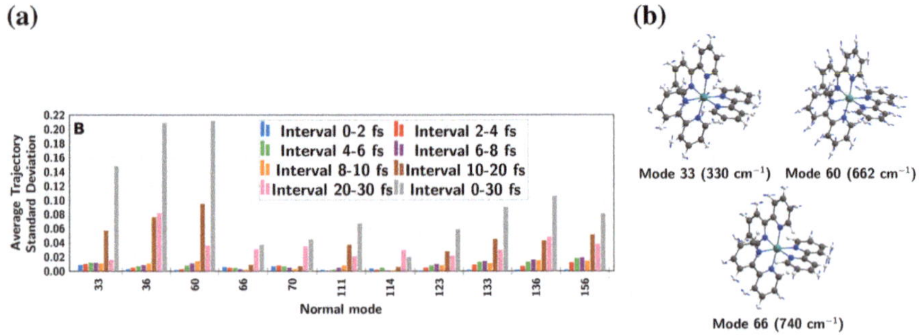

Fig. 6 **a** Total standard deviation of relevant normal modes of $[Ru(bpy)_3]^{2+}$ for an averaged trajectory at different time intervals. **b** Most relevant normal modes selected from the standard deviation of the averaged trajectory (Fig. 6a). Reprinted with permission from [69]. Copyright 2018 American chemical society

tifying the normal modes that show more significant changes over time, see, e.g., modes 33, 60, and 66 in Fig. 6b; and thus, they correspond to the modes that influence most the ISC processes. Importantly, these modes are low-frequency ones and involve hydrogen bond bending, but also Ru–N bond stretching in some cases. The motions in the nitrogen and ruthenium atoms are thus the most important ones promoting ISC through ES mixing.

3.3 Anti-Kasha Emissions in Ru(II) Complexes: Kinetic Control of Photoluminescence or Solvent Effects?

The next computational work [99] is chosen to highlight the often intricate nature of the emissive processes of TMCs, leading in some cases to, e.g., dual photoluminescence and/or anti-Kasha emissions (i.e., emission from higher-lying ESs). In addition, the interplay between two extrinsic factors, i.e., temperature and environment, was concomitantly assessed in this study.

Two different anti-Kasha emissions are possible: those arising between thermally equilibrated ES (case A in Fig. 7a) and those resulting from non-thermally equilibrated ESs (case B in Fig. 7a). The thermally-equilibrated case is characterized by: (i) rather small $ES_n - ES_1$ energetic differences, and thus thermal equilibrium between the ES is at play; and (ii) absence of excitation wavelength-dependent emissive properties. Conversely, a non-thermally equilibrated scenario is typified by: (i) large $ES_n - ES_1$ energy gaps, so that photon emission from ES_n is faster than the sluggish $ES_n \rightarrow ES_1$ IC or ISC processes, and (ii) excitation wavelength-dependent emissive properties. In Fig. 7b are depicted the two heteroleptic $[Ru(H)(CO)(N^\wedge N)(tpp)_2]^+$ complexes (tpp = triphenylphosphine) under study in [99]. In the case of **1**, $[Ru(H)(CO)(dmphen)(tpp)_2]^+$ (dmphen = 5,6-dimethyl-1, 10-phenanthroline), temperature-dependent emissive properties were experimentally observed [100]. More in details, an emission switch from solely ^{3}MLCT emission at 298 K to dual photoluminescence at 77 K was measured. This ^{3}MLCT state and the higher-lying ^{3}LC state are responsible for the two emissive bands at 77 K. Conversely, for **2**, $[Ru(H)(CO)(bpy)(tpp)_2]^+$, only ^{3}MLCT emission was observed regardless of the temperature. Two possible origins of the anti-Kasha emissions in **1** arise, namely: (i) different temperature-dependent solvent relaxation effects for the ^{3}MLCT and ^{3}LC states or (ii) a kinetically controlled scenario of photoluminescence. The solvent hypothesis relies on the fact that both (i) solute geometry relaxation in the ES and (ii) ES density-dependent relaxation of the solvent polarization occur at 298 K before the actual photon emission takes place. Conversely, at 77 K, the glassy conformation of the media hinders the solvent molecules to reorganize themselves to adapt to the solute charge distribution in the electronic ES. Therefore, if the ^{3}MLCT and ^{3}LC excitations involve different density rearrangements, it could be the case that the ^{3}MLCT state is the lowest lying ES (Kasha state) at 298 K but not at 77 K. The second hypothesis (kinetically controlled scenario of photoluminescence) relies on the fact that thermal equilibrium between the involved ES could be possible at 298 K, but not at 77 K.

The two above hypotheses were examined in [99]. In a first step, the geometries of the relevant stationary points of **1**, i.e., the two ^{3}ES minima and the TS connecting both minima, were optimized with UB3LYP. In Fig. 8a, the schematic Jablonski diagram is shown while Fig. 8b contains the spin density distribution at the ^{3}MLCT and ^{3}LC minima. The ^{3}MLCT state is the lowest adiabatic emissive state, i.e., the Kasha state, being 4.0 kcal/mol below the ^{3}LC state. In addition, there is a TS connecting both minima, which is found 1.8 kcal/mol above the ^{3}LC minimum. The $\alpha_{P\text{-}Ru\text{-}P}$ bite angle at the ^{3}LC minimum (169°) is much larger than at the ^{3}MLCT minimum (139°). Therefore, due to the steric hindrance between the two $-PPh_3$ units at the ^{3}MLCT minimum, one of the phenyl rings is forced to twist at this geometry (see Fig. 8b). The imaginary mode (31i cm^{-1}) of the TS connecting both minima mainly involves the torsion of this phenyl ring. The photodeactivation channel of **1** can be summarized as follows. Photoexcitation to the singlet manifold (S_n) leads to the population of the triplet manifold (T_m) with almost unity yield, as very efficient

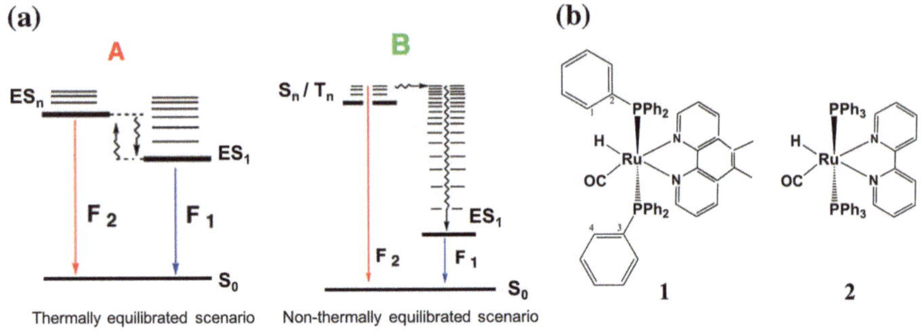

Fig. 7 **a** Thermally equilibrated (A) and non-thermally equilibrated (B) anti-Kasha scenarios. **b** Chemical structures of complexes 1–2. Figure **b** is reproduced from [99] by permission of the PCCP Owner Societies

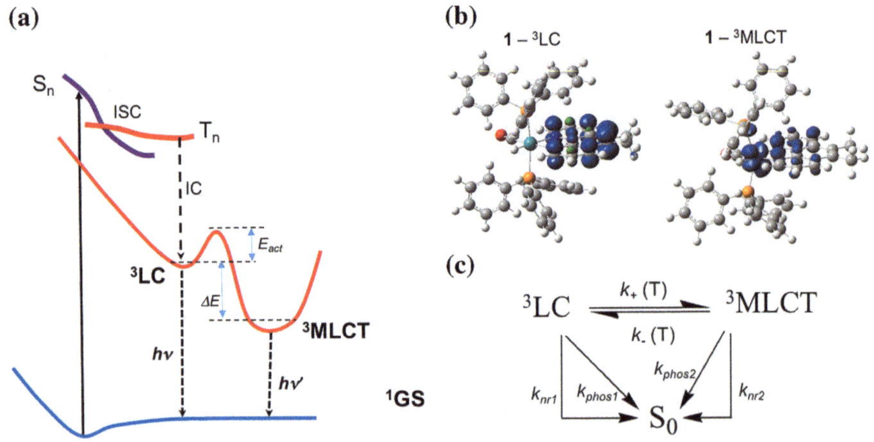

Fig. 8 **a** Schematic Jablonski diagram of complex 1. **b** Spin density distributions of the ^{3}LC and ^{3}MLCT states of complex 1. **c** Kinetic model for the decay kinetics of complex 1. Reproduced from [99] by permission of the PCCP owner societies

and ultrafast ISC processes occur in these type of Ru(II) complexes (we recall to Sect. 3.2). Then, after ISC processes, the well of the ^{3}LC minimum will preferentially be populated in a first stage, since this geometry is closer to the FC region. Once IC and vibrational relaxation to the well of the ^{3}LC minimum are completed, the system will be able to surpass the rather small $^3LC \rightarrow {}^3MLCT$ activation barrier (1.8 kcal/mol) if enough thermal energy is available. As the $^3MLCT \rightarrow {}^3LC$ back reaction possesses a much larger activation barrier (6.8 kcal/mol), the $^3LC \rightarrow {}^3MLCT$ process would be reversible (i.e., thermally equilibrated ESs) or irreversible depending on the given temperature.

ΔSCF-B3LYP and TD-B3LYP calculations were performed to get insights into the emissive characteristics of **1**. To determine phosphorescence rates, QR TD-B3LYP calculations were performed at the ^{3}MLCT and ^{3}LC optimized geometries.

The phosphorescence emission maxima were computed on the basis of ΔSCF-PCM-B3LYP calculations, and then compared to the vertical singlet–triplet PCM-TD-DFT excitations for emission. Both types of calculations supported the assignment of the ES involved in the experimentally observed emission at 77 and 298 K. Thus, ΔSCF-PCM-B3LYP emission maxima values are 1.59 and 2.12 eV for the ^{3}MLCT and ^{3}LC states, respectively, in good agreement with the observed blue shift when cooling down the samples. To assess the influence of ES density-dependent relaxation effects of the solvent polarization, a state-specific PCM approach was used in combination with the TD-DFT calculations, namely the perturbative corrected linear response (cLR)-PCM model [44]. In such a scheme, the dynamic component of the solvent polarization rearranges to equilibrate with the ES charge density of the solute, which is taken into account using the one-particle TD-DFT density matrix. cLR-PCM-TD-DFT calculations are therefore better suited than their LR counterparts to compute emission energies. The cLR-PCM-TD-DFT values are slightly red-shifted (by ca. 0.2–0.3 eV) for both the ^{3}MLCT and ^{3}LC emissions with respect to their corresponding LR-PCM-TD-DFT values, and thus, the ^{3}MLCT state is always the lowest adiabatic ES regardless of the solvation scheme used. Therefore, a putative destabilization of the ^{3}MLCT state at 77 K due to the lack of solvent relaxation effects at this temperature is not responsible of the blue shift observed experimentally. Indeed, the ^{3}MLCT and ^{3}LC dipole moments are rather similar between them, thus hinting at a limited impact of solvent relaxation effects.

Once this hypothesis is disregarded, the second hypothesis (kinetically controlled scenario of photoluminescence) was analyzed. In Fig. 8c, the kinetic model including all possible ES decay channels for **1** is shown. The ^{3}LC-^{3}MLCT equilibrium is controlled by the forward $k_+(T)$ and reverse $k_-(T)$ rate constants, which can be expressed in an Arrhenius form. At 77 K, a $k_+/k_- = 2.3 \times 10^{11}$ value is computed, hence indicating that the $^3\mathrm{LC} \rightarrow {}^3\mathrm{MLCT}$ process is irreversible. Thus, ^{3}LC-like emission is predominantly observed at this temperature (but certainly coexisting with ^{3}MLCT-like emission), because k_{phos1} competes with k_{nr1} and k_+ (77 K). Additionally, the ^{3}MLCT-like emission is concomitantly observed due to the confluence of two factors: (i) $k_{\mathrm{phos2}} \gg k_{\mathrm{phos1}}$ (7.6×10^2 and $2.2\ \mathrm{s}^{-1}$ are the computed values, respectively) and (ii) a non-negligible k_+ (77 K) value. Conversely, at 298 K, a $k_+/k_- = 863$ value is obtained, which implies that a thermal equilibrium between the ^{3}LC and ^{3}MLCT states is likely. Consequently, one can assess whether dual photoluminescence or solely emission from the higher or lower ES will arise with the following expression

$$\frac{\phi(^3\mathrm{MLCT})}{\phi(^3\mathrm{LC})} = \frac{k_r(^3\mathrm{MLCT})}{k_r(^3\mathrm{LC})} e^{\left[\frac{-\Delta E(^3\mathrm{LC}-^3\mathrm{MLCT})}{k_B T}\right]}, \tag{8}$$

where the ^{3}MLCT/^{3}LC ratio of quantum yields is determined by their radiative rates, the energy gap, and the $k_B T$ factor [101]. If the ratio is below 10^{-2}, only ^{3}LC emission will take place, while if this value is above 10^{+2} implies solely emission from the ^{3}MLCT state. Finally, values between these two thresholds lead to dual emissions. The computed ratio is substantially above $>10^{+2}$, pointing to sole ^{3}MLCT emission at 298 K. This is because: (i) $k_{\mathrm{phos2}} \gg k_{\mathrm{phos1}}$ and, in addition, the ^{3}MLCT state is the lowest adiabatic ES. In conclusion, the switch from dual photoluminescence to sole ^{3}MLCT-like emission from 298 to 77 K roots on the possibility to attain thermally equilibrated ES at room temperature.

4 Conclusion and Perspectives

One can easily recognize that the photophysical and photochemical properties of TMCs are more chameleonic than those of organic compounds [101]. The significant experimental and theoretical activity that has emerged in this field during the past years has permitted to underline the complexity of the ES dynamics in TMCs. As a consequence of the gained understanding, early empirical concepts developed for the rationalization of spectral features, lifetimes, ES kinetics, and photoluminescence efficiencies are nowadays a matter of debate in the community. Indeed, the use of simple qualitative rules, such as, El-Sayed- [102] and Kasha- [8] rules, as well as the energy gap law [31], should cautiously be done for TMCs, as these empirical laws can often lead to erroneous predictions of the preferred photochemical outcomes. Importantly, all these new insights would have not been possible without the strong synergy between experimental and theoretical and computational efforts. Throughout this chapter, the advances and the numerous difficulties that are still present in the field have been highlighted and will not be repeated here. Obviously, the forthcoming theoretical/computational developments in quantum chemical methods for the ES, ES decay rate formalisms, and ES reaction dynamic methods should tackle the current deficiencies to model the photodeactivation channels of TMCs.

References

1. Daniel C, Gourlaouen C (2017) Chemical bonding alteration upon electronic excitation in transition metal complexes. Coord Chem Rev 344:131–149
2. Daniel C (2015) Photochemistry and photophysics of transition metal complexes: quantum chemistry. Coord Chem Rev 282–283:19–32
3. González L, Escudero D, Serrano-Andrés L (2012) Progress and challenges in the calculation of electronic excited states. Chem Phys Chem 13:28–51
4. Stufkens DJ, Vlček A (1998) Ligand-dependent excited state behavior of Re(I) and Ru(II) carbonyl-diimine complexes. Coord Chem Rev 177:127–179
5. Mai S, Plasser F, Dorn J, Fumanal M, Daniel C, González L (2018) Quantitative wave function analysis for excited states of transition metal complexes. Coord Chem Rev 361:74–97
6. Le Bahers T, Adamo C, Ciofini I (2011) A qualitative index of spatial extent in charge-transfer excitations. J Chem Theory Comput 7:2498–2506
7. Jacquemin D, Le Bahers T, Adamo C, Ciofini I (2012) What is the "best" atomic charge model to describe through-space charge-transfer excitations? Phys Chem Chem Phys 14:5383–5388
8. Kasha M (1950) Characterization of electronic transitions in complex molecules. Discuss Faraday Soc 9:14–19
9. Penfold TJ, Gindensperger E, Daniel C, Marian CM (2018) Spin-vibronic mechanism for intersystem crossing. Chem Rev 118:6975–7025

10. Baryshnikov G, Minaev B, Ågren H (2017) Theory and calculation of the phosphorescence phenomenon. Chem Rev 117:6500–6537
11. Uoyama H, Goushi K, Shizu K, Nomura H, Adachi C (2012) Highly efficient organic light-emitting diodes from delayed fluorescence. Nature 492:234–238
12. Sousa C, Alías M, Domingo A, de Graaf C (2018) Deactivation of excited states in transition metal complexes: insight from computational chemistry. Chem Eur J. https://doi.org/10.1002/chem.201801990
13. Chergui M (2015) Ultrafast photophysics of transition metal complexes. Acc Chem Res 48:801–808
14. Chergui M (2012) On the interplay between charge, spin and structural dynamics in transition metal complexes. Dalton Trans 41:13022–13029
15. Tang KC, Liu KL, Chen IC (2004) Rapid intersystem crossin in highly phosphorescent iridium complexes. Chem Phys Lett 386:437–441
16. Cannizzo A, Blanco-Rodríguez AM, El Nahhas A, Szebera J, Zalis S, Vlček A, Chergui M (2008) Femtosecond fluorescence and intersystem crossing in Rhenium(I) carbonyl-bipyridine complexes. J Am Chem Soc 130:8967–8974
17. Cannizzo A, van Mourik F, Gawelda W, Zgrablic G, Bressler C, Chergui M (2006) Broadband femtosecond fluorescence spectroscopy of $[Ru(bpy)_3]^{2+}$. Angew Chem Int Ed 45:3174–3176
18. Damrauer NH, Cerullo G, Yeh A, Boussie TR, Shank CV, McCusker JK (1997) Femtosecond dynamics of excited-state evolution in $[Ru(bpy)_3]^{2+}$. Science 275:54–57
19. van der Veen RM, Cannizzo A, van Mourik F, Vlček A, Chergui M (2011) Vibrational relaxation and intersystem crossing of binuclear metal complexes in solution. J Am Chem Soc 133:305–315
20. Kukura P, McCamant DW, Mathies RA (2007) Femtosecond stimulated raman spectroscopy. Annu Rev Phys Chem 58:461–488
21. Yoon S, Kukura P, Stuart CM, Mathies RA (2006) Direct observation of the ultrafast intersystem crossing in tris (2, 2-bipyridine) Ruthenium(II) using femtosecond stimulated raman spectroscopy. Mol Phys 104:1275–1282
22. Smeigh AL, Creelman M, Mathies RA, McCusker JK (2008) Femtosecond time-resolved optical and raman spectroscopy of photoinduced spin crossover: temporal resolution of low-to-high spin optical switching. J Am Chem Soc 130:14105–14107
23. Chergui M, Zewail AH (2009) Electron and X-ray methods of ultrafast structural dynamics: advances and applications. Chem Phys Chem 10:28–43
24. Zhang W, Alonso-Mori R, Bergmann U, Bressler C, Chollet M, Galler A, Gawelda W, Hadt RG, Hartsock RW, Kroll T, Kjaer KS, Kubicek K, Lemke HT, Liang HW, Meyer DA, Nielsen MM, Purser C, Robinson JS, Solomon EI, Sun Z, Sokaras D, van Driel TB, Vanko G, Weng TC, Zhu D, Gaffney KJ (2014) Tracking excited-state charge and spin dynamics in iron coordination complexes. Nature 509:345–348
25. Jonas DM (2003) Two-dimensional femtosecond spectroscopy. Annu Rev Phys Chem 54:425–463
26. Carbery WP, Verma A, Turner DB (2017) Spin-orbit coupling drives femtosecond nonadiabatic dynamics in a transition metal compound. J Phys Chem Lett 8:1315–1322
27. Heitz MC, Finger K, Daniel C (1997) Photochemistry of organometallics: quantum chemistry and photodissociation dynamics. Coord Chem Rev 159:171–193
28. Ribbing C, Daniel C (1994) Spin-orbit coupled excited states in transition metal complexes: a configuration interaction treatment of $HCo(CO)_4$. J Chem Phys 100:6591
29. Casida ME, Huix-Rotllant M (2012) Progress in time-dependent density-functional theory. Annu Rev Phys Chem 63:287–323
30. Escudero D, Laurent A, Jacquemin D (2017) Time-dependent density functional theory: a tool to explore excited states. In: Leszczynski J, Kaczmarek-Kedziera A, Puzyn T, Papadopoulos M, Reis HK, Shukla M (eds) Handbook of computational chemistry. Springer, Cham, pp 1–43
31. Englman R, Jortner J (1970) The energy gap law for radiationless transitions in large molecules. Mol Phys 18:145–164

32. Yarkony DR (2012) Nonadiabatic quantum chemistry—past, present and future. Chem Rev 112:481–498
33. Wagenknecht PS, Ford PC (2011) Metal centered ligand field excited states: their roles in the design and performance of transition metal based photochemical molecular devices. Coord Chem Rev 255:591–616
34. Durham B, Caspar JV, Nagle JK, Meyer TJ (1982) Photochemistry of tris (2, 2′-bipyridine) Ruthenium(2+) ion. J Am Chem Soc 104:4803–4810
35. Sajoto T, Djurovich PI, Tamayo AB, Oxgaard J, Goddard WA III, Thompson ME (2009) Temperature dependence of blue phosphorescent cyclometalated Ir(III) complexes. J Am Chem Soc 131:9813–9822
36. Escudero D (2016) Quantitative prediction of photoluminescence quantum yields of phosphors from first principles. Chem Sci 7:1262–1267
37. Zhang X, Jacquemin D, Peng Q, Shuai Z, Escudero D (2018) General approach to compute phosphorescent OLED efficiency. J Phys Chem C 122:6340–6347
38. Mai S, Marquetand P, González L (2015) A general method to describe intersystem crossing dynamics in trajectory surface hopping. Int J Quantum Chem 115:1215–1231
39. Cui G, Thiel W (2014) Generalized trajectory surface-hopping method for internal conversion and intersystem crossing. J Chem Phys 141:124101
40. Crespo-Otero R, Barbatti M (2018) Recent advances and perspectives on nonadiabatic mixed quantum-classical dynamics. Chem Rev 118:7026–7068
41. Morzan UN, Alonso de Armiño DJ, Foglia NO, Ramírez F, González Lebrero MC, Scherlis DA, Estrín DA (2018) Spectroscopy in complex environments from QM–MM Simulations. Chem Rev 118:4071–4113
42. Mennucci B (2012) Polarizable continuum model. WIREs Comput Mol Sci 2:386–404
43. Barboza Formiga AL, Vancoillie S, Pierloot K (2013) Electronic spectra of N-heterocyclic pentacyanoferrate(II) complexes in different solvents, studied by multiconfigurational perturbation theory. Inorg Chem 52:10653–10663
44. Caricato M, Mennucci B, Tomasi J, Ingrosso F, Cammi R, Corni S, Scalmani G (2006) Formation and relaxation of excited states in solution: a new time dependent polarizable continuum model based on time dependent density functional theory. J Phys Chem 124:124520
45. Improta R, Barone V, Scalmani G, Frisch MJ (2006) A state-specific polarizable continuum model time dependent density functional theory method for excited state calculations in solution. J Phys Chem 125:054103
46. Sisto A, Glowacki DR, Martinez TD (2014) Ab initio nonadiabatic dynamics of multichromophore complexes: a scalable graphical-processing-unit-accelerated exciton framework. Acc Chem Res 47:2857–2866
47. Curutchet C, Muñoz-Losa A, Monti S, Kongsted J, Scholes GD, Mennucci B (2009) Electronic energy transfer in condensed phase studied by a polarizable QM/MM model. J Chem Theory Comput 5:1838–1848
48. Jacob CR, Neugebauer J (2014) Subsystem density-functional theory. WIREs Comput Mol Sci 4:325–362
49. Andersson K, Malmqvist PA, Roos BO (1992) Second-order perturbation theory with a complete active space self-consistent field reference function. J Chem Phys 96:1218
50. Malmqvist PA, Pierloot K, Shahi ARM, Cramer JC, Gagliardi L (2008) The restricted active space followed by second-order perturbation theory method: theory and application to the study of CuO_2 and Cu_2O_2 systems. J Chem Phys 128:204109
51. Pierloot K (2011) Transition metals compounds: outstanding challenges for multiconfigurational methods. Int J Quantum Chem 111:3291–3301
52. Radoń M, Drablik G (2018) Spin states and other ligand-field states of aqua complexes revisited with multireference ab Initio calculations including solvation effects. J Chem Theory Comput 14:4010–4027
53. Stein CJ, Reiher M (2016) Automated selection of active orbital spaces. J Chem Theory Comput 12:1760–1771

54. Bao JJ, Dong SS, Gagliardi L, Truhlar DG (2018) Automatic selection of an active space for calculating electronic excitation spectra by MS-CASPT2 or MC-PDFT. J Chem Theory Comput 14:2017–2025
55. Wouters S, Bogaerts T, van der Voort P, van Speybroeck V, van Neck D (2014) DMRG-SCF study of the singlet, triplet, and quintet states of oxo-Mn (salen). J Chem Phys 140:241103
56. Marti KH, Reiher M (2011) New electron correlation theories for transition metal chemistry. Phys Chem Chem Phys 13:6750–6759
57. Chan GKL, Sharma S (2011) The density matrix renormalization group in quantum chemistry. Annu Rev Phys Chem 62:465–481
58. Li Manni G, Smart SD, Alavi A (2016) Combining the complete active space self-consistent field method and the full configuration interaction quantum monte carlo within a super-CI framework, with application to challenging metal-porphyrins. J Chem Theory Comput 12:1245–1258
59. Phung QM, Wouters S, Pierloot K (2016) Cumulant approximated second-order perturbation theory based on the density matrix renormalization group for transition metal complexes: a benchmark study. J Chem Theory Comput 12:4352–4361
60. Freitag L, Knecht S, Angeli C, Reiher M (2017) Multireference perturbation theory with cholesky decomposition for the density matrix renormalization group. J Chem Theory Comput 13:451–459
61. Ziegler T, Seth M, Krykunov M, Autschbach J, Wang F (2009) On the relation between time-dependent and variational density functional theory approaches for the determination of excitation energies and transition moments. J Chem Phys 130:154102
62. Ziegler T (1991) Approximate density functional theory as a practical tool in molecular energetics and dynamics. Chem Rev 91:651–667
63. Escudero D, Thiel W (2014) Assessing the density functional theory-based multireference configuration interaction (DFT/MRCI) method for transition metal complexes. J Chem Phys 140:194105
64. Latouche C, Skouteris D, Palazzetti F, Barone V (2015) TD-DFT benchmark on inorganic Pt(II) and Ir(III) complexes. J Chem Theory Comput 11:3281–3289
65. Niehaus TA, Hofbeck T, Yersin H (2015) Charge-transfer excited states in phosphorescent organo-transition metal compounds: a difficult case for time dependent density functional theory? RSC Adv 5:63318–63329
66. Le Bahers T, Brémond E, Ciofini I, Adamo C (2014) The nature of vertical excited states of dyes containing metals for DSSC applications: insights from TD-DFT and density based indexes. Phys Chem Chem Phys 16:14435–14444
67. Fumanal M, Daniel C (2016) Description of excited states in $[Re(Imidazole)(CO)_3(Phen)]^+$ including solvent and spin-orbit coupling effects: Density functional theory versus multiconfigurational wavefunction approach. J Comput Chem 37:2454–2466
68. Georgieva I, Aquino AJA, Trendafilova N, Santos PS, Lischka H (2010) Solvatochromic and ionochromic effects of Iron(II) bis(1, 10-phenanthroline) dicyano: a theoretical study. Inorg Chem 49:1634–1646
69. Atkins AJ, González L (2017) Trajectory surface-hopping dynamics including intersystem crossing in $[Ru(bpy)_3]^{2+}$. J Phys Chem Lett 8:3840–3845
70. Atkins AJ, Talotta F, Freitag L, Boggio-Pasqua M, González L (2017) Assessing excited state energy gaps with time-dependent density functional theory on Ru(II) complexes. J Chem Theory and Comput 13:4123–4145
71. Gagliardi L, Truhlar DG, Li Manni G, Carlson RK, Hoyer CE, Bao JL (2017) Multiconfiguration pair-density functional theory: a new way to treat strongly correlated systems. Acc Chem Res 50:66–73
72. Sharkas K, Savin A, Jensen HJA, Toulouse J (2012) A multiconfigurational hybrid density-functional theory. J Chem Phys 137:044104
73. Grimme S, Waletzke M (1999) A combination of Kohn-Sham density functional theory and multi-reference configuration interaction methods. J Chem Phys 111:5645

74. Lyskov I, Kleinschmidt M, Marian CM (2016) Redesign of the DFT/MRCI hamiltonian. J Chem Phys 144:034104
75. Marian CM (2012) Spin–orbit coupling and intersystem crossing in molecules. WIREs Comput Mol Sci 2:187–203
76. Pauli W (1927) Zur quantenmechanik des magnetischen elektrons. Z Phys 43:601–623
77. Park JW, Shiozaki T (2017) Analytical derivative coupling for multistate CASPT2 theory. J Chem Theory Comput 13:2561–2570
78. Sand AM, Hoyer CE, Sharkas K, Kidder KM, Lindh R, Truhlar DG, Gagliardi L (2018) Analytic gradients for complete active space pair-density functional theory. J Chem Theory Comput 14:126–138
79. Harvey JN, Aschi M, Schwarz H, Koch W (1998) The singlet and triplet state of phenyl cation. a hybrid approach for locating minimum energy crossing points between non-interacting potential energy surfaces. Theor Chem Acc 99:95–99
80. Heully JL, Alary F, Boggio-Pasqua M (2009) Spin-orbit effects on the photophysical properties of $[Ru(bpy)_3]^{2+}$. J Chem Phys 131:184308
81. Maeda S, Ohno K, Morokuma K (2009) Automated global mapping of minimal energy points on seams of crossing by the anharmonic downward distortion following method: a case study of H2CO. J Phys Chem A 113:1704–1710
82. Maeda S, Taketsugu T, Ohno K, Morokuma K (2015) From roaming atoms to hopping surfaces: mapping out global reaction routes in photochemistry. J Am Chem Soc 137:3433–3445
83. Harabuchi Y, Eng J, Gindensperger E, Taketsugu T, Maeda S, Daniel C (2016) Exploring the mechanism of ultrafast intersystem crossing in rhenium(I) carbonyl bipyridine halide complexes: key vibrational modes and spin-vibronic quantum dynamics. J Chem Theory Comput 12:2335–2345
84. Niu Y, Peng P, Deng C, Gao X, Shuai Z (2010) Theory of excited state decays and optical spectra: application to polyatomic molecules. J Phys Chem A 114:7817–7831
85. Minaev B, Baryshnikov G, Ågren H (2014) Principles of phosphorescent organic light emitting devices. Phys Chem Chem Phys 16:1719–1758
86. Minaev B (1999) The singlet-triplet absorption and photodissociation of the HOCl, HOBr, and HOY molecules calculated by the MCSCF quadratic response method. J Phys Chem A 103:7294–7309
87. Mori K, Goumans TPM, van Lenthe E, Wang F (2014) Predicting phosphorescent lifetimes and zero-field splitting of organometallic complexes with time-dependent density functional theory including spin-orbit coupling. Phys Chem Chem Phys 16:14523–14530
88. Peng Q, Niu Y, Shi Q, Gao X, Shuai Z (2013) Correlation function formalism for triplet excited state decay: combined spin-orbit and nonadiabatic couplings. J Chem Theory Comput 9:1132–1143
89. Etinski M, Tatchen J, Marian CM (2017) Time-dependent approaches for the calculation of intersystem crossing rates. J Chem Phys 134:154105
90. Kleinschmidt M, van Wüllen C, Marian CM (2015) Intersystem-crossing and phosphorescence rates in fac-Ir(III)$(ppy)_3$: a theoretical study involving multi-reference configuration interaction wavefunctions. J Chem Phys 142:094301
91. Sousa C, de Graaf C, Rudavskyi A, Broer R, Tatchen J, Etinski M, Marian CM (2013) Ultrafast deactivation mechanism of the excited singlet in the light-induced spin crossover of [Fe (2, 2-bipyridine$)_3]^{2+}$. Chem Eur J 19:17541–17551
92. Beck MH, Jäckle A, Worth GA, Meyer HD (2000) The multi-configurational time-dependent hartree approach: a highly efficient algorithm for propagating wavepackets. Phys Rep 324:1–105
93. Fumanal M, Gindensperger E, Daniel C (2018) Ultrafast intersystem crossing vs internal conversion in α-diimine transition metal complexes: quantum evidence. J Phys Chem Lett 9:5189–5195
94. Tully JC (1990) Molecular dynamics with electronic transitions. J Chem Phys 93:1061–1071
95. Richter M, Marquetand P, González-Vázquez J, Sola I, González L (2011) SHARC: ab initio molecular dynamics with surface hopping in the adiabatic representation including arbitrary couplings. J Chem Theory Comput 7:1253–1258

96. Mai S, Marquetand P, González L (2018) Nonadiabatic dynamics: the SHARC approach. WIREs Comput Mol Sci. https://doi.org/10.1002/wcms.1370
97. Tavernelli I, Curchod B, Rothlisberger U (2011) Nonadiabatic molecular dynamics with solvent effects: a LR-TDDFT QM/MM study of Ruthenium(II) tris (Bipyridine) in water. Chem Phys 391:101–109
98. Liu XY, Zhang YH, Fang WH, Cui G (2018) Early-time excited-state relaxation dynamics of iridium compounds: distinct roles of electron and hole transfer. J Phys Chem A 122:5518–5532
99. Jacquemin D, Escudero D (2018) Thermal equilibration between excited states or solvent effects: unveiling the origins of anomalous emissions in heteroleptic Ru(II) complexes. Phys Chem Chem Phys 20:11559–11563
100. Kamecka A, Muszynska W, Kapturkiewicz A (2017) Luminescence properties of heteroleptic $[Ru(H)(CO)(N\hat{}N)(tpp)_2]^+$ complexes: comparison with their $[Os(H)(CO)(N\hat{}N)(tpp)_2]^+$ analogues. J Lumin 192:842–852
101. Escudero D, Jacquemin D (2015) Computational insights into the photodeactivation dynamics of phosphors for OLEDs: a perspective. Dalton Trans 44:8346–8355
102. El-Sayed M (1963) Spin orbit coupling and the radiationless processes in nitrogen heterocyclics. J Chem Phys 38:2834–2838

Mechanism and Kinetics in Homogeneous Catalysis: A Computational Viewpoint

Jeremy N. Harvey

Abstract The use of computational methods based on electronic structure theory and statistical mechanics to study reaction mechanisms and kinetics in homogeneous catalysis, especially organometallic catalysis and organocatalysis, is reviewed. The chapter focuses mostly on examples from the authors' own group, published over the last two decades, and discusses progress and remaining challenges. It is argued that while it plays a valuable role in mechanistic studies, computation is not yet able to replace experimental studies.

1 Introduction

Quantum chemical methods play an ever-greater role in investigating chemical reaction mechanisms. In this chapter, I will discuss some important aspects associated with this type of modeling, with a specific focus on organic and organometallic chemistry, including homogeneous catalytic reactions. This is an area in which my group has been working essentially since the beginning of my independent research career, yet unlike for some other fields in which we have worked, I have not previously written any reviews on this topic. As well as drawing on examples of our own work from various periods, I will also discuss work from other groups and attempt to address what I feel are some of the main challenges to this type of work at the current time. Of necessity, this will also require looking back at how well various studies carried out previously have stood up to the test of time.

J. N. Harvey (✉)
Department of Chemistry and Division of Quantum Chemistry and Physical Chemistry, KU Leuven, Celestijnenlaan 200F, 3001 Leuven, Belgium
e-mail: Jeremy.harvey@kuleuven.be

E. Broclawik et al. (eds.), *Transition Metals in Coordination Environments*, Challenges and Advances in Computational Chemistry and Physics 29,
https://doi.org/10.1007/978-3-030-11714-6_10

2 Chemical Reaction Mechanisms

Ever since the progress of chemical theories meant that the notion of 'chemical reaction mechanism' was a recognizable concept, it has been clear that understanding the mechanisms of chemical reactions provides a powerful tool that can be used to help in the development of synthetically more attractive variations. Accordingly, many techniques have been developed to study reaction mechanisms. One of the most powerful approaches is based on determining the reaction kinetics, while varying experimental conditions such as temperature, initial concentrations of reagents, additives or catalysts, pressure, and perhaps also the structure of the reagents through the modification of substituents. This enormous field of endeavor leads to most of the modern understanding of organic and organometallic reaction mechanisms as it is described in textbooks and used by researchers to help imagine new reactions or reaction conditions.

A huge number of examples of the power of this technique could be given and would take this chapter into areas that are not directly relevant. One particular family of studies will however be mentioned, to underline just how much information can be obtained by such techniques: the work done by the groups of Jencks [1] and others concerning the hydrolysis of carbonyl compounds such as amides. The broad mechanism of this reaction was established early on to involve nucleophilic attack by water or hydroxide ion on the carbonyl carbon, leading to a tetrahedral intermediate, which, perhaps after some adjustment of its protonation, can then collapse with loss of the amine moiety to yield a carboxylate salt. Studying the kinetics for this reaction under one particular set of conditions can tell you how fast the reaction proceeds, but *detailed* study of the kinetics can reveal much more. For example, it is possible to identify that a given 'general' acid catalyst is involved in transferring a proton to the amide carbonyl oxygen at the rate-determining transition state (TS) of the reaction, the extent to which this proton transfer has taken place, the fact that upon using a different acid catalyst, the rate-determining step changes to another step in the mechanism, and far more besides. All of this based only on kinetic studies, without the need for any direct experimental detection of the transition states to reaction!

The name given to the field most often associated with such studies is 'physical organic chemistry,' which correctly identifies that early work of this type usually addressed organic reactions. However, essentially the same techniques can also be applied to organometallic and inorganic chemistry. Indeed, much of our modern understanding of organometallic reactivity comes from similar studies. One famous example of this type of insight concerned the hydrogenation of alkenes by a rhodium complex, which was shown to proceed not through the relatively abundant metal–alkene complex that can be detected experimentally under reaction conditions, but instead from an isomeric and more reactive form [2].

As a student, like many people of my and preceding generations, I was taught about reaction mechanisms largely based upon studies of this type. However, in the intervening period, theoretical and computational methods have come to adopt a greater and greater role in examining reaction mechanisms. Theory in its various forms has always played a considerable role in developing our mechanistic understanding. And the use of quantum chemistry for modeling molecular electronic structure and exploring reactive potential energy surfaces is a practice with an already long track record that predates both of the experimental studies mentioned above. However, the balance between experiment and computation has firmly tipped toward the latter in recent years.

The move toward computation as a source of mechanistic information is this author's view partly due to a relative drop in prestige for experimental studies, especially when set against the difficulty of such studies. Given the enormous progress that physical organic chemistry made in understanding reactivity, more and more reactions, even 'new' synthetic procedures, can now be understood at least qualitatively by analogy to well-understood previous reactions. Also, the remaining relatively poorly understood reactions are nowadays, more often than not, much more challenging to investigate, due to the involvement of complex reaction conditions or of unstable reagents, or due to the unavailability or complexity of suitable analytical methods to assess reaction progress on the timescale of reaction. This does not mean that high-quality experimental studies of mechanism have ceased to be carried out. To again take just one example, concerning a reaction my group has been interested in, the Suzuki–Miyaura cross-coupling reaction, considerable insight has been provided in the last years by exquisite studies of two key steps, boronic ester activation [3] and transmetallation [4, 5]. However, the proportion of mechanistic studies using experimental methods has dropped considerably. Indeed, the work just cited [3, 5] includes significant contributions from computation also.

Accompanying the drop in the reward:risk ratio for experimental studies, there has been a huge rise in the applicability and power of computational methods. This can in turn be traced to improvements in computer hardware, to new and more accurate approximate theoretical methods, and to more efficient computer implementations of existing methods. Combined, these changes have led to an enormous change in the relative cost of computational studies in recent years. In the past, studies typically addressed model compounds in which most substituents in the experimentally studied compounds were removed, solvent and counterion effects were ignored, and very approximate techniques such as Hartree–Fock were applied. Nowadays, most synthetically relevant transformations are of a size such that the whole system containing reactant molecules, catalyst, and key solvent molecules contains less than a few hundred atoms and can be directly studied using methods such as density functional theory using relatively modest commodity computing resources.

One further reason that can be suggested for the rise in computational methods concerns the way in which different methods convey their results. Most experimental techniques provide *indirect* information concerning structure, energy, and bonding of key intermediates and TSs. In order to convert such insights into concepts that can be used to assist reaction design, synthetic chemists need to invest considerable effort, perhaps through the means of constructing three-dimensional molecular 'models' of the system of interest. In contrast, computational methods provide at least the illusion of *direct* access to these properties, with modern graphical rendering methods seeming to allow the researchers to immerse themselves into the atomistic-level realm within which molecules come together to react. The benefit in terms of supporting researchers' intuitions is considerable.

All of these factors have led to an explosion in the number of scientists studying reaction mechanisms computationally, with an accordingly large rise in the number of publications. Like most changes, the huge benefits brought about by this shift in emphasis within mechanistic studies carry some problems with them also. These negative aspects show up in a variety of ways, as does the awareness that the field has of these difficulties. Here, I would like to cite a withering criticism of computational studies published by the editors of a major journal in the field of catalysis, as part of an editorial about 'ethical conduct.' In their perspective, the authors write:

> A second example is that researchers often overly rely on computation in interpreting a reaction mechanism. The conclusions derived from an unrealistic assumption are far from truth, and confuse and mislead the community. It is worse than nothing. The problem is not the quality of the calculations, but arises from the abuse of a powerful tool by inexperienced researchers. We all know that organic synthesis with wrong starting materials will never result in the correct target product. We should appreciate solid experimental evidence more than frivolous computations. This prevailing trend is harmful to the community [6].

As can be seen, the authors explicitly argue both that computational studies can be wrong and misleading and that experimental studies should be preferred. Such an attitude can be expected to arise whenever a field such as mechanistic studies undergoes such a large shift in its research methodology in a relatively short period. As such, some of the language in the above extract can appear exaggerated or beside the point. Nevertheless, in my view, the arguments expressed are valuable and find many echoes in contemporary evaluations of mechanistic computational chemistry. In the following sections, I will attempt to explore some of the issues that lead to such attitudes, and how the field can respond to these challenges in future.

3 Reactivity Studies in Organic and Organometallic Chemistry

The work in our group on mechanistic organic and organometallic chemistry was initiated shortly after I started to work as an independent researcher in computational chemistry in 1999, and was facilitated by the fact that my Ph.D. studies were in the field of experimental organic chemistry, with a focus on reaction mechanisms involving organosulfur species [7, 8]. The first organic [9] and organometallic [10] studies carried out in my group were both published in 2002. I had not initially intended to perform such work, but it became obvious that there was a large demand for such studies and that the existing methods had the scope to provide significant mechanistic input. Already at the time, many experimental groups were performing their own computational studies, and the key density functional theory (DFT) and continuum solvation (or polarizable continuum model, PCM) techniques that still form the core of computational studies were available. This led to a perception that theoretical work in mechanistic organic chemistry was a very 'applied' subfield of computational chemistry, with little in the way of theoretical challenge. Given our work in the area in the intervening years, I now believe that the perception that this was a 'mature' subfield was incorrect in a number of important ways, and these will form some of the themes for this study.

3.1 Sulfur Ylide Epoxidation

The first study [9] concerned reaction of a sulfur ylide R_2S–CH–R′ with an aldehyde R″–CHO to form an epoxide R′–CH(O)CH–R″. The impetus for the study came from a desire to gain a better understanding of the factors governing the diastereoselectivity of the reaction (whether *cis* or *trans* or epoxides are formed) and its enantioselectivity (which enantiomer of the epoxide is favored when using a given chiral R_2S group in the ylide reagent). The broad mechanism of the reaction was already clear by analogy to known chemistry and from previous studies: Initial nucleophilic addition of the ylide carbon atom to the aldehyde forms a zwitterionic intermediate (or betaine), after which intramolecular nucleophilic substitution with the negatively charged oxygen atom of the betaine as nucleophile and the dialkyl sulfonium R_2S group as leaving group forms the epoxide (Fig. 1).

However, prior experimental work had led to confusing evidence concerning which of these two steps was rate-limiting, and which led to the observed preference for the *trans* diastereoisomer of the epoxide. This motivated our computational study, which was carried out using the popular B3LYP variant [11] of DFT, together with a continuum (PCM) model of the acetonitrile solvent used in the experiments. The calculations led to the potential energy surfaces shown in Fig. 2. These energy surfaces suggested a surprising interpretation of the experimental observations: First of all, the second step from Fig. 1, ring formation, appeared to actually be very facile, with a low barrier, for both diastereoisomers. In contrast, nucleophilic addition proceeded with a significant barrier and occurred preferentially through 'cisoid' addition TSs, in which the developing negative charge on the carbonyl oxygen atom was relatively close to the positive charge on the sulfonium sulfur atom. The initially formed betaines were accordingly also in cisoid conformeric form, whereas ring formation requires instead a transoid conformation in which the departing sulfonium group is roughly *anti* to the alkoxide oxygen. Forming such conformation requires passing

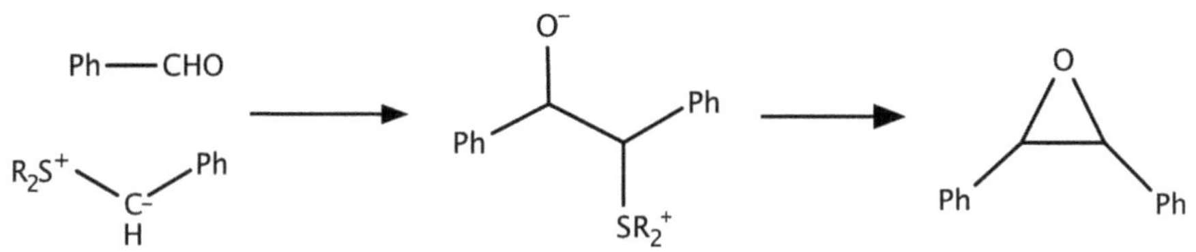

Fig. 1 Overall mechanism of sulfur ylide-mediated epoxidation of aldehydes

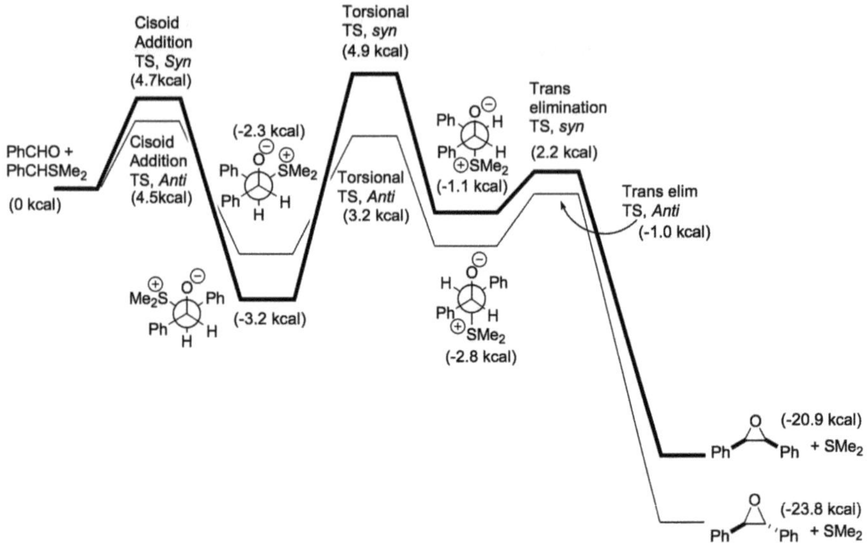

Fig. 2 Potential energy surfaces (B3LYP/6-311+G**//PCM) for sulfur ylide epoxidation. Reprinted with permission from [9], Copyright (2002) American Chemical Society

over a torsional TS, and to our surprise, this TS was found to the highest along the whole reaction profile in the case of the cis epoxide formation. This observation was able to rationalize the already mentioned confusing experimental observations and furthermore suggested a more general model that could account also for observed patterns of enantioselectivity [12].

Looking back at this project [9], some aspects are striking in terms of the methodology used. First, we used geometry optimization including the effects of solvation as treated by a continuum model, which was not yet common at the time, but has in the meantime essentially become completely standard, at least for charged or polar systems. Next, we used DFT for geometry optimization, as is also fairly typical—but we used the standard B3LYP functional [11] for this purpose. B3LYP was developed and parameterized for smaller molecules, and only later was it realized that it can lead to very significant errors for larger systems due to the neglect or at least inaccurate treatment of dispersion interactions. This will become a theme later in this chapter. Finally, we reported *potential energies* (or, more precisely, gas-phase potential energies augmented by a PCM estimate of the solvation free energy taking into account self-consistent relaxation of the gas-phase electron density and structure with respect to the continuum model of solvent) for the different stationary points, rather than free energies.

It has nowadays become completely standard to calculate statistical mechanics corrections in order to obtain standard free energies for species studied in mechanistic studies of reactivity. The reason for this is that free energies relate much more closely to experimental observables—equilibrium constants and rate constants—than do potential energies. Yet at the time, including these corrections had not yet become mandatory in computational chemistry. The reasons for this are manifold. First, calculating the required vibrational frequencies was and remains computationally demanding for larger systems (the model treated in our study, while of very modest size by today's standards, was not completely trivial at the time). Next, the accuracy of vibrational frequencies computed for systems whose structure was optimized with PCM solvation was not always very high, due to various numerical issues. Finally, there was some doubt about the need to include 'ideal-gas-like' free energy corrections for systems in solution. In retrospect, these doubts were incorrect, but they were quite widespread, and can perhaps be attributed to the fact that in many cases, free energy corrections to the potential energy roughly match in magnitude—but with opposite sign—the dispersion corrections needed to improve DFT energies. The fact that these two apparently very different effects roughly match in magnitude is associated with the fact that both dispersion effects and entropy effects tend to be large when dealing with processes in which two fragments come together to form a larger species (or the reverse of course). Hence for unimolecular processes, both effects tend to be small, whereas they become large for addition steps (like the first step in Figs. 1 and 2), and fortuitously roughly match in magnitude.

Given these many differences in the theoretical protocol used in [9] and the current state of the art, it is natural to wonder to what extent the conclusions reached are still applicable. In Table 1, the electronic energies from reference [9] are compared to new results generated for this review, including a 'best estimate' of the relative free energies at a level of theory similar to that used by us in a recent study of another reaction [16].

As can be seen in the table, including the corrections for entropy and for dispersion does indeed have a very large impact on relative energies. For example, the betaine complexes are stabilized by 3–5 kcal mol^{-1} compared to the earlier study. The CCSD(T) calculations match the B3LYP-D3 results quite well, though the cisoid betaines are somewhat more stable relative to reactants with the CCSD(T) approach. Another difference concerns the effect of the free energy corrections. As expected, these destabilize all the species relative to reactants, which again makes sense since one is bringing two separate molecules together to form one complex. As mentioned above, then, the dispersion and free energy corrections are in opposite directions, but the free energy correction is in this case somewhat larger, so the relative free energy of the TSs compared to reactants is somewhat higher than the published [9] relative potential energies with B3LYP.

In terms of the qualitative picture emerging from the calculations, the comforting outcome is that the new calculations remain broadly in favor of the previous conclusion: The deciding factor making the *trans* epoxide product dominate is the high energy and free energy of the *syn* torsional TS, which is confirmed to be rate-limiting.

As a final note on this reaction, it is worth noting that carrying out the new calculations was greatly facilitated by the fact that the original paper [9] contained optimized structures for all species in the Supporting Information, so that this data was available long after the original work had been done. In the past, quantum chemical studies often included full structural data for all optimized species in Figures, and various

Table 1 Relative energies and free energies of key species in Fig. 2 based on 2002 B3LYP calculations [9] and new B3LYP-D3 and DLPNO-CCSD(T) calculations (unpublished, 2018)

Species	E(B3LYP)	E(B3LYP-D3)	E(CCSD(T))	G(CCSD(T))
Reactants	0.0	0.0	0.0	0.0
Syn Add TS	4.68	−2.98	−2.63	10.17
Anti Add TS	4.50	−2.01	−2.45	11.30
Syn cisoid betaine	−3.19	−8.35	−14.03	1.01
Anti cisoid betaine	−2.27	−5.08	−9.27	5.49
Syn Torsional TS	4.92	1.24	−2.04	13.99
Anti Torsional TS	3.24	−0.37	−4.06	11.88
Syn Elimination TS	−0.08[a]	1.38	−1.18	12.17
Anti Elimination TS	−1.03	−0.96	−1.89	11.92

E(B3LYP): B3LYP/6-311+G**(CH_3CN)//B3LYP/6-31+G*(CH_3CN) electronic energies including PCM solvation using the Jaguar Poisson–Boltzmann approach, from [9]; **E(B3LYP-D3)**: electronic energies including PCM solvation using the SMD parameterization in Gaussian and the D3 empirical dispersion correction [13], B3LYP-D3/6-311++G(3df, p)(CH_3CN) single-point energies at B3LYP-D3/6-31+G(d)(CH_3CN) structures; **E(CCSD(T))**, single-point electronic energies at the DLPNO-CCSD(T) level of theory [14] at the B3LYP-D3/6-31+G(d)(CH_3CN) structures. Vacuum DLPNO-CCSD(T) calculations [14] using the aug-cc-pVDZ and aug-cc-pVTZ basis sets were carried out and basis set extrapolation to the complete basis set (CBS) limit was performed using the approach of Truhlar [15], with the B3LYP/6-31+G(d) solvation free energy included; and **G(CCSD(T))**: the E(CCSD(T)) values were complemented by zero-point, thermal and entropic corrections (298 K) derived from B3LYP-D3/6-31+G(d)(CH_3CN) vibrational frequency calculations

[a]This value was misreported as +2.2 kcal mol^{-1} in tables and figures in the main text of [9], though the total energies included in the Supporting Information was correct, and the value shown here is based on that correct value

repositories to collect results of quantum chemical computations were set up. With increasing computer power leading to more and more optimized structures for larger and larger species being generated in typical studies, these old ways of sharing data became increasingly impossible to perform. By and large, though, the key data can be shared via publishing optimized structures as Cartesian coordinates in the Supporting Information of papers (repeating a calculation for a given structure is usually relatively straightforward). Nowadays, there is an increasing move to encourage or mandate open sharing of data generated in scientific studies, and norms for how to do this in quantum chemistry continue to evolve. The example discussed here suggests that accessible archiving of Cartesian coordinates—at the very least—is strongly beneficial and should be the norm in quantum chemistry, though as computers continue to increase in power, the volume of data generated now often exceeds what can reasonably be included in print even in Supporting Information, and other types of archiving are developing to cope with this [17].

3.2 Ketone Hydrogenation by Ruthenium Hydrides

The second case study discussed here is the hydrogenation of ketones by ruthenium diamine diphosphine complexes [10], on which topic our first computational organometallics mechanisms paper was published, also in 2002. This project arose again from collaboration with an experimental group, with the aim being to provide independent structural and bonding insight into a key intermediate, as well as clues to the detailed mechanism. The schematic mechanism proposed in the study is shown in Fig. 3, where the amidoamine intermediate ($Ru(PPh_3)_2H(NH_2CMe_2CMe_2NH)$) (at the bottom right of the figure) could be isolated by the experimental group and characterized by X-ray crystallography. Kinetics for catalytic hydrogenation of acetophenone in benzene were determined both for the case of the Ph_3P-containing catalyst and for a related one in which the two Ph_3P phosphines were replaced by BINAP and showed second-order rate laws, with the Rate $= k \times p(H_2) \times [Ru]$.

Our calculations [10] first of all focused on the novel amidoamine intermediates, which display unusual distorted trigonal bipyramidal structures that were well reproduced by B3LYP structural optimization. Furthermore, we also computed a potential energy surface (Fig. 4) with the energies and structures of intermediates and transition state for the model reaction involving catalyst $(H_3P)_2RuH_2(NH_2CH_2CH_2NH_2)$ (**6′** in Fig. 4) and acetone. As in the previous section, we decided to compute only potential energies rather than free energies, and in this case, given the very nonpolar solvent benzene used in the experiments, continuum solvent was omitted. A hydrogenation transition state for the hypothesized concerted hydride and proton delivery from respectively ruthenium and nitrogen to respectively carbon and oxygen was

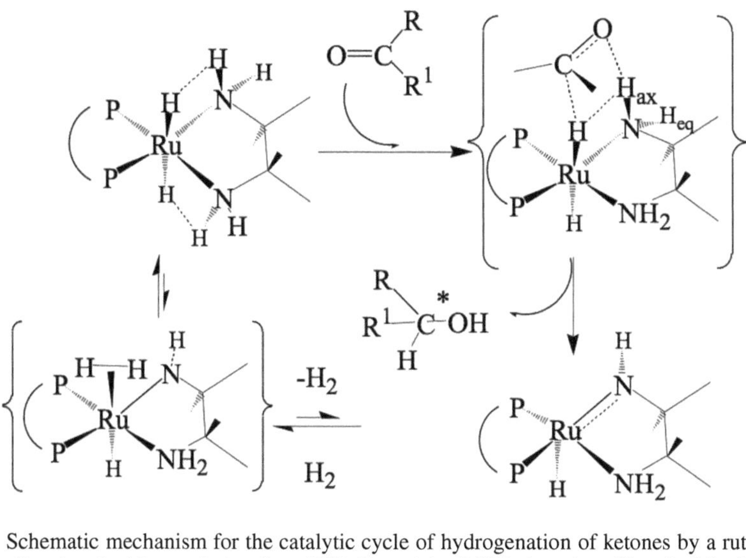

Fig. 3 Schematic mechanism for the catalytic cycle of hydrogenation of ketones by a ruthenium catalyst. Reprinted with permission from [10], Copyright (2002) American Chemical Society

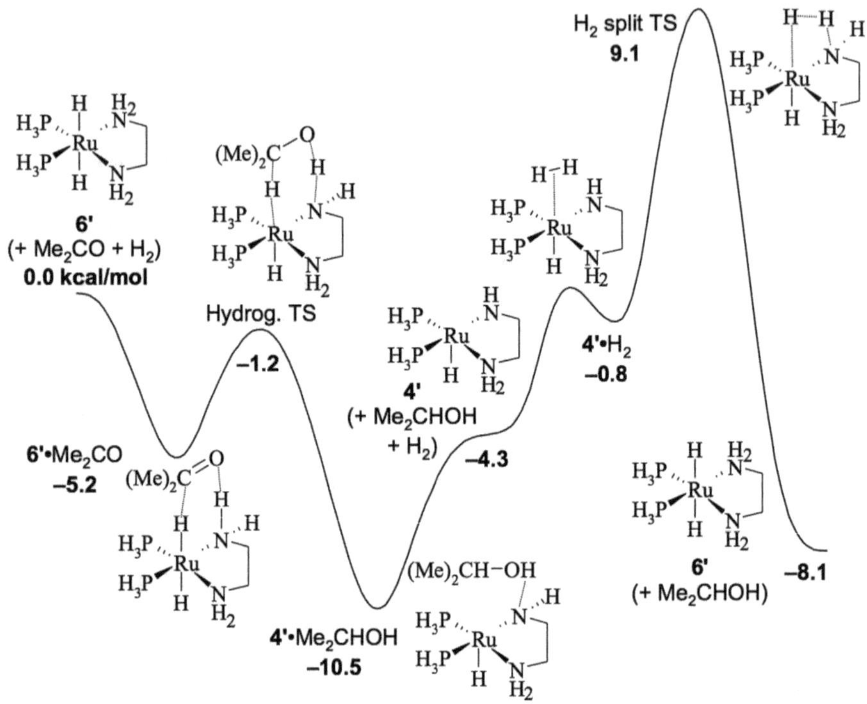

Fig. 4 Potential energy surface for hydrogenation of acetone by a ruthenium diamine dihydride complex. Reprinted with permission from [10], Copyright (2002) American Chemical Society

readily located, and found to lie low in energy, while a higher-energy H_2 splitting TS led to regeneration of the dihydridoruthenium species **6′** from the amidoamine intermediate **4′**, and had a higher barrier. Given that the H_2 splitting TS was so much higher in energy than that for hydrogenation, the hydrogen activation step was concluded to be turnover-limiting, consistent with the observed reaction order.

In a subsequent study [18], we studied the reactivity in the full system, obtained hydrogenation and H_2-splitting TSs in the case of Ru(S-BINAP)H_2(S,S-cydn), where S,S-cydn is the chiral diamine 1,2-diaminocyclohexane, and related species. These calculations—again mostly performed using B3LYP—provided insight into the origin of enantioselectivity of hydrogenation, and further details concerning the mechanism. The reason for highlighting this study here is however different: A referee of our work pointed out that our calculations, as described in the first submitted manuscript version, did not account for the observed selectivity of the hydrogenation, once possible reversibility of the ketone hydrogenation step was taken into account. The issue can be understood with reference to Fig. 4, which does not differ in this respect from the potential energy surface for the 'real' system nor from the free energy surface. Basically, the amidoamine species **4** can react either with hydrogen to regenerate the diamine species, or with the alcohol product to regenerate the ketone.

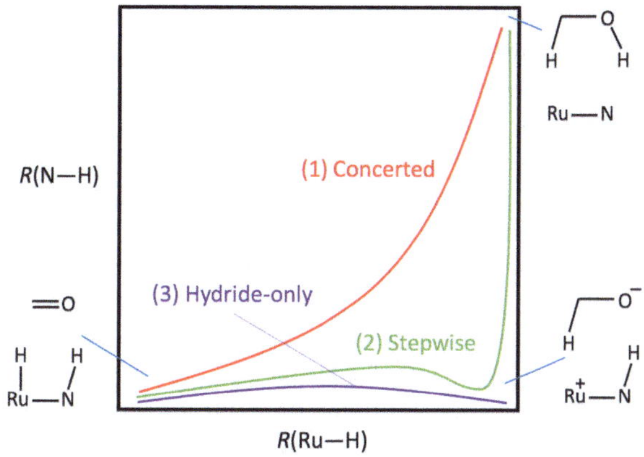

Fig. 5 Schematic illustration of concerted (1), stepwise (2), and hydride transfer only (3) mechanisms in ruthenium hydride ammine hydrogenation of ketones

Since the barrier to the latter process is lower, then hydrogenation of ketone should be reversible, so that even if the initial hydrogenation is enantioselective, product should be racemized under reaction conditions, and no enantioselectivity should be observed.

This referee was completely right—we had neglected to spot this point, and in our revised manuscript, we acknowledged this problem. One possible explanation is that our calculations were not accurate enough and that more accurate electronic energies as well as treatment of entropic effects would have improved the situation. However, the discrepancy was fairly large (the lowest hydrogenation TS was almost 14 kcal mol^{-1} lower in energy than the hydrogen splitting TS), and the expected entropic effect was unlikely to be large, as both steps were bimolecular. An alternative explanation was more plausible: There was already evidence that under protic conditions, a different mechanism for H_2 splitting and indeed for hydrogenation can occur. In this variant, the amidoamine becomes protonated on nitrogen, yielding a cationic five-coordinate ruthenium species, which itself can coordinate H_2, which can then be deprotonated, to yield the diamine dihydride complex. In fact, the amidoamine species need not even be formed: Instead of concerted hydride and proton transfer to the ketone, simple hydride transfer can occur, forming an alkoxide that is then protonated by solvent, and the cationic five-coordinate species mentioned, which can then activate dihydrogen as discussed above (see Fig. 5).

In a more recent computational study [19], support for these pathways has been provided from a thorough examination of different possible routes. Also a molecular dynamics approach using an ab initio DFT potential energy surface for a related Ru-catalyzed reaction showed that these ionic variants tend to dominate in protic solvents [20].

The reason for highlighting this problem here is twofold. First, it shows that it is possible to derive the incorrect kinetic conclusions from a calculated potential energy surface. In this case, considering the uncertainties relating to the calculated potential energy surfaces, and the neglect of entropic effects, the barriers for each individual reaction step seemed to be consistent with observed reactivity. However, the *relative* height of the two barriers was not consistent with observed behavior. Why did we miss this point? A first explanation is simply that beyond a certain mechanistic complexity, kinetic analysis based only on computed free energies or indeed potential energies becomes difficult to perform and errors are made. This system was not highly complex, so other people such as the referee could readily detect the problem. Indeed, having learned from this case and others, we would probably no longer make this mistake for a system of this relative simplicity. Still, it is worthy of note that errors of this type can occur and are expected to become more frequent as complexity increases. A second explanation is that our focus in the second study [19] was enantioselectivity, which we simply assumed *must* be due to the hydrogenation TSs, and accordingly paid too little attention to other points.

A second issue highlighted by this mistake is that chemical reaction mechanisms are complicated, in ways that are sometimes hard to capture in computational studies. For ketone hydrogenation by diamine dihydride ruthenium complexes, our first study [10] had focused on understanding reactivity and kinetics in a non-chiral variant of the reaction, carried out in a nonpolar solvent. Additionally, in that study, the amidoamine intermediate shown on the top right in Fig. 5 (and bottom right in Fig. 3) had been isolated and shown to be catalytically competent. The experimental rate law was compatible with hydrogen splitting being turnover-limiting (although the experimental activation enthalpy was somewhat smaller than the calculated one). All of this suggests that the mechanism suggested in that paper, and shown in Figs. 3 and 4, is actually broadly correct *under the reaction conditions used there* (broadly correct in that H_2-splitting is indeed turnover-limiting, though the too-high barrier may mean that the detailed mechanism for this step needs to be refined). In contrast, with protic solvents, the variant (3) shown in Fig. 5—which is likely to be very unfavorable under non-protic, nonpolar conditions—may dominate. The free energy landscape underlying Fig. 5 is relatively flat, and its shape is sensitive to the environment. Hence, mechanistic conclusions based on one set of reaction conditions may not apply under all conditions.

3.3 *Mechanism Discovery:* Cis–Trans *Isomerization of Alkenes*

In 2011, we published a study of a deceptively simple reaction: reversible *cis–trans* isomerization of an alkene [21]; see Fig. 6. This isomerization was discovered by our experimental collaborators in the group of Prof. Guy Lloyd-Jones, using the simple Pd(II) catalyst $PdCl_2$ in solvents such as acetonitrile. The experimental studies showed that this reaction could partly be caused by small amounts of palladium hydride impurities formed under the reaction conditions, but conditions were found where this side route could be inhibited, yet isomerization continued to be observed. The question was, therefore, how does Pd(II) cause isomerization? Clearly, the alkene could coordinate to Pd, which would weaken the double-bond character of the alkene

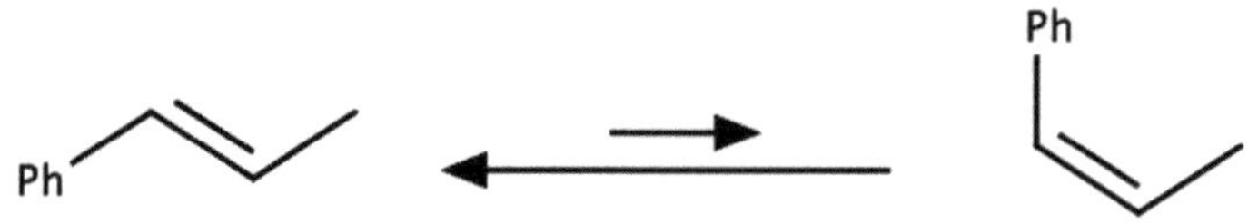

Fig. 6 Alkene *cis–trans* isomerization

C–C bond, but equally clearly, this would occur within the framework of the standard Dewar–Chatt–Duncanson picture for alkene complexes, which requires a structure in which both carbon atoms interact with the Pd center, thereby leading to partial Pd–C bonds and partial three-membered ring character, which would have the effect of prohibiting rotation and isomerization.

We examined this computationally, studying a variety of possible mechanisms involving Pd(II) alkene complexes. One possibility we considered is that the Pd could carry out an electrophilic addition to the alkene to form a β-palladium carbocation species Pd–C–C$^+$ that would be less stable than the normal alkene complex, but might be low enough in energy to form transiently, and then undergo rotation around the C–C bond. Another possibility was that a nucleophile in the medium—solvent or an impurity such as water—could add to one of the carbons, forming a species Pd–C–C–Nu that could undergo rotation then loss of the nucleophile. This would require that approach of the nucleophile and its departure would occur from opposite faces of the Pd–alkene complex. Indeed, such pathways could be located with DFT calculations. However, all isolated TSs lay much too high in energy to be able to account for the observed reactivity. Strenuous tests were made to benchmark the DFT method used to explore the potential energy surface (B3LYP with a flexible basis set for optimization, followed by single-point energy correction using the Grimme D3 dispersion correction), and in so far as we could judge, inaccuracy in the theoretical method was much smaller than the mismatch in calculated barrier and observed reactivity, which was larger than 10 kcal mol^{-1} for all mechanisms and computational approaches considered.

Lengthy discussions ensued with the experimental collaborator, until a breakthrough came from the observation that under some conditions, the experimental reaction kinetics showed a reaction order larger than 1 with respect to palladium catalyst. As the reaction order was intermediate (smaller than 2, larger than 1), the implication of this was not clear. It is known that $PdCl_2$ species can form dimers with two bridging chlorides and two external chlorides, and such species were considered computationally. However, coordination at one palladium center in such complexes led to reaction mechanisms and barriers that were barely different from those found with the monomeric catalyst model. Upon continued encouragement from Prof. Lloyd-Jones, further mechanistic variants were considered, until finally a complicated bimetallic mechanism could be identified that was able to account for the experimental observations (Fig. 7). In this mechanism, the aforementioned palladium dimer (species **8** in Fig. 7) could undergo ring opening, to form an alkene complex **9** with only one bridging chloride, and this could undergo an intramolecular

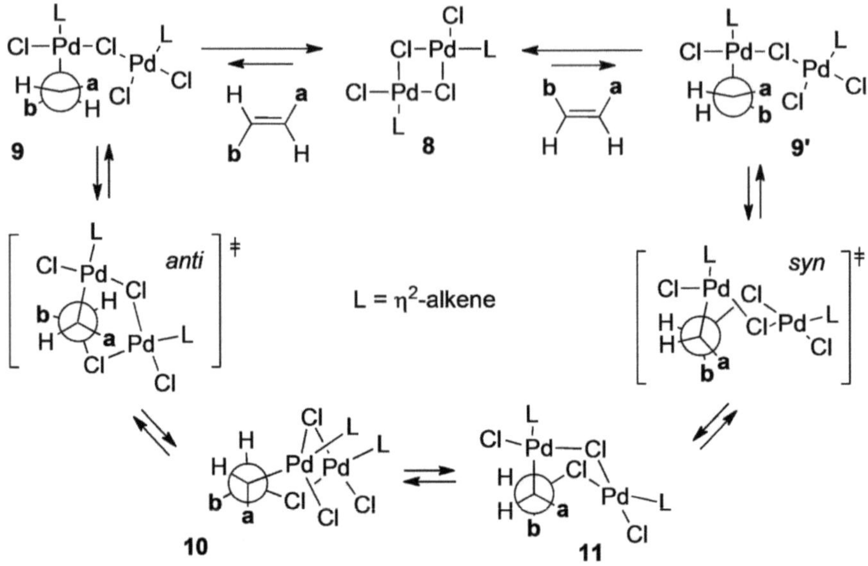

Fig. 7 Bimetallic mechanism for isomerization of a generic alkene CHa=CHb by a dimeric palladium complex, species 8. Reprinted with permission from [21], copyright John Wiley and Sons (2011)

cyclization step, in which a free chloride ligand on the *other* palladium center carried out a nucleophilic addition on one of the alkene carbons, to form a six-membered ring complex **10**, which was able to undergo conformational change involving rotation around the erstwhile carbon–carbon double bond, before ring opening.

This mechanism, by taking account of the initial alkene coordination and the equilibrium between monomeric and dimeric palladium species, was able to reproduce the observed kinetics perfectly, and the fitted elementary rate constants were in good agreement with the calculated relative free energies, with errors of 2–4 kcal mol^{-1} that were well within the estimated accuracy of the computational approach.

The reason for highlighting this study here is because the nature of the difficulty encountered when seeking the correct reactivity model in our work [21] was a very clear form of a common problem in computational studies: Our initial way of casting the problem to be solved had led us to a misconception about the microscopic nature of the reaction: We assumed that the catalyst was a $PdCl_2$ center. This made it very hard to make the leap of imagination that the true catalyst was the dimer, and in fact this leap was only made following an express invitation from the experimental colleague to consider dimers. In retrospect, this was perhaps an obvious step to take, since it was known that palladium chlorides form dimers. Still, in retrospect many things are obvious. Even after switching to dimers, there was some work to do to identify the correct mechanism, but this was in many ways an easier part of the work than was the step to change the model to a dimeric palladium species.

The problem encountered here can be expressed in a more formal way: The correct mechanism simply *cannot be represented* on the potential energy surface of the [Pd, Cl_2, ligand, alkene] model system (the ligand was acetonitrile solvent) initially built. No amount of exploration of the potential energy surface of this model could possibly generate the correct mechanism. Extension to variants such as [Pd, Cl_2, ligand, alkene, nucleophile] was tested, but the extension to [Pd_2, Cl_4, $ligand_n$, alkene] was only made after the experimental evidence supporting dimers was found. As it happens, the initial calculations were performed on a model system with two palladium atoms and *one* additional solvent ligand ($n = 1$), and this model too is unable to account for the reactivity, but optimization of one candidate structure did hint that perhaps a species with $n = 2$, as shown in Fig. 7, might be favorable, and then calculations with that model confirmed this hint, leading to the correct mechanism. Note that the $n = 1$ potential energy surface is a lower-dimensional manifold on the $n = 2$ surface, and the monomeric palladium surface is a lower-dimensional manifold of the dimeric surface, so species and TSs with effectively *lower* stoichiometry can be discovered when exploring the surface of the species with higher stoichiometry, but in both cases the reverse is not true.

The impossibility to represent the correct reaction mechanism in a given model system is a major challenge for computational mechanistic studies, since there does not appear to be a systematic yet affordable theory-based approach to generate the correct model stoichiometry. Systematic approaches for exploring the potential energy surface for a given assumed model stoichiometry are undergoing major expansion [22] and are already very computationally demanding. Approaches requiring addition of additional fragments to the starting reaction model can also be formulated, but on the one hand, they do usually require setting a *maximum* stoichiometry initially, and on the other hand, they rapidly lead to exponentially increasing computing costs. Model building remains the area where the computational chemist's chemical intuition (or rather, that of their experimental collaborators, etc) remains essential in many cases. To close this section, I will point out that had we been unlucky, and had the mismatch between the calculated free energy barrier for the initially studied mechanisms and the experimentally observed rates been somewhat smaller, then we might have been led to publish this incorrect mechanism.

3.4 Morita–Baylis–Hillman Reaction

Our fourth case study is a reaction mechanism that we have been interested in for some years now, the coupling of an acrylic ester with an aldehyde with catalysis by an amine to form a new carbon–carbon bond. This particular variant of the Morita–Baylis–Hillmann reaction has received a lot of synthetic interest, with synthetic chemists being interested in developing catalysts and conditions that maximize reactivity, and optionally that allow enantioselective formation of one or other of the mirror-image products. An experimental colleague initiated the collaboration, with the view of obtaining insight into the structure of key rate-limiting TSs so as to be

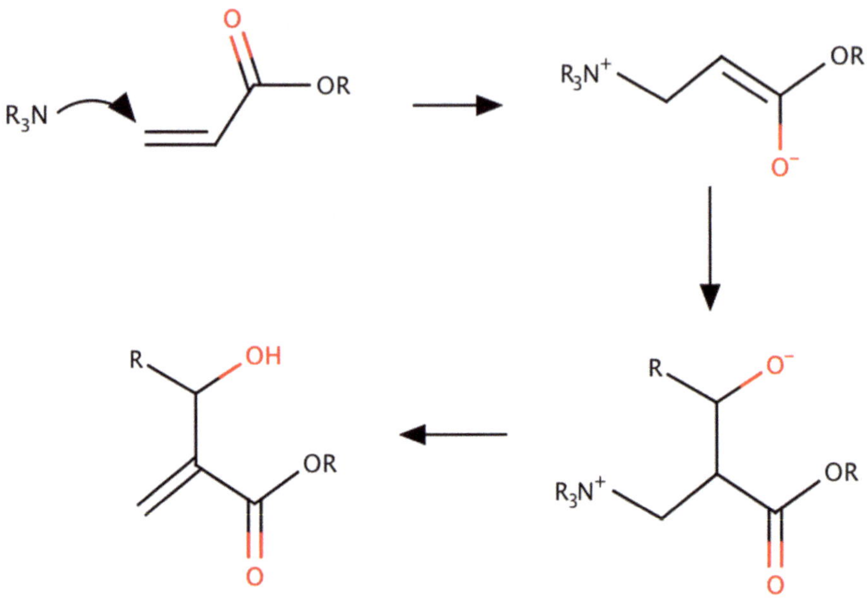

Fig. 8 Generic mechanism of the Morita–Baylis–Hillman reaction

able to assist with development of an enantioselective version of the reaction. The broad features of the reaction mechanism were already known and could indeed be guessed fairly easily by an experienced mechanistic organic chemist. Nucleophilic addition of the amine catalyst to the acrylic ester forms a zwitterionic enolate, which can itself undergo nucleophilic addition to the aldehyde, with proton shuffling and loss of amine-generating product (Fig. 8).

Our first study of this reaction [22] used a similar methodology to that of [9]: the B3LYP flavor of DFT together with a medium-sized basis and the Poisson–Boltzmann form of PCM to treat solvation. Methyl acrylate was the chosen substrate, together with benzaldehyde, and trimethylamine was the model used for the catalyst. Based on experimental evidence that the proton shuffling could be facilitated by a second equivalent of aldehyde or by the alcohol group of the product, we included in some calculations a second equivalent of benzaldehyde, and a methanol molecule to represent the protic product. Only potential energies (together with the solvation free energy from PCM) were included.

The pathways explored are summarized in Fig. 9, with key calculated relative energies also shown. The initial nucleophilic addition step is endothermic and is followed by endothermic addition to aldehyde. In the case where no protic product or solvent was considered to be present, this was followed by addition of a second aldehyde to form a deprotonated hemiacetal intermediate, which could transfer a proton from the α position of the acrylate moiety to ultimately generate product. In contrast, in the presence of the alcohol (methanol) molecule, this molecule could

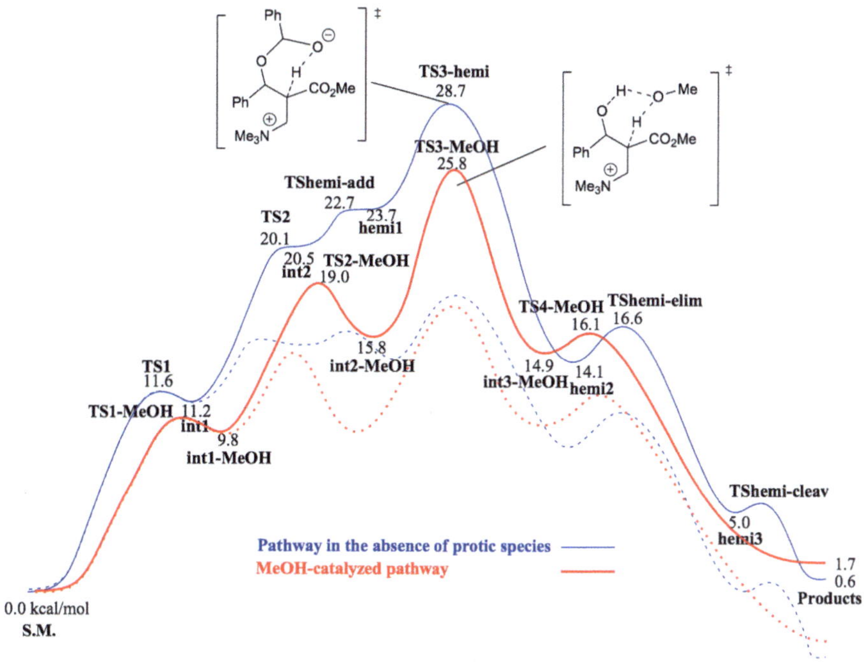

Fig. 9 Potential energy surfaces for the Morita–Baylis–Hillman reaction, based on B3LYP and a continuum solvent model. Reproduced with permission from [22], Copyright (2007) American Chemical Society

lead to a 'proton shuttle' pathway moving the same α-proton to the oxygen atom, above a TS of similar height to that modeled in the hemiacetal pathway.

This was the first DFT study of the mechanism of the Morita–Baylis–Hillmann reaction to work on a (nearly) realistic model system, at the DFT level, and including PCM solvation (a model of THF, the non-protic solvent sometimes used in experimental studies). Also, we made efforts to calibrate the B3LYP functional by performing a set of correlated ab initio calculations on a model system. This was done using the G3MP2 composite method and seemed to indicate reasonable accuracy for B3LYP. Finally, a number of possible side reactions and mechanistic variants were studied, including, under the protic conditions, proton transfer from solvent to the anionic center in the key zwitterionic intermediate. Finally, by obtaining detailed atomistic models of intermediates and TSs, we were able to obtain insight into the challenge in carrying out enantioselective versions of the reaction: We found multiple close-lying TSs with different relative arrangement of the amine and the forming chiral centers. Since the use of a chiral amine catalyst is the obvious way to try to perform an enantioselective variant of the reaction, this is clearly a problematic observation, since it suggests that selectivity will be counteracted by the existence of close-lying conformers and diastereoisomers en route to products. For all these reasons, we felt emboldened to include the word 'accurate' in the title of our paper [22].

In retrospect, this was unwise, since while some of the mechanistic conclusions drawn from the study were indeed accurate (such as the observations about enantioselectivity), and some features of our study made it more likely to be accurate than previous studies, some of the underlying calculated results were very far from being accurate, as became clear in following years.

First, we decided to go beyond potential energy surfaces and compute free energies, somewhat like has been reported here in Table 1. In fact, hand-waving estimated of free energies of activation had been included in a footnote in [22] and hinted that the chosen theoretical protocol overestimated the barrier to reaction. This was emphatically confirmed upon calculating free energies in a new study [23]. The calculated free energy of the rate-limiting TS using the same DFT functional as in the original study, B3LYP, was above 50 kcal mol^{-1} relative to reactants, and thereby well above the estimated free energy barrier based on experimentally observed reactivity, which was of the order of 25 kcal mol^{-1}.

The new study simultaneously suggested a resolution of this disagreement for the case of the non-protic reaction conditions, involving a second equivalent of aldehyde. A first effect leading to barrier lowering is to change the model of the nucleophile from trimethylamine to one of the catalysts that is actually used experimentally, quinuclidine. This is a much better nucleophile, and this change lowers the barrier by a few kcal mol^{-1}. A much more important effect was observed based on local coupled-cluster ab initio correlated calculations, and on dispersion-corrected B3LYP (using the same type of Grimme correction already mentioned before in this study): It appears that B3LYP *hugely* overestimates the relative potential energy of the key TSs. This is because the Morita–Baylis–Hillman reaction in this variant involves bringing *four* moderately large organic molecules (one tertiary amine, one acrylate ester, and two aldehydes) together to form a single species. The dispersion effect of roughly 5 kcal mol^{-1} shown in Table 1 is for bringing *two* species together, i.e., forming one new intermolecular interface. Here, one forms six such interfaces (assuming each fragment is in contact with each other), or three, assuming a one-dimensional arrangement—in any case, a much larger effect is expected and indeed found.

The second paper [23] did not go on to consider the reaction under protic conditions, partly due to the large amount of work needed to perform, and critically assess, the local coupled-cluster calculations for the large system. A second reason was that the situation in the presence of the alcohol product was considered to be more complex, so that continuum solvent models were feared to provide a less accurate model of the solvation effects. We did not at all consider reaction in the presence of a large amount of alcohol, as when the reaction is performed in such solvents. Reaction in alcohols is not strikingly faster than in polar aprotic solvents including relatively apolar solvents such as THF or dichloromethane, so we considered that except for the additional possibility for shuffling shown in Fig. 9, we did not expect the mechanism to be hugely different in alcohols.

Following both of these papers, a very thorough experimental study [24] was reported of the Morita–Baylis–Hillman reaction, considering a slightly different vari-

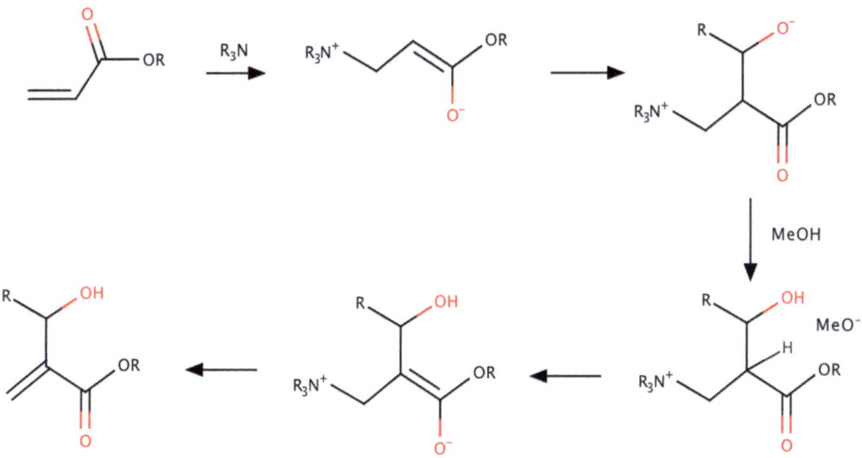

Fig. 10 Experimentally supported mechanism for the Morita–Baylis–Hillman reaction, as demonstrated in [24]

ant: with *p*-nitrobenzaldehyde in place of benzaldehyde, and carried out in methanol solvent, in the presence of the DABCO amine catalyst (which is similar to the already mentioned quinuclidine). This paper used a series of experiments to assess the free energy of a number of intermediates and TSs in Fig. 8, as well as for some additional species. As well as reporting these results, the authors reported their own computational results, using the B3LYP density functional as in our earlier study [22] along with some dispersion-corrected methods. Based on their new computational results, and those reported by others such as our study [22], the authors argued forcefully that computational methods were not yet mature enough to provide valuable mechanistic information about such reactions. Their conclusions [24] state for example that '[I]t is not clear to us that any significant accurate information that was not already apparent from experiment either has been, or could have been, reliably garnered purely from computations.'

The main reasons for this conclusion can be paraphrased as follows:

- Conventional organic mechanistic theories are sufficient to predict the mechanism in some detail—by identifying the key steps, and indeed estimating their thermodynamics and kinetics.
- The previously published and new computations led to results that were severely inaccurate in relation to experiment, and furthermore these results were very sensitive to computational choices such as the treatment of dispersion or of entropic effects.
- *The* correct mechanism, sketched in Fig. 10, had either not been considered or incorrectly ruled out by computational studies.

These arguments are indeed quite strong and do indeed emphasize that truly *predictive* studies of reaction mechanisms are much harder to carry out than is usually acknowledged in the computational community. Post hoc rationalizations of experimental observations are much more common. The suggestion that basic organic chemistry knowledge is already enough to identify the mechanism has some merit, and indeed I already noted near Fig. 8 that the broad features of the mechanism are obvious from basic knowledge of organic chemistry, with the key discussion concerning the way in which the proton shuffling occurs. Also, it has been noted that the calculated energies in our first study, derived from B3LYP, were in error by more than 20 kcal mol^{-1} for some species, which is a huge error. For these reasons, the criticisms of the previous computational studies were fully warranted.

However, it should be noted that *the* mechanism as shown in Fig. 10 cannot in fact be the *only* mechanism for the reaction, since it explicitly requires the presence of a proton-donating alcohol solvent, while the reaction can be performed in non-protic solvents, or using a neat mixture of the (non-protic) reactants. Hence, another mechanism—the two-aldehyde mechanism mentioned above and considered in both of our computational studies [22, 23]—must be favored under some conditions (it is not completely clear under which conditions which of the mechanisms is favored). It should also be noted that our own work had in fact never set out to predict the mechanism in pure alcohol solvent (the alcohol that we did consider was the product, although it was modeled as methanol). Finally, we note that the conclusions surrounding enantioselectivity are much less easy to derive simply using paper and pencil, and knowledge of organic reaction mechanisms.

While retrodiction is much less satisfying than prediction, the publication of [24] did spur us on (together with another colleague, Ragahavan Sunoj, who had also previously published on the Morita–Baylis–Hillman reaction) to revisit the reaction and attempt to examine computationally the mechanism for the precise variant of the reaction studied in the new study [24], i.e., with DABCO catalyst, methyl acrylate, and *p*-nitrobenzaldehyde, in methanol. Using local coupled-cluster methods, the free energy surface emerging from this new study [16] is shown in Fig. 11. As can be seen our calculated free energies are in fair agreement with experiment, with a maximum error of just under 5 kcal mol^{-1}, in line with the expected accuracy of the methods used.

Our new study [16] also aimed to examine the *origin* of the errors in earlier studies. Some of these are obvious: The poor description of dispersion in B3LYP has already been mentioned several times in this chapter and leads to hugely inaccurate energies for this reaction. It has also already been mentioned that some of the 'error' in some of the studies comes from the fact that the mechanism of the Morita–Baylis–Hillman reaction is different under different reaction conditions. We also showed that changes in the solvent model (various continuum models or a free energy perturbation method with molecular mechanics), or in the treatment of the entropy arising from soft vibrational modes, or changing from one more accurate dispersion-corrected DFT functional to another or to coupled-cluster theory, are enough to account for changes in relative free energies of several kcal mol^{-1}. The Morita–Baylis–Hillman reaction involves very large changes in the solvation free energy, so it may represent a more challenging target for accurate computational studies than some other reactions involving less change in polarity along the reaction path.

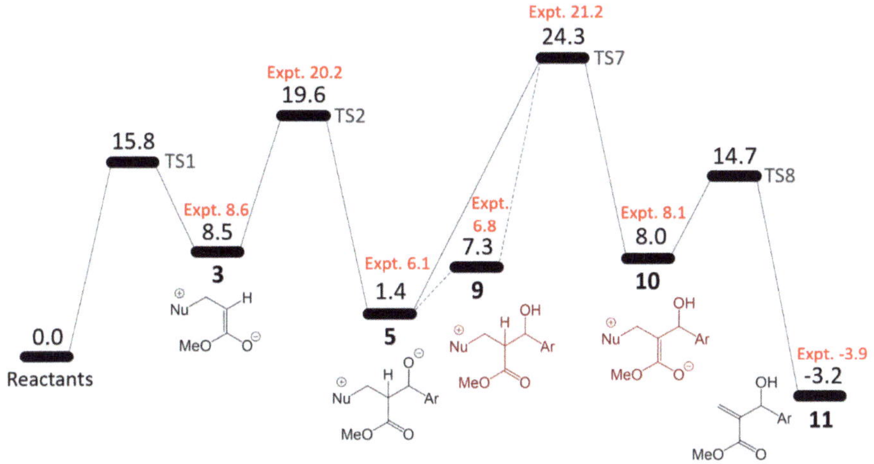

Fig. 11 Free energies (in kcal mol^{-1}) relative to reactants (p-nitrobenzaldehyde and methyl acrylate, with catalyst Nu = DABCO), calculated at the CCSD(T) level and including mixed continuum and explicit treatment of solvent. Reproduced from [16] with permission from the PCCP Owner Societies

One important factor was the handling of the proton-transfer step with solvent shown as the rightmost step in Fig. 10. This step had actually been considered in [22], but had been found there to be endothermic by 19.2 kcal mol^{-1}, whereas experimentally the standard free energy change is only +0.7 kcal mol^{-1} [24]. Part of this very large difference is due to the fact that the +19.2 kcal mol^{-1} is a potential energy difference, not a free energy. Another difference is that the calculation was performed in continuum THF solvent, not the more polar methanol. A more important factor, though, is that the reaction was modeled as involving the formation of a naked methoxide anion, stabilized only by the continuum model, while the reactant was modeled as involving the hydrogen-bonded complex between the zwitterionic intermediate and methanol solvent. It is known [see, e.g., 25]—but was not sufficiently appreciated by us at the time of writing [22] that 'direct' calculation of the difference in pK_A between two acids—which is effectively what our calculation at the time amounted to—can lead to *very* large errors due to imbalances in the treatment of solvation. Indeed, in our newer study [16], proton transfer for this step was found to be very unfavorable when treating the methoxide anion produced as an isolated species. Upon including 'microsolvation' through which this ion was surrounded by methanol molecules, much better accuracy was obtained.

Another important factor is the treatment of entropy. As already mentioned in the context of Table 1, when bringing two molecules together to form an intermediate or TS, there is a large entropy effect on the free energy. Here, one is bringing four molecules together, so the effect is very large. In our original paper [22], we had suggested that the effect as calculated using gas-phase-like statistical mechanics, which is mainly driven by the Sackur–Tetrode equation for the translational entropy, might be exaggerated. This suggestion was motivated by several studies [26] arguing that experimental entropies of solvation are quite negative, so that solutes have a lower entropy than suggested by the Sackur–Tetrode equation. These negative entropies of solvation are indeed frequently observed experimentally, but their existence does not mean that one should not use a version of the Sackur–Tetrode equation to compute entropies hence free energies in solution. Indeed, the negative entropy of solvation should be accounted for by the use of the continuum solvent model, which is parameterized to reproduce the experimental free energy of solvation, including both its enthalpic and entropic factors. Seeking to make a separate adjustment of the solvent-phase entropy would lead to a double-counting of solvation entropies. In our new study [16], the 'full' Sackur–Tetrode entropy is included, and the good agreement with experiment vindicates the present argument. A recent study [27] has compared various ways to compute solvent-phase entropies and also argues against making substantial ad hoc adjustments to the Sackur–Tetrode approach.

4 Conclusions

The above examples show that computational chemistry can make important contributions to understanding reaction mechanisms in organic and organometallic chemistry. However, the examples described show that obtaining accurate results is much more challenging than had been appreciated when I started to carry out work in this area. The chapter outlines many of the pitfalls that we fell into along the way—or that we avoided due to good luck, and the vigilance of referees or experimental collaborators, rather than anything else.

What are some of the key lessons learned? First is that electronic structure theory remains challenging. It is to be hoped that there will be no future discovery of unexpected but very large errors in a commonly used method, such as the discovery of the magnitude of the error made by using non-dispersion-adjusted functionals such as B3LYP that occurred over the period covered by the research described in this chapter. In retrospect, the discovery of the importance and magnitude of this error for organic and organometallic mechanistic chemistry must be considered to be enormously chastening for the field. It occurred due to the fact that the chemical space where functionals such as B3LYP had been widely benchmarked—largely, a space occupied only by small molecules—ceased to overlap with the regions of chemical space where it was being used, as more powerful computers enabled application to larger and larger species. This observation invites caution for the future, as computers continue to expand in capacity, so more and more systems of more and more varied properties are studied. My view is that the future will contain more chastening experiences where a given method is discovered to be much less accurate than expected for a problem to which it has applied and where results have been published.

A second observation and recommendation relates to the fact that while calculating potential energy surfaces and free energies is definitely challenging, deducing the kinetic and thermodynamic conclusions of these calculations can also be surprisingly demanding. In our study of ruthenium-catalyzed hydrogenation of ketones [18], we made a mistake in our analysis of this type. This was pointed out to us by a referee prior to publication, fortunately, and could be remedied by more careful thinking. I argue that such errors occur quite frequently and that beyond a certain level of mechanistic complexity, authors would be well advised to carry out explicit kinetic simulation [28] of the network of reaction steps they have modeled, in order to check that the emerging behavior meshes with that which they had expected.

A third recommendation is that when studying details of reaction mechanisms, one should pay great attention to the fact that most interesting chemical transformations do not have one single mechanism, but rather exist on a mechanistic manifold, with the preferred route or rate-limiting step sometimes changing when substrates or reaction conditions are changed.

A fourth point is that the ‘unknown unknowns’ in mechanistic exploration will continue to tax our ingenuity as computational chemists and invite us to think carefully about possible mechanisms. The unknown unknowns here refer to species or transition states that we simply have not thought about and that perhaps cannot even be represented while using the atomistic model that we have chosen to use to study the problem at hand. Occasionally, as in our study of palladium-catalyzed alkene isomerization, these unknown unknowns, once they become known, will turn out to lie low in free energy and to play an important role in the reaction.

One more conclusion emerging from this work is that statistical mechanics plays an important role in mechanistic organic and organometallic chemistry. Because the corresponding calculations typically performed, using simple ‘ideal-gas’ statistical mechanics with the rigid-rotor and harmonic oscillator approximations and the Sackur–Tetrode equation for translational entropy, are very undemanding computationally, in comparison with the electronic structure calculations, there may be a tendency to overlook their importance and to reach incorrect conclusions about their accuracy. For example, we ourselves argued at one point [22] that adjustments to solution-phase free energies may need to be applied, based on an incorrect analysis of the theoretical framework. In fact, more recent work makes it clear that this suggestion is seriously incorrect and that the ‘standard’ ideal-gas-type expressions are reasonably correct.

As an overall conclusion taken from this work, I would like to argue that the computing accurate free energy changes, rate constants and mechanisms for organic and organometallic chemistry is still highly challenging. There has been huge progress, so that computation can now provide valuable assistance in studying mechanisms, and indeed in some cases, near-quantitative conclusions can be obtained. However, for the foreseeable future at least, it does not appear likely that computational chemistry will become predictive enough to be able to completely replace experimental approaches for studying mechanisms.

References

1. Jencks WP (1976) Enforced general acid-base catalysis of complex reactions and its limitations. Acc Chem Res 9:425–432. https://doi.org/10.1021/ar50108a001
2. Landis CR, Halpern J (1987) Asymmetric hydrogenation of methyl-(*Z*)-α-acetamidocinnamate catalyzed by {1,2-bis(phenyl-*o*-anisyl)phosphino)ethan}rhodium(I): kinetics, mechanism, and origin of enantioselection. J Am Chem Soc 109:1746–1754. https://doi.org/10.1021/ja00240a025
3. Gonzalez JA, Ogba OM, Morehouse GF, Rosson N, Houk KN, Leach AG, Cheong PHY, Burke MD, Lloyd-Jones GC (2016) MIDA boronates are hydrolysed fast and slow by two different mechanisms. Nat Chem 8:1067–1075. https://doi.org/10.1038/nchem.2571
4. Thomas AA, Denmark SE (2016) Pre-transmetalation intermediates in the Suzuki-Miyaura reaction revealed: the missing link. Science 352:329–332. https://doi.org/10.1126/science.aad6981
5. Thomas AA, Wang H, Zahrt AF, Denmark SE (2017) Structural, kinetic, and computational characterization of the elusive Arylpalladium(II)boronate complexes in the Suzuki–Miyaura reaction. J Am Chem Soc 139:3805–3821. https://doi.org/10.1021/jacs.6b13384
6. Noyori R, Richmond JP (2013) Ethical conduct in chemical research and publishing. Adv Synth Catal 355:3–8. https://doi.org/10.1002/adsc.201201128
7. Harvey J, Brichard MH, Viehe HG (1993) Electrophilic ene-type reactions of phenyl vinyl sulfoxide with alkenes. J Chem Soc Perkin Trans 1:2275–2280. https://doi.org/10.1039/P19930002275
8. Harvey JN, Viehe HG (1995) 3-thio-claisen rearrangement of the allyl vinyl sulfonium ion. J Chem Soc Chem Commun 2345–2346. https://doi.org/10.1039/c39950002345
9. Aggarwal VK, Harvey JN, Richardson J (2002) Unravelling the mechanism of epoxide formation from sulfur ylids and aldehydes. J Am Chem Soc 124:5747–5756. https://doi.org/10.1021/ja025633n
10. Abdur-Rashid K, Clapham S, Hadzovic A, Harvey JN, Lough AJ, Morris RH (2002) Mechanism of the hydrogenation of ketones catalyzed by trans-dihydrido(diamine)ruthenium(II) complexes. J Am Chem Soc 124:15104–15118. https://doi.org/10.1021/ja016817p
11. Becke AD (1993) Density-functional thermochemistry. III. The role of exact exchange. J Chem Phys 98:5648–5652. https://doi.org/10.1063/1.464913
12. Aggarwal VK, Richardson J (2003) The complexity of catalysis: origins of enantio- and diastereocontrol in sulfur ylide mediated epoxidation reactions. Chem Commun 2644–2651. https://doi.org/10.1039/b304625g
13. Grimme S, Antony J, Ehrlich S, Krieg H (2010) A consistent and accurate ab initio parametrization of density functional dispersion correction (DFT-D) for the 94 elements H-Pu. J Chem Phys 132:154104. https://doi.org/10.1063/1.3382344
14. Riplinger C, Pinski P, Becker U, Valeev EF, Neese F (2016) Sparse maps—a systematic infrastructure for reduced-scaling electronic structure methods. II. Linear scaling domain based pair natural orbital coupled cluster theory. J Chem Phys 144:024109. https://doi.org/10.1063/1.4939030
15. Truhlar DG (1998) Basis-set extrapolation. Chem Phys Lett 194:45–48. https://doi.org/10.1016/S0009-2614(98)00866-5
16. Liu Z, Patel C, Harvey JN, Sunoj RB (2017) Mechanism and reactivity in the Morita–Baylis–Hillman reaction: the challenge of accurate computations. Phys Chem Chem Phys 19:30647–30657. https://doi.org/10.1039/c7cp06508f
17. For one example of a modern approach to archving quantum chemistry results, see the ioChem-BD project. https://www.iochem-bd.org/
18. Leyssens T, Peeters D, Harvey JN (2008) Origin of enantioselective hydrogenation of ketones by RuH_2(diphosphine)(diamine) catalysts: a theoretical study. Organometallics 27:1514–1523. https://doi.org/10.1021/om700940m

19. Dub PA, Henson NJ, Martin RL, Gordon JC (2014) Unravelling the mechanism of the asymmetric hydrogenation of acetophenone by [RuX_2(diphosphine)(1,2-diamine)] catalysts. J Am Chem Soc 136:3505–3521. https://doi.org/10.1021/ja411374j
20. Pavlova A, Meijer EJ (2012) Understanding the role of water in aqueous ruthenium-catalyzed transfer hydrogenation of ketones. ChemPhysChem 13:3492–3496. https://doi.org/10.1002/cphc.201200454
21. Tan EHP, Lloyd-Jones GC, Harvey JN, Lennox AJJ, Mills BM. [$(RCN)_2PdCl_2$]-catalyzed alkene isomerization: selective inhibition of migration reveals a non-hydride pathway for E/Z interconversion. Angew Chem Int Ed 50:9602–9606. https://doi.org/10.1002/anie.201103947
22. Robiette R, Aggarwal VK, Harvey JN (2007) Mechanism of the Morita-Baylis-Hillman reaction: a computational investigation. J Am Chem Soc 129:15513–15525. https://doi.org/10.1021/ja0717865
23. Harvey JN (2010) Ab initio transition state theory for polar reactions in solution. Faraday Discuss 145:487–505. https://doi.org/10.1039/b907340j
24. Plata RE, Singleton DA (2015) A case study of the mechanism of alcohol-mediated Morita Baylis–Hillman reactions. The importance of experimental observations. J Am Chem Soc 137:3811–3826. https://doi.org/10.1021/ja5111392
25. Ho J, Coote ML (2010) A universal approach for continuum solvent pK_a calculations: are we there yet? Theor Chem Acc 125:3–21. https://doi.org/10.1007/s00214-009-0667-0
26. Lau JKC, Deubel DV (2006) Hydrolysis of the anticancer drug cisplatin: pitfalls in the interpretation of quantum chemical calculations. J Chem Theory Comput 2:103–106. https://doi.org/10.1021/ct050229a
27. Besora M, Vidossich P, Lledos A, Ujaque G, Maseras F (2018) Calculation of reaction free energies in solution: a comparison of current approaches. J Phys Chem A 122:1392–1399. https://doi.org/10.1021/acs.jpca.7b11580
28. Besora M, Maseras F (2018) Microkinetic modeling in homogeneous catalysis. WIREs Comput Mol Sci 8 (in press). https://doi.org/10.1002/wcms.1372

Computational Modelling of Structure and Catalytic Properties of Silica-Supported Group VI Transition Metal Oxide Species

Jarosław Handzlik

Abstract Chromium, molybdenum and tungsten oxides supported on amorphous silica are catalysts for many reactions, including large-scale industrial processes. Although these systems have been extensively studied for many years, there are still a few unresolved issues, concerning mainly the nature of the active sites and mechanisms of their formation. Computational studies, using cluster or periodic models to represent the catalyst surface, are helpful in interpretation of spectroscopic data and can provide complementary information about the catalytic process. In this chapter, such computational works on CrO_x/SiO_2, MoO_x/SiO_2 and WO_x/SiO_2 systems are presented. It is seen that coordination environment of the transition metal, determined also by local surface properties, is a key factor influencing catalytic activity of the surface metal species. This results in complex structure–activity relationships. While a great progress has been achieved in modelling of these systems, from simple clusters to advanced periodic slabs, theoretical determination of complex reaction mechanisms using surface models with representative distribution of metal sites is still a challenge for computational catalysis.

1 Introduction

Group VI transition metal oxides supported on amorphous silica are effective catalysts for many important reactions, including large-scale industrial processes like ethene polymerization or alkene metathesis. These ill-defined systems have been extensively studied for many years using both experimental and theoretical methods. Nevertheless, the structure of the surface chromium, molybdenum and tungsten oxide species on silica has been a subject under debate for a long time, and sometimes contradictory conclusions, based on spectroscopic data, were drawn. Moreover, the nature of the active sites, usually surface organometallic species, as well as the way

J. Handzlik (✉)
Faculty of Chemical Engineering and Technology, Cracow University of Technology, ul. Warszawska 24, 31-155 Kraków, Poland
e-mail: jhandz@pk.edu.pl

E. Broclawik et al. (eds.), *Transition Metals in Coordination Environments*, Challenges and Advances in Computational Chemistry and Physics 29,
https://doi.org/10.1007/978-3-030-11714-6_11

how they are formed, is even less understood. On the other hand, knowledge about the catalyst structure at the atomic level is extremely important for studying the catalytic reaction mechanisms and necessary in effective designing of new catalysts. Coordination environment of the transition metal is a key factor influencing the catalytic properties of the surface metal species. Especially, the local properties of the support, which can be considered as a multidentate ligand coordinated by the metal centre, can dramatically affect the reactivity. However, establishing the structure–activity relationships for silica-supported metal oxide catalysts is still a challenging task, mainly due to heterogeneity of the metal species on amorphous surface and low fraction of the active sites formed in situ.

Computational modelling, taking advantage of growing computer power, enables for development of advanced models representing surface metal species. Density functional theory (DFT) methods, offering a good balance between the accuracy and the cost, are nowadays commonly used in the field of computational catalysis. Theoretical investigations are helpful in interpretation of spectroscopic data concerning supported transition metal oxide species. Quantum chemistry methods, especially when combined with advanced models of the surface, can also provide complementary information about the catalytic system, not accessible at this moment by experimental techniques. Better understanding the nature of the active sites and their precursors, as well as mechanisms of the catalytic reactions and the role of the amorphous surface, is possible owing to computations. Many examples of such computational studies, concerning silica-supported chromium, molybdenum and tungsten oxide systems, are presented in this chapter.

2 Surface Modelling

An adequate model representing surface is crucial for realistic description of the supported metal oxide systems. Especially, modelling amorphous materials with inhomogeneous distribution of surface metal sites is challenging. There are two main approaches to simulate surface of solid, cluster approximation [1–6] and periodic slab models [1–5, 7–10]. A cluster model is a finite fragment cut off from the solid, with dangling bonds usually saturated by hydrogen atoms. Consequently, long-range interactions are neglected, which can be justified for covalent amorphous materials like silica. Small clusters, often used in the past due to limitations of computer power, cannot reproduce the complexity of the amorphous surface, although they are handy for efficient computations of reaction mechanisms. Nowadays, it is possible to employ very big cluster models, containing dozens or even hundreds of atoms [3, 4]. To reduce the computational cost or enlarge the model, partitioning into layers treated at different levels of theory can be done [11]. The active site and its close vicinity are then calculated at a higher level of theory (*inner layer*), while the rest of the system (*outer layer*) is described by a computationally less demanding method. An example of such hybrid method is the ONIOM partitioning scheme [11] (Fig. 1).

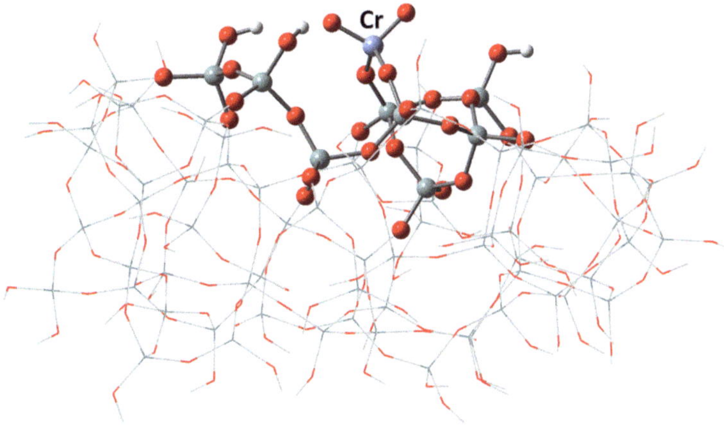

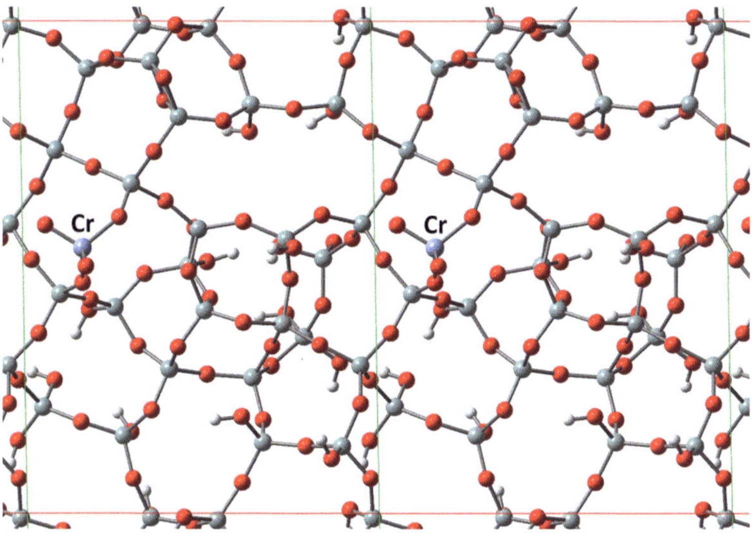

Fig. 1 Example cluster (top) and periodic [4] (bottom) models of Cr(VI) oxide species on silica. In the first case, the two-layer ONIOM method is applied (the wireframe part of the model represents the outer layer)

In the case of the periodic models, the surface of the solid is represented by a slab inside the unit cell that is repeated using periodic boundary conditions (Fig. 1) [1]. For realistic description of the catalyst surface, especially for amorphous materials, the unit cell should be large enough to avoid artificial periodicity and lateral interactions between a surface metal site or adsorbed reactant and its periodic images. Several advanced slab models of amorphous silica were developed from DFT calculations [12–16], and some of them were used in computational studies presented below.

3 CrO_x/SiO_2 System

Chromia–silica system is well known as the Phillips catalyst for ethene polymerization, discovered in 1950s [17–20]. Approximately 40–50% of the worldwide production of high-density polyethylene is related to this catalyst [17]. A unique property of the Phillips catalyst is the fact that no cocatalyst is necessary to achieve the catalytic activity, in contrast to other transition metal systems used for ethene polymerization, such as Ziegler–Natta and metallocene catalysts. Silica-supported chromium oxide systems are also effective in other catalytic reactions, like dehydrogenation of alkanes [21], oxidative dehydrogenation of hydrocarbons in presence of oxygen [22, 23] or carbon dioxide [24, 25], and various selective oxidation reactions [20, 26–28].

3.1 Structure of Surface Chromium Oxide Species—Experimental Data

Many techniques, including UV-vis, Raman, XANES (X-ray absorption near-edge structure), EXAFS (extended X-ray absorption fine structure), EPR (electron paramagnetic resonance), XPS (X-ray photoelectron spectroscopy) and IR (infrared) spectroscopy, have been used to characterize chromium oxide species on silica. After high-temperature calcination of chromia–silica systems, well-dispersed surface Cr(VI) oxide species are mainly present [17–26, 28–36], although small amounts of Cr(V) and Cr(III) species were also detected, as well as Cr_2O_3 clusters at higher Cr loadings [17, 19–24, 26, 32, 34]. The nature of the Cr(VI) species on silica has been discussed for many years. Monomeric, dimeric and polymeric Cr(VI) species were often postulated, mainly based on the UV-vis DRS (diffuse reflectance spectroscopy) data [19, 20, 24]. On the other hand, Raman spectroscopy, UV-vis, XANES and EXAFS studies conclusively demonstrated that the monomeric form is the major or even the only surface Cr(VI) oxide species [19, 23, 26, 29–36]. It is described as a tetrahedral dioxo species with two Cr–O–Si linkages (Fig. 2a). Minor five-coordinate monooxo Cr(VI) species, bonded to the silica surface by four Cr–O–Si linkages (Fig. 2b), was also suggested [32–34]; however, this proposal has been recently questioned [35, 36].

Even more complex and problematic is nature of reduced chromium species, which seems to be strongly dependent on the reduction conditions. If CO is used as the reducing agent, Cr(VI) species are almost selectively converted to Cr(II) [5, 17–20, 30, 37–39], whereas reduction with H_2 results in Cr(III) [34, 37] or both Cr(III) and Cr(II) [5, 24] oxidation states. The presence of water during reduction favours formation of Cr(III) species [5]. When the calcinated Phillips catalyst is contacted with ethene, Cr(VI) is reduced to Cr(II) and/or Cr(III) [17–19, 34, 40, 41].

According to the UV-vis DRS studies, there are three distinct reduced chromium oxide species on silica: pseudo-tetrahedral Cr(II), pseudo-octahedral Cr(II) and pseudo-octahedral Cr(III) [20, 24, 42]. In addition to the surface reduced chromium

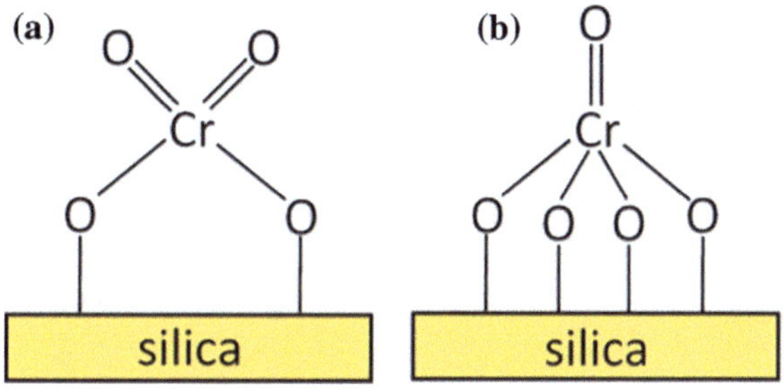

Fig. 2 Proposed structures for monomeric Cr(VI) oxide species on silica

species, Cr_2O_3 clusters can be present [5, 20, 24]. Cr(III) and Cr(II) oxidation states were also detected by XPS [19, 28, 43]. Based on IR spectroscopy data concerning CO chemisorption on the silica-supported Cr(II) species, three families of the Cr(II) sites can be distinguished, characterized by a different degree of coordinative unsaturation [19, 20, 30, 44]. Isolated Cr(V) species, dispersed Cr(III) species and Cr(III) clusters in the CrO_x/SiO_2 systems can be observed by EPR [20, 23, 24, 39, 45–47]. Although intermediate Cr(IV) oxidation state might be also expected during the reduction of the surface Cr(VI) species with two-electron reducing agents, like CO or H_2, there are very few reports suggesting that Cr(IV) species are indeed experimentally observed [38, 39].

3.2 *Structure of Surface Chromium Species—Computational Modelling*

Taking into account a complex molecular picture of the CrO_x/SiO_2 catalyst and amorphous character of the silica support, computational modelling of this system is a difficult task even today. In the past, only small cluster models of the chromium species on silica were possible to apply for efficient computations. Nevertheless, even simple models, if reasonable constructed, can be helpful in interpretation of spectroscopic data.

Espelid and Børve [48] used F-terminated clusters containing 2–3 Si atoms to calculate d–d transition energies and intensities for monomeric Cr(II) and Cr(III) species, and dimeric Cr(II) species. The transition energies, computed from second-order, multireference many-body perturbation theory (CASPT2 method), were in agreement with the experimental UV-vis DRS data reported for reduced chromia–silica systems. On this basis, the assignment of the experimental bands to specific coordination and oxidation states of chromium was proposed. Mainly the same cluster models were applied to investigate at the DFT level the structure, stability and vibrational properties of surface Cr(II) and Cr(III) species after CO adsorption [49]. Damin et al. [50] computationally studied interaction between Cr(II) species on silica and probe molecules (CO, N_2) employing the model of Espelid and Børve.

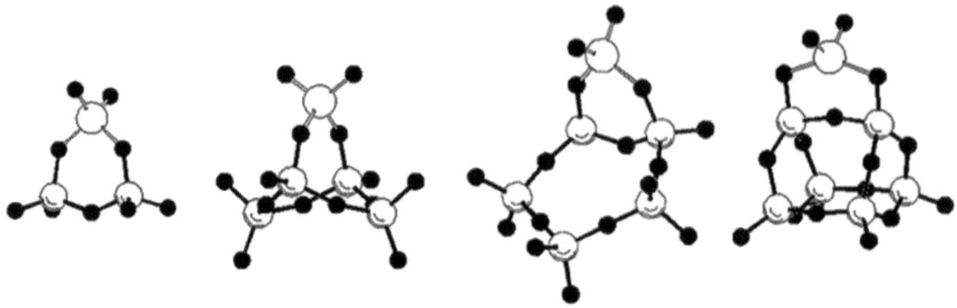

Fig. 3 Cluster models of monomeric dioxo Cr(VI) species on silica. Adapted from [29] with permission from the PCCP Owner Societies

In addition to F-termination of the cluster, other terminations (H, OH) were also examined. Interestingly, by varying the percentage of Hartree–Fock exchange in hybrid DFT functionals, they found that its increase to 35–40% allows for a better agreement with the experimental IR data, compared to the B3LYP method (20% of Hartree–Fock exchange).

In combined experimental and theoretical (B3LYP) studies, Dines and Inglis [29] used cluster models ranging in complexity (Fig. 3) to obtain geometrical parameters of the monomeric dioxo Cr(VI) species on silica and enable the assignment of the measured Raman spectra. It was concluded that only monomeric chromium species are present in the Cr(VI)/SiO_2 system at low Cr loading. Recently, the electronic structure of isolated Cr(VI) species on silica was comprehensively investigated using two complementary techniques: UV-vis absorption and fluorescence spectroscopy [35]. The observed electronic absorption spectrum was compared with simulated spectra obtained from TD-DFT (B3LYP) calculations for cluster models of the dioxo Cr(VI) species containing various chromasiloxane ring structures, built based on the models used by Dines and Inglis. Additionally, the five-coordinate monooxo Cr(VI) species was considered. The obtained results provided no evidence for the presence of the monooxo Cr(VI) sites in significant amounts.

Small clusters were also applied to model surface Cr(VI) species which were formed after CrO_2Cl_2 grafting on silica and characterized with IR, XANES and EXAFS spectroscopy [51, 52]. The energetic effects for possible routes for the formation of the surface Cr(VI) sites were calculated with the B3LYP functional. It was concluded that fraction of Cr(VI) species with strained chromasiloxane ring depends on silica dehydroxylation temperature and is related to the catalytic activity in ethene polymerization. In next combined experimental and computational work on this system [53], reduced surface Cr(II) species were modelled by a chromasiloxane ring embedded in a large cluster (37 Si atoms) cut off from the β-cristobalite structure. The ONIOM method and hybrid B3PW91 functional with increased percentage of Hartree–Fock exchange (40%) were used to study adsorption of CO on the Cr(II) sites. They were coordinated by either one or two siloxane ligands, in agreement with EXAFS results. The former, three-coordinate Cr(II) species were proposed to

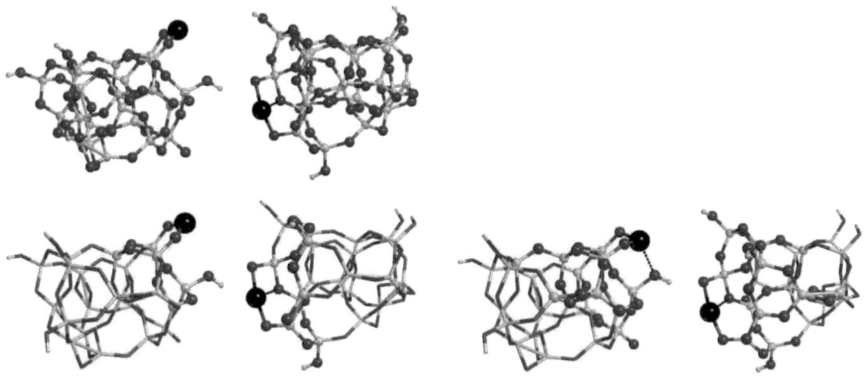

Fig. 4 Cluster models of monomeric Cr(II) species on silica (black sphere: Cr atom). For models at the bottom the ONIOM scheme with differently defined inner layer was applied. Adapted from [44], Copyright (2015), with permission from Elsevier

be relevant for ethene polymerization activity. Based on these results, smaller cluster models of Cr(II) monocarbonyl, Cr(II) dicarbonyl and Cr(III) vinyl species were further built and used in DFT calculations to aid in vibrational assignments in recent experimental studies on the initiation mechanism for the Phillips ethene polymerization catalyst [39, 47].

The ONIOM partitioning scheme was applied for clusters containing 22 Si atoms to model Cr(II) species in the Phillips catalyst (Fig. 4) and examine CO adsorption on these sites [44]. High-level calculations were performed with the long-range corrected hybrid ωB97X-D functional, including empirical dispersion corrections. Additionally, electronic transitions were computed at the TD-DFT level for smaller models, limited to the inner layer. Two differently located Cr(II) sites were modelled. The first one is more coordinatively unsaturated and leads to stable dicarbonyl species upon CO adsorption. In the second one, chromium strongly interacts with a siloxane bridge and monocarbonyl species is rather formed. The models were validated by comparison of the computational results with the experimental IR and UV-vis spectroscopy data for the Cr(II)/SiO_2 catalyst with a very low Cr loading. It was concluded that both kinds of the Cr(II) sites coexist in the real system and their relative ratio depends on the activation temperature. Similar computational approach was recently used for modelling Cr(VI) and Cr(II) species on silica and was validated by comparison of the simulated and experimental XANES spectra [41]. Simulation of the XANES spectrum for a model of Cr(II) species with methylformate and two ethene molecules in the coordination sphere supported the proposal based on operando spectroscopic investigations of the Phillips catalyst that sixfold coordinated Cr(II) sites, in interaction with the oxygenated by-products, are mainly involved in ethene polymerization.

Isolated monooxo and dioxo Cr(VI) species on silica under dehydrated conditions were modelled using medium-size (15 Si atoms) clusters derived from the β-cristobalite structure [54]. Geometries of the whole systems were fully optimized with the PW91 and TPSS functionals to account for the amorphous nature of silica. On the basis of test calculations, the PW91 method was found to be most accurate, among many density functionals, in predicting relative energies of chromium oxo compounds. It was concluded that the relative stability of the monooxo Cr(VI) species, compared to the dioxo one, is much lower than the relative stability of the monooxo Mo(VI) species in the molybdena–silica system, which was earlier computationally studied using analogous models [55].

High-resolution ^{1}H MAS NMR and ^{29}Si CP/MAS NMR spectroscopy was combined with DFT (B3LYP) computations employing polyhedral oligomeric silsesquioxane (POSS) models to investigate the role of various silanol groups in formation of surface Cr(VI) species [56]. It was suggested that monomeric dioxo Cr(VI) species are most favourable located on pairs of a single and an adjacent geminal silanols accompanied with a geminal silanol left. Replacing a pair of geminal silanols by the chromium species is not preferred energetically. The calculated enthalpies of the grafting reactions indicate also higher preference for monomeric Cr(VI) species, compared to Cr(VI) dimers.

Although cluster models of various complexity, representing CrO_x/SiO_2 catalyst, were successfully applied in many computational works, especially those combined with spectroscopic studies, they are usually more or less arbitrarily constructed. To develop more advanced models, being able to reproduce heterogeneity of the surface chromium oxide species, realistic and well-validated models of amorphous silica [12–16] are necessary. Such a model, developed and validated by Tielens et al. [13], representing hydroxylated surface (5.8 OH nm^{-2}) in the unit cell including 120 atoms ($Si_{27}O_{54}\cdot 13\ H_2O$), was employed in systematic computational studies of the $Cr(VI)/SiO_2$ system [7]. Mono-, di-, tri- and tetragrafted monomeric chromium(VI) species at different degrees of hydration were modelled (Fig. 5). Their formation can be regarded as grafting H_2CrO_4 unit with dehydroxylation of surface silanols. Different types of silanols were involved in grafting reactions: isolated (Si–OH), vicinal (HO–Si–O–Si–OH), geminal (HO–Si–OH) and nonvicinal (two Si–OH groups not directly connected). The relative stabilities of the surface chromium species in a wide range of temperatures were determined from the atomistic thermodynamic approach, based on the calculated energies (PBE functional) for grafting reactions. The monografted hydroxy dioxo Cr species was predicted to be most stable at lower temperatures. An increase of temperature favours digrafted dioxo Cr species (Fig. 2a) and then tetragrafted monooxo Cr species (Fig. 2b). The comparison of the calculated Cr=O stretching frequencies for the surface chromium species with experimental Raman spectra indicated the digrafted dioxo Cr species as the most representative for the $Cr(VI)/SiO_2$ system.

A large number of periodic and cluster models of the SiO_2 surface were applied in extensive DFT (PW91) investigations of chromium(VI) oxide species on partly dehydroxylated silica [4]. The periodic model of Tielens et al. [7, 13] was properly modified to achieve the initial surface with 3.1 OH nm^{-2}, and, 2.4–1.3 OH nm^{-2} after grafting the Cr species, which roughly corresponds to the conditions of the catalyst thermal treatment before the reaction [17–19, 30, 57, 58]. Similar to the original model, the chromium coverage was about 0.4 atoms nm^{-2}, in agreement

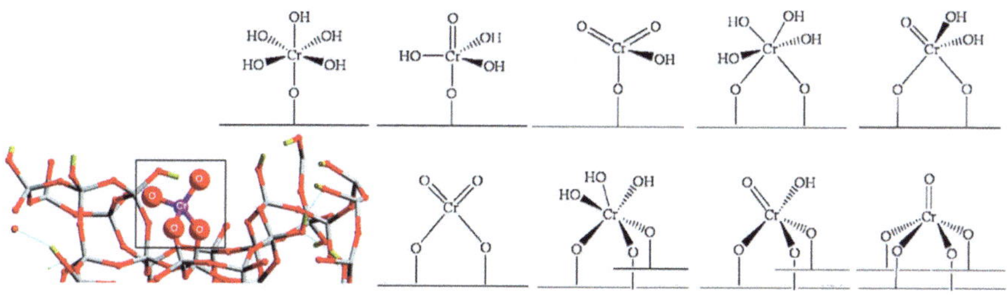

Fig. 5 Modelled Cr(VI) oxide species on hydroxylated silica surface and periodic model of digrafted dioxo Cr(VI) species. Adapted with permission from [7]. Copyright (2012) American Chemical Society

with typical Cr loadings for the Phillips catalyst [17–19, 30]. Based on the periodic models, a set of cluster models (33 Si atoms) was also built. Additionally, cluster models (21, 24, 26 and 72 Si atoms) derived from another structure of amorphous silica were developed, as well as periodic (24 Si atoms in the unit cell) and cluster (63 and 97 Si atoms) models based on the β-cristobalite structure (Fig. 6). With a variety of models, relative stabilities of the tetrahedral dioxo Cr(VI) species and five-coordinate monooxo Cr(VI) species (Fig. 2) were determined. The relative energies were shown to depend on the location of the Cr site on the surface and the structure of the model. More flexible amorphous models allow for easier formation of the four Cr–O–Si linkages of the monooxo Cr(VI) species, compared to the models based on the crystalline structure. Nevertheless, all the approaches led to the same general conclusion that the dioxo Cr(VI) species are more stable than the monooxo species. This is consistent with the above-mentioned experimental results [35, 36]. It was again noticed that the energetic preference for the dioxo Cr(VI) species over the monooxo Cr(VI) species is stronger than in the case of the analogous models for the Mo(VI) species on silica [3]. The vibrational frequency analysis for the variously modelled surface Cr(VI) species, including simulation of the isotopic effect, allowed for a detailed interpretation of the experimental Raman spectra for the Cr(VI)/SiO_2 system [4]. It was predicted that the asymmetric O=Cr=O stretching mode for the dioxo species and the Cr=O stretching mode for the monooxo species can overlap.

By considering in the next computational studies [5] dimeric Cr(VI) species on silica, often postulated in the literature [19, 20, 24, 42], it was predicted that they are less stable than the monomeric Cr(VI) species. The structures of various reduced chromium oxide species, both monomeric (Fig. 7) and dimeric (Fig. 8), were also calculated. The use of advanced periodic and cluster models of amorphous silica, taken from the previous work [4], allowed to deal with the heterogeneity of variously located reduced Cr species. For instance, three-coordinate Cr(II) species interacting with siloxane ligand, which was proposed to be relevant in ethene polymerization [39, 44, 53], was directly obtained after geometry optimization without any arbitrarily construction of the model. Therefore, such models can be helpful in investigations of structure–activity relationships for the Phillips catalyst. On the basis of the vibrational frequency analysis, some new assignments for reported Raman spectroscopy data

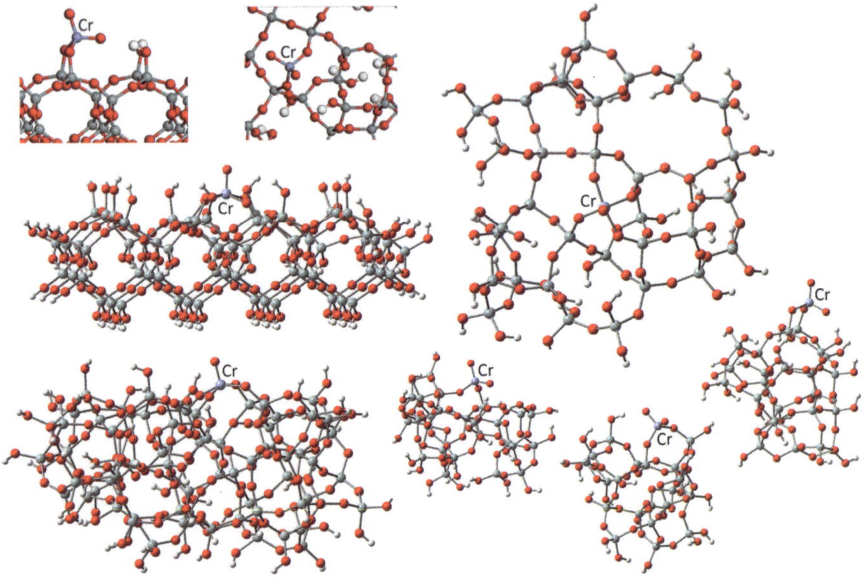

Fig. 6 Example periodic and cluster models of monomeric Cr(VI) oxide species on silica. Adapted with permission from [4]. Copyright (2013) American Chemical Society

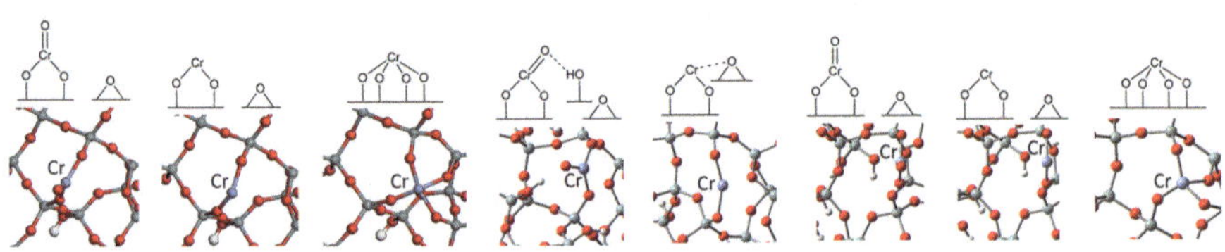

Fig. 7 Example periodic and cluster models of monomeric Cr(IV) and Cr(II) oxide species on silica. Adapted from [5], Copyright (2016), with permission from Elsevier

concerning reduced CrO_x/SiO_2 systems were proposed. The energetic effects of complex redox paths, leading to various Cr(II) and Cr(III) sites, were computed using the PW91 functional. In addition to the DFT calculations, in situ UV-vis DRS studies of the reduction of CrO_x/SBA-1 system with CO and H_2 were carried out. It was observed that the presence of water favours the formation of Cr(III) species, what can be explained, in accordance with the DFT results, by reoxidation of Cr(II) species and hydrolysis of the Cr–O–Si bonds, finally leading to Cr_2O_3 generation. Using the TD-DFT (ωB97X-D) approach, excitation energies for the surface Cr species were calculated. For a given oxidation state of chromium, they depend significantly on the structure of the Cr species and its interaction with the surface siloxane bridges and silanol groups.

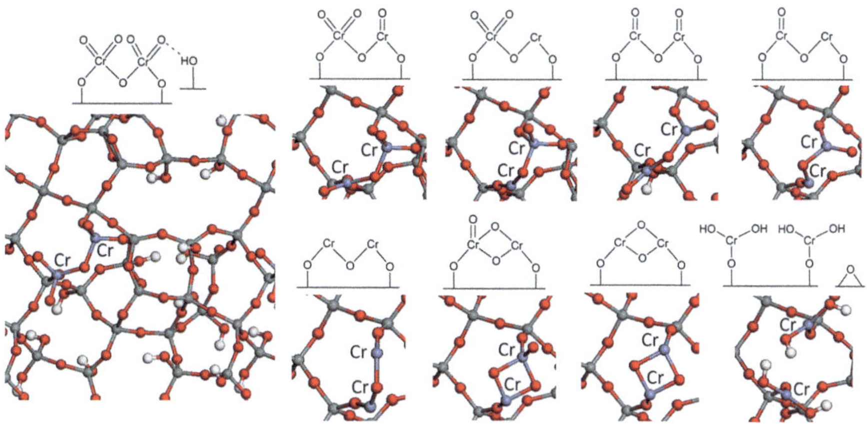

Fig. 8 Example periodic models of dimeric chromium oxide species on silica. Adapted from [5], Copyright (2016), with permission from Elsevier

3.3 *Catalytic Activity—Computational Studies*

Many theoretical investigations addressed mechanisms of the reactions catalysed by the CrO_x/SiO_2 system. The size of the models was usually limited to reduce computational cost of exploring many potential reaction pathways. In DFT (BP86) studies of ethane dehydrogenation over Cr(III) sites on silica, cluster models constructed either in ad hoc manner or systematically, starting from low-index surfaces of α- and β-cristobalite, were used [59, 60]. To examine the accuracy of the cluster approximation, hybrid QM/MM periodic calculations were also done. Mechanism of acetylene and methylacetylene cyclotrimerization catalysed by Cr(II)/SiO_2 system was investigated with a simple cluster model (2 Si atoms) representing chromasiloxane ring [61, 62]. On the basis of benchmark calculations of triplet–quintet energy gaps performed for many density functionals, taking the CASPT2 value as the reference, the B3PW91 functional with Hartree–Fock exchange increased to 28% was selected for the mechanistic calculations [61].

Despite many decades of investigations, the nature of the active sites in the Phillips catalyst, as well as the mechanism of their formation, is still not well recognized and raise much controversy in the literature. CrO_x/SiO_2 does not require any cocatalyst as a source of alkyl group to form the polymer chain via the Cossee–Arlman insertion. Instead, ethene must somehow react with surface chromium species to generate the initiating ligand. A lot of computational works concerning the CrO_x/SiO_2 system were focused on this issue.

First comprehensive computational studies on the mechanism of ethene polymerization over the Phillips catalyst were performed by Espelid and Børve [63–65]. They compared several initiation and propagation routes using the BP86 functional and small cluster models representing monomeric and dimeric reduced chromium species on silica. It was concluded that various chromium(IV) organic species, being potential products of reaction between monomeric Cr(II) site and ethene, are not effective in chain propagation [63]. However, easily generated Cr(IV) chromacyclopentane intermediate can lead, after ethylene insertion, to experimentally observed [66] 1-hexene formation. The latter is a product of intramolecular β-H transfer, which competes with the ring expansion, and the kinetic preference depends on the Cr site strain. On the other hand, Cr(III) alkyl site was predicted to be the active propagating species, although the mechanism of its formation was not explained [63]. Instead, generation of propagating Cr(IV) alkyl site through hydrogen transfer from a surface silanol to Cr(II) species was later proposed [64]. Another initiation route postulated involves dimeric Cr(II) species which might react with ethene to form cyclic dimeric Cr(III) propagating species without external sources of hydrogen [65].

Various initiation pathways for the Phillips catalyst, starting from Cr(II) site modelled by chromasiloxane ring (2 Si atoms), were calculated with the B3LYP functional [67]. The key intermediate was Cr(IV) chromacyclopentane species. According to their experimental results [43, 68], the authors took into account the presence of formaldehyde, which can be formed during reduction of the surface Cr(VI) species by ethene. They found that adsorption of one formaldehyde molecule on the Cr site favours ethene dimerization to 1-butene through Cr(IV) hydride intermediate and metathesis to form propene/ethene through a Cr(IV) chromacyclobutane intermediate. This proposal might explain experimentally observed production of propene and butene during the induction period of ethene polymerization over the Phillips catalyst [43, 68]. On the other hand, after formaldehyde desorption, a ring expansion pathway to Cr(IV) chromacycloheptane and subsequent one-step reductive elimination of 1-hexene were predicted to be kinetically preferred.

Conley et al. [69, 70] showed that well-defined Cr(III)/SiO_2 catalysts are active in ethene polymerization. They proposed that heterolytic C–H bond activation of ethene forms the first Cr–C bond, which was supported by computational (B3LYP-D3) studies [70]. Two F-terminated cluster models were constructed (Fig. 9), representing major tri-coordinate Cr(III) species and minor species with Cr(III) coordinated to an additional siloxane bridge, which was confirmed by comparison of experimental and computed vibrational shifts after CO adsorption. The calculated reaction pathways included initiation by hydrogen transfer from ethene to generate $(\equiv SiO)_2(\equiv SiOH)$Cr(III)-vinyl species, subsequent propagation according to the Cossee–Arlman mechanism and less kinetically favoured termination via reverse hydrogen transfer to chromium alkyl. Coordination of two ethene molecules to the Cr(III) site facilitates the C–H bond activation, but it is prevented if the additional siloxane ligand is present. On this basis, it was suggested that Cr(III) sites coordinating siloxane bridge are inactive in polymerization. The same computational approach was used for investigations of propane dehydrogenation over the Cr(III)/SiO_2 system [70].

The C–H bond activation mechanism was further re-examined by Fong et al. [6], who used the same cluster model representing Cr(III) species and concluded that the initiation step is too slow. In addition, they found a termination route which is faster than propagation, so only oligomers would be formed. Similar results were obtained for the analogous mechanism on Cr(II) site represented by the chromasiloxane cluster

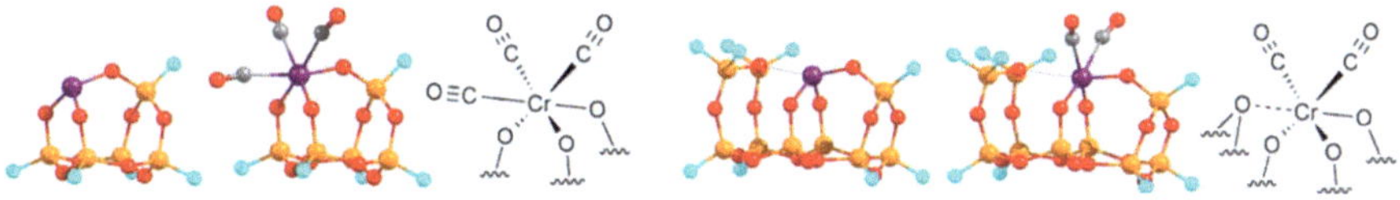

Fig. 9 Cluster models of Cr(III) oxide species on silica and the corresponding CO adducts. Adapted with permission from [70]. Copyright (2015) American Chemical Society

(2 Si atoms). Other routes considered for the Cr(II) site, i.e., chromacycle ring expansion, oxachromacycle ring expansion and carbene mechanism, were also excluded because of too high Gibbs energy barriers calculated (ωB97X-D functional) for the initiation or propagation steps. Instead, $(\equiv SiO)_2Cr(III)$-alkyl species without vinyl end were proposed as the active sites; however, the mechanism of their formation remained unknown.

A large number of potential initiation (Fig. 10), propagation and termination mechanisms for ethene polymerization involving surface Cr(II), Cr(III) and Cr(V) sites on silica were computationally investigated on the PBE0-D3 level [71], applying cluster models obtained from the β-cristobalite or amorphous silica [4, 13] structure (Fig. 11). It was predicted that Cr(II) oxachromacycle ring expansion is a more effective polymerization mechanism than the routes via Cr(II) chromacycle or (≡SiO)(≡SiOH)Cr(II)-vinyl site, mainly because of the kinetic preference for propagation, compared to the termination step. The mechanisms involving Cr(III) oxachromacycle [40, 71] or $(\equiv SiO)_2(\equiv SiOH)Cr(III)$-vinyl species would be less kinetically accessible than the corresponding routes for the Cr(II) sites. However, the calculated barriers might strongly depend on a silica model used, i.e. on a more realistic description of the local coordination environment. It was also shown that $(\equiv SiO)_2Cr(III)$-OH species can transform into $(\equiv SiO)_2Cr(III)\text{-}CH{=}CH_2$ [40, 71], which enables rapid propagation, kinetically favoured over the possible termination steps. The proposed mechanism was consistent with operando spectroscopy studies [40], which indicated the Cr(III) vinyl site as the active reaction intermediate during ethylene polymerization over the CrO_x/SiO_2 catalyst. Trying to explain how Cr(III) sites might be generated from Cr(II) species, most likely formed in the Phillips catalyst after Cr(VI) reduction by ethene, it was suggested that defect sites on the silica surface can play a role [71].

Structure–activity relationships for well-defined $Cr(III)/SiO_2$ catalyst were studied with periodic DFT (PBE) approach [10], based on the amorphous silica model [15] with 372 atoms in the unit cell and a surface silanol density of 1.1 OH nm^{-2}. The models of surface Cr(III) species were constructed by substituting ≡SiOH fragments with Cr (Fig. 12). Reactivity of five resulting $(\equiv SiO)_3Cr(III)$ sites of different location and strain were compared for the C–H bond activation and oxachromacycle ring expansion mechanisms. It was found that both routes are facilitated by strained Cr(III) species and propagation is more accessible kinetically than termination. Strained sites favour the oxachromacycle pathway of ethene polymerization, whereas both mechanisms can compete in the case of less strained and thereby less

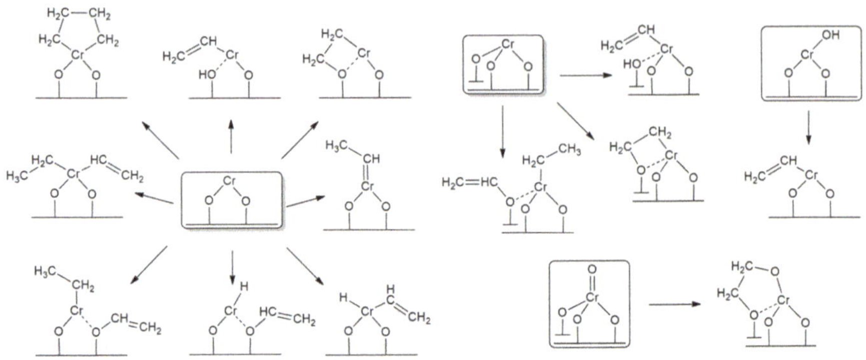

Fig. 10 Potential initiation routes for ethene polymerization involving Cr(II), Cr(III) and Cr(V) oxide species on silica. Adapted from [71], Copyright (2017), with permission from Elsevier

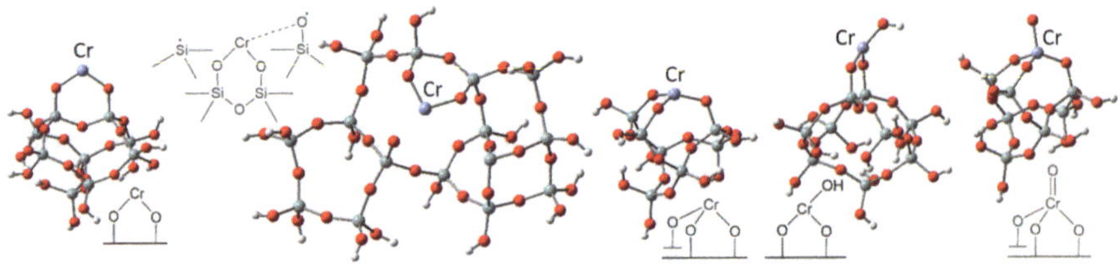

Fig. 11 Cluster models of Cr(II), Cr(III) and Cr(V) oxide species on silica. Adapted from [71], Copyright (2017), with permission from Elsevier

active Cr(III) species. On this basis, the authors suggested that the distribution of active sites with different local environments accounts for the broad distribution of polymer chains. In the next work of this group [72], the oxachromacycle ring expansion and C–H bond activation mechanisms were theoretically examined for propene polymerization, which allowed to explain experimentally observed different reactivity of the well-defined Cr(III)/SiO_2 system towards ethene and propene.

An influence of the coordination environment of the surface chromium species on their reactivity was also pointed in the next computational work of Fong et al. [73]. By applying cluster models, the ωB97X-D functional and variational transition state theory, they examined a one-electron initiation mechanism for ethene polymerization over the Phillips catalyst, according to Kissin and Brandolini proposal [74]. In the first step, various organo-Cr(IV) sites might be formed by oxidative addition of ethene to surface Cr(II) species. Intramolecular hydrogen transfer in Cr(II) bis(ethene) complex would lead to $(\equiv SiO)_2Cr(IV)(CH_2–CH_3)(CH=CH_2)$ species. Subsequent Cr–C bond homolysis gives active $(\equiv SiO)_2Cr(III)(CH=CH_2)$ site and ethyl radical, in agreement with the observed generation of organic radicals [47] and Cr(III) vinyl species [40, 47] during the initiation stage of ethene polymerization. The studied effect of siloxane coordination to the Cr site suggested that the proposed mechanism might be effective if the siloxane ligands were hemilabile, allowing

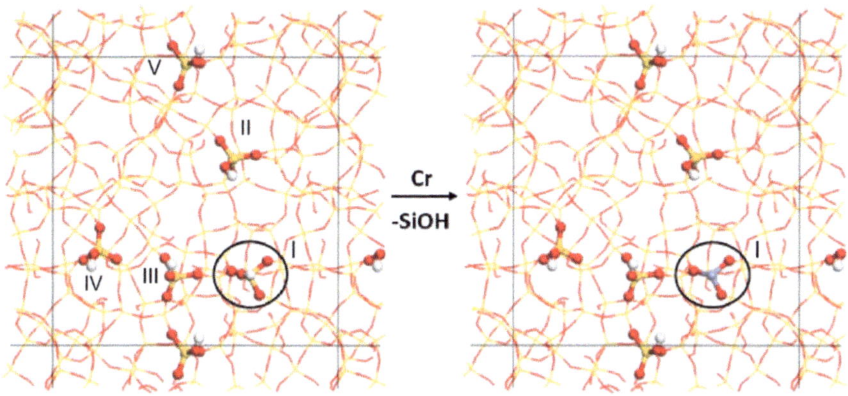

Fig. 12 Construction of Cr(III)/SiO_2 periodic model by substitution of one of the five SiOH groups in the unit cell with Cr. Reprinted from [10], Copyright (2017), with permission from Elsevier

access to lower and higher coordination states during the first and second initiation steps, respectively (Fig. 13). However, to confirm such a scenario, a more advanced model would be required, reproducing the flexibility of the silica framework. Because Cr(IV) chromacyclopentane site is the most kinetically accessible product of oxidative addition of ethene to Cr(II), an alternative Cr–C bond homolysis mechanism was examined [75]. According to this proposal, Cr(IV) chromacyclopentane species undergoes homolysis to generate tethered butyl radical which can attach to an adjacent Cr(II) site forming an n-butyl bridge. Then, ethene insertion might occur at both of the resulting Cr(III) alkyl sites. Although this mechanism does not require a change in Cr coordination number during the reaction, it needs neighbouring Cr sites instead.

In most industrial polymerization processes by the Phillips technology, the initial Cr(VI) species are reduced by ethene; however, the reaction mechanisms for the reduction stage are not well recognized. This issue was addressed in computational (PBE0-D3) studies using cluster models developed from the β-cristobalite structure [76]. It was shown that the reduction mechanism strongly depends on the coordination of the surface Cr(VI) sites. In the case of the major four-coordinate dioxo Cr(VI) species, the most kinetically favoured reduction pathway involves both oxo ligands and leads to the formation of Cr(II) site and two formaldehyde molecules. Potential minor five-coordinate monooxo Cr(VI) species would be reduced to a very stable cyclic Cr(IV) site, being rather a spectator species than a reactive intermediate. Reduction of the dioxo and monooxo Cr(VI) species by CO also leads to Cr(II) and Cr(IV) sites, respectively. To explain experimentally observed CO_2 release after treating of the Cr(VI)/SiO_2 system with ethene at higher temperatures [40], the mechanism of formaldehyde oxidation to carbon oxides and water was proposed. An alternative conversion of formaldehyde to methyl formate over Cr(II) sites, as suggested from experimental investigations [41], was examined too; however, the calculated pathways are less kinetically accessible. On the other hand, transforma-

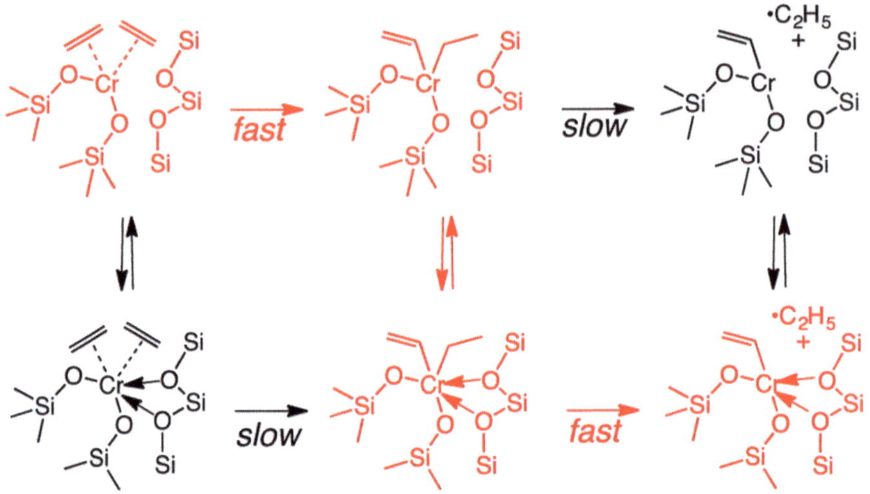

Fig. 13 Proposed initiation mechanism for ethene polymerization over the Phillips catalyst, requiring hemilabile siloxane ligands. Reprinted with permission from [73]. Copyright (2016) American Chemical Society

tion of formaldehyde to various surface intermediates was predicted to occur easily, which might explain why formaldehyde is usually not detected in gas phase during the early stage of ethene polymerization.

4 MoO_x/SiO_2 System

Molybdenum oxide supported on amorphous silica is a popular catalyst which exhibits activity in a variety of reactions, including alkene metathesis [77–85], selective oxidation of methane [86–88], alkenes [22, 89, 90] and alcohols [27, 91–93], as well as, oxidation of ammonia [94]. The active sites or their precursors are usually highly dispersed surface molybdenum species, often proposed to be monomeric species [78, 81–85, 88].

4.1 Structure of Surface Molybdenum Oxide Species—Experimental Data

The structure of the silica-supported Mo(VI) species under dehydrated conditions have been discussed for many years, on the basis of UV-vis, Raman, XANES, NEXAFS, EXAFS and IR spectroscopy results [32, 33, 81, 84, 85, 87–90, 92, 95–104]. Today, it is rather established that the isolated surface Mo(VI) sites are present as

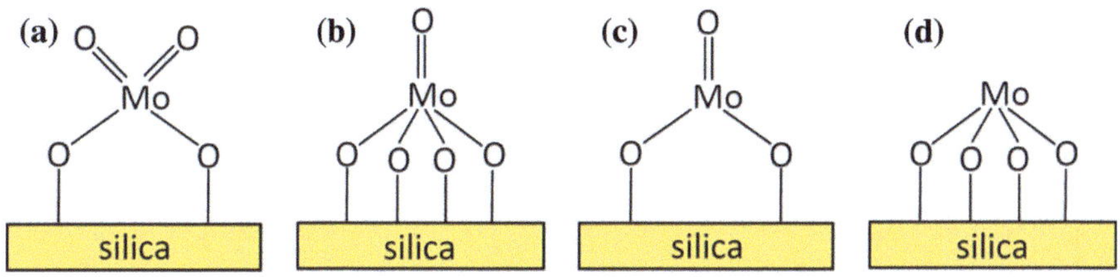

Fig. 14 Proposed structures for monomeric Mo(VI) and Mo(IV) oxide species on silica

major four-coordinate dioxo Mo(VI) species (Fig. 14a) [32, 33, 81, 84, 89, 90, 97, 99–101, 104]. Minor five-coordinate monooxo Mo(VI) species can also exist (Fig. 14b). Some authors also proposed dimeric/oligomeric Mo(VI) oxide species [85, 102, 103] or clusters [85], besides monomers, at relatively low Mo loadings. At higher Mo content, crystalline MoO_3 phase is additionally observed [32, 84, 104].

Considering the reduced MoO_x/SiO_2 catalyst, two or three distinct Mo(V) species on silica were observed with EPR after reduction of the Mo(VI)/SiO_2 system by H_2 [79, 105] or its thermal treatment in inert gas [79]. XPS measurements indicated the existence of Mo(VI), Mo(V) and Mo(IV) states after more severe reduction of the catalyst by H_2 [79]. The presence of Mo(IV) species in H_2-reduced MoO_x/SiO_2 system was supported by XANES studies [88, 98]. Effective generation of surface Mo(IV) species by photoreduction of molybdena-silica catalysts in CO was also reported [77, 78]. The nature of the Mo(IV) species might depend on the precursor structure (Fig. 14).

4.2 *Structure of Surface Molybdenum Species—Computational Modelling*

Similar to modelling chromia–silica catalysts, rather simple cluster models were used in the past to represent molybdenum oxide species on silica. Radhakrishnan et al. [97] combined NEXAFS, EXAFS and Raman spectroscopy studies of the Mo(VI)/SiO_2 system with Hartree–Fock calculations using small cluster models (1–2 Si atoms). They found that tetrahedral dioxo and distorted octahedral Mo(VI) species, the latter modelled as five-coordinate monooxo species, can be present on the silica surface. On the basis of DFT (B3LYP) calculations with slightly larger cluster models, Chempath et al. [106] proposed that surface dioxo and monooxo Mo(VI) species (Fig. 14a, b) can be in equilibrium with each other. However, by comparing XANES and EXAFS spectra simulated for the Mo(VI) and Mo(IV) models with the corresponding experimental spectra, they concluded that isolated Mo(VI) sites on silica are present as dioxo species, and Mo(IV) sites exist as three-coordinate monooxo species (Fig. 14c). This proposal was additionally confirmed by vibrational frequency analysis, referred to experimental Raman data at the time. In contrast,

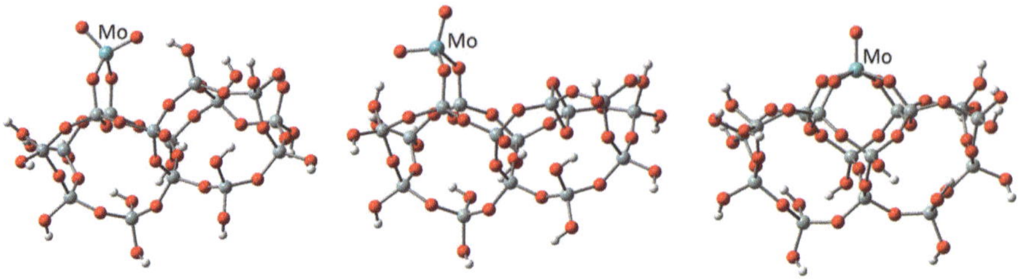

Fig. 15 Cluster models of Mo(VI) oxide species on silica. Adapted from [55], Copyright (2009), with permission from Elsevier

the results of B3LYP cluster studies employing the POSS model to represent silica suggested that the vibrational assignments for the surface Mo(VI) species might be reverse than those originally proposed from the experiment [32, 33] and, therefore, the monooxo Mo(VI) species might be dominant [107]. The POSS-based cluster models were also applied in combined computational (RPBE) and experimental investigations of MoO_x/SBA-15 system [101]. By comparison of the calculated and experimental NEXAFS spectra, it was concluded that the tetrahedral dioxo Mo(VI) species is the major species, while the pentahedrally coordinated monooxo Mo(VI) species exists in small quantities if at all.

Relative energies of the dioxo and monooxo Mo(VI) species on silica, represented by medium-size (15 Si atoms) cluster models (Fig. 15) derived from the β-cristobalite structure, were computed with various density functionals, selected on the basis of test calculations for molybdenum oxo compounds [55]. Although the monooxo species was predicted to be more stable, the theoretically determined Mo=O stretching frequencies confirmed the vibrational assignments [32, 33] indicating that the dioxo species is dominant.

In comprehensive DFT (PW91) studies of the MoO_x/SiO_2 system [3], a variety of advanced cluster models (15–97 Si atoms) were developed from the β-cristobalite and amorphous silica structure (Fig. 16). The two-layer ONIOM partitioning scheme was adopted to the largest models. In addition, periodic models based on the β-cristobalite framework were applied. It was found that relative energies of the Mo(VI) species depend on their location and a silica model used. If the local silica structure is flexible enough to facilitate formation of four Mo–O–Si linkages, the monooxo Mo(VI) species (Fig. 14b) are thermodynamically preferred under dehydrated conditions. Most locations, however, due to geometrical constraints on the surface, are favourable for the dioxo Mo(VI) species (Fig. 14a), which therefore should be in majority. This conclusion was supported by the vibrational frequency analysis, consistent with the proposed assignments of the experimental Raman bands [32, 33]. Reduction of the surface Mo(VI) species to Mo(IV) sites by H_2 was calculated as a highly endergonic process, in contrast to the exergonic reduction of Cr(VI) to Cr(IV) [5], which is in agreement with much worse reducibility of molybdena–silica catalysts, compared to chromia–silica systems. It was predicted that relative stabilities of the Mo(IV) species (Fig. 14c, d) are determined by the relative energies of the corresponding

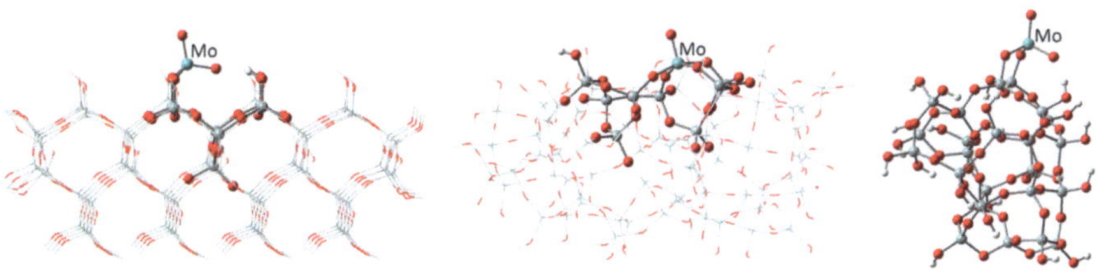

Fig. 16 Example cluster models of Mo(VI) oxide species on silica. In the first two cases, the ONIOM method is applied. Adapted with permission from [3]. Copyright (2012) American Chemical Society

Mo(VI) precursors; hence, three-coordinate monooxo Mo(IV) species (Fig. 14c) are most likely formed during Mo(VI) reduction with two-electron reducing agents.

Dioxo functionality of the major monomeric Mo(VI) species on silica was also confirmed by DFT (PBE) atomistic thermodynamic approach combined with vibrational frequency analysis [8]. Analogous to the previous work on CrO_x/SiO_2 system [7], mono-, di, tri- and tetragrafted monomeric molybdenum(VI) species at different degrees of hydration were considered using the slab model of hydroxylated silica surface [13]. The Mo grafting site was investigated systematically for the type of silanol (isolated, vicinal, geminal, nonvicinal or in a nest) accessible on the surface, as well as its effect on hydrogen bond formation with the Mo species and its stabilization. Based on calculated surface Gibbs energies and comparison between calculated and experimental vibrational frequencies, the digrafted dioxo Mo(VI) species (Fig. 14a) was predicted to be the dominant over a wide range of temperatures. Only at low temperatures, a monografted hydroxy dioxo Mo(VI) species would be more preferred thermodynamically. The tetragrafted monooxo Mo(VI) species (Fig. 14b) is more stable than the digrafted dioxo species, but it requires a nest of at least four silanol groups for grafting, so its presence is merely due to the distribution of the silanol density on the surface. This conclusion is consistent with the postulated geometrical constrains on the dehydroxylated silica surface, which hinder the formation of the tetragrafted monooxo Mo(VI) species [3].

4.3 Catalytic Activity—Computational Studies

Both dioxo and monooxo isolated Mo(VI) species were considered as the active sites or their precursors in computational studies of reactions catalysed by the MoO_x/SiO_2 system. Based on B3LYP cluster calculations, two mechanisms were proposed for methane oxidation to formaldehyde, depending on which Mo(VI) species, the dioxo or monooxo one, is assumed as the active site [108]. It was shown that the pathway involving the dioxo Mo(VI) species and reduced monooxo Mo(IV) species (Fig. 14a, c) is in a better agreement with experimental kinetic data. The mechanisms of methanol oxidation to formaldehyde over the dioxo and monooxo Mo(VI) sites were also found to be different each other and only in the latter case cleavage of the Mo–O–Si linkage is predicted [107].

Similar to other supported metal oxide catalysts for alkene metathesis, active metal alkylidene sites in the MoO_x/SiO_2 system are generated in situ from the surface Mo oxide species upon contact with alkene [82–85]. Among several initiation mechanisms considered in the literature, reduction of Mo(VI) to Mo(IV) with propene, followed by 1,2-hydrogen shift, was recently proposed, based on experimental results [82, 83]. It was also reported that high-temperature activation of the MoO_x/SiO_2 catalyst under an alkene-containing atmosphere is highly efficient for low-temperature alkene metathesis [85]. However, these issues were not addressed in computational works, although the propagation mechanism for alkene metathesis was studied in detail.

In a series of papers [109–111], DFT (B3LYP) studies of alkene metathesis over MoO_x/SiO_2 catalyst were reported. The Mo(VI) and Mo(IV) alkylidene species (Fig. 17) were represented by various cluster models (7–49 Si atoms) derived from the β-cristobalite structure. The largest models were developed using the two-layer ONIOM partitioning scheme [110, 111]. By exploring the pathways of ethene metathesis over tetrahedral Mo(VI) methylidene sites, it was found that a cycloreversion step, involving trigonal bipyramidal (TBP) molybdacyclobutane intermediate, is more favoured kinetically than a side transformation to a more stable square pyramidal (SP) molybdacyclobutane [109], in contrast to the theoretical results reported for molybdena–alumina systems [112–115]. The TBP $\rightarrow$ SP pseudorotation was proposed to be a reversible deactivation route. The comparison of the computed and experimental [78] C–H stretching frequencies suggested that the experimentally observed molybdacyclobutane species were the stable SP sites, not the reactive TBP intermediates. It was also shown that metathesis activity of the Mo(VI) alkylidene sites strongly depends on their location on the silica surface, similar to Mo(VI) alkylidene species on γ-alumina [113–116]. The electronic and geometrical structure analysis indicated that the main factor affecting the reactivity is the geometry of the Mo site, determined by surface constrains [109, 111]. Especially, the more advanced models (Fig. 18) enabled investigations of the relationship between metathesis activity of the tetrahedral Mo(VI) methylidene species (Fig. 17a) and the local structure of partially dehydroxylated silica surface [111]. Five-coordinate Mo(VI) alkylidene species (Fig. 17b) and Mo(IV) alkylidene species (Fig. 17c) were also examined, and it was concluded that these sites are inactive in alkene metathesis, in contrast to the four-coordinate Mo(VI) alkylidene sites, further additionally studied with the paired interacting orbitals (PIO) method [117]. Thus, it was shown [109, 111] that the metathesis activity of the Mo alkylidene sites is determined by the coordination environment of molybdenum, its oxidation state and the geometry of the Mo species, which is influenced by the local structure of the silica surface. The latter factor reflects heterogeneity of the active Mo sites.

Another strategy to account for inhomogeneous distribution of active sites on amorphous supports and investigate the structure–reactivity relationship was proposed by Goldsmith et al. [118]. They developed an algorithm which generates a representative set of small cluster models allowing to relate the activation energy

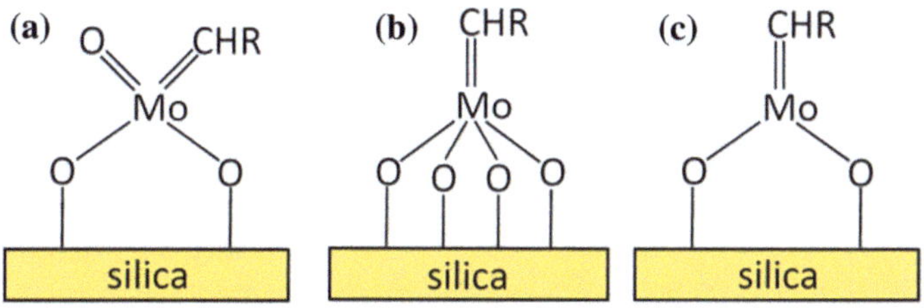

Fig. 17 Potential structures of Mo(VI) and Mo(IV) alkylidene species on silica

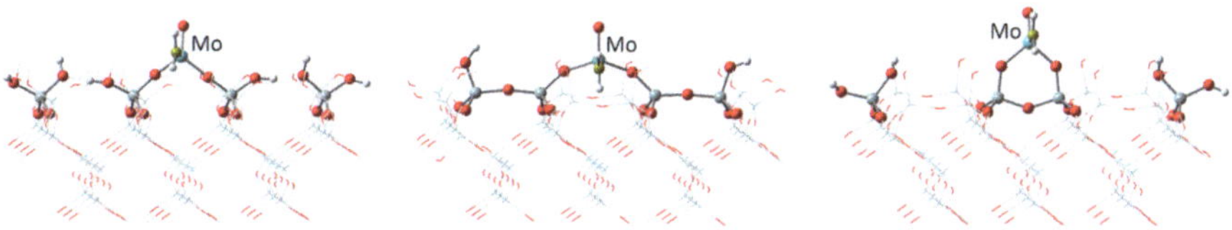

Fig. 18 Example models representing tetrahedral Mo(VI) methylidene species on SiO_2 which differ significantly each other in predicted metathesis activity. Adapted with permission from [111]. Copyright (2007) American Chemical Society

to the geometry and energy of the active site. By applying this method to model the above-mentioned TBP → SP molybdacyclobutane pseudorotation, they found a clear relationship between the Si–Si distance and the activation barrier. The same reaction was later studied by Ewing et al. [119] who used an advanced periodic model of amorphous silica (250–270 atoms in the unit cell) [14], considering 15 different surface structures with silanol density of 4, 3.2 and 2.3 OH nm^{-2}. They developed an approach for generating a large number of isolated metal atom sites on amorphous surface (Fig. 19). It was shown that the local structure of silica in the vicinity of the Mo site affects significantly the reaction and activation energy. A very complex structure–energy relationship was revealed, which results from various effects, including the Mo site interactions with neighbouring silanols and rotation-induced relaxation of the surface, which are not accounted for in small cluster models.

5 WO_x/SiO_2 System

Silica-supported tungsten oxide is mainly used as the industrial catalyst for metathesis of light alkenes [84, 85, 120–123]. In the past, ethene and butene were produced from propene in the Phillips triolefin process. Nowadays, due to a world shortage of propene, the reverse process, known as olefins conversion technology (OCT), is carried out on a large scale. The WO_x/SiO_2 system is also active in other catalytic reactions, including selective oxidation of methane [124], butene [22], styrene [125] and methanol [27], as well as, photocatalytic water splitting [126]. Similar to

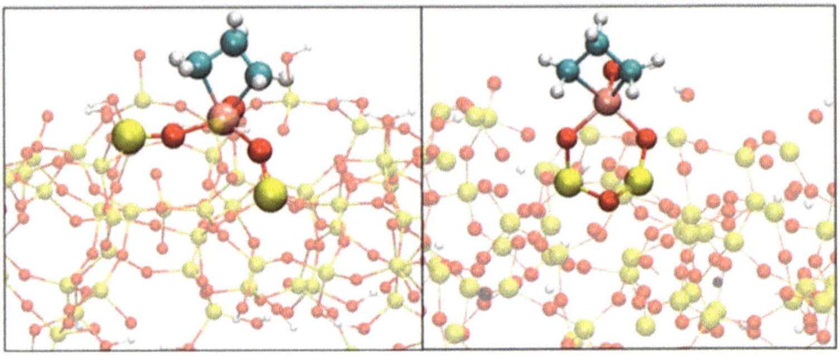

Fig. 19 Example models of SP molybdacyclobutane species differently located on the silica surface. Adapted with permission from [119]. Copyright (2016) American Chemical Society

CrO_x/SiO_2 and MoO_x/SiO_2, the catalytic activity of WO_x/SiO_2 is often attributed to well-dispersed surface metal oxide species [84, 85, 121–123, 126].

The structure of the surface W(VI) species in oxidized WO_x/SiO_2 systems under dehydrated conditions was determined mainly from in situ Raman, UV-vis, XANES and EXAFS studies [32, 33, 84, 100, 121]. It was postulated that isolated four-coordinate dioxo W(VI) species and five-coordinate monooxo W(VI) species coexist on the silica surface and their ratio is a function of temperature. On the other hand, surface oligomeric W(VI) species were sometimes proposed [127, 128]. At higher tungsten loadings, crystalline WO_3 nanoparticles are also present [32, 84, 121, 127, 128].

Following the previous computational works on silica-supported chromium(VI) [7] and molybdenum(VI) [8] oxide species, Guesmi et al. [9] performed DFT (PBE) studies of isolated tungsten(VI) oxide species on hydroxylated silica surface represented by the slab model of Tielens et al. [13]. Thermodynamic stability of mono-, di, tri- and tetragrafted monomeric tungsten(VI) species at different degrees of hydration was determined for a wide range of temperatures (Fig. 20). It was predicted that hydrated species with W–OH groups might be present only at very low temperatures. In more typical conditions, five-coordinate monooxo W(VI) species was found to be most stable; however, this tetragrafted species is statistically disfavoured due to the low probability to have four silanols, required for grafting, in one nest. For this reason, the digrafted dioxo W(VI) species is expected to dominate over the more thermodynamically preferred monooxo species. The presence of both dioxo and monooxo species was additionally confirmed by the computed W=O stretching frequencies, being close to the corresponding experimental Raman data. Finally, a comparison between the tungsten oxide species and other group VI metal oxide species on silica confirmed a greater similarity with the Mo species than with the Cr species.

In recent experimental works [85, 121–123], activation of the WO_x/SiO_2 catalyst for alkene metathesis was extensively investigated. It was postulated that the isolated dioxo W(VI) species are the active site precursors. Similar to the MoO_x/SiO_2 system,

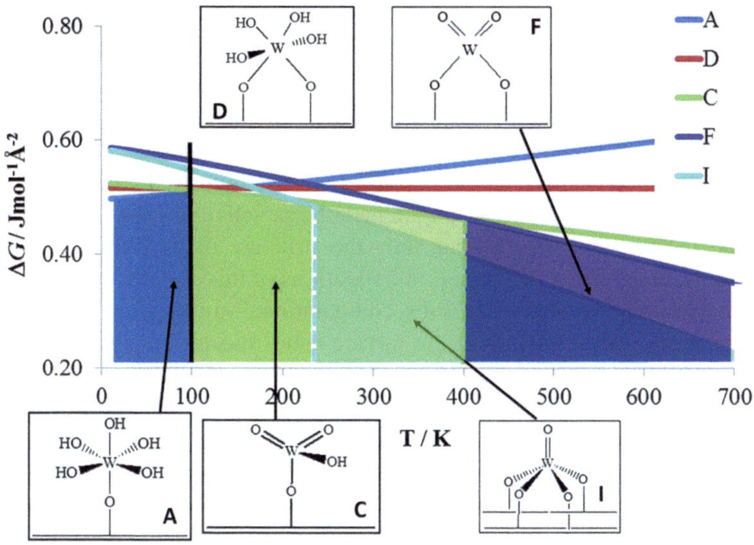

Fig. 20 Phase diagram (surface energy vs. temperature) showing the stability ranges for various W(VI) oxide species on silica [9]—adapted by permission of The Royal Society of Chemistry

the high-temperature activation in the presence of alkene was found to be very effective [85]. It was also proposed that three distinct sites of different metathesis activity are generated during the catalyst activation with alkenes [122]. Surprisingly, computational studies on the mechanism of alkene metathesis catalysed by the WO_x/SiO_2 system were not reported.

6 Concluding Remarks

From the computational works presented in this chapter, it is evident that a great progress has been achieved in the field of modelling silica-supported group VI metal oxide systems. Whereas only simple and more or less arbitrarily constructed cluster models could be used in the past, the most advanced today's slab models contain hundreds of atoms in the unit cell and are able to account for the heterogeneity of the surface metal species.

A number of theoretical works concerned the CrO_x/SiO_2 system, mainly as the Phillips catalyst for ethene polymerization. They addressed the nature of the oxidized and reduced chromium oxide species on silica, as well as polymerization mechanisms, especially formation of the active sites. It seems that the latter subject still requires further investigations, although many new important results, both experimental and theoretical, were reported recently.

Modelling of the molybdenum oxide species on silica allowed to support the experimental proposals about their structure. The MoO_x/SiO_2 system was also computationally studied as the catalyst for selective oxidation reactions and alkene metathesis. However, the still unknown mechanism of generation of the metathesis active sites in this system was not examined, and theoretical studies on this issue would be desired.

Although the WO_x/SiO_2 system is an important industrial catalyst for alkene metathesis, computational works on the metathesis activity of the surface tungsten oxide species on silica are lacking, only the structure of the W(VI) species was modelled. Hence, further theoretical investigations of this system are also needed.

Combined with experimental results, computational studies often enabled for better recognizing of the structure of the surface metal species on silica. On the other hand, many mechanisms of the catalytic reactions, especially the formation of the active sites from the metal oxide precursors, have not been understood well enough yet. Applying advanced surface models with representative distribution of metal sites to determine complex reaction mechanisms and structure–activity relationships seems to be a challenge for future works. Even more challenging might be an application of ab initio molecular dynamics methods to study catalytic reactions over the silica-supported metal oxide systems.

References

1. Sautet P, Delbecq F (2010) Catalysis and surface organometallic chemistry: a view from theory and simulations. Chem Rev 110:1788–1806
2. Handzlik J, Kurleto K (2013) Theoretical investigations of heterogeneous olefin metathesis catalysts. Curr Org Chem 17:2796–2813
3. Handzlik J, Ogonowski J (2012) Structure of isolated molybdenum(VI) and molybdenum(IV) oxide species on silica: periodic and cluster DFT studies. J Phys Chem C 116:5571–5584
4. Handzlik J, Grybos R, Tielens F (2013) Structure of monomeric chromium(VI) oxide species supported on silica: periodic and cluster DFT studies. J Phys Chem C 117:8138–8149
5. Gierada M, Michorczyk P, Tielens F, Handzlik J (2016) Reduction of chromia-silica catalysts: a molecular picture. J Catal 340:122–135
6. Fong A, Yuan Y, Ivry SL, Scott SL, Peters B (2015) Computational kinetic discrimination of ethylene polymerization mechanisms for the Phillips (Cr/SiO_2) catalyst. ACS Catal 5:3360–3374
7. Guesmi H, Tielens F (2012) Chromium oxide species supported on silica: a representative periodic DFT model. J Phys Chem C 116:994–1001
8. Guesmi H, Grybos R, Handzlik J, Tielens F (2014) Characterization of molybdenum monomeric oxide species supported on hydroxylated silica: a DFT study. Phys Chem Chem Phys 16:18253–18260
9. Guesmi H, Grybos R, Handzlik J, Tielens F (2016) Characterization of tungsten monomeric oxide species supported on hydroxylated silica; a DFT study. RSC Adv 6:39424–39432
10. Floryan L, Borosy AP, Núñez-Zarur F, Comas-Vives A, Copéret C (2017) Strain effect and dual initiation pathway in Cr^{III}/SiO_2 polymerization catalysts from amorphous periodic models. J Catal 346:50–56
11. Dapprich S, Komáromi I, Byun KS, Morokuma K, Frisch MJ (1999) A new ONIOM implementation in Gaussian98. Part I. The calculation of energies, gradients, vibrational frequencies and electric field derivatives. J Mol Struct THEOCHEM 461–462:1–21

12. Ugliengo P, Sodupe M, Musso F, Bush IJ, Orlando R, Dovesi R (2008) Realistic models of hydroxylated amorphous silica surfaces and MCM-41 mesoporous material simulated by large-scale periodic B3LYP calculations. Adv Mater 20:4579–4583
13. Tielens F, Gervais C, Lambert JF, Mauri F, Costa D (2008) Ab initio study of the hydroxylated surface of amorphous silica: a representative model. Chem Mater 20:3336–3344
14. Ewing CS, Bhavsar S, Veser G, McCarthy JJ, Johnson JK (2014) Accurate amorphous silica surface models from first-principles thermodynamics of surface dehydroxylation. Langmuir 30:5133–5141
15. Comas-Vives A (2016) Amorphous SiO_2 surface models: energetics of the dehydroxylation process, strain, ab initio atomistic thermodynamics and IR spectroscopic signatures. Phys Chem Chem Phys 18:7475–7482
16. Gierada M, Petit I, Handzlik J, Tielens F (2016) Hydration in silica based mesoporous materials: a DFT model. Phys Chem Chem Phys 18:32962–32972
17. McDaniel MP (2010) A review of the Phillips supported chromium catalyst and its commercial use for ethylene polymerization. In: Advances in Catalysis, vol. 53, pp. 123–606
18. McDaniel M (2017) Manipulating polymerization chemistry of Cr/silica catalysts through calcination. Appl Catal A Gen 542:392–410
19. Groppo E, Lamberti C, Bordiga S, Spoto G, Zecchina A (2005) The structure of active centers and the ethylene polymerization mechanism on the Cr/SiO_2 catalyst: a frontier for the characterization methods. Chem Rev 105:115–183
20. Weckhuysen BM, Wachs IE, Schoonheydt RA (1996) Surface chemistry and spectroscopy of chromium in inorganic oxides. Chem Rev 96:3327–3349
21. Hakuli A, Harlin ME, Backman LB, Krause AOI (1999) Dehydrogenation of i-butane on CrO_x/SiO_2 catalysts. J Catal 184:349–356
22. Ramani NC, Sullivan DL, Ekerdt JG, Jehng J-M, Wachs IE (1998) Selective oxidation of 1-butene over silica-supported Cr(VI), Mo(VI), and W(VI) oxides. J Catal 176:143–154
23. Cherian M, Rao MS, Hirt AM, Wachs IE, Deo G (2002) Oxidative dehydrogenation of propane over supported chromia catalysts: influence of oxide supports and chromia loading. J Catal 211:482–495
24. Michorczyk P, Pietrzyk P, Ogonowski J (2012) Preparation and characterization of SBA-1-supported chromium oxide catalysts for CO_2 assisted dehydrogenation of propane. Microporous Mesoporous Mater 161:56–66
25. Botavina MA, Agafonov YA, Gaidai NA, Groppo E, Cortés Corberán V, Lapidus AL, Martra G (2016) Towards efficient catalysts for the oxidative dehydrogenation of propane in the presence of CO_2: Cr/SiO_2 systems prepared by direct hydrothermal synthesis. Catal Sci Technol 6:840–850
26. Kim DS, Tatibouet J-M, Wachs IE (1992) Surface structure and reactivity of CrO_3/SiO_2 catalysts. J Catal 136:209–221
27. Jehng J-M, Hu H, Gao X, Wachs IE (1996) The dynamic states of silica-supported metal oxide catalysts during methanol oxidation. Catal Today 28:335–350
28. Liotta LF, Venezia AM, Pantaleo G, Deganello G, Gruttadauria M, Noto R (2004) Chromia on silica and zirconia oxides as recyclable oxidizing system: structural and surface characterization of the active chromium species for oxidation reaction. Catal Today 91–92:231–236
29. Dines TJ, Inglis S (2003) Raman spectroscopic study of supported chromium(VI) oxide catalysts. Phys Chem Chem Phys 5:1320–1328
30. Groppo E, Damin A, Bonino F, Zecchina A, Bordiga S, Lamberti C (2005) New strategies in the Raman study of the Cr/SiO_2 Phillips catalyst: observation of molecular adducts on Cr(II) sites. Chem Mater 17:2019–2027
31. Moisii C, Deguns EW, Lita A, Callahan SD, van de Burgt LJ, Magana D, Stiegman AE (2006) Coordination environment and vibrational spectroscopy of Cr(VI) sites supported on amorphous silica. Chem Mater 18:3965–3975
32. Lee EL, Wachs IE (2007) In situ spectroscopic investigation of the molecular and electronic structures of SiO_2 supported surface metal oxides. J Phys Chem C 111:14410–14425

33. Lee EL, Wachs IE (2008) In situ Raman spectroscopy of SiO_2-supported transition metal oxide catalysts: an isotopic $^{18}O-^{16}O$ exchange study. J Phys Chem C 112:6487–6498
34. Chakrabarti A, Wachs IE (2015) The nature of surface CrO_x sites on SiO_2 in different environments. Catal Lett 145:985–994
35. Peek NM, Jeffcoat DB, Moisii C, van de Burgt L, Profeta S, Scott SL, Stiegman AE (2018) Reassessment of the electronic structure of Cr(VI) sites supported on amorphous silica and implications for Cr coordination number. J Phys Chem C 122:4349–4358
36. Moisii C, Jeffcoat D, Peek N, van de Burgt L, Scott SL, Stiegman AE (2018) Do monooxo sites exist in silica-supported Cr(VI) materials? Reassessment of the resonance Raman spectra. J Phys Chem C 122:17149–17160
37. Gaspar AB, Martins RL, Schmal M, Dieguez LC (2001) Characterization of Cr^{2+} and ethylene polymerization on Cr/SiO_2 catalysts. J Mol Catal A Chem 169:105–112
38. Groppo E, Damin A, Otero Arean C, Zecchina A (2011) Enhancing the initial rate of polymerisation of the reduced Phillips catalyst by one order of magnitude. Chem Eur J 17:11110–11114
39. Brown C, Krzystek J, Achey R, Lita A, Fu R, Meulenberg RW, Polinski M, Peek N, Wang Y, van de Burgt LJ, Profeta S, Stiegman AE, Scott SL (2015) Mechanism of initiation in the Phillips ethylene polymerization catalyst: redox processes leading to the active site. ACS Catal 5:5574–5583
40. Chakrabarti A, Gierada M, Handzlik J, Wachs IE (2016) Operando molecular spectroscopy during ethylene polymerization by supported CrO_x/SiO_2 catalysts: active sites, reaction intermediates, and structure-activity relationship. Top Catal 59:725–739
41. Barzan C, Piovano A, Braglia L, Martino GA, Lamberti C, Bordiga S, Groppo E (2017) Ligands make the difference! Molecular insights into Cr^{VI}/SiO_2 Phillips catalyst during ethylene polymerization. J Am Chem Soc 139:17064–17073
42. Weckhuysen BM, De Ridder LM, Schoonheydt RA (1993) A quantitative diffuse reflectance spectroscopy study of supported chromium catalysts. J Phys Chem 97:4756–4763
43. Liu B, Nakatani H, Terano M (2002) New aspects of the induction period of ethene polymerization using Phillips CrO_x/SiO_2 catalyst probed by XPS, TPD and EPMA. J Mol Catal A Chem 184:387–398
44. Budnyk A, Damin A, Groppo E, Zecchina A, Bordiga S (2015) Effect of surface hydroxylation on the catalytic activity of a $Cr(II)/SiO_2$ model system of Phillips catalyst. J Catal 324:79–87
45. Weckhuysen BM, De Ridder LM, Grobet PJ, Schoonheydt RA (1995) Redox behavior and dispersion of supported chromium catalysts. J Phys Chem 99:320–326
46. Weckhuysen BM, Schoonheydt RA, Mabbs FE, Collison D (1996) Electron paramagnetic resonance of heterogeneous chromium catalysts. J Chem Soc Faraday Trans 92:2431–2436
47. Brown C, Lita A, Tao Y, Peek N, Crosswhite M, Mileham M, Krzystek J, Achey R, Fu R, Bindra JK, Polinski M, Wang Y, van de Burgt LJ, Jeffcoat D, Profeta S, Stiegman AE, Scott SL (2017) Mechanism of initiation in the Phillips ethylene polymerization catalyst: ethylene activation by Cr(II) and the structure of the resulting active site. ACS Catal 7:7442–7455
48. Espelid Ø, Børve KJ (2001) Theoretical analysis of d-d transitions for the reduced Cr/silica system. Catal Lett 75:49–54
49. Espelid Ø, Børve KJ (2002) Theoretical analysis of CO adsorption on the reduced Cr/silica system. J Catal 205:177–190
50. Damin A, Vitillo JG, Ricchiardi G, Bordiga S, Lamberti C, Groppo E, Zecchina A (2009) Modeling CO and N_2 adsorption at Cr surface species of Phillips catalyst by hybrid density functionals: effect of Hartree-Fock exchange percentage. J Phys Chem A 113:14261–14269
51. Demmelmaier CA, White RE, van Bokhoven JA, Scott SL (2008) Nature of $\equiv SiOCrO_2Cl$ and $(\equiv SiO)_2CrO_2$ sites prepared by grafting CrO_2Cl_2 onto silica. J Phys Chem C 112:6439–6449
52. Demmelmaier CA, White RE, van Bokhoven JA, Scott SL (2009) Evidence for a chromasiloxane ring size effect in Phillips (Cr/SiO_2) polymerization catalysts. J Catal 262:44–56
53. Zhong L, Lee M-Y, Liu Z, Wanglee Y-J, Liu B, Scott SL (2012) Spectroscopic and structural characterization of $Cr(II)/SiO_2$ active site precursors in model Phillips polymerization catalysts. J Catal 293:1–12

54. Handzlik J, Kurleto K (2013) Assessment of density functional methods for thermochemistry of chromium oxo compounds and their application in a study of chromia-silica system. Chem Phys Lett 561–562:87–91
55. Handzlik J (2009) DFT study of molybdena-silica system—a selection of density functionals based on their performance in thermochemistry of molybdenum compounds. Chem Phys Lett 469:140–144
56. Cheng R, Liu X, Fang Y, Terano M, Liu B (2017) High-resolution ^{29}Si CP/MAS solid state NMR spectroscopy and DFT investigation on the role of geminal and single silanols in grafting chromium species over Phillips Cr/silica catalyst. Appl Catal A Gen 543:26–33
57. Zhuravlev LT (2000) The surface chemistry of amorphous silica. Zhuravlev model. Colloids Surfaces A Physicochem Eng Asp 173:1–38
58. Ek S, Root A, Peussa M, Niinistö L (2001) Determination of the hydroxyl group content in silica by thermogravimetry and a comparison with ^{1}H MAS NMR results. Thermochim Acta 379:201–212
59. Lillehaug S, Børve KJ, Sierka M, Sauer J (2004) Catalytic dehydrogenation of ethane over mononuclear Cr(III) surface sites on silica. Part I. C-H activation by σ-bond metathesis. J Phys Org Chem 17:990–1006
60. Lillehaug S, Jensen VR, Børve KJ (2006) Catalytic dehydrogenation of ethane over mononuclear Cr(III)-silica surface sites. Part 2: C-H activation by oxidative addition. J Phys Org Chem 19:25–33
61. Liu Z, Cheng R, He X, Wu X, Liu B (2012) DFT functional benchmarking on the energy splitting of chromium spin states and mechanistic study of acetylene cyclotrimerization over the Phillips Cr(II)/silica catalyst. J Phys Chem A 116:7538–7549
62. Liu Z, Cheng R, He X, Liu B (2013) Reactivity and regioselectivity of methylacetylene cyclotrimerization over the Phillips Cr/silica catalyst: a DFT study. ACS Catal 3:1172–1183
63. Espelid Ø, Børve KJ (2000) Theoretical models of ethylene polymerization over a mononuclear chromium(II)/silica site. J Catal 195:125–139
64. Espelid Ø, Børve KJ (2002) Molecular-level insight into Cr/silica Phillips-type catalysts: polymerization-active mononuclear chromium sites. J Catal 205:366–374
65. Espelid Ø, Børve KJ (2002) Molecular-level insight into Cr/silica Phillips-type catalysts: polymerization-active dinuclear chromium sites. J Catal 206:331–338
66. Potter KC, Beckerle CW, Jentoft FC, Schwerdtfeger E, McDaniel MP (2016) Reduction of the Phillips catalyst by various olefins: stoichiometry, thermochemistry, reaction products and polymerization activity. J Catal 344:657–668
67. Zhong L, Liu Z, Cheng R, Tang S, Qiu P, He X, Terano M, Liu B (2012) Active site transformation during the induction period of ethylene polymerization over the Phillips CrO_x/SiO_2 catalyst. ChemCatChem 4:872–881
68. Liu B, Nakatani H, Terano M (2003) Mechanistic implications of the unprecedented transformations of ethene into propene and butene over Phillips CrO_x/SiO_2 catalyst during induction period. J Mol Catal A: Chem 201:189–197
69. Conley MP, Delley MF, Siddiqi G, Lapadula G, Norsic S, Monteil V, Safonova OV, Copéret C (2014) Polymerization of ethylene by silica-supported dinuclear Cr^{III} sites through an initiation step involving C-H bond activation. Angew Chem Int Ed 53:1872–1876
70. Conley MP, Delley MF, Núñez-Zarur F, Comas-Vives A, Copéret C (2015) Heterolytic activation of C-H bonds on Cr^{III}-O surface sites is a key step in catalytic polymerization of ethylene and dehydrogenation of propane. Inorg Chem 54:5065–5078
71. Gierada M, Handzlik J (2017) Active sites formation and their transformations during ethylene polymerization by the Phillips CrO_x/SiO_2 catalyst. J Catal 352:314–328
72. Delley MF, Praveen CS, Borosy AP, Núñez-Zarur F, Comas-Vives A, Copéret C (2017) Olefin polymerization on $Cr(III)/SiO_2$: mechanistic insights from the differences in reactivity between ethene and propene. J Catal 354:223–230
73. Fong A, Peters B, Scott SL (2016) One-electron-redox activation of the reduced Phillips polymerization catalyst, via alkylchromium(IV) homolysis: a computational assessment. ACS Catal 6:6073–6085

74. Kissin YV, Brandolini AJ (2008) Chemistry of olefin polymerization reactions with chromium-based catalysts. J Polym Sci Part A Polym Chem 46:5330–5347
75. Fong A, Vandervelden C, Scott SL, Peters B (2018) Computational support for Phillips catalyst initiation via Cr-C bond homolysis in a chromacyclopentane site. ACS Catal 8:1728–1733
76. Gierada M, Handzlik J (2018) Computational insights into reduction of the Phillips CrO_x/SiO_2 catalyst by ethylene and CO. J Catal 359:261–271
77. Shelimov BN, Elev IV, Kazansky VB (1986) Use of photoreduction for activation of silica-molybdena catalysts for propylene metathesis: comparison with thermal reduction. J Catal 98:70–81
78. Vikulov KA, Elev IV, Shelimov BN, Kazansky VB (1989) IR and UV-vis spectroscopic studies of the stable $Mo{=}CH_2$ carbene complexes over photoreduced silica-molybdena catalysts with chemisorbed cyclopropane, and their role in olefin metathesis reactions. J Mol Catal 55:126–145
79. Zhang B, Liu N, Lin Q, Jin D (1991) The effects of Mo oxidation states on olefin metathesis. J Mol Catal 65:15–28
80. Handzlik J, Ogonowski J, Stoch J, Mikołajczyk M, Michorczyk P (2006) Properties and metathesis activity of molybdena-alumina, molybdena-silica-alumina and molybdena-silica catalysts—a comparative study. Appl Catal A Gen 312:213–219
81. Balcar H, Mishra D, Marceau E, Carrier X, Žilková N, Bastl Z (2009) Molybdenum oxide catalysts for metathesis of higher 1-alkenes via supporting MoO_2(acetylacetonate)$_2$ and MoO_2(glycolate)$_2$ on SBA-15 mesoporous molecular sieves. Appl Catal A Gen 359:129–135
82. Amakawa K, Wrabetz S, Kröhnert J, Tzolova-Müller G, Schlögl R, Trunschke A (2012) In situ generation of active sites in olefin metathesis. J Am Chem Soc 134:11462–11473
83. Amakawa K, Kröhnert J, Wrabetz S, Frank B, Hemmann F, Jäger C, Schlögl R, Trunschke A (2015) Active sites in olefin metathesis over supported molybdena catalysts. ChemCatChem 7:4059–4065
84. Lwin S, Wachs IE (2014) Olefin metathesis by supported metal oxide catalysts. ACS Catal 4:2505–2520
85. Ding K, Gulec A, Johnson AM, Drake TL, Wu W, Lin Y, Weitz E, Marks LD, Stair PC (2016) Highly efficient activation, regeneration, and active site identification of oxide-based olefin metathesis catalysts. ACS Catal 6:5740–5746
86. Banares MA, Fierro JLG, Moffat JB (1993) The partial oxidation of methane on MoO_3/SiO_2 catalysts: influence of the molybdenum content and type of oxidant. J Catal 142:406–417
87. Ohler N, Bell AT (2005) Selective oxidation of methane over MoO_x/SiO_2: isolation of the kinetics of reactions occurring in the gas phase and on the surfaces of SiO_2 and MoO_x. J Catal 231:115–130
88. Ohler N, Bell AT (2006) Study of the elementary processes involved in the selective oxidation of methane over MoO_x/SiO_2. J Phys Chem B 110:2700–2709
89. Thielemann JP, Hess C (2012) Structure of silica-supported molybdenum oxide studied by in situ spectroscopy under reactive and non-reactive conditions. J Catal 288:124–126
90. Thielemann JP, Hess C (2013) Monitoring silica supported molybdenum oxide catalysts at work: a Raman spectroscopic study. ChemPhysChem 14:441–447
91. Ono T, Anpo M, Kubokawa Y (1986) Catalytic activity and structure of MoO_3 highly dispersed on SiO_2. J Phys Chem 90:4780–4784
92. Banares MA, Hu HC, Wachs IE (1994) Molybdena on silica catalysts: role of preparation methods on the structure-selectivity properties for the oxidation of methanol. J Catal 150:407–420
93. Zhang W, Desikan A, Oyama ST (1995) Effect of support in ethanol oxidation on molybdenum oxide. J Phys Chem 99:14468–14476
94. Biermann JJP, Janssen FJJG, Ross JRH (1992) Nitrogen containing species as intermediates in the oxidation of ammonia over silica supported molybdena catalysts. Appl Catal A Gen 86:165–179
95. Hu H, Wachs IE, Bare SR (1995) Surface structures of supported molybdenum oxide catalysts: characterization by Raman and Mo L^3-edge XANES. J Phys Chem 99:10897–10910

96. Takenaka S, Tanaka T, Funabiki T, Yoshida S (1998) Structures of molybdenum species in silica-supported molybdenum oxide and alkali-ion-modified silica-supported molybdenum oxide. J Phys Chem B 102:2960–2969
97. Radhakrishnan R, Reed C, Oyama ST, Seman M, Kondo JN, Domen K, Ohminami Y, Asakura K (2001) Variability in the structure of supported MoO_3 catalysts: studies using Raman and X-ray absorption spectroscopy with ab initio calculations. J Phys Chem B 105:8519–8530
98. Ohler N, Bell AT (2005) A study of the redox properties of MoO_x/SiO_2. J Phys Chem B 109:23419–23429
99. Tian H, Roberts CA, Wachs IE (2010) Molecular structural determination of molybdena in different environments: aqueous solutions, bulk mixed oxides, and supported MoO_3 catalysts. J Phys Chem C 114:14110–14120
100. Wachs IE, Roberts CA (2010) Monitoring surface metal oxide catalytic active sites with Raman spectroscopy. Chem Soc Rev 39:5002–5017
101. Guo CS, Hermann K, Hävecker M, Thielemann JP, Kube P, Gregoriades LJ, Trunschke A, Sauer J, Schlögl R (2011) Structural analysis of silica-supported molybdena based on X-ray spectroscopy: quantum theory and experiment. J Phys Chem C 115:15449–15458
102. Thielemann JP, Kröhnert J, Hess C (2010) Nitric oxide adsorption and oxidation on SBA-15 supported molybdenum oxide: a transmission IR study. J Phys Chem C 114:17092–17098
103. Thielemann JP, Ressler T, Walter A, Tzolova-Müller G, Hess C (2011) Structure of molybdenum oxide supported on silica SBA-15 studied by Raman, UV-Vis and X-ray absorption spectroscopy. Appl Catal A Gen 399:28–34
104. Amakawa K, Sun L, Guo C, Hävecker M, Kube P, Wachs IE, Lwin S, Frenkel AI, Patlolla A, Hermann K, Schlögl R, Trunschke A (2013) How strain affects the reactivity of surface metal oxide catalysts. Angew Chem Int Ed 52:13553–13557
105. Louis C, Che M (1987) EPR investigation of the coordination sphere of Mo^{5+} ions on thermally reduced silica-supported molybdenum catalysts prepared by the grafting method. J Phys Chem 91:2875–2883
106. Chempath S, Zhang Y, Bell AT (2007) DFT studies of the structure and vibrational spectra of isolated molybdena species supported on silica. J Phys Chem C 111:1291–1298
107. Gregoriades LJ, Döbler J, Sauer J (2010) Oxidation of methanol to formaldehyde on silica-supported molybdena: density functional theory study on models of mononuclear sites. J Phys Chem C 114:2967–2979
108. Chempath S, Bell AT (2007) A DFT study of the mechanism and kinetics of methane oxidation to formaldehyde occurring on silica-supported molybdena. J Catal 247:119–126
109. Handzlik J (2005) Metathesis activity and properties of Mo-alkylidene sites differently located on silica. A density functional theory study. J Phys Chem B 109:20794–20804
110. Handzlik J (2007) Application of the ONIOM (QM/QM) method in the study of molybdena–silica system active in olefin metathesis. Int J Quantum Chem 107:2111–2119
111. Handzlik J (2007) Theoretical investigations of isolated Mo(VI) and Mo(IV) centers of a molybdena-silica catalyst for olefin metathesis. J Phys Chem C 111:9337–9348
112. Handzlik J, Ogonowski J (2002) DFT study of ethene metathesis proceeding on monomeric Mo^{VI} centres of MoO_3/Al_2O_3 catalyst. The role of the molybdacyclobutane intermediate. J Mol Catal A: Chem 184:371–377
113. Handzlik J (2004) Metathesis activity of monomeric Mo-methylidene centres on (1 0 0) and (1 1 0)C surfaces of γ-Al_2O_3—a comparative DFT study. Surf Sci 562:101–112
114. Handzlik J, Ogonowski J, Tokarz-Sobieraj R (2005) Dependence of metathesis activity of Mo-methylidene sites on their location on (1 0 0) γ-Al_2O_3—a theoretical study. Catal Today 101:163–173
115. Handzlik J (2007) Properties and metathesis activity of monomeric and dimeric Mo centres variously located on γ-alumina—a DFT study. Surf Sci 601:2054–2065
116. Handzlik J, Sautet P (2008) Active sites of olefin metathesis on molybdena-alumina system: a periodic DFT study. J Catal 256:1–14
117. Handzlik J, Czernecki M, Shiga A, Śliwa P (2012) Paired interacting orbitals (PIO) study of Mo/SiO_2 and Mo/HZSM-5 catalysts for olefin metathesis. Comput Theor Chem 991:174–181

118. Goldsmith BR, Sanderson ED, Bean D, Peters B (2013) Isolated catalyst sites on amorphous supports: a systematic algorithm for understanding heterogeneities in structure and reactivity. J Chem Phys 138:204105
119. Ewing CS, Bagusetty A, Patriarca EG, Lambrecht DS, Veser G, Johnson JK (2016) Impact of support interactions for single-atom molybdenum catalysts on amorphous silica. Ind Eng Chem Res 55:12350–12357
120. Mol JC (2004) Industrial applications of olefin metathesis. J Mol Catal A Chem 213:39–45
121. Lwin S, Li Y, Frenkel AI, Wachs IE (2016) Nature of WO_x sites on SiO_2 and their molecular structure-reactivity/selectivity relationships for propylene metathesis. ACS Catal 6:3061–3071
122. Lwin S, Wachs IE (2017) Catalyst activation and kinetics for propylene metathesis by supported WO_x/SiO_2 catalysts. ACS Catal 7:573–580
123. Howell JG, Li Y-P, Bell AT (2016) Propene metathesis over supported tungsten oxide catalysts: a study of active site formation. ACS Catal 6:7728–7738
124. de Lucas A, Valverde JL, Cañizares P, Rodriguez L (1999) Partial oxidation of methane to formaldehyde over W/SiO_2 catalysts. Appl Catal A Gen 184:143–152
125. Adam F, Iqbal A (2011) The liquid phase oxidation of styrene with tungsten modified silica as a catalyst. Chem Eng J 171:1379–1386
126. Liu G, Wang X, Wang X, Han H, Li C (2012) Photocatalytic H_2 and O_2 evolution over tungsten oxide dispersed on silica. J Catal 293:61–66
127. Ross-Medgaarden EI, Wachs IE (2007) Structural determination of bulk and surface tungsten oxides with UV-vis diffuse reflectance spectroscopy and Raman spectroscopy. J Phys Chem C 111:15089–15099
128. Chauvin J, Thomas K, Clet G, Houalla M (2015) Comparative influence of surface tungstate species and bulk amorphous WO_3 particles on the acidity and catalytic activity of tungsten oxide supported on silica. J Phys Chem C 119:12345–12355

Catalytic Properties of Selected Transition Metal Oxides—Computational Studies

Witold Piskorz and Filip Zasada

Abstract This chapter is the review of the computational methods applied to the transition metal oxides most abundant in heterogeneous catalysis and is focused on the influence of the environment on the transition metal cation properties. The shortcomings of the most commonly used DFT level of theory are discussed, and its extensions towards more realistic environment are presented. The modern reactive force-field methods are also mentioned. The embedding schemes most commonly found in the quantum-chemical or classical description of the heterogeneous processes are discussed. The errors stemming from the non-completeness of the basis function, i.e. the basis set superposition error, found in the calculations with atomic basis, and the Pulay stress, occurring in the planewave calculations, together with remedies, are briefly described. It is shown that in all discussed systems, i.e. CeO_2, TiO_2, ZrO_2, zeolites, *d*-electron metal spinels, and V_2O_5, the appropriately applied Hubbard DFT GGA+U methods are successful for the compromise between computational cost and resultant accuracy. The much more time-consuming hybrid functionals give slightly more accurate results and, moreover, are more universal in the sense that they do not need calibration against experiment contrary to DFT+U where the Hubbard correction needs to be carefully selected for modelling particular properties.

1 Introduction

Generally, the *coordination environment* term is considered as in the coordination between ligand and central ion in the complex molecular compounds in, e.g., solutions in homogeneous catalysis [1, 2]. This term can be, however, applied also to the solid state—the periodic (PBC, extended) systems, which can be locally, and in the

W. P. dedicates this work to his mother on the occasion of her round jubilee.

W. Piskorz (✉) · F. Zasada
Faculty of Chemistry, Jagiellonian University, Kraków, Poland
e-mail: wpiskorz@chemia.uj.edu.pl

E. Broclawik et al. (eds.), *Transition Metals in Coordination Environments*,
Challenges and Advances in Computational Chemistry and Physics 29,
https://doi.org/10.1007/978-3-030-11714-6_12

simplified, qualitative, manner, described through the crystal field theory [3] or the ligand field theory [4] as well. Indeed, the present chapter is devoted to the extended systems, particularly to the challenges of the application of computational methods to the description of such systems. Such approach is particularly useful in qualitative and intuition-aimed description of the *d*-electron metal oxides (transition metal oxides, TMO) like titania, zirconia, or spinels, to name just a few. The *f*-electron systems (rare earths, RE) are most abundantly represented by ceria.

The coordination environment of the surface atoms is among the three pivotal aspects in understanding the chemistry of the metal oxides surface as pointed out by Barteau [5], the other being the redox properties of the oxide and the oxidation state of the surface.

The local coordination environment can be used for study of the redox or protic properties in heterogeneous catalytic processes. Focusing on the importance of the coordination environment can also lead to the description of the charge transfer processes (polaron theory) in such oxides.

In this chapter, the successes and challenges of the computational studies on selected oxides, ceria, titania, zirconia, zeolites, *d*-electron spinels, and vanadia, are reviewed to give the reader the general insight as for the applicability of certain levels of theory and the possible quantities that can be calculated.

2 Methods

2.1 The DFT and Other Quantum-Chemical Methods

Due to the size of the models in the heterogeneous processes, the computational methods of choice belong to the large DFT [6, 7] family. They comprise both computationally cheaper local density approximation, LDA [7, 8] or generalised gradient approximation, GGA [9, 10], levels of theory or, more demanding, hybrid functionals, i.e. those with admixture of the Hartree–Fock exchange term in the functional, e.g. the widely found B3LYP functional [11] ("classical" in organic chemistry) or, more physically grounded, modern hybrid functionals like PBE0, HSE06.

The LDA functionals, despite their functional dependence only on the electron density and stemming from the homogeneous electron gas model, surprisingly well describe chemical systems due to the partial cancellation of exchange and Coulombic correlation effects [12], and the reasonable reproduction of the spherical average of the exchange–correlation hole [13] (even though the approximate and exact holes differ qualitatively). For detailed discussion of the exchange and correlation approximation in the LDA approximations for the atomic systems, see, e.g., the monograph [14].

In the Hartree–Fock method, which can be regarded as a special case of the DFT—with exact exchange term but with complete neglect of Coulombic correlation—the self-interaction energy is cancelled out by the exchange term. The uncompensated

Coulombic repulsion in the pure DFT functionals tends to delocalise the unpaired electron in the radicals [15] and also affects the electronic structure of the narrow band semiconductors [16].

The hybrid functionals are preferable in the systems with strong correlation at the expense of the necessity of calculation of the Hartree–Fock four-centre integrals and hence the significant rise of the computational cost. For such systems, the non-hybrid functionals have difficulty in the description of the properties stemming explicitly from electron Coulombic correlation, like spin density localisation or the band gap width. Such weakness leads to, e.g., metallisation of the Mott insulators [17, 18]. The hybrid functionals cancel out some of the self-interaction, i.e. effect of the interaction of given electron with the electron density to which the given electron contributes. The self-interaction problem is particularly pronounced in the d- or f-electron systems, which are of special importance in the solid-state catalysis. The post-Hartree–Fock methods, like CCSD(T) often regarded as the high-accuracy method of choice can, at the time of writing, be applied to relatively small clusters (vide infra) only.

Generally, one of the methods of simplification of the extraordinarily difficult problem of application of quantum mechanics to the chemical problems is the simplification of the Hamiltonian: instead of the system with many electron, many nuclei interactions, the simplified Hamiltonian with only selected degrees of freedom, appropriate for the description of the desired class of problems is applied. For instance, the Hubbard model is particularly applicable in the systems with strong electronic correlation.

The Hubbard DFT+U level of theory [19–21] can be applied for the strongly correlated d- or f-electron system of the same size as in the case of pure functionals. The DFT+U method accounts for the correlation effects arising from the strong on-site Coulomb repulsion and exchange interactions and, in the formulation of Dudarev et al. [22], incorporates the (semi-empirical) parameter $U_{\text{eff}} = U - J$, where U is a parameter describing the energy increase for an extra electron on a particular site and J is a parameter representing the screened exchange energy.

If possible, the computationally very challenging, but giving excellent accuracy, methods can be used. These are the methods based on the analysis of the poles of the Green's function for the system with screened Coulombic kernel, derived from many-body perturbation theory (MBPT), i.e. the $G_{(0)}W_{(0)}$ methods. For detailed discussion of GW methods, see, e.g., [23, 24].

Recently, a plethora of computational studies was published on reducible oxides—both on bulk and surfaces—with use of DFT+U [25, 26], hybrid DFT [27, 28], or GW method [29]. In these articles, it is shown that the known shortcomings of pure DFT (e.g. metallisation of isolating ceria) can be significantly improved thus correcting the underestimated band gap. The same roots has the issue of the charge or spin density localisation at the defects [30].

The DFT-KS is the theory of the ground state so the other methods, beyond the DFT, must be used to precisely describe the excited states. One of the most successful and widely used is the time-dependent density functional theory (TD-DFT), founded by the Runge–Gross theorem, the analogue to the Hohenberg-Kohn theorem, but concerning the evolving many-body system [31, 32]; according to it, there exists a unique mapping between the time-dependent external potential of a system and its time-dependent density. Hence, instead of the wavefunction of $3N$ variables (excluding spin), the three-dimensional electron density can be concerned as a fabric for the determination of all observables of the system. Based on the linear response theory, the spectra and other information on the excited states to be derived from TD-DFT [33].

The other computationally demanding effect, the London dispersion interactions, can be parameterised (based on either experimental or very accurate quantum-chemical calculations) and form an additive term, essentially force field-like, to the Hamiltonian. Such semi-empirical approach is coded as DFT+D and is computationally as cheap as plain DFT [34], while the accuracy is close to the high quality of very costly benchmark CCSD(T) calculations.

For the Grimme's D2 [35, 36] and D3 [37–39] versions of DFT+D, the dispersion term has the simple and computationally fast form of single-atomic parameters, what implies, besides their accuracy, the strength of these methods. Hence, the calculation of the dispersion energy is independent of the electronic structure and its computational cost is limited to the iteration of the atomic pairs. This DFT+D approach is possible due to the mainly two-atomic nature of London dispersion interactions, namely the potential of interaction between two atoms of given elements is a function of that elements only and is independent of the chemical environment of that atoms. Moreover, such diatomic term is a simple function of the atomic scalar terms. Once the atomic terms are parameterised, the atom–atom contribution can be calculated instantly. Alternative scheme, e.g., by Dion et al. [40–42], called the van der Waals density functional (vdW-DF) method, is based directly on the electron density. It comprises the non-local term accounting in an approximate way for the non-local electron correlation effects. Comparing to the local functional, it is an improvement even though it is obtained quite simply, using the double space integration. Its accuracy is improved for the systems with pronounced dispersion effects, it tends to be, however, inferior than the GGA functionals for the systems with hydrogen bonds [43, 44].

2.2 *The Classical Mechanics—Force Fields*

The classical force-field (FF) dynamics is a versatile tool for the description of physical processes like diffusion, sorption, solution, or conformational analysis. Due to little computational cost, FF is capable to describe the systems with millions of atoms for the timescale as long as nanoseconds. In its classical formulation, it cannot, however, describe the making or breaking of chemical bonds. The step forward in the improvement of description of "chemical" properties is the reactive force-field (ReaxFF) level of theory [45], where the possibility of bond breaking and making is based on the "bond order" analysis which in turn depends on the interatomic distances. The interatomic potential parameters are obtained either semi-empirically, or by means of quantum-chemical electronic structure calculations

[46]. The parameterisation for given element is transferable, i.e. parameters set is independent of the chemical environment of given atom.

2.3 Other Issues

2.3.1 Embedding Schemes

Embedding of the small, "active" system, described by high-accuracy, high-cost level of theory in the large environment, treated with moderate-accuracy, computationally cheap method is a concept frequently realised in practice. Typically, the high-level method belongs to QC, e.g. DFT or post-Hartree–Fock method, while the environment ("host") is described by the molecular mechanics (MM) based on the force-field, or the semi-empirical method. Such arrangement is called the QM/MM scheme, or the Integrated Molecular Orbital-Molecular Mechanics (IMOMM [47]) method. This approach evolved also into high-level-QM/low-level-QM (IMOMO [48], Integrated Molecular Orbital-Molecular Orbital method), or even into the generalised, multilayer ONIOM [49] (N-layered Integrated MO and MM) scheme. The introduction of the potential calculated in the originally periodic molecular mechanics as the external potential (through the hydrogen atom linkers) for the quantum-chemical calculations is called QM-Pot [50].

2.3.2 Basis Sets

The most commonly used types of basis set in the solid-state calculations are the plane waves (and augmented PW) and atom-centred analytical (Gaussian [51, 52], Slater-type [53, 54]), or numerical [55], basis sets (linear combination of atomic orbitals, LCAO). In the planewave world, the localised orbitals (the Wannier functions [56, 57]), being the counterpart of localised molecular orbitals, are also regarded as a useful canvas to describe the atomic properties and are often used in the interpretation of the planewave calculation results.

To reduce the complexity, the pseudopotentials, or furthermore, to increase the accuracy, the augmentation of the plane waves by the atom-centred wavefunctions (WF), are often introduced. The replacement of rapidly oscillating wavefunctions by the smooth pseudo wavefunctions (moreover, still preserving the way to calculate all-electron properties; PAW [58]) allows for further reduction of the number of the basis set.

For the sake of completeness, the real-space grid (generally orthorhombic) representation of the wavefunction ought to be mentioned, found in some computational codes. For the PAW method, the grids of different density can be used [59]. The real-space grid approach removes the bottleneck of the PW approach—the interprocessor communications in the fast Fourier transform (FFT) routines.

Due to the non-completeness of the basis set, the following errors occur: for the localised basis sets, it is the basis set superposition error (BSSE) while for the planewave basis sets, the error named the Pulay stress. In both cases, the basis set changes when the set of atom changes (former case) or when the unit cell size or shape changes (latter case). Generally, both errors are lowered when the size of the basis set increases. For the avoidance of the Pulay stress during the optimisation of the cell volume, the series of calculations for the varied cell size are performed (with the same cut-off energy hence with different number of plane waves) and the resultant energy *versus* size of the cell function is fitted to some equation of state, typically to the Birch–Murnaghan [60] EOS.

2.3.3 Computed Observables and Other Quantities

Having chosen the computational method, a plethora of observables and other properties can be calculated, e.g. SCF-derived properties like the band structure and density of electronic states (DOS), or the optimisation of the cell parameters and atom positions, miscellaneous spectroscopic properties—interpretation of XAFS, population (in the PBC world, the dominating are Bader [61, 62] and DDEC [63–65]) and bond order (DDEC) analyses, etc. Besides static parameters, the dynamic properties can be calculated including the phase stability [66, 67], thermal properties, and the Raman or dipole infrared spectroscopies.

In the studies of the solid interfaces, the scanning tunnelling microscopy (STM) is one of the most important techniques and the most common theory used to simulate the STM images is the Tersoff–Hamann theory [68], (TH). Its conceptual simplicity is based on the *s*-wave approximation, namely the replacement of the electronic structure of the tip by single orbital of spherical symmetry (i.e. orbital *s*). Thanks to this approximation, the tunnelling current is proportional to the substrate local density of states integrated over the range of voltage given by the bias voltage. The drawback of TH method is the neglect of the real electronic structure of the tip (and its modification induced by sticking of miscellaneous species to the tip during experiment) and the limited resolution. Its main advantage over the more accurate Bardeen approximation is speed: the change of simulated current ("microscope settings") changes the image at an instant, and the integration itself is also relatively fast. On the other hand, the Bardeen formalism [69], originally derived from Oppenheimer's scattering theory, is used for the cases when the failure of TH is expected, like in [70], where the Hofer's bSKAN [71] implementation of Bardeen theory was used together with VASP code.

2.3.4 Atomistic Thermodynamics

The computational modelling of the nanocrystal morphology is most commonly realised by the Wulff construction based on the (free) energy calculations at the (most commonly) DFT+U level of theory. The Wulff [72] construction is based on the theorem that the minimum surface energy of the convex polyhedron is achieved when distances r_{hkl} from the centre of the polyhedron to the (hkl) surface with energy of γ_{hkl} are held by relationship: $\gamma_{\mathrm{hkl}}/r_{\mathrm{hkl}} = \mathrm{const}; \ \forall hkl$. The values of the surface free energy are usually calculated *via* the well-established atomistic thermodynamic modelling [73, 74].

2.3.5 Solvation Effects

The solid/liquid interfaces can be modelled *via* DFT augmented with solution effects (DFTsol) [75, 76] or on the classical force-field level of theory. The solvation effects are more abundantly implemented in the molecular codes (PCM and, more recently, COSMO) than in the PBC programs [76]; the agreement between computational results and the experimental data is very good, up to a few hundredths of eV in the solvation energy.

Among the PBC codes, the VASPsol [76] extension to VASP belongs to the most popular. Besides the energetics, the DFTsol extension of the DFT reproduces very well the electrochemical properties [77] of the systems in thermodynamic equilibrium with the liquid (e.g. the potential of zero charge, PZC).

3 Systems: Oxides

Out of the coordination environments for the TMs, definitely the oxide ligands are the most common. Most of the industrial catalytic processes occurring on the surface of transition metal oxide solids are the redox processes or the protic processes. In the former, both TM cationic sites and the anionic oxide sites can act as the redox centres, hence the number of elementary steps can be significant. Therefore, for the detailed study thereof the simplified preliminary models can be used, like non-redox metal oxide (e.g. alkaline earth oxides [78, 79]), where only oxide redox centres exist. In both cases, the oxidiser atoms/molecules with surface ions atoms form the reactive oxygen species (ROS) which are the by-products of the catalytic process.

3.1 Reducible and Non-reducible Oxides

Typically, the “reducible oxides”, RO, term concerns the oxides with relatively low energy of O liberation [80], with band gaps <3 eV, and denotes the oxides of *d*-electron metals: Ti, V, Fe, Co, Hf, Zr, Mn, W, Ni, and *f*-electron: Ce, Pr, and Sm. The distinguishing between reducible and non-reducible oxides is, however, non-rigorous and is often dependent on the particular reaction, especially on the redox potential of the reaction environment. For example, Paier et al. [81] reasonably define the reducible oxides as those prone to reduction in the “catalytic” conditions, i.e. for the O_2 partial pressure higher than *ca.* 10^{-14} bar (ultra high vacuum lower limit) and the temperatures up to *ca.* 1000 °C.

In most cases, the oxygen atom transfer from the catalyst to the adsorbed molecule, along the Mars–van Krevelen (MvK [82]) mechanism, leads to the oxygen vacancy formation. In order to close the catalytic cycle and restore the initial stoichiometry of the catalyst, the reaction must be carried out under oxygen (or other oxidising agent, e.g. N_2O) pressure so that gas-phase oxygen species can interact with the surface, dissociate, and eventually refill the vacancy. To assess the origin of the oxidiser—to distinguish between the solid O donor or the gas-phase O donor (in other words, to scrutinise the mechanism to tell the MvK from the reactive surface oxygen species-based one)—the isotopic measurements are performed [83].

The reducible oxides can form bulk materials, nanocrystals, and nanostructured or functionalised surfaces. They are versatile systems whose unique properties stem from the wealth of the oxidation states of metal cations and from easy transformation between them [84].

The reducible oxides tend to fulfil the Grasselli's [85] canonical properties ("seven pillars") of a well-performing and efficient redox catalyst, namely as originally enumerated: "lattice oxygen, metal–oxygen bond strength, host structure, redox, multifunctionality of active sites, site isolation, and phase cooperation". One postulate, however, seems to be worth adding: the mutual electronic communication of the sites which allows for the catalytic cycle to close in the larger scale, both temporary and spatially, thus facilitating the redox properties in the more flexible way. In other words, the catalytic cycle does not necessarily have to close upon *each and every* catalytic act ("differentially") but only on average, "integrally". One site can be, e.g., an electron donor for more than one reacting molecule while the needed electrons can be supplied by other sites and delivered through the bulk by, e.g., polaron mechanism. Such electron shuttling requires, however, the cooperation of the conductive bulk system.

3.1.1 Ceria CeO_2

Although cerium is the lanthanide metal, it will be discussed together with transition metals since its oxide's redox chemistry is actually very close to that of TM oxides.

Ceria, CeO_2, is one of the most abundant redox metal oxide catalysts and definitely the most common lanthanide oxide in the redox catalysis [86] for many reasons. Ceria acts as an oxygen reservoir in both simple and composite catalysts [87, 88] in automotive technology (the most successful, ceria-utilising three-way catalysts, TWC [89, 90]), water-gas shift, WGS [91, 92] ($CO + H_2O \rightarrow CO_2 + H_2$), in fuel cells technology [93, 94], and also in organic catalysis [95]. It has also a prominent usage in the petrochemical industry in hydrocarbon reforming [96] and oxidation [97].

Its reduction and reoxidation during catalytic process are crucial in pure ceria, which was devoted a number of applied [98] and fundamental experimental [99, 100] and theoretical [81, 101–103] studies, in doped ceria [104], and in metal–ceria composite system catalysis. Ceria, possessing a complex electronic structure [105, 106], is particularly interesting due to its unique properties: pronounced structure sensitivity [107, 108], exploited by the shape-controlled synthesis [109], easy oxygen transport and storage [88, 110], acid–base properties [111]. The oxygen transport phenomena include the facile creation, healing, and diffusion of oxygen vacancies, and is pronounced especially at the ceria surfaces. The wide applicability of ceria originates from its ability to the fast and repeatedly undergoing the catalytic redox cycle (i.e. high oxygen storage capability) for the low energy of forming, diffusion, and healing of the oxygen vacancies. Ceria is also a perspective catalyst for the two-step water splitting process yielding hydrogen [112].

Ceria is also a common reactive support for other metal oxides like vanadia in the oxidative dehydrogenation reactions of alkanes to produce alkenes [113]. In some systems, however, ceria plays a role of inactive support [114].

Structure

Ceria in normal conditions crystallises in the fluorite structure ($Fm\bar{3}m$). In the cubic arrangement of the equivalent O^{2-} anions, the Ce^{4+} cations occupy tetrahedral positions around the O^{2-} sites. The mixed valence ground state of the Ce^{4+} once was believed to be deduced from the XPS spectra [115, 116] (partial charge transfer from the valence O^{2-} $2p$ orbital to the virtual $4f^0$ state of cerium cation), it is, however, not the only interpretation, see, e.g., Wuilloud et al. [117], who say in their abstract that "A mixed valence can be definitely excluded in CeO_2", supporting their reasoning by the many-body calculations. Based on the reflectance measurements, there is a possibility of small occupation of $4f^1$ configuration [118]. The experimental band gap width is 3.00 eV [117].

Even though the interactions between O^{2-} and Ce^{4+} have some covalent component, for the sake of simplicity ceria is often regarded and a fully ionic compound, even in the case of surface ions. The low-index faces of ceria (Fig. 1) have stability in the following order: $E_{(111)} > E_{(110)} > E_{(100)}$ [105, 119]. The non-polar $CeO_2(111)$

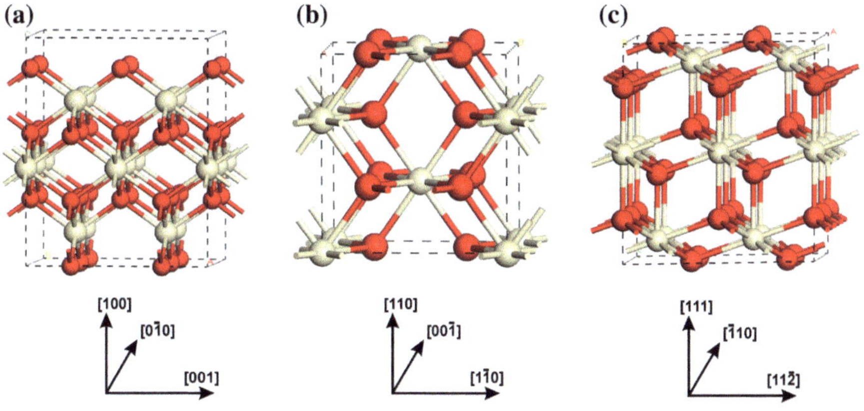

Fig. 1 Unrelaxed exposed facets of ceria. **a** $(2 \times 2)3L$ (100); **b** $(1 \times 1)4L$ (110); **c** $(\sqrt{3} \times 2)3L$ (111)

surface undergoes negligible relaxation [120], while the reconstruction of the other low-index surfaces, (110) [121] and (100) [122], is significant. These surfaces differ in their reactivity in oxidation (the most active: (110) and (100) [123]) and hydrogenation processes (the most active: (111)). Thus, the understanding of the morphology of the nanograins is crucial in the prediction of the catalytic properties.

It was also shown by Fronzi et al. [124] that, when moving from the oxygen-rich to oxygen-lean environment, the most stable surface is in the following sequence: the stoichiometric $CeO_2(111)$, the $CeO_2(111)$ surface with subsurface O vacancies, and, eventually, the $CeO_2(111)$: Ce-terminated surface.

Morphology

The ceria lattice parameters have been successfully calculated using either pure GGA (PBE), GGA+U, or hybrid functionals (HSE06) with discrepancies, comparing to experimental values [125, 126], in the range of *ca.* +1.1%, +1.5%, or −0.4%, respectively [27]. The computational values of bulk properties of CeO_2 and Ce_2O_3 are also summarised in the article by Fronzi et al. [124]. According to them, the calculated cerium sesquioxide Ce_2O_3 lattice constant a for LDA and GGA functionals is 3.86 Å and 3.77 Å, respectively, what is close to the results of Da Silva et al. [27] (3.83 Å and 3.77 Å respectively), who also report that the hybrid DFT and LDA+U give comparable lattice constants to those obtained by pure DFT, while application of GGA+U worsens the lattice parameters (too large a and c values).

The issue of the equilibrium morphology of ceria nanograins was studied computationally by many researchers, e.g. ref. [123, 124, 127, 128], to mention a few.

Although theoretical investigations on stoichiometric bulk ceria and gas–surface reactions are readily available, there are very scarce computational analyses about the similar systematic studies of moderately cerium oxide particles of catalytic relevance. Chen et al. employed simulated annealing method to find the structures of $(CeO_2)_n$ ($n = 1, \ldots, 5$) with global minimum potential energy, and refined obtained structures with DFT simulation showing that the coordination numbers of Ce and O atoms are very different from those in bulk CeO_2 material [129]. By combination of DFT-based modelling and spectroscopic studies, Burrow et al. revealed that the cerium oxide gas-phase clusters show structural motifs reminiscent of the bulk ceria, in contrast to TMI oxide clusters [130]. Theoretical study of Wu et al. indicates that the $Ce_nO_{2n+1}^-$ clusters contain oxygen-centred radicals ($O^{-\bullet}$) and the nature of the spin density distributions within the clusters controls the experimentally observed size-dependent reactivity. The reactivity of the oxygen-deficient cerium oxide cluster ions, $Ce_nO_m^+$ ($n = 2, \ldots, 10$, $m \leqslant 2n$), was investigated using DFT method, and it was shown that CeO_2 NPs are able to extract oxygen atoms from CO, CO_2, NO, N_2O, and O_2 in the gas phase [131]. It was also revealed that oxygen transfer reactions may be explained in terms of the energy balance between the bond dissociation energy of the oxygen-containing molecule and the oxygen affinity of the oxygen-deficient cerium oxide cluster.

The comparison of the performance of pure GGA (PBE), PBE+U(4.5), hybrid HSE06, and van der Waals functional vdW-DF+U on the adsorption of water on the (111) surface of ceria is found in the article by Fernández-Torre et al. [132] They considered a few associative two dissociative geometries of adsorbed waters and found, based on their own results and those reviewed from the literature, that the inclusion of the van der Waals interactions increases (to the absolute value) the binding energy of water by *ca.* 0.18 eV per molecule but do not change the water overall binding behaviour. They also show that water adsorbs on top of a Ce^{4+} atom either as a molecule (forming one hydrogen bond with the surface O atom) or as a hydroxyl pair, and these states differ very slightly in energy (in agreement with [133], where PW91+U(5.0) is used) so both forms—associated and dissociated—can coexist.

More involved and unbiased approach, namely the ab initio-MD (AIMD), within the DFT+U(7.0) level of theory (PBE functional, pseudopotentials, double-ζ Gaussian basis set with auxiliary planewave basis set, Γ-point integration of irreducible Brillouin zone), was used by Ren et al. [128] who studied three low-index surfaces of ceria, (111), (110), and (100), in contact with vapour and liquid water. They conclude, based on the Wulff construction, that in the vapour phase at lower T two low-index partially hydroxylated surfaces, (111) and (100), dominate and the contribution of (100) is lowered for the nanograins at higher temperatures, when the grain becomes dehydrated. In the aqueous conditions, however, the hydroxylated (111) surface is dominant.

Formation of O Vacancy

The energy of formation of O vacancy on the surface is relevantly lower than in the bulk [102] and is significantly dependent on the exposed face, the lowest is for the (110) face [105]. The structure of the surface vacancy, crucial in the catalytic processes, is non-trivial. For years it was believed that the localisation of the electrons emerging from the formation of the oxygen vacancy, i.e., the position of the reduced Ce^{3+} cations, was next to the vacancy. As was shown by, e.g., Ganduglia-Pirovano et al. [101], there are multiple structures of almost identical stabilities (differences within the range of a few tenths of eV) in the first to fourth coordination sphere and in the first to second cationic layer [81]. Such mobile O vacancies can be formed by doping with aliovalent cations [134], by exposition to the oxygen-lean conditions at high temperatures [135–137], or by chemical reduction [138].

The electronic state of Ce in the reduced CeO_2 surface has been devoted a number of experimental articles. The evidences for the existence of Ce^{3+} in the surface or intrafacial region of ceria have been obtained by means of XPS, (HR) EELS [139–141], IR [111, 142], and EPR [143]. The experiments show that in the reduced ceria the electrons reducing Ce^{4+} to Ce^{3+} are localised at the $4f$ orbitals. The morphology of the defects have been studied by STM, and it was shown that the oxygen defects created thermally on the (111) surface are either isolated or form the aggregates [121, 122, 144]. The computational studies of the surface of reduced CeO_2 clearly show the failure of the GGA-PBE or PW91 [103] functional which erroneously yield a metallic ground state of defected ceria, so do the LDA functionals [145]. The hybrid functionals tend to overestimate the CeO_2 band gap by 45% (PBE0) and 15% (HSE), comparing to experiments [27]. Very reasonable is the pragmatic, albeit non-universal, DFT+U approach as show by Nolan et al. [105, 146] and Fabris et al. [145], where the calculations with standard atomic orbitals and the Wannier localised functions were carried out for reduced (defect concentration of $1/4$) ceria (100), (110), and (111) surfaces yielding the well agreed surface reconstruction. The obtained geometries did not significantly depend on the particular variant of the DFT+U method, contrary to the defect formation energies, and it can be inferred that the Ce ions closest to the vacancy are repelled from it, while the O anions move towards the defect site [145]. The value of U, however, needs to be adjusted accordingly to the required property. For example, to obtain the energy of CO adsorption on CeO_2 the value of U was found to be 2 eV [147], while for the proper reproduction of the electronic structure of CeO_2 the value of $U = 4.5$–6.0 eV is required [27, 148, 149].

The use of hybrid functional, e.g. HSE, significantly improves the results [81] for the reduction of the self-interaction error, which leads to the exaggeration of electron delocalisation. The electrons occupy two localised Ce 4f states adjacent to the vacancy and, in the energy terms, *ca.* 1 eV below the conduction band within the band gap. According to, e.g. Paier et al. [81], the hybrid functional can be used for validation of the results obtained by PBE+U and gives results generally closer to the experiment than PBE+U. Adding the dispersion effects, DFT+D, improves the results shifting them towards experimental values by 0.12–0.15 eV. Nonetheless, the deviation from the experiment for the O vacancy formation was still quite high, *ca.* 1.7 eV for HSE+D or 2.2 eV for PBE+U+D. For the low concentrations of defects, i.e. for the calculations on the model of supercell $4 \times 4 \times 4$, the HSE functional underestimates the defect formation by *ca.* 0.5 eV with respect to the best experimental results available (4.2 ± 0.3 eV [150]), while PBE+U(4.5) underestimates them by *ca.* 1.4 eV.

O Diffusion

Another important parameter in the description of redox properties of oxides is the oxygen diffusion. For CeO_2, the DFT+U calculations [151] gave the energy barrier of 51 kJ/mol, very close to the experimental value of 50 ± 16 kJ/mol for defect concentration of 4%. The electronic conductivity, also contributing to the ceria conductivity, has been modelled by small polarons. Panhans et al. [150] in their experiments at fixed $p(O_2)$ found the conductivity to be n-type with resultant activation energy of $1/2 \Delta H_d(1/2 O_2) + \Delta H^{\sharp}_{\text{polaron}} = 2.57 \pm 0.33$ eV, where $\Delta H_d(1/2 O_2)$ is the defect formation energy, and $\Delta H^{\sharp}_{\text{polaron}}$ is the energy barrier for polaron hopping. The $\Delta H^{\sharp}_{\text{polaron}}$ value was found to be 0.48 eV, very close to the experimental value of 0.40 eV [152].

O Chemisorption

The process of chemisorption of O_2 was studied both computationally and experimentally by the Raman spectroscopy [153], which can distinguish between reactive oxygen species of different O-O bond length and thus different stretching frequency, i.e. superoxo ($O_2^{\bullet-}$) and peroxo (O_2^{2-}). The former can be also identified by the EPR spectroscopy [153]. The mostly dispersive nature of the species adsorption on the pristine surface can be deduced from the work of Huang et al. where the adsorption energy obtained for the PBE+U method was negligible [154], in line with the fact that the dispersion effects were not included in the computational method. The adsorption on the defected (111) surface, however, is exoenergetic by 1.72 eV and the peroxy group is formed: $[2Ce^{3+},\ V_O] + O_2 \rightarrow [2Ce^{4+},\ O_2^{2-}]$—this is the process of healing of oxygen defects upon reoxidation. The detachment of the O atom from the peroxy group costs *ca.* 2.6 eV, twice less than the removal of oxygen from pristine (111) surface (not counting the O_2 association energy) [81]. On the other hand, Hu et al. [155] reported the formation of superoxo group (confirmed by EPR [143]), thus only one Ce^{3+} ion is reoxidised: $[2Ce^{3+},\ V_O] + O_2 \rightarrow [Ce^{4+},\ Ce^{3+},\ O_2^{\bullet-}]$.

Catalysis

Among the processes catalysed by ceria, the most important industrially are the water-gas shift (WGS, $CO + H_2O \rightarrow CO_2 + H_2$), the selective NO_x reduction (SCR) by ammonia, methanation of CO_2 (Sabatier reaction, $CO_2 + 4\,H_2 \rightarrow CH_4 + 2\,H_2O$, which is the combination of the reversed water-gas shift reaction, $H_2 + CO_2 \rightarrow CO + H_2O$, and a CO methanation, $3\,H_2 + CO \rightarrow CH_4 + H_2O$ [156]).

The WGS process is industrially used for the production of hydrogen and for the CO removal from the feed stream for fuel cells [157] or ammonia synthesis [158]. In the computational study of Bruix [159] on the Pt nanoparticles on ceria, the following parameters were set: PW91+U(4.0) functional, the energy cut-off of 415 eV. The slab of three CeO_2 (nine atomic) layers and a 3×4 surface cell was used as a model. The separating vacuum layer was 15 Å thick. The atoms in the bottom O–Ce–O layer were fixed at the bulk positions during the geometry optimisation. The conclusion of the computational part of Bruix et al. article [159] is that ceria influences the electronic structure of the small noble metal (Pt) particles in such a way that the O-H bond dissociation is facilitated to such extent that when moving from Pt(111) surface to the Pt_8 clusters on CeO_2(111), the dissociation of water becomes exoenergetic ($\Delta E_{reac} = -0.24$ to -0.32 eV), what is beneficial comparing to other WGS catalytic systems (e.g. for Pt(111) $\Delta E_{reac} = +0.65$ eV, for Pt_{79} cluster $\Delta E_{reac} = -0.03$ eV). This property was attributed to the exposition of low-coordinative Pt atoms and to the high flexibility of the nanoparticle.

The similar conclusion about non-sufficient activity of Pt was taken by Flaherty et al. [160], who modelled computationally the WGS on Pt(111) surface concluding that CO dissociates *via* two pathways, the Boudouard reaction (disproportionation

$2\,CO \rightarrow C + CO_2$) and through a COH intermediate. Nucleation and subsequent oligomerisation of carbon deactivate the catalyst.

In the review by Tang et al. [161], there are discussed three important factors governing initial steps of the NH_3-SCR process, namely the interaction of process reactants—NO, O_2, and NH_3—with ceria. The presence of O_2 relevantly increases the efficiency of NO reduction, although in the absence of O_2 the reaction still takes place. The prerequisite step for the reduction of NO by ammonia catalysed by ceria is the adsorption which is facilitated by the surface energy rise due to the presence of defects like kinks, steps, or vacancies. This was confirmed by the DFT+U calculations by Yang et al. [162] and shown that the adsorption energy of NO on reduced ceria is *ca.* 10 times higher, then on stoichiometric ceria. The DFT+U calculations [163] of NO on reduced $CeO_2(110)$ revealed the preference of NO to adsorb on the surface oxygen vacancy. The subsequent NO_2 formation–dissociation cycles lead to the NO diffusion on the ceria surface. Two NO molecules adsorbed at neighbour O vacancies can form a N_2O_2 dimer, eventually liberating N_2, what was confirmed by Nolan [164] who studied computationally also the CO adsorption. The adsorption of NO on non-stoichiometric ceria easily leads to the NO dissociation due to the strong interaction between NO and ceria, what was confirmed by Luo et al. [165] who found that during the thermoprogrammed desorption of NO the new species: N_2, N_2O, and O_2 were detected.

The NO_2 adsorption on (100), (110), and (111) surfaces of defected ceria leads to the embedding of single O into the vacancy, to bending of the O–N–O and to elongation of $N{-}O_{vac}$ bond (by *ca.* 10%) with accompanying partial reoxidation of Ce^{3+}, as modelled computationally by Nolan et al. [166]. The adsorption energies were found: -2.40 eV for the (111) surface,-2.25 eV for the (110) surface, and -2.32 eV for the (100) surface.

One of the main motivations of the carbon dioxide reduction by hydrogen is the facilitation of the remote product transport using the conventional pipeline system, what is virtually impossible in the case of hydrogen. Such process is economically efficient with source of cheap hydrogen other than the fossil hydrocarbons, e.g. from the biomass, or, perspectively, from water electrolysis. The CO_2 reduction process is also interesting from the fundamental point of view, as a way to study the utilisation of CO_2 yielding other products of industrial importance.

It is commonly assumed that the CO_2 methanation follows two steps: the first one is the reduction of CO_2 to CO and further conversion to alkanes along the Fischer-Tropsch process. The CO_2 reduction process was modelled computationally by, e.g., Sharma et al. [167] (Ru-doped ceria), who studied their systems with LDA+U(5.5) approach and concluded that LDA+U gives better agreement with experiment than GGA+U [25, 26] as for the unit cell parameter and the energetics of the reaction $Ce_2O_3 + {}^1\!/_2O_2 \rightarrow 2\,CeO_2$. Sharma et al. [167] conclude that ruthenium atoms tend to stay at (or migrate to, if allowed, i.e. if well equilibrated) the surface region and the catalytic CO_2 methanation over Ru-doped ceria does not include the Fischer-Tropsch step. The authors also point out the application of the Sabatier principle (they call it the "moderation principle"), namely the substitutional doping by the replacement of the host cations by the lower valency ones (e.g. La-doped ceria) weakens the binding of oxygen and thus facilitates the formation of O vacancy and so enhances the catalytic activity *via* the MvK mechanism, but when the donation of O is too high, then the process becomes stoichiometric instead of catalytic as the healing of O vacancies is too slow.

3.1.2 Titania TiO_2

The TiO_2 systems are among the most popular photocatalysts [168–170] of many types of processes, e.g. methane forming from CO_2 and H_2O [169], which is a highly desirable perspective process in the days of environmental concern by lowering the CO_2 emission and forming fuels with use of solar energy. Among the numerous system, e.g. TiO_2, CdS, ZrO_2, ZnO, and MgO, TiO_2 seems the most convenient concerning oxidation properties, charge transport properties, stability, non-toxicity, and cost [171]. Another highly industrially demanded processes, belonging to so-called green chemistry, include the splitting of water (WS) into hydrogen and oxygen with use of the solar energy (Grätzel's dye-sensitised solar cells, DSSC [172–174]), air/water purification devices, or self-cleaning windows and façades. The photocatalytic oxidation of pollutants is also of high environmental concern [168]. The position of the valence and conduction bands of TiO_2 allows for the oxidation of water at the (semiconductor) photoanode and the reduction of water at the cathode leading to the water splitting into H_2 and O_2 (see Fig. 2). The wide band gap, however,

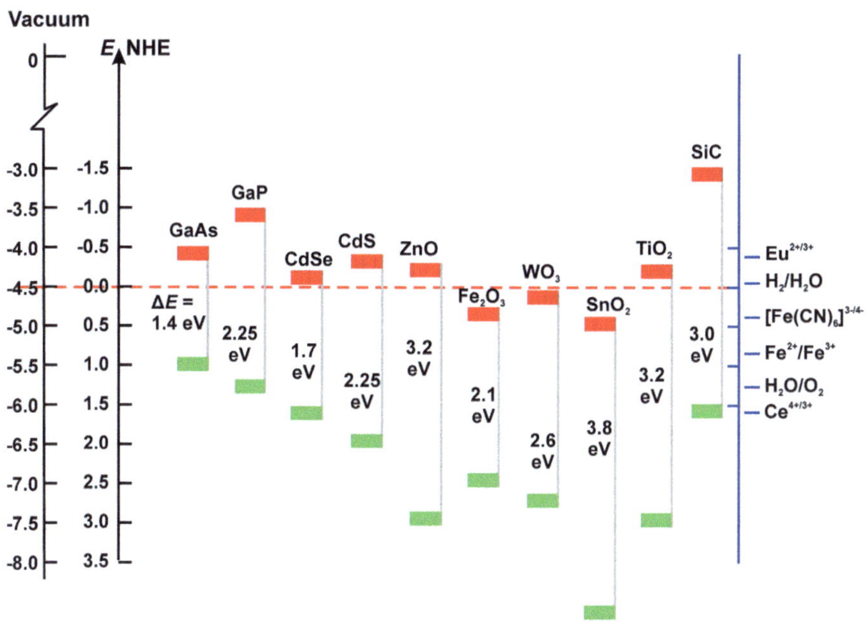

Fig. 2 Band positions of several semiconductors in contact with aqueous electrolyte at pH 1. Reproduced from [175], with permission

limits the absorption of light by TiO_2 to the UV part of the spectrum dramatically lowering its efficiency towards solar light. On the other hand, the semiconductors with narrow bad gap are prone to the photocorrosion. The solution of this deadlock is the sensitisation of the semiconductor by the agent able to absorb visible light and inject carriers into the substrate.

Although there is wealth of experimental articles on TiO_2, there is not many of them on computational studies.

The number of various titania catalysts was studied by Anpo et al. [176] in the form of: highly dispersed, powdered, and the monocrystal. They found the important dependence of the kind of catalyst and the CO_2/H_2O ratio, and temperature on its efficiency and selectivity. In the case of rutile monocrystal, the higher selectivity towards methanol and methane was observed for (100) than for (110) facet. The reactants and intermediates were studied by high-resolution electron energy loss spectroscopy. The difference in activity between (100) and (110) is attributed to the different Ti/O surface atomic ratio, and the more spacey (100) surface allows for the contact with CO_2 and H_2O molecules, and also have higher reductive potential in the excited state, facilitating the reduction of CO_2 towards CH_4. Besides, the TiO_2 polymorphs, rutile and anatase, belong to the most studied single crystal systems [177].

The computational description of defected TiO_2 requires the usage of the hybrid functionals or, when the computational cost thereof is prohibitively high, the DFT+U functional.

Since the zero-dimensional (0-D) TiO_2 NPs are the most fundamental nanostructures of TiO_2 which can be utilised as starting points for more complex materials with more specified and improved performance, they belong to the group of the most investigated solid-state materials of the past 20 years. In these investigations, the DFT (and more advanced QChM) cluster calculations are of indisputable importance. In this context, numerous experimental and theoretical studies investigated isolated titanium oxide clusters to correlate their structures and properties with those of the bulk phases [178–181]. In particular, the Ti_nO_{2n} and Ti_nO_{2n+1} clusters were found to be the most stable neutral ones, while Ti_nO_{2n-1} and Ti_nO_{2n-2} clusters were formed by fragmentation [182]. Zhai et al. [183] have probed the electronic structure and band gap evolution of titanium dioxide clusters, $(TiO_2)_n$ ($n = 1, \ldots, 10$), using photoelectron spectroscopy (PES) comparing the results with available theoretical data [184] showing that they are valuable to verify future high-level theoretical calculations. The possibility of antiferromagnetic stabilisation in the titania clusters was predicted (thanks to the evolutionary algorithm USPEX and DFT+U calculations) within a high-symmetry geometric structure of the bare cube-shape Ti_8O_{12} cluster, in excellent agreement with experimental results [185]. For such TiO_2 NP, unique chemical bonding was described where electrons of Ti atoms interacting in antiferromagnetic fashion to lower the total energy of the system. Illas et al. investigated the electronic band gap of $(TiO_2)_n$ nanoparticles ($n = 1, \ldots, 20$) using a relativistic all-electron description, within the G_0W_0 method [186]. It was shown that high-level G_0W_0 approach may be successfully implemented to study the electronic band gap of realistic size nanoparticles at an affordable computational cost giving results that are directly related with those from photoemission spectroscopy. The results revealed that systematic correlation exists between band gaps, ionisation potentials, and electron affinities of TiO_2 nanoparticles. In the paper by Berardo et al., a wide range of many-body (G_0W_0, qsGW, EA/IP-EOM-CCSD) and DFT (B3LYP, PBE) methods were used to study the ionisation potentials and electron affinities (vertical and adiabatic) together with fundamental gap and exciton binding energy values of bare and hydroxylated TiO_2 nanoclusters [187]. Berardo et al. describe how the clusters electronic properties change as a function of size and hydroxylation and compare the performance and predictions of the different classes of methods as well.

The importance of the reduced Ti^{3+} species in the electric conductivity [177], photocatalysis, photochemical water splitting, photochemical oxidation of pollutants, and the technology of dye-sensitised solar cells is well recognised in the literature. The presence of the (paramagnetic) Ti^{3+} cation, whose presence supports the proposed mechanism, in reduced rutile was proved via the EPR spectroscopy (the *g*-tensor typical for the Ti $3d^1$ state [188]).

Structure and Morphology

The most common polymorphs of TiO_2 are rutile and anatase (the third one—brookite—is much less popular) [177]. Their crystal structures are sketched in Fig. 3, where the distorted octahedral coordination of Ti^{4+} is clearly visible. Of these two, anatase monocrystals are very difficult to obtain and hence this polymorph is very scarcely discussed in surface science articles; it is, however, more active photocatalytically and often found in the industrial colloidal TiO_2.

The thermodynamically most stable, and hence the most intensively investigated, including computational studies, rutile surface is (110), see Fig. 4 (left), which exposes the plane built of O_{3c} and Ti_{5c}, while the fully coordinated Ti_{6c} are bonded

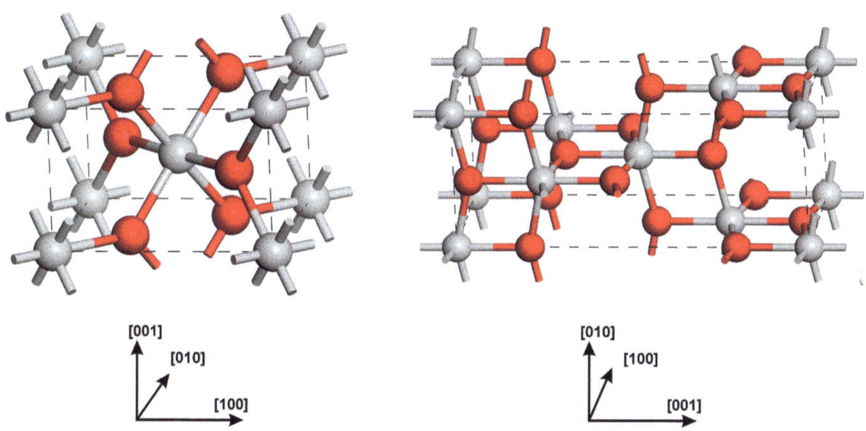

Fig. 3 Crystal structures of bulk rutile (left) and anatase (right). Red spheres: O, grey spheres: Ti

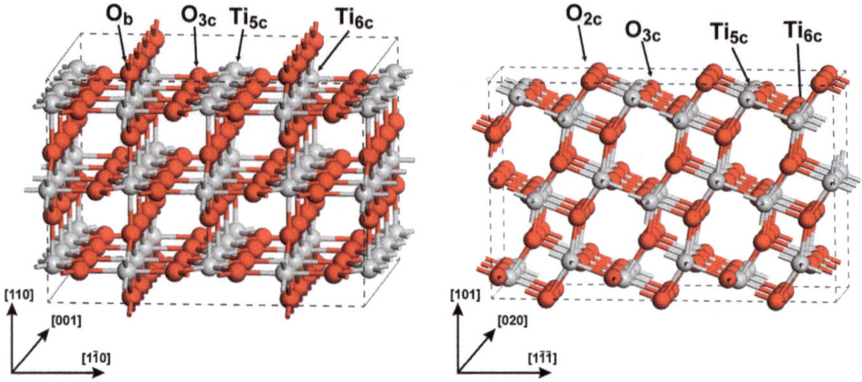

Fig. 4 Structure of the rutile (110) surface (left) and anatase (101) surface (right) of their most stable terminations. Red spheres: O, grey spheres: Ti

to exposed, bridging, doubly-coordinated oxygen atoms (O_b) forming a ridge. The coordinatively unsaturated O_b and Ti_{5c} atoms are reactive sites on this surface. The relaxation of rutile (110) is relatively low, mainly in the normal direction, and affects mostly O_b. The anatase (101) terrace face exposes unsaturated Ti_{5c} and O_{2c} which are expected to be the most reactive.

The other surfaces, less stable than (110), but also found in research, are (001) and (100).

As Labat has shown [189], studying TiO_2 with different Hamiltonians, the Hartree–Fock and DFT, with both LDA-VWN, GGA-PBE, and hybrid functional, that the excellent agreement with experiment for the band structure and binding energy was achieved for both B3LYP and PBE0 functionals. The best geometrical structure was obtained at the PBE0 level. Labat used both all-electron Gaussian and PAW [58] basis sets.

Water Adsorption

The issue of water adsorption, e.g. the nature of adsorbed species, is crucial for catalysis. For example, many organic processes are catalysed by protons which form hydroxyl groups at the surface (vide infra).

Recently, Futera et al. [190] studied the rutile (110) and anatase (101) water interface with use of the ReaxFF simulations. They studied the dynamics of hydrogen bonds by Luzar–Chandler model used to predict the mean lifetime of the hydrogen bonds and conclude noting the spontaneous dissociation of water on both rutile and anatase surfaces with OH group remaining essentially on Ti_{5c} while H^+ shifting to the nearest doubly-coordinated oxygen atom. The polarisation of the surface (voltage drop from -2 to 6 V/Å) forces the water molecules in the interfacial region (*ca.* 6.5 Å) to order. Formation of such ordered structure was also confirmed by the change of the diffusion coefficient, namely from $2.28 \cdot 10^{-9}\ m^2\,s^{-1}$ in the internal of the water layer to the value of $4.67 \cdot 10^{-11}\ m^2\,s^{-1}$ for rutile and $6.48 \cdot 10^{-11}\ m^2\,s^{-1}$ for anatase.

The issue of water adsorption and the question whether H_2O adsorbs associatively or dissociatively is a matter of long dispute. As early as in 1995, Goniakowski and Gillian [191] showed computationally that both forms are energetically favourable. On the other hand, the embedded cluster Hartree–Fock studies [192] showed preferential occurrence of associative mechanism and argue that the contribution of dissociated forms postulated by planewave DFT studies is due to the overestimation of the hydrogen bonds in the dissociated forms by planewave DFT [193]. This is in line with conclusions of the Car–Parrinello simulations of Langel et al. [194], who found that spontaneous dissociation of water occurs on the O vacancy of (100) surface and does not at the (110) surface.

The experimental results suggest, however, that water adsorbs mainly associatively and the dissociation occurs at low coverages, what can be associated with surface defects. This conclusion is in line with recent STM experiment, which present the ability of O vacancies on anatase (101) surface to dissociate water yielding bridging hydroxyls [195]. The dominant role of O_b vacancies in rutile (110) was also supported by Pang et al. [196] The analogous conclusion on high stability of bridging hydroxyls on the rutile (110) surface is expressed in the article of Wendt et al. [197], who studied the rutile system both experimentally (STM) and computationally, and in the article of Morgan et al. [198]

Morgan et al. [198] also reviewed the former attempts to describe computationally the rutile (110) surface. They note that hybrid functionals improve the description of the band structure (an hence the bad gap width) [199], comparing to pure DFT, and they show that the hybrid B3LYP calculations localise defect states where two electrons are observed on Ti^{3+}_{6c} and Ti^{3+}_{5c}. Other article, by Bredow and Pacchioni [200], shows that for B3LYP in the cluster model of the reduced titania (110) surface although the excess charge is localised on the Ti^{3+}_{6c} and Ti^{3+}_{5c}, but as many as four sites are engaged, not just two. For HF-LYP functional, the spread of the excess charge is qualitatively different, is localised on two Ti^{3+}_{5c} centres, adjacent to the defect. The authors of [198], based on their own work, attribute the B3LYP behaviour to the fact that the amount of the HF exchange component has been fitted to the results of calculations for the first and second rows of elements and for heavier elements changing of the HF part could lead to more accurate results. The usage of GGA+U($U \geqslant 4.20$) can reproduce the charge localisation on two Ti^{3+}_{5c} centres (see Fig. 5), and generally the experimental band state.

Oxygen Vacancy Formation

Morgan et al. also discussed also the vacancy formation energy obtained by themselves ($E(O_{vac}) = 3.66$ eV for GGA+U(4.20)) and by Wu et al. [201] and by Rasmussen et al. [202] (3.52 and 3.03 eV for different supercells).

Fig. 5 View along the [001] direction of the isosurface of the charge density associated with the band gap state. The contour level is 0.06 electrons $Å^{-3}$. Reproduced from [198] with permission

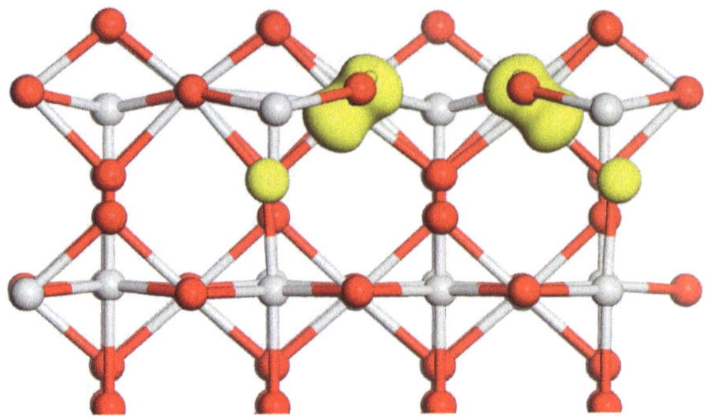

The more scarcely discussed in literature anatase (101) surface has also been studied by Nadeem et al. [195] and their investigations suggest that the surface vacancies dissociate water and form bridging OH groups.

The healing of oxygen vacancies was also studied in the above-mentioned works of Langel et al. [194] and Wendt et al. [197], who found that O_2 dissociates in vacancy.

Catalysis/Photocatalysis

The study of photocatalysis on TiO_2 dates back to 1979, when Honda et al. [203] showed the ability to reduce photocatalytically carbon dioxide to organic compounds, e.g. formic acid and aldehyde, methane and methanol by the suspension of TiO_2 assisted by non-oxide n-type semiconductor (to red-shift the absorption band of TiO_2) particles in water [204]. The Cu-TiO_2 ($\leqslant$5 wt%) [205] system was also used yielding methane and ethylene at the absence of oxygenated products (CH_3OH or CH_2O). However, Anpo et al. [176] report the formation of CH_3OH on copper-loaded (0.3−1.0 wt%) fine-grained titania and the less efficiency towards CH_4. They attribute the photocatalytic activity to the Cu^+ species, confirmed by XPS, on TiO_2.

The issue of shifting of rutile or anatase light absorption from the UV range to the visible light has been crucial from the point of view of solar photocatalysis or for light-driven reactions for years. In its pristine form, rutile absorbs as little as 4% of the solar light due to its high intrinsic band gap (3.0 eV) [206]. In this purpose, doping with non-metallic atoms, e.g. nitrogen, seems very efficient [206–208], although there are also reports on carbon and sulphur which, being anionic, can contribute to the states supposed to be localised above the valence band maximum which in turn can increase the photoactivation in the range of visible light. For the case of C, Di Valentin et al. [209] computationally (PBE functional, Car–Parrinello approach for geometry optimisation, a $2 \times 2 \times 3$ cell for rutile, a $2\sqrt{2} \times 2\sqrt{2} \times 1$ cell for anatase) shown that for low C concentrations and in oxygen-lean conditions C substitutes O in the lattice and O vacancies are formed, while in high O concentration C substitutes Ti and also occupy interstitials. The multidoping effect was also observed, apparently stemming from the interspecies redox. Carbon doping of rutile, both O-substitutional or interstitial, forms the localised occupied states in the band gap (shift of absorption edge up to 2 eV) [209]. The Ti-substitution in rutile does not form the localised states in the band gap and (in opposition to anatase) does not narrow the band gap.

Nitrogen in most cases adopts the substitutional positions in TiO_2, in rare cases it can, however, be present in interstitials [208, 210]. The substitutional N-doping does not lead to any multi-atom effects [209, 210]. The nitrogen ($2p$) states lie a few hundreds of meV above the valence band top.

Self-organisation

The issue of self-organisation of organic adsorbates on metallic (coinage metals) surfaces yielding graphene nanoribbons has been extensively studied [211, 212]. However, the rutile (110) surface turned out to be also very perspective and better suited substrate for the bottom-up assembly of such structures. The process of self-organisation on the substrate of functionalised semiconductor surfaces, e.g. TM oxides, can be much more tunable, for example by hydration and thus decorating by protons, than on metal surfaces. On the other hand, the TMO have interesting optical, photo- and electrochemical properties which can be tuned, e.g. sensitised, by the adsorbates. Kolmer et al. [213] demonstrated that the presence of surface hydroxyl groups (protons, particularly) is crucial in the polymerisation of aryl halide precursors on the (2×1) reconstructed rutile $TiO_2(011)$ surface ([213], Fig. 1C). The moderate concentration of hydroxyl leads to the formation of long molecular oligomers. Increasing the hydroxyl coverage of the surface results in the formation of shorter oligomers and the hydroxyl-free surface suppresses the polymerisation utterly. Kolmer et al. used 10,10′-dibromo-9,9′-bianthryl (DBBA, see, e.g., [213], Fig. 1) as a precursor and performed both experimental and computational studies of the reaction thermodynamics and also the STM microscopy imaging ([213], Fig. 1, right panel). These studies were based on PW91+D functional. The reaction route on rutile differs distinctively from the route on metallic substrates: the most important difference stems from their very different chemical nature and properties emerging thereof. The concerted process of early C–C bond formation together with late C–Br bond cleavage followed by the multistep proton-assisted coupling was found to be the easiest (the lowest activation energy, modelled computationally, below 1.95 eV) among the several conceivable routes. The heterolytically formed protons, specific to the metal oxides, transferred from surface hydroxyl groups to DBBA admolecules (1.39 eV), facilitate the coupling, the subsequent migration of protons on the aromatic framework is easy (0.65–1.13 eV), and then preferred attachment to the Br-bound carbon atom occurs. The subsequent relevant weakening of the C–Br bond facilitates the C–C bond formation and release of Br_2 molecules (0.82 eV).

In the articles of Zasada et al. [214], the issue of terephthalic anhydride monolayer on rutile (110) surface was studied computationally ($Ti_{48}O_{96}$ supercell; PW91+D and PBE+D; $2 \times 2 \times 1$ sampling of IBZ; 400 eV planewave cut-off energy). For the high coverage limit, several packing domains were tested and similar adsorption energies were found suggesting possible coexistence of such domains. It was also shown that for different STM settings, qualitatively different (with different motif, dimeric or monomeric) images are obtained.

In [215], the possibility of formation of surface attached metal-organic frameworks (SURMOF) was noted and the HR-TEM image was simulated. The Zn-MOF-5 bulk model was built based on the X-ray diffraction study of Li et al. [216]: the unit cell was cubic ($Fm\bar{3}m$ space group) with $Zn_{32}O_{104}C_{192}H_{96}$ stoichiometry, and the initial lattice constant of 25.86 Å. The almost commensurate matching of rutile and Zn-MOF-5 was noted; hence, the mutual mismatch of the supercell as a function of the number of MOF unit cells was discussed. The MOF cell was found much softer than rutile cell (calculated bulk moduli of 18.41 GPa and 216.00 GPa, respectively, both values are close to experimental data). The adhesion of Zn-MOF-5 layer to rutile substrate (-0.32 eV nm^{-2}) was found only possible when dispersion forces (addition of -0.39 eV nm^{-2}) are included in the computations to overcome the stress ($+0.69$ eV nm^{-2}) imposed by interfacial strain ($\varepsilon[001] = 0.31\%$ and $\varepsilon[1\bar{1}0] = 2.86\%$). The HR-TEM simulations were performed with use of the JEMS software [217, 218].

The other surface-assisted formation of self-organised structures adsorbed on the rutile surface, both the dispersion forces driven [70] and covalent bonded structures [213, 219], were noticed, the latter stemming from the polycondensation catalysed by protons adsorbed on the rutile surface in the form of hydroxyl groups. The former case, i.e. the monolayer formed densely due to the intermolecular and molecule-surface attraction concerned the 3,4,9,10-perylenetetracarboxylic dianhydride (PTCDA) on TiO_2-(110) surface [70]. The STM images for the low-coverage case were modelled within the Tersoff–Hamann approximation, and it can be seen that the PTCDA molecules adsorb in the intact state. For the monolayer, several adsorption modes were tested and, given the experimentally (LEED and STM) found packing—$c(6 \times 2)$, one structure with significant bending of PTCDA molecules along the protruding O_b ridges (see [70] Fig. 6) was found. The modelling of the STM in agreement with experiment was possible only when the Bardeen theory simulation was performed (see [70], Fig. 6d, e).

The self-organisation of morphological effects, i.e. the formation of nanoripples by Ar^+ beams in UHV conditions and the reversible reorientation of them by 90° during temperature change, was observed with STM and modelled at GGA level by Kolmer et al. [188] Several channels of diffusion were postulated, and the calculation of the energy barriers leads to the conclusion that diffusion of Ti adatoms alone, characterised by the barrier of 1.14 eV along the [001] direction ("*in-channel*"), and ever higher in the other directions (2.79 eV for "*cross-channel*", 3.25 eV across the [001] step), cannot be responsible for the mass movement required for the formation of ripples. The assistance of the mobile oxygen surface species (diffusion barrier of 0.42 eV) dramatically lowers the energy barriers (e.g. 0.71 eV for concerted "*in-channel*" hop of Ti(O)). For the lowest $T = 150$ K, the ripples formation is ruled mainly by the erosion of the ascending step edges as the diffusion processes are virtually hindered ([188], Fig. 3). For slightly higher temperature, 300 K, the assisted diffusion of Ti(O) becomes intensive for "*in-channel*" and the "*cross [1–11]*" modes while keeps hindered for "*cross-channel*" and the "*cross [001]*". The coexistence of the ion erosion in the direction perpendicular to the [001] steps with the mass transport along the [001] direction results in the formation of the anisotropic structure. For $T > 400$ K the diffusion of Ti^{3+} (formed upon surface O_2 desorption: $2O^{2-} + 2Ti^{4+} \rightarrow O_{2(g)} + 2Ti^{3+}$) into the bulk takes place. In these conditions, the organisation of the ripples is no longer governed by diffusion but rather by sputtering along the direction normal to [001] steps. For higher $T = 720$ K, the excess surface Ti diffuse into the bulk and the stoichiometry of the $TiO_2(110)$ surface is restored what make the ripples long and, as for lower T, well ordered along the direction of ion beam projection.

3.1.3 Zirconia ZrO_2

Zirconium oxide belongs to the family of oxides with high dielectric constant ("high-K") [220]. This issue is crucial for the modern microelectronics, namely in the MOS technology where the insulating oxide layer must not be too thin to prevent tunnelling charge leakage, yet providing required capacitance. The most common materials are ZrO_2, HfO_2, and their mixtures with SiO_2 ($K = 3.9$). Their disadvantage is the low stability (metastability) of amorphous phase, contrary to SiO_2, which allows for the formation of the interfaces with miscellaneous substrates yielding little electrical defects. The metastable phase can be stabilised by alloying with, e.g. Si, Al, or N.

The other application of ZrO_2 is catalysis, both as an active phase or active phase component [221, 222] or as a substrate [223, 224], due to its thermal and chemical stability, high refractive index, low thermal conductivity, hardness, and remarkable oxygen ion conduction [225–227]. Zirconia is a well-known solid acid catalyst, particularly when mixed with titania [228], in the isomerisation, oxidation, dehydrogenation, dehydration, alkylation and dealkylation, SCR and photocatalytic processes, as well as in the manufacturing of hydrocarbon gas sensors (see, e.g., the references in [228]).

Structure and Morphology

The properties of three most important ambient pressure phases, the fluorite cubic structure c-ZrO_2 (O_h^5, $Fm\overline{3}m$ space group, see Fig. 6a, stable at $T > 2370\,^\circ$C), the tetragonal phase t-ZrO_2 (D_{4h}^{15}, $P4_2/nmc$, see Fig. 6b, stable for T from $1170\,^\circ$C to $2370\,^\circ$C), and the room temperature stable baddeleyite monoclinic m-ZrO_2 (C_{2h}^5, $P2_1/c$, see Fig. 6c, $T < 1170\,^\circ$C) phase, were studied computationally by Vanderbilt et al. [220] and Zhao et al. [229] who used LDA functional, while Jomard et al. [230] used the GGA functional.

The high-T cubic phase can be stabilised by doping by MgO, CaO, or Y_2O_3 [225, 226, 231]. The tetragonal phase is also stable in the RT when it forms nanograins with size up to *ca.* 20 nm [232], what indicates that the impact of the crystal size should be accounted for in the case of nanometric size. The stability and low-temperature phase transition of monoclinic, tetragonal, cubic, orthorhombic-I ($Pbca$)

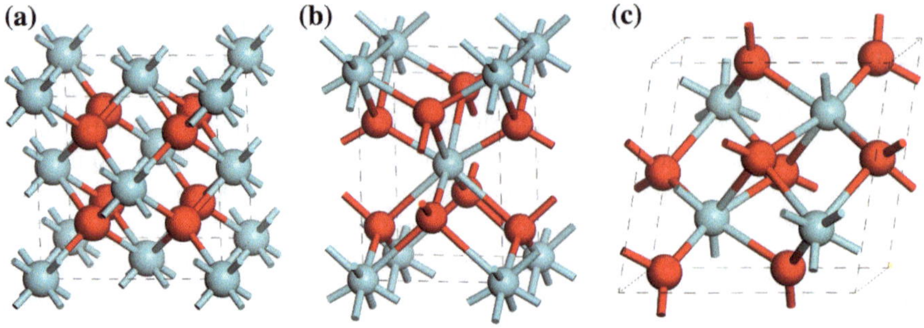

Fig. 6 Structures of zirconia polymorphs: **a** cubic, **b** tetragonal, **c** monoclinic. Colour coding: oxygen ions—red spheres, zirconium ions—grey spheres

and orthorhombic-II (cotunnite) phases of ZrO_2 and HfO_2 were studied computationally (LDA and GGA) by Jaffe et al. [233], who found that GGA yields the results agreeing with experiment. They found also that the ionicity decreases from monoclinic to orthorhombic-II phase. The non-periodic, cluster approach study was performed by Li and Dixon [234] using the CCSD(T) and DFT levels of theory. The B3LYP and BP86 functionals were used, the aug-cc-pVxZ basis set for O and aug-cc-pVxZ-PP basis set with pseudopotentials for Zr and Hf (where $x = \mathrm{D}$ for geometry optimisation and vibrational analysis while $x = \mathrm{T}$ for single point calculations). The $(MO_2)_n$ clusters (M = Zr, Hf; n = 1–4) and their anions were studied, and it was found that DFT properly predicted the conformers for neutral clusters and some of the anionic clusters. The electron excitation energies were calculated by TD-DFT, EOM-CCSD, and CCSD(T), and the latter gave the best results. The important finding is that the band gap depends strongly on the cluster structure and hence can be tuned to match the visible range of the light, an issue crucial for photocatalysis, see Fig. 7.

More recently, the dependence of the DFT functional, namely the amount of the Hartree–Fock exact exchange, on the band gap width, structure, and stability of several wide gap semiconductors, including TiO_2 and ZrO_2, also confronted against the GW results, has been studied by Gerosa et al. [235]. They concluded that the dielectric-dependent hybrid functional reproduces well both ground-state properties (lattice constant, reaction energy, phase stability) and excited-state properties (band gap width). Gerosa et al. tested also the results obtained with LCAO and PW basis sets.

Water Adsorption

The issue of adsorption of water on the ZrO_2 surface is very important since water can form both dissociative and associative moieties which both modify the nature of the active sites [236] and also severely influence the stability of the ZrO_2 polymorphs thus, e.g., delaying the phase transition [237].

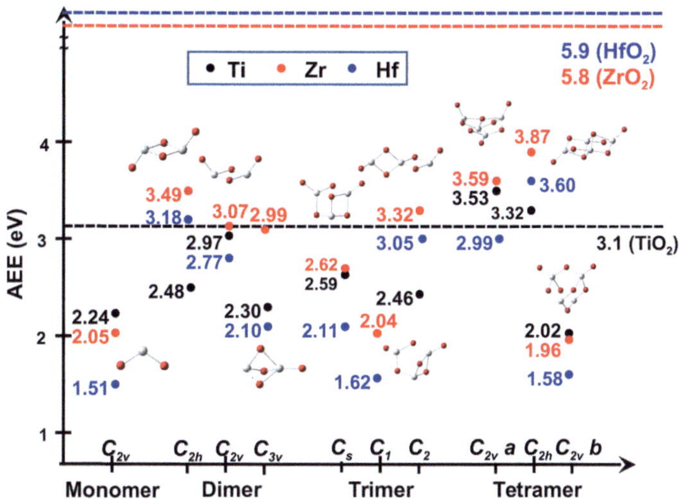

Fig. 7 Adiabatic energy gaps in eV for the low-energy conformers of the ground singlet states of $(MO_2)_n$ (n = 1–4) clusters for M = Ti (black), Zr (red), and Hf (blue) calculated at the CCSD(T)/aD//B3LYP/aD level. The experimental values of the bulk metal oxide band gaps are shown. Reprinted with permission from [234]. Copyright 2010 American Chemical Society

To study the phenomenon of ZrO_2 nanograin hydration, the ab initio morphology of bare and hydrated polymorphs of ZrO_2, tetragonal [238] and monoclinic [239], were systematically studied computationally at the PW91 level of theory yielding very good agreement with HR-TEM imaging (see Fig. 8c). The obtained surface Gibbs free energies were used to construct the Wulff solid. The hydration was modelled *via* the multisite Langmuir isotherm, which was also reformulated and fitted to the empirical Fowler–Guggenheim equation, which in turn allowed for the assessment of the average lateral interaction energy. The influence of the nanograin size on its stability was accounted for through the balance between the bulk stress tensor and the surface energy (Fig. 8b and [238], Fig. 8). If only the diagonal, hydrostatic term is taken into account, the surface energy is connected with internal pressure *via* the Laplace–Young equation. The contribution to the surface energy introduced by the edges and corners was negligible, as shown by, e.g., Barnard et al. [240] The equilibrium morphology change during T change is presented in Fig. 8a for monoclinic zirconia, while for tetragonal ZrO_2, see [238], Fig. 7.

3.1.4 Zeolites

The specific place in the world of TM cations in their coordination environment is occupied by zeolites, where TM cations can be guests, either isolated or forming clusters, introduced in the structure of the p-electronic oxide which, in turn, can significantly change their properties.

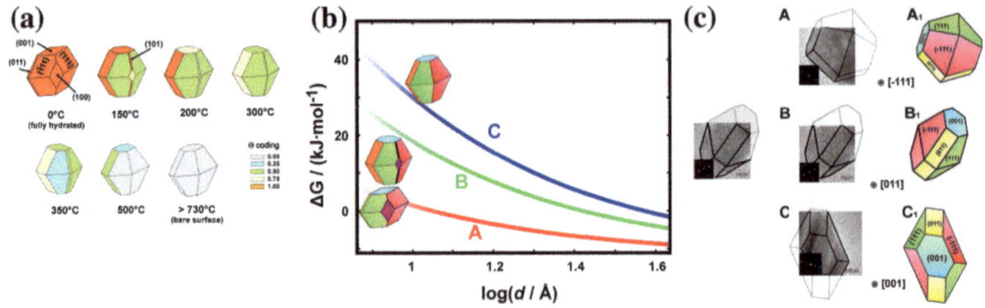

Fig. 8 **a** Wulff shapes of m-ZrO_2 as a function of temperature for $p_{H_2O} = 0.01$ atm; **b** size dependence of the Gibbs free energy of the m-ZrO_2 nanocrystals for (A) fully hydroxylated ($T < 95\,°C$), (B) half-hydroxylated ($\Theta = 0.5$, $T \sim 300\,°C$), and (C) bare surface ($T > 750\,°C$); **c** TEM pictures of the monoclinic zirconia nanocrystals (A–C) together with the calculated Wulff shapes (A_1–C_1), oriented along [−111], [011], and [001] directions to align with the observed 2D images. Adapted with permission from Piskorz et al. [239]. Copyright (2011) American Chemical Society

Structure

According to Liebau et al. [241] zeolites can be classified to the family of (micro)porous tectosilicates, i.e. forming 3D-frameworks of interconnected tetrahedral units of $[SiO_4]^{-4}$ and, optionally, $[AlO_4]^{-5}$, which can be classified as secondary building units (SBU). To satisfy the electroneutrality of the framework, each $[AlO_4]^{-5}$ unit, having an extra negative charge comparing to the $[SiO_4]^{-4}$ tetrahedron, must be counterbalanced by the single charge of the cation, e.g. a proton or single valence cation, or by the n-valence cation, shared by n of $[AlO_4]^{-5}$ units. Due to the repulsion of the formally single negative aluminium tetrahedra, they do not accommodate adjacent positions but must be separated by $[SiO_4]^{-4}$ units, what is known as the Löwenstein rule. The question of localisation of aluminium oxide units in frameworks is a matter of plethora of studies, both experimental [242–248] and computational [242, 249–251]. The issue of zeolite structures is comprehensively reviewed in [252]. The Si positions in the frameworks are standardised, e.g. see Fig. 9. The family of zeolitic structures is numerous; currently, 229 different zeolite structures are known.

The TM cations are responsible for the redox properties of the zeolitic systems, and this property is most frequently accompanied by the presence of acidic centres thus forming the bifunctional catalysts. The TM ions can occupy both the framework positions (isomorphic substitution), see, e.g., [253–255], or the extraframework positions (ion exchangeable sites) [251, 252]. Due to the easier accessibility, the latter case is most commonly found in catalysis.

The availability of the micropore systems for the reacting molecules can be increased by forming the system of mesopores (2–50 nm) [256] or hierarchisation of zeolites [257]. This important issue, however, will not be discussed here for sake of conciseness, nor will be the issue of non-TM zeolites application.

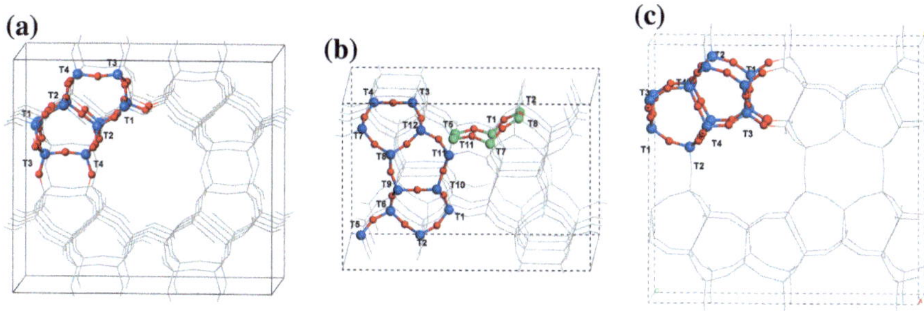

Fig. 9 Annotated Si positions in framework of **a** MOR, **b** MFI, **c** FER; blue spheres—the non-equivalent Si positions, red spheres—O atoms, green spheres—commonly regarded set of Si atoms exposed to large channels

Catalytic Activity

The TM zeolites are active in, e.g., the CH_4 to CH_3OH direct conversion [258], nitrogen oxide decomposition [259, 260], H–H bond activation [249, 261, 262], processes involving alkane/arene like dehydrogenation [263] and aromatisation of light alkanes [264, 265], and also in alkylation [266] and selective oxidation [267] processes.

Zeolites can accommodate not only the bare cations [262] or small clusters (e.g. mono-(μ-oxo)dicupric [268]), but also the complexes as large as Ru(III) benzimidazole or 2-ethyl Ru(III) benzimidazole. Selvaraj et al. [269] studied the benzimidazole complexes for their pharmaceutical importance as the biomimetic system with the imidazole ring system, and used the hybrid, B3LYP functional. Starting from the optimised geometry, the TD-DFT calculations were performed. As Selvaraj et al. conclude: "Upon encapsulation of free complex into Zeolite Y, reduces the HOMO-LUMO gap, global hardness and increases the softness values of the complexes, resulting in a higher catalytic ability" in the perspective catalytic cycle of phenol oxidation.

In the purpose of resolution of the issue of the electron density flow channels, the computational tool, the Natural Orbitals for the Chemical Valence (NOCV [270, 271]) analysis, combined by the energy decomposition approach (Extended Transition State, ETS) of Ziegler and Rauk [272], has been successfully applied [262] for the H_2 interaction with ZSM-5 system with d^{10} cations: Zn^{2+}, Ag^+, Cu^+, and Cd^{2+}. The aim was to devise a novel "geometric" descriptor, Δr^{NOCV}, dependent on the NOCV electronic channels: σ-donation, polarisation, and d_π backdonation. It was found that backdonation process has stronger influence on the adsorbed hydrogen molecule activation than the one connected with depopulation of bonding orbitals. The Turbomole package on the B3LYP level of theory with def2-TZVP basis set was used. The NOCV analysis was done with use of *natorbs* program [273]. In this article, the proposed pathway for the H_2 storage process was also studied.

NO Activation

The activation of NO by Cu(I) sites in zeolites is a matter of long-lasting studies, since its discovery by Iwamoto et al. [259, 260], both experimental and computational [274–281]. The GGA and hybrid functionals (PBE0, HSE03, HSE06, B3LYP) are most abundantly used, and the hybrid ONIOM scheme is also found. The hybrid functionals are slightly more accurate in predicting geometric parameters and excitation energies, and give higher density localisation than GGA. The binding energies of cations to the framework, the band gap width for pure siliceous zeolite, and the exchange splitting for extraframework cations is also higher for the hybrid functionals [281]. Adsorption energies for CO and NO in metal-exchanged chabazite are linearly correlated with the stability of the cation location, and the values obtained with hybrid functionals are closer to experiment than obtained with GGA [281]. The same conclusion holds also for bond length and stretching frequency relationship. The frequency values calculated with GGA for Cu(I)-chabazite are more accurate, but, on the other hand, the red-shift values are better for hybrids. Contrary to GGA, hybrid functionals correctly predict a blue-shift of the CO stretching mode upon adsorption, due to the more accurate prediction of the HOMO-LUMO gap of the molecule and the band gap of the zeolitic host [281]. The exaggeration of the exchange splitting and hence to the too wide HOMO-LUMO gaps in hybrids lead, however, to the less accurate description of the system with high spins. The use of hybrid functionals also overestimates the wavenumbers which can be corrected, in pragmatic although not elegant way, by scaling [281].

The Fe-ZSM-5 zeolite, exhibiting catalytic activity in NO decomposition, has been also studied computationally, e.g., by Heyden et al. [282] (B3LYP functional, results comparable to the cited calculations at the CASSCF-MCQDPT2 level of theory), who studied as many as 46 different surface species and 63 elementary reactions. Based on the Landau–Zener theory, they concluded that the correction stemming from the spin inversion probability was smaller than the immanent DFT error.

The influence of two zeolite models of different size, the T1 ($[Al(OH)_4]^-$) and the M7 (six silica tetrahedra and one aluminium oxide tetrahedron) structures, on the $Cu^{+/2+}$ cation properties in NO activation was studied in the article of Kozyra and Piskorz [283]. There the NOCV analysis was used with the following fragment selection: (T1/M7-$Cu^{+/2+}$) and (NO) to elucidate the influence of the zeolite-cation electron donation/backdonation on the cation-NO charge flow. Such analysis has been done before for the interaction of Ag and Cu sites with ethene, ethyne, formaldehyde [284], benzene [285], and NO [286, 287] (Turbomole, B3LYP functional, def2-TZVP basis set, QM/MM scheme in QM-Pot [288] code). The geometric, energetic, and spectroscopic (EPR) studies of TMI at ZSM-5 were also performed by Pietrzyk et al. [289] They used the VWN functional (with correction of BPW91) and M5 and Z6 cluster models of framework. The following TM ions were used: Mo^{5+} (configuration d^1, term ^{2}D), Fe^{3+}, Mn^{2+}, Cr^+ (d^5, ^{6}S), Fe^{2+} (d^6, ^{5}D), Co^{2+} (d^7, ^{4}F), Ni^{2+} (d^8, ^{3}F), Ni^+, Cu^{2+} (d^9, ^{2}D), and Cu^+, Zn^{2+} (d^{10}, ^{1}S). The authors of [289] conclude that essentially two mechanisms of N_2O decomposition can be distinguished: the electron transfer, when the transient N_2O^- is formed and the $N_2{-}O^-$ bond is cleaved, and the oxygen atom transfer, facilitated by the spin catalysis process. The activation of NO molecules can follow through either the oxidative (nitrosonium) or the reductive (nitroside) complex.

Methane Activation

The other catalytic issue, commonly represented in literature, is the methane activation on zeolitic systems. The review, augmented with own DFT+*D* calculations, on methane activation on TH-exchanged zeolites was written by Kulkarni et al. [258] They summarise the impact of TM cation, zeolite framework, and active site motif on the calculated energy barrier of C–H bond activation.

The computational study of Benco et al. [249], concerning the extraframework Zn^{2+} Lewis site, revealed the relative stability of Al substituting Si in non-equivalent positions of mordenite and the stability of the extraframework Zn^{2+} cation (double Al substitution). It was found that the most favourable Al substitution is for the T2 and T3 sites, the least is for T4 site, and the preferential location of Zn^{2+} is in the small ring (5MR, 6MR); see Fig. 10.

The issue of partial methane oxidation covers the wide range of zeolitic systems, beginning with Fe/ZSM-5 [290, 291], Cu/ZSM-5, FeCu/ZSM-5 [292], Fe/FER [293], and Cu/MOR [268].

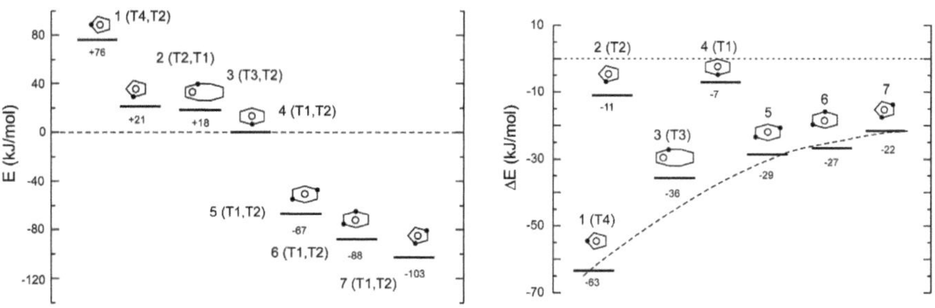

Fig. 10 Left: The stability of Zn-MOR structures for large Al-Al distances (structures 1–4) and for short Al-Al distances (structures 5–7). Symmetries of Al sites are given in parentheses. In structures 1–4, the Zn^{2+} cation is connected to the Al site indicated first. Full (empty) circles in the sketch of the zeolite structure show the location of the framework Al (extraframework Zn) atom. Right: Dissociation energies of H_2 on Zn-MOR. For all configurations compared in $\Delta E = E^{tot}_{diss} - E^{tot}_{ads}$ is displayed. The negative values show that dissociative adsorption stabilises all configurations irrespective of the position of the Zn^{2+} cation. Reprinted with permission from [249]. Copyright 2005 American Chemical Society

Hydrocarbon Transformation

The hydrocarbon transformation on zeolites was studied computationally for, e.g. process of ethyne cyclotrimerisation ([294], geometry optimisation and vibrational analysis with B3LYP functional; simulation of ^{13}C NMR spectra by B3LYP functional, GIAO, TZVP basis set with ECP). The article of Archipov et al. [295] aims at the understanding of benzene sorption in CuHY zeolite as a key step in the oxidation of benzene to phenol, using the QM-Pot approach. They conclude that benzene molecule η^2-coordinated to copper; after preferential adsorption of Cu to SIII the coordination to framework oxygens is reduced, although the differentiation between two adsorption sites, SII and SIII, is difficult based on the IR spectrum. The π-backdonation from the Cu^+ $3d$ is more dominating than p-π donation from the benzene molecule and the oxidation state of copper changes from Cu(II) to Cu(I). Benzene adsorbs both at Brønsted sites and at Cu centres.

Sorption/Diffusion

Besides the reactivity, the theory and modelling of diffusion in zeolites have gained sound attention in literature [296–300]. Such properties, either static or time-dependent, require an enormous number of the energy evaluations so, as such, need the very fast methods, typically the classical force field. The simulations help resolving non-trivial phenomena like non-Langmuir adsorption isotherms, adsorption/desorption hysteresis loops, or faster diffusivity of certain substituted hydrocarbons [298].

3.1.5 Spinels of *d*-Electron Metals

Spinel oxides belong to scientifically and technologically most attractive oxides with remarkable record of widespread applications in heterogeneous catalysis [301–303], energy storage and conversion [304, 305] or sensor devices [306, 307]. In particular, cobalt spinel oxides derivatives exhibit high activity in low-temperature decomposition of nitrous oxide [308–311], and $CoNi_2O_4$ was reported to be highly active in CH_4 combustion. Mixed Co–Fe–Al spinels, with optical band gaps between 1.6 and 2.0 eV, generate strong p-type photocurrent for photoelectrochemical splitting of water, whereas $ZnCo_2O_4$ [312] or $CoMn_2O_4$ [313] has a good cycling capacity in the Li^+ charge–discharge processes.

Structure and Morphology

Spinels form a wide family of oxides which crystallise in the cubic crystal system of $Fd3m$ (or $F\bar{4}3m$ [314]) space group and are characterised by the general AB_2O_4 formula. The oxygen anions form the face-centred cubic lattice, where the tetrahedral

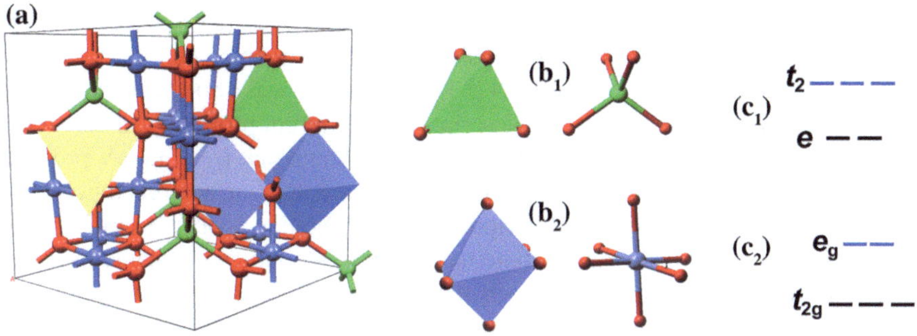

Fig. 11 Cubic 8 Co_3O_4 unit cell (**a**) of spinel together with TM cations in tetrahedral (**b**$_1$) and distorted octahedral (**b**$_2$) coordination environment. The crystal field splitting imposed by *ted* (**c**$_1$) and *oct* (**c**$_2$) environment

(*ted*) position can be occupied by, e.g., divalent "A" ions (Mg, Fe, Ni, Mn, Co, and Zn), whereas the octahedral (*oct*) position—by trivalent "B" ions (Co, Al, Fe, Cr, or Mn)—such way of site occupation defines the normal spinel ($[A]^{ted}[B_2]^{oct}O_4$); see Fig. 11. The inverse spinel structure is defined by formula: $[B]^{ted}[AB]^{oct}O_4$.

Other combinations incorporating di-, tri-, or tetravalent cations, including Co, Mn, Zn, Fe, Cr, Ti, and Si, are also possible; thus, the spinel structure can be regarded as versatile matrix being able to accommodate a wide range of metal cations and oxidation states [315, 316] what implies their high catalytic activity in the number of processes. Moreover, spinel properties can be precisely tuned being still within the same generic structure.

It should not be overlooked that in the quantum-chemical modelling, the flexibility of the substitution of the transition metal cations by the electronically hard main group cations (e.g. $Co^{3+} \leftrightarrow Al^{3+}$ or $Co^{2+} \leftrightarrow Mg^{2+}$) in the spinel structure, due to the similar ionic radii of the appropriate counterparts, allows for the enforcement of the electron density distribution among ions thus allowing for the introduction of the desired oxidation states (formally the localisation of the electronic defect—electron or hole) for selected cations. Such process can be done in a polaronic way.

Computational Issues

Taking into account the moderate size of spinel conventional unit cell (56 ions: $A_8B_{16}O_{32}$) most of the bulk-related calculations is conducted within periodic approach, with DFT-based methods. Calculations may be performed for as-yet-hypothetical structures what can be useful in the prediction of novel phases that could be obtained by driving the system through a structural transition [317].

Most of the existing and hypothetical spinel-type oxides, however, exhibit complex atomic and magnetic structure which is a challenge for the standard LDA and GGA-DFT methods, and a typical problem is the error arising from the inappropriate description of the Coulomb on-site repulsion [318]. In the case of spinels, the choice of the Hubbard U value has been systematically investigated [319–322], and it has been shown that computationally expensive hybrid functionals exhibit little advantage over the properly selected GGA+U [323, 324]. It was shown, however, that the U parameter is not universal for description of different properties of calculated TMO systems. In particular case of Co_3O_4, it was shown that $U = 3.0$ eV provides a better overall description of the electronic structure and surface reactivity while $U = 5.9$ eV is better suited for description of the magnetic properties [320]. Generally, LDA calculations underestimate the spinel oxide lattice constants by 1–2% relative to the experimental values, whereas GGA functionals overestimate lattice parameters [325]. LDA describes the total energy of bulk oxides rather properly, however, fails severely when employed for bond energy calculations leading to errors of several tens of per cent [326]. The GGA functionals generally yield much better bond energies, surface energies, and adsorption energies. It has been reported [327] that LSDA overestimates exchange interactions, possibly due to the overestimating of *p-d* hybridisation [317, 328] whereas GGA functionals generally give better values [329]. Since the value of U is usually fitted to other observables such as band gap or the atomic magnetic moments, the computed exchange interactions can diverge significantly from exact values [317].

Contrary to pure DFT, hybrid functionals have shown excellent results [330, 331]. While in pure HF calculations, the band gaps are overestimated, a hybrid approach yields reasonable band gaps. More involved methods for an improved treatment of electron correlation, e.g. the Møller–Plesset (MP2) expansion, quantum Monte Carlo (QMC) approaches, or dynamical mean-field theory (DMFT) are currently too computationally demanding to be generally applied to complex-structured oxides and are found in practice only in cluster calculations [332].

The mere calculations of total energy of the unit cell systems can be used to distinguish between the normal ($[A]^{ted}[B_2]^{oct}O_4, \lambda = 0$), inverse ($[B]^{ted}[AB]^{oct}O_4, \lambda = 1$), or mixed ($[A_{1-\lambda}B_\lambda]^{ted}[A_\lambda B_{2-\lambda}]^{oct}O_4$, $0 < \lambda < 1$) spinel structure. The degree of inversion, λ, is governed by several factors, e.g. the cation radius ratio, Coulomb interactions between the cations, and crystal field effects of the octahedral site preference energy of cations [333]. The total energy calculations allow for assessing the magnetic ordering of the paramagnetic centres, e.g. antiferromagnetic ordering in cobalt spinel [334].

It was shown that DFT-based methods are able to reproduce the partially inverse $MgAl_2O_4$ spinel Raman spectra with high accuracy, and to solve some of the uncertainties in peak attributions [335]. In case of cubic $LiMn_2O_4$, the calculations helped to understand transformation between cubic and orthorhombic structures [336]. The first-principles calculations may also provide insight into influence of pressure on spinel oxide structural. In such way, the prediction of structural and phase transitions upon pressure was studied in case of $ZnAl_2O_4$ and $ZnGa_2O_4$ with the excellent reproduction of experimental trends [337]. As reported by Price et al., the sequence of four transitions with increasing pressure was predicted for the cubic spinel phase of calcium ferrite [338]. The more involved characteristic of bulk spinel oxides such as detailed electronic properties, optical properties, or superconductivity, which involve excited-state properties or high accuracy of the description of electronic correlation are usually beyond the validity of density functional theory and the methods like GW are used.

The fact that distinct surface planes of oxides differs in concentration and coordination of the exposed ions (active sites) is in line with the general observation that the catalytic activity of the oxide catalysts depends on the grain size and surface morphological features which together influence the nature of the active sites exposed by the spinel samples [339, 340]. The {110} facets of Co_3O_4 nanocrystals have higher catalytic activity for CO oxidation than the {100} and {111}, which is reasoned by the high concentration of octahedral Co^{3+} sites [341]. Contrary, for the ORR reaction, the predominant {111} and {100} surfaces exhibit higher catalytic activity attributed to the exposition of Co^{2+} cations [342]. On the other hand, in the case of methane combustion, the catalytic performance of the Co_3O_4 is believed to be related to the surface energy [343].

Owing to present state of computational chemistry, fairly accurate first-principles modelling of surface structure and energetics made the theoretical predictions of the resultant morphology for faceted nanocrystals possible [344]. By employing the Wulff construction along with the ab initio thermodynamics, the influence of chemical environment can also be taken into account [345]. The mixed cobalt spinels were studied computationally, and the predicted nanocrystal shapes were in good agreement with experiment [334, 346]. Such environmentally independent morphology prediction may be treated as the first-order approximation of ambient conditions and, *via* the first-principles thermodynamics, the inclusion of entropic corrections and polarisable continuum (DFTsol) leads to more realistic shape modelling. Such approach was recently employed for cobalt oxide which high catalytic performance in redox processes attributed usually to the presence of reactive oxygen species (ROS) and to lability of lattice oxygen. The latter can be released with formation of surface oxygen vacancies even under mild conditions [347, 348].

In this context, the influence of oxygen rich and lean conditions on different terminations of cobalt spinel most abundant surfaces was studied [323, 349–351]. For most abundant (100) surface, both reactive oxygen species [350, 352] and oxygen vacancies [349, 353] were studied by means of DFT combined with ab initio thermodynamics revealing that three principal states of the spinel surface in function of temperature correspond to oxygen adsorption ($T < 350-400\,°C$), bare surface region ($350-400\,°C$ to $550-700\,°C$), and oxygen release region ($T > 550-700\,°C$).

Recently, similar approach was employed for (111) surface, revealing that the ROS species and vacancies temperature stabilities region are shifted towards higher temperatures in this case [354]. As for (110) surface in contact with gaseous O_2, Selloni et al. presented a comparative study of and using DFT+U calculations with different U values, discussing influence of Hubbard parameter on structural parameters, electronic properties, and surface energetics [320]. Surface energy calculations indicate that the (110)-A termination is more stable in a wide range of the oxygen chemical potential [355], in agreement with surface science experiments [345, 356].

Catalytic Activity

Due to their unique properties in redox catalysis, spinels are often regarded as low-cost and stable substitutes of noble metals.

As in any TM oxides, the defects in the spinel-type oxides are crucial in the electrocatalysis and photocatalysis activity [357]. It is well accepted that creation of oxygen vacancies can improve the electron conductivity of oxides [358] and that such defects are involved in catalytic activity in MvK mechanism [359]. In this context, it is beneficial that for many spinel-type oxides the concentration of oxygen vacancy may be controlled easily by O_2 partial pressure or the synthesis temperature [360, 361]. Non-intrinsic defects are often caused by external dopants such as substitutions of A and B sites with different cations, thus the cation-doped spinel oxides can be expressed as $A_{1-x}A'_xB_{2\text{-}y}B'_yO_4$ ($0 \leqslant {}^x/_y \leqslant 1$). The possibility of feasible non-intrinsic defect introduction provides the powerful tool for improvements of the catalyst performance by bulk modification (doping with alien cations) [362, 363] or by tuning the surface properties of the catalyst with alkali promoters [364, 365].

To address the phenomenon of CO oxidation at temperature as low as −77 °C, Wang et al. [366] studied computationally the commonly exposed faces of Co_3O_4 and concluded that the MvK mechanism was preferred and Co^{3+} was the active site. Furthermore, the high reactivity of low-coordination oxygen ions was attributed to their weak binding to the substrate lattice [367]. The MvK mechanism was also assessed computationally (DFT, PBE functional) in the CO oxidation by N_2O over Co_3O_4 (110) surface [368]. To explain spectacular activity of Ni-doped cobalt spinel in CH_4 combustion, Hu et al. investigated a complete catalytic cycle for methane combustion on the Co_3O_4(110) surface and compared it with that on the Co_3O_4(100) surface on the basis of first-principles calculations [369]. The conditions of Langmuir–Hinshelwood and MvK mechanisms in CH_4 oxidation over cobalt spinel were shown computationally, and the continuous Co partial reduction adapts to the catalyst redox state to the current thermodynamic conditions [359]. It was shown [370] that in the methane-to-methanol process, the surface Co–O pairs were the active sites, where the two ions provide a synergistic effect for the first C–H bond activation to yield surface Co–CH_3 and O–H species. Methanol oxidation on cobalt spinel (110) and (100) surfaces was studied by means of spin-polarised density functional theory with the GGA+U framework [371], and the results indicate that CH_3OH can adsorb to surface lattice oxygen atom (O_{2f}/O_{3f}) to form Co–O bond directly, and the adsorption of CH_3OH and its decomposition products on (110)-B is more stable than on (111)-B, which means CH_3OH prefers Co^{3+} rather than Co^{2+}.

Adsorption and oxidation of NH_3 were studied on $ZnFe_2O_4$ [372] and Co_3O_4 [373, 374] spinels. In the first case, it was shown (PW91 functional) that NH_3 molecule preferred to adsorb on the surface Zn atom over the spinel $ZnFe_2O_4$(110) surface with adsorption energy of 203.125 kJ/mol. Such stable configuration of H_3N@Zn corresponded to high activation degree of NH_3 with pronounced electron transfer (−0.157 e) and two N–H bonds elongated significantly. For cobalt spinel, the successive dehydrogenation of ammonia on (110)-B termination was studied by means of spin-polarised DFT within the PBE+U (U_{eff} = 3.3 eV) framework and the calculated energy barriers of NH_3, NH_2, and NH dehydrogenation are in the range of 29–67 kJ mol^{-1}, indicating a high activity of the surface towards dehydrogenation and that the surface reactivity is attributed to the presence of a low-coordinated surface lattice site.

The activation of gas-phase oxygen was addressed by modelling of the isotopic oxygen exchange reaction ($^{18}O_2 + ^{16}O_2 \rightarrow 2\,^{16}O^{18}O$) over cobalt spinel (100) surface in time- and temperature-resolved catalytic experiments [352]. Overall process was divided into four elementary steps, and the corresponding reaction barriers (both electronic and entropic contributions) were calculated by means of periodic GGA+U and atomistic thermodynamics. The mobility of lattice oxygen (crucial in MvK mechanism) was also studied [349, 354], showing that DFT+U calculated vacancy formation energies conjoined with first-principles thermodynamics are able to reproduce results of TPD-O_2 experiment over cobalt spinel [83]. In this context, it was also reported that oxygen vacancy formation is generally much easier on $NiCo_2O_4$(001) than on Co_3O_4(001) surfaces, suggesting that the former oxide may be a better catalyst for oxidation reactions based on the MvK mechanism [375].

Low-cost catalytic decomposition of N_2O into N_2 and O_2 is a subject of intensive investigations due to its harmful impact on the environment [256, 376], and the kinetic of N_2O decomposition over oxide material containing transition metal ions has been extensively investigated to find the correlation between the electronic structure of the catalysts (characterised essentially by their electron donor properties) and their reactivity [377–379]. The most promising performance was observed so far for cobalt spinel-based catalysts [308, 380]. The energetics of the postulated elementary steps (N_2O adsorption, N_2O activation through dissociative electron or oxygen atom transfer, adoxygen surface diffusion and recombination and dioxygen desorption) was calculated using a cluster approach [310]. It was concluded the N_2O activation and the formation of dioxygen are energetically most demanding steps, whereas the barrier for the oxygen surface diffusion was distinctly smaller. The effect of potassium promotion on *de*N_2O activity of various 3*d* electron spinels (Mn_3O_4, Fe_3O_4, Co_3O_4) [381] was investigated in terms of periodic DFT-*r*PBE molecular modelling of a surface dipole model ($K^{\delta+}$–$O_{surf}^{\delta-}$) whose formation weakened the interfacial potential and increased the Fermi energy leading to the decrease of the calculated spinel work function by 0.45 eV (very close to the experimental $\Delta\Phi = 0.5$ eV).

To better understand the hydrogenation mechanism, Lu et al. [382] modelled the reaction steps using DFT+U+D. They have demonstrated that H_2 dissociation is a complicated two-step process, followed by the surface diffusion of H adatom. The calculations reveal that the presence of the surface oxygen vacancy facilitates the stepwise hydrogenation of ethylene (drop of activation energy from 1.19 to 0.47 eV) due to the weaker bond strength of formed OH group.

Spinel-type oxides are promising also in the ORR process because they have excellent electrical conductivity originating from the electron hopping between different valence states of metals in octahedral sites, and in this way, they provide active metal centres for O_2 adsorption and activation in ORR.

Recently, Si et al. [383] performed DFT-PW91 calculations to determine the structure and energy of O_2, HOO*, O*, HO*, HOOH, H_2OO, and CH_3OH molecules adsorbed on $Mn_2AlO_4(001)$ and on Pt(001). The results together with electrochemical measurements reveal that the ORR on the Mn_2AlO_4 catalyst is a four-electron process and although Mn_2AlO_4 is less active than Pt-based catalyst, the former has much better methanol tolerance. The similar computational approach was used to study the bifunctional (both OER and ORR active) spinel-type MFe_2O_4 oxides (M = Co, Mn, Ni) showing that the ORR on MFe_2O_4 is a direct four-electron pathway and that $CoFe_2O_4$ exhibits the best ORR/OER activity among tested oxides [384].

To conclude, an easy control of parameters such as composition, structure, morphology, and valence state made the spinel-type oxides a suitable catalysts for numerous reactions, particularly, the NO_x reduction [385], CO oxidation [386], CO_2 reduction [387], hydrogen evolution reaction [388, 389], ORR and OER [333], NH_3 oxidation [390], formaldehyde oxidation [391], methane combustion [392], alcohols oxidation [393, 394], H_2O_2 decomposition [395], urea oxidation [396], NH_4ClO_4 decomposition [397], and methylene blue degradation [398].

3.1.6 Vanadia V_2O_5

Vanadium oxides belong to an important class of materials with diverse temperature-dependent electronic, magnetic, and catalytic properties [399–401]. The significant electron–lattice interactions and electron–electron correlations which control most of the V_xO_y properties [402] together with the structure-related characteristic have stimulated vanadia development for a large number of applications, particularly as a smart material for energy, sensors, optoelectronics, and catalysis. The large variety of stable and metastable types of vanadium oxide exist; however, thermodynamic consideration reveals that relevant phases, in terms of stability and formation likelihood, are V_2O_3, V_7O_{13}, V_6O_{13}, V_2O_5, and VO_2 [402]. It is also proved that depending on the ambient conditions and temperature, the phase transformations between these oxides and metal-to-insulator transitions of some vanadia phases can occur [399]. Apparently, such complex structural and electronic transformations may be responsible for behaviour of vanadia-based systems in surface science applications.

Despite the vast number of possible vanadium oxides, V_2O_5 has been paid special attention and is nowadays the most studied V_xO_y system, not only in terms of catalytic applications (which are discussed in more detail below). V_2O_5 is well-established chromogenic material, which is able to respond to external stimuli such as electromagnetic radiation, temperature, and electrical charge [403, 404]. In electrochemical applications, V_2O_5 is used for fabrication of hybrid electrodes for supercapacitors [405, 406], Na-based pseudocapacitors [407], and alkali batteries with high capacity and cyclability [408]. The layered structure (see below) of V_2O_5 gives rise to its highly anisotropic transport properties used for the development of thin film transistors and related electronics [409]. The room temperature surface reactivity of vanadium pentoxide is responsible not only for its sensing properties reported for ethanol [410] and ammonia [411] molecules, but also for significant photocatalytic activity [412, 413].

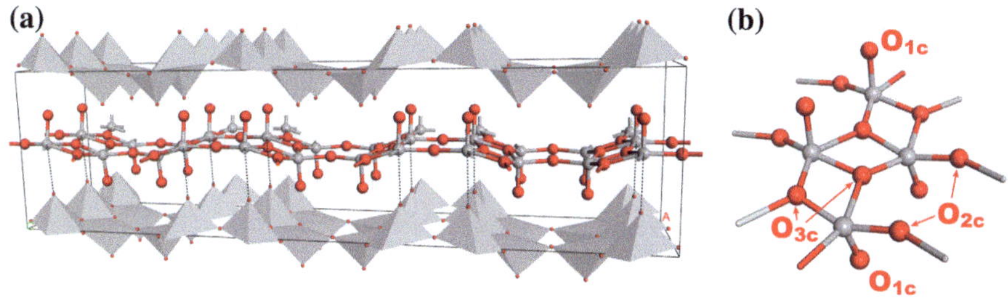

Fig. 12 Structure of V_2O_5

Structure

The vanadium pentoxide belongs to the family of the layered vanadium oxides, known as the Wadsley phases, which comprises also VO_3, V_3O_7, V_4O_9, and V_6O_{13} with general formula V_nO_{2n+1} with $n = 1, \ldots, 6$ [402]. The V_2O_5 itself features eleven structures, among which the most stable is the orthorhombic lattice crystallising in the $Pmnm$ symmetry (space group no. 59, Schoenflies symbol: D_{2h}^{13}) with the unit cell of the following parameters: $a = 11.512$ Å, $b = 3.564$ Å, $c = 4.368$ Å, crystal density = 3.4 g/cm³ [414], containing two stoichiometric units ($2 \cdot V_2O_5$).

Its layered structure is shown in a supercell (see Fig. 12a) where the middle layer is represented by the ball and stick, whereas the top and bottom layers are shown in polyhedral representation. The layer is built up from alternating up-up-down-down motifs of distorted [VO_5] square pyramids which share both corners (motifs pointing in the same direction) and edges (motifs pointing in opposite direction) with neighbouring units. The bounding between layer is realised by the van der Waals' and electrostatic interactions (presented as dashed line), which define a quasi-octahedral coordination environment for the V atoms upon taking into consideration a long $V \cdots O$ bond with the distant oxygen atom in the next layer.

The detailed coordination of vanadium is better revealed in Fig. 12b, where the structure fragment is presented. Vanadium forms one bond perpendicular to the (001) plane and four V–O bonds in the (001) plane. The VO_5 subunits, sharing edges, are mirror-inverted resulting from the V^{5+}–V^{5+} repulsive interaction. In this way in bulk V_2O_5, there are three structurally different oxygen centres, terminal (vanadyl) oxygen, O_{1c}, coordinated to V atom through a rather short bond ($d_{V\text{–}O} = 1.58$ Å) and bridging oxygen atoms, O_{2c} and O_{3c}, two and three coordinated, respectively, and characterised by V–O distances ranging between 1.78 and 2.01 Å.

It is generally accepted that under ambient conditions, hydrated vanadia clusters are present on oxide supports whereas under dehydrated conditions, the supported vanadia phases consist of isolated and polymeric surface VO_4 species [415]. The monomeric vanadyl species are likely present on the catalyst surface at low vanadium loadings, while dimeric and polymeric vanadium oxides and crystalline V_2O_5 can be formed on the surface at relatively high vanadium coverages. It is also well known that the most beneficial, in terms of catalytic activity, is a close to monolayer coverage where all the support is covered by dispersed oxide species and also some V_2O_5 nanoparticles can be present [415, 416].

Such insights into the V_2O_5 status in supported catalytic systems allow for identifying the active phases, thus enable the molecular engineering and tuning by controlling the number of catalytic active sites on the support and their specific activity [417–419].

Catalytic Activity

The catalytic properties of vanadia and vanadia-based materials are mainly selective oxidation reactions with the generally accepted “nucleophilic” mechanism in which lattice oxygen participates in the reaction by being incorporated into the organic species [420]. The resulting oxygen vacancies may be refilled by dioxygen following the MvK mechanism, or the defect concentration exceeds a critical value, resulting in the vanadia surface topology change [421, 422]. Thus, supported vanadium oxide catalysts have found extensive applications as oxidation catalysts in the chemical, petroleum, and environmental industries [415]. An SiO_2-supported vanadia catalyse the oxidation of sulphur dioxide to sulphur trioxide in the production of sulphuric acid [423, 424]. Supported V_2O_5/TiO_2 system, consisting of a 2D surface vanadium oxide phase strongly interacting with the TiO_2 support, represents a new class of catalytic materials [425, 426] with large number of applications. It is used for selective oxidation of o-xylene to phthalic anhydride, selective oxidation of alkyl pyridines to nicotinic acid and selective catalytic reduction (SCR) of NO_x emissions with ammonia or urea [426]. Commercial V_2O_5-WO_3/TiO_2 shows good catalytic activity because of its highly stable activity for NO_x removal and low sensitivity to SO_2 poisoning [427, 428]. It has also been found to activate oxidative destruction of chlorinated hydrocarbons [429] and oxidation of elemental mercury to mercury oxides that are easier to trap. The V_2O_5/Al_2O_3 catalysts were applied for oxidative dehydrogenation of propane to propene [430]. Vanadium-based catalysts supported on sulphated or tungstated ZrO_2 were used for SCR of NO showing improved potassium poisoning resistance compared to traditional titania-supported system [431]. Thorough understanding of the catalyst active phase on the molecular level may be achieved only by cooperation of surface science experiments and molecular modelling techniques.

Computational Issues

In the work of Kempf et al. [432], the periodic structure of V_2O_5 has been optimised with employment of Hartree–Fock method, with good agreement for the structural parameters and for the vanadyl bond stretching force constant, but with much too low binding energy calculated. The latter parameter may be improved by the employment of the GGA-based DFT [433]. In the LDA calculation [434], the resulting bulk structure was well reproduced—the atomic coordinates agree up to 0.01 Å, and the unit cell parameters are underestimated by $\sim$1.25% only. Sauer and Vyboishchikov reported the BP86/D(T)ZVP results for bulk V_2O_5 which reasonably reproduce both cell parameters (within 2%) and atomic coordinates (within 2 pm) within the V_2O_5 layer [435]. The interlayer V–O distance, however, was overestimated by 10 pm, which may stem from its intermolecular character, and a proper description may therefore be a challenge for DFT calculations. The strength of the interaction between the individual layers in the crystal is very low, a few kJ/mol per V_2O_5 unit [433, 435].

For V_2O_5, the natural cleavage plane is (001) surface characterised by the lowest surface energy thus highest stability comparing to other facets of V_2O_5. It was studied by means of periodic methods and the structural and electronic properties of the (001) surface, calculated as single-layer slab, are close to those of the bulk [433, 434] in good agreement with recent experimental results [436, 437]. It was also reported that the vanadia surface is mostly ionic with the vanadyl bond V–O1c (for labelling, see Fig. 12b) being mostly covalent and the atomic charges on oxygen atoms decrease with their coordination: O3c > O2c > O1c for both bulk and (001) surface. This may suggest a large reactivity of the threefold coordinated oxygen. On the other hand, the pDOS shows that vanadyl oxygen levels are much closer to the Fermi level than those of O2c or O3c. Therefore, a large reactivity can be expected for O1 site [433]. A metal-to-insulator transition of the V_2O_5(001) surface was studied by DFT and Monte Carlo simulations corroborated with in situ STM and band gap mapping, derived from scanning tunnelling spectroscopy data. The transition restricted to the surface layers was observed at 350–400 K and occurred anisotropically across the surface with the formation of vanadyl oxygen vacancies preferentially growing along the vanadyl rows. Further heating to 800 K leads to irreversible surface reduction which proceeds sequentially $V_2O_5(001) \rightarrow V_6O_{13}(001) \rightarrow V_2O_3(0001)$, as observed by high-resolution STM [399]. Witko et al. reported theoretical results for (010) and (100) which are characterised by stronger relaxation effect than (001) facet. Calculated surface energies were used to obtain a Wulff shape which highly resembles the equilibrium shape of real V_2O_5 observed by SEM. It was concluded that even both (100) and (010) surfaces constitute only 15.5% of the total surface area, they cannot be excluded from the catalytic processes' studies.

Interface Modelling

The issue of vanadia dispersion on support was addressed by gas-phase cluster calculations by Sauer et al. [438], where the neutral $(V_2O_5)_n$ clusters ($n = 1, \ldots, 12$) were studied by DFT-GGA calculations and it was shown that gas-phase clusters are very different from the layer structure of the solid bulk since vanadium is fourfold coordinated and threefold coordinated oxygen is avoided. For small clusters, the doubly-bridged $OV–O_2–VO_2$ structure is observed, whereas for larger models the polyhedral oligomeric species with vanadyl groups at the corners and V–O–V bonds at the edges are expected. Similar geometrical parameters for the vanadium clusters were also reported by Calatayud et al. [439] (B3LYP) together with topological analysis of the Electron Localisation Function. The latter method was used to understand the nature of bonding which showed that larger clusters are energetically favourable rather than the smaller V_xO_y species and that the V–O bond can be considered as mostly ionic.

The thermodynamic stability of dimeric V_2O_5 species on anatase (100) and (001) perfect surfaces at low coverage was studied by Calatayud et al. [440] with conclusion that relaxation of the outermost surface layers leads to stabilisation of the covered slabs, especially for the (001) surface, and seems necessary for the correct description of the adsorption. The comparison of the electronic properties of the model active centres for silica- and titania-supported vanadium catalysts was reported by Avdeev et al. [441], and the results reveal that the weak interface bond is realised on SiO_2 whereas for TiO_2 the strong bonding is expected. The latter results from the structural correspondence of TiO_2 and VO_x which promotes the formation of polymeric oxovanadium groups on the carrier surface by the isomorphic substitution of Ti for V. Additionally, the TD-DFT method was used to calculate the energies of the excited states and optical spectra of the adsorbates which enabled for the identification of the frequencies typical for monomeric and dimeric oxovanadium-supported species. It was revealed that such amorphous catalyst exhibits increased acidic power compared to the pure V_2O_5 and that the V ions embedded in the lower layer are of great importance in this context. The possible weak and strong bonding between TiO_2 support and V_2O_5 monolayer were also investigated by ab initio methods [442]. In the former case, the vanadia monolayer maintains its crystallographic structure; however, the weak interaction is strong enough to induce a charge redistribution responsible for vanadium acidity enhancement. In the second model, the presence of the anatase support alters significantly the structure of the V_2O_5 monolayer, resulting in a much higher epitaxy. It was also observed that reactivity of the oxygen sites in H adsorption for the different catalyst models correlates well with the frontier orbital levels of the metal oxide. The stability of reduced and reoxidated VO_x species (following the MvK cycle) on the reconstructed TiO_2(001) surface has also been theoretically investigated [443]. It was revealed that the low binding energy of vanadyl oxygen species is related to the structure relaxation of the surface VO_x species within a monolayer and that the stable peroxide species V(O–O) can be formed on the fully oxidised VO_x/TiO_2 surfaces. Taking into account that functional groups V=O and V–O–V and their reduced forms V–OH and V–OH–V are the main components of the active sites in VO_x/TiO_2 catalysts [444, 445], Avdeev and Tapilin studied computationally the effect of water on the electronic structure of active sites of supported vanadium oxide catalyst [446]. They show that H_2O dissociates readily on the reduced vanadium sites forming the stable surface OH groups which are then reoxidated by gas-phase O_2 yielding hydroxylated monolayer. The calculated hydroxylation–hydration reaction pathway reveals that the lattice oxygen of the surface vanadia species is exchanged with water oxygen rather than with O_2.

More involved V_2O_5 support system of ultrathin SiO_2/Mo(112) film was described theoretically in the paper of Todorova et al. [447] where low-coverage differently anchored vanadia species (monomers and dimers) and clusters have been investigated by means of first-principles thermodynamics. It was revealed that at low V chemical potentials, monomeric species are stable and with increasing V or

O_2 chemical potentials, dimeric VO_x species and then large clusters (V_6O_{15}) are expected. The characteristic feature of all these stable species is the presence of surface vanadyl groups. Importantly, all vibrational frequencies observed for the vanadia/SiO_2/Mo(112) model system were reproduced theoretically for the V_4O_{10} cluster anchored to the silica surface by two V–O(2)–Si interface bonds.

Vanadium oxide catalysts are frequently used in selective oxidation reactions, and it is well established that their catalytic properties depend strongly on their ability to provide surface oxygen as a reactant following the MvK mechanism. Since V_2O_5 possesses two kinds of V–O–V bonds and a vanadyl V=O bond, several authors have tried to relate experimentally the activity and/or selectivity to specific bonds of the oxide [448–450]. However, despite extended structural and spectroscopic studies, the role of bond types was not elucidated by experimental approach. In this context, the theoretical investigation of V_2O_5 lattice oxygen stability was reported.

Hermann et al. [420] reported the cluster studies of the different oxygen vacancies at the V_2O_5 (010) surface, showing that removal of oxygen is energetically demanding while the pre-adsorption of hydrogen can facilitate this process substantially. The oxygen extraction leads to the local relaxation affecting interlayer binding which may lead to major reconstruction of the surface as proposed from experiments [451]. The electronic structure analysis reveals that the VO formation results in chemical reduction of neighbouring vanadium centres, which is expressed by increased vanadium 3*d* states occupation. The effect of the support for vanadia-based catalysts in terms of the V=O bond dissociation energetic was also tested [452], and it was shown that both polyhedral oligomeric vanadia species on silica and isolated vanadia species on α-Al_2O_3 have similar V=O bond dissociation energies (250 to 300 kJ mol^{-1}). These findings explain why the vanadium oxide particles supported on Al_2O_3 films are easily covered with vanadyl groups even at UHV conditions [453]. The computational comparison of the reduced states of different phases of V_2O_5, γ-V_2O_5(001) and α-V_2O_5(001), was performed for wide range of defect concentrations, (1/6 ML to full ML of defects), with the main conclusion that in reducing environment the α-V_2O_5(001) surface would be fully reduced, whereas the γ-V_2O_5-(001) surface is only partially reduced [454] and, still exposing the surface vanadyl oxygen groups, may be more reactive in such conditions.

The formation of oxygen vacancies in H_2-rich environment was also studied [455], and it was revealed that O_2 does not adsorb on the fully oxidised surface of the catalyst but reoxidises oxygen vacancies. It was also shown that nucleophilic oxygen $O^{2-}_{(s)}$ is present above 550 K on the catalyst surface and is probably responsible for the propene formation in the propane oxidative dehydrogenation (ODH) reaction. The formation of oxygen vacancies on less abundant low-index surfaces of V_2O_5 was studied by Goclon et al. [456], and it was concluded that although in equilibrium only about 15% of the surface area of a crystallite consists of (100) and (010) facets, these surfaces will make considerable contribution to the activity of V_2O_5 as an oxidation catalyst.

NO_x *Catalytic Reduction by* NH_3

The molecular aspects of vanadia-based catalysts performance in selective reduction of NO by NH_3 were also studied by quantum-chemical modelling. Using V_2O_9H model cluster, Gilardoni et al. postulated that the reaction is initiated by NH_3 adsorption on a Brønsted site with formation of NH_4^+ surface cation [457]. Afterwards, gaseous NO reacts with such activated ammonia yielding NH_2NO intermediate and then the reaction products N_2 and H_2O. The computational results agree with isotopic labelling studies giving support to a dual-site Eley-Rideal-type mechanism. Quite similar conclusions were reported by Soyer et al. following their $V_2O_9H_8$ cluster simulations performed at B3LYP/6–31G** level [458].

The influence of vanadia loading in V_2O_5/TiO_2 catalyst on NH_3 activation was studied theoretically by Anstrom et al. [459] with conclusion that vanadia species formed at low loadings (monomeric clusters) are less active than oligomeric clusters expected for higher V_2O_5 coverages. It was reasoned by the fact that Brønsted acid sites may be formed by the transfer of hydrogen atoms from Ti–OH groups to vanadia oxide oligomeric moieties, and that NH_3 adsorption is facilitated by presence of neighbouring vanadyl V=O bonds. The adsorption of NO on pure and Mo defected (001)-V_2O_5 surface (in both pristine and reduced state) was also studied [460]. The detailed analysis of electronic structure revealed that substituting a V atom with Mo leads to a reduction of the adjacent metal atom, practically without changing the electronic properties of the oxygen atoms. Furthermore, it was shown that the NO adsorption on the reduced surfaces of V_2O_5 and $V_{2-x}Mo_xO_5$ is exothermic and that the effect of the molybdenum presence for energetic stabilisation of adsorbed NO is not distinct. The role of Brønsted acidic centres in NH_3 activation on other low-index vanadium pentoxide surfaces was also studied [461]. It was shown that in all cases, the hydrogen bonds play a major role in this process and that the active sites for ammonia activation in the SCR mechanism are localised not only at saturated (010) sites but also at the unsaturated (001) and (100) V_2O_5 surfaces.

Based on the earlier work on adsorption, diffusion, and reaction of the different surface species, Gruber [462] and Hermann proposed detailed theoretical description of the elementary steps of the catalytic NO_x reduction with NH_3 [462]. In this paper, geometric and energetic details and reaction paths at Brønsted sites of the perfect surface as well as at Lewis sites of the reduced surface were evaluated using extended cluster models. The results suggest that Brønsted and Lewis reaction mechanisms differ in many aspects such as overall barrier energy, number of elementary steps, and reaction intermediates.

The DFT calculations (B3LYP) were also used to complement experimental findings for the dehydrogenation of CH_3OH mediated by VO^+ [463]. The reaction stationary points for both triplet and singlet potential energy surface (PES) have been considered, with conclusion that initial C–H bond activation is the rate determining step in both cases. Interestingly, the reported mechanism cannot be considered as an VO-induced rearrangement of a CH_3OH cation [464], because an electron transfer in the adduct $(CH_3OH)VO^+$ from the methanol moiety to VO^+ does not occur to a notable extent. The mechanism of the CH_3OH to CH_2O conversion on

$V_2O_3(0001)$ and $V_2O_5(001)$ surfaces was analysed by Göbke et al. [465] revealing that a methoxy layer is formed after CH_3OH adsorption and the abstracted hydrogen is stabilised on V=O groups with formation of hydroxy species. It was shown that in case of $V_2O_3(0001)$, H_2O desorption produces additional defects effectively doubling the methoxy coverage. For $V_2O_5(001)$, the reactions for CH_3OH and H_2O formation compete for the hydroxy species that stabilise the surface methoxy groups by production of a hydrogen deficiency.

4 Conclusions

The issue of coordination environment in the solid-state world is dominated by oxides. In the catalytic processes, the redox properties are defined mostly by the presence and the environment of TM cations in various coordination, like in the bulk or at the surface, owing to the wealth of their oxidation states, relatively close in energy.

Even though the above discussed systems do not exhaust the wide variety of TM in heterogeneous catalysis, this chapter aims to review the most successful computational methods used to describe their unique properties.

Since more involved methods (e.g. post-HF or GW methods and, for the larger systems, even hybrid DFT functionals) are computationally very demanding, the Hubbard DFT+U method is a good compromise nowadays. It should be, however, carefully parameterised, taking into account the system and even the property to be addressed.

Acknowledgements The present work was in part funded by the National Science Centre (grant 2016/23/B/ST4/00088). Part of the calculations were performed in the Cyfronet PL-Grid supercomputer centre in Kraków.

References

1. Hartwig JF (2010) Organotransition metal chemistry: from bonding to catalysis. University Science Books, Sausalito
2. Fajardo J, Peters JC (2017) J Am Chem Soc 139(45):16105. https://doi.org/10.1021/jacs.7b10204
3. Bethe H (1929) Annalen der Physik 395(2):133. https://doi.org/10.1002/andp.19293950202
4. Miessler GL, Fischer PJ, Tarr DA, (2014) Inorg Chem. Pearson
5. Barteau MA (1996) Chem Rev 96(4):1413. https://doi.org/10.1021/cr950222t
6. Hohenberg P, Kohn W (1964) Phys Rev 136(3B):B864 (1964). https://link.aps.org/doi/10.1103/PhysRev.136.B864
7. Kohn W, Sham LJ (1965) Phys Rev 140(4A):A1133. https://doi.org/10.1103/PhysRev.140.A1133. https://link.aps.org/doi/10.1103/PhysRev.140.A1133
8. Ceperley DM, Alder BJ (1980) Phys Rev Lett 45(7):566. http://escholarship.org/uc/item/2d7023jm.pdf

9. Perdew JP, Burke K, Ernzerhof M (1996) Phys Rev Lett 77(18):3865. https://doi.org/10.1103/PhysRevLett.77.3865
10. Hammer B, Hansen LB, Nørskov JK (1999) Phys Rev B **59**(11):7413. https://link.aps.org/doi/10.1103/PhysRevB.59.7413
11. Becke AD (1993) J Chem Phys 98(7):5648. https://doi.org/10.1063/1.464913https://doi.org/10.1063/1.464913
12. Hood RQ, Chou M, Williamson A, Rajagopal G, Needs R (1998) Phys Rev B Condens Matter Mater Phys 57(15):8972. https://doi.org/10.1103/PhysRevB.57.8972. https://link.aps.org/doi/10.1103/PhysRevB.57.8972
13. Jones RO (2006) In: Grotendorst J, Blügel, Marx D (eds) Computational nanoscience: do it yourself! (John von Neumann Institute for Computing, Jülich 2006), Chap 31, pp 45–70. https://doi.org/10.1007/978-3-642-86105-5
14. Dreizler RM, Gross EKU (1990) Density functional theory. Springer. https://doi.org/10.1007/978-3-642-86105-5. https://link.springer.com/book/10.1007/978-3-642-86105-5
15. Sodupe M, Bertran J, Rodriguez-Santiago L, Baerends EJ (1999) J Phys Chem A 103(1):166. https://doi.org/10.1021/jp983195u
16. de I, Moreira PR, Illas F, Martin RL (2002) Phys Rev B Condens Matter Mater Phys **65**(15):1551021. https://doi.org/10.1103/PhysRevB.65.155102. https://link.aps.org/doi/10.1103/PhysRevB.65.155102
17. Austin IG, Mott NF (1970) Science 168(3927):71. https://doi.org/10.1126/science.168.3927.71. http://www.sciencemag.org/cgi/doi/10.1126/science.168.3927.71
18. Himmetoglu B, Floris A, De Gironcoli S, Cococcioni M (2014) Int J Quantum Chem 114(1):14. http://doi.wiley.com/10.1002/qua.24521
19. Anisimov VI, Zaanen J, Andersen OK (1991) Phys Rev B **44**(3):943. https://link.aps.org/doi/10.1103/PhysRevB.44.943
20. Anisimov VI, Aryasetiawan F, Liechtenstein A (1997) J Phys Condens Matter 9(4):767. https://doi.org/10.1088/0953-8984/9/4/002. http://stacks.iop.org/0953-8984/9/i=4/a=002?key=crossref.fb949d5eca1b7900b10858389e3ab0c7
21. Cococcioni M (2012) In: Pavarini E, Koch E, Anders F, Jarrell M (eds) Correlated electrons: from models to materials modeling and simulation, Chap. 4. Forschungszentrum, Juelich, p. 40. http://www.cond-mat.de/events/correl12
22. Dudarev SL, Botton GA, Savrasov SY, Humphreys CJ, Sutton AP (1998) Phys Rev B 57(3):1505. https://doi.org/10.1103/PhysRevB.57.1505. https://link.aps.org/doi/10.1103/PhysRevB.57.1505
23. Faber C, Boulanger P, Duchemin I, Attaccalite C, Blase X (2013) J Chem Phys 139(19):194308. https://doi.org/10.1063/1.4830236. https://doi.org/10.1063/1.4830236
24. Aryasetiawan F, Gunnarsson O (1998) Rep Prog Phys 61(3):237 (1998). http://stacks.iop.org/0034-4885/61/i=3/a=002
25. Loschen C, Carrasco J, Neyman KM, Illas F (2007) Phys Rev B 75(3):35115. https://doi.org/10.1103/PhysRevB.75.035115. http://reichling.physik.uos.de/download_paper.php?paper=PhysRevB75p035115(2007)_Loschen.pdf%5Cn%3CGotoISI%3E://000243895400048
26. Loschen C, Carrasco J, Neyman KM, Illas F (2011) Phys Rev B 84:199906. https://doi.org/10.1103/PhysRevB.75.035115
27. Da Silva JL, Ganduglia-Pirovano MV, Sauer J, Bayer V, Kresse G (2007) Phys Rev B Condens Matter Mater Phys 75(4):45121. https://doi.org/10.1103/PhysRevB.75.045121. https://link.aps.org/doi/10.1103/PhysRevB.75.045121
28. Ganduglia-Pirovano MV, Da Silva JLF, Sauer J (2009) Phys Rev Lett 102(2):026101. https://doi.org/10.1103/PhysRevLett.102.026101. https://link.aps.org/doi/10.1103/PhysRevLett.102.026101
29. Jiang H, Gomez-Abal RI, Rinke P, Scheffler M (2009) Phys Rev Lett 102(12):126403. https://doi.org/10.1103/PhysRevLett.102.126403. http://link.aps.org/doi/10.1103/PhysRevLett.102.126403

30. Pacchioni G (2015) In: Jupille J, Thornton G (eds) Defects at oxide surfaces. Chap 1.3 Proble, Springer series in surface sciences, vol 58. Springer International Publishing, Cham, p 472. https://doi.org/10.1007/978-3-319-14367-5. https://link.springer.com/book/10.1007/978-3-319-14367-5
31. Runge E, Gross EKU (1984) Phys Rev Lett 52(12):997. https://doi.org/10.1103/PhysRevLett.52.997. https://link.aps.org/doi/10.1103/PhysRevLett.52.997
32. Marques MAL, Ullrich CA, Nogueira F, Rubio A, Burke K, Gross EKU, (2006) Time-dependent density functional theory. Lecture notes in physics, vol 706. Springer, Heidelberg. https://doi.org/10.1007/b11767107. http://link.springer.com/10.1007/b11767107
33. Casida ME, Huix-Rotllant M (2015) Springer International Publishing, Cham pp 1–60. https://doi.org/10.1007/128_2015_632. https://link.springer.com/chapter/10.1007/128_2015_632
34. Wu X, Vargas MC, Nayak S (2001) J Chem Phys 115(19):8748. https://doi.org/10.1063/1.1412004. https://doi.org/10.1063/1.1412004
35. Grimme S (2004) J Computat Chem 25(12):1463. https://doi.org/10.1002/jcc.20078. http://doi.wiley.com/10.1002/jcc.20078
36. Grimme S (2006) J Computat Chem 27(15):1787. https://doi.org/10.1002/jcc.20495. http://doi.wiley.com/10.1002/jcc.20495
37. Grimme S, Antony J, Ehrlich S, Krieg H (2010) J Chem Phys 132(15):154104. https://doi.org/10.1063/1.3382344. https://doi.org/10.1063/1.3382344
38. Smith DGA, Burns LA, Patkowski K, Sherrill CD (2016) J Phys Chem Lett 7(12):2197. https://doi.org/10.1021/acs.jpclett.6b00780. https://doi.org/10.1021/acs.jpclett.6b00780
39. Reckien W, Janetzko F, Peintinger MF, Bredow T (2012) J Comput Chem 33(25):2023. https://doi.org/10.1002/jcc.23037. https://onlinelibrary.wiley.com/doi/10.1002/jcc.23037
40. Dion M, Rydberg H, Schröder E, Langreth DC, Lundqvist BI (2004) Phys Rev Lett 92(24):246401. https://doi.org/10.1103/PhysRevLett.92.246401. https://link.aps.org/doi/10.1103/PhysRevLett.92.246401
41. Dion M, Rydberg H, Schröder E, Langreth DC, Lundqvist BI (2005) Phys Rev Lett **95**(10), 109902. https://link.aps.org/doi/10.1103/PhysRevLett.95.109902
42. Klimes J, Bowler DR, Michaelides A (2011) Phys Rev B Condens Matter Mater Phys 83(19):195131. https://doi.org/10.1103/PhysRevB.83.195131. https://link.aps.org/doi/10.1103/PhysRevB.83.195131
43. Santra B, Michaelides A, Fuchs M, Tkatchenko A, Filippi C, Scheffler M (2008) J Chem Phys 129(19):194111 https://doi.org/10.1063/1.3012573. https://doi.org/10.1063/1.3012573
44. Gulans A, Puska M, Nieminen R (2009) Phys Rev B 79(20):201105. https://doi.org/10.1103/PhysRevB.79.201105. https://link.aps.org/doi/10.1103/PhysRevB.79.201105
45. Senftle TP, Hong S, Islam MM, Kylasa SB, Zheng Y, Shin YK, Junkermeier C, Engel-Herbert R, Janik MJ, Aktulga HM, Verstraelen T, Grama A, van Duin ACT (2016) NPJ Comput Mater 2(1):15011. https://doi.org/10.1038/npjcompumats.2015.11. http://www.nature.com/articles/npjcompumats201511
46. Lid S, Köppen S, Colombi Ciacchi L (2017) Comput Mater Sci 140:307. https://doi.org/10.1016/j.commatsci.2017.09.003. https://www.sciencedirect.com/science/article/pii/S0927025617304731
47. Maseras F, Morokuma K (1995) J Comput Chem 16(9):1170. https://doi.org/10.1002/jcc.540160911. http://doi.wiley.com/10.1002/jcc.540160911
48. Humbel S, Sieber S, Morokuma K (1998) J Chem Phys 105(5):1959. https://doi.org/10.1063/1.472065. https://aip.scitation.org/doi/abs/10.1063/1.472065
49. Svensson M, Humbel S, Froese RDJ, Matsubara T, Sieber S, Morokuma K (1996) J Phys Chem 100(50):19357. https://doi.org/10.1021/jp962071j. http://pubs.acs.org/doi/abs/10.1021/jp962071j
50. Sauer J, Sierka M (2000) J Comput Chem 21(16):1470. https://doi.org/10.1002/1096-987X(200012)21:16<1470::AID-JCC5>3.0.CO;2-L. http://doi.wiley.com/10.1002/1096-987X%28200012%2921%3A16%3C1470%3A%3AAID-JCC5%3E3.0.CO%3B2-L
51. Boys SF (1950) Proc R Soc Lond Ser A Math Phys Sci **200**(1063):542 LP. http://rspa.royalsocietypublishing.org/content/200/1063/542.abstract

52. Huzinaga S (1984) Gaussian basis sets for molecular calculations. Elsevier, Amsterdam
53. Slater JC (1930) Phys Rev 36(1):57. https://doi.org/10.1103/PhysRev.36.57. https://link.aps.org/doi/10.1103/PhysRev.36.57
54. te Velde G, Bickelhaupt FM, Baerends EJ, Fonseca Guerra C, van Gisbergen SJ, Snijders JG, Ziegler T (2001) J Comput Chem 22(9):931. https://doi.org/10.1002/jcc.1056. http://doi.wiley.com/10.1002/jcc.1056
55. Delley B (2000) J Chem Phys 113(18):7756. https://doi.org/10.1063/1.1316015. https://doi.org/10.1063/1.1316015
56. Wannier GH (1937) Phys Rev 52(x):191. https://doi.org/10.1103/PhysRev.52.191. https://link.aps.org/doi/10.1103/PhysRev.52.191
57. Mostofi AA, Yates JR, Pizzi G, Lee YS, Souza I, Vanderbilt D, Marzari N (2014) Comput Phys Commun 185(8):2309. https://doi.org/10.1016/j.cpc.2014.05.003. https://www.sciencedirect.com/science/article/pii/S001046551400157X?via%3Dihub
58. Blöchl PE (1994) Phys Rev B 50(24):17953. https://doi.org/10.1103/PhysRevB.50.17953. https://link.aps.org/doi/10.1103/PhysRevB.50.17953
59. Mortensen JJ, Hansen LB, Jacobsen KW (2005) Phys Rev B 71(3):035109. https://doi.org/10.1103/PhysRevB.71.035109. https://link.aps.org/doi/10.1103/PhysRevB.71.035109
60. Murnaghan FD (1944) Proc Natl Acad Sci 30(9):244. https://doi.org/10.1073/pnas.30.9.244. http://www.pnas.org/cgi/doi/10.1073/pnas.30.9.244
61. Bader RFW (1990) Atoms in molecules—a quantum theory. Oxford University Press, Oxford
62. Henkelman G, Arnaldsson A, Jónsson H (2006) Comput Mater Sci 36(3):354. https://doi.org/10.1016/J.COMMATSCI.2005.04.010. https://www.sciencedirect.com/science/article/pii/S0927025605001849
63. Manz TA, Limas NG (2016) RSC Adv 6(53):47771. https://doi.org/10.1039/C6RA04656H. http://dx.doi.org/10.1039/C6RA04656H
64. Limas NG, Manz TA (2016) RSC Adv 6(51):45727. https://doi.org/10.1039/C6RA05507A. http://dx.doi.org/10.1039/C6RA05507A
65. Manz TA (2017) RSC Adv 7(72):45552. https://doi.org/10.1039/C7RA07400J. http://xlink.rsc.org/?DOI=C7RA07400J
66. Kerisit S, Rosso KM, Yang Z, Liu J (2010) J Phys Chem C 114(44):19096. https://doi.org/10.1021/jp103809s. http://pubs.acs.org/doi/10.1021/jp103809s
67. Arsentev M, Hammouri M, Kovalko N, Kalinina M, Petrov A (2017) Comput Mater Sci 140:181. https://doi.org/10.1016/J.COMMATSCI.2017.08.045. https://www.sciencedirect.com/science/article/pii/S0927025617304676
68. Tersoff J, Hamann DR (1985) Phys Rev B 31(2):805. https://doi.org/10.1103/PhysRevB.31.805. https://link.aps.org/doi/10.1103/PhysRevB.31.805
69. Bardeen J (1961) Phys Rev Lett 6(2):57. https://doi.org/10.1103/PhysRevLett.6.57. https://link.aps.org/doi/10.1103/PhysRevLett.6.57
70. Godlewski S, Tekiel A, Piskorz W, Zasada F, Prauzner-Bechcicki JS, Sojka Z, Szymonski M (2012) ACS Nano 6(10):8536. https://doi.org/10.1021/nn303546m
71. Hofer WA, Foster AS, Shluger AL (2003) Rev Mod Phys 75(4):1287. https://doi.org/10.1103/RevModPhys.75.1287. https://link.aps.org/doi/10.1103/RevModPhys.75.1287
72. Wulff G (1901) Zeitschrift für Krystallographie und Mineralogie 34(5/6):449
73. Reuter K, Scheffler M (2003) Phys Rev B Condens Matter Mater Phys 68(4):045407. https://doi.org/10.1103/PhysRevB.68.045407. https://link.aps.org/doi/10.1103/PhysRevB.68.045407
74. Geysermans P, Finocchi F, Goniakowski J, Hacquart R, Jupille J (2009) Phys Chem Chem Phys 11(13):2228. https://doi.org/10.1039/b812376d. http://xlink.rsc.org/?DOI=b812376d
75. Zhang CH, Chen B, Jin Y, Sun DB (2017) J Phys Chem Solids 110:129. https://doi.org/10.1016/j.jpcs.2017.06.006. https://www.sciencedirect.com/science/article/pii/S0022369717300525
76. Mathew K, Sundararaman R, Letchworth-Weaver K, Arias TA, Hennig RG (2014) J Chem Phys 140(8):84106. https://doi.org/10.1063/1.4865107. https://doi.org/10.1063/1.4865107

77. Letchworth-Weaver K, Arias TA (2012) Phys Rev B Condens Matter Mater Phys 86(7):75140. https://doi.org/10.1103/PhysRevB.86.075140. https://link.aps.org/doi/10.1103/PhysRevB.86.075140
78. Piskorz W, Zasada F, Stelmachowski P, Diwald O, Kotarba A, Sojka Z (2011) J Phys Chem C 115(45):22451. https://doi.org/10.1021/jp2070826
79. Piskorz W, Zasada F, Stelmachowski P, Kotarba A, Sojka Z (2013) J Phys Chem C 117(36):18488. https://doi.org/10.1021/jp405459g
80. Puigdollers AR, Schlexer P, Tosoni S, Pacchioni G (2017) ACS Catal 7(10):6493. https://doi.org/10.1021/acscatal.7b01913. https://doi.org/10.1021/acscatal.7b01913
81. Paier J, Penschke C, Sauer J (2013) Chem Rev 113(6):3949. https://doi.org/10.1021/cr3004949. https://doi.org/10.1021/cr3004949
82. Mars P, van Krevelen D (1954) Chem Eng Sci 3:41. https://doi.org/10.1016/S0009-2509(54)80005-4. https://www.sciencedirect.com/science/article/pii/S0009250954800054
83. Zasada F, Janas J, Piskorz W, Sojka Z (2017) Res Chem Intermed 43(5):2865. https://doi.org/10.1007/s11164-016-2798-y
84. Trovarelli A, Fornasiero P (2013) Catalysis by ceria and related materials, 2nd edn. Imperial College Press, London
85. Grasselli RK (2002) Topics Catal 21(1):79. https://doi.org/10.1023/A:1020556131984. https://doi.org/10.1023/A:1020556131984
86. Huang W, Gao Y (2014) Catal Sci Technol 4(11):3772. https://doi.org/10.1039/C4CY00679H. http://dx.doi.org/10.1039/C4CY00679H
87. Trovarelli A (1996) Catal Rev 38(4):439. https://doi.org/10.1080/01614949608006464. http://www.tandfonline.com/doi/abs/10.1080/01614949608006464
88. Yao HC, Yao YF (1984) J Catal 86(2):254. https://doi.org/10.1016/0021-9517(84)90371-3
89. Gandhi HS, Graham GW, McCabe RW (2003) J Catal 216(1–2):433. https://doi.org/10.1016/S0021-9517(02)00067-2. https://www.sciencedirect.com/science/article/pii/S0021951702000672?via%3Dihub
90. Mullins DR (2015) Surf Sci Rep 70(1):42. https://doi.org/10.1016/j.surfrep.2014.12.001. https://www.sciencedirect.com/science/article/pii/S016757291400034X#bib4
91. Fu Q, Saltsburg H, Flytzani-Stephanopoulos M (2003) Science 301(5635):935. https://doi.org/10.1126/science.1085721
92. Deng W, Flytzani-Stephanopoulos M (2006) Angew Chem Int Edn 45(14):2285. https://doi.org/10.1002/anie.200503220. http://dx.doi.org/10.1002/anie.200503220
93. Yu EH, Wang X, Liu XT, Li L (2013) In: Liang ZX, Zhao TS (eds) Catalysts for alcohol-fuelled direct oxidation fuel cells. Energy and environment series. The Royal Society of Chemistry, Cambridge, pp 227–249. https://doi.org/10.1039/9781849734783. http://dx.doi.org/10.1039/9781849734783
94. Park S, Vohs JM, Gorte RJ (2000) Nature 404(6775):265. https://doi.org/10.1038/35005040
95. Delgado JJ, del Río E, Chen X, Blanco G, Pintado JM, Bernal S, Calvino JJ (2013) In: Trovarelli A, Fornasiero P 2nd (edn) Catalysis by ceria and related materials. Imperial College Press, pp 47–138
96. Farrauto R, Hwang S, Shore L, Ruettinger W, Lampert J, Giroux T, Liu Y, Ilinich O (2003) Annu Rev Mater Res 33(1):1. https://doi.org/10.1146/annurev.matsci.33.022802.091348. http://www.annualreviews.org/doi/10.1146/annurev.matsci.33.022802.091348
97. Heck RM, Farrauto RJ (2001) Appl Catal A Gen 221(1–2):443. https://doi.org/10.1016/S0926-860X(01)00818-3. https://www.sciencedirect.com/science/article/pii/S0926860X01008183
98. Capdevila-Cortada M, Vilé G, Teschner D, Pérez-Ramírez J, López N (2016) Appl Catal B Environ 197:299. https://doi.org/10.1016/j.apcatb.2016.02.035. https://www.sciencedirect.com/science/article/pii/S0926337316301242
99. Trovarelli A (2002) Catalysis by ceria and related materials. Imperial College Press
100. Xiong YP, Kishimoto H, Yamaji K, Yoshinaga M, Horita T, Brito ME, Yokokawa H (2011) Solid State Ion 192(1):476. https://doi.org/10.1016/j.ssi.2010.07.017. https://www.sciencedirect.com/science/article/pii/S0167273810004157

101. Ganduglia-Pirovano MV, Hofmann A, Sauer J (2007) Surface Sci Rep 62(6):219. https://doi.org/10.1016/j.surfrep.2007.03.002. https://www.sciencedirect.com/science/article/pii/S0167572907000295
102. Sayle TXT, Parker SC, Catlow CRA (1994) Surface Sci 316(3):329. https://doi.org/10.1016/0039-6028(94)91225-4. https://www.sciencedirect.com/science/article/pii/0039602894912254
103. Yang Z, Woo TK, Baudin M, Hermansson K (2004) J Chem Phys 120(16):7741. https://doi.org/10.1063/1.1688316. http://aip.scitation.org/doi/10.1063/1.1688316
104. Mogensen M, Sammes NM, Tompsett GA (2000) Solid State Ion 129(1):63. https://doi.org/10.1016/S0167-2738(99)00318-5
105. Nolan M, Parker SC, Watson GW (2005) Surf Sci 595(1–3):223. https://doi.org/10.1016/j.susc.2005.08.015. https://www.sciencedirect.com/science/article/pii/S003960280500926X?via%3Dihub
106. Esch F, Fabris S, Zhou L, Montini T, Africh C, Fornasiero P, Comelli G, Rosei R (2005) Science 309(5735):752 LP. http://science.sciencemag.org/content/309/5735/752.abstract
107. Aneggi E, Llorca J, Boaro M, Trovarelli A (2005) J Catal 234(1):88. https://doi.org/10.1016/J.JCAT.2005.06.008. https://www.sciencedirect.com/science/article/pii/S0021951705002435
108. Désaunay T, Bonura G, Chiodo V, Freni S, Couzinié JP, Bourgon J, Ringuedé A, Labat F, Adamo C, Cassir M (2013) J Catal 297:193. https://doi.org/10.1016/J.JCAT.2012.10.011. https://www.sciencedirect.com/science/article/pii/S0021951712003247
109. Mai HX, Sun LD, Zhang YW, Si R, Feng W, Zhang HP, Liu HC, Yan CH (2005) J Phys Chem B 109(51):24380. https://doi.org/10.1021/jp055584b. https://doi.org/10.1021/jp055584b
110. Skorodumova NV, Simak SI, Lundqvist BI, Abrikosov IA, Johansson B (2002) Phys Rev Lett 89(16):166601/1. https://doi.org/10.1103/PhysRevLett.89.166601. https://link.aps.org/doi/10.1103/PhysRevLett.89.166601
111. Binet C, Daturi M, Lavalley JC (1999) Catal Today 50(2):207. https://doi.org/10.1016/S0920-5861(98)00504-5. https://www.sciencedirect.com/science/article/pii/S0920586198005045?via%3Dihub
112. Abanades S, Legal A, Cordier A, Peraudeau G, Flamant G, Julbe A (2010) J Mater Sci 45(15):4163. https://doi.org/10.1007/s10853-010-4506-4. https://doi.org/10.1007/s10853-010-4506-4
113. Weckhuysen BM, Keller DE (2003) Catal Today **78**(1-4 SPEC):25. https://doi.org/10.1016/S0920-5861(02)00323-1. https://www.sciencedirect.com/science/article/pii/S0920586102003231?via%3Dihub
114. Berner U, Schierbaum K, Jones G, Wincott P, Haq S, Thornton G (2000) Surf Sci 467(1–3):201. https://doi.org/10.1016/S0039-6028(00)00770-6. https://www.sciencedirect.com/science/article/pii/S0039602800007706
115. Fujimori A (1983) Phys Rev B 28(4):2281. https://doi.org/10.1103/PhysRevB.28.2281. https://link.aps.org/doi/10.1103/PhysRevB.28.2281
116. Kotani A, Mizuta H, Jo T, Parlebas J (1985) Solid State Commun 53(9):805. https://doi.org/10.1016/0038-1098(85)90223-6. https://www.sciencedirect.com/science/article/pii/0038109885902236?via%3Dihub
117. Wuilloud E, Delley B, Schneider WD, Baer Y (1984) Phys Rev Lett 53(2):202. https://doi.org/10.1103/PhysRevLett.53.202. https://link.aps.org/doi/10.1103/PhysRevLett.53.202
118. Marabelli F, Wachter P (1987) Phys Rev B **36**(2):1238. http://link.aps.org/doi/10.1103/PhysRevB.36.1238
119. Jiang Y, Adams JB, van Schilfgaarde M (2005) J Chem Phys 123(6):064701. https://doi.org/10.1063/1.1949189. http://aip.scitation.org/doi/10.1063/1.1949189
120. Nörenberg H, Briggs G (1999) Surf Sci 424(2–3):L352. https://doi.org/10.1016/S0039-6028(99)00212-5. https://www.sciencedirect.com/science/article/pii/S0039602899002125
121. Nörenberg H, Briggs G (1999) Surf Sci 433–435:127. https://doi.org/10.1016/S0039-6028(99)00070-9. https://www.sciencedirect.com/science/article/pii/S0039602899000709
122. Nörenberg H, Harding J (2001) Surf Sci 477(1):17. https://doi.org/10.1016/S0039-6028(01)00700-2. https://www.sciencedirect.com/science/article/pii/S0039602801007002

123. Trovarelli A, Llorca J (2017) ACS Catal 7(7):4716. https://doi.org/10.1021/acscatal.7b01246. http://pubs.acs.org/doi/10.1021/acscatal.7b01246
124. Fronzi M, Soon A, Delley B, Traversa E, Stampfl C (2009) J Chem Phys 131(10):104701. https://doi.org/10.1063/1.3191784. https://aip.scitation.org/doi/abs/10.1063/1.3191784
125. Gerward L, Olsen JS (1993) Powder Diffraction 8(02):127. https://doi.org/10.1017/S0885715600017966. http://www.journals.cambridge.org/abstract_S0885715600017966
126. Gerward L, Staun Olsen J, Petit L, Vaitheeswaran G, Kanchana V, Svane A (2005) J Alloys Compd 400(1–2):56. https://doi.org/10.1016/J.JALLCOM.2005.04.008. https://www.sciencedirect.com/science/article/pii/S0925838805003403
127. Kaneko K, Inoke K, Freitag B, Hungria AB, Midgley PA, Hansen TW, Zhang J, Ohara S, Adschiri T (2007) Nano Lett 7(2):421. https://doi.org/10.1021/nl062677b. https://pubs.acs.org/doi/abs/10.1021/nl062677bpubs.acs.org/doi/abs/10.1021/nl062677b
128. Ren Z, Liu N, Chen B, Li J, Mei D (2018) J Phys Chem C 122(9):4828. https://doi.org/10.1021/acs.jpcc.7b10208. http://pubs.acs.org/doi/10.1021/acs.jpcc.7b10208
129. Chen C, Chen HL, Weng MH, Ju SP, Chang JG, Chang CS (2008) Chin J Catal 29(11):1117. https://doi.org/10.1016/S1872-2067(09)60011-5. http://linkinghub.elsevier.com/retrieve/pii/S1872206709600115
130. Burow AM, Wende T, Sierka M, Włodarczyk R, Sauer J, Claes P, Jiang L, Meijer G, Lievens P, Asmis KR (2011) Phys Chem Chem Phys 13(43):19393. https://doi.org/10.1039/c1cp22129a. http://xlink.rsc.org/?DOI=c1cp22129a
131. Nagata T, Miyajima K, Hardy RA, Metha GF, Mafuné F (2015) J Phys Chem A 119(22):5545. https://doi.org/10.1021/acs.jpca.5b02816. http://pubs.acs.org/doi/10.1021/acs.jpca.5b02816
132. Fernández-Torre D, Kośmider K, Carrasco J, Ganduglia-Pirovano MV, Pérez R (2012) J Phys Chem C 116(25):13584. https://doi.org/10.1021/jp212605g. http://pubs.acs.org/doi/10.1021/jp212605g
133. Marrocchelli D, Yildiz B (2012) J Phys Chem C 116(3):2411. https://doi.org/10.1021/jp205573v. http://pubs.acs.org/doi/10.1021/jp205573v
134. Yoshida H, Deguchi H, Miura K, Horiuchi M, Inagaki T (2001) Solid State Ion 140(3–4):191. https://doi.org/10.1016/S0167-2738(01)00854-2. https://www.sciencedirect.com/science/article/pii/S0167273801008542
135. Ricken M, Nölting J, Riess I (1984) J Solid State Chem 54(1):89. 10.1016/0022-4596(84)90135-X. https://www.sciencedirect.com/science/article/pii/002245968490135X
136. Kümmerle E, Heger G (1999) J Solid State Chem 147(2):485. https://doi.org/10.1006/JSSC.1999.8403. https://www.sciencedirect.com/science/article/pii/S0022459699984037
137. Hull S, Norberg S, Ahmed I, Eriksson S, Marrocchelli D, Madden P (2009) J Solid State Chem 182(10):2815. https://doi.org/10.1016/J.JSSC.2009.07.044. https://www.sciencedirect.com/science/article/pii/S0022459609003545
138. Perrichon V, Laachir A, Bergeret G, Frety R, Tournayan L, Touret O (1994) J Chem Soc Faraday Trans 90(5):773. https://doi.org/10.1039/FT9949000773. http://dx.doi.org/10.1039/FT9949000773
139. Romeo M, Bak K, El Fallah J, Le Normand F, Hilaire L (1993) Surf Interf Anal 20(6):508. https://doi.org/10.1002/sia.740200604. http://doi.wiley.com/10.1002/sia.740200604
140. Pfau A, Schierbaum K (1994) Surf Sci 321(1–2):71. https://doi.org/10.1016/0039-6028(94)90027-2. https://www.sciencedirect.com/science/article/pii/0039602894900272
141. Mullins D, Overbury S, Huntley D (1998) Surf Sci 409(2):307. https://doi.org/10.1016/S0039-6028(98)00257-X. https://www.sciencedirect.com/science/article/pii/S003960289800257X
142. Binet C, Badri A, Lavalley JC (1994) J Phys Chem 98(25):6392. https://doi.org/10.1021/j100076a025. http://pubs.acs.org/doi/abs/10.1021/j100076a025
143. Soria J, Martínez-Arias A, Conesa JC (1995) J Chem Soc Faraday Trans 91(11):1669. https://doi.org/10.1039/FT9959101669. http://xlink.rsc.org/?DOI=FT9959101669
144. Namai Y, Fukui KI, Iwasawa Y (2003) J Phys Chem B 107(42):11666. https://doi.org/10.1021/jp030142q. http://pubs.acs.org/doi/abs/10.1021/jp030142q

145. Fabris S, Vicario G, Balducci G, de Gironcoli S, Baroni S (2005) J Phys Chem B 109(48):22860. https://doi.org/10.1021/jp0511698. http://pubs.acs.org/doi/abs/10.1021/jp0511698
146. Nolan M, Grigoleit S, Sayle DC, Parker SC, Watson GW (2005) Surf Sci 576(1–3):217. https://doi.org/10.1016/J.SUSC.2004.12.016. https://www.sciencedirect.com/science/article/pii/S0039602804015651
147. Huang M, Fabris S (2008) J Phys Chem C 112(23):8643. https://doi.org/10.1021/jp709898r
148. Migani A, Vayssilov GN, Bromley ST, Illas F, Neyman KM (2010) J Mater Chem 20(46):10535. https://doi.org/10.1039/C0JM01908A. http://dx.doi.org/10.1039/C0JM01908A
149. Castleton CWM, Kullgren J, Hermansson K (2007) J Chem Phys 127:244704. https://doi.org/10.1063/1.2800015
150. Panhans MA, Blumenthal RN (1993) Solid State Ion 60(4):279. https://doi.org/10.1016/0167-2738(93)90006-O. https://www.sciencedirect.com/science/article/abs/pii/016727389390006O
151. Nolan M, Fearon J, Watson G (2006) Solid State Ion 177(35–36):3069. https://doi.org/10.1016/j.ssi.2006.07.045. http://linkinghub.elsevier.com/retrieve/pii/S0167273806004528
152. Tuller HL, Nowick AS (1979) J Electrochem Soc 126(2):209. https://doi.org/10.1149/1.2129007. http://jes.ecsdl.org/cgi/doi/10.1149/1.2129007
153. Choi YM, Abernathy H, Chen HT, Lin MC, Liu M (2006) Chem Phys Chem 7(9):1957. https://doi.org/10.1002/cphc.200600190. http://doi.wiley.com/10.1002/cphc.200600190
154. Huang M, Fabris S (2007) Phys Rev B Condensed Matter Materials Physics 75(8):081404. https://doi.org/10.1103/PhysRevB.75.081404. https://link.aps.org/doi/10.1103/PhysRevB.75.081404
155. Li HY, Wang HF, Gong XQ, Guo YL, Guo Y, Lu G, Hu P (2009) Phys Rev B 79(19):193401. https://doi.org/10.1103/PhysRevB.79.193401. https://link.aps.org/doi/10.1103/PhysRevB.79.193401
156. Stangeland K, Kalai D, Yu Z (2017) Energy Procedia 105:2022. https://doi.org/10.1016/J.EGYPRO.2017.03.577. https://www.sciencedirect.com/science/article/pii/S187661021730629X
157. Burch R (2006) Phys Chem Chem Phys 8(47):5483. https://doi.org/10.1039/B607837K. http://dx.doi.org/10.1039/B607837K
158. Hinrichsen KO, Kochloefl K, Muhler M (2008) In: Ertl G, Knözinger H, Schüth F, Weitkamp J Handbook of heterogeneous catalysis, 2nd edn, Chap. 13.12. Wiley-VCH Verlag GmbH& Co. KGaA, Weinheim, Germany, pp 2905–2920
159. Bruix A, Rodriguez JA, Ramírez PJ, Senanayake SD, Evans J, Park JB, Stacchiola D, Liu P, Hrbek J, Illas F (2012) J Am Chem Soc 134(21):8968. https://doi.org/10.1021/ja302070k. http://pubs.acs.org/doi/10.1021/ja302070k
160. Flaherty DW, Yu WY, Pozun ZD, Henkelman G, Mullins CB (2011) J Catal 282(2):278. https://doi.org/10.1016/J.JCAT.2011.06.024. https://www.sciencedirect.com/science/article/pii/S0021951711002223
161. Tang C, Zhang H, Dong L (2016) Catal Sci Technol 6(5):1248. https://doi.org/10.1039/C5CY01487E. http://xlink.rsc.org/?DOI=C5CY01487E
162. Yang Z, Woo TK, Hermansson K (2006) Surf Sci 600(22):4953. https://doi.org/10.1016/J.SUSC.2006.08.018. https://www.sciencedirect.com/science/article/pii/S0039602806008867?via%3Dihub
163. Zhang J, Gong XQ, Lu G (2014) Phys Chem Chem Phys 16(32):16904. https://doi.org/10.1039/C4CP02235A. http://xlink.rsc.org/?DOI=C4CP02235A
164. Nolan M (2009) J Phys Chem C 113(6):2425. https://doi.org/10.1021/jp809292u. http://pubs.acs.org/doi/10.1021/jp809292u
165. Luo MF, Zhong YJ, Zhu B, Yuan XX, Zheng XM (1997) Appl Surf Sci **115**(2):185. https://doi.org/10.1016/S0169-4332(97)80203-6. https://www.sciencedirect.com/science/article/pii/S0169433297802036?via%3Dihub

166. Nolan M, Parker SC, Watson GW (2006) J Phys Chem B 110(5):2256. https://doi.org/10.1021/jp055624b. https://pubs.acs.org/doi/abs/10.1021/jp055624bpubs.acs.org/doi/abs/10.1021/jp055624b
167. Sharma S, Hu Z, Zhang P, McFarland EW, Metiu H (2011) J Catal 278(2):297. https://doi.org/10.1016/J.JCAT.2010.12.015. https://www.sciencedirect.com/science/article/pii/S0021951710004446
168. Qi K, Zasada F, Piskorz W, Indyka P, Gryboś J, Trochowski M, Buchalska M, Kobielusz M, Macyk W, Sojka Z (2016) J Phys Chem C 120(10):5442. https://doi.org/10.1021/acs.jpcc.5b10983
169. Varghese OK, Paulose M, LaTempa TJ, Grimes CA (2010) Nano Lett 10(2):750. https://doi.org/10.1021/nl903742x. http://pubs.acs.org/doi/abs/10.1021/nl903742x
170. Kisch H (2014) Semiconductor photocatalysis: principles and applications. Wiley-VCH Verlag GmbH & Co. KGaA, Weinheim, Germany. https://doi.org/10.1002/9783527673315. http://doi.wiley.com/10.1002/9783527673315
171. Li Y, Wang WN, Zhan Z, Woo MH, Wu CY, Biswas P (2010) Appl Catal B Environ 100(1–2):386. https://doi.org/10.1016/j.apcatb.2010.08.015. https://www.sciencedirect.com/science/article/pii/S0926337310003681
172. Grätzel M (2003) J Photochem Photobiol C Photochem Rev 4(2):145. https://doi.org/10.1016/S1389-5567(03)00026-1. https://www.sciencedirect.com/science/article/pii/S1389556703000261?via%3Dihub
173. Gong J, Sumathy K, Qiao Q, Zhou Z (2017) Renewable and sustainable energy reviews 68:234. https://doi.org/10.1016/j.rser.2016.09.097. https://www.sciencedirect.com/science/article/pii/S1364032116305883
174. Richhariya G, Kumar A, Tekasakul P, Gupta B (2017) Renewable and sustainable energy reviews 69:705. https://doi.org/10.1016/J.RSER.2016.11.198. https://www.sciencedirect.com/science/article/pii/S1364032116309571
175. Grätzel M (2001) Nature 414:338. https://doi.org/10.1038/35104607. https://doi.org/10.0.4.14/35104607
176. Anpo M, Yamashita H, Ichihashi Y, Ehara S (1995) J Electroanal Chem 396(1–2):21. https://doi.org/10.1016/0022-0728(95)04141-A. https://www.sciencedirect.com/science/article/pii/002207289504141A?via%3Dihub
177. Diebold U (2003) Surf Sci Rep 48(5–8):53. https://doi.org/10.1016/S0167-5729(02)00100-0. http://linkinghub.elsevier.com/retrieve/pii/S0167572902001000
178. Hagfeldt A, Lunell S, Sieghbahn HOG (1994) Int J Quantum Chem 49(2):97. https://doi.org/10.1002/qua.560490205. http://doi.wiley.com/10.1002/qua.560490205
179. Huang CN, Bow JS, Zheng Y, Chen SY, Ho NJ, Shen P (2010) Nanoscale Res Lett 5(6):972. https://doi.org/10.1007/s11671-010-9591-4. http://www.nanoscalereslett.com/content/5/6/972
180. Tsipis AC, Tsipis CA (1999) Phys Chem Chem Phys 1(18):4453. https://doi.org/10.1039/a904269e. http://xlink.rsc.org/?DOI=a904269e
181. Albaret T, Finocchi F, Noguera C (1999) Faraday Discuss 114:285. https://doi.org/10.1039/a903066b. http://xlink.rsc.org/?DOI=a903066b
182. Matsuda Y, Bernstein ER (2005) J Phys Chem A 109(2):314. https://doi.org/10.1021/jp040670h. http://pubs.acs.org/doi/abs/10.1021/jp040670h
183. Zhai HJ, Wang LS (2007) J Am Chem Soc 129(10):3022. https://doi.org/10.1021/ja068601z. http://pubs.acs.org/doi/abs/10.1021/ja068601z
184. Albaret T, Finocchi F, Noguera C (2000) J Chem Phys 113(6):2238. https://doi.org/10.1063/1.482038. http://aip.scitation.org/doi/10.1063/1.482038
185. Yu X, Oganov AR, Popov IA, Qian G, Boldyrev AI (2016) Angew Chem Int Edn 55(5):1699. https://doi.org/10.1002/anie.201508439. http://doi.wiley.com/10.1002/anie.201508439
186. Morales-García Á, Valero R, Illas F (2017) J Chem Theory Comput 13(8):3746. https://doi.org/10.1021/acs.jctc.7b00308. http://pubs.acs.org/doi/10.1021/acs.jctc.7b00308
187. Berardo E, Kaplan F, Bhaskaran-Nair K, Shelton WA, van Setten MJ, Kowalski K, Zwijnenburg MA (2017) J Chem Theory Comput 13(8):3814. https://doi.org/10.1021/acs.jctc.7b00538. http://pubs.acs.org/doi/10.1021/acs.jctc.7b00538

188. Kolmer M, Ahmad Zebari AA, Goryl M, Buatier De Mongeot F, Zasada F, Piskorz W, Pietrzyk P, Sojka Z, Krok F, Szymonski M Phys Rev B Condens Matter Mater Phys 88(19). https://doi.org/10.1103/PhysRevB.88.195427
189. Labat F, Baranek P, Domain C, Minot C, Adamo C (2007) J Chem Phys 126(15):154703. https://doi.org/10.1063/1.2717168. http://aip.scitation.org/doi/10.1063/1.2717168
190. Futera Z, English NJ (2017) J Phys Chem C 121(12):6701. https://doi.org/10.1021/acs.jpcc.6b12803. http://pubs.acs.org/doi/10.1021/acs.jpcc.6b12803
191. Goniakowski J, Gillan MJ (1995) Surf Sci 350(1–3):145. https://doi.org/10.1016/0039-6028(95)01252-4. http://arxiv.org/abs/mtrl-th/9508009%0Adx.doi.org/10.1016/0039-6028(95)01252-4
192. Stefanovich EV, Truong TN (1999) Chem Phys Lett 299(6):623. https://doi.org/10.1016/S0009-2614(98)01295-0. https://www.sciencedirect.com/science/article/pii/S0009261498012950
193. Bandura AV, Sykes DG, Shapovalov V, Troung TN, Kubicki JD, Evarestov RA (2004) J Phys Chem B 108(23):7844. https://doi.org/10.1021/jp037141i. https://pubs.acs.org/doi/abs/10.1021/jp037141ipubs.acs.org/doi/abs/10.1021/jp037141i
194. Langel W (2002) Surf Sci 496(1–2):141. https://doi.org/10.1016/S0039-6028(01)01606-5. https://www.sciencedirect.com/science/article/pii/S0039602801016065
195. Nadeem IM, Harrison GT, Wilson A, Pang CL Zegenhagen J Thornton G (2017) J Phys Chem B 122(2). https://doi.org/10.1021/acs.jpcb.7b06955. http://pubs.acs.org/doi/abs/10.1021/acs.jpcb.7b06955
196. Pang CL, Lindsay R, Thornton G (2008) Chem Soc Rev 37(10):2328. https://doi.org/10.1039/b719085a. http://xlink.rsc.org/?DOI=b719085a
197. Wendt S, Schaub R, Matthiesen J, Vestergaard EK, Wahlström E, Rasmussen MD, Thostrup P, Molina LM, Lægsgaard E, Stensgaard I, Hammer B, Besenbacher F (2005) Surf Sci 598(1–3):226. https://doi.org/10.1016/j.susc.2005.08.041. https://www.sciencedirect.com/science/article/pii/S0039602805010629?via%3Dihub
198. Morgan BJ, Watson GW (2007) Surf Sci 601(21):5034. https://doi.org/10.1016/j.susc.2007.08.025. https://www.sciencedirect.com/science/article/pii/S0039602807008758?via%3Dihub
199. Muscat J, Wander A, Harrison N (2001) Chem Phys Lett 342(3–4):397. https://doi.org/10.1016/S0009-2614(01)00616-9. https://www.sciencedirect.com/science/article/pii/S0009261401006169
200. Bredow T, Pacchioni G (2002) Chem Phys Lett 355(5–6):417. https://doi.org/10.1016/S0009-2614(02)00259-2. https://www.sciencedirect.com/science/article/pii/S0009261402002592
201. Wu X, Selloni A, Nayak SK (2004) J Chem Phys 120(9):4512. https://doi.org/10.1063/1.1636725. http://aip.scitation.org/doi/10.1063/1.1636725
202. Rasmussen MD, Molina LM, Hammer B (2004) J Chem Phys 120(2):988. https://doi.org/10.1063/1.1631922. http://aip.scitation.org/doi/10.1063/1.1631922
203. Inoue T, Fujishima A, Konishi S, Honda K (1979) Nature 277(5698):637. https://doi.org/10.1038/277637a0. http://www.nature.com/articles/277637a0
204. Halmann M, Ulman M, Aurian-Blajeni B (1983) Solar Energy 31(4):429. https://doi.org/10.1016/0038-092X(83)90145-7. https://www.sciencedirect.com/science/article/pii/0038092X83901457?via%3Dihub
205. Adachi K, Ohta K, Mizuno T (1994) Solar Energy 53(2):187. https://doi.org/10.1016/0038-092X(94)90480-4. https://www.sciencedirect.com/science/article/pii/0038092X94904804
206. Diwald O, Thompson TL, Zubkov T, Walck SD, Yates JT (2004) J Phys Chem B 108(19):6004. https://doi.org/10.1021/jp031267y. http://pubs.acs.org/doi/abs/10.1021/jp031267y
207. Asahi R, Morikawa T, Ohwaki T, Aoki K, Taga Y (2001) Science 293(5528):269. https://doi.org/10.1126/science.1061051. http://www.ncbi.nlm.nih.gov/pubmed/11452117
208. Sakthivel S, Janczarek M, Kisch H (2004) J Phys Chem B 108(50):19384. https://doi.org/10.1021/jp046857q. https://pubs.acs.org/doi/abs/10.1021/jp046857q
209. Di Valentin C, Pacchioni G, Selloni A (2005) Chem Mater 17(26):6656. https://doi.org/10.1021/cm051921h. https://pubs.acs.org/doi/abs/10.1021/cm051921h

210. Di Valentin C, Pacchioni G, Selloni A, Livraghi S, Giamello E (2005) J Phys Chem B 109(23):11414. https://doi.org/10.1021/jp051756t. https://pubs.acs.org/doi/abs/10.1021/jp051756t
211. Grill L, Dyer M, Lafferentz L, Persson M, Peters MV, Hecht S (2007) Nature Nanotechnol 2(11):687. https://doi.org/10.1038/nnano.2007.346. http://www.nature.com/articles/nnano.2007.346
212. Cai J, Ruffieux P, Jaafar R, Bieri M, Braun T, Blankenburg S, Muoth M, Seitsonen AP, Saleh M, Feng X, Müllen K, Fasel R (2010) Nature 466(7305):470. https://doi.org/10.1038/nature09211. http://www.nature.com/articles/nature09211
213. Kolmer M, Ahmad Zebari AA, Prauzner-Bechcicki JS, Piskorz W, Zasada F, Godlewski S, Such B, Sojka Z, Szymonski M (2013) Angew Chem Int Edn 52(39):10300. https://doi.org/10.1002/anie.201303657
214. Zasada F, Piskorz W, Godlewski S, Prauzner-Bechcicki JS, Tekiel A, Budzioch J, Cyganik P, Szymonski M, Sojka Z (2011) J Phys Chem C 115(10):4134. https://doi.org/10.1021/jp111014r
215. Zasada F, Piskorz W, Gryboś J, Sojka Z (2014) J Phys Chem C 118(17):8971. https://doi.org/10.1021/jp412756a
216. Li H, Eddaoudi M, O'Keeffe M, Yaghi OM (1999) Nature 402(6759):276. https://doi.org/10.1038/46248. http://www.nature.com/doifinder/10.1038/46248
217. Stadelmann PA (1987) Ultramicroscopy 21(2):131. https://doi.org/10.1016/0304-3991(87)90080-5. https://www.sciencedirect.com/science/article/pii/0304399187900805
218. Stadelmann PA. JEMS-SAAS. http://www.jems-saas.ch/
219. Kolmer M, Zuzak R, Ahmad Zebari AA, Godlewski S, Prauzner-Bechcicki JS, Piskorz W, Zasada F, Sojka Z, Bléger D, Hecht S, Szymonski M (2015) Chem Commun **51**(56):11276. https://doi.org/10.1039/C5CC02989A. http://xlink.rsc.org/?DOI=C5CC02989A
220. Vanderbilt D, Zhao X, Ceresoli D (2005) Thin Solid Films 486(1–2):125. https://doi.org/10.1016/j.tsf.2004.11.232. https://www.sciencedirect.com/science/article/pii/S0040609004019066?via%3Dihub
221. Jung KT, Bell AT (2002) Catal Lett 80(1–2):63. https://doi.org/10.1023/A:1015326726898. http://link.springer.com/10.1023/A:1015326726898
222. Matter PH, Braden DJ, Ozkan US (2004) J Catal 223(2):340. https://doi.org/10.1016/j.jcat.2004.01.031. https://www.sciencedirect.com/science/article/pii/S0021951704000533
223. Adamski A, Sojka Z, Dyrek K, Che M, Wendt G, Albrecht S (1999) Langmuir 15(18):5733. https://doi.org/10.1021/la981431m. https://pubs.acs.org/doi/abs/10.1021/la981431m
224. Korhonen ST, Airaksinen SM, Krause AOI (2006) Catal Today 112(1–4):37. https://doi.org/10.1016/j.cattod.2005.11.053. http://linkinghub.elsevier.com/retrieve/pii/S0920586105008710
225. Zavodinsky VG, Chibisov AN (2006) J Phys Conf Ser 29:173. https://doi.org/10.1088/1742-6596/29/1/033. http://stacks.iop.org/1742-6596/29/i=1/a=033?key=crossref.fc6f5dcadf81965d0d034b0436efe5a2
226. Durandurdu M (2009) Epl 88(6):66001. https://doi.org/10.1209/0295-5075/88/66001. http://stacks.iop.org/0295-5075/88/i=6/a=66001?key=crossref.d10a35965ddce7bab98cdc40aaaa3d69
227. NETL (2004) Seventh Edition Fuel Cell Handbook. Tech. rep., National Energy Technology Laboratory, Pittsburgh, PA, and Morgantown, WV. https://doi.org/10.2172/834188. http://www.osti.gov/servlets/purl/834188-H0AaAO/native/
228. Reddy BM, Khan A (2005) Catal Rev 47(2):257. https://doi.org/10.1081/CR-200057488. http://www.tandfonline.com/doi/abs/10.1081/CR-200057488
229. Zhao X, Vanderbilt D (2002) Phys Rev B Condens Matter Mater Phys 65(7):1. https://doi.org/10.1103/PhysRevB.65.075105. https://link.aps.org/doi/10.1103/PhysRevB.65.075105
230. Jomard G, Petit T, Pasturel A, Magaud L, Kresse G, Hafner J (1999) Phys Rev B 59(6):4044. https://doi.org/10.1103/PhysRevB.59.4044. https://link.aps.org/doi/10.1103/PhysRevB.59.4044

231. Zhang L, Zhu L, Virkar AV (2016) J Power Sources 302:98. https://doi.org/10.1016/j.jpowsour.2015.10.026. https://www.sciencedirect.com/science/article/pii/S0378775315304134
232. Garvie RC (1965) J Phys Chem 69(4):1238. https://doi.org/10.1021/j100888a024. http://pubs.acs.org/doi/abs/10.1021/j100888a024
233. Jaffe JE, Bachorz RA, Gutowski M (2005) Phys Rev B Condens Matter Mater Phys 72(14):144107. https://doi.org/10.1103/PhysRevB.72.144107. https://link.aps.org/doi/10.1103/PhysRevB.72.144107
234. Li S, Dixon DA (2010) J Phys Chem A 114(7):2665. https://doi.org/10.1021/jp910310j. http://pubs.acs.org/doi/abs/10.1021/jp910310j
235. Gerosa M, Bottani CE, Caramella L, Onida G, Di Valentin C, Pacchioni G (2015) Phys Rev B Condens Matter Materials Phys 91(15):155201. 10.1103/PhysRevB.91.155201. https://link.aps.org/doi/10.1103/PhysRevB.91.155201
236. Zhu J, van Ommen JG, Lefferts L (2006) Catal Today 117(1–3):163. https://doi.org/10.1016/J.CATTOD.2006.05.046. https://www.sciencedirect.com/science/article/pii/S0920586106002550
237. Boulc'h F, Dessemond L, Djurado E (2004) J Eur Ceram Soc 24(6):1181. https://doi.org/10.1016/S0955-2219(03)00563-6. https://www.sciencedirect.com/science/article/pii/S0955221903005636
238. Piskorz W, Gryboś J, Zasada F, Zapała P, Cristol S, Paul JF, Sojka Z (2012) J Phys Chem C 116(36):19307. https://doi.org/10.1021/jp3050059
239. Piskorz W, Gryboś J, Zasada F, Cristol S, Paul JF, Adamski A, Sojka Z (2011) J Phys Chem C 115(49):24274. https://doi.org/10.1021/jp2086335
240. Barnard AS, Curtiss LA (2005) Rev Adv Mater Sci 10(2):105
241. Liebau F, Gies H, Gunawardane RP, Marler B (1986) Zeolites 6(5):373. https://doi.org/10.1016/0144-2449(86)90065-5. https://www.sciencedirect.com/science/article/pii/0144244986900655
242. Akporiaye DE, Dahl IM, Mostad HB, Wendelbo R (1996) J Phys Chem 100(10):4148. https://doi.org/10.1021/jp952189k. http://pubs.acs.org/doi/abs/10.1021/jp952189k
243. Olson DH, Khosrovani N, Peters AW, Toby BH (2000) J Phys Chem B 104(20):4844. https://doi.org/10.1021/jp000417r. http://pubs.acs.org/doi/abs/10.1021/jp000417r
244. Lu B, Kanai T, Oumi Y, Sano T (2007) J Porous Mater 14(1):89. https://doi.org/10.1007/s10934-006-9012-3. http://link.springer.com/10.1007/s10934-006-9012-3
245. Sazama P, Tabor E, Klein P, Wichterlova B, Sklenak S, Mokrzycki L, Pashkkova V, Ogura M, Dedecek J (2016) J Catal 333:102. https://doi.org/10.1016/j.jcat.2015.10.010. https://www.sciencedirect.com/science/article/pii/S0021951715003383
246. Di Iorio JR, Gounder R (2016) Chem Mater 28(7):2236. https://doi.org/10.1021/acs.chemmater.6b00181. http://pubs.acs.org/doi/10.1021/acs.chemmater.6b00181
247. Klinowski J (1984) Prog Nucl Magn Reson Spectrosc 16:237. https://doi.org/10.1016/0079-6565(84)80007-2. https://www.sciencedirect.com/science/article/pii/0079656584800072?via%3Dihub
248. Majda D, Paz FA, Friedrichs D, Foster MD, Simperler A, Bell RG, Klinowski J (2008) J Phys Chem C 112(4):1040. https://doi.org/10.1021/jp0760354. https://pubs.acs.org/doi/abs/10.1021/jp0760354
249. Benco L, Bucko T, Hafner J, Toulhoat H (2005) J Phys Chem B 109(43):20361. https://doi.org/10.1021/jp0530597. https://pubs.acs.org/doi/abs/10.1021/jp0530597
250. García-Pérez E, Dubbeldam D, Liu B, Smit B, Calero S (2007) Angew Chem Int Edn 46(1–2):276. https://doi.org/10.1002/anie.200603136. http://doi.wiley.com/10.1002/anie.200603136
251. Meeprasert J, Kungwan N, Jungsuttiwong S, Namuangruk S (2014) Microporous Mesoporous Mater 195:227. https://doi.org/10.1016/j.micromeso.2014.04.038. https://www.sciencedirect.com/science/article/pii/S1387181114002248?via%3Dihub
252. Xu R, Pang W, Yu J, Huo Q, Chen J (2007) Chem zeolites and related porous materials synthesis and structure. Wiley, Asia. https://doi.org/10.1002/9780470822371.

https://www.wiley.com/en-pl/Chemistry+of+Zeolites+and+Related+Porous+Materials:+Synthesis+and+Structure-p-9780470822333
253. Kowalak S, Stawiński K, Makowiak A (2001) Microporous Mesoporous Mater 44–45:283. https://doi.org/10.1016/S1387-1811(01)00194-9. https://www.sciencedirect.com/science/article/pii/S1387181101001949?via%3Dihub
254. Nogier JP, Millot Y, Man PP, Shishido T, Che M, Dzwigaj S (2009) J Phys Chem C 113(12):4885. https://doi.org/10.1021/jp8099829. http://pubs.acs.org/doi/10.1021/jp8099829
255. Trejda M, Ziolek M, Millot Y, Chalupka K, Che M, Dzwigaj S (2011) J Catal 281(1):169. https://doi.org/10.1016/j.jcat.2011.04.013. https://www.sciencedirect.com/science/article/pii/S0021951711001357?via%3Dihub
256. Pérez-Ramírez J, Christensen CH, Egeblad K, Christensen CH, Groen JC (2008) Chem Soc Rev 37(11):2530. https://doi.org/10.1039/b809030k. http://xlink.rsc.org/?DOI=b809030k
257. Roth WJ, Nachtigall P, Morris RE, Čejka J (2014) Chem Rev 114(9):4807. https://doi.org/10.1021/cr400600f. http://pubs.acs.org/doi/10.1021/cr400600f
258. Kulkarni AR, Zhao ZJ, Siahrostami S, Nørskov JK, Studt F (2018) Catal Sci Technol 8(1):114. https://doi.org/10.1039/C7CY01229B. http://xlink.rsc.org/?DOI=C7CY01229B
259. Iwamoto M, Yokoo S, Sakai K, Kagawa S (1981) J Chem Soc Faraday Trans 1 77(7):1629. https://doi.org/10.1039/f19817701629. http://xlink.rsc.org/?DOI=f19817701629
260. Iwamoto M, Furukawa H, Mine Y, Uemura F, Mikuriya SI, Kagawa S (1986) J Chem Soc Chem Commun 0(16):1272. https://doi.org/10.1039/c39860001272. http://xlink.rsc.org/?DOI=c39860001272
261. Oda A, Torigoe H, Itadani A, Ohkubo T, Yumura T, Kobayashi H, Kuroda Y (2012) Angew Chem Int Edn 51(31):7719. https://doi.org/10.1002/anie.201201000. http://doi.wiley.com/10.1002/anie.201201000
262. Kozyra P, Piskorz W (2016) Phys Chem Chem Phys 18(18):12592. https://doi.org/10.1039/C5CP05493A. http://xlink.rsc.org/?DOI=C5CP05493A
263. Stepanov AG, Arzumanov SS, Gabrienko AA, Parmon VN, Ivanova II, Freude D (2008) Chem Phys Chem 9(17):2559. https://doi.org/10.1002/cphc.200800569. http://doi.wiley.com/10.1002/cphc.200800569
264. Biscardi JA, Meitzner GD, Iglesia E (1998) J Catal 179(1):192. https://doi.org/10.1006/jcat.1998.2177. https://www.sciencedirect.com/science/article/pii/S0021951798921777
265. Niu X, Gao J, Miao Q, Dong M, Wang G, Fan W, Qin Z, Wang J (2014) Microporous Mesoporous Mater 197:252. https://doi.org/10.1016/j.micromeso.2014.06.027. https://www.sciencedirect.com/science/article/pii/S138718111400345X
266. Wang X, Xu J, Qi G, Li B, Wang C, Deng F (2013) J Phys Chem C 117(8):4018. https://doi.org/10.1021/jp310872a. http://pubs.acs.org/doi/10.1021/jp310872a
267. Smeets PJ, Woertink JS, Sels BF, Solomon EI, Schoonheydt RA (2010) Inorg Chem 49(8):3573. https://doi.org/10.1021/ic901814f. http://pubs.acs.org/doi/abs/10.1021/ic901814f
268. Woertink JS, Smeets PJ, Groothaert MH, Vance MA, Sels BF, Schoonheydt RA, Solomon EI (2009) Proc Natl Acad Sci USA 106(45):18908. https://doi.org/10.1073/pnas.0910461106. http://www.ncbi.nlm.nih.gov/pubmed/19864626, www.pubmedcentral.nih.gov/articlerender.fcgi?artid=PMC2776445
269. Selvaraj T, Rajalingam R, Balasubramanian V (2018) Appl Surf Sci 434:781. https://doi.org/10.1016/j.apsusc.2017.11.011. https://www.sciencedirect.com/science/article/pii/S0169433217332282?via%3Dihub
270. Mitoraj M, Michalak A (2007) J Mol Model 13(2):347. https://doi.org/10.1007/s00894-006-0149-4. http://link.springer.com/10.1007/s00894-006-0149-4
271. Michalak A, Mitoraj M, Ziegler T (2008) J Phys Chem A 112(9):1933. https://doi.org/10.1021/jp075460u. https://pubs.acs.org/doi/abs/10.1021/jp075460u
272. Ziegler T, Rauk A (1979) Inorg Chem 18(6):1558. https://doi.org/10.1021/ic50196a034. http://pubs.acs.org/doi/abs/10.1021/ic50196a034

273. M. Radoń. Natorbs (v. 0.3)—utility for computing natural (spin) orbitals and natural orbitals for chemical valence. http://www.chemia.uj.edu.pl/~mradon/natorbs
274. Pietrzyk P, Sojka Z, Dzwigaj S, Che M (2007) J Am Chem Soc 129(46):14174. https://doi.org/10.1021/ja076689q. https://pubs.acs.org/doi/abs/10.1021/ja076689q
275. Pietrzyk P, Piskorz W, Sojka Z, Broclawik E (2003) J Phys Chem B 107(25):6105. https://doi.org/10.1021/jp034173x
276. Broclawik E, Datka J, Gil B, Piskorz W, Kozyra P (2000) Topics Catal 11(1–4):335. https://doi.org/10.1023/a:1027235511555. http://link.springer.com/10.1023/A:1027235511555
277. Uzunova EL, Göltl F, Kresse G, Hafner J (2009) J Phys Chem C 113(13):5274. https://doi.org/10.1021/jp809927k. http://pubs.acs.org/doi/10.1021/jp809927k
278. Davidová M, Nachtigallová D, Nachtigall P, Sauer J (2004) J Phys Chem B 108(36):13674. https://doi.org/10.1021/jp0478007. https://pubs.acs.org/doi/abs/10.1021/jp0478007
279. Rejmak P, Broclawik E, Góra-Marek K, Radoń M, Datka J (2008) J Phys Chem C 112(46):17998. https://doi.org/10.1021/jp8042686. http://pubs.acs.org/doi/10.1021/jp8042686
280. Izquierdo R, Rodríguez LJ, Añez R, Sierraalta A (2011) J Mol Catal A Chem 348(1–2):55. https://doi.org/10.1016/j.molcata.2011.07.018. https://www.sciencedirect.com/science/article/pii/S1381116911003189
281. Göltl F, Hafner J (2012) J Chem Phys 136(6):64501. https://doi.org/10.1063/1.3676410. http://aip.scitation.org/doi/10.1063/1.3676410
282. Heyden A, Peters B, Bell AT, Keil FJ (2005) J Phys Chem B 109(5):1857. https://doi.org/10.1021/jp040549a. https://pubs.acs.org/doi/abs/10.1021/jp040549a
283. Kozyra P, Piskorz W (2015) Phys Chem Chem Phys PCCP 17(20):13267. https://doi.org/10.1039/c4cp05932h. http://www.ncbi.nlm.nih.gov/pubmed/25920323
284. Broclawik E, Załucka J, Kozyra P, Mitoraj M, Datka J (2010) J Phys Chem C 114(21):9808. https://doi.org/10.1021/jp1002676. http://pubs.acs.org/doi/10.1021/jp1002676
285. Kozyra P, Załucka J, Mitoraj M, Brocławik E, Datka J (2008) Catal Lett 126(3–4):241. https://doi.org/10.1007/s10562-008-9620-4. http://link.springer.com/10.1007/s10562-008-9620-4
286. Kozyra P, Radon M, Datka J, Broclawik E (2012) Struct Chem 23(5):1349. https://doi.org/10.1007/s11224-012-0050-y. http://link.springer.com/10.1007/s11224-012-0050-y
287. Radoń M, Kozyra P, Stępniewski A, Datka J, Broclawik E (2013) Can J Chem 91(7):538. https://doi.org/10.1139/cjc-2012-0536. http://www.nrcresearchpress.com/doi/abs/10.1139/cjc-2012-0536
288. Sierka M, Sauer J (2000) J Chem Phys 112(16):6983. https://doi.org/10.1063/1.481296. https://aip.scitation.org/doi/10.1063/1.481296
289. Pietrzyk P, Zasada F, Piskorz W, Kotarba A, Sojka Z (2007) Catal Today 119(1–4):219. https://doi.org/10.1016/j.cattod.2006.08.054. https://www.sciencedirect.com/science/article/pii/S0920586106005256?via%3Dihub
290. Dubkov K, Ovanesyan N, Shteinman A, Starokon E, Panov G (2002) J Catal 207(2):341. https://doi.org/10.1006/JCAT.2002.3552. https://www.sciencedirect.com/science/article/pii/S0021951702935529?via%3Dihub
291. Starokon EV, Parfenov MV, Arzumanov SS, Pirutko LV, Stepanov AG, Panov GI (2013) J Catal 300:47. https://doi.org/10.1016/J.JCAT.2012.12.030. https://www.sciencedirect.com/science/article/pii/S002195171200423X
292. Xu J, Armstrong RD, Shaw G, Dummer NF, Freakley SJ, Taylor SH, Hutchings GJ (2016) Catal Today 270:93. https://doi.org/10.1016/J.CATTOD.2015.09.011. https://www.sciencedirect.com/science/article/pii/S0920586115005659
293. Park KS, Kim JH, Park SH, Moon DJ, Roh HS, Chung CH, Um SH, Choi JH, Bae JW (2017) J Mol Catal A Chem 426:130. https://doi.org/10.1016/J.MOLCATA.2016.11.008. https://www.sciencedirect.com/science/article/pii/S1381116916304824
294. Kletnieks PW, Liang AJ, Craciun R, Ehresinann JO, Marcus DM, Bhirud VA, Klaric MM, Hayman MJ, Guenther DR, Bagatchenko OP, Dixon DA, Gates BC, Haw JF (2007) Chem Eur J 13(26):7294. https://doi.org/10.1002/chem.200700721. http://doi.wiley.com/10.1002/chem.200700721

295. Archipov T, Santra S, Ene AB, Stoll H, Rauhut G, Roduner E (2009) J Phys Chem C 113(10):4107. https://doi.org/10.1021/jp805976a. http://pubs.acs.org/doi/10.1021/jp805976a
296. Xiao J, Wei J (1992) Chem Eng Sci 47(5):1123. https://doi.org/10.1016/0009-2509(92)80236-6. https://www.sciencedirect.com/science/article/pii/0009250992802366
297. Xiao J, Wei J (1992) Chem Eng Sci 47(5):1143. https://doi.org/10.1016/0009-2509(92)80237-7. https://www.sciencedirect.com/science/article/pii/0009250992802377
298. Song L, Sun Z, Duan L, Gui J, McDougall GS (2007) Microporous Mesoporous Mater 104(1–3):115 https://doi.org/10.1016/j.micromeso.2007.01.015. https://www.sciencedirect.com/science/article/pii/S1387181107000418
299. Gonçalves CV, Cardoso D (2008) Microporous Mesoporous Mater 116(1–3):352. https://doi.org/10.1016/j.micromeso.2008.04.022. https://www.sciencedirect.com/science/article/pii/S138718110800200X
300. Abdelrasoul A, Zhang H, Cheng CH, Doan H (2017) Microporous Mesoporous Mater 242:294. https://doi.org/10.1016/j.micromeso.2017.01.038. https://www.sciencedirect.com/science/article/pii/S1387181117300380
301. Yoshino H, Ohnishi CH, Hosokawa S, Wada K, Inoue M (2011) J Mater Sci 46(3):797. https://doi.org/10.1007/s10853-010-4818-4. http://link.springer.com/10.1007/s10853-010-4818-4
302. Wilczkowska E, Krawczyk K, Petryk J, Sobczak JW, Kaszkur Z (2010) Appl Catal A Gen 389(1–2):165. https://doi.org/10.1016/j.apcata.2010.09.016. http://linkinghub.elsevier.com/retrieve/pii/S0926860X10006708
303. Amrousse R, Tsutsumi A, Bachar A, Lahcene D (2013) Appl Catal A Gen 450:253. https://doi.org/10.1016/j.apcata.2012.10.036. http://linkinghub.elsevier.com/retrieve/pii/S0926860X12006849
304. Li WY, Xu LN, Chen J (2005) Adv Funct Mater 15(5):851. https://doi.org/10.1002/adfm.200400429. http://doi.wiley.com/10.1002/adfm.200400429
305. Woodhouse M, Herman GS, Parkinson BA (2005) Chem Mater 17(17):4318. https://doi.org/10.1021/cm050546q. http://pubs.acs.org/doi/abs/10.1021/cm050546q
306. Wang G, Liu H, Horvat J, Wang B, Qiao S, Park J, Ahn H (2010) Chem A Eur J 16(36):11020. https://doi.org/10.1002/chem.201000562. http://doi.wiley.com/10.1002/chem.201000562
307. Fu L, Liu Z, Liu Y, Han B, Hu P, Cao L, Zhu D (2005) Adv Mater 17(2):217. https://doi.org/10.1002/adma.200400833. http://doi.wiley.com/10.1002/adma.200400833
308. Zasada F, Stelmachowski P, Maniak G, Paul JF, Kotarba A, Sojka Z (2009) Catal Lett 127(1–2):126. https://doi.org/10.1007/s10562-008-9655-6. http://link.springer.com/10.1007/s10562-008-9655-6
309. Stelmachowski P, Zasada F, Maniak G, Granger P, Inger M, Wilk M, Kotarba A, Sojka Z (2009) Catal Lett 130(3–4):637. https://doi.org/10.1007/s10562-009-0014-z. http://link.springer.com/10.1007/s10562-009-0014-z
310. Piskorz W, Zasada F, Stelmachowski P, Kotarba A, Sojka Z (2008) Catal Today 137(2–4):418. https://doi.org/10.1016/j.cattod.2008.02.027
311. Karásková K, Obalová L, Jirátová K, Kovanda F (2010) Chem Eng J 160(2):480. https://doi.org/10.1016/j.cej.2010.03.058. http://linkinghub.elsevier.com/retrieve/pii/S1385894710002949
312. Qiu Y, Yang S, Deng H, Jin L, Li W (2010) J Mater Chem 20(21):4439. https://doi.org/10.1039/c0jm00101e. http://xlink.rsc.org/?DOI=c0jm00101e
313. Zhou L, Zhao D, Lou XW (2012) Adv Mater 24(6):745. https://doi.org/10.1002/adma.201104407. http://doi.wiley.com/10.1002/adma.201104407
314. Fleet ME (1986) J Solid State Chemi 62(1):75. https://doi.org/10.1016/0022-4596(86)90218-5. http://linkinghub.elsevier.com/retrieve/pii/0022459686902185
315. Sickafus KE, Wills JM, Grimes NW (2004) J Am Ceram Soc 82(12):3279. https://doi.org/10.1111/j.1151-2916.1999.tb02241.x. http://doi.wiley.com/10.1111/j.1151-2916.1999.tb02241.x
316. Paudel TR, Zakutayev A, Lany S, D'Avezac M, Zunger A (2011) Adv Funct Mater 21(23):4493. https://doi.org/10.1002/adfm.201101469. http://doi.wiley.com/10.1002/adfm.201101469

317. Rabe KM (2010) Annu Rev Condens Matter Phys 1(1):211. https://doi.org/10.1146/annurev-conmatphys-070909-103932. http://www.annualreviews.org/doi/10.1146/annurev-conmatphys-070909-103932
318. Cohen AJ, Mori-Sanchez P, Yang W (2008) Science 321(5890):792. https://doi.org/10.1126/science.1158722. http://www.sciencemag.org/cgi/doi/10.1126/science.1158722
319. Qiao L, Xiao HY, Meyer HM, Sun JN, Rouleau CM, Puretzky AA, Geohegan DB, Ivanov IN, Yoon M, Weber WJ, Biegalski MD (2013) J Mater Chem C 1(31):4628. https://doi.org/10.1039/c3tc30861h. http://xlink.rsc.org/?DOI=c3tc30861h
320. Selcuk S, Selloni A (2015) J Phys Chem C 119(18):9973. https://doi.org/10.1021/acs.jpcc.5b02298. http://pubs.acs.org/doi/abs/10.1021/acs.jpcc.5b02298
321. Wang L, Maxisch T, Ceder G (2006) Phys Rev B Conden Matter Mater Phys 73(19):195107. https://doi.org/10.1103/PhysRevB.73.195107. https://link.aps.org/doi/10.1103/PhysRevB.73.195107
322. O'Brien CJ, Rák Z, Brenner DW (2013) J Phys Condens Matter 25(44):445008. https://doi.org/10.1088/0953-8984/25/44/445008. http://stacks.iop.org/0953-8984/25/i=44/a=445008?key=crossref.24ee6c1a77b9ba202f63e7419ccf1d0e
323. Montoya A, Haynes BS (2011) Chem Phys Lett 502(1–3):63. https://doi.org/10.1016/j.cplett.2010.12.015. http://linkinghub.elsevier.com/retrieve/pii/S000926141001599X
324. Chen J, Wu X, Selloni A (2011) Phys Rev B 83(24):245204. https://doi.org/10.1103/PhysRevB.83.245204. https://link.aps.org/doi/10.1103/PhysRevB.83.245204
325. Wu Z, Cohen RE, Singh DJ (2004) Phys Rev B 70(10):104112. https://doi.org/10.1103/PhysRevB.70.104112. https://link.aps.org/doi/10.1103/PhysRevB.70.104112
326. Koch W, Holthausen MC (2001) A chemist's guide to density functional theory. Wiley-VCH Verlag GmbH, Weinheim. FRG. https://doi.org/10.1002/3527600043. http://doi.wiley.com/10.1002/3527600043
327. Koo HJ, Whangbo MH (2008) Inorg Chem 47(11):4779. https://doi.org/10.1021/ic800216j. http://pubs.acs.org/doi/abs/10.1021/ic800216j
328. Wei SH, Zunger A (1993) Phys Rev B 48(9):6111. https://doi.org/10.1103/PhysRevB.48.6111. https://link.aps.org/doi/10.1103/PhysRevB.48.6111
329. Rodriguez-Fortea A, Llunell M, Alemany P, Canadell E (2009) Inorg Chem 48(13):5779. https://doi.org/10.1021/ic900105u. http://pubs.acs.org/doi/abs/10.1021/ic900105u
330. Piskunov S, Kotomin EA, Heifets E, Maier J, Eglitis RI, Borstel G (2005) Surf Sci 575(1–2):75. https://doi.org/10.1016/j.susc.2004.11.008. http://linkinghub.elsevier.com/retrieve/pii/S0039602804014414
331. Wahl R, Vogtenhuber D, Kresse G (2008) Phys Rev B Condens Matter Mater Phys 78(10):104116. https://doi.org/10.1103/PhysRevB.78.104116. https://link.aps.org/doi/10.1103/PhysRevB.78.104116
332. Kvyatkovskii OE, Karadag F, Mamedov A, Zakharov GA (2004) Phys Solid State 46(9):1717. https://doi.org/10.1134/1.1799192. http://link.springer.com/10.1134/1.1799192
333. Zhao Q, Yan Z, Chen C, Chen J (2017) Chem Rev 117(15):10121. https://doi.org/10.1021/acs.chemrev.7b00051. http://pubs.acs.org/doi/10.1021/acs.chemrev.7b00051
334. Zasada F, Piskorz W, Stelmachowski P, Kotarba A, Paul JF, Płociński T, Kurzydłowski KJ, Sojka Z (2011) J Phys Chem C 115(14):6423. https://doi.org/10.1021/jp200581s
335. Lazzeri M, Thibaudeau P (2006) Phys Rev B Condens Matter Mater Phys 74(14):140301. https://doi.org/10.1103/PhysRevB.74.140301. https://link.aps.org/doi/10.1103/PhysRevB.74.140301
336. Fang CM, de Wijs GA, Loong CK, de With G (2007) J Mater Chem 17(46):4908. https://doi.org/10.1039/b706814j. http://xlink.rsc.org/?DOI=b706814j
337. López-Moreno S, Romero AH, Rodríguez-Hernandez P, Muñoz A (2009) High Press Res 29(4):573. https://doi.org/10.1080/08957950903474635. http://www.tandfonline.com/doi/abs/10.1080/08957950903474635
338. Ono S, Brodholt JP, Price GD (2008) Phys Chem Miner 35(7):381. https://doi.org/10.1007/s00269-008-0231-9. http://link.springer.com/10.1007/s00269-008-0231-9

339. Ding Y, Xu L, Chen C, Shen X, Suib SL (2008) J Phys Chem C 112(22):8177. https://doi.org/10.1021/jp0773839. http://pubs.acs.org/doi/10.1021/jp0773839
340. Feng J, Zeng HC (2003) Chem Mater 15(14):2829. https://doi.org/10.1021/cm020940d. http://pubs.acs.org/doi/abs/10.1021/cm020940d
341. Xie X, Li Y, Liu ZQ, Haruta M, Shen W (2009) Nature 458(7239):746. https://doi.org/10.1038/nature07877. http://www.nature.com/articles/nature07877
342. Xiao J, Kuang Q, Yang S, Xiao F, Wang S, Guo L (2013) Sci Rep 3(1):2300. https://doi.org/10.1038/srep02300. http://www.nature.com/articles/srep02300
343. Hu L, Peng Q, Li Y (2008) J Am Chem Soc 130(48):16136. https://doi.org/10.1021/ja806400e. http://pubs.acs.org/doi/abs/10.1021/ja806400e
344. Yu X, Huo CF, Li YW, Wang J, Jiao H (2012) Surf Sci 606(9–10):872. https://doi.org/10.1016/j.susc.2012.02.003. http://linkinghub.elsevier.com/retrieve/pii/S0039602812000544
345. Zasada F, Piskorz W, Cristol S, Paul JF, Kotarba A, Sojka Z (2010) J Phys Chem C 114(50):22245. https://doi.org/10.1021/jp109264b
346. Zasada F, Gryboś J, Indyka P, Piskorz W, Kaczmarczyk J, Sojka Z (2014) J Phys Chem C 118(33):19085. https://doi.org/10.1021/jp503737p. http://pubs.acs.org/doi/abs/10.1021/jp503737p
347. Takita Y, Tashiro T, Saito Y, Hori F (1986) J Catal 97(1):25. https://doi.org/10.1016/0021-9517(86)90033-3. http://linkinghub.elsevier.com/retrieve/pii/0021951786900333
348. Maniak G, Stelmachowski P, Kotarba A, Sojka Z, Rico-Pérez V, Bueno-López A (2013) Appl Catal B Environ 136–137:302. https://doi.org/10.1016/j.apcatb.2013.01.068. http://linkinghub.elsevier.com/retrieve/pii/S0926337313001045
349. Zasada F, Piskorz W, Sojka Z (2015) J Phys Chem C 119:33. https://doi.org/10.1021/acs.jpcc.5b05136
350. Zasada F, Piskorz W, Janas J, Gryboś J, Indyka P, Sojka Z (2015) ACS Catal 5(11):6879. https://doi.org/10.1021/acscatal.5b01900
351. Hashim AH, Zayed AOH, Zain SM, Lee VS, Said SM (2018) Appl Surf Sci 427:1090. https://doi.org/10.1016/j.apsusc.2017.09.085. http://linkinghub.elsevier.com/retrieve/pii/S0169433217327344
352. Zasada F, Piskorz W, Janas J, Budiyanto E, Sojka Z (2017) J Phys Chem C 121(43):24128. https://doi.org/10.1021/acs.jpcc.7b09597
353. Shojaee K, Montoya A, Haynes BS (2013) Comput Mater Sci 72:15. https://doi.org/10.1016/j.commatsci.2013.02.001. http://linkinghub.elsevier.com/retrieve/pii/S0927025613000505
354. Zasada F, Gryboś J, Piskorz W, Sojka Z (2018) J Phys Chem C 122:5. https://doi.org/10.1021/acs.jpcc.7b11869
355. Chen J, Selloni A (2012) Phys Rev B Condens Matter Mater Phys 85(8):085306. https://doi.org/10.1103/PhysRevB.85.085306. https://link.aps.org/doi/10.1103/PhysRevB.85.085306
356. Vaz C, Wang HQ, Ahn C, Henrich V, Baykara M, Schwendemann T, Pilet N, Albers B, Schwarz U, Zhang L, Zhu Y, Wang J, Altman E (2009) Surf Sci 603(2):291. https://doi.org/10.1016/j.susc.2008.11.022. http://linkinghub.elsevier.com/retrieve/pii/S0039602808008054
357. Chen D, Chen C, Baiyee ZM, Shao Z, Ciucci F (2015) Chem Rev 115(18):9869. https://doi.org/10.1021/acs.chemrev.5b00073. http://pubs.acs.org/doi/10.1021/acs.chemrev.5b00073
358. Uchaker E, Cao G (2015) Chem Asian J 10(8):1608. https://doi.org/10.1002/asia.201500401. http://doi.wiley.com/10.1002/asia.201500401
359. Zasada F, Janas J, Piskorz W, Gorczyńska M, Sojka Z (2017) ACS Catal 7(4):2853. https://doi.org/10.1021/acscatal.6b03139
360. Uusi-Esko K, Rautama EL, Laitinen M, Sajavaara T, Karppinen M (2010) Chem Mater 22(23):6297. https://doi.org/10.1021/cm102003y. http://pubs.acs.org/doi/abs/10.1021/cm102003y
361. Kaczmarczyk J, Zasada F, Janas J, Indyka P, Piskorz W, Kotarba A, Sojka Z (2016) ACS Catal 6(2):1235. https://doi.org/10.1021/acscatal.5b02642
362. Omata K, Takada T, Kasahara S, Yamada M (1996) Appl Catal A Gen 146(2):255. https://doi.org/10.1016/S0926-860X(96)00151-2. http://linkinghub.elsevier.com/retrieve/pii/S0926860X96001512

363. Yan L, Ren T, Wang X, Gao Q, Ji D, Suo J (2003) Catal Commun 4(10):505. https://doi.org/10.1016/S1566-7367(03)00131-6. http://linkinghub.elsevier.com/retrieve/pii/S1566736703001316
364. Ohnishi C, Asano K, Iwamoto S, Chikama K, Inoue M (2007) Catal Today 120(2):145. https://doi.org/10.1016/j.cattod.2006.07.042. http://linkinghub.elsevier.com/retrieve/pii/S0920586106004731
365. Asano K, Ohnishi C, Iwamoto S, Shioya Y, Inoue M (2008) Appl Catal B Environ 78(3–4):242. https://doi.org/10.1016/j.apcatb.2007.09.016. http://linkinghub.elsevier.com/retrieve/pii/S0926337307002901
366. Wang HF, Kavanagh R, Guo YL, Guo Y, Lu G, Hu P (2012) J Catal 296:110. https://doi.org/10.1016/j.jcat.2012.09.005. http://linkinghub.elsevier.com/retrieve/pii/S0021951712002849
367. Wang YG, Yang XF, Li J (2016) Chin J Catal 37(1):193. https://doi.org/10.1016/S1872-2067(15)60969-X. http://linkinghub.elsevier.com/retrieve/pii/S187220671560969X
368. Xu XL, Yang E, Li JQ, Li Y, Chen WK (2009) Chemcatchem 1(3):384. https://doi.org/10.1002/cctc.200900115
369. Hu W, Lan J, Guo Y, Cao XM, Hu P (2016) ACS Catal 6(8):5508. https://doi.org/10.1021/acscatal.6b01080. http://pubs.acs.org/doi/10.1021/acscatal.6b01080
370. Wang Y, Yang X, Hu L, Li Y, Li J (2014) Chin J Catal 35(4):462. https://doi.org/10.1016/S1872-2067(14)60043-7. http://linkinghub.elsevier.com/retrieve/pii/S1872206714600437
371. Lv CQ, Liu C, Wang GC (2014) Catal Commun 45:83. https://doi.org/10.1016/j.catcom.2013.10.039. http://linkinghub.elsevier.com/retrieve/pii/S156673671300410X
372. Zou CY, Ji W, Shen Z, Tang Q, Fan M (2018) Appl Surf Sci 442:778. https://doi.org/10.1016/j.apsusc.2018.02.037. http://linkinghub.elsevier.com/retrieve/pii/S0169433218303805
373. Shojaee K, Haynes BS, Montoya A (2014) Appl Surf Sci 316(1):355. https://doi.org/10.1016/j.apsusc.2014.08.021. http://linkinghub.elsevier.com/retrieve/pii/S016943321401753X
374. Shojaee K, Haynes BS, Montoya A (2017) Proc Combust Inst 36(3):4365. https://doi.org/10.1016/j.proci.2016.06.100. http://linkinghub.elsevier.com/retrieve/pii/S1540748916301584
375. Shi X, Bernasek SL, Selloni A (2017) J Phys Chem C 121(7):3929. https://doi.org/10.1021/acs.jpcc.6b12005. http://pubs.acs.org/doi/abs/10.1021/acs.jpcc.6b12005
376. Kondratenko EV, Pérez-Ramírez J (2007) Catal Today 121(3–4):197. https://doi.org/10.1016/j.cattod.2006.08.066. http://linkinghub.elsevier.com/retrieve/pii/S092058610600589X
377. Yakovlev AL, Zhidomirov GM, van Santen RA (2001) Catal Lett 75(1/2):45. https://doi.org/10.1023/A:1016692419859. http://link.springer.com/10.1023/A:1016692419859
378. Drago RS, Jurczyk K, Kob N (1997) Appl Catal B Environ 13(1):69. https://doi.org/10.1016/S0926-3373(96)00088-4. http://linkinghub.elsevier.com/retrieve/pii/S0926337396000884
379. Scagnelli A, Valentin CD, Pacchioni G (2006) Surf Sci 600(2):386. https://doi.org/10.1016/j.susc.2005.10.048. http://linkinghub.elsevier.com/retrieve/pii/S0039602805012264
380. Stelmachowski P, Maniak G, Kotarba A, Sojka Z (2009) Catal Commun 10(7):1062. https://doi.org/10.1016/j.catcom.2008.12.057. http://linkinghub.elsevier.com/retrieve/pii/S1566736709000120
381. Maniak G, Stelmachowski P, Zasada F, Piskorz W, Kotarba A, Sojka Z (2011) Catal Today 176(1):369. https://doi.org/10.1016/j.cattod.2010.11.043
382. Lu J, Song J, Niu H, Pan L, Zhang X, Wang L, Zou JJ (2016) Appl Surf Sci 371:61. https://doi.org/10.1016/j.apsusc.2016.02.209. http://linkinghub.elsevier.com/retrieve/pii/S0169433216303968
383. Si C, Wang Y, Zhang J, Gao H, Lv L, Han L, Zhang Z (2016) Nano Energy 23:105. https://doi.org/10.1016/j.nanoen.2016.03.012. http://linkinghub.elsevier.com/retrieve/pii/S2211285516300234
384. Si C, Zhang Y, Zhang C, Gao H, Ma W, Lv L, Zhang Z (2017) Electrochim Acta 245:829. https://doi.org/10.1016/j.electacta.2017.06.029. http://linkinghub.elsevier.com/retrieve/pii/S0013468617312677
385. Obalová L, Karásková K, Jirátová K, Kovanda F (2009) Appl Catal B Environ 90(1–2):132. https://doi.org/10.1016/j.apcatb.2009.03.002. http://linkinghub.elsevier.com/retrieve/pii/S0926337309000939

386. PalDey S, Gedevanishvili S, Zhang W, Rasouli F (2005) Appl Catal B Environ 56(3):241. https://doi.org/10.1016/j.apcatb.2004.09.013. http://linkinghub.elsevier.com/retrieve/pii/S0926337304005569
387. Wang S, Ding Z, Wang X (2015) Chem Commun 51(8):1517. https://doi.org/10.1039/C4CC07225A. http://xlink.rsc.org/?DOI=C4CC07225A
388. Walter MG, Warren EL, McKone JR, Boettcher SW, Mi Q, Santori EA, Lewis NS (2010) Chem Rev 110(11):6446. https://doi.org/10.1021/cr1002326. http://pubs.acs.org/doi/abs/10.1021/cr1002326
389. Kang D, Kim TW, Kubota SR, Cardiel AC, Cha HG, Choi KS (2015) Chem Rev 115(23):12839. https://doi.org/10.1021/acs.chemrev.5b00498. http://pubs.acs.org/doi/10.1021/acs.chemrev.5b00498
390. Martin A, Luck F, Armbruster U, Patria L, Radnik J, Schneider M (2005) Topics Catal 33(1–4):155. https://doi.org/10.1007/s11244-005-2522-4. http://link.springer.com/10.1007/s11244-005-2522-4
391. Liu P, He H, Wei G, Liang X, Qi F, Tan F, Tan W, Zhu J, Zhu R (2016) Appl Catal B Environ 182:476. https://doi.org/10.1016/j.apcatb.2015.09.055. http://linkinghub.elsevier.com/retrieve/pii/S0926337315301818
392. Chen J, Shi W, Yang S, Arandiyan H, Li J (2011) J Phys Chem C 115(35):17400. https://doi.org/10.1021/jp202958b. http://pubs.acs.org/doi/10.1021/jp202958b
393. Anu Prathap M, Srivastava R (2013) Nano Energy 2(5):1046. https://doi.org/10.1016/j.nanoen.2013.04.003. http://linkinghub.elsevier.com/retrieve/pii/S2211285513000645
394. Sun S, Zhou Y, Hu B, Zhang Q, Xu ZJ (2016) J Electrochem Soc 163(2):H99. https://doi.org/10.1149/2.0761602jes. http://jes.ecsdl.org/lookup/doi/10.1149/2.0761602jes
395. Zhu X, Zhao H, Niu X, Liu T, Shi L, Lan M (2016) Microchim Acta 183(8):2431. https://doi.org/10.1007/s00604-016-1887-3. http://link.springer.com/10.1007/s00604-016-1887-3
396. Periyasamy S, Subramanian P, Levi E, Aurbach D, Gedanken A, Schechter A (2016) ACS Appl Mater Interf 8(19):12176. https://doi.org/10.1021/acsami.6b02491. http://pubs.acs.org/doi/10.1021/acsami.6b02491
397. Alizadeh-Gheshlaghi E, Shaabani B, Khodayari A, Azizian-Kalandaragh Y, Rahimi R (2012) Powder Technol 217:330. https://doi.org/10.1016/j.powtec.2011.10.045. http://linkinghub.elsevier.com/retrieve/pii/S0032591011005882
398. Cui B, Lin H, Liu YZ, Li JB, Sun P, Zhao XC, Liu CJ (2009) J Phys Chem C 113(32):14083. https://doi.org/10.1021/jp900028t. https://pubs.acs.org/doi/10.1021/jp900028t
399. Blum RP, Niehus H, Hucho C, Fortrie R, Ganduglia-Pirovano MV, Sauer J, Shaikhutdinov S, Freund HJ (2007) Phys Rev Lett 99(22):226103. https://doi.org/10.1103/PhysRevLett.99.226103. https://link.aps.org/doi/10.1103/PhysRevLett.99.226103
400. Haber J, Witko M, Tokarz R (1997) Appl Catal A Gen 157(1–2):3. https://doi.org/10.1016/S0926-860X(97)00017-3. http://linkinghub.elsevier.com/retrieve/pii/S0926860X97000173
401. Surnev S, Ramsey MG, Netzer FP (2003) Prog Surf Sci 73(4–8):117. https://doi.org/10.1016/j.progsurf.2003.09.001. http://linkinghub.elsevier.com/retrieve/pii/S0079681603000947
402. Bahlawane N, Lenoble D (2014) Chem Vapor Depos 20(7-8-9):299. https://doi.org/10.1002/cvde.201400057. http://doi.wiley.com/10.1002/cvde.201400057
403. Gavrilyuk A, Tritthart U, Gey W (2011) Phys Chem Chem Phys 13(20):9490. https://doi.org/10.1039/c0cp02201b. http://xlink.rsc.org/?DOI=c0cp02201b
404. Lamsal C, Ravindra NM (2013) J Mater Sci 48(18):6341. https://doi.org/10.1007/s10853-013-7433-3. http://link.springer.com/10.1007/s10853-013-7433-3
405. Yang J, Lan T, Liu J, Song Y, Wei M (2013) Electrochim Acta 105:489. https://doi.org/10.1016/j.electacta.2013.05.023. http://linkinghub.elsevier.com/retrieve/pii/S0013468613009274
406. Perera SD, Rudolph M, Mariano RG, Nijem N, Ferraris JP, Chabal YJ, Balkus KJ (2013) Nano Energy 2(5):966. https://doi.org/10.1016/j.nanoen.2013.03.018. http://linkinghub.elsevier.com/retrieve/pii/S2211285513000530
407. Yeager MP, Du W, Bishop B, Sullivan M, Xu W, Su D, Senanayake SD, Hanson J, Teng X (2013) Chem Sus Chem 6(12):2231. https://doi.org/10.1002/cssc.201300480. http://doi.wiley.com/10.1002/cssc.201300480

408. Su D, Wang G (2013) ACS Nano 7(12):11218. https://doi.org/10.1021/nn405014d. http://pubs.acs.org/doi/10.1021/nn405014d
409. Glynn C, Thompson D, Paez J, Collins G, Benavente E, Lavayen V, Yutronic N, Holmes JD, González G, O'Dwyer C (2013) J Mater Chem C 1(36):5675. https://doi.org/10.1039/c3tc31104j. http://xlink.rsc.org/?DOI=c3tc31104j
410. Chebout K, Iratni A, Bouremana A, Sam S, Keffous A, Gabouze N (2013) Solid State Ion 253:164. https://doi.org/10.1016/j.ssi.2013.09.055. http://linkinghub.elsevier.com/retrieve/pii/S0167273813004694
411. Modafferi V, Panzera G, Donato A, Antonucci P, Cannilla C, Donato N, Spadaro D, Neri G (2012) Sens Actuators B Chem 163(1):61. https://doi.org/10.1016/j.snb.2012.01.007. http://linkinghub.elsevier.com/retrieve/pii/S0925400512000305
412. Sobolev VI, Danilevich EV, Koltunov KY (2013) Kinet Catal 54(6):730. https://doi.org/10.1134/S0023158413060128. http://link.springer.com/10.1134/S0023158413060128
413. Abazari R, Sanati S, Saghatforoush LA (2014) Chem Eng J 236:82. https://doi.org/10.1016/j.cej.2013.09.056. http://linkinghub.elsevier.com/retrieve/pii/S1385894713012497
414. Enjalbert R, Galy J (1986) Acta Crystallogr Sect C Cryst Struct Commun 42(11):1467. https://doi.org/10.1107/S0108270186091825. http://scripts.iucr.org/cgi-bin/paper?S0108270186091825
415. Wachs IE (2013) Dalton Trans 42(33):11762. https://doi.org/10.1039/c3dt50692d. http://xlink.rsc.org/?DOI=c3dt50692d
416. Carrero CA, Schloegl R, Wachs IE, Schomaecker R (2014) ACS Catal 4(10):3357. https://doi.org/10.1021/cs5003417. http://pubs.acs.org/doi/10.1021/cs5003417
417. Guerrero-Pérez MO (2017) Catal Today 285:226. https://doi.org/10.1016/j.cattod.2017.01.037. https://www.sciencedirect.com/science/article/pii/S0920586117300378
418. Wachs IE (2005) Catal Today 100(1–2):79. https://doi.org/10.1016/j.cattod.2004.12.019. http://linkinghub.elsevier.com/retrieve/pii/S0920586104008077
419. Guerrero-Pérez MO, Wachs IE (2016) Catal Today 277:201. https://doi.org/10.1016/j.cattod.2016.09.031. http://linkinghub.elsevier.com/retrieve/pii/S0920586116305867
420. Hermann K, Witko M, Druzinic R, Tokarz R (2001) Appl Phys A Mater Sci Process 72(4):429. https://doi.org/10.1007/s003390100756. http://link.springer.com/10.1007/s003390100756
421. Haber J (1992) in new developments. In: Ruiz P, Delmon B (eds) Selective oxidation by heterogeneous catalysis. Elsevier Science Publishers B.V, Amsterdam, pp 279–304
422. Tilley RJD (1983) Surface properties and catalysis by non-metals. D. Reidel Publishing Company, Dordrecht/Boston/Lancaster. https://doi.org/10.1007/978-94-009-7160-8. http://link.springer.com/10.1007/978-94-009-7160-8
423. Bartholomew CH, Farrauto RJ (2010) Fundamentals of industrial catalytic processes 2nd edn. Wiley Inc., Hoboken, NJ, USA. https://doi.org/10.1002/9780471730071. http://doi.wiley.com/10.1002/9780471730071
424. Schlosser EG (1971) Angewandte Chemie 83(20):812. https://doi.org/10.1002/ange.19710832023. http://doi.wiley.com/10.1002/ange.19710832023
425. Wachs IE, Weckhuysen BM (1997) Appl Catal A Gen 157(1–2):67. https://doi.org/10.1016/S0926-860X(97)00021-5. http://linkinghub.elsevier.com/retrieve/pii/S0926860X97000215
426. Weissermel K, Arpe HJ (1978) Industrial organic chemistry: important raw materials and intermediates. Verlag Chemie, Weinheim, New York
427. Kong M, Liu Q, Wang X, Ren S, Yang J, Zhao D, Xi W, Yao L (2015) Catal Commun 72:121. https://doi.org/10.1016/j.catcom.2015.09.029. http://linkinghub.elsevier.com/retrieve/pii/S156673671530090X
428. Yu Y, He C, Chen J, Yin L, Qiu T, Meng X (2013) Catal Commun 39:78. https://doi.org/10.1016/j.catcom.2013.05.010ShortCommunication. http://linkinghub.elsevier.com/retrieve/pii/S1566736713001775
429. Lomnicki S, Lichtenberger J, Xu Z, Waters M, Kosman J, Amiridis MD (2003) Appl Catal B Environ 46(1):105. https://doi.org/10.1016/S0926-3373(03)00215-7. http://linkinghub.elsevier.com/retrieve/pii/S0926337303002157

430. Høj M, Jensen AD, Grunwaldt JD (2013) Appl Catal A: Gen 451:207. https://doi.org/10.1016/j.apcata.2012.09.037. http://linkinghub.elsevier.com/retrieve/pii/S0926860X12006308
431. Due-Hansen J, Boghosian S, Kustov A, Fristrup P, Tsilomelekis G, Ståhl K, Christensen CH, Fehrmann R (2007) J Catal 251(2):459. https://doi.org/10.1016/j.jcat.2007.07.016. http://linkinghub.elsevier.com/retrieve/pii/S0021951707002862
432. Kempf JY, Silvi B, Dietrich A, Catlow CRA, Maigret B (1993) Chem Mater 5(5):641. https://doi.org/10.1021/cm00029a011. http://pubs.acs.org/doi/abs/10.1021/cm00029a011
433. Yin X, Fahmi A, Endou A, Miura R, Gunji I, Yamauchi R, Kubo M, Chatterjee A, Miyamoto A (1998) Appl Surf Sci 130–132:539. https://doi.org/10.1016/S0169-4332(98)00111-1. http://linkinghub.elsevier.com/retrieve/pii/S0169433298001111
434. Chakrabarti A, Hermann K, Druzinic R, Witko M, Wagner F, Petersen M (1999) Phys Rev B 59(16):10583. https://doi.org/10.1103/PhysRevB.59.10583. https://link.aps.org/doi/10.1103/PhysRevB.59.10583
435. Vyboishchikov SF, Sauer J (2001) J Phys Chem A 105(37):8588. https://doi.org/10.1021/jp012294w. http://pubs.acs.org/doi/abs/10.1021/jp012294w
436. Costa AD, Mathieu C, Barbaux Y, Poelman H, Dalmai-Vennik G, Fiermans L (1997) Surf Sci 370(2–3):339. https://doi.org/10.1016/S0039-6028(96)00956-9. http://linkinghub.elsevier.com/retrieve/pii/S0039602896009569
437. Goschke RA, Vey K, Maier M, Walter U, Goering E, Klemm M, Horn S (1996) Surf Sci 348(3):305. https://doi.org/10.1016/0039-6028(95)00998-1. http://linkinghub.elsevier.com/retrieve/pii/0039602895009981
438. Asmis KR, Santambrogio G, Brümmer M, Sauer J (2005) Angew Chem Int Edn 44(20):3122. https://doi.org/10.1002/anie.200462894. http://doi.wiley.com/10.1002/anie.200462894
439. Calatayud M, Andrés J, Beltrán A (2001) J Phys Chem A 105(42):9760. https://doi.org/10.1021/jp011535x. http://pubs.acs.org/doi/abs/10.1021/jp011535x
440. Calatayud M, Mguig B, Minot C (2003) Surf Sci 526(3):297. https://doi.org/10.1016/S0039-6028(02)02673-0. http://linkinghub.elsevier.com/retrieve/pii/S0039602802026730
441. Avdeev VI, Zhidomirov GM (2005) J Struct Chem 46(4):577. https://doi.org/10.1007/s10947-006-0174-2. http://link.springer.com/10.1007/s10947-006-0174-2
442. Alexopoulos K, Hejduk P, Witko M, Reyniers MF, Marin GB (2010) J Phys Chem C 114(7):3115. https://doi.org/10.1021/jp910685z. http://pubs.acs.org/doi/10.1021/jp910685z
443. Avdeev VI, Tapilin VM (2009) J Phys Chem C 113(33):14941. https://doi.org/10.1021/jp904211a. http://pubs.acs.org/doi/10.1021/jp904211a
444. Deo G, Wachs I, Haber J (1994) Crit Rev Surf Chem 4:141
445. Coudurier G, Védrine J (2000) Catal Today 56(4):415. https://doi.org/10.1016/S0920-5861(99)00301-6. http://linkinghub.elsevier.com/retrieve/pii/S0920586199003016
446. Avdeev VI, Tapilin VM (2010) J Phys Chem C 114(8):3609. https://doi.org/10.1021/jp911145c. http://pubs.acs.org/doi/10.1021/jp911145c
447. Todorova TK, Döbler J, Sierka M, Sauer J (2009) J Phys Chem C 113(19):8336. https://doi.org/10.1021/jp811387z. http://pubs.acs.org/doi/10.1021/jp811387z
448. Inomata M, Miyamoto A, Murakami Y (1980) J Catal 62(1):140. https://doi.org/10.1016/0021-9517(80)90429-7. http://linkinghub.elsevier.com/retrieve/pii/0021951780904297
449. Volta JC, Portefaix JL (1985) Appl Catal 18(1):1. https://doi.org/10.1016/S0166-9834(00)80296-1. http://linkinghub.elsevier.com/retrieve/pii/S0166983400802961
450. Ono T, Numata H (1997) J Mol Catal A Chem 116(3):421. https://doi.org/10.1016/S1381-1169(96)00425-6. http://linkinghub.elsevier.com/retrieve/pii/S1381116996004256
451. Bielanski A, Haber J (1990) Oxygen in catalysis. Marcel Dekker, New York, NY
452. Sauer J, Döbler J (2004) Dalton Trans (19):3116. https://doi.org/10.1039/B402873B. http://xlink.rsc.org/?DOI=B402873B
453. Magg N, Giorgi JB, Schroeder T, Bäumer M, Freund HJ (2002) J Phys Chem B 106(34):8756. https://doi.org/10.1021/jp0204556. http://pubs.acs.org/doi/abs/10.1021/jp0204556
454. Ganduglia-Pirovano MV, Sauer J (2005) J Phys Chem B 109(1):374. https://doi.org/10.1021/jp046233k. http://dx.doi.org/10.1021/jp046233k

455. Słoczyński J, Grabowski R, Kozłowska A, Tokarz-Sobieraj R, Witko M (2007) J Mol Catal A Chem 277(1–2):27. https://doi.org/10.1016/j.molcata.2007.07.022. http://linkinghub.elsevier.com/retrieve/pii/S1381116907004190
456. Goclon J, Grybos R, Witko M, Hafner J (2009) Phys Rev B 79(7):075439. https://doi.org/10.1103/PhysRevB.79.075439. https://link.aps.org/doi/10.1103/PhysRevB.79.075439
457. Gilardoni F, Weber J, Baiker A (1997) J Phys Chem A 101(34):6069. https://doi.org/10.1021/jp9701606. http://pubs.acs.org/doi/abs/10.1021/jp9701606
458. Soyer S, Uzun A, Senkan S, Onal I (2006) Catal Today 118(3–4 SPEC. ISS.), 268. https://doi.org/10.1016/j.cattod.2006.07.033. http://linkinghub.elsevier.com/retrieve/pii/S0920586106004433
459. Anstrom M, Dumesic J, Topsøe NY (2002) Catal Lett 78(1–4):281. https://doi.org/10.1023/A:1014996215585
460. Kornelak P, Michalak A, Najbar M (2005) Catal Today 101(2):175. https://doi.org/10.1016/j.cattod.2005.01.015. http://linkinghub.elsevier.com/retrieve/pii/S0920586105000167
461. Szaleniec M, Drzewiecka-Matuszek A, Witko M, Hejduk P (2013) J Mol Model 19(10):4487. https://doi.org/10.1007/s00894-013-1951-4. http://link.springer.com/10.1007/s00894-013-1951-4
462. Gruber M, Hermann K (2013) J Chem Phys 139(24):244701. https://doi.org/10.1063/1.4849556. http://aip.scitation.org/doi/10.1063/1.4849556
463. Engeser M, Schröder D, Schwarz H (2005) Chem Eur J 11(20):5975. https://doi.org/10.1002/chem.200401352. http://doi.wiley.com/10.1002/chem.200401352
464. Gauld JW, Radom L (1997) J Am Chem Soc 119(41):9831. https://doi.org/10.1021/ja970785h. http://pubs.acs.org/doi/abs/10.1021/ja970785h
465. Göbke D, Romanyshyn Y, Guimond S, Sturm JM, Kuhlenbeck H, Döbler J, Reinhardt U, Ganduglia-Pirovano MV, Sauer J, Freund HJ (2009) Angew Chem Int Edn 48(20):3695. https://doi.org/10.1002/anie.200805618. http://doi.wiley.com/10.1002/anie.200805618

Molecular Electrochemistry of Coordination Compounds—A Correlation Between Quantum Chemical Calculations and Experiment

Piotr P. Romańczyk and Stefan S. Kurek

Abstract This contribution shows how molecular electrochemistry may benefit from the application of DFT methods combined with implicit solvent models. The progress in quantum chemical calculations, including efficient solvation models, has brought about the development of effective computational protocols that allow accurate (to 0.05 V) reproduction of experimental redox potentials of mono- and dinuclear complexes, including electrocatalytically relevant systems and mixed-valence compounds. These calculations may also help to understand how electronic and structural factors, modulated by the changes in both first and second coordination spheres, and the local environment (dielectric medium and specific interactions), govern the ability of transition metal complexes to undergo electron transfer (ET) processes. Understanding the principles that lie behind it is of great importance in redox chemistry and catalysis, and biological systems. After a brief introduction to modelling approaches and discussion of challenges for calibration of computational protocols based on comparison with experimental data, a number of noteworthy case studies are given. Specifically, the determination of ferrocenium/ferrocene absolute potentials in solvents commonly used in electrochemistry is discussed, the redox behaviour of Cu and Fe systems affected by H-bonding, followed by the presentation of intriguing properties of mono- and bimetallic Mo/W scorpionates. Particularly, electrochemical communication between metal centres and a baffling (auto)catalytic dehalogenation triggered by ET through a $C{-}H\cdots O_{alkoxide}$ hydrogen bond, the mechanism of which was unravelled owing to the application of dispersion-corrected DFT calculations, are highlighted.

P. P. Romańczyk (✉) · S. S. Kurek
Faculty of Chemical Engineering and Technology, Molecular Electrochemistry Group,
Cracow University of Technology, ul. Warszawska 24 31-155, Kraków, Poland
e-mail: piotrom@chemia.pk.edu.pl

S. S. Kurek
e-mail: skurek@chemia.pk.edu.pl

E. Broclawik et al. (eds.), *Transition Metals in Coordination Environments*,
Challenges and Advances in Computational Chemistry and Physics 29,
https://doi.org/10.1007/978-3-030-11714-6_13

1 Introduction

Molecular electrochemistry addresses redox behaviour and transformation of molecules initiated by electron transfer (ET). Understanding the electronic and structural factors affected by the changes in both first and second coordination spheres of a metal–ligand system, and the effects of local environment (dielectric medium and specific interactions) that govern the ability of transition metal complexes to undergo ET processes, is of great importance in organic and inorganic redox chemistry and catalysis, in artificial and biological systems. In this contribution, we focus on computational protocols that use DFT combined with continuum solvation models allowing accurate (to 0.05 V) prediction of redox potentials. After brief introduction to modelling approaches, we discuss selected examples of mono- and dinuclear complexes, including ferrocene—an internal redox potential standard in non-aqueous electrochemistry, (bio)catalytically relevant Mo, W, Cu or Fe systems, and mixed-valent (MV) compounds with interesting and unexpected electronic properties, for which results of DFT computations successfully match experimental properties. The chapter also covers the second coordination sphere and H-bonding effects on reduction potential and the problem of electronic coupling in MV compounds assessed based on electrochemical measurements. The last part of the chapter describes ET-triggered bond-breaking processes and elucidation of possible reaction pathways with the use of computations. Case studies on Mo and W nitrosyl alkoxide scorpionates are reviewed to show how molecular electrocatalysis of polychlorinated alkanes dehalogenation and its inhibition can be controlled by non-covalent interactions, like C−H···O hydrogen and/or halogen bonding, and dispersion forces, bringing about activation of polyhalogenated alkanes and giving rise to enhanced rates of electron transfer. Revealing the relationship of described systems with some enzyme-mimicking activity is significant for the rational designing of new electrocatalysts.

2 DFT Modelling of Redox Potentials

Redox potential, a characteristic property of a given metal–ligand system, can be controlled by introducing changes in the first and second coordination spheres, and the local environment—solvent, ions and molecules present in the solution. The prediction of accurate redox potentials is important in elucidating catalytic reaction mechanisms, catalyst designing, and in modelling of active sites in metalloenzymes. The use of the approach based on the DFT combined with implicit solvent models that started with the 2002 Friesner and Baik paper [1] has recently been reviewed [2, 3], and here, it will only be briefly outlined, followed by the discussion of selected examples involving transition metal systems.

Scheme 1 Thermodynamic cycle commonly used in calculation of $\Delta_{ea}G^{\circ}_{(soln)}$. An alternative approach is to calculate $\Delta_{ea}G^{\circ}_{(soln)}$ directly within continuum solvation model, see text and [7]

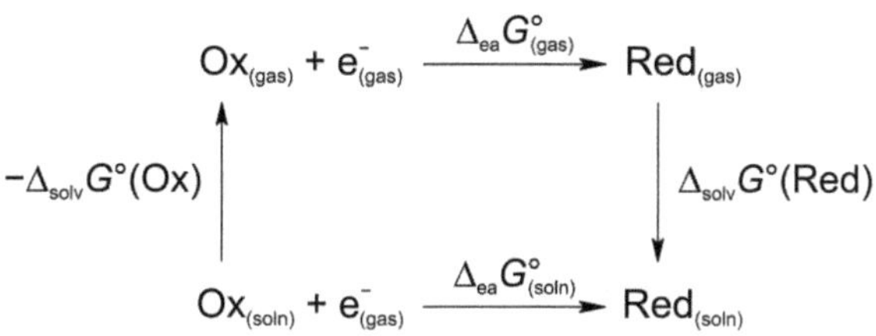

The standard redox potential (E°) of an Ox/Red couple is related to the standard Gibbs energy of an electron attachment in solution, i.e. $Ox_{(soln)} + ne^{-}_{(gas)} \rightarrow Red_{(soln)}$, by (1):

$$E^{\circ}_{Ox/Red} = -\Delta_{ea}G^{\circ}_{(soln)}/nF - E^{\circ}_{abs}(\text{ref}) \tag{1}$$

where F is Faraday's constant and n is the number of transferred electrons (all over this chapter, n is assumed to be 1), $E^{\circ}_{(abs)}$ is the absolute potential of the reference used, usually standard hydrogen electrode (SHE; recommended values are 4.281 V [4] 4.32 V [5], or 4.44 V [6]),[1] saturated calomel electrode (SCE; 0.241 V vs. SHE), Ag/AgCl (in 3 M NaCl 0.209 V vs. SHE) or Fc^{+}/Fc (vide infra). $\Delta_{ea}G^{\circ}_{(soln)}$ can be calculated using a thermodynamic cycle (see Scheme 1), and expressing $\Delta_{ea}G^{\circ}_{(gas)}$ as the sum of electron affinity, and thermal and entropic contributions, from the equation:

$$\Delta_{ea}G^{\circ}_{(soln)} = -E_{ea,gas} + \Delta\Delta G^{1atm}_{0\rightarrow 298K,gas} + \Delta\Delta G^{1atm\rightarrow 1M} + \Delta_{solv}G^{\circ}(\text{Red}) - \Delta_{solv}G^{\circ}(\text{Ox}) \tag{2}$$

In the above equation, $E_{ea,gas}$ ($=-\Delta H_0$) is the adiabatic electron affinity of the oxidised form, i.e. the negative of the enthalpy change at 0 K for the reduction, the $\Delta\Delta G^{1atm}_{0\rightarrow 298K,gas}$ term is a contribution to the Gibbs energy of heating from 0 to 298 K, $\Delta\Delta G^{1atm\rightarrow 1M}$ is a correction for the standard state change from 1 atm to 1 mol L^{-1} (1.89 kcal mol^{-1} at 298 K) that cancels out for processes with the same number of molecules in the initial and final states, and $\Delta_{solv}G^{\circ}(\text{Red}) - \Delta_{solv}G^{\circ}(\text{Ox})$ is the difference between the solvation free energies of reduced and oxidised forms.

Thus, to get E°, one needs to combine the gas-phase energetics, calculated at a reasonably high level of theory, with thermodynamic contributions derived from

[1]Differences between $E^{\circ}_{(abs)}$(SHE) values result from different values of the hydration free energy of H^{+} and the $H^{+}_{(gas)}$ free energy of formation. The value of 4.44 V based on experimental data at 298.15 K refers to the outer potential of the phase (Volta potential), whereas the computed values of 4.28 or 4.32 V correspond to the inner potential of the phase (Galvani potential), which is exactly what the QC results refer to. The IUPAC-recommended experimental value, (4.44 ± 0.02) V, is not very appropriate; it must differ by the value of surface potential of water from the computed data. The surface potential is not directly measurable and must be calculated. Its values range between 0.16 and 0.14 V, resulting in the E_{abs}(SHE) equal to 4.28 or 4.32 V. A future work may bring a new value, but not differing more than by 40 mV from the published so far.

frequency calculation commonly done in the harmonic approximation (for the gas-phase or solution-optimised structure),[2] and $\Delta_{solv}G°$. The solvation free energies can be received at a low computational cost employing an implicit solvent model (ISM), when specific interactions, i.e. hydrogen bonding, solvent molecule coordination, can be neglected. Otherwise, there is a need for taking into account one or a few explicit solvent molecules to form a solvent–solute cluster embedded in a continuous environment [8], in water; however, explicit solvation, e.g. within a QM/MM framework, may be required [9]. The commonly used continuum solvation models are polarisable continuum models (PCMs) [10], the solvation model based on solute density (SMD) [11] (which uses the IEF-PCM algorithm for bulk electrostatics), the conductor-like screening model (COSMO), and COSMO for realistic solvation COSMO-RS (statistical thermodynamics for the molecular surface interactions, COSMO polarisation charge densities used) [12]. $\Delta_{solv}G°$ may be computed using, e.g. B3LYP/6-31+G(d,p) (COSMO-RS requires the use of BP/TZP), in accordance with the level of theory applied in parametrisation of the solvation model; nevertheless, the results are not largely dependent on the model chemistry, which delivers a reasonably accurate electronic density for a solute [3]. Nowadays, continuum solvation models allow obtaining reliable redox potentials with mean unsigned error (MUE) even of ca. 0.1 V; for organic solutes in acetonitrile using SMD Guerard and Arey [13] obtained MUE equal to 0.13 V, in water it is higher. Note that we frequently take advantage of error compensation between $E_{ea,gas}$ and $\Delta_{solv}G°$ (and other effects). In case structural changes upon solvation are not very large, using gas-phase optimised geometries in both phases is justified and recommended [2].

2.1 Problems with Calibration Based on Comparison with Experimental Data

Potentials calculated with DFT/ISM can be directly compared with voltammetrically determined $E_{1/2}$ potentials[3] for reversible or quasi-reversible processes, however, provided well-calibrated[4] computational protocols (method/basis set, solvation model)

---

[2]If the liquid and gas-phase solute structures differ appreciably, using partition functions computed in solution is a correct approach, see [83].

[3]$E_{1/2}$ denotes a mean of voltammetric anodic and cathodic peak potentials. $E_{1/2} = E_f$ (E_f is formal potential) if the diffusion coefficients of the oxidised (Ox) and reduced (Red) forms are equal. E_f potential refers to the situation, when the concentration ratio of [Ox] to [Red] at the electrode surface is equal to one, which obviously is not true in the case of more complex systems, like, e.g. proton-coupled reactions. Then, knowing the electrode reaction, and the concentration of that other participating species, like protons, the relation between the E_f and E° (based on concentrations) can easily be found.

[4]Calibrated against the best available experimental data or highly accurate QM methods, such as G3(MP2)-RAD or explicitly correlated coupled cluster theory CCSD(T)-F12 calculations, which could now be applied even for midsized metal complexes, if only single reference quantum methods are applicable.

for a given class of compounds have been established. This methodology, however, does not account for a couple of factors that may also affect the redox potential.

One of these factors is the ion energy lowering caused by ionic strength, induced by base electrolyte (also called supporting electrolyte; it is used in large excess to ensure diffusion-controlled transport of electroactive species, thus making migration in the electric field negligible). As the following example shows, this stabilisation energy is important for non-polar solvents. In these solvents, it comes to the association of the base electrolyte and this should be taken into account, when calculating the ionic strength. For n-Bu_4NPF_6 in dichloromethane, the fraction of free ions in 0.1 M solution in CH_2Cl_2 is equal to only 7.7% [14]. The resulting ionic strength would cause Fc^+/Fc potential shift to a value lower by 53 mV, as estimated from the extended Debye–Hückel law. For more polar solvents, the shift would be much smaller, even, if assumed that there would not come to any association in these polar solvents. Another approach is measuring the variation in the dielectric constant (relative permittivity) value effected by the presence of electrolytes. Using dyes that show known pronounced solvatochromism, it is possible to determine the dielectric constant based on their fluorescence measurements. It appeared that for n-Bu_4NBF_4 solutions in dichloromethane, its dielectric constant increases substantially from 8.93 for neat solvent to 14.1 for 0.1 M concentration, and to 18.0 for 0.2 M solution [15], which lowers the oxidation potential of ferrocene accordingly.

Another effect that leads to redox potential shift owing to stabilisation of ions is ion-pairing. For highly charged species, like $[Ru(bpy)_3]^{2+/3+}$, ion-pairing effect becomes significant attaining almost 500 mV in CH_2Cl_2, however, this effect is drastically reduced to below 50 mV in polar MeCN, as shown by Batista group [16] by including explicit BF_4^- counterions in modelling. Ion-pairing effect may be relatively small for molecules like ferrocene in 1,2-dichloroethane containing 0.1 M hexafluorophosphate or perchlorate, in which case the measured ion-pairing effect is equal to ca. 30 mV [17], which means indistinguishable from the simple ionic strength effect described above. However, in the case of ferrocenes with electron-donating or withdrawing substituents, interacting more specifically with counter ions, this may bring about the shift of almost 200 mV [17].

It is still a common practice of quoting the redox potentials of processes proceeding in non-aqueous systems as they were measured, usually versus aqueous reference electrodes, like SCE or Ag/AgCl electrodes. This practice should be discouraged, as the potential measured in this way contains liquid junction potential occurring at the boundary between the non-aqueous phase in the cell and the aqueous solution in the reference electrode. This potential of the magnitude from tens to even a few hundred mV depends not only on the solvents, but is a function of the types and concentrations of electrolytes in both phases [18, 19]. For instance, liquid junction potential between 0.1 M solution of tetraethylammonium picrate in DMF and a saturated aqueous KCl solution in SCE is 172 mV, and when the solvent is acetonitrile, it is equal to 93 mV [18]. In most cases, the higher potential is at the organic phase side, making the potentials measured versus SCE higher by these values. Despite some newly published procedures that enable the experimental determination of the Gibbs energy of transfer of single ions between solvents [20], it is by far more convenient to use an internal standard, of which ferrocene is the most popular. Its oxidation potential obviously depends on the solvent used, and in accurate calculations, it should be remembered that it may be affected by the type and concentration of the base electrolyte, which has to be used in excess in voltammetric techniques.

Recently, it has been reported that using microporous glass frits, like Vycor glass, to separate aqueous and non-aqueous solutions, may be a source of an additional error exceeding 50 mV in the measured potential [21]. However, it concerns only solutions of low ionic strength. For base electrolyte concentrations used typically in cyclic voltammetry (0.1 M) and concentrated aqueous solutions in reference electrodes, this effect is fortunately close to zero.

2.2 *Absolute Potential of Fc^+/Fc*

Ferrocene has been used as an internal standard in non-aqueous electrochemistry since long, and in 1984 IUPAC recommended [22, 23] to report measured potentials versus Fc^+/Fc, then assumed to be a solvent-independent redox system. This was based on a virtually constant difference between $E_{1/2}$ of ferrocene and *bis*(biphenyl)chromium(I) tetraphenylborate measured in 0.1 M tetraethyl-/tetrabutylammonium perchlorate solutions [22]. Following the above IUPAC recommendation and another one defining the E°_{abs}(SHE) equal to (4.44 ± 0.02) V [6] (see footnote 1 on page 3), and combining it with the only known value of E°(Fc^+/Fc) in aqueous solutions measured as 0.400 V versus SHE [24], it was assumed that the universal absolute potential, $E^{\circ}_{abs}\left(Fc^+/Fc\right)$, the same for all solvents, is equal to 4.84 V. In fact, as it will be shown below, the oxidation potential of Fc depends on the solvent used. Ferrocene undergoes reversible 1e oxidation. Owing to the fact that geometry differences between neutral and oxidised forms are negligible, inner reorganisation energy is very low. This results in a perfect 59 mV value of the cyclic voltammetric anodic-to-cathodic peak separation, as expected for a reversible or Nernstian electrode process. In this case, it means fast ET, the rate of electrode process controlled by diffusion, and at each point, the concentration ratio [Ox]/[Red] at the electrode surface is equal to that calculated from the Nernst equation for the electrode potential, provided ohmic potential drop (iR) is zero).

As mentioned above, the 1984 IUPAC recommendation was based on the assumption of the Fc^+/Fc absolute redox potential being the same in all solvents, which soon appeared to be only a very crude approximation. The development of efficient quantum chemical calculations and accurate solvent models allowed the calculation of the Fc^+/Fc potential in various solvents. Thus, Namazian et al. [25] carried out high-level ab initio calculations to obtain reliable values of the absolute reduction potentials of Fc^+/Fc in MeCN, 1,2-dichloroethane (DCE), and dimethylsulphoxide (DMSO). They combined the G3(MP2)-RAD-Full-TZ gas-phase energetics (on top of LanL2TZf/6-31G(d) geometries), regarded as "chemically accurate", with solvation Gibbs energies from two implicit solvent models, SMD and COSMO-RS. The authors have also established that employing PCM and CPCM models results in positive $\Delta_{solv}G°(Fc)$ in MeCN, which is obviously physically unrealistic, because Fc is well soluble in this solvent.

Romańczyk et al. [8] obtained reliable values of $E°(Fc^+/Fc)$ from computations at the SMD/B3LYP-D2/def2-QZVPPD//B3LYP/LanL2TZf/6-31G(d) level of theory. This method provides virtually the same gas-phase energetics for ionisation of Fc as does the G3(MP2)-RAD-Full-TZ level. Table 1 compares $E°_{(abs)}$ of Fc^+/Fc couple, computed using the methodologies described above, for other non-aqueous solvents commonly used in electrochemistry, and water, for which no such data were available. Note that the values obtained by Namazian et al. [25] (numbers marked with footnote c Table 1) include the correction for the Gibbs energy of free electrons (−0.038 eV) based on Fermi–Dirac statistics, whereas $E°$s computed by us (footnotes a and b) follow the "ion convention" (see [3] for explanation of thermochemical conventions concerning the thermodynamics of the electron). We found a linear dependence between ferrocene oxidation potentials and the reciprocal of solvent dielectric constants only in the case of the SMD solvation; the obtained linear regression equation reads $E°_{abs} = 4.910 + 2.091/\varepsilon$ (for 8 solvents), which may be employed in the estimation of this potential in other solvents. Obviously, in the case of solvation models for which the bulk electrostatic contribution to $\Delta_{solv}G°$ is calculated from the Poisson equation (like SMD) or the generalised Born approximation (e.g. SM12 [26]), a linear relationship between the difference $\Delta_{solv}G°(Red) - \Delta_{solv}G°(Ox)$ (and $E°$) with respect to $1/\varepsilon$ is expected. It should be noted that the $E°_{(abs)}(Fc^+/Fc)$ values based on the SMD and COSMO-RS solvation energies are very similar only for polar solvents, whereas for non-polar ones they differ appreciably (even by 0.25 V). The COSMO-RS derived $E°_{(abs)}$s in DCE and DMSO solvents are not consistent with experimental data: higher $E°_{(abs)}$s in solvents of lower polarity (see Table 1) and the measured formal potential in DCE (or CH_2Cl_2) being of an order of 0.1 V higher than that in DMSO [27, 28]. Moreover, Krishtalik [29] has demonstrated that for metallocenes, like ferrocene and cobaltocene, the dielectric response energy contribution to the solvation energy can be calculated with the Born equation applying the same effective radii for all the aprotic solvents.

Knowing the absolute potentials of the Fc^+/Fc pair in a given solvent makes it possible to present the computed redox potentials for other redox processes in the ferrocenium/ferrocene scale, i.e. versus Fc^+/Fc, and compare them with the experimentally determined values in the same scale. It is also possible, to convert thus computed potentials to the commonly applied SCE scale by using the experimental formal potential of Fc^+/Fc in a given solvent measured versus SCE [31], by simply adding it to the value calculated versus Fc^+/Fc.

$$E_{comp}(\text{vs. SCE}) = E_{comp}\left(\text{vs. Fc}^+/\text{Fc}\right) + E°\left(\text{Fc}^+/\text{Fc vs. SCE}\right) \tag{3}$$

Please note that the potential versus SCE also depends on the type and concentration of supporting electrolyte used. Reference [31] cites a couple of examples of this variation. The differences due to using different base electrolytes in the same solvents may reach ca. 40 mV, but we know from our own experience that changing the concentration of the base electrolyte in dichloromethane from 0.1 to 0.2 M

Table 1 Absolute reduction potentials of Fc^+/Fc couple in non-aqueous solvents commonly used in electrochemistry, and water. The values from two implicit solvent models, SMD or COSMO-RS, are compared

Solvent	ε	$E^{\circ}_{(abs)}$		
		SMD	COSMO-RS	Exp.
Tetrahydrofuran	7.43	5.192[a]	4.959 (4.945)[b]	
CH_2Cl_2	8.93	5.143[a]	4.971 (4.966)[b]	
1,2-dichloroethane	10.13	5.116[a] 5.102[c]	4.974 (4.971)[b] 4.927[c]	5.08[d] 5.01[e]
PhCN	25.59	4.991[a]	4.968 (4.976)[b]	
MeOH	32.61	4.974[a]	4.974 (4.986)[b]	
MeCN	35.69	4.967[a] 4.964[c]	4.988 (5.026)[b] 4.988[c]	4.98[f]
N,*N*-dimethylformamide	37.22	4.966[a]	4.971 (4.986)[b]	
DMSO	46.83	4.954[a] 4.952[c]	4.992 (5.027)[b] 5.043[c]	
Water	78.36	4.937[a]	4.931 (4.927)[b]	4.84[g]

[a]Computed at the SMD/B3LYP-D2/def2-QZVPPD//B3LYP/LanL2TZf/6-31G(d) level; data from [8] or this work. The potential in tetrahydrofuran given in Table is our recently obtained revised value

[b]Our results obtained using the COSMO-RS module with Klamt or ADF combi2005 (in parentheses) parameters as implemented in ADF 2017.105 program package [30] and gas-phase energetics calculated at the same level as in footnote a

[c]Calculated using G3(MP2)-RAD-Full-TZ gas-phase energies (on top of B3LYP/LanL2TZf/6-31G(d) geometries) combined with SMD or COSMO-RS (Klamt parameters using ADF 2008.01 software) solvation free energies [25]. These values include the correction for the Gibbs energy of the electron (−0.038 eV)

[d]Obtained from formal redox potential of Fc^+/Fc in DCE (assumed to be 0.64 V vs. SHE) and by taking 4.44 V as the absolute potential of SHE

[e]From [25]. Value derived by taking Fc ionisation energy equal to experimental value of 6.71 eV that it is significantly lower than that from G3(MP2)-RAD-Full-TZ calculations (7.047 eV)

[f]From [25]. Obtained using $E_{1/2}$ of Fc^+/Fc in MeCN versus SCE (0.38 V [31]) and 4.60 V for SCE in MeCN. The latter value is a result of a combination of SHE potential assumed to be 4.36, and 0.24 V versus SHE for SCE

[g]Obtained from $E^{\circ}(Fc^+/Fc)$ in aqueous solutions (0.400 V vs. SHE) and by taking 4.44 V as E°_{abs}(SHE); see footnote 1 for comment on this value

may bring about larger deviations. So it is always better to rely on an internal standard, like ferrocene. Conversion of calculated absolute potentials to the fundamental electrochemical scale versus SHE or SCE in aqueous systems is also possible by subtracting E°_{abs}(SHE), equal to 4.281 V [4] (or 4.32 V [5]). By further subtracting E(SCE vs. SHE) = 0.241 V, we get the potential versus SCE. Note that the value of absolute potential of SCE widely cited after [1], 4.1888 V, is obviously wrong, as it was calculated by erroneously assuming the potential of SCE being negative (!), −0.2412 V vs. SHE. For the correct value see footnote f to Table 1.

2.3 *Molybdenum and Tungsten Scorpionates*

The 16*e* Mo and W nitrosyl scorpionates (a common name for tris(pyrazol-1-yl)borate complexes), stable in solid and in solution, show well-defined reversible reduction potentials (and in some cases also reversible 1e oxidation processes), readily controllable in the enormous range of 2200 mV by selecting suitable co-ligands [32]. Their *catalytic* properties in dehalogenation of some chloroalkanes were detected only recently (vide infra), whereas the closely related oxo-analogues were utilised in enzyme modelling a long time ago [33, 34].

The redox potential of the nitrosyl, oxo- and oxothio-$Mo^{II/I}$ and $W^{II/I}$ pairs in the catalytically relevant alkoxide scorpionates can be accurately reproduced (to 0.05 V) by DFT calculations at the IEF-PCM/B3LYP(-D2)/LACV3P+/6-311++G(2d,2p) level on top of the IEF-PCM/B3LYP/LANL2DZ/6-31G(d,p) or, when more precise geometries are needed, COSMO/B3LYP/def2-TZVPP structures. A triple-ζ basis set containing a diffuse d function on metal is essential in description of the low Mo^I or W^I oxidation state. Amongst the computationally investigated examples of this class of complexes are the $[Mo(NO)Tp^{Me2}(OMe)_2]$ complex (Tp^{Me2} = hydrotris(3,5-dimethylpyrazol-1-yl)borate), which undergoes reduction at $E_{1/2} = -1.84$ V ($E^\circ_{calc} = -1.82$ V)[5] as well as oxidation at +0.91 V [35], and tungsten analogues [36] depicted in Fig. 1. As can be seen, in all these cases, the calculated potentials match the experimental values very accurately, and ν_{NO} frequency is reproduced well for the W^{II} complex. Interestingly, the difference between the SOMO energies for the Mo and W dimethoxy scorpionates (0.58 eV) agrees nicely with the difference between their $E_{1/2}$ values for reduction (0.56 V). The characteristic feature of these dimethoxy Mo and W complexes is that the reduction is an evidently slower process with $\Delta E_p = 250$–122 mV, an effect of a relatively high inner-sphere reorganisation energy due to the appreciable elongation of the $M{-}O_{alk}$ bond (by 0.12 Å) and a decrease in $M{-}O_{alk}{-}C$ angle (thus, a lesser degree of $p\pi_O$-$d\pi_{Mo}$ charge donation occurs), whereas upon oxidation the structural changes are more subtle (Δd $M{-}O_{alk} = 0.05$ Å) and ΔE_p drops to 82–68 mV.

The potentials obtained in the DFT modelling of the nitrosyl Mo and W species in CH_2Cl_2 were received with the use of 4.84 V as $E^\circ_{abs}(Fc^+/Fc)$ (see above), which deserves a comment. Obviously, this value is not likely to be the absolute potential of Fc in CH_2Cl_2 (vide supra), and the remarkable accuracy (to 0.05 V) achieved for a series of $\{M^{II/I}(NO)(Tp^{Me2})\}^{2+/1+}$ alkoxides [35–37] is due to a systematic error of 0.3 V of the B3LYP functional (and/or the IEF-PCM solvation) for these complexes. Intriguingly, it seems that this is not the case for oxo-Mo^V species discussed below, i.e. where the same methodology as for the nitrosyl species was used but a value of 5.14 V for $E_{abs}(Fc)$ yielded excellent agreement with experiment. In contrast to the accurate B3LYP results, the non-hybrid functionals BP86 and OLYP underestimate the reduction potentials for this class of Mo–NO complexes, yielding too cathodic values by even hundreds of mV.

[5] The computed potential given in [35] (−1.80 V) was obtained from calculations done in a slightly differently contracted basis set (LANL2DZ) for Mo.

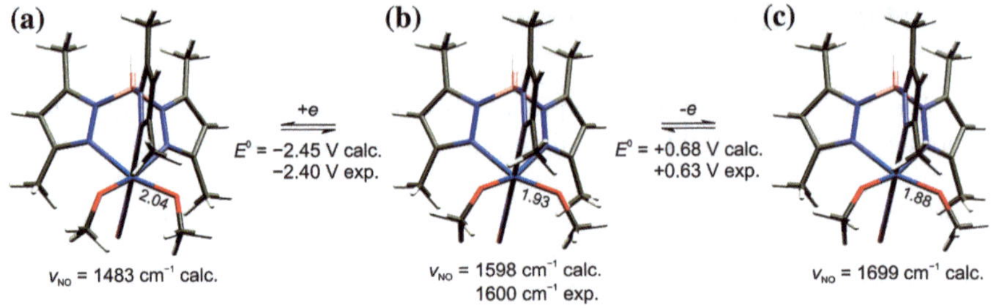

Fig. 1 Optimised geometries of neutral, one-electron reduced and one-electron oxidised forms of [W(NO)Tp^{Me2}(OMe)$_2$] with standard redox potentials and ν_{NO} frequencies obtained from IEF-PCM/B3LYP/LACV3P+/6-311++G(2d,2p)//COSMO/B3LYP/def2-TZVPP calculations in CH_2Cl_2; W-O$_{alkoxide}$ distances in Å

The methodology described just above for nitrosyl alkoxy scorpionates may also be successfully applied in redox potential modelling of oxo-Mo$^{VI/V}$ and cis-oxothio-Mo$^{VI/V}$ centres present in many Mo enzymes involved in oxygen or sulphur atom transfer reactions; scorpionate complexes capable to stabilise these centres have long been serving as excellent synthetic models of these enzymes [34]. Figure 2 shows two examples of such systems: [Mo^V(O)Tp^{Me2}(OMe)$_2$], electron transfer properties of which have been studied by Vannucci et al. [38], and [Tp^{iPr}-Mo^{VI}OS(OC$_6$H$_4$sec-Bu-2)] synthesised and characterised by Doonan et al. [39]. The latter species constitute a model of the active site in xanthine oxidase in sulphur atom transfer (SAT) reactions [40]. The Mo$^{VI/V}$ redox potentials calculated by us [37] for both complexes match excellently the experimental $E_{1/2}$ values (recorded in CH_2Cl_2 or MeCN) as well as E° value computed by Vannucci et al. (at the OPTX/TZP level with COSMO solvation effects) for the [Mo^V(O)Tp^{Me2}(OMe)$_2$] complex [38].

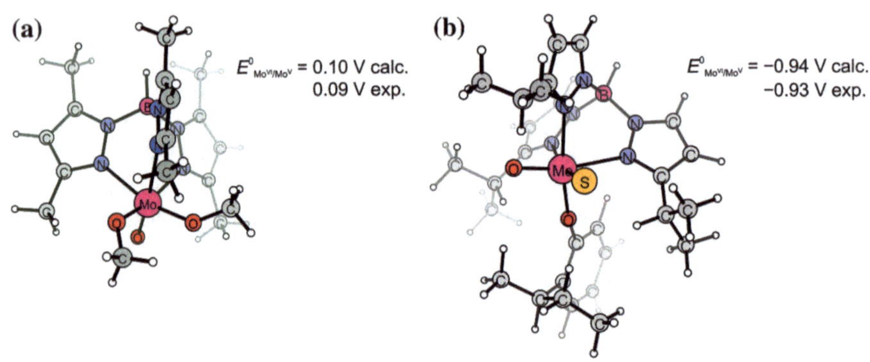

Fig. 2 Optimal geometries of **a** [Mo^V(O)Tp^{Me2}(OMe)$_2$] and **b** [Tp^{iPr}Mo^{VI}OS(OC$_6$H$_4$sec-Bu-2)] complexes in CH_2Cl_2 and MeCN, respectively, with standard redox potentials of Mo^{VI}/Mo^V; calculations at IEF-PCM/B3LYP/LACV3P+/6-311++G(2d,2p)//IEF-PCM/B3LYP/LANL2DZ/6-31G(d,p) level

2.4 *Effect of Second Coordination Sphere and H-Bonding Changes on Reduction Potential*

The environment surrounding the transition metal complex beyond the primary ligand may have a profound effect on its redox potential, particularly in biological systems [41]. Of particular significance may be varying the geometric structural parameters and non-covalent interactions in the secondary coordination sphere, like hydrogen, halogen, chalcogen and similar types of bonding with primary ligands. They may be pronounced specifically in enzymes, but there are more and more examples reported in simpler systems.

Glaser and co-workers [42] reported on the influence of hydrogen bonds on the reduction potential of Cu^{II} in an extended salen-like framework. They compared $E_{1/2}$ values of an unsubstituted $[\mathbf{L^{H}}Cu^{II}]$ complex with its analogue, $[(H_2\mathbf{L^{OH}})Cu^{II}]$, possessing OH groups H-bonded with the coordinated aryl-oxygen donor atoms (Fig. 3a). A positive shift of 270 mV manifests destabilisation of Cu^{II} oxidation state in $[(H_2L^{OH})Cu^{II}]$. The calculated (at the BH-LYP/TZV(p,d) level) shift is slightly higher (400 mV) compared with the experiment. The conclusive evidence that the effect is due to H-bonding and not to the presence of electron-donating OH substituents in the naphthalene ring was obtained from DFT calculations performed for a model complex, $[(H_2L^{OH,rot})Cu^{II}]$, with the OH groups rotated by 180° (which precludes the formation of H-bonding), showing that the $[(H_2L^{OH,rot})Cu^{II}]$ complex is reduced at a potential 180 mV *more cathodic* than $[L^{H}Cu^{II}]$ (i.e. stabilisation of the Cu^{II} state).

Yang et al. [43] using photoelectron spectroscopy and theoretical calculations probed the effect of a single H-bond on the redox potential of the $[Fe_4S_4]^{2+/3+}$ couple in cubane-like complexes being iron–sulphur proteins active site models (Fig. 3b). They have concluded that one OH···S hydrogen bond

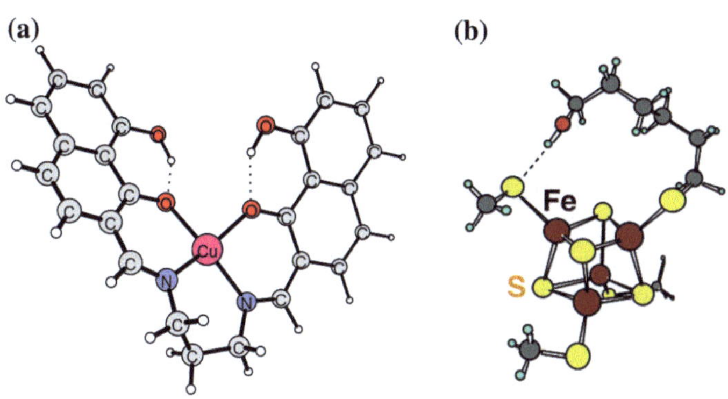

Fig. 3 **a** Optimal geometry of $[(H_2\mathbf{L^{OH}})Cu^{II}]$ (unsubstituted $[\mathbf{L^{H}}Cu^{II}]$ complex contains H atoms instead of OH groups) at BH-LYP/TZV(p,d) level [42], **b** B3LYP/6-31G** optimised structure of the H-bonded conformation of $[Fe_4S_4(SCH_3)_3(SC_6H_{12}OH)]^{2-}$. Reprinted with permission from [43]. Copyright 2004 American Chemical Society

to the terminal S-ligand increases the oxidation potential by ~130 mV. However, the main alkylene chain orientation change, accompanied by the Fe–S–C–C dihedral angle variation, induced by the H-bond formation, cannot be excluded, which also may affect the reduction potential.

We note that such geometrical effects can considerably alter the reduction potential, examples of which are salen-type Cu(II) complexes (an effect of 160 mV) [44] and a chelate $[W(NO)(Tp^{Me2})O(CH_2)_4O]$ complex which undergoes reduction at a potential 180 mV more anodic than its diethoxy analogue [45]. The redox potential shift in the latter example is excellently reproduced by DFT calculations (at the level of theory pointed out in Fig. 2) [37].

2.5 *Further Relevant Reports*

Recent literature provides a wealth of new successful applications of quantum chemical calculations that reveal how DFT/ISM approaches can be employed to design new complexes with specific redox potentials, interesting examples of which we shall refer to only briefly. Bím et al. [46] provided an example of how computational electrochemistry can be used to resolve the interpretation of electrochemical data obtained for experimentally elusive ferryl species, like $[(N4Py)Fe^{IV}O]$ (N4Py = *N*,*N*-bis(2-pyridylmethyl)-bis(2-pyridyl)methylamine), by using a protocol based on B3LYP-D3 energetics and COSMO-RS solvation, first created and calibrated for accurate evaluation of one-electron reduction potentials of a set of diverse mononuclear non-haem iron redox couples. Rudshteyn et al. [47] explained using a combined electrochemical and DFT study (SMD/B3LYP/LANL2DZ/6-311G*), how redox potentials may be affected by thermodynamic and structural factors in the case of tungsten–alkylidyne $W(CR)L_4X$ complexes (L is a neutral ligand, and X is an anionic ligand), which are of interest owing to their application as photoreductants and in the H_2 activation. In a review of the recent progress in the quantum chemical modelling of homogeneous water oxidation catalysis, Liao and Siegbahn [48] also discussed the DFT/ISM approach, including the B3LYP-D3, B3LYP*-D3 and M06 series of functionals that are commonly applied in this field, to calculate redox potentials. Knowing them enables constructing catalytic cycles and hence revealing the reaction mechanisms. Some DFT functionals like MN12-L with Wertz's correction and TPSS-D3 with the IEF-PCM/scaled-UAKS implicit solvation model may attain the accuracy (RMSD = 154 mV with respect to experiment) of CCSD(T) methods for specific classes of complexes, as demonstrated by Šebesta et al. [49] in the evaluation of reduction potentials for the Pt(IV) complexes: tetraplatin, satraplatin and its derivatives, well known for their anticancer activity.

3 Electrochemical Communication Across Saturated Dioxoalkylene and Oxo Bridges in Dimolybdenum Scorpionates

Electron transfer between donor–acceptor pairs (D-A) is a process of great importance in chemical, physical and biological systems, particularly metalloproteins, where covalent σ- and π-bonds, H-bonds [50, 51] and through-space tunnelling often assist ET [52–55]. The investigation of transition metal complexes in mixed-valence states, of which the Creutz–Taube ion ($Ru^{II}Ru^{III}$) is the best-known example [56], reveals the correlation between the magnitude, range and character of D···A electronic interactions and the structure of the bridging ligand, such as its length, substitution pattern, geometric/electronic structure [57, 58], and the surrounding medium (solvent, base electrolyte [59]). Moreover, the relationship between the localisation/delocalisation of an unpaired electron and the relative orientation of the local D and A coordinate systems may confer the properties of the molecular switches and rheostats on complexes [60–62]. According to the Robin and Day seminal paper [63], mixed-valence systems are divided into three main classes corresponding to the degree of electronic coupling between D and A along the bridge. In compounds belonging to class I, there is no coupling between diabatic redox states and the odd electron is fully localised. Conversely, the charge in class III molecules is totally delocalised over both redox centres (a single ground-state minimum on the free-energy surface without ET barrier), and systems in-between fit into class II (a double single ground-state minimum). This has been discussed in detail in review papers [52, 58] in the light of Marcus theory [64], with implications for the properties of mixed-valence complexes.

Metal–metal electronic interactions in a mixed-valence state are manifested in electrochemical studies. For instance, in the voltammogram, the measure of electrochemical interactions is differences in reduction (or oxidation) potentials, $\Delta_{\text{red/ox}}E_{1/2}$, between both metal centres. The $\Delta_{\text{red/ox}}E_{1/2}$ difference is observed when the change in electron density on one metal centre (caused by the initial reduction or oxidation) affects the second centre by a combination of *through-space* (Coulombic) and *through-bond* (inductive or resonance contribution) interactions. Thermodynamic stability of mixed-valence ions is quantified by the comproportionation constant K_c, for the equilibrium between isovalent and mixed-valence forms:

$$\{\text{M-Q-M}\}^n + \{\text{M-Q-M}\}^{n-2} \rightleftharpoons 2\{\text{M-Q-M}\}^{n-1} \tag{4}$$

The value of K_c is related to the splitting of redox potentials ($\Delta E°$) by: $K_c = e^{F\Delta E°/(RT)}$ (it can range from the statistical value of 4 to more than 10^{24} ($\Delta E_{1/2} =$ 1440 mV) as reported for the Creutz–Taube analogue $[\{Mo(NO)(Tp^{Me2})Br\}_2(\mu\text{-}pz)]^-$ [65]). A practical consequence of this equilibrium is that only for sufficiently large electrochemical interactions ($\Delta E_{1/2} > 200$ mV) it is possible to selectively reduce or oxidise (electrochemically or chemically) just one redox centre.

Note that directly correlating K_c, for systems with two identical redox sites, with the degree of electronic coupling between the diabatic states for electron donor and acceptor (H_{ab} parameter) without other evidences (spectroscopic, QC calculations) for the extent of an extra-electron delocalisation may be wrong. However, it is routinely employed in the literature (see, e.g., [66, 67]). Only a qualitative relationship may exist, and, as demonstrated in an excellent review by Winter [68], a large redox potential splitting is not a necessary condition for appreciable degree of ground-state delocalisation in the MV state. Below, we will show two examples of entirely valence-trapped $Mo^{I}Mo^{II}$ systems (class I in Robin–Day classification) despite it showing substantial to very large potential splittings for two consecutive 1e reductions, demonstrating suitability of DFT modelling of their properties.

Mixed-valence complexes based on di-$\{Mo^{II/I}(NO)\}^{2+/1+}$ fragments stabilised by κ^3-hydrotris(3,5-dimethylpyrazol-1-yl)borate) (=scorpionate), discussed above, demonstrate a wide range of behaviour, from fully delocalised to valence-trapped. Their amazing electronic and magnetic properties have been described in the literature, mainly of those with π-coupled bridges [32, 57]. Generally, much less work has been devoted to species (this class and other classes) containing saturated alkylene ligand systems. Aside from the examples below, we should mention here a study by Ding et al. [69] who analysed the extent of delocalisation (H_{ab}) in Fc-bridge-Fc systems with various bridging ligands, including saturated C–C single bonds, by conventional and constrained [70] DFT methods. The latter approach consists in constructing charge-localised states by imposing an external constraint, i.e. adding an effective potential to the Hamiltonian. They concluded that only constrained DFT is capable to predict quantitatively the strength of electronic coupling for systems, which are incorrectly described by conventional DFT as delocalised owing to the significant self-interaction error in common approximate DFT functionals.

Let us now return to dimolybdenum species. The separation of the reduction potentials, $\Delta_{red}E_{1/2}$, for the complexes based on di-$\{Mo^{II}(NO)Tp^{Me2}\}^{2+}$ fragments incorporating single or double $[OCH_2(CZ_2)_nCH_2O]^{2-}$ bridges ($n = 0$–3, Z = H or F) ranges between 70 and 360 mV, which reflects modest to substantial electrochemical interactions transmitted through the σ-framework and corresponds to the K_c value up to *ca.* 10^6 [71, 72]. This can be compared with complexes with unsaturated bridges allowing a strong electronic coupling, like the $[(PY5Me_2)_2V_2(\mu\text{-}5,6\text{-dmbzim})]^{3+}$ cation ($\Delta_{ox}E_{1/2} = 376$ mV in MeCN [73]) and the Creutz−Taube ion ($\Delta_{ox}E_{1/2} =$ 390 mV in water [56]),[6] or with Fc-CH_2-CH_2-Fc for which $\Delta_{ox}E_{1/2}$ (in CH_2Cl_2) is unmeasurable [74], which agrees well with the small H_{ab} value predicted by constrained DFT for the 1e oxidised form of the latter complex [69].

The redox potential splitting ($\Delta_{red}E_{1/2}$) measured for a di-Mo^{II} complex with a single ethane-1,2-diolate bridge equals 310 mV in CH_2Cl_2, and the IR spectrum of the $Mo^{I}Mo^{II}$ form shows two NO stretching vibrations instead of one, as in the neutral $Mo^{II}Mo^{II}$, pointing to valence-trapped behaviour [75]. As can be seen in Fig. 4, both electrochemical and spectral features are excellently reproduced by computations. Again, the IEF-PCM/B3LYP/LACV3P+/6-311++G(2d,2p) level of

[6] Note, however, that for these species, $\Delta E_{1/2}$ might be larger in a less polar solvent.

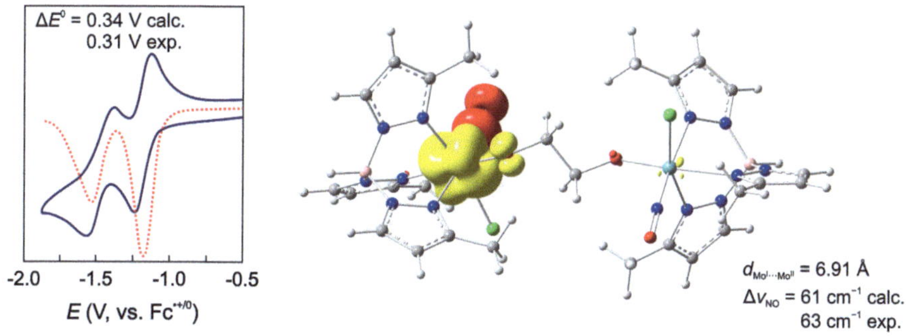

Fig. 4 Left panel: Cyclic and differential pulse voltammograms in CH_2Cl_2 for the $\{Mo(NO)Tp^{Me2}Cl\}_2(\mu\text{-}OCH_2CH_2O)$ dinuclear complex. The redox potential calculated at the IEF-PCM/B3LYP/LACV3P+/6-311++G(2d,2p) level for the B3LYP/LANL2DZpd geometries. Right panel: Spin density contour plot (0.005 au) for Mo^{II}/Mo^{I} mixed-valence form together with the difference in ν_{NO} frequencies for Mo^{II} and Mo^{I} centres and $Mo^{II}\cdots Mo^{I}$ distance

theory, appropriate for the monometallic Mo and W systems described above, is performing well also for a full model of the $\{Mo(NO)Tp^{Me2}Cl\}_2(\mu\text{-}OCH_2CH_2O)$ dinuclear complex (in [75], a simplified model has been used, in which outer Me groups are replaced by H atoms).

Interestingly, the doubly bridged analogue, *anti*-$[\{Mo(NO)(Tp^{Me2})(\mu\text{-}OCH_2CH_2O)\}_2]$, turned out to be insoluble in solvents other than chloroform, which points to specific interactions between the solvent and the solute. In attempting to measure metal–metal interactions, we saw how nicely the complex catalysed $CHCl_3$ reduction, which is the subject of next subsection.

The goal of the study described by Romańczyk et al. in [62] was to answer the following two fundamental questions: (i) how does a relative orientation of $\{Mo(NO)\}^{2+/3+}$ groups affect the degree of delocalisation of an unpaired electron in the μ-oxo complex, (ii) why, in an optimal geometry, the two NO ligands are mutually orthogonal. The system selected for this study—a binuclear nitrosyl complex $[(HO)(Tp^{Me2})(ON)MoOMo(NO)(Tp^{Me2})(OH)]$ (Fig. 5)—exhibits extremely interesting electronic properties. This compound in the $\{16e{:}17e\}^-$ valence state (after $1e$ reduction) is, amongst the symmetrically substituted binuclear complexes with a short μ-oxo bridge, a rare and unexpected example of the total valence trapping of an unpaired electron on one metal centre, in both EPR and IR timescales [76]. Thus, the very strong electrochemical interactions determined for this complex with $\Delta_{red}E_{1/2}$ (>1000 mV in CH_2Cl_2, comproportionation constant $K_c > 10^{16}$) do not arise from a delocalisation of an unpaired electron via the MoOMo bridge (i.e. through-bond effects), but from the through-space metal–metal interactions (Class I).

DFT:B3LYP modelling at the IEF-PCM(CH_2Cl_2)/B3LYP/def2-TZVPP//B3LYP/LanL2DZ level was carried out for a crystallographically characterised complex $[\{Mo(NO)(Tp^{Me2})(OH)\}_2(\mu\text{-}O)]$ (Fig. 5) and its singly reduced form: $[Tp^{Me2}]^-$, with truncated one outer CH_3 per a pyrazolyl group. Geometry optimisation provided a structure with perpendicular NO groups, i.e.

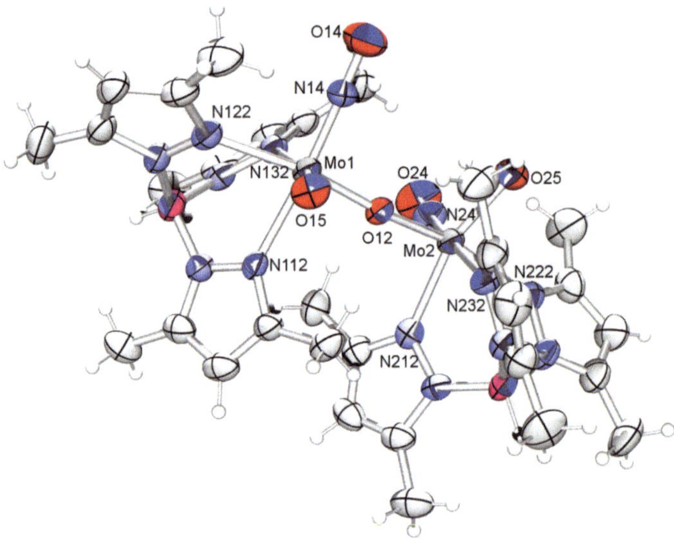

Fig. 5 Molecular structure of [{Mo(NO)(Tp^{Me2})(OH)}$_2$(μ-O)] (Tp^{Me2} = hydrotris(3,5-dimethyl-pyrazol-1-yl)borate) complex determined by X-ray diffraction. The Mo···Mo distance is 3.79 Å. Reprinted with permission from [62]. Copyright 2010 American Chemical Society

with a torsion angle (O)N(14)–Mo1···Mo2–N(O) of *ca.* 90°, resembling that obtained from the X-ray experiment. The validity of the model was confirmed by the agreement of the calculated ν_{NO} frequencies and the redox potentials (derived from the total free energy of an electron attachment in solution, $\Delta_{ea}G°_{(soln)}$) with the experimental data. The lowest unoccupied orbital is the d_{xy} orbital on one of the Mo centres (Fig. 6) (whereby the z-axis is determined by the Mo—N—O arrangement, and the x-axis by the Mo···Mo direction), which accepts an electron after reduction giving the $\{16e{:}17e\}^-$ mixed-valence state. For this electron to be able to delocalise onto the second centre, the d_{xy} orbitals on both centres should overlap via the oxygen p orbital, which is not possible, when NO groups, and consequently the local z axes, are mutually perpendicular. Therefore, the excess electron is trapped (Fig. 7a). It is important to note that the strongly twisted equilibrium geometry is a distinctive feature of other nitrosylmolybdenum μ-oxo complexes as well (also before reduction).[7] Their oxo-analogues, containing two $\{Mo(=O)\}^{3+}$ moieties that lie in one plane (or are only slightly twisted) [77], have entirely different properties. The orientation of NO groups in the equilibrium structure, unfavourable for the intercentre coupling, is a result of a destabilising effect of $(d, \pi_x{}^*)_b$ pair repulsion (dominating contribution of $d_{xz,Mo}$), which forces the molecule to twist and prohibits

[7] Similar geometry of the {(ON)MoOMo(NO)} moiety has also been found in structurally characterised complexes with non-Tp ligands, e.g. [{Mo^{II}(NO)('S_4')}$_2$(μ-O)] ('S_4' = 2,3,8,9-dibenzo-1,4,7,10-tetrathiadecane(2−) or [Cp*Mo(NO)R]$_2$(μ-O) (Cp* = η^5-C_5Me_5, R = CH_2SiMe_3 or Me). References are given in [62].

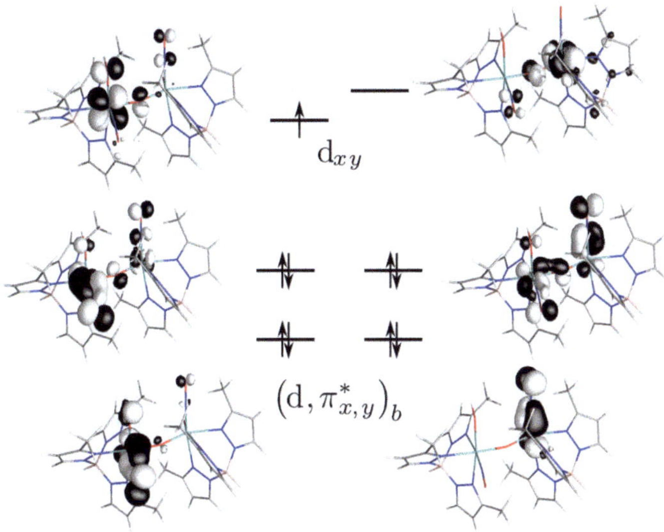

Fig. 6 Important occupied molecular orbitals (B3LYP/def2-TZVPP, isovalue 0.05 au) for optimal geometry of $[\{Mo(NO)(Tp^{3\text{-}Me})(OH)\}_2(\mu\text{-}O)]^-$ (Mo^IMo^{II}). In the analogous μ-O complex comprising $\{Mo^V(=O)\}^{3+}/\{Mo^{VI}(=O)\}^{4+}$ pair, the d_{xz} and d_{yz} orbitals are empty and strongly antibonding. Reprinted with permission from [62]. Copyright 2010 American Chemical Society

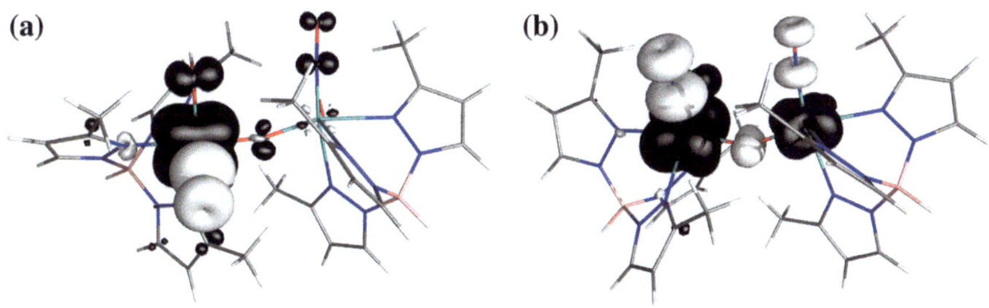

Fig. 7 Spin density plots for **a** equilibrium structure of $[\{Mo(NO)(Tp^{Me})(OH)\}_2(\mu\text{-}O)]^-$ (angle (O)N–Mo1···Mo2–N(O) = 98.9°) and **b** maximum electronic delocalisation structure (angle (O)N–Mo1···Mo2–N(O) = 23°). Contour plot for isovalue 0.002 au. Reprinted with permission from [62]. Copyright 2010 American Chemical Society

electron delocalisation (see Fig. 6). In the case of $\{Mo^V(O)\}^{3+}/\{Mo^{VI}(O)\}^{4+}$ pair, the d_{xz} orbitals are unoccupied and the additional repulsion between the electron pairs does not occur; thus, the μ-oxo complexes with $\{Mo(O)\}^{3+/4+}$ cores may easily adopt *syn* or *anti* conformations, enabling electron delocalisation.

Calculations have demonstrated that maximum delocalisation of an unpaired electron (via a hole-transfer mechanism), measured as a spin population on the second 16*e* Mo centre in the model complex $[\{Mo(NO)(Tp^{Me})X\}_2(\mu\text{-}O)]^-$, was achieved (at an energetic cost of 16.5 kcal/mol) for a small torsion angle of NO groups (Fig. 7b).

DFT predicts that an electron exchange along the Mo–O–Mo array could therefore be triggered by exciting the bridge torsional mode (characteristic of a molecular rheostat), using a high-boiling solvent or by irradiation at *ca.* 50–200 cm^{-1} (note that low-frequency modes are often underestimated owing to anharmonicity effects), which is also suggested by the temperature-dependent EPR spectra recorded for $[\{Mo(NO)(Tp^{Me2})X\}_2(\mu\text{-}O)]^-$ (X = Cl or O_2CPh) complexes [76]. It was demonstrated that the steric hindrance (Tp^{Me}–Tp^{Me} repulsion) and the weak hydrogen bonds, i.e. OH···ON, OH···OH and C–H···$O_{(NO/OH)}$, had no substantial effect on the molecular geometry.

4 Electrocatalytic Reductive Dehalogenation Driven by Non-covalent Interactions—Binding and Activation in Mo/W-Alkoxide System

The Lewis acid–Lewis base interactions, for example the C–H···A (A = O, N or π acceptor) hydrogen bonds supported by halogen interactions, and the London dispersion forces, are responsible for the formation of receptor–substrate systems, which are crucial in molecular recognition and (bio)catalysis. It has been shown [35, 36, 78] that these interactions may also play a significant role in electrocatalytic reductive dehalogenation processes.

Electroreduction of halogenated hydrocarbons is a slow process due to a high barrier for the electron transfer. There are only a few classes of transition metal complexes showing *catalytic* activity in the reduction of these compounds. The examples include Co, Ni and Fe tetrapyrrole macrocycles (as active sites in vitamin B_{12} [79], Factor F_{430} [80] and cytochrome P450 [81]) and 1*e* reduced forms of $\{M^{II}(NO)(Tp^{Me2})\}^{2+}$ (M = Mo, W) alkoxide complexes. In our earlier studies (e.g. [45]) we have shown that the latter compounds, having a more cathodic potential $E_{1/2}$ than −1.77 V, effectively catalyse the $CHCl_3$ electroreduction. The untypical character of the cyclic voltammetry curve (the sudden, almost vertical drop-down, a half-peak width, $E_{p/2} - E_p$, of just a few mV, see Fig. 8b) points to an activated process, occurring rapidly, according to a new mechanism, which has not been described in the literature so far. Moreover, it has been indicated that the addition of alcohols inhibits the catalysis. The presence of the Cl_3C–H···$O_{alkoxide}$ interactions in the crystal structures of $CHCl_3$ solvates of the Mo alkoxide complexes suggested a key role of a step involving an activation of the C–H bond in $CHCl_3$ placed inside the cavity formed by the two pz rings of the Tp ligand. It was therefore important to answer the following questions: Can such adducts exist in solution? and What is their fate after the Mo^{II} reduction?

The aim of the electrochemical and theoretical studies conducted by our group was to elucidate the mechanism of the chloroform [35] and pentachloroethane [36] electrocatalytic dehalogenation in the presence of Mo^I/W^I alkoxide complexes, and to explain the inhibition mechanism (Fig. 8). By using voltammetric methods, it was

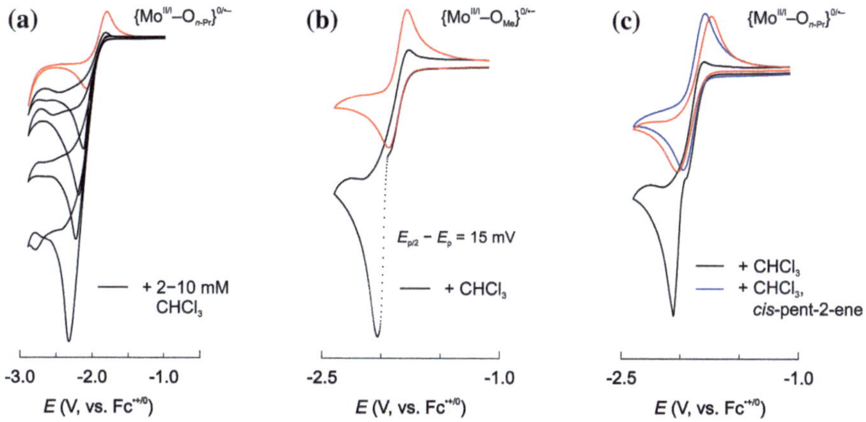

Fig. 8 Chloroform reduction in the presence of $[Mo(NO)(Tp^{Me2})(OR)_2]$ (R = Me or *n*-Pr) in CH_2Cl_2 on Pt. **a** The effect of increasing amounts of $CHCl_3$. The reduction occurs at a potential favourably shifted by ca. 1 V in comparison with the uncatalysed process. **b** Illustration of sudden process acceleration (points sampled every 10 ms) upon addition of $CHCl_3$. Note that for typical voltammetric wave the $E_p - E_{p/2}$ difference is 56.5 mV. **c** Inhibitive effect of adding *cis*-2-pentene; the addition of alcohols has a similar effect

shown that the initial 16*e* bis-alkoxide complex is fully recovered after the reduction of chloroform. Moreover, the combination of gas–liquid chromatography (GLC) analysis and electrochemical methods enabled the detection of dehalogenation products: CH_2Cl_2 (the absence of any C_2 compounds), Cl^- anions and transiently formed $:CCl_2$. The effect of the presence of *cis*-2-pentene on the $CHCl_3$ electroreduction was also investigated. The addition of an olefin, as well as alcohols, inhibits the catalysis, which points to the radical/carbene character of the process.

The DFT-D modelling (with Grimme's D3 dispersion correction [82]) was performed with the hybrid B3LYP functional and the triple-ζ quality basis set (def2-TZVPP) in the CH_2Cl_2 (COSMO model). The applied exchange-correlation functional correctly predicted the geometries for the class $\{Mo(NO)(Tp^{Me2})\}$ complexes (as shown above), also for the $\{Mo^{II}{-}O_{alk}\}^0\cdots HCCl_3$ adduct (cf. the crystal structure of a Mo^{II}-alkoxy scorpionate showing $CHCl_3$ molecules interacting with the alkoxy groups, see reference 20 in [35]), and enabled obtaining the accurate Mo/W redox potentials (vide supra).

Our studies described in [35, 36, 78] allowed showing that the formation of the adduct with the $C{-}H\cdots O$ bonding between the polychloroalkane containing an acidic hydrogen atom in the molecule and the catalyst ($\{M^I{-}O_{alk}\}^{\bullet -}$) serving also as an electron donor facilitates dissociative electron transfer and initiates selective dehalogenation (Fig. 9). The most important role here play the close and prolonged contact between the electron donor and the chloroalkane molecule, increasing the probability of ET (*proximity effect*) and transition state stabilisation. In the same conditions, polychloroalkanes and alkenes containing no H atoms in the molecule, for instance CCl_4

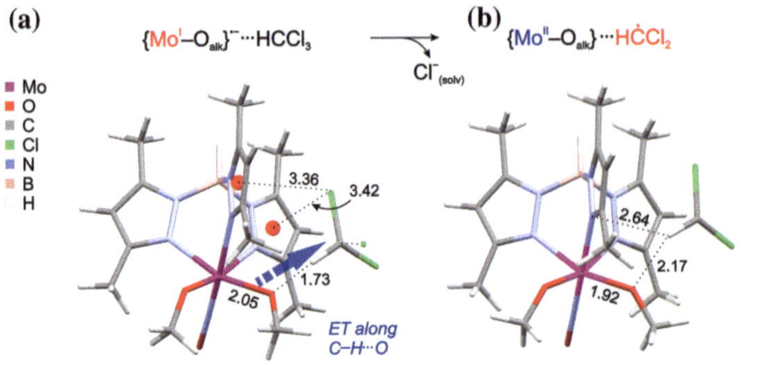

Fig. 9 **a** DFT-D optimised geometries (COSMO/B3LYP-D3/def2-TZVPP) of $\{Mo^I{-}O_{alk}\}^{\bullet-}\cdots HCCl_3$ adduct just before ET coupled with the C−Cl bond cleavage (C−Cl bond elongated to 2.20 Å). Note the substantial shortening of the C−H···O$_{alk}$ and Cl···π distances (in the equilibrium geometry the distances are: 1.82 and 3.43 Å). **b** The product of dissociative electron transfer with $Cl^-_{(solvated)}$ moved to infinity, i.e. $\{Mo^{II}{-}O_{alk}\}\cdots HCCl_2^{\bullet}$. The adduct is unstable and $CHCl_2^{\bullet}$ may be easily substituted by a solvent or $CHCl_3$ molecule. Reprinted from [35] with permission from the PCCP Owner Societies

and C_2Cl_4, do not undergo reduction. It is noteworthy that the electron transfer in the Mo^I adduct is not the only possible scenario that may be postulated. The possibility of the alkoxy ligand dissociation or its protonation by $CHCl_3$ was also considered; however, these hypotheses were excluded on the energetic basis and on the basis of an inconsistency with experiment (the complex was fully recoverable).

It is important to note that the H-bond in the $\{Mo^I(NO)(Tp^{Me2})(O_{alkoxide})\}^{\bullet-}\cdots HCCl_3$ adduct is one of the strongest known hydrogen bonds of the C−H···O type, which is reflected by the geometry ($d_{H\cdots O}$ = 1.82 Å and nearly linear $\theta_{C-H\cdots O}$ angle equal to 172.2°), a large $\Delta\nu_{(C-H)}$ red-shift of 380 cm^{-1}, and a noticeable reorganisation of electronic density along the H-bond axis as evidenced by natural orbitals for chemical valence (NOCV) method (Fig. 10).

The thermodynamic stability of the $Mo^{II/I}$ adducts and their tungsten analogues was assessed on the basis of the calculated standard binding Gibbs energies in solution; for the $\{Mo^I{-}O_{alk})\}^{\bullet-}\cdots HCCl_3$ adduct $\Delta_{bind}G^\circ = -1.6$ kcal mol^{-1}. Additionally, the binding energies determined at the DFT-D3 level were compared with those obtained from ordinary DFT (Fig. 11), and it was concluded that dispersion significantly contributes to the binding energy [78], and including it in modelling is mandatory to explain the formation of these electrocatalytically relevant adducts.

The Gibbs energy of the intramolecular dissociative ET ($\Delta_{DET}G^\circ$) from Mo^I to $CHCl_3$ (Fig. 9) is expressed as:

$$\Delta_{DET}G^\circ = -F\left(E^\circ_{CHCl_3/\,CHCl_2^\bullet + Cl^-} - E^\circ_{\{Mo^{II/I}-O_{alk}\}^{0/\bullet-}}\right) + \Delta\Delta_{bind}G^\circ_{CHCl_2^\bullet/\,CHCl_3} \quad (5)$$

where E°s denote reduction potentials of $CHCl_3$ and a Mo^{II} complex, and the last term is the difference between free binding energies of $CHCl_2^\bullet$ and $CHCl_3$ with the

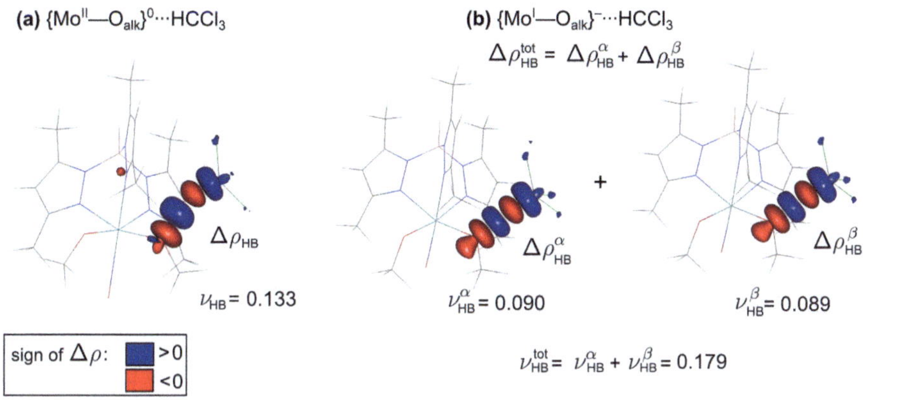

Fig. 10 NOCV analysis for the $\{Mo^{II/I}-O_{alk}\}^{0/\bullet-}$ adducts with $CHCl_3$ reveal a characteristic pattern of H-bonding (HB). The principal contributions to the deformation density ($\Delta\rho_{HB}$) are shown for 16e Mo^{II} (**a**) and 17e Mo^{I} $HCCl_3$ (**b**) adducts (with respect to $\{Mo^{II/I}-O_{alk}\}^{0/\bullet-}$ and $HCCl_3$ as fragments) obtained from a pair of NOCVs with the largest eigenvalue (ν_{HB}). For an open-shell $\{Mo^{I}-O_{alk})\}^{\bullet-}\cdots HCCl_3$ species there are two (qualitatively identical) deformation densities and eigenvalues for spin-up (α) and spin-down (β) electrons, respectively, obtained from spin-unrestricted calculations. Contour plots correspond to $5 \cdot 10^{-4}$ bohr^{-3} isovalue. Reprinted from [35] with permission from the PCCP Owner Societies

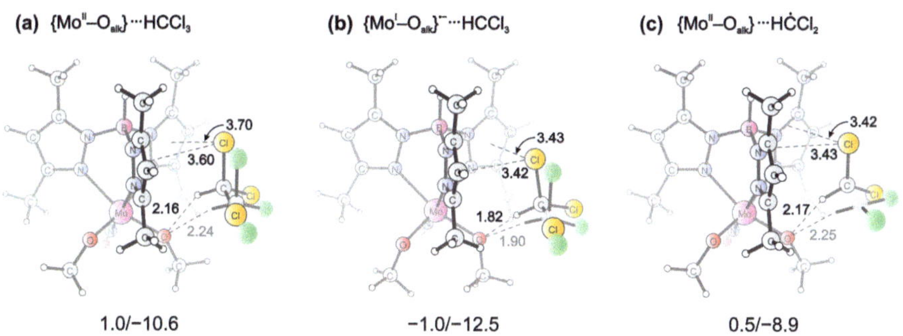

Fig. 11 Comparison of DFT-D3 (sharp, Cl atoms yellow) and DFT (diffused, Cl atoms green) optimised geometries for $\{Mo^{II/I}-O_{alk}\}^{0/\bullet-}$ adducts with $CHCl_3$ (**a** and **b**) and $CHCl_2^{\bullet}$ radical (**c**) in CH_2Cl_2 (COSMO). Dotted lines show $C-H\cdots O_{alkoxide}$ bonding and $C-Cl\cdots\pi_{pyrazolyl}$ dispersive interactions; distances in Å. Numbers at the bottom indicate binding energies (DFT/DFT-D3) in kcal/mol. Reprinted from [78], Copyright (2013), with permission from Elsevier

$Mo^{II/I}$ centre. The potential of the $Mo^{II/I}$ couple may be tuned by choosing appropriate alkoxide ligands; note that $E°$ for $CHCl_3$ was obtained from CCSD(T) calculations to overcome the problem of C–Cl bond energy, which is highly underestimated by DFT. Obviously, $\Delta_{DET}G°$ depends on the dielectric constant of the solvent playing an essential role in stabilising the Cl^- anion that is eventually formed. Importantly, the calculations indicated that $\Delta_{DET}G°$ is negative for Mo alkoxides with $E° \leq$ ca.

−1.6 V versus Fc, which excellently agrees with the experimental finding that the catalysis proceeds with complexes having a more cathodic potential than −1.77 V.

The proposed mechanisms of $CHCl_3$ and pentachloroethane cathodic dehalogenation in the presence of Mo/W alkoxide complexes, which are fully consistent with the experiment (cyclic voltammetry, GLC analyses of the products of the electrocatalytic reduction, the simulation of the voltammogram, the inhibitory effects) and with the DFT calculations, are presented in Fig. 12. The plausible reaction paths of further dehalogenation steps were determined on the basis of reaction energies and activation energies of the consecutive reactions with the participation of $CHCl_3$ or C_2HCl_5 and the products of their degradation.

In the case of chloroform, the process has an *autocatalytic* (reflected in the shape of the cyclic voltammetry curve, Fig. 8b), radical carbene character. After triggering, the reaction proceeds as a series of consecutive alternating electron and proton transfers, direct or $Mo^{II/I}$-mediated ($CHCl_2^{\cdot}$ and $:CCl_2$ may be reduced by Mo^I).

Moreover, the mechanism of the inhibition was elucidated and in the case of alcohols consists in the competitive, stronger binding of the alcohol molecule to the active site (see Fig. 13a), whereas for olefins, the addition of the $CHCl_2^{\cdot}$ radicals to the C=C bonds occurs easily (the binding energy of the olefin to Mo^I is too low to block the $CHCl_3$ binding). It is important to note that the autocatalytic reaction can be ruled out in *N*,*N*-dimethylformamide, since this solvent reacts with $:CCl_2$—that is why the studies were conducted in dichloromethane.

It was demonstrated that in the case of pentachloroethane, the first step, i.e. the intramolecular electron transfer assisted by the C−H···O bond, requires the application of the alkoxide complex exhibiting a more cathodic potential than that of the $Mo^{II/I}$ pair, i.e. $\{W^I{-}O_{alkoxide}\}^{\bullet-}$. In the experiment, the chelate complex,

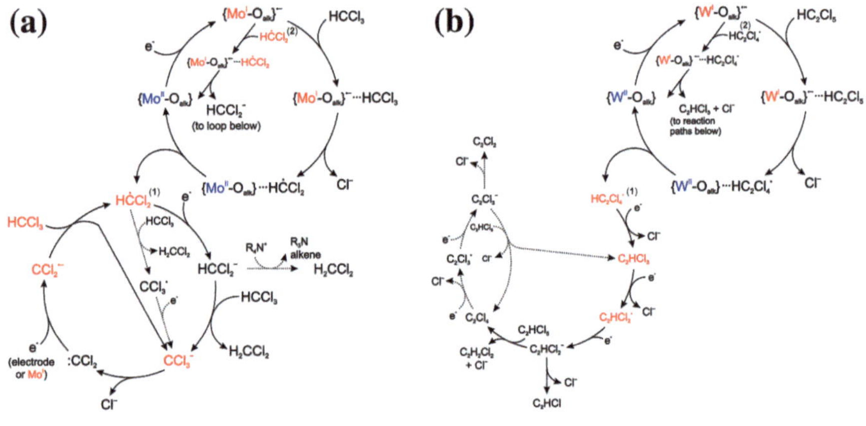

Fig. 12 Mechanism of cathodic dehalogenation of **a** chloroform, with key branching in the organic loop, **b** pentachloroethane in the presence of Mo/W alkoxides. Dotted lines denote reactions which are (**a**) less likely to occur (slower) or (**b**) feasible, but at potentials significantly more cathodic than the $W^{II/I}$ redox potential. (**a**) Reproduced from [35] with permission from the PCCP Owner Societies, and (**b**) reprinted from [36], Copyright (2014), with permission from Elsevier

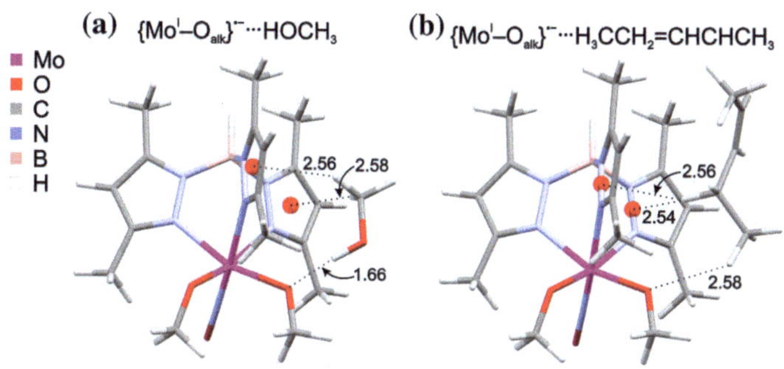

Fig. 13 DFT-D optimised geometries (distances in Å) of $\{Mo^I{-}O_{alk}\}^{\bullet-}$ interacting with CH_3OH (**a**) and cis-2-pentene (**b**). Dotted lines show O–H···O or C–H···O bonding and C–H···π_{pz} dispersive interactions; only the shortest contacts are drawn. Methanol binds with the Mo^I site slightly stronger than $CHCl_3$; $\Delta_{bind}G^\circ$ is equal -2.2 kcal mol^{-1} for CH_3OH and -1.6 kcal mol^{-1} for $CHCl_3$, conversely, for the alkene $\Delta_{bind}G^\circ$ is positive (2.3 kcal mol^{-1}). Reprinted from [35] with permission from the PCCP Owner Societies

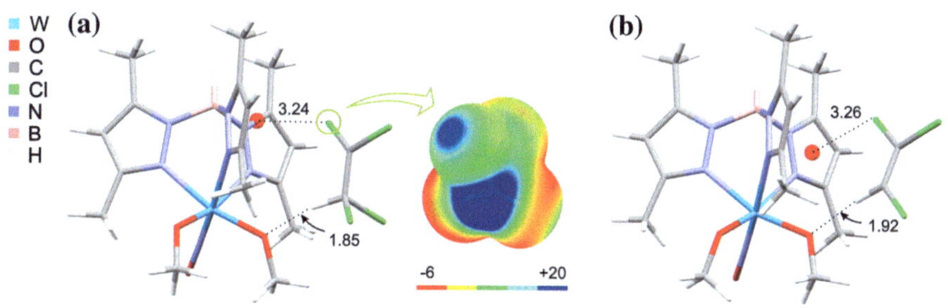

Fig. 14 Intermediates of pentachloroethane reduction in the presence of $W^{II/I}$. DFT-D optimised geometries (COSMO/B3LYP-D3/def2-TZVPP) for **a** $\{W^I{-}O_{alk}\}\cdots HC_2Cl_4^\bullet$, and **b** $\{W^I{-}O_{alk}\}^{\bullet-}\cdots HC_2Cl_3$ formed upon intramolecular dissociative ET into $C_2HCl_4^\bullet$. Dotted lines show C–H···$O_{alkoxide}$ bonding and C–Cl···$\pi_{pyrazolyl}$ interactions, which in (**a**) exhibit a weak halogen bonding character (distances in Å). Electrostatic potential for $C_2HCl_4^\bullet$ on the 0.001 a.u. isosurface of the total B3LYP/6-311++G(2d,2p) electronic density is shown; the potential energies are presented in the -6 (or less) to +20 kcal mol^{-1} (or more) range. Reprinted from [36], Copyright (2014), with permission from Elsevier

$[W(NO)(Tp^{Me2})O(CH_2)_4O]$, was utilised, whereas the $[W(NO)(Tp^{Me2})(OMe)_2]$ was used in the calculations. The initial product of the first stage of the C_2HCl_5 dehalogenation—$C_2HCl_4^\bullet$—may also interact with the W^I complex (even stronger than C_2HCl_5), due to its ability to form, apart from the H-bond, the C–Cl···$\pi_{pyrazolyl}$ halogen bonding (Fig. 14a). The reduction of $C_2HCl_4^\bullet$ radical, by W^I or at electrode, yields in one step C_2HCl_3 and Cl^-.

However, analogously to the systems containing cytochrome P450, the process is suppressed at the trichloroethylene stage. The stability of the $\{W^I{-}O_{alk}\}^{\bullet-}\cdots HC_2Cl_3$ adduct (Fig. 14b) is much lower than that of its analogue with C_2HCl_5 ($\Delta_{bind}G^\circ$ = +0.1 kcal mol^{-1} for C_2HCl_3 and -1.2 kcal mol^{-1} for C_2HCl_5), which correlates with a much weaker catalytic effect (see Fig. 1d in [36]). The reduction product of trichloroethylene is *cis*-$C_2HCl_2^{\bullet}\cdots Cl^-$, which is also true for the intramolecular ET within the W^I adduct. If an effective $H^{\bullet}$ donor is not present in the system, the *cis*-$C_2HCl_2^{\bullet}$ radical is rapidly reduced to the corresponding carbanion.

The DFT calculations performed by us made it possible to elucidate why chloroform can be reduced by the Mo^I, whereas pentachloroethane cannot undergo this reaction in identical conditions, despite the fact that the $E°$ potentials for both chloroalkanes are similar. In the case of $\{Mo^I{-}O_{alk}\}^{\bullet-}\cdots HC_2Cl_5$ adduct, the distance between the electron donor and acceptor ($O_{alkoxide}\cdots C{-}Cl_{acceptor}$) is *ca.* 1 Å higher than in the analogous $CHCl_3$ adduct, which results in the need to apply a higher overpotential (W^{II}/W^I). The unique *autocatalytic* character of the $CHCl_3$ dehalogenation was also explained. It results from the substantial differences in the reactivity of the $CHCl_3$ and C_2HCl_5 degradation products; the branching in the "organic loop" is observed only in the case of chloroform (cf. Fig. 12a, b).

Importantly, effects similar to those recognised for the presented catalytic system $\{M^I{-}O_{alk}\}^{\bullet-}$ (M = Mo or W), i.e. the proximity and stabilisation effects, and competitive inhibition, operate in enzymatic catalysis and originate in the Lewis acid–Lewis base interactions and London dispersion forces. The obtained results (described above for chloroform and pentachloroethane dehalogenation in the presence of Mo/W complexes, and also presented in the paper [78] for the porphyrin adducts) enabled a general conclusion to be drawn concerning the features that a molecule should exhibit in order to be a good redox electrocatalyst. Amongst them, the most essential appears to be an anion radical form with an appropriate redox potential, rather than dianion, exhibiting rigid structure with a spacious SOMO, showing strong dispersive interactions enabling the formation of an adduct enhanced by H– or halogen bonding, preferably in an environment of a low-dielectric constant.

5 Concluding Remarks

This chapter demonstrates how molecular electrochemistry may benefit from the application of density functional theory methods combined with implicit solvent models. From the pioneering work of Baik and Friesner [1], a significant progress has been made giving rise to chemically accurate predictions of the redox potential of catalytically relevant transition metal complexes, including also bioinorganic systems. These achievements enabled showing how the thermodynamics of key electron transfer steps in these processes is controlled by metal–ligand interactions, perturbations in the second coordination sphere, the local environment and a variety of non-covalent interactions. Most pivotal factor for this has been the development of novel theoretical models, particularly, efficient solvation models (still remaining a big challenge), dispersion-corrected functionals, and more accurate correlated methods, not to mention an enormous growth in computational power. Crucial to this and future success is the calibration of the computational DFT protocol based on either reliable experimental data or proven benchmark computational methods.

Calibration of computational methods based on comparison with experimental data mounts also challenges. Measurements require some specific conditions to be met, and which may affect the measured values, that are not necessarily given in data tables. Voltammetry needs the presence of relatively high concentrations of a base electrolyte that influences the determined E° due to ionic strength (or alternatively variation of dielectric constant) and ion-pairing effects. Both effects are less important in polar solvents. The potential shift direction always results from the stabilisation of ions; e.g., if a cation (or a more positive cation) is the product of oxidation, the potential will be less positive. In non-polar media, like chlorinated aliphatics (extremely non-polar environments, like aliphatic hydrocarbons are not used in electrochemistry) the potential shift is generally moderate (53 mV for Fc^{+}/Fc in 0.1 M n-Bu_4NPF_6 in dichloromethane). Ion-pairing brings about the same effect for species like ferrocene; however, it may become significant for highly charged species (500 mV for $[Ru(bpy)_3]^{2+/3+}$ in CH_2Cl_2). This effect can be successfully modelled by including explicit counterions.

Care should be taken, when using redox potentials in non-aqueous solutions measured against popular aqueous reference electrodes and not versus internal standards. Conversion parameters may be found in the literature, but it should be remembered that the liquid junction potential occurring between the aqueous and non-aqueous phases that adds a contribution to the measured value, is affected by a couple of factors, not only the solvent types, but also the types and concentrations of electrolytes on both sides. Thus, it is always better to rely on an internal standard, like ferrocene. The reliable Fc^{+}/Fc absolute potentials in solvents commonly employed in electrochemistry can be calculated by quantum chemical methods using SMD or COSMO-RS solvation energies, and the SMD solvation model provided values that are much more consistent with experiment for low-polarity solvents.

It was shown here that it is possible to obtain outstanding accuracy in the prediction of redox potentials for monometallic and bimetallic Mo/W scorpionates in non-aqueous solvents, clarify unexpected properties of mixed-valence bimetallic complexes incorporating dioxoalkylene or oxo bridges, enigmatic dehalogenation catalysis and demonstrate how H-bonding may affect the redox behaviour of Cu and Fe systems. The presented examples of entirely valence-trapped $Mo^{I}Mo^{II}$ systems, despite showing medium to large electrochemical interactions (the splitting of redox potentials, $\Delta E_{1/2}$), contradict the common practice of correlating $\Delta E_{1/2}$ with the degree of electronic coupling. DFT-D calculations allowed solving of the mechanism of (auto)catalytic chloroform or pentachloroethane dehalogenation triggered by a dissociative electron transfer through a charge-assisted $C{-}H\cdots O$ hydrogen bond and the explanation of the catalysis inhibition. It has been proved that these interactions, crucial for enzymatic activity, may play a decisive role not only in substrate recognition (e.g. undesirable xenobiotics), but also in electron transfer. The obtained results could be useful in designing the new electrocatalysts and sensors.

Acknowledgements We thank Prof. Ewa Brocławik (Polish Academy of Sciences), Dr. Mariusz Radoń (Jagiellonian University) and Klemens Noga (Academic Computer Center CYFRONET) for a long-standing collaboration and very valuable, fruitful discussions. Support for this work from the PL-Grid Infrastructure is gratefully acknowledged.

References

1. Baik M-H, Friesner RA (2002) Computing redox potentials in solution: density functional theory as a tool for rational design of redox agents. J Phys Chem A 106:7407–7412. https://doi.org/10.1021/jp025853n
2. Marenich AV, Ho J, Coote ML, Cramer CJ, Truhlar DG (2014) Computational electrochemistry: prediction of liquid-phase reduction potentials. Phys Chem Chem Phys 16:15068–15106. https://doi.org/10.1039/C4CP01572J
3. Ho J, Coote ML, Cramer CJ, Truhlar DG (2015) Theoretical calculation of reduction potentials. In: Hammerich O, Speiser B (eds) Organic electrochemistry, 5th edn. CRC Press, Boca Raton, pp 229–259
4. Isse AA, Gennaro A (2010) Absolute potential of the standard hydrogen electrode and the problem of interconversion of potentials in different solvents. J Phys Chem B 114:7894–7899. https://doi.org/10.1021/jp100402x
5. Hofer TS, Hünenberger PH (2018) Absolute proton hydration free energy, surface potential of water, and redox potential of the hydrogen electrode from first principles: QM/MM MD free-energy simulations of sodium and potassium hydration. J Chem Phys 148:222814. https://doi.org/10.1063/1.5000799
6. Trasatti S (1986) The absolute electrode potential: an explanatory note. Pure Appl Chem 58:955–966. https://doi.org/10.1351/pac198658070955
7. Ho J (2015) Are thermodynamic cycles necessary for continuum solvent calculation of pKas and reduction potentials? Phys Chem Chem Phys 17:2859–2868. https://doi.org/10.1039/C4CP04538F
8. Romańczyk PP, Rotko G, Kurek SS (2014) The redox potential of the phenyl radical/anion couple and the effect thereon of the lithium cation: a computational study. Electrochem Commun 48:21–23. https://doi.org/10.1016/j.elecom.2014.08.003
9. Wang LP, Van Voorhis T (2012) A polarizable QM/MM explicit solvent model for computational electrochemistry in water. J Chem Theory Comput 8:610–617. https://doi.org/10.1021/ct200340x
10. Tomasi J, Persico M (1994) Molecular interactions in solution: an overview of methods based on continuous distributions of the solvent. Chem Rev 94:2027–2094. https://doi.org/10.1021/cr00031a013
11. Marenich AV, Cramer CJ, Truhlar DG (2009) Universal solvation model based on solute electron density and on a continuum model of the solvent defined by the bulk dielectric constant and atomic surface tensions. J Phys Chem B 113:6378–6396. https://doi.org/10.1021/jp810292n
12. Klamt A (2011) The COSMO and COSMO-RS solvation models. WIREs Comput Mol Sci 1:699–709. https://doi.org/10.1002/wcms.56
13. Guerard JJ, Arey JS (2013) Critical evaluation of implicit solvent models for predicting aqueous oxidation potentials of neutral organic compounds. J Chem Theory Comput 9:5046–5058. https://doi.org/10.1021/ct4004433
14. Tsierkezos NG, Philippopoulos AI (2009) Studies of ion solvation and ion association of n-tetrabutylammonium hexafluorophosphate and n-tetrabutylammonium tetraphenylborate in various solvent. Fluid Phase Equilib 277:20–28. https://doi.org/10.1016/j.fluid.2008.11.004
15. Bao D, Millare B, Xia W, Steyer BG, Gerasimenko AA, Ferreira A, Contreras A, Vullev VI (2009) Electrochemical oxidation of ferrocene: a strong dependence on the concentration of the supporting electrolyte for nonpolar solvents. J Phys Chem A 113:1259–1267. https://doi.org/10.1021/jp809105f

16. Konezny SJ, Doherty MD, Luca OR, Crabtree RH, Soloveichik GL, Batista VS (2012) Reduction of systematic uncertainty in DFT redox potentials of transition-metal complexes. J Phys Chem C 116:6349–6356. https://doi.org/10.1021/jp300485t
17. Shi Y, Fang L, Li X, Qu L, Shao H (2015) The observation of ion-pairing effect based on substituent effect of ferrocene derivatives. J Electroanal Chem 757:258–262. https://doi.org/10.1016/j.jelechem.2015.10.003
18. Diggle JW, Parker AJ (1974) Liquid junction potential in electrochemical cells involving a dissimilar solvent junction. Aust J Chem 27:1617–1621. https://doi.org/10.1071/CH9741617
19. Izutsu K (2011) Liquid junction potential between electrolyte solutions in different solvents. Anal Sci 27:685–694. https://doi.org/10.2116/analsci.27.685
20. Radtke V, Ermantraut A, Himmel D, Koslowski T, Leito I, Krossing I (2018) The Ideal ionic liquid salt bridge for the direct determination of gibbs energies of transfer of single ions, part I: the concept. Angew Chem Int Ed 57:2344–2347. https://doi.org/10.1002/anie.201707333
21. Mousavi MPS, Saba SA, Anderson EL, Hillmyer MA, Bühlmann P (2016) Avoiding errors in electrochemical measurements: effect of frit material on the performance of reference electrodes with porous frit junctions. Anal Chem 88:8706–8713. https://doi.org/10.1021/acs.analchem.6b02025
22. Gritzner G, Kuta J (1984) Recommendations on reporting electrode potentials in nonaqueous solvents. Pure Appl Chem 56:461–466. https://doi.org/10.1351/pac198456040466
23. Gritzner G (2013) Reference redox systems in nonaqueous systems and the relation of electrode potentials in nonaqueous and mixed solvents to standard potential in water. In: Inzelt G, Lewenstam A, Scholz F (eds) Handbook of reference electrodes. Springer, Berlin Heidelberg, pp 25–31. https://doi.org/10.1007/978-3-642-36188-3_2
24. Koepp H-M, Wendt H, Strehlow H (1960) Der Vergleich der Spannungsreihen in verschiedenen Solventien. II. Z Elektrochem 64:483–491. https://doi.org/10.1002/bbpc.19600640406
25. Namazian M, Lin CY, Coote ML (2010) Benchmark calculations of absolute reduction potential of ferricinium/ferrocene couple in nonaqueous solutions. J Chem Theory Comput 6:2721–2725. https://doi.org/10.1021/ct1003252
26. Marenich AV, Cramer CJ, Truhlar DG (2013) Generalized born solvation model SM12. J Chem Theory Comput 9:609–620. https://doi.org/10.1021/ct300900e
27. Noviandri I, Brown KN, Fleming DS, Gulyas PT, Lay PA, Masters AF, Phillips L (1999) The decamethylferrocenium/decamethylferrocene redox couple: a superior redox standard to the ferrocenium/ferrocene redox couple for studying solvent effects on the thermodynamics of electron transfer. J Phys Chem B 103:6713–6722. https://doi.org/10.1021/jp991381+
28. Tsierkezos NG (2007) Cyclic voltammetric studies of ferrocene in nonaqueous solvents in the temperature range from 248.15 to 298.15 K. J Solut Chem 36:289–302. https://doi.org/10.1007/s10953-006-9119-9
29. Krishtalik LI (2008) Solvation energy of metallocenes pilot ions: Beyond the Born approximation. Electrochim Acta 53:3722–3733. https://doi.org/10.1016/j.electacta.2007.10.026
30. ADF Program System, release 2017.105; Scientific computing & modelling NV, Vrije Universiteit: Amsterdam, The Netherlands
31. Connelly NG, Geiger WE (1996) Chemical redox agents for organometallic chemistry. Chem Rev 96:877–910. https://doi.org/10.1021/cr940053x
32. McCleverty JA, Ward MD, Jones CJ (2001) Redox-activity in complexes with Mo(NO) and Mo(O) cores: facts and consequences. Comments Inorg Chem 22:293–326. https://doi.org/10.1080/02603590108050873
33. Enemark JH, Cooney JJA, Wang JJ, Holm RH (2004) Synthetic analogues and reaction systems relevant to the molybdenum and tungsten oxotransferases. Chem Rev 104:1175–1200. https://doi.org/10.1021/cr020609d
34. Young CG (2016) Scorpionate complexes as models for molybdenum enzymes. Eur J Inorg Chem 2016:2357–2376. https://doi.org/10.1002/ejic.201501387
35. Romańczyk PP, Radon M, Noga K, Kurek SS (2013) Autocatalytic cathodic dehalogenation triggered by dissociative electron transfer through a C−H···O hydrogen bond. Phys Chem Chem Phys 15:17522–17536. https://doi.org/10.1039/C3CP52933A

36. Romańczyk PP, Rotko G, Noga K, Radoń M, Andryianau G, Kurek SS (2014) The effect of C−H···O bonding and Cl···π interactions in electrocatalytic dehalogenation of C_2 chlorides containing an acidic hydrogen. Electrochim Acta 140:497–504. https://doi.org/10.1016/j.electacta.2014.04.175
37. Romańczyk PP, Kurek SS (2019) Towards the accurate prediction of redox potentials of catalytically relevant nitrosyl, oxo- and oxothio-Mo and W scorpionates. A computational study. Manuscript in preparation
38. Vannucci AK, Snyder RA, Gruhn NE, Lichtenberger DL, Enemark JH (2009) New insights into solvolysis and reorganization energy from gas-phase, electrochemical, and theoretical studies of oxo-Tp*Mo^V molecules. Inorg Chem 48:8856–8862. https://doi.org/10.1021/ic9011058
39. Doonan CJ, Nielsen DJ, Smith PD, White JM, George GN, Young CG (2006) Models for the molybdenum hydroxylases: synthesis, characterization and reactivity of *cis*-oxosulfido-Mo(VI) complexes. J Am Chem Soc 128:305–316. https://doi.org/10.1021/ja056109u
40. Donahue JP (2006) Thermodynamic scales for sulfur atom transfer and oxo-for-sulfido exchange reactions. Chem Rev 106:4747–4783. https://doi.org/10.1021/cr050044w
41. Hosseinzadeh P, Lu Y (2016) Design and fine-tuning redox potentials of metalloproteins involved in electron transfer in bioenergetics. Biochim Biophys Acta 1857:557–581. https://doi.org/10.1016/j.bbabio.2015.08.006
42. Glaser T, Liratzis I, Kataeva O, Fröhlich R, Piacenz M, Grimme S (2006) Direct influence of hydrogen-bonding on the reduction potential of a Cu^{II} center. Chem Commun 2006:1024–1026. https://doi.org/10.1039/B517340J
43. Yang X, Niu S, Ichiye T, Wang LS (2004) Direct measurement of the hydrogen-bonding effect on the intrinsic redox potentials of [4Fe−4S] cubane complexes. J Am Chem Soc 126:15790–15794. https://doi.org/10.1021/ja045709a
44. Hirotsu M, Kuwamura N, Kinoshita I, Kojima M, Yoshikawa Y, Ueno K (2009) Steric, geometrical and solvent effects on redox potentials in salen-type copper(II) complexes. Dalton Trans 2009:7678–7683. https://doi.org/10.1039/B906013H
45. Włodarczyk AJ, Romańczyk PP (2010) Redox and catalytic properties of alkoxides based on tris(3,5-dimethylpyrazolyl)borato tungsten nitrosyls. Inorg Chim Acta 363:813–816. https://doi.org/10.1016/j.ica.2009.11.034
46. Bím D, Rulíšek L, Srnec M (2018) Computational electrochemistry as a reliable probe of experimentally elusive mononuclear nonheme iron species. J Phys Chem C 122:10773–10782. https://doi.org/10.1021/acs.jpcc.8b02698
47. Rudshteyn B, Vibbert HB, May R, Wasserman E, Warnke I, Hopkins MD, Batista VS (2017) Thermodynamic and structural factors that influence the redox potentials of tungsten–alkylidyne complexes. ACS Catal 7:6134–6143. https://doi.org/10.1021/acscatal.7b01636
48. Liao RZ, Siegbahn PEM (2017) Quantum chemical modeling of homogeneous water oxidation catalysis. Chemsuschem 10:4236–4263. https://doi.org/10.1002/cssc.201701374
49. Šebesta F, Baxová K, Burda JV (2018) Redox potentials for tetraplatin, satraplatin, its derivatives, and ascorbic acid: a computational study. Inorg Chem 57:951–962. https://doi.org/10.1021/acs.inorgchem.7b01894
50. de Rege PJ, Williams SA, Therien MJ (1995) Direct evaluation of electronic coupling mediated by hydrogen bonds: implications for biological electron transfer. Science 269:1409–1413. https://doi.org/10.1126/science.7660123
51. Porter TM, Heim GP, Kubiak CP (2018) Stable mixed-valent complexes formed by electron delocalization across hydrogen bonds of pyrimidinone-linked metal clusters. J Am Chem Soc 140:12756–12759. https://doi.org/10.1021/jacs.8b09273
52. Parthey M, Kaupp M (2014) Quantum-chemical insights into mixed-valence systems: within and beyond the Robin-Day scheme. Chem Soc Rev 43:5067–5088. https://doi.org/10.1039/C3CS60481K
53. Bu D, Xiong Y, Tan YN, Meng M, Low PJ, Kuang DB, Liu CY (2018) Understanding the charge transport properties of redox active metal-organic conjugated wires. Chem Sci 9:3438–3450. https://doi.org/10.1039/C7SC04727D

54. Zhang J, Kuznetsov AM, Medvedev IG, Chi Q, Albrecht T, Jensen PS, Ulstrup J (2008) Single-molecule electron transfer in electrochemical environments. Chem Rev 108:2737–2791
55. Blumberger J (2015) Recent advances in the theory and molecular simulation of biological electron transfer reactions. Chem Rev 115:11191–11238. https://doi.org/10.1021/acs.chemrev.5b00298
56. Creutz C, Taube H (1973) Binuclear complexes of ruthenium ammines. J Am Chem Soc 95:1086–1094. https://doi.org/10.1021/ja00785a016
57. McCleverty JA, Ward MD (1998) The role of bridging ligands in controlling electronic and magnetic properties in polynuclear complexes. Acc Chem Res 31:842–851. https://doi.org/10.1021/ar970315s
58. D'Alessandro DM, Keene FR (2006) Current trends and future challenges in the experimental, theoretical and computational analysis of intervalence charge transfer (IVCT) transitions. Chem Soc Rev 35:424–440. https://doi.org/10.1039/B514590M
59. Barrière F, Geiger WE (2006) Use of weakly coordinating anions to develop an integrated approach to the tuning of $\Delta E_{1/2}$ values by medium effects. J Am Chem Soc 128:3980–3989. https://doi.org/10.1021/ja058171x
60. Chisholm MH, Feil F, Hadad CM, Patmore NJ (2005) Electronically coupled MM quadruply-bonded complexes (M = Mo or W) employing functionalized terephthalate bridges: toward molecular rheostats and switches. J Am Chem Soc 127:18150–18158. https://doi.org/10.1021/ja0550982
61. Fox MA, Le Guennic B, Roberts RL, Brue DA, Yufit DS, Howard JAK, Manca G, Halet JF, Hartl F, Low PJ (2011) Simultaneous bridge-localized and mixed-valence character in diruthenium radical cations featuring diethynylaromatic bridging ligands 133:18433–18446. https://doi.org/10.1021/ja207827m
62. Romańczyk PP, Noga K, Włodarczyk AJ, Nitek W, Broclawik E (2010) Torsionally controlled electronic coupling in mixed-valence oxodimolybdenum nitrosyl scorpionates—a DFT study. Inorg Chem 49:7676–7684. https://doi.org/10.1021/ic100024h
63. Robin MB, Day P (1968) Mixed valence chemistry-a survey and classification. Adv Inorg Chem Radiochem 10:247–422. https://doi.org/10.1016/S0065-2792(08)60179-X
64. Marcus RA, Sutin N (1985) Electron transfers in chemistry and biology. Biochim Biophys Acta 811:265–322. https://doi.org/10.1016/0304-4173(85)90014-X
65. Wlodarczyk A, Doyle GA, Maher JP, McCleverty JA, Ward MD (1997) Characterisation of two different mixed-valence states in a strongly interacting dimolybdenum complex. Chem Commun 1997:769–770. https://doi.org/10.1039/A608157F
66. Zanello P (2003) Inorganic electrochemistry: theory, practice and application. Royal Society of Chemistry, Oxford
67. Ventura K, Smith MB, Prat JR, Echegoyen LE, Villagrán D (2017) Introducing students to inner sphere electron transfer concepts through electrochemistry studies in diferrocene mixed-valence systems. J Chem Educ 94:526–529. https://doi.org/10.1021/acs.jchemed.6b00642
68. Winter RF (2014) Half-wave potential splittings $\Delta E_{1/2}$ as a measure of electronic coupling in mixed-valent systems: triumphs and defeats. Organometallics 33:4517–4536. https://doi.org/10.1021/om500029x
69. Ding F, Wang H, Wu Q, Van Voorhis T, Chen S, Konopelski JP (2010) Computational study of bridge-assisted intervalence electron transfer. J Phys Chem A 114:6039–6046. https://doi.org/10.1021/jp912049p
70. Kaduk B, Kowalczyk T, Van Voorhis T (2012) Constrained density functional theory. Chem Rev 112:321–370. https://doi.org/10.1021/cr200148b
71. Włodarczyk AJ, Romańczyk PP, Kurek SS, Nitek W, McCleverty JA (2008) Electrochemical interactions in binuclear molybdenum and tungsten nitrosyl complexes incorporating saturated n-alkanediolate bridging ligands. Crystal structures of $[Mo(NO)(Tp^{Me2})I\{O(CH_2)_2OCOCH_3\}]$ and $[W(NO)(Tp^{Me2})I_2]$ $0.5I_2$ $0.5C_6H_5CH_3$. Polyhedron 27:783–796. https://doi.org/10.1016/j.poly.2007.11.008
72. Włodarczyk AJ, Romańczyk PP (2009) Molybdenum doubly bridged and chelate nitrosyl complexes incorporating saturated n-alkanediolate ligands. Inorg Chim Acta 362:4635–4638. https://doi.org/10.1016/j.ica.2009.03.032

73. Bechlars B, D'Alessandro DM, Jenkins DM, Iavarone AT, Glover SD, Kubiak CP, Long JR (2010) High-spin ground states via electron delocalization in mixed-valence imidazolate-bridged divanadium complexes. Nat Chem 2:362–368. https://doi.org/10.1038/nchem.585
74. Ferguson G, Glidewell C, Opromolla G, Zakaria CM, Zanello P (1996) The redox behaviour of some bis-ferrocenyl compounds: crystal and molecular structures of diferrocenylmethane and diferrocenylmethanol. J Organomet Chem 517:183–190. https://doi.org/10.1016/0022-328X(96)06199-2
75. Noga K, Romańczyk PP, Włodarczyk AJ, Broclawik E (2008) Theoretical modeling of electrochemical interactions in bimetallic molybdenum nitrosyl complexes incorporating saturated bridges. Polyhedron 27:2819–2824. https://doi.org/10.1016/j.poly.2008.06.019
76. Wlodarczyk A, Coles SJ, Hursthouse MB, Abdul KM, Lieberman HF (1997) Symmetrically- and asymmetrically-substituted oxo-bridged binuclear molybdenum nitrosyls: synthetic, electrochemical, spectroscopic and structural studies. J Chem Soc Dalton Trans 1997:2921–2930. https://doi.org/10.1039/A704179I
77. Millar AJ, Doonan CJ, Laughlin LJ, Tiekink ERT, Young CG (2002) Atom transfer chemistry and electrochemical behavior of Mo(VI) and Mo(V) trispyrazolylborate complexes: new mononuclear and dinuclear species. Inorg Chim Acta 337:393–406. https://doi.org/10.1016/S0020-1693(02)01105-2
78. Romańczyk PP, Noga K, Radoń M, Rotko G, Kurek SS (2013) On the role of noncovalent interactions in electrocatalysis. Two cases of mediated reductive dehalogenation. Electrochim Acta 110:619–627. https://doi.org/10.1016/j.electacta.2013.05.006
79. Liao RZ, Chen SL, Siegbahn PEM (2015) Which oxidation state initiates dehalogenation in the B12-dependent enzyme NpRdhA: Co^{II}, Co^{I}, or Co^{0}? ACS Catal 5:7350–7358. https://doi.org/10.1021/acscatal.5b01502
80. Szatkowski L, Hall MB (2016) Dehalogenation of chloroalkanes by nickel(I) porphyrin derivatives, a computational study. Dalton Trans 45:16869–16877. https://doi.org/10.1039/C6DT02632J
81. Li XX, Zheng QC, Wang Y, Zhang HX (2014) Theoretical insights into the reductive metabolism of CCl_4 by cytochrome P450 enzymes and the CCl_4-dependent suicidal inactivation of P450. Dalton Trans 43:14833–14840. https://doi.org/10.1039/C4DT02065K
82. Grimme S, Antony J, Ehrlich S, Krieg H (2010) A consistent and accurate ab initio parametrization of density functional dispersion correction (DFT-D) for the 94 elements H-Pu. J Chem Phys 132:154104. https://doi.org/10.1063/1.3382344
83. Ribeiro RF, Marenich AV, Cramer CJ, Truhlar DG (2011) Use of solution-phase vibrational frequencies in continuum models for the free energy of solvation. J Phys Chem B 115:14556–14562. https://doi.org/10.1021/jp205508z

The Quest for Accurate Theoretical Models of Metalloenzymes: An Aid to Experiment

Matthew G. Quesne and Sam P. de Visser

Abstract Enzymes are versatile oxidants in Nature that catalyze a range of reactions very efficiently. Experimental studies on the mechanism of enzymes are sometimes difficult due to the short lifetime of catalytic cycle intermediates. Theoretical modeling can assist and guide experiment and elucidate mechanisms for fast reaction pathways. Two key computational approaches are in the literature, namely quantum mechanics/molecular mechanics (QM/MM) on complete enzyme structures and QM cluster models on active site structures only. These two approaches are reviewed here. We give examples where the QM cluster approach worked well and, for instance, enabled the bioengineering of an enzyme to change its functionality. In addition, several examples are given, where QM cluster models were insufficient and full QM/MM structures were needed to establish regio-, chemo-, and stereoselectivities.

1 Introduction

Metalloenzymes are extremely diverse and ubiquitous family of biocatalysts that use earth abundant and often redox-active metal co-factors to vastly increase their turnover frequencies [1–4]. However, the development of accurate models for such catalysts has proven particularly challenging; owing to both the difficulties involved in the realistic treatment of transition metals and the complex three-dimensional structures of these large catalysts. It was not until the addition of generalized gradient approximations to density functional theory (DFT) methods in the 1990s [5] that the study of these types of catalysts really came of age and many examples started being reported of studies of enzymatic reaction mechanisms [6–9]. During the

M. G. Quesne (✉)
School of Chemistry, Cardiff University, Main Building, Park Place, Cardiff CF10 3AT, UK
e-mail: quesnem@cardiff.ac.uk

S. P. de Visser
The Manchester Institute of Biotechnology and School of Chemical Engineering and Analytical Science, The University of Manchester, 131 Princess Street, Manchester M1 7DN, UK
e-mail: sam.devisser@manchester.ac.uk

E. Broclawik et al. (eds.), *Transition Metals in Coordination Environments*,
Challenges and Advances in Computational Chemistry and Physics 29,
https://doi.org/10.1007/978-3-030-11714-6_14

stratospheric expansion of modeling techniques in the decades that followed, three main broad strategies emerged, namely: quantum mechanical cluster approach, quantum mechanics/molecular mechanics (QM/MM), and the study of biomimetic model systems. Since a subsequent chapter in this book will discuss biomimetic models, here we will focus specifically on the QM/MM and QM cluster model techniques on enzymes only.

QM cluster models focus on the active site region of the enzyme and study the active site alone, i.e., the active site and its immediate surrounding but without the rest of the protein and solvent. These cluster models can range in size from several dozen to hundreds of atoms, but are treated with high-accuracy computational methods such as wave function or DFT approaches. In general, all models typically include the (metal) co-factor and its first coordination sphere as well as the substrate(s). In addition, often close-lying hydrophylic residues that donate hydrogen bonds or that are involved in salt-bridges or π-stacking interactions are included. Often these cluster models contain geometric constraints that force residues (or part of them) to stay in a specific orientation, which may lead to convergence problems of the calculation. However, they do contain the coordination features of the central metal ion as well as a part of or the complete substrate-binding pocket. Therefore, they should mimic substrate approach and catalysis as good as a full enzyme model. One of the first applications of the QM cluster technique to metalloenzymes was reported in 1997 and focused on the enzyme methane monooxygenase [10].

QM/MM protocols split the metalloenzyme into both a small active core region, which is to be treated using QM methods, and apply classical mechanical techniques to everything else. Figure 1 shows an example of prolyl-4-hydroxylase [11, 12], where the QM/MM structure is split into a QM region (in the amber circle) and an MM region (in the green circle). A major difficulty with QM/MM involves obtaining a good description of the border between the QM and MM region and the interaction of the MM with the QM. Several reviews that give a very good overview of the early applications of QM/MM techniques are available [13–16]. While the underlying concept of the QM/MM method has been around since the 1970s [17], it was not until the aforementioned improvement in the QM methodology that the results of these studies became widely applied to metalloenzymes [18]. Therefore, the first full QM/MM study into a transition metal-dependent enzyme was undertaken in 2000 on galactose oxidase [19].

There are several very detailed reviews into the various advantages and drawbacks of QM/MM and cluster model techniques, with a focus on metalloenzymes [8, 9, 20, 21]. Therefore, in this chapter we intend to highlight some of the latest developments in both techniques as well as to describe some of the most useful applications of each, using several case studies.

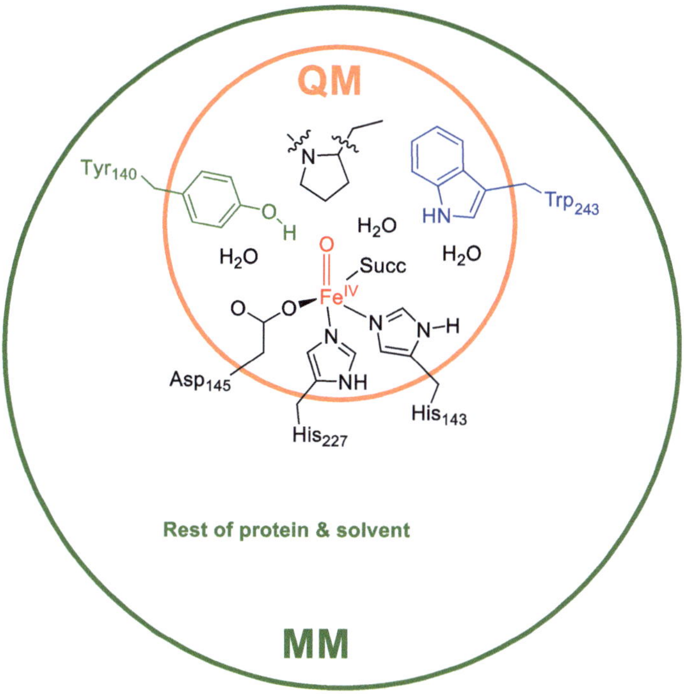

Fig. 1 Splitting a solvated enzyme structure into a QM and MM region

2 Methodology

2.1 Cluster Models

Reluctance to accept reactivity studies conducted on cluster models, which by design only focus on a small portion of the enzyme-substrate complex, arose due to concerns around the absence of long-range polarization and dispersion contributions from the neglected protein and solvent environments [22]. Many of these initial concerns were mitigated with a study into a manganese catalase cluster that concluded these effects to be very small (1–3 kcal mol^{-1}) [23]. However, subsequent structure convergence studies, performed using QM cluster models ranging in size from 27 to 220 atoms, showed that changes in the size of the model could affect the energetics of the reaction pathway by as much as 14 kcal mol^{-1} [24, 25]. Therefore, the selection of the model is of utmost importance in cluster model calculations, although the same rule applies to the choice of the QM region in QM/MM.

Adding an implicit solvent to a QM cluster model, necessitates creating a cavity around the system alongside defining a probe radius of the solvent [26]. As discussed previously, such techniques can prove insufficient for metalloenzymes that control reactivity by controlling the solvation of their active site regions [27] or that use substrate position to compartmentalize their reactivity [28]. However, in the view of these authors such objections should be system-specific and when done correctly (and on appropriate systems), cluster model techniques remain powerful tools for studying enzymatic systems in general and metalloenzymes in particular.

2.2 Quantum Mechanics/Molecular Mechanics (QM/MM)

QM/MM studies begin with selecting heavy atom coordinates obtained from X-ray crystallography [29] and deposited at the protein databank. It is often impossible to crystallize a protein-substrate complex; therefore, competitive inhibitors are often used to bind to an active site in a manner that closely mimics the natural substrate or alternatively the metal is replaced by an unreactive metal such as Zn(II). As a consequence, QM/MM calculations involve a detailed setup procedure starting from a protein databank file (pdb) [29] that will need to be carefully analyzed and often modified and/or corrected to create an appropriate starting structure for the QM/MM calculations. After such a structure is obtained, some heavy atom coordinates need to be modified into those of the starting structure, for instance, the user may want to change the resting state crystal structure into a QM/MM model of the active oxidant of the enzyme. This could for instance, mean replacing an iron(III)-water complex with an iron(IV)-oxo structure [28]. Next, hydrogen atoms must be re-added, since the X-ray scattering ability of their single electron is not great enough to be resolved. Several protonation procedures exist but a popular one utilizes a web server a common Web service for proteins uses the PROPKA technique [30]. This is typically done at an estimated pH value of approximately 7 [30], which is the active pH of most proteins. Crystallization techniques necessitate the removal of water from the protein suspension, and this combined with difficulties in resolving their positions leads to a reduced amount of crystal waters, present in the electron density maps. Therefore, there is a need to re-solvate the metalloenzyme using a water sphere or box centered around the enzyme/substrate system. After the solvation step, a molecular dynamics (MD) simulation is run, with pressure being exerted from the periphery of the system to force water molecules into the protein. Biomolecular force fields that have been extensively parameterized for use on biopolymers, such as AMBER [31], CHARMM [32], and GROMOS [33], are highly recommended. For a detailed overview of how to set up a QM/MM calculation and the difficulties and challenges encountered, a tutorial review on the subject is available in the literature [21]. As described below, one of the real advantages in this type of setup relates to the explicit solvation approach; whereby, a systematic solvation protocol allows explicit water molecules into the active site and so well-prepared QM/MM models often replicate the specific polarity of the metalloenzyme more accurately.

Once the system is solvated, several different QM/MM protocols can be followed, whereby the complete system is split into two regions as explained in Fig. 1 above: a QM region for the inner core of the protein and the MM region for the rest. The interaction between these two regions is important for the electronic description of the

QM region, which is particularly relevant as often there are chemical bonds between the two regions. Often these bonds are capped and the QM region is calculated using hydrogen-link atoms [15].

Two main approaches are available in the literature for the interaction of the MM atoms with the QM atoms, namely mechanical embedding [34] and electrostatic embedding protocols [18]. Mechanical embedding protocols consider the electrostatic interaction between initial charges in the MM region and QM regions; however, only with an electrostatic methodology are MM charges incorporated into the QM Hamiltonian explicitly. Convergence studies on QM/MM models, with over 40 different QM regions, showed that the absolute mean deviation increased from 1.7 kcal mol^{-1} using a mechanical embedding protocol to around 5 kcal mol^{-1} when electronic embedding was considered [35]. Therefore, while electrostatic embedding may be considered more accurate, when using such a protocol the choice of QM region is extremely important.

The interaction of the QM and MM region to give the overall QM/MM energies is calculated through either a subtractive or additive protocol. Two layer ONIOM (QM:MM) is an example of a subtractive QM/MM protocol, whereby, linker atoms (L) are used to cap only the inner (I) subsystems and the whole system is calculated using the MM force field [36]. The QM region plus linker atoms is calculated both at the QM and MM level and therefore the energy of the MM inner subsystem ($E_{MM}(I + L)$) needs to be subtracted from the energy of the inner subsystem calculated with the QM methodology ($E_{QM}(I + L)$), (1).

$$E_{QM/MM}^{Whole} = E_{MM}^{Whole} + E_{QM}(I + L) - E_{MM}(I + L) \tag{1}$$

Alternatively, an additive QM/MM approach can be employed, where the MM calculation is only performed on the outer subsystem [37]. The QM calculation is performed in much the same way as for the subtractive scheme; however, for this protocol an additional QM/MM coupling term must also be included to describe the various bonding, van der Waals and electrostatic interactions between the two subsystems.

$$E_{QM/MM}^{Whole} = E_{MM}^{Enviroment} + E_{QM}(I + L) + E_{QMMM}^{Coupling} \tag{2}$$

While there are many different types of subtractive and additive schemes as well as embedding techniques, it is also true that with careful setup both protocols should give identical results [38].

3 Developments in QM Cluster Calculations

This section presents an overview of the latest developments in QM cluster model techniques. Several brief case studies are used that highlight specific breakthrough in this field.

3.1 Co-factor Free Dioxygenase Reaction Mechanism

Our first example concerns one of the most abundant groups of metalloenzymes, namely the dioxygenases that generally use transition metals to catalyze the activation of molecular oxygen in otherwise spin forbidden reactions [39–42]. However, a small subgroup of this superfamily is able to direct this triplet to singlet conversion without the incorporation of a co-factor [43]. One of the earliest examples of these types of catalysts was (1H)-3-hydroxy-4-oxoquinaldine 2,4-dioxygenase (HOD), which catalyzes oxygen-dependent *N*-heteroaromatic ring cleavage, in the (1H)-3 hydroxy-4-oxoquinaldine (QND) substrate [44]. Two recent papers used combined experimental DFT cluster models and QM/MM techniques to explore how this co-factor-free enzyme is able to replicate metalloenzyme functionality without incorporating a metal or flavin co-factor [45, 46].

Cluster models of various sizes incorporated the same active site residues in both studies (Fig. 2). The smaller model (shown in the blue circle) includes the substrate and truncated models of His_{251}, Asp_{126}, Ser_{101} and Trp_{160} as well as the backbone of Trp_{36}. The larger model retains all these atoms and also incorporates the imidazole groups of His_{38}, His_{100} and His_{102} as well as the three water molecules that were deemed to be within interaction distance of several QM residues. Importantly, the O_2 molecule was only included in the latter study [45]; while the former used some QM/MM calculations to validate the cluster model-derived reaction mechanism [46].

The first paper focused on the initial reaction step of HOD catalysis and explored the effect of amino acid substitution using both experimental, cluster model and QM/MM techniques [46]. This combined approach highlights the importance of the His/Asp dyad; whereby, it was determined that His_{251} was not basic enough to catalyze substrate deprotonation alone. Site-specific mutations produced the variants D126A and H251A that targeted each residue in this dyad separately. These mutants were then replicated theoretically by removing a carboxylate and an imidazole groups from the wild-type (WT) cluster models. These cluster models utilized the hybrid DFT functional UB3LYP [5, 47] in combination with the basis set 6−31G(3d,p), as implemented in the Gaussian software package [36]. Additionally, QM/MM models were constructed and optimized with a combination of the CHARMM27 force field [32] and UB3LYP, as implemented in DL_POLY [48] and TURBOMOLE [49], respectively. The protocol used here is identical to that discussed in more depth in Sect. 4; therefore, here we will only state that the protein databank (PDB) file 2WJ4, a solvent sphere of 35Å and the basis set SV(P) was used in this study.

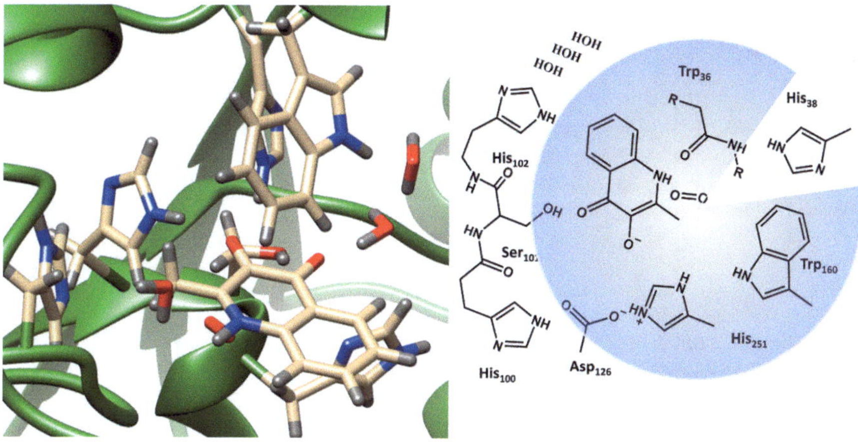

Fig. 2 Structure of the minimal (blue circle) and large (all residues) DFT models, as modified from the protein crystal structure 2WJ4

It is evident that the initial optimizations using these two techniques produce dramatically different substrate orientations, as can be seen from the bond lengths displayed in Fig. 3. All the cluster models show substantial substrate migration away from Trp_{36} and Ser_{101}, with the exact positioning of the substrate being very model dependent. In contrast, when using the QM/MM approach the substrate position is barely affected by site mutations and in each case QND remains in hydrogen-bonding distance to Trp_{36} and Ser_{101}. Importantly, spontaneous hydrogen transfer is only seen with the wild-type model, with the QND–OH species being the local minima in both the D126A and H251A clusters. This computational observation is in excellent agreement with experiment; whereby, when stop-flow measurements were used to assess the deprotonation rate constant (k_H), the values were between 5–40-fold lower in D126A, (depending on pH) then the WT and activity was too low to measure in the H251A variant.

The very substantial movement of the QND substrate is perhaps unsurprising since it is positioned at the surface of the cluster model and is not surrounded by many other residues. It is possible that either the use of a very large cluster model or the addition of restraints on QND could give better structural agreement with experiment. However, constraining the substrate atoms could have led to unphysically high barriers due to QND re-orientation and extensive bond rearrangement, during the reaction path, and an even large QM region would be too computationally expensive. Despite the limitations of the cluster models in accurately simulating substrate orientation, these systems did prove sufficient by providing compelling electronic explanations for the observed loss of catalytic activity between the different variants. Table 1 indicates that only cluster models of the wild-type showed exothermic deprotonation of QND, with the hydrogen atom abstraction disfavored in the other two variants. These energies relate to the sum of the bond dissociation energy of (QND)O–H and

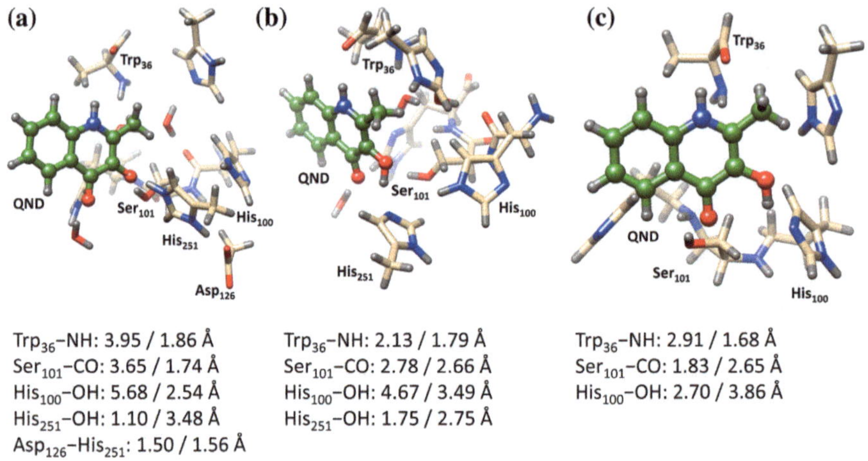

Fig. 3 Comparative geometries between DFT cluster model and QM/MM optimized geometries for HOD and its variants. **a** WT HOD, **b** D1216A, and **c** H251A systems optimized at B3LYP//cluster and B3LYP//QMMM level of theory. Distances between certain active site residues and QND atoms were taken DFT optimized (left) and QM/MM optimized (right) systems

Table 1 QND deprotonation by His residues, using active site models of different HOD variants

	$\Delta E + ZPE_a$	ΔG
WT	−0.9	−0.8
D126A	13.1	14.0
H251A	28.0	27.8

$_a$Zero-point (ZPE) and Gibbs free energies for QND deprotonation using UB3LYP in combination with the 6−31G(3d,p) basis set. All energies are given in kcal mol^{-1}.

the bond association energy of (His)N–H, with the His_{100} substituting His_{251}, as the hydrogen atom acceptor, in the H251A mutant model. A more detailed look into the thermodynamic driving forces of this reaction showed that hydrogen bonding of the histidine to the aspartic acid residue gives an increase of its proton affinity by 12 kcal mol^{-1} predominantly by increasing the electron affinity of the transfer residue. This value was obtained by calculating the difference in free energy, gained by transferring a proton from QND to either His_{251} alone or the His_{251}–Asp_{126} dimer. Importantly, in this specific example the substantially poor description of substrate orientation appears not to have been that important in the usefulness of the models because the Trp_{36}(CO)–QND(NH) hydrogen bond only effected the hydrogen transfer energies by ~2 kcal mol^{-1}.

The second study focused on the rate-limiting oxygen-dependent step of the reaction mechanism and concluded (somewhat unexpectedly) that it proceeded via an oxygen bond triplet intermediate [46]. Scheme 1 highlights the differences in the proposed and calculated reaction mechanism and shows that the calculated potential

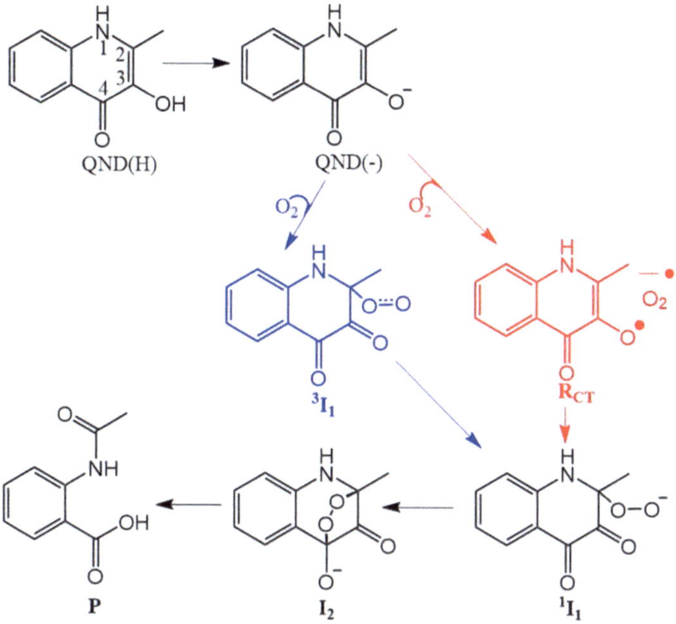

Scheme 1 Potential reaction mechanism(s) for the catalytic cycle of HOD. Both the proposed (red) and calculated (blue) first intermediates are shown

energy surface deviates with an initial oxygen-bound triplet intermediate (shown in blue). The original basis for the long-range electron transfer hypothesis came from spin-trapping experiments [50]; however, the calculated spin densities seem to indicate that the observed radical is more likely associated with the transformation from QND(−) to $^3\mathbf{I}_1$, which is accompanied by an elongation of the C2–C3 bond and a rehybridization from sp^2 to sp^3. Furthermore, transient-state stop-flow experiments, which accompanied these calculations, were unable to detect the signature of the proposed $\mathbf{R}_{CT}$ species.

3.2 *Bioengineering of S-para-Hydroxymandalate Synthase into R-para-Hydroxymandalate Synthase*

QM cluster models can provide useful insight into the active sites of metalloenzymes, their structure and reactivity. In a combined experimental and computational approach, the mechanism of S-p-hydroxymandalate synthase was investigated [51], which is a non-heme iron dioxygenase that utilizes molecular oxygen and converts p-hydroxyphenylpyruvate enantioselectively into S-p-hydroxymandalate. Using a QM cluster model, the key step in the reaction mechanism, namely hydrogen atom abstraction by an iron(IV)-oxo species (Fig. 4) was investigated.

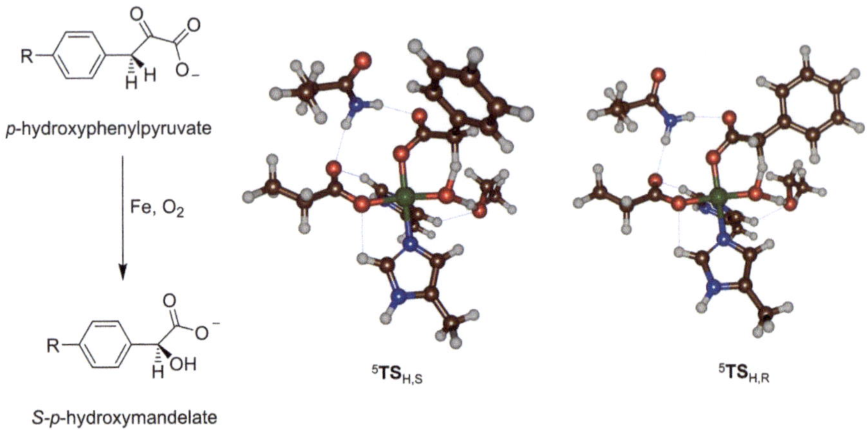

Fig. 4 Reaction catalyzed by S-p-hydroxymandalate synthase and QM cluster optimized transition states for hydrogen atom abstraction of the pro-S (left) and pro-R (right) hydrogen atoms by an iron(IV)-oxo species

The cluster model included the first coordination sphere of protein residues, namely the side chains of Gln_{325} and Thr_{234} as well as the substrate. Subsequently, the hydrogen atom abstraction and OH rebound of both hydrogen atoms from the α-position of the substrate was investigated with DFT methods. These two pathways give the pro-S and pro-R hydrogen atom abstraction transition states ($^5\mathbf{TS}_{H,S}$ and $^5\mathbf{TS}_{H,R}$) by an iron(IV)-oxo species in the quintet spin state. After the transition states, as seen before [52–55], the calculations relax to a radical intermediate prior to an OH rebound to form the S- and R-hydroxymandalate products.

Subsequently, these QM cluster structures of the transition states, for pro-S versus pro-R hydrogen atom abstraction, were inserted into the crystal structure coordinates of the protein. Based on the pro-R transition state structure, the computational modeling suggested to engineering the protein with three mutations for the biosynthesis of R-p-hydroxymandalate, namely the Ser221Met/Val223Phe/Tyr359Ala variant. When this triple mutant was synthesized experimentally, it indeed showed >96% R-product over S-products as proposed by the modeling [51].

4 Developments in QM/MM Calculations

This section is intended to provide a short but comprehensive overview of the latest developments in QM/MM methodology with some recent examples from our groups. Several brief case studies that summarize recent work in the field are used to highlight specific exciting breakthrough protocols.

4.1 Explicit Solvent Effects Captured with QM/MM Models

Another extremely diverse superfamily of oxygenases are the heme-dependent metalloenzymes, the cytochrome P450s, which are represented by over 20,000 unique sequences and are extremely well conserved throughout the natural world [6, 56–60]. Humans also possess a very large number of P450 homologs localized mainly in liver hepatocytes, where they are responsible for the metabolism of harmful xenobiotics, such as drug molecules [61] but also phthalates from cosmetics products [62]; however, with this family of biocatalysts being so functionally diverse they also catalyze other pathways, including hormone biosynthesis [63–65]. Such a massively diverse functionality is hard to correlate to the large amount of structural conservation seen throughout the P450 superfamily of metalloenzymes. However, much of the small amount of structural differences are localized close to the enzyme active site, which amplifies the effects of such changes. Several members of this superfamily such as $P450_{cam}$ possess relatively tight-binding pockets, whereas those that catalyze the oxidation of large lipid structures, such as $P450_{BM3}$ may have very open substrate-binding regions [66–69]. There has been a large amount of biotechnological interest in many P450 isozymes owing primarily to their much superior product regio/stereo—selectivity, when compared to more traditional catalysts used by the pharmaceutical industry [70, 71].

This superdiverse family of biocatalysts conserve many core structural as well as spectroscopic features, chiefly an absorbance feature at 450 nm, which is observed once CO binds the iron(III) co-factor of their heme co-enzyme [72]. Figure 5 shows the active site of $P450_{CAM}$ and illuminates the other main feature conserved throughout all P450s, namely the thiolate linkage to iron that connects the heme co-factor to the protein via a cysteinate residue [73–75]. In the case of $P450_{CAM,}$ Cys_{357} occupies this coordination position. Figure 5 represents the substrate-bound 5-coordinate iron(III) complex that after O_2 binding and two protonation steps is converted into an iron(IV)-oxo heme cation radical species called Compound I (Cpd I) [76–78]. Theoretical models have been able to predict the likelihood of competitive doublet and quartet spin states that bifurcate the reaction mechanisms catalyzed by these enzymes and leads to the possible spin-state selective product composition [78–82] (see Scheme 2). Cpd I is responsible for an initial hydrogen atom transfer (HAT) step, with the potential energy surfaces of each spin state then deviating; one spin state gives low rebound barriers (low-spin), while the other spin state (high-spin) has significant barriers that implicate a finite lifetime of the radical intermediate leading to rearrangement pathways and by-products of side reactions [80, 83].

Many computational studies on P450 systems have been performed and generally they reproduce experimental findings well [84–90]. Although, these small DFT cluster models had proved useful in justifying a possible mechanism to explain the presence of the two distinct and competitive product pathways seen in experiment, they were insufficient when it came to describe whether these differences were caused by some adaptation in the protein environments or by some innate properties of the substrate(s) themselves. Because of this, our first example in this section concerns

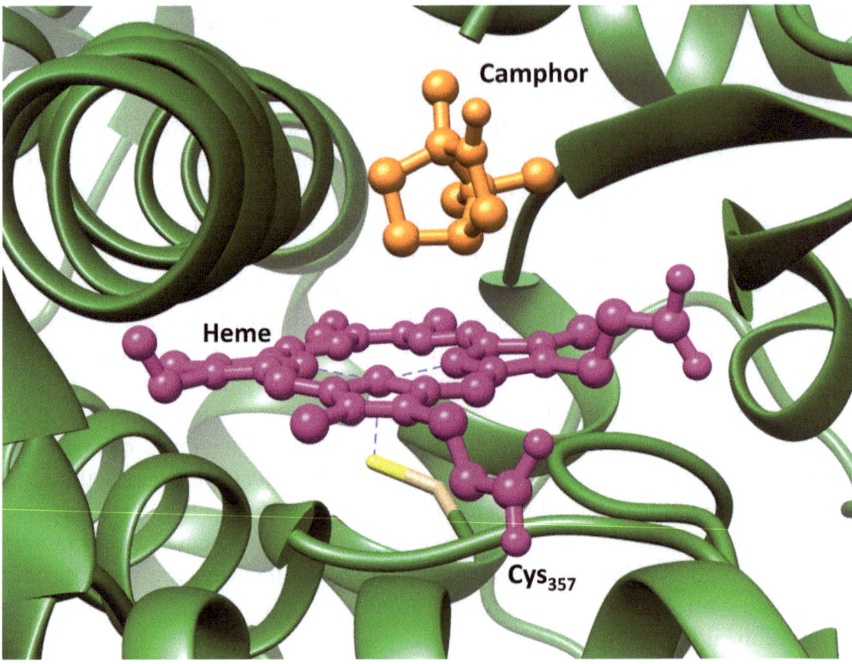

Fig. 5 Active site region of $P450_{cam}$, from PDB 2CPP with heme in purple and substrate (camphor) in amber

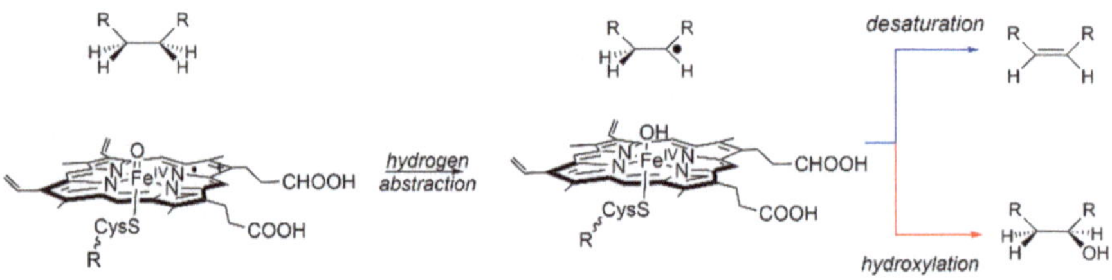

Scheme 2 Alternative hydroxylation and desaturation pathways

work that used a combination of: DFT, QM/MM, and valence bond (VB) modeling techniques to help answer this fundamental question [61].

In order to quantify the effect of substrate driven chemoselectivity, small DFT models of the first coordination sphere of the Fe(IV)=O center of Cpd I were used to study its reaction mechanism(s) with: valpronic acid (VA), ethane (ET), and dihydroanthracene (DHA) [61]. In vitro enzymatic studies [91] have shown that P450 catalyzed oxidation of ethane promotes exclusive alcohol formation, while DHA is desaturated to form anthracene and VA can go through either the hydroxyl rebound or desaturation pathways. The minimal DFT cluster models did predict the experimentally observed exclusive alcohol formation; however, when the VA and DHA substrates were used the calculations predicted both possible pathways to be bar-

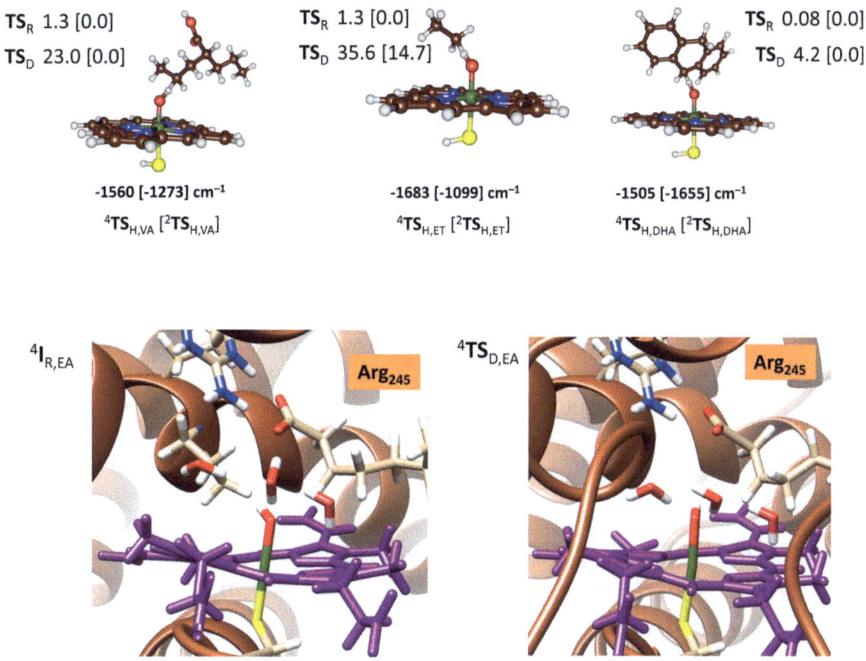

Fig. 6 QM cluster (top) and QM/MM (bottom) models of substrate oxidation by P450 Cpd I. (Top) Hydrogen atom abstraction geometries and ΔG for VA, ET, and DHA. Hydrogen abstraction (**TS**$_H$), OH rebound (**TS**$_R$), and desaturation (**TS**$_D$) barriers shown. (Bottom) optimized QM/MM geometries of the HAT intermediate ($^4I_{R,EA}$) and decarboxylation transition state ($^4TS_{D,EA}$), modeled using OleT$_{JE}$

rierless in the doublet spin state, with the quartet spin states dominated by alcohol production (see Fig. 6). The fact that the same reaction profiles in VA and DHA leads experimentally in one case to a mix of product formation and in the other purely desaturation, meaning that product specificity does not correlated to the kinetically controlled reactivity of the first coordination sphere of compound I and the substrate but instead must at least in part be influenced by the larger protein environment.

Subsequently, QM/MM models of P450 OleT$_{JE}$ were created (lower panel of Fig. 6), which is a P450 peroxygenase that performs the decarboxylation of long-chain fatty acids. In particular, the regioselectivity for desaturation versus α-hydroxylation was investigated for eicosanoic acid (EA) as a substrate [92].

The QM/MM model utilized a crystal structure of the enzyme/substrate complex [93] and modified it to form a substrate-bound Cpd I [94, 95] active species before proceeding along the reaction mechanism, using a well-established protocol [28, 96]. In this study, the inclusion of the protein environment completely alters the preferred reaction mechanism (from that seen in the DFT cluster models). Whereby, using the QM/MM protocol the preferred spin state changes from a doublet to a quartet and the decarboxylation barrier (^{4}**TS**$_D$) reduces from 17.8 kcal mol^{-1} (in the cluster model) to 5.1 kcal mol^{-1}. More importantly, the relative preference of the pathways flips with the hydroxyl rebound barrier ($^4\mathbf{TS_R}$) increasing from 5.1 kcal mol^{-1} to 6.6 kcal mol^{-1}, in the QM/MM model. This reversal of the barrier ordering leads to a preference for olefin production that can be at least partly understood when examining the $^4\mathbf{I_R}$ and $^4\mathbf{TS_R}$ structure shown in Fig. 6. As mentioned in previous sections, the ability of a well-designed QM/MM protocol to replicate an accurately solvated active site is one of the largest advantages of this type of technique, above cluster models. The water network surrounding the iron(IV)-hydroxo of $^4\mathbf{I_{EA}}$ forms a bridge to the guanidine groups of Arg_{245}, which in turn was found to make the rotation of hydroxyl group more difficult and thereby reversing the chemoselectivity by pushing $^4\mathbf{TS_{R,EA}}$ above $^4\mathbf{TS_{D,EA}}$. The authors subsequently highlight the fact that $P450_{cam}$ has a very tight-binding pocket that is likely to exclude water and therefore prevent formation of analogous hydrogen-bonding networks and retain its preference for forming the alcohol.

4.2 *QM/MM Techniques Are Often Essential for Replicating Minor Structural Anomalies that Lead to Large Changes in Reactivity*

Our second example in this section concerns AlkB, which is an α-ketoglutarate (αKG)-dependent metalloenzyme that utilizes dioxygen and with the help of α-ketoglutarate forms a high-valent iron(IV)-oxo intermediate, which transfers its oxygen atom to the substrate [97–100]. AlkB is a member of another highly conserved group of iron oxygenases that are responsible for the demethylation of methylated DNA bases [101–103]. DNA bases are methylated in response to chemical damage and if left unrepaired would eventually result in a halt in DNA replication by preventing Watson–Crick base pair formation [104–106]. AlkB is the proposed mechanism based on experimental studies of taurine/α-ketoglutarate dioxygenase and computational modeling (mostly using QM cluster methods) is shown in Scheme 3. The resting state is an iron(II) held to the protein via linkages with two histidine and one carboxylate (Asp or Glu) ligands. Three water molecules complete the octahedral coordination environment of the resting state. It is envisaged that an initial O_2 binding leads to the formation of an iron(III)-superoxo structure that attacks the β-carbon of αKG, which is followed by decarboxylation to form succinate and CO_2 and an iron(IV)-oxo species [52, 107]. The latter reacts with substrate by hydrogen atom abstraction and hydroxyl rebound steps, catalyzed by the resulting ferryl-oxo intermediate species [101–103, 108].

However, while the crystal structure of the AlkB enzyme [109] confirms the presence of a standard 2—His/1—Asp iron-binding motif, the oxygen coordination site is surprisingly not proximal to the substrate (see Fig. 7), but orthogonal to it. Therefore, a computational study was undertaken using a combination of QM/MM and QM cluster techniques to assess the validity of the mechanism in Scheme 3

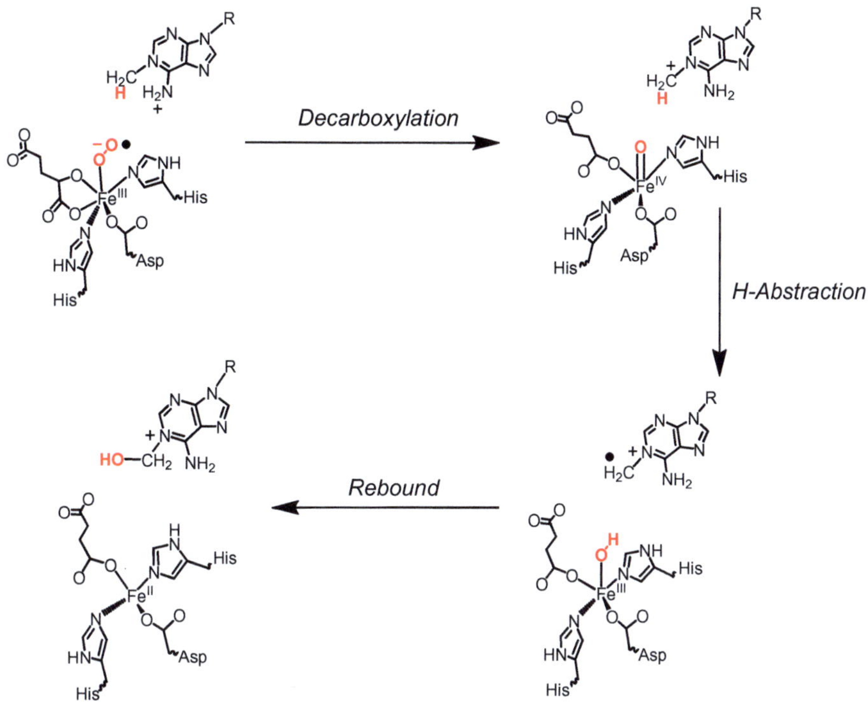

Scheme 3 Typical substrate hydroxylation mechanism catalyzed by the non-heme iron oxygenase superfamily

[28]. Although the authors did originally attempt to place oxygen in a more proximal position (*trans* to His_{187}), they found that stereochemical clashes made such a position unfeasible. Therefore, an Fe(IV)oxo species with the oxygen in the free coordination position was created and an initial oxygen atom rotation was undertaken, to move the oxygen from the position *trans* to His—131 (**R**) to the one *trans* to His—187 (**R'**).

The QM/MM methodology used had been carefully benchmarked previously [110, 111] and starts from the 3I20 crystal structure coordinates, which were further modified as seen in Fig. 7 [109]. The QM region included an iron(IV)-oxo group, truncated His_{131} His_{187} and Asp_{133} amino acids, an acetate for the succinate and the N^1-methyladenine substrate. The protein was protonated using PROPKA and solvated using the CHARMM software package [112]. A large molecular dynamics simulation was then run to provide several low-energy snapshots for the QM/MM calculations.

The authors benchmarked their snapshots and chosen density functional methods by using B3LYP, B3LYP*, BP86, and M06 [5, 47, 113, 114]. They report that all methods and snapshots produced the same spin-state ordering and reaction specificity, and therefore focus on the B3LYP energetics of the snapshot taken at 500

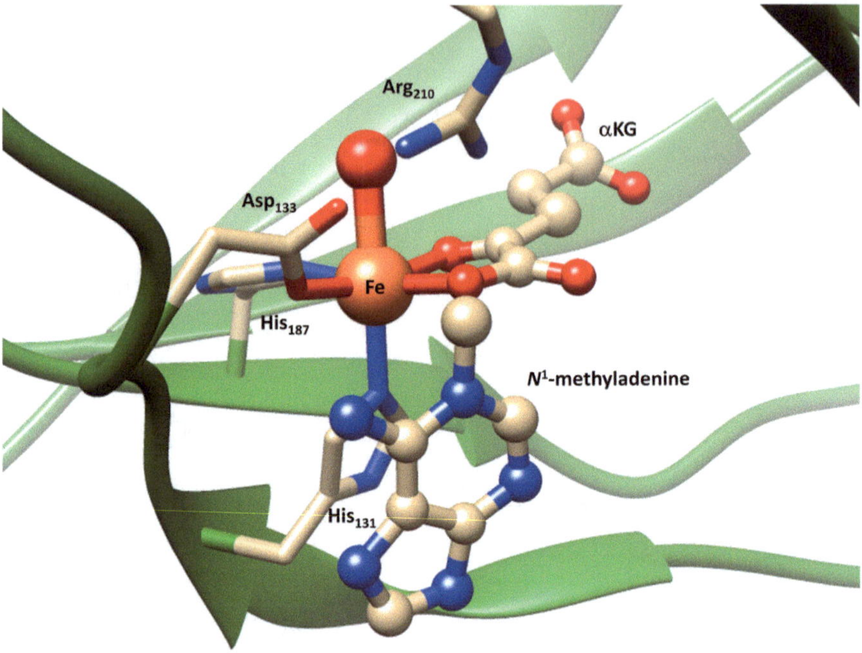

Fig. 7 Active site region of AlkB, from PDB: 3I20. An oxygen atom replaces water, *trans* to His_{131}

picoseconds. On the dominant quintet surface, the isomerization barrier (^{5}**TS**$_I$) from ^{5}**R** to ^{5}**R'** is $\Delta E + ZPE = 9.0$ kcal mol^{-1} lead to an isomerization product (**R'**) of $\Delta E + ZPE = -6.0$ kcal mol^{-1}, see Fig. 8. Although work on a related iron halogenase (SyrB2) [113] showed a slightly higher isomerization barrier of 13.4 kcal mol^{-1}, the authors note that in that case the increase in barrier height was likely due to the presence of a halide as well as an oxygen rotation [113]. Electronically, σ^*_{x2-y2} is singly occupied in ^{5}**R**, whereas the energy of this orbital is elevated in ^{5}**R'** and the σ^*_{z2} orbital occupied. The increased stability of ^{5}**R** is because the HOMO (σ^*_{z2}) orbital in ^{5}**R'** has smaller amounts of antibonding interactions than those of the σ^*_{x2-y2} orbital of ^{5}**R**. The LUMO (σ^*_{z2}) in ^{5}**R** has localized electron density dominated by antibonding interaction along the Fe–O bond, whereas the σ^*_{x2-y2} orbital of ^{5}**R'** shows large amounts of Asp_{133}, His_{187} and succinate involvement. These differences result in a decrease in spin density from 0.75 to 0.69 as ^{5}**R** rotates to ^{5}**R'**, and strongly implies that this rotation would lead to a slightly less efficient oxidant. It is clear that the protein environment assists in this rotation with hydrogen bonds between Arg_{210} and the oxo group assisting in stabilizing the rotational transition state. The authors go on to propose that the enzyme might use the less powerful oxidant to separate the radical superoxo activation and hydrogen atom abstraction steps. To strengthen this argument, cluster models were hydrogen atom abstraction of $\Delta E + ZPE = 32.8$ kcal mol^{-1} where calculated for the iron(III)-superoxo and while this indicates that the iron(III)-superoxo is a less efficient oxidant then **R'**, with a HAT

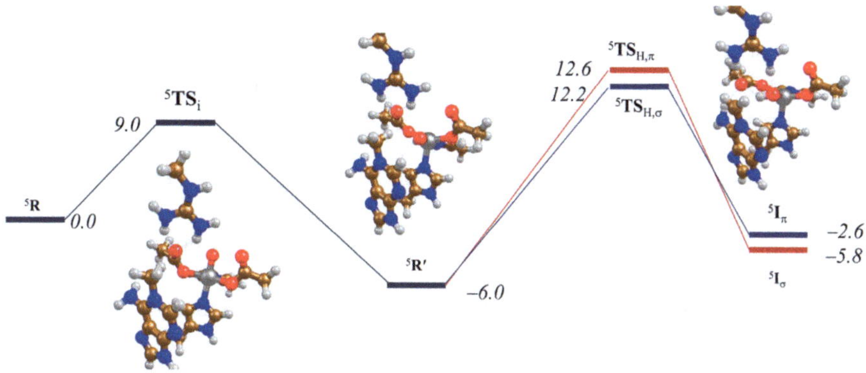

Fig. 8 Potential energy surface for the rotation of an oxygen group toward an active confirmation, as calculated using QM/MM. Zero-point corrected energies obtained at UB3LYP/B2//UB3LYP/B1 are given in kcal/mol^{-1}

barrier of 23.4 kcal mol^{-1}, any amount of hydrogen atom abstraction by this species could prove devastating for catalysis (see Fig. 8).

Minimal cluster models of the iron's first coordination sphere were also created to investigate the HAT barriers for: N^1-methylguanine (27.9 kcal mol^{-1}), N^1-methyladenine (23.4 kcal mol^{-1}), N^3-methylcytosine (6.5 kcal mol^{-1}), and N^3-methythymine (10.9 kcal mol^{-1}). These results are in excellent agreement with previous cluster model studies that found a HAT barrier of 20.9 kcal mol^{-1} for the AlkB demethylation of adenine [115]. However, both these studies report barriers that could indicate turnover frequencies that are too slow for efficient biocatalytic reactions at room temperature in biosystems. A common drawback of using QM cluster models is the unrealistically large degrees of freedom associated with the movement of the substrate, which in turn can lead to unphysically large barriers due to overstabilization of the reactant species. Another important indicator that the variations of barrier heights might be unphysical came from DFT calculation on the relative strengths of the C–H (BDE$_{CH}$) bond in each substrate. By subtracting the combined energies of the substrate radicals and a hydrogen atom from the energy of their protonated equivalents, it was assessed that BDE$_{CH}$ values for each substrate fall within 5 kcal mol^{-1} of each other. Therefore, the large variation in barrier heights could not be substrate driven and indeed detailed analysis of the cluster model systems indicated that multiple hydrogen-bonding interactions with various oxygens were primarily responsible for this phenomenon.

Using the QM/MM protocol, two competitive hydrogen atom transfer channels were found with the $^5\sigma$-pathway transferred an alpha electron into the virtual σ^*_{x2-y2} orbital forming ($^5\mathbf{I}_\sigma$) with $\pi^{*\uparrow}_{xz}\ \pi^{*\uparrow}_{yz}\ \pi^{*\uparrow}_{xy}\ \sigma^{*\uparrow}_{z2}\ \sigma^{*\uparrow}_{x2-y2}\ \phi^{\downarrow}_{Sub}$ configuration, and the $^5\pi$-pathway transferring a beta electron into the partially occupied π^*_{xz} orbital forming ($^5\mathbf{I}_\pi$) $\pi^{*2}_{xz}\ \pi^{*\uparrow}_{yz}\ \pi^{*\uparrow}_{xy}\ \sigma^{*\uparrow}_{z2}\ \sigma^{*0}_{x2-y2}\ \phi^{\uparrow}_{Sub}$ configuration. Usually, non-heme iron(IV)-oxo metalloenzymes favor the $^5\sigma$-pathway because the optimal angle of substrate approach for the correct orbital overlap is approximately 180°, which usually results in less stereochemical clashes than seen in the optimal 120° of the π-pathway and also leads to the maximum level of exchange interactions between unpaired electrons on Fe(III) [52, 113, 116–121]. The σ-pathway for hydrogen-transfer barrier ($^5\mathbf{TS}_{H,\sigma}$) included a Fe–O–C angle of 141.4° which is obviously considerable smaller than the optimal; while the angle of substrate approach seen in $^5\mathbf{TS}_{H\pi}$ (131.4°) is considerably more favorable for π hydrogen abstraction. The potential energy surfaces produced by both pathways (from $^5\mathbf{R}$ to $^5\mathbf{I}$) are shown in Fig. 8 and illustrate that both $^5\mathbf{TS}_{H,\sigma}$ and $^5\mathbf{TS}_{H,\pi}$ have comparable energies: 18.2 versus 18.6 kcal mol^{-1}, which is considerably more realistic then the value of over 20 kcal mol^{-1} found using cluster models.

The final step of the reaction mechanism involves a rebound of the hydroxyl group back onto the radical substrate intermediate, with both $^5\mathbf{I}_\sigma$ and $^5\mathbf{I}_\pi$ forming strong hydrogen-bonding interaction between carboxylate group of Asp_{133} and the Fe(III) – OH. This then leads to relatively high rebound barriers of >13.1 kcal mol^{-1}, which are very far away from the 0.6 kcal mol^{-1} found with the DFT cluster model calculations [115].

5 Conclusion

Computational studies on non-heme and heme monoxygenases and dioxygenases are useful for the understanding of metalloenzymes and in particularly for providing insights into the chemical and physical properties of short-lived and highly reactive intermediates. The key to these studies remains the choice of the model, which should be a good representation of the real system. Two approaches have been used, namely QM models and QM/MM techniques. The latter take the structure of the full enzyme with a solvent layer into consideration whereas the former only include key features of the first and second coordination sphere. We have shown several examples where small active site models using the QM cluster technique give viable results that compare well with experiment. For systems, however, where substrate binding is tight and or a network of hydrogen-bonding interactions exist, a complete QM/MM approach may be more appropriate.

Acknowledgements This work is partially supported as part of an EPSRC low carbon fuels grant (EP/N009533/1).

References

1. Finkelstein J (2009) Metalloproteins Nat 460:813–813. https://doi.org/10.1038/460813a
2. Lu Y, Yeung N, Sieracki N, Marshall NM (2009) Design of functional metalloproteins. Nature 460:855–862. https://doi.org/10.1038/nature08304

3. Siegbahn PE, Blomberg MR (2010) Quantum chemical studies of proton-coupled electron transfer in metalloenzymes. Chem Rev 110:7040–7061. https://doi.org/10.1021/cr100070p
4. Valdez CE, Smith QA, Nechay MR, Alexandrova AN (2014) Mysteries of metals in metalloenzymes. Acc Chem Res 47:3110–3117. https://doi.org/10.1021/ar500227u
5. Becke AD (1993) Density-functional thermochemistry. III. The role of exact exchange. Chem Phys 98:5648–5652
6. Meunier B, De Visser SP, Shaik S (2004) Mechanism of oxidation reactions catalyzed by cytochrome P450 enzymes. Chem Rev 104:3947–3980
7. Li D, Wang Y, Han K (2012) Recent density functional theory model calculations of drug metabolism by cytochrome P450. Coord Chem Rev 256:1137–1150
8. Blomberg MRA, Borowski T, Himo F et al (2014) Quantum chemical studies of mechanisms for metalloenzymes. Chem Rev 114:3601–3658. https://doi.org/10.1021/cr400388t
9. de Visser SP, Quesne MG, Martin B et al (2014) Computational modelling of oxygenation processes in enzymes and biomimetic model complexes. Chem Commun 50:262–282. https://doi.org/10.1039/c3cc47148a
10. Siegbahn EM, Crabtree RH (1997) Mechanism of C−H activation by diiron methane monooxygenases: quantum chemical studies. J Am Chem Soc 119:3103–3113. https://doi.org/10.1021/JA963939M
11. Timmins A, Saint-André M, de Visser SP (2017) Understanding how Prolyl-4-hydroxylase structure steers a Ferryl Oxidant toward scission of a strong C-H Bond. J Am Chem Soc 139:9855–9866. https://doi.org/10.1021/jacs.7b02839
12. Timmins A, de Visser SP (2017) How are substrate binding and catalysis affected by mutating Glu127 and Arg161 in Prolyl-4-hydroxylase? A QM/MM and MD study. Front Chem 5:94. https://doi.org/10.3389/fchem.2017.00094
13. Gao J, Truhlar DG (2002) Quantum mechanical methods for enzyme kinetics. Annu Rev Phys Chem 53:467–505
14. Senn HM, Thiel W (2007) QM/MM studies of enzymes. Curr Opin Chem Biol 11:182–187
15. Senn HM, Thiel W (2009) QM/MM methods for biomolecular systems. Angew Chemie Int Ed 48:1198–1229. https://doi.org/10.1002/anie.200802019
16. van der Kamp MW, Mulholland AJ (2013) Combined quantum mechanics/molecular mechanics (QM/MM) methods in computational enzymology. Biochemistry 52:2708–2728
17. Warshel A, Levitt M (1976) Theoretical studies of enzymic reactions: dielectric, electrostatic and steric stabilization of the carbonium ion in the reaction of lysozyme. J Mol Biol 103:227–249. DOI: https://doi.org/0022-2836(76)90311-9 [pii]
18. Field MJ, Bash PA, Karplus M (1990) A combined quantum mechanical and molecular mechanical potential for molecular dynamics simulations. J Comput Chem 11:700–733. https://doi.org/10.1002/jcc.540110605
19. Rothlisberger U, Carloni P, Doclo K, Parrinello M (2000) A comparative study of galactose oxidase and active site analogs based on QM/MM Car-Parrinello simulations. J Biol Inorg Chem 5:236–250. https://doi.org/10.1007/s007750050368
20. Borowski T, Quesne M, Szaleniec M (2015) QM and QM/MM methods compared: Case studies on reaction mechanisms of metalloenzymes. In: Advances in protein chemistry and structural biology. Academic Press Inc, pp 187–224
21. Quesne MG, Borowski T, De Visser SP (2016) Quantum mechanics/molecular mechanics modeling of enzymatic processes: caveats and breakthroughs. Chem—A Eur J 22:2562–2581. https://doi.org/10.1002/chem.201503802
22. Mulholland AJ, Grant GH, Richards WG (1993) Computer modeling of enzyme catalyzed reaction-mechanisms. Protein Eng 6:133–147
23. Siegbahn PEM (2001) Modeling aspects of mechanisms for reactions catalyzed by metalloenzymes. J Comput Chem 22:1634–1645. https://doi.org/10.1002/jcc.1119
24. Sevastik R, Himo F (2007) Quantum chemical modeling of enzymatic reactions: the case of 4-oxalocrotonate tautomerase. Bioorg Chem 35:444–457
25. Georgieva P, Himo F (2010) Quantum chemical modeling of enzymatic reactions: the case of histone lysine methyltransferase. J Comput Chem 31:1707–1714

26. Cramer CJ, Truhlar DG (1992) AM1-SM2 and PM3-SM3 parameterized SCF solvation models for free energies in aqueous solution. J Comput Aided Mol Des 6:629–666. https://doi.org/10.1007/BF00126219
27. Faponle AS, Quesne MG, de Visser SP (2016) Origin of the regioselective fatty-acid hydroxylation versus decarboxylation by a cytochrome P450 peroxygenase: what drives the reaction to Biofuel production? Chemistry 22:5478–5483. https://doi.org/10.1002/chem.201600739
28. Quesne MG, Latifi R, Gonzalez-Ovalle LE et al (2014) Quantum mechanics/molecular mechanics study on the oxygen binding and substrate hydroxylation step in AlkB repair enzymes. Chem—A Eur J 20:435–446. https://doi.org/10.1002/chem.201303282
29. Berman HM, Westbrook J, Feng Z et al (2000) The protein data bank nucleic acids research. Nucl Acids Res 28:235–242
30. Dolinsky TJ, Czodrowski P, Li H et al (2007) PDB2PQR: expanding and upgrading automated preparation of biomolecular structures for molecular simulations. Nucl Acids Res 35:W522–W525
31. Wang J, Wolf RM, Caldwell JW et al (2004) Development and testing of a general amber force field. J Comput Chem 25:1157–1174. https://doi.org/10.1002/jcc.20035
32. Brooks BR, Brooks CL 3rd, Mackerell ADJ et al (2009) CHARMM: the biomolecular simulation program. J Comput Chem 30:1545–1614. https://doi.org/10.1002/jcc.21287
33. Oostenbrink C, Villa A, Mark AE, Van Gunsteren WF (2004) A biomolecular force field based on the free enthalpy of hydration and solvation: the GROMOS force-field parameter sets 53A5 and 53A6. J Comput Chem 25:1656–1676
34. Maseras F, Morokuma K (1995) Imomm: A new ab initio + molecular mechanics geometry optimization scheme of equilibrium structures and transition states. J Comp Chem 16:1170–1179
35. Hu L, Söderhjelm P, Ryde U (2011) On the convergence of QM/MM energies. J Chem Theory Comput 7:761–777
36. Frisch MJ, Trucks GW, Schlegel HB et al (2009) Gaussian 09, Revision B.01. Gaussian 09, Revis. B.01, Gaussian, Inc, Wallingford CT
37. Sherwood P, de Vries AH, Guest MF et al (2003) QUASI: a general purpose implementation of the QM/MM approach and its application to problems in catalysis. J Mol Struct THEOCHEM 632:1–28
38. Cao L, Ryde U (2018) On the difference between additive and subtractive QM/MM calculations. Front Chem 6:89. https://doi.org/10.3389/fchem.2018.00089
39. Solomon EI, Brunold TC, Davis MI et al (2005) Oxygenases: mechanisms and structural motifs for O-2 activation. Chem Rev 105:550–555. https://doi.org/10.1021/cr040653o
40. Bugg TDH, Ramaswamy S (2008) Non-heme iron-dependent dioxygenases: unravelling catalytic mechanisms for complex enzymatic oxidations. Curr Opin Chem Biol 12:134–140. https://doi.org/10.1016/j.cbpa.2007.12.007
41. Quesne MG, Faponle AS, Goldberg DP, Visser SPD (2015) Catalytic function and mechanism of heme and nonheme iron(IV)-Oxo complexes in nature
42. de Visser SP (2018) Mechanistic insight on the activity and substrate selectivity of nonheme iron dioxygenases. Chem Rec. https://doi.org/10.1002/tcr.201800033
43. Fetzner S, Steiner RA (2010) Cofactor-independent oxidases and oxygenases. Appl Microbiol Biotechnol 86:791–804. https://doi.org/10.1007/s00253-010-2455-0
44. Bauer I, Max N, Fetzner S, Lingens F (1996) 2,4-dioxygenases catalyzing n-heterocyclic-ring cleavage and formation of carbon monoxide. purification and some properties of 1H-3-hydroxy-4-oxoquinaldine 2,4-dioxygenase from arthrobacter sp. Ru61a and comparison with 1H-3-hydroxy-4-oxoquinoline 2,4-dioxygenase from pseudomonas putida 33/1. Eur J Biochem 240:576–583. https://doi.org/10.1111/j.1432-1033.1996.0576h.x
45. Hernández-Ortega A, Quesne MG, Bui S et al (2014) Origin of the proton-transfer step in the cofactor-free 1-H-3-hydroxy-4-oxoquinaldine 2,4-dioxygenase: effect of the basicity of an active site His residue. J Biol Chem 289:8620–8632. https://doi.org/10.1074/jbc.M113.543033

46. Hernández-Ortega A, Quesne MG, Bui S et al (2015) Catalytic mechanism of cofactor-free dioxygenases and how they circumvent spin-forbidden oxygenation of their substrates. J Am Chem Soc 137:7474–7487. https://doi.org/10.1021/jacs.5b03836
47. Lee C, Yang W, Parr RG (1988) Development of the Colle-Salvetti correlation-energy formula into a functional of the electron density. Phys Rev B 37:785
48. Smith W, Yong CW, Rodger PM (2002) DL_POLY: application to molecular simulation. Mol Simul 28:385–471
49. Ahlrichs R, Bar M, Haser M et al (1989) Chem Phys Lett 162:165. Curr version see. https://www.turbomole.com
50. Thierbach S, Bui N, Zapp J et al (2014) Substrate-Assisted O2 activation in a cofactor-independent dioxygenase. Chem Biol 21:217–225. https://doi.org/10.1016/J.CHEMBIOL.2013.11.013
51. Pratter SM, Konstantinovics C, Di Giuro CML et al (2013) inversion of enantioselectivity of a mononuclear non-heme iron (II)-dependent hydroxylase by tuning the interplay of metal-center geometry and protein structure. Angew Chemie 125:9859–9863
52. de Visser SP (2006) Propene activation by the oxo-iron active species of taurine/alpha-ketoglutarate dioxygenase (TauD) enzyme. How does the catalysis compare to heme-enzymes? J Am Chem Soc 128:9813–9824. https://doi.org/10.1021/ja061581g
53. Latifi R, Bagherzadeh M, de Visser SP (2009) Origin of the correlation of the rate constant of substrate hydroxylation by nonheme iron(IV)-oxo complexes with the bond-dissociation energy of the C-H bond of the substrate. Chemistry (Easton) 15:6651–6662. https://doi.org/10.1002/chem.200900211
54. Karamzadeh B, Kumar D, Sastry GN, de Visser SP (2010) Steric factors override thermodynamic driving force in regioselectivity of proline hydroxylation by prolyl-4-hydroxylase enzymes. J Phys Chem A 114:13234–13243. https://doi.org/10.1021/jp1089855
55. Faponle AS, Seebeck FP, de Visser SP (2017) Sulfoxide synthase versus cysteine dioxygenase reactivity in a nonheme iron enzyme. J Am Chem Soc 139:9259–9270. https://doi.org/10.1021/jacs.7b04251
56. de Montellano PRO (2010) Hydrocarbon hydroxylation by cytochrome P450 enzymes. Chem Rev 110:932
57. de Montellano PRO (2004) Cytochrome P450: structure, mechanism, and biochemistry. Springer
58. Denisov IG, Makris TM, Sligar SG, Schlichting I (2005) Structure and chemistry of cytochrome P450. Chem Rev 105:2253–2277. https://doi.org/10.1021/cr0307143
59. Groves JT (2003) The bioinorganic chemistry of iron in oxygenases and supramolecular assemblies. Proc Natl Acad Sci 100:3569–3574. https://doi.org/10.1073/pnas.0830019100
60. Kadish KM, Smith KM, Guilard R (2010) Handbook of porphyrin science. World Scientific
61. Ji L, Faponle AS, Quesne MG et al (2015) Drug metabolism by cytochrome P450 enzymes: what distinguishes the pathways leading to substrate hydroxylation over desaturation? Chem—A Eur J 21:9083–9092. https://doi.org/10.1002/chem.201500329
62. Reinhard FC, de Visser S (2017) Biodegradation of cosmetics products: a computational study of cytochrome P450 metabolism of phthalates. Inorganics 5:77. https://doi.org/10.3390/inorganics5040077
63. Guengerich FP (2001) Common and uncommon cytochrome P450 reactions related to metabolism and chemical toxicity. Chem Res Toxicol 14:611–650
64. Munro AW, Girvan HM, McLean KJ (2007) Variations on a (t) heme–novel mechanisms, redox partners and catalytic functions in the cytochrome P450 superfamily. Nat Prod Rep 24:585–609
65. Posner GH, O'Neill PM (2004) Knowledge of the proposed chemical mechanism of action and cytochrome P450 metabolism of antimalarial trioxanes like artemisinin allows rational design of new antimalarial peroxides. Acc Chem Res 37:397–404
66. Atkins WM, Sligar SG (1987) Metabolic switching in cyctochrome P-450cam: deuterium isotope effects on regiospecificity and the monooxygenase/oxidase ratio. J Am Chem Soc 109:3754–3760

67. Gelb MH, Heimbrook DC, Malkonen P, Sligar SG (1982) Stereochemistry and deuterium isotope effects in camphor hydroxylation by the cytochrome P450cam monooxygenase system. Biochemistry 21:370–377
68. Davydov DR, Hui Bon Hoa G, Peterson JA (1999) Dynamics of protein-bound water in the heme domain of P450BM3 studied by high-pressure spectroscopy: comparison with P450cam and P450 2B4. Biochemistry 38:751–761
69. Ruettinger RT, Wen L-P, Fulco AJ (1989) Coding nucleotide, 5'regulatory, and deduced amino acid sequences of P-450BM-3, a single peptide cytochrome P-450: NADPH-P-450 reductase from Bacillus megaterium. J Biol Chem 264:10987–10995
70. Grogan G (2011) Cytochromes P450: exploiting diversity and enabling application as biocatalysts. Curr Opin Chem Biol 15:241–248
71. O'Reilly E, Köhler V, Flitsch SL, Turner NJ (2011) Cytochromes P450 as useful biocatalysts: addressing the limitations. Chem Commun 47:2490–2501
72. Omura T, Sato R (1962) A new cytochrome in liver microsomes. J Biol Chem 237:PC1375-PC1376
73. Auclair K, Moënne-Loccoz P, de Montellano PR (2001) Roles of the proximal heme thiolate ligand in cytochrome P450cam. J Am Chem Soc 123:4877–4885
74. Poulos TL, Finzel BC, Gunsalus IC et al (1985) The 2.6-A crystal structure of Pseudomonas putida cytochrome P-450. J Biol Chem 260:16122–16130
75. Schlichting I, Berendzen J, Chu K et al (2000) The catalytic pathway of cytochrome P450cam at atomic resolution. Science (80-) 287:1615–1622
76. Rittle J, Green MT (2010) Cytochrome P450 compound I: capture, characterization, and CH bond activation kinetics. Science (80-) 330:933–937
77. de Visser SP, Shaik S, Sharma PK et al (2003) Active species of horseradish peroxidase (HRP) and cytochrome P450: two electronic chameleons. J Am Chem Soc 125:15779–15788. https://doi.org/10.1021/ja0380906
78. Shaik S, Kumar D, de Visser SP et al (2005) Theoretical perspective on the structure and mechanism of cytochrome P450 enzymes. Chem Rev 105:2279–2328. https://doi.org/10.1021/cr030722j
79. Kamachi T, Yoshizawa K (2003) A theoretical study on the mechanism of camphor hydroxylation by compound I of cytochrome P450. J Am Chem Soc 125:4652–4661
80. de Visser SP, Ogliaro F, Shaik S (2001) How does ethene inactivate cytochrome P450 en route to its epoxidation? A density functional study the research is supported in part by the ISF and in part by the ministry of science, culture, and sport. F.O. acknowledges the European Union for a Marie Cu. Angew Chem Int Ed Engl 40:2871–2874. DOI: https://doi.org/10.1002/1521-3773(20010803)40:15<2871::AID-ANIE2871>3.0.CO;2-R [pii]
81. Kumar D, de Visser SP, Sharma PK et al (2004) Radical clock substrates, their C-H hydroxylation mechanism by cytochrome P450, and other reactivity patterns: what does theory reveal about the clocks' behavior? J Am Chem Soc 126:1907–1920. https://doi.org/10.1021/ja039439s
82. Quesne MG, Senthilnathan D, Singh D et al (2016) Origin of the enhanced reactivity of μ-nitrido-bridged diiron(IV)-Oxo porphyrinoid complexes over cytochrome P450 compound i. ACS Catal 6:2230–2243. https://doi.org/10.1021/acscatal.5b02720
83. de Visser SP, Porro CS, Quesne MG et al (2013) Overview on theoretical studies discriminating the two-oxidant versus two-state-reactivity models for substrate monoxygenation by cytochrome p450 enzymes. Curr Top Med Chem 13:2218–2232. https://doi.org/10.2174/15680266113136660155
84. Forkert P-G, Lee RP (1997) Metabolism of ethyl carbamate by pulmonary cytochrome P450 and carboxylesterase isozymes: involvement of CYP2E1 and hydrolase A. Toxicol Appl Pharmacol 146:245–254
85. Gunes A, Bilir E, Zengil H et al (2007) Inhibitory effect of valproic acid on cytochrome P450 2C9 activity in epilepsy patients. Basic Clin Pharmacol Toxicol 100:383–386
86. Lee RP, Parkinson A, Forkert PG (1998) Isozyme-selective metabolism of ethyl carbamate by cytochrome P450 (CYP2E1) and carboxylesterase (hydrolase A) enzymes in murine liver microsomes. Drug Metab Dispos 26:60–65

87. Loch JM, Potter J, Bachmann KA (1995) The influence of anesthetic agents on rat hepatic cytochromes P450 in vivo. Pharmacology 50:146–153
88. Rettie AE, Rettenmeier AW, Howald WN, Baillie TA (1987) Cytochrome P-450–catalyzed formation of delta 4-VPA, a toxic metabolite of valproic acid. Science (80-) 235:890–893
89. Sadeque AJM, Fisher MB, Korzekwa KR et al (1997) Human CYP2C9 and CYP2A6 mediate formation of the hepatotoxin 4-ene-valproic acid. J Pharmacol Exp Ther 283:698–703
90. Wen X, Wang JS, Kivisto KT et al (2001) In vitro evaluation of valproic acid as an inhibitor of human cytochrome P450 isoforms: preferential inhibition of cytochrome P450 2C9 (CYP2C9). Br J Clin Pharmacol Engl 547–553
91. Groves JT, McClusky GA (1976) Aliphatic hydroxylation via oxygen rebound. Oxygen transfer catalyzed by iron. J Am Chem Soc 98:859–861
92. Rude MA, Baron TS, Brubaker S et al (2011) Terminal olefin (1-alkene) biosynthesis by a novel P450 fatty acid decarboxylase from Jeotgalicoccus species. Appl Environ Microbiol 77:1718–1727
93. Belcher J, McLean KJ, Matthews S et al (2014) Structure and biochemical properties of the alkene producing cytochrome P450 OleTJE (CYP152L1) from the Jeotgalicoccus sp. 8456 bacterium. J Biol Chem 289:6535–6550
94. Porro CS, Sutcliffe MJ, de Visser SP (2009) Quantum mechanics/molecular mechanics studies on the sulfoxidation of dimethyl sulfide by compound i and compound 0 of cytochrome P450: which is the better oxidant?â€. J Phys Chem A 113:11635–11642. https://doi.org/10.1021/jp9023926
95. Postils V, Saint-André M, Timmins A et al (2018) Quantum mechanics/molecular mechanics studies on the relative reactivities of compound I and II in cytochrome P450 enzymes. Int J Mol Sci 19. https://doi.org/10.3390/ijms19071974
96. Kumar D, Thiel W, de Visser SP (2011) Theoretical study on the mechanism of the oxygen activation process in cysteine dioxygenase enzymes. J Am Chem Soc 133:3869–3882. https://doi.org/10.1021/ja107514f
97. Fu Y, Dai Q, Zhang W et al (2010) The AlkB domain of mammalian ABH8 catalyzes hydroxylation of 5-Methoxycarbonylmethyluridine at the wobble position of tRNA. Angew Chemie 122:9069–9072
98. Giri NC, Sun H, Chen H et al (2011) X-ray absorption spectroscopy structural investigation of early intermediates in the mechanism of DNA repair by human ABH2. Biochemistry 50:5067–5076
99. Li D, Delaney JC, Page CM et al (2012) Exocyclic carbons adjacent to the N 6 of adenine are targets for oxidation by the escherichia coli adaptive response protein AlkB. J Am Chem Soc 134:8896–8901
100. Yu B, Hunt JF (2009) Enzymological and structural studies of the mechanism of promiscuous substrate recognition by the oxidative DNA repair enzyme AlkB. Proc Natl Acad Sci USA 106:14315–14320. DOI: https://doi.org/0812938106 [pii]. DOI: https://doi.org/10.1073/pnas.0812938106
101. Falnes PO, Johansen RF, Seeberg E (2002) AlkB-mediated oxidative demethylation reverses DNA damage in Escherichia coli. Nature 419:178–182. DOI: https://doi.org/10.1038/nature01048 [pii]
102. Trewick SC, Henshaw TF, Hausinger RP et al (2002) Oxidative demethylation by Escherichia coli AlkB directly reverts DNA base damage. Nature 419:174–178. https://doi.org/10.1038/nature00908
103. Yi C, Yang CG, He C (2009) A non-heme iron-mediated chemical demethylation in DNA and RNA. Acc Chem Res 42:519–529. DOI: https://doi.org/10.1021/ar800178j [pii]. DOI: https://doi.org/10.1021/ar800178j
104. Ladwein KI, Jung M (2011) Oxidized cytosine metabolites offer a fresh perspective for active DNA demethylation. Angew Chemie Int Ed 50:12143–12145
105. Mishina Y, Duguid EM, He C (2006) Direct reversal of DNA alkylation damage. Chem Rev 106:215–232. https://doi.org/10.1021/cr0404702

106. Mishina Y, He C (2006) Oxidative dealkylation DNA repair mediated by the mononuclear non-heme iron AlkB proteins. J Inorg Biochem 100:670–678
107. Borowski T, Bassan A, Siegbahn PEM (2004) Mechanism of dioxygen activation in 2-oxoglutarate-dependent enzymes: a hybrid DFT study. Chem—A Eur J 10:1031–1041. https://doi.org/10.1002/chem.200305306
108. O'Brien PJ (2006) Catalytic promiscuity and the divergent evolution of DNA repair enzymes. Chem Rev 106:720–752. https://doi.org/10.1021/cr040481v
109. Hunt JF, Yu B, Edstrom WC et al (2006) Crystal structures of catalytic complexes of the oxidative DNA/RNA repair enzyme AlkB. Nature 439:879–884. https://doi.org/10.1038/nature04561
110. Godfrey E, Porro CS, de Visser SP (2008) Comparative quantum mechanics/molecular mechanics (QM/MM) and density functional theory calculations on the oxo-iron species of taurine/alpha-ketoglutarate dioxygenase. J Phys Chem A 112:2464–2468. https://doi.org/10.1021/jp710999v
111. Quesne MG, Ward RA, de Visser SP (2013) Cysteine protease inhibition by nitrile-based inhibitors: a computational study. Front Chem 1:39. https://doi.org/10.3389/fchem.2013.00039
112. Brooks BR, Bruccoleri RE, Olafson BD et al (1983) CHARMM: A program for macromolecular energy, minimization, and dynamics calculations. J Comput Chem 4:187–217
113. Borowski T, Noack H, Radoń M et al (2010) Mechanism of selective halogenation by SyrB2: a computational study. J Am Chem Soc 132:12887–12898. https://doi.org/10.1021/ja101877a
114. Reiher M, Salomon O, Artur Hess B (2001) Reparameterization of hybrid functionals based on energy differences of states of different multiplicity. Theor Chem Acc Theory Comput Model (Theoretica Chim Acta) 107:48–55
115. Liu H, Llano J, Gauld JW (2009) A DFT study of nucleobase dealkylation by the DNA repair enzyme AlkB. J Phys Chem B 113:4887–4898. https://doi.org/10.1021/jp810715t
116. Sahu S, Widger LR, Quesne MG et al (2013) Secondary coordination sphere influence on the reactivity of nonheme iron(II) complexes: an experimental and DFT approach. J Am Chem Soc 135:10590–10593. https://doi.org/10.1021/ja402688t
117. Wong SD, Bell CB, Liu LV et al (2011) Nuclear resonance vibrational spectroscopy on the FeIV = OS = 2 Non-Heme site in TMG3tren: experimentally calibrated insights into reactivity. Angew Chemie Int Ed 50:3215–3218
118. Ye S, Neese F (2011) Nonheme oxo-iron(IV) intermediates form an oxyl radical upon approaching the C-H bond activation transition state. Proc Natl Acad Sci USA 108:1228–1233. DOI: https://doi.org/1008411108 [pii]. DOI: https://doi.org/10.1073/pnas.1008411108
119. Yang T, Quesne MG, Neu HM et al (2016) Singlet versus Triplet Reactivity in an Mn(V)-Oxo Species: Testing Theoretical Predictions Against Experimental Evidence. J Am Chem Soc 138:12375–12386. DOI: https://doi.org/10.1021/jacs.6b05027
120. Cantú Reinhard FG, de Visser SP (2017) Oxygen atom transfer using an iron(IV)-Oxo embedded in a tetracyclic N-Heterocyclic Carbene system: how does the reactivity compare to cytochrome P450 compound I? Chem—A Eur J 23:2935–2944. https://doi.org/10.1002/chem.201605505
121. Shaik S, Chen H, Janardanan D (2011) Exchange-enhanced reactivity in bond activation by metal–oxo enzymes and synthetic reagents. Nat Chem 3:19–27. https://doi.org/10.1038/nchem.943

Applications of Computational Chemistry to Selected Problems of Transition-Metal Catalysis in Biological and Nonbiological Systems

Hajime Hirao

Abstract The chemistry of transition-metal-containing systems is highly complex and diverse and thus lends itself to careful computational investigation. Indeed, computational chemistry can play fundamentally important roles in elucidating the catalytic mechanisms of such systems, by offering information about short-lived intermediates and transition states as well as factors that determine catalytic properties, which is not easily attained by experimental means. A quantum mechanical description of a targeted catalytic system could be difficult or unfeasible in many circumstances, especially when large systems such as metalloenzymes and coordination polymers are studied. Nevertheless, valuable insights can still be gained from hybrid computational techniques that allow concrete realizations of extensive reaction pathway analyses. This chapter gives a brief overview of some of our recent attempts to study the structure and activity of transition-metal-containing systems varying in size using several computational approaches.

1 Introduction

The catalytic roles of transition metals in a broad range of chemical and biological processes have attracted the interest of many researchers over the past decades. Many attempts have been made to understand the factors that underlie the catalytic properties of transition-metal complexes. However, numerous variations exist in ligands used to form complexes and oxidation and spin states of transition metals, giving rise to diverse reactivity patterns that may not easily be generalized. In addition, ligands often play noninnocent roles in catalysis by getting oxidized or reduced, thereby adding complexity to the problem. We believe that computational chemistry has many useful roles to play in teasing out catalytic mechanisms of such systems.

H. Hirao (✉)
Department of Chemistry, City University of Hong Kong, Tat Chee Avenue, Kowloon, Hong Kong, China
e-mail: hhirao@cityu.edu.hk

E. Broclawik et al. (eds.), *Transition Metals in Coordination Environments*, Challenges and Advances in Computational Chemistry and Physics 29,
https://doi.org/10.1007/978-3-030-11714-6_15

The diversity of transition-metal complexes and the prominent role of computational chemistry in mechanistic studies can be seen, for example, in iron-containing enzymes and their synthetic analogues in bioinorganic chemistry. In the case of the cytochrome P450 enzymes (P450s), the central iron ion is coordinated by a deprotonated protoporphyrin IX ligand in the equatorial sites and by a cysteinate ligand in the proximal site [1, 2]. It is known that various intermediates are formed in the catalytic cycle of P450s, in which the oxidation state of the iron can vary between +2 and +4. The +4 state taken by the oxoiron(IV) porphyrin π-cation radical species called compound I (Cpd I) is of particular interest in the context of catalysis, because Cpd I can effect difficult chemical transformations such as C–H activation of alkane substrates via the rebound mechanism [3]. Other types of heme enzymes also exist wherein an amino acid residue other than cysteinate is used as the proximal ligand [4, 5]. Furthermore, there are a number of nonheme iron enzymes in nature that use aspartate, glutamate, histidine, or other anionic groups as ligands [6, 7]. As in the case of heme enzymes, an oxoiron(IV) species is used for C–H activation in some cases [8], and other types of active species have also been identified [9, 10]. The intriguing but elusive nature of active species in heme and nonheme iron enzymes has prompted chemists to prepare synthetic analogues of enzymatic active species, and substantial progress has been made in this area [11–15]. Computational chemistry has contributed significantly to understanding the reactivity patterns of iron enzymes and synthetic analogues [16–24].

In this chapter, we discuss some of our recent attempts to study the structure and catalytic properties of transition-metal-containing systems of different sizes, such as homogeneous transition-metal complexes, metalloenzymes, and metal–organic frameworks (MOFs). Kohn–Sham density-functional theory (DFT) is usually the first choice for describing the quantum mechanical nature of these systems, yet we also sometimes employ hybrid techniques such as quantum mechanics and quantum mechanics (QM/QM') and quantum mechanics and molecular mechanics (QM/MM) to deal with large molecular systems. Application of these techniques to specific chemical reactions often yields unexpected insights.

2 Computational Methods for Studying Catalysis in Various Systems

The computational treatment of transition-metal-containing systems is challenging in several respects. One of the major challenges pertains to high computational costs of quantum mechanical calculations. Correlated ab initio molecular orbital (MO) methods are usually too demanding for reaction pathway analyses of such systems. As such, DFT in the Kohn–Sham formalism is often used for this purpose [25–38]. Many exchange–correlation functionals have been developed so far to approximate the exchange–correlation energy in the Kohn–Sham DFT. Which functional performs best for a given purpose has been (and will continue to be) a matter of intense debate, and many validation studies have been conducted [39–42]. These studies have shown that no functional is fully reliable, which is due to the challenges associated with accurately describing dispersion, self-interaction correction, dynamical correlation, and nondynamical correlation [28]. Nevertheless, B3LYP has traditionally been one of the most popular functionals in the chemistry community [43, 44]. Whereas criticism against B3LYP is sometimes harsh, relatively good performance of B3LYP has also been pointed out for many applications. For example, an important question in relation to the studies of transition-metal complexes is how well DFT works in predicting the relative stability of different spin states. Verma et al. have recently performed benchmark calculations on 14 iron complexes using 20 exchange–correlation functionals [45]. In their benchmark study, PW6B95 predicted the correct ground spin states for all 14 complexes examined. B3LYP, MPW1B95, and MN15 also performed well, producing correct answers for 13 complexes. This number decreased to 12 when the OPBE, OLYP, and M06-L functionals were used. The use of a larger basis set allows the calculated energy to be closer to the convergent result for a given functional and reduces the occurrence of poor descriptions of electronic distributions, which is therefore recommended. However, in practice, large basis sets slow down DFT computation significantly, and geometry is not as sensitive to the basis set used as energy is. Therefore, geometry optimization is typically performed with a small basis set (e.g., of double-ς quality), which is followed by single-point energy evaluation with a larger basis set. It should also be noted that larger basis sets do not necessarily guarantee higher accuracy of energy values because of the approximate nature of exchange–correlation functionals. In many cases, effective-core potential basis sets are employed for transition metals. Semiempirical (or semiclassical) dispersion corrections such as DFT-D3 are often tacked onto the DFT energy to improve the description of noncovalent interactions between fragments in the system [46].

When the target system is not very large and contains no more than ~150 atoms (with the number depending on the available computers), DFT can easily be applied to the system in its entirety (Fig. 1a). Today, the majority of DFT calculations that probe the mechanisms of homogeneous catalytic reactions employ this kind approach.

However, the system would contain more atoms than can be handled by this DFT approach. One possible solution to this problem is to use the QM/QM' method in which two different levels of quantum mechanics are used in combination (Fig. 1b). QM and QM' denote higher- and lower-level QM methods, respectively. For example, a smaller basis set may be used for the QM' calculation than for the QM calculation, which can reduce the computational cost compared with the case of full QM calculation. Alternatively, semiempirical MO or density-functional tight-binding (DFTB) methods may be used for the lower-level QM' calculation. Any other combination of two (or even more) different levels of QM methods is also possible.

The use of QM' in QM/QM' permits computational treatments of systems that are larger than what can be treated by pure QM. However, QM/QM' is still too expensive when studying enzymes that contain thousands of atoms. In such cases, it is advantageous to use the hybrid QM/MM method (Fig. 1c) [47–49], because MM is computationally much less costly than QM'. When studying enzymatic systems, DFT is usually used for the QM part of QM/MM, whereas force-field parameters that

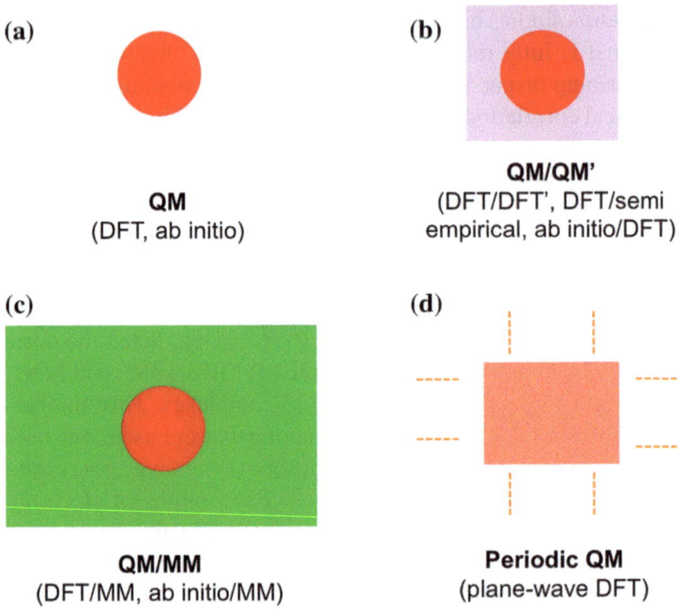

Fig. 1 Schematic illustrations of various computational methods that are useful for computational studies of transition-metal-containing systems

have been well established in the area of biological simulation are employed for the MM description. There are several different types of QM/MM methods, especially in terms of their coupling and embedding schemes. Figure 2 gives a concise summary of these methods. As for how the QM and MM calculations are coupled, the additive scheme is regarded as an orthodox approach that divides the system into QM- and MM-described sub-regions. The total energy is calculated basically as the sum of the QM, MM, and QM–MM interaction energies. By contrast, the subtractive scheme is adopted in the ONIOM method, and here the total energy is extrapolated from three different calculations [50–52]. Both coupling schemes are similar as far as produced energy trends are concerned. However, the subtractive method always uses "complete" molecular fragments for energy evaluations, and this has the advantage that one can easily combine any number of different QM methods or MM methods. Although the same kinds of combinations are essentially possible in the additive scheme, "incomplete" molecular fragments such as the green-colored one in the top-left part of Fig. 2a must be prepared in this case, which makes the procedure to define sub-regions technically cumbersome, especially when more than two sub-regions are defined. In addition, QM' calculations on such an incomplete fragment may also pose another technical problem concerning how the boundary atoms in the QM' region are treated. As such, QM/QM' calculations are executed more straightforwardly with the subtractive scheme. In QM/MM, there are also several embedding schemes, but

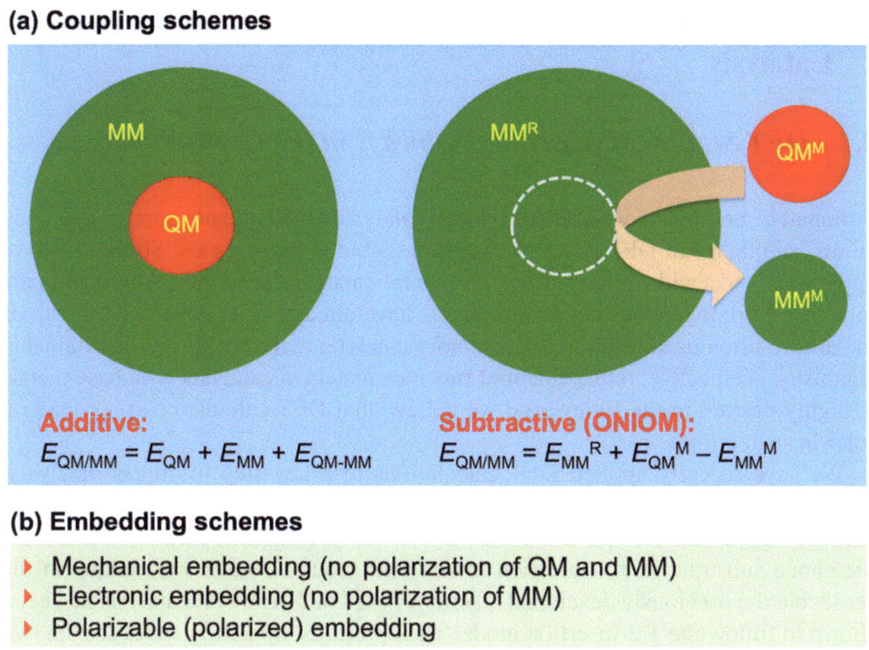

Fig. 2 Summary of QM/MM methods

in most cases, the electronic embedding or the mechanical-embedding scheme is used.

Solid materials like transition-metal oxides contain quite a large number of atoms, but exploiting the fact that they have periodic structures, quantum mechanical treatments of such systems are still possible with periodic DFT (Fig. 1d). Periodic DFT calculations are executed mostly with plane-wave basis sets and exchange–correlation functionals based on the generalized gradient approximation (GGA). As MOFs also possess periodic structures, they can be treated with periodic DFT. However, MOFs are not as densely packed as metal-oxide materials, and thus, the unit cell tends to be large. In addition, convergence problems are occasionally encountered. Because hybrid functionals have been relatively well validated for use in reaction pathway analyses and widely used in the chemistry community, we prefer to use such functionals when studying MOF catalysis. However, in the periodic DFT, the use of hybrid functionals is not as practical as in the atomic-orbital DFT. GGA functionals also tend to overstabilize low spin states of transition metals. For these reasons, we have recently tried using QM/MM with a hybrid functional for studying chemical reactions occurring in MOFs.

3 Studies of Transition-Metal Complexes for Homogeneous Catalysis

3.1 DFT Studies of Reaction Pathways in the Ground State

Fortunately, homogeneous transition-metal catalysts usually do not contain too many atoms, and then full DFT descriptions are possible in many cases. So far, we have applied DFT to a wide range of transition-metal-catalyzed reactions. Although many impressive homogeneous catalytic systems have already been developed, precious metals are often used to induce satisfactory catalytic activities. From a sustainable chemistry perspective, replacement of precious metals in catalysts with base metals is highly desired in the future, and we believe that DFT calculations play integral roles in such efforts.

We have recently applied DFT calculations to a key step in copper-catalyzed asymmetric addition of organoborons to enones [53]. Our DFT calculations showed that the ligand–Cu–Ph species, which is formed during the reaction, reacts with the enone substrate in 1,4-insertion mode. This reaction mode was unique in the sense that the previously developed reactions based on Rh and Pd catalysts had been shown to follow the 1,2-insertion mode. The optimized transition states for the two possible pathways leading to *R* and *S* products were not equally stable (Fig. 3). Thus, the transition state for the *R*-product formation can form hydrogen bonds between aromatic C–H bonds of the phosphoramidite and the oxygen atom of the chalcone more favorably than the transition state for the other pathway, which leads to greater stability of the former transition state.

DFT calculations were also applied to another kind of Cu-catalyzed reaction for the synthesis of 1,4-diaryl-1*H*-imidazoles [54]. Detailed experimental and theoretical studies were conducted to examine several possible mechanisms in detail. Eventually, a mechanism was proposed that involves a tandem two-step process and the formation of an *N*-arylformimidate intermediate (Fig. 4).

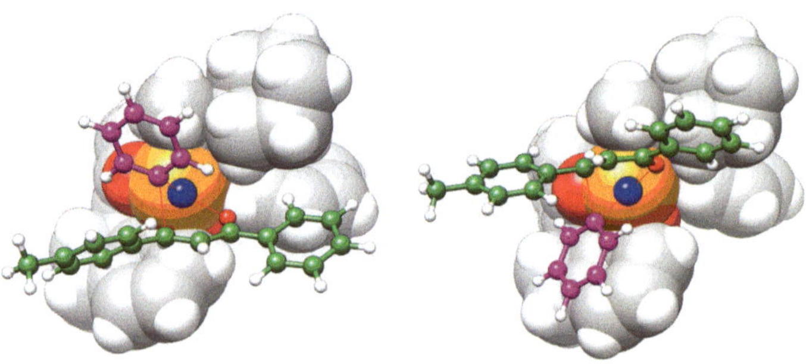

Fig. 3 DFT-optimized transition states for the 1,4-insertion of the ligated phenylcopper(I) species to *p*-methylchalcone (Reprinted with permission from [53]. Copyright 2016 American Chemical Society)

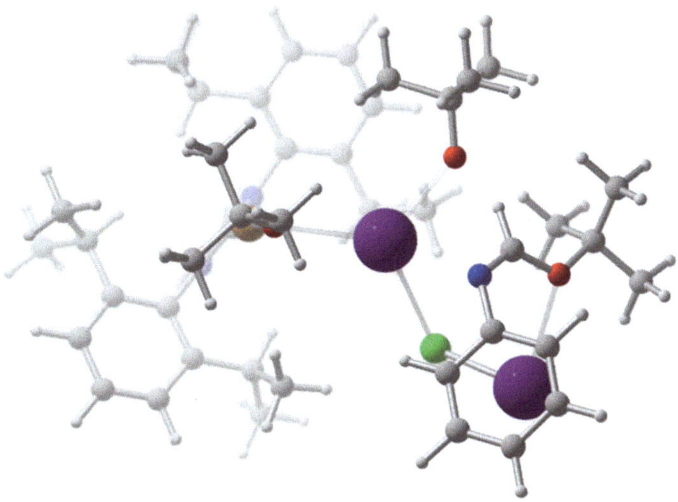

Fig. 4 *N*-arylformimidate formed during the reaction

3.2 DFT Studies of Reaction Pathways in Excited States

Gazi et al. used a vanadium(V)-oxo complex (Fig. 5) for photocatalytic C–C bond cleavage of a (*β*-O-4) lignin model compound under visible light irradiation [55, 56]. Lignin is one of the major constituents of abundantly available lignocellulosic biomass [57, 58], but as compared with the other two major constituents, namely cellulose and hemicelluloses, lignin is notoriously resistant to various kinds of degradation processes, thus making its selective depolymerization under ambient conditions difficult. Lignin is known to have several typical linkages in its structure, and *β*-O-4 is the most abundant one [59, 60]. The development of methods for efficient lignin depolymerization is highly desired because it could reduce the reliance on fossil fuels in synthesizing aromatic compounds. In addition, if visible light is efficiently utilized, fossil fuels will not have to be used to drive reactions. The use of vanadium

Fig. 5 Vanadium(V)-oxo complex examined

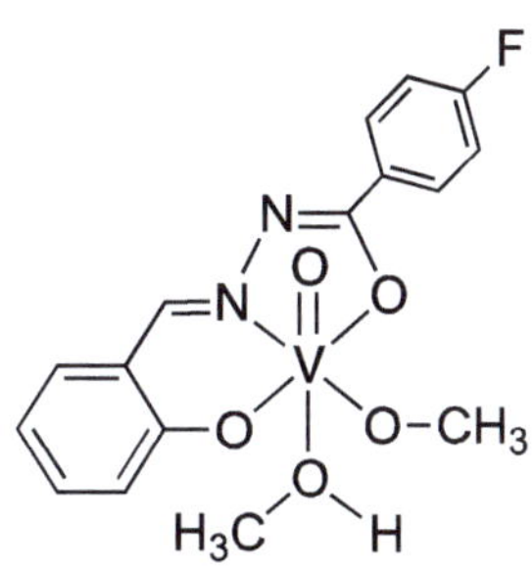

is interesting in that harvesting light usually requires a less abundant metal, typically ruthenium or iridium.

DFT calculations were used to understand the mechanism of this photocatalytic C–C bond cleavage reaction in detail (Fig. 6) [57]. As time-dependent DFT and correlated ab initio MO methods are deemed too expensive for describing the entire reaction pathways in the photo-excited states, ordinary DFT calculations were performed to model excited states. It turns out that the first excited singlet state can be described reasonably well with the broken-symmetry DFT method. Also, the triplet excited state can be described by simply specifying the spin state as triplet. The calculations showed that the β-O-4 substrate coordinating to the equatorial site of the vanadium-oxo complex cannot thermally undergo C–C bond cleavage. However, the low-lying excited state in the triplet (T_1) state, which corresponds to a ligand-to-metal charge transfer (LMCT) excited state, has a low barrier to homolytic C–C bond cleavage. The excited state therefore makes the reaction possible. In this LMCT transition, an electron is transferred from the hydrazine benzohydroxamate ligand to the vanadium(V) center, to form an oxidized ligand and a vanadium(IV) state. After an electronic excitation to a higher-lying excited state, intramolecular relaxation, internal conversion, and intersystem crossing may occur to access the LMCT-type T_1 state, in which homolytic C–C bond cleavage is a facile process. The low barrier obtained here is attributed to the fact that one of the two unpaired electrons, which is produced upon homolytic C–C bond cleavage, is efficiently absorbed into the low-lying orbital of the oxidized ligand (Fig. 7). Thus, as a result of the C–C bond cleavage, aryl formate (one of the major products) and a radical species are left behind. The latter species will rapidly combine with an O_2 molecule that is present in the system under aerobic conditions. Subsequently, the alkyl peroxy radical binds to the vanadium center, and the formal oxidation state of vanadium reverts back to +5. In the second step, aryl formate is formed (Fig. 8). After all, the oxidation state of the vanadium center is restored to +5, so that the vanadium complex can play a catalytic role by effecting C–C cleavage in many catalytic rounds. The attachment of electron-withdrawing groups to the hydrazine benzohydroxamate led to the improvement of the catalytic activity; this can be attributed to the stabilization of the ligand orbital, which enhances the process illustrated in Fig. 7 [56].

In the photocatalytic reaction under anaerobic conditions, only a small amount (3%) of the aryl aldehyde product was obtained in experiment. The formation of the aldehyde product can be understood from the fact that the initial LMCT-driven homolytic C–C bond cleavage process to form the aldehyde (the process from **RC** to **Int1** in Fig. 6) is possible even in the absence of O_2. As for the low yield of the aldehyde product, it indicates that the vanadium complex is consumed rapidly, instead of acting as a catalyst. According to our DFT analysis (Fig. 9), the recovery of the vanadium(V) state under anaerobic conditions is possible after a second LMCT, but the subsequent barrier is higher than that for the initial C–C bond cleavage process. Thus, the LMCT-driven C–C bond cleavage is favored over the LMCT-driven vanadium(V) recovery, which results in the rapid depletion of the catalytically active vanadium(V) species via preferential C–C bond cleavage. The pathways obtained for the reaction under anaerobic conditions theoretically lead to aryl formate. However,

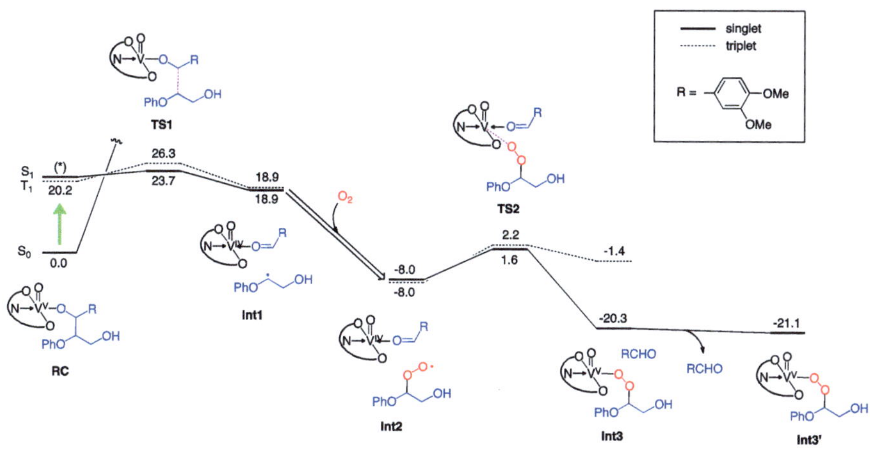

Fig. 6 First half of the photocatalytic C–C bond reaction of a lignin model compound that occurs with a vanadium-oxo complex in aerobic conditions. The B3LYP(SCRF)6-311+G(d,p)//6-31G* method was used, and the energies are presented in kcal/mol unit. Adapted from [55] with permission from The Royal Society of Chemistry

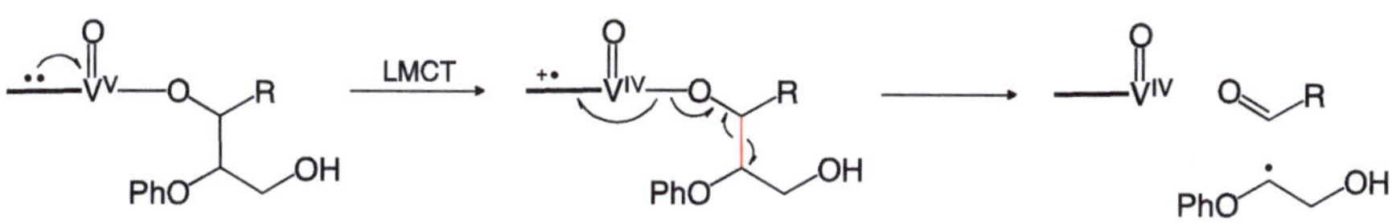

Fig. 7 Schematic illustration of the LMCT-driven C–C bond cleavage reaction

no yield of aryl formate can be explained by the higher barriers in these pathways than that for the initial C–C bond cleavage and the rapid consumption of the vanadium(V)-oxo species.

3.3 *QM/QM' Studies of Reactions Catalyzed by Transition-Metal Catalysis Having Large Ligands*

ONIOM-based QM/QM' calculations were applied to nickel-catalyzed transfer hydrogenation reactions [59, 60]. To obtain chiral alkyl amines that are omnipresent in drugs and agrochemicals in enantioselective manners, bulky ligands were used experimentally. The QM and QM' methods employed were the B3LYP and M06-L methods, respectively, and a smaller basis set was used for QM'. The reason we can expect some speedup in QM/QM' is that QM methods such as DFT do not scale linearly, which makes accurate QM calculations on larger models very time-consuming. Thus, even though three separate calculations on the real and model systems are needed (for QM'^{R}, QM^{M}, and QM'^{M}, see Fig. 1a, the top-right figure)

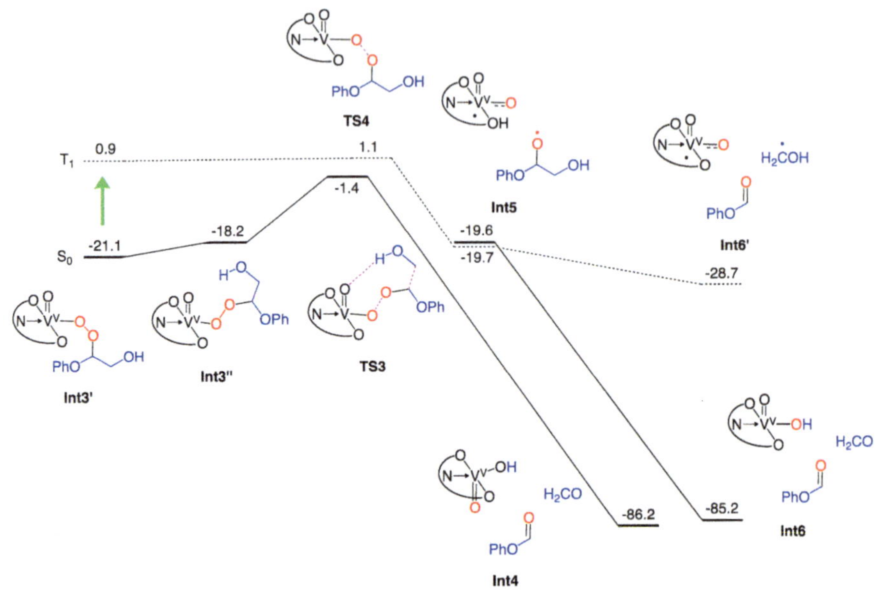

Fig. 8 DFT-derived pathway (in kcal/mol) for the second half of the reaction. Adapted from [55] with permission from The Royal Society of Chemistry

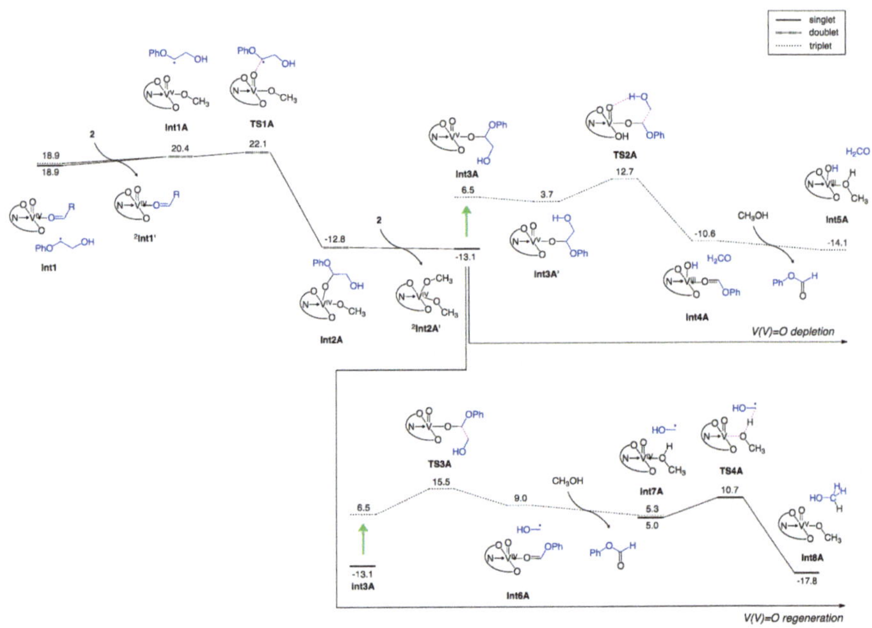

Fig. 9 DFT-derived reaction pathway (in kcal/mol) for the reaction after the initial C–C bond cleavage step. Reprinted from [55] with permission from The Royal Society of Chemistry

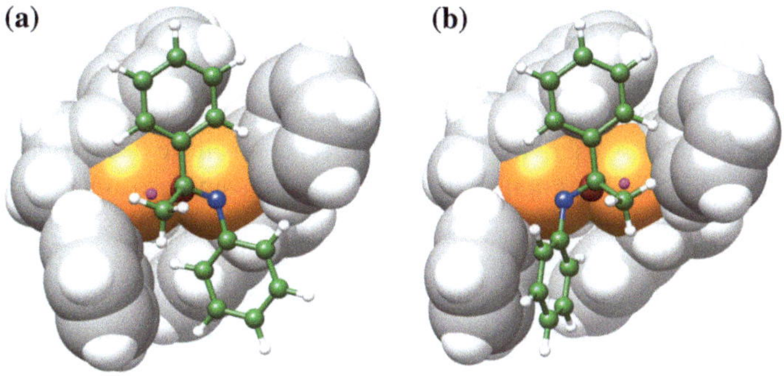

Fig. 10 Two possible transition states, **a** TS-*S* and **b** TS-*R*, for the hydride insertion step from the (hydride)nickel(II) species into the *N*-phenylketimine substrate (Reprinted with permission from [60]. Copyright 2016 Wiley)

in the QM/QM' calculation, the omission of a high-level QM calculation on the real model (QM^R) results in faster execution of total energy evaluation. In the application of QM/QM' to the nickel-catalyzed enantioselective reductive amination of ketones, reaction pathways leading to two possible enantiomers were determined. Close inspection of the two major transition states in the hydride transfer step (Fig. 10) showed that the substrate in the favored TS-*S* does not undergo severe steric distortion in the pocket of the catalyst. By contrast, in the other transition state (TS-*R*), the coplanarity of the *N*-phenyl ring with the C=N bond of the imine cannot be maintained, thus leading to some destabilization of this pathway. Extensive ONIOM pathway analyses were performed to examine which of the oxidation states nickel, +2 or +1, is used for the reaction, and the results argued for the use of the Ni(II) state.

4 Studies of Metalloenzymes

Despite the vitals roles metalloenzymes play in many physiological processes, the mechanisms underlying enzymatic reactions are not entirely clear in many cases. A major reason is that short-lived intermediates of metalloenzymes cannot easily be trapped and characterized experimentally. Although enzymes generally consist of thousands of atoms, which can hardly be described by full DFT calculations, enzymatic reactions are typically local phenomena around a particular active site, which permits the use of truncated active models for studying their reaction mechanisms. Nevertheless, protein surroundings often have large impacts on the reactivity and selectivity of enzymatic reactions, and such effects can be described effectively by MM methods. Hence, DFT and QM/MM are two major approaches to the computational studies of chemical reactions occurring in metalloenzymes.

4.1 Studies of Heme Enzymes

For example, P450s use the short-lived Cpd I species for oxidation reactions, and the nature of Cpd I and its reactivity in metabolic reactions have been widely explored theoretically [16–18]. Other than metabolism, mechanism-based inactivation (MBI) of P450s is a problem of practical importance that presents a serious concern in virtually all drug discovery projects. Using truncated small models and DFT calculations, we have investigated the mechanisms of several MBI processes [61–65]. These calculations provided valuable insights into how Cpd I reacts with inhibitor molecules to cause P450 inhibition.

P450s also sometimes play essential roles in synthesizing signaling molecules. For example, CYP19A1 (aromatase) is known to catalyze the androgen-to-estrogen conversion in three major reaction steps. Although the first and second steps are already well understood, there has been controversy over whether the active species in the third step is Cpd I or ferric peroxide (Fig. 11a, b). Until recently, many people had believed that ferric peroxide should be the active species because experiments supported it. In this ferric peroxide mechanism, one of the two oxygen atoms from O_2 is incorporated into the formic acid product (Fig. 11c). However, recent ^{18}O-labeling experiments could not find the O_2-derived oxygen in formic acid [66], which indicated that the Cpd I mechanism rather than the ferric peroxide mechanism is operative. We performed QM/MM calculations to answer the question of how Cpd I could affect the third step of the aromatase reaction [67]. The QM/MM enzyme model was built from a crystal structure of human CYP19A1 (PDB code 3EQM) [68]. Based on the QM/MM results, we proposed a novel mechanism in which Cpd I abstracts a hydrogen atom from the O–H bond of the *gem*-diol form of the substrate to yield the aromatized product. The QM/MM calculations also suggested that Cpd I can abstract the hydrogen atom of the C–H bond of *gem*-diol or aldehyde to form a carboxylic acid product, although the fate of the O_2-derived oxygen was different in these reactions, which explained how and why the carboxylic acid was obtained in the experimental condition.

In our studies of P450s, we are also interested in how different groups within enzymes interact with each other. Thus, we have tried to identify the origins of interactions with the aid of various methods [69–71]. Interaction energies between fragments were decomposed into electrostatic and several other terms.

4.2 Studies of Nonheme Enzymes

We have been applying the DFT and QM/MM methods to many other metalloenzymes [72–77]. Most recently, we have studied the mechanism of CO oxidation catalyzed by Mo–Cu carbon monoxide dehydrogenase (Mo–Cu CODH) using the ONIOM QM/MM method [78]. The QM/MM enzyme model was built from a crystal structure (PDB code 1N5W) [79]. Over the past two decades, the mechanism of the

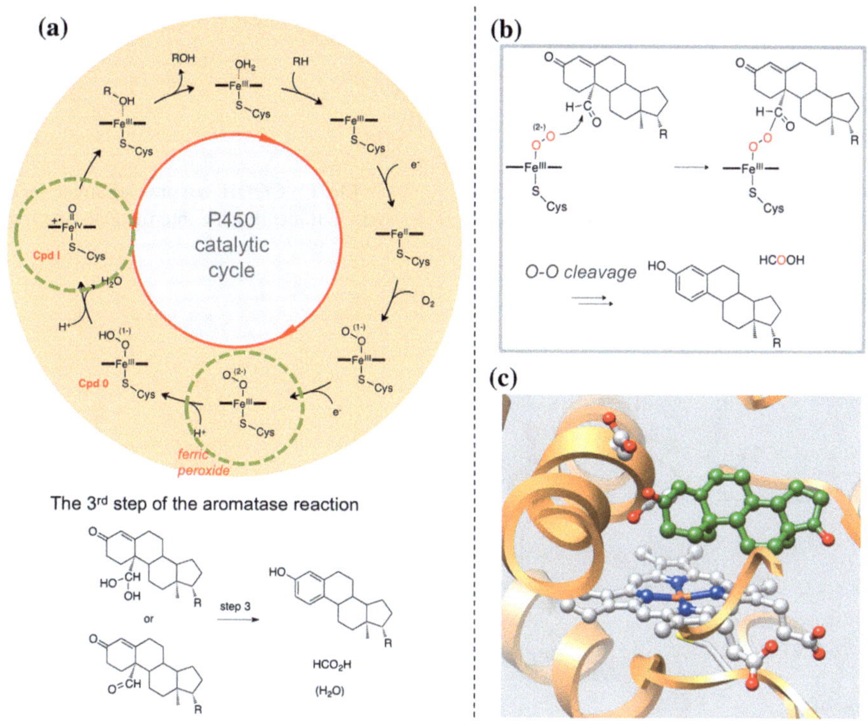

Fig. 11 **a** General catalytic cycle of P450s with the two possible active species in the third step of the aromatase reaction highlighted. **b** The ferric peroxide mechanism. **c** X-ray structure of human aromatase (PDB code 3EQM)

CO oxidation reaction in Mo–Cu CODH has been an intriguing but controversial issue, and thus, the mechanism of enzyme catalysis has not been completely clear. Several lines of evidence pointed to the involvement of an Mo^{VI}=O species in the oxidation of CO, where the oxo ligand is situated in the equatorial site (Fig. 12a). One important clue to understanding how CO is oxidized was the crystal structure obtained for the enzyme in complex with *n*-butylisocyanide, which also undergoes a reaction in Mo–Cu CODH. In the crystal structure, a thiocarbamate complex that looks like the one in Fig. 12b was observed, which implied that an analogous thiocarbonate intermediate (Fig. 12c) may also be formed in the reaction of CO. Before our studies, there had been a few DFT studies of this enzyme using small active-site models [80–83]. One important conclusion drawn from the previous DFT studies was that CO can be oxidized without any problem in the enzyme, but the release of the produced CO_2 from thiocarbonate should be very difficult.

Using QM/MM, we first examined what kind of hydrogen-bonding pattern is stable for the Mo^{VI}=O active species. Our QM/MM exploration yielded two different geometries (**Int1A** and **Int1B** in Fig. 13). We infer that **Int1B**, in which the oxo group is hydrogen bonded to the neutral Glu763 residue, is initially formed in the formation

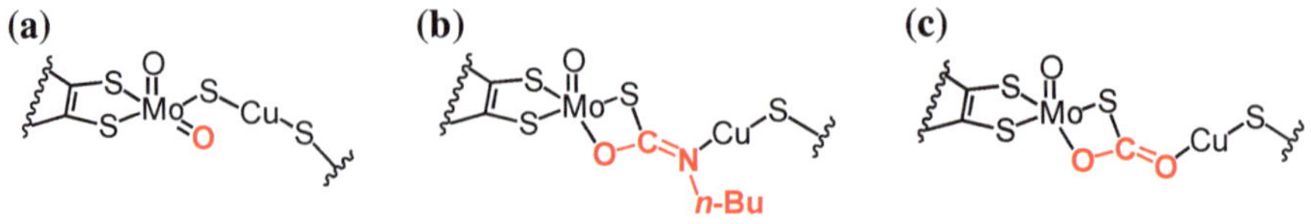

Fig. 12 Schematic illustrations of **a** the active species of Mo–Cu CODH, **b** a thiocarbamate intermediate formed in the reaction of *n*-butylisocyanide, and **c** a thiocarbonate intermediate that may be formed in the reaction of CO

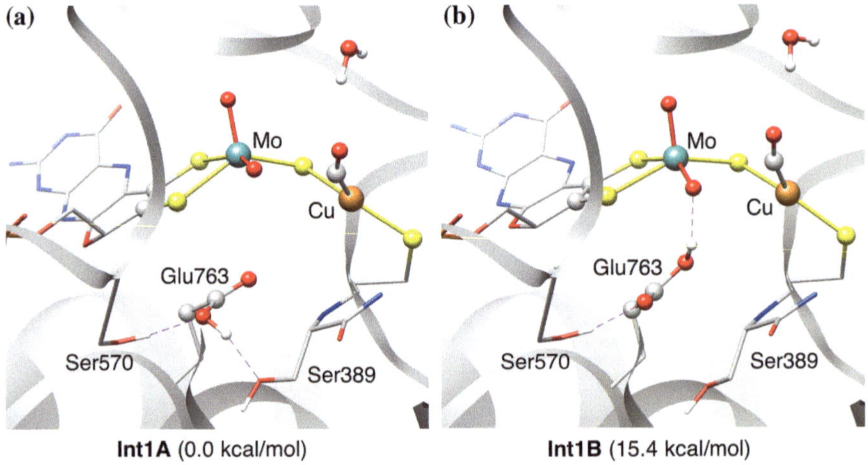

Fig. 13 Two possible geometries of the Mo^{VI}=O species and their relative free energies in kcal/mol (Reprinted with permission from [78]. Copyright 2018 Royal Society of Chemistry)

of the Mo^{VI}=O species. This is because the equatorial oxo unit should be derived from a water molecule, which will eventually get deprotonated with Glu763 used as a proton acceptor. However, **Int1A** was shown to be much more stable than **Int1B**, which indicates that **Int1A** is readily formed and used for the subsequent oxidation reactions.

Figure 14 shows the reaction energy profile determined by QM/MM calculations. Consistent with the hypothesis that the reaction involves the formation of a thiocarbonate intermediate, the corresponding thiocarbonate intermediate (**Int4A**) can be formed in an exothermic manner without surpassing high barriers. We could not locate any transition state for the direct release of CO_2 from **Int4**. However, we found that CO_2 can be released relatively easily from the previously formed **Int3A** species via **TS4A**. Inspection of the relative energies of the species on the reaction pathway shows that **Int3A** is accessible from **Int4A** and thus it is possible for **Int4A** to release CO_2 via **Int3A**. The net barrier for the CO_2 release was only 12.9 kcal/mol, which did not agree well with the previous suggestions from DFT calculations that the CO_2 release is not a facile process.

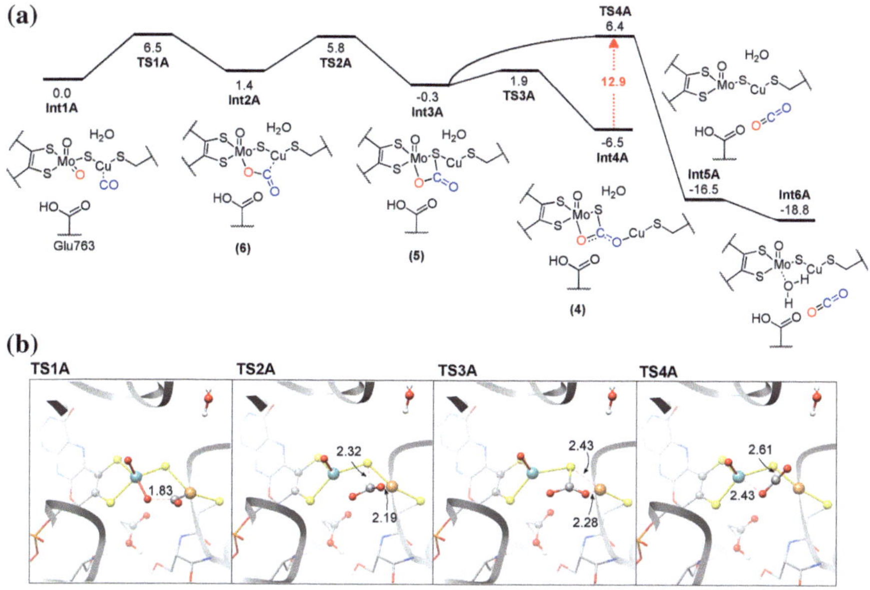

Fig. 14 **a** QM/MM-derived free energy profile (in kcal/mol) for the CO oxidation in Mo–Cu CODH and **b** the geometries of optimized transition states. Key distances are shown in Å unit (Reprinted with permission from [78]. Copyright 2018 Royal Society of Chemistry)

This result initially made us suspect that the protein environmental effect exerted to the reaction is very large. However, an additional energy decomposition analysis of the QM/MM-calculated energy gap between **Int4A** and **TS4A** did not corroborate this hypothesis. Therefore, we decided to perform DFT calculations on our own. Even in the DFT calculations, we did not obtain a high barrier for the CO_2 release. From these results, we concluded that the problematic CO_2 release step in the reaction in Mo–Cu CODH is actually not kinetically difficult even without the protein environment. However, we noticed that the reaction is calculated as more exothermic by QM/MM than by DFT, which suggests that the protein environment still has an important role to play in this reaction by making it thermodynamically more favored.

We also performed QM/MM calculations on the reaction of *n*-butylisocyanide. In this reaction, a distinctly different energy profile was obtained. Thus, although a thiocarbamate intermediate can easily be formed, this intermediate is so stable that *n*-butylisocyanate cannot be released from the active-site cluster, which is consistent with the fact that the thiocarbamate complex was observed in the previous X-ray study. According to our results, thiocarbonate is also formed during the reaction of CO, but it is converted easily to the more stable product state.

By combining all data from QM/MM calculations, we proposed a mechanism of the Mo–Cu CODH catalyzed CO oxidation, which consists of three major steps

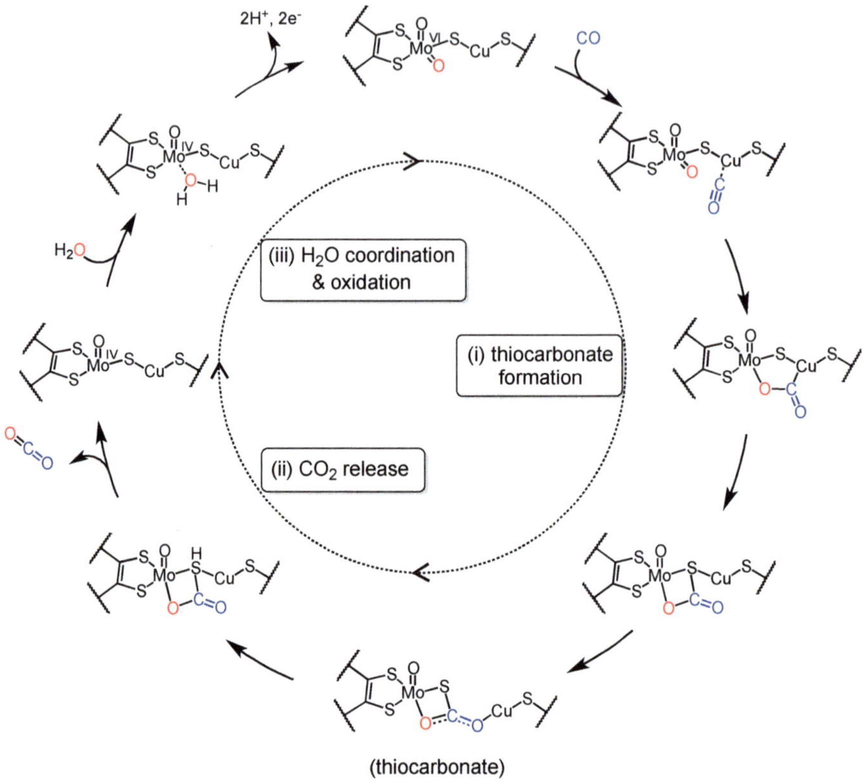

Fig. 15 Our proposed mechanism of the CO oxidation catalyzed by Mo–Cu CODH (Reprinted with permission from [78]. Copyright 2018 Royal Society of Chemistry)

(Fig. 15): (i) thiocarbonate formation, (ii) CO_2 release, and (iii) H_2O coordination and oxidation.

5 Studies of MOFs

5.1 QM/MM Studies of MOFs

We have recently made new attempts to study chemical reactions occurring in MOFs using QM/MM and MM [84–88]. MOFs are similar to metalloenzymes in the sense that both have metal ions and organic groups as building blocks, although their structures are fundamentally different. QM/MM has been applied to a number of reactions that occur in metalloenzymes and other enzymes, but by comparison, applications of QM/MM to MOF-catalyzed reactions had been few and far between.

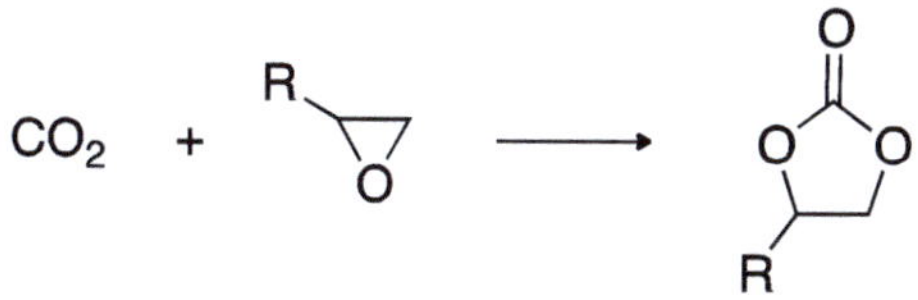

Fig. 16 Cycloaddition of CO_2 to epoxides

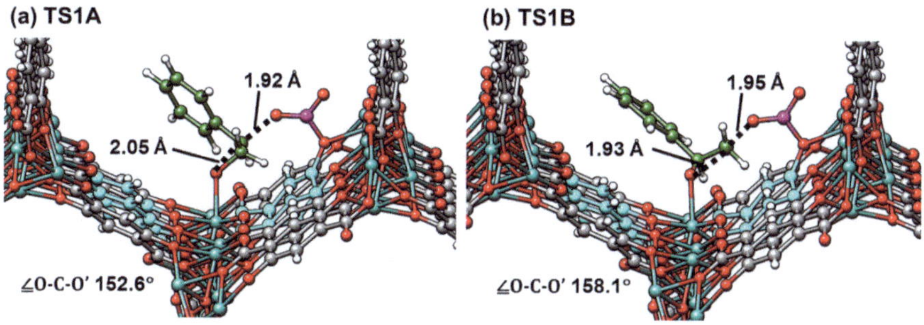

Fig. 17 Transition states for the first step (ring opening) in the reaction catalyzed by Co-MOF-74 (Reprinted with permission from [87]. Copyright 2018 American Chemical Society)

Most recently, we have studied MOF-catalyzed cycloaddition reactions of CO_2 with an epoxide substrate using QM/MM (Fig. 16) [87]. Over the years, the CO_2 coupling reaction has attracted considerable attention in relation to carbon capture and utilization (CCU), and various types of catalysts for this reaction have been developed [89]. In recent years, an increasing number of reports have come out that demonstrated that MOFs can catalyze the cycloaddition reaction [90]. For example, Co-MOF-74 was shown to catalyze the cycloaddition of CO_2 to styrene oxide [91]. CO_2 is basically a very inert molecule, and thus, it does not have a sufficient nucleophilicity to open the epoxide ring. As such, many of the catalytic reactions developed thus far employ a cocatalyst like a quaternary ammonium salt, and the halide ion included in the cocatalyst is responsible for the ring opening. Interestingly, in the case of the reaction catalyzed by Co-MOF-74, a cocatalyst was not needed in forming the 4-phenyl-1,3-dioxolan-2-one (styrene carbonate) product, which should be beneficial in the context of easier product separation.

Our QM/MM study identified the oxygen atom from the dobdc^{4-} (2,5-dioxido-1,4-benzenedicarboxylate) linker as the Lewis base site (Fig. 17). The Lewis base site activates CO_2 to make it sufficiently nucleophilic. The activated CO_2 can now attack one of the two carbon atoms in the epoxide moiety of styrene. The attack of CO_2 on the carbon atom that is not substituted by phenyl was found to have a lower barrier, which is probably due to steric effects.

Although the ring-opening process may give the impression that the substrate can gain larger flexibility, the QM/MM analysis showed that the substrate structure after ring opening is actually tight, pointing to the difficulty in its ring closure to directly form the styrene carbonate (Fig. 18). Such ring closure requires the cleavage of either

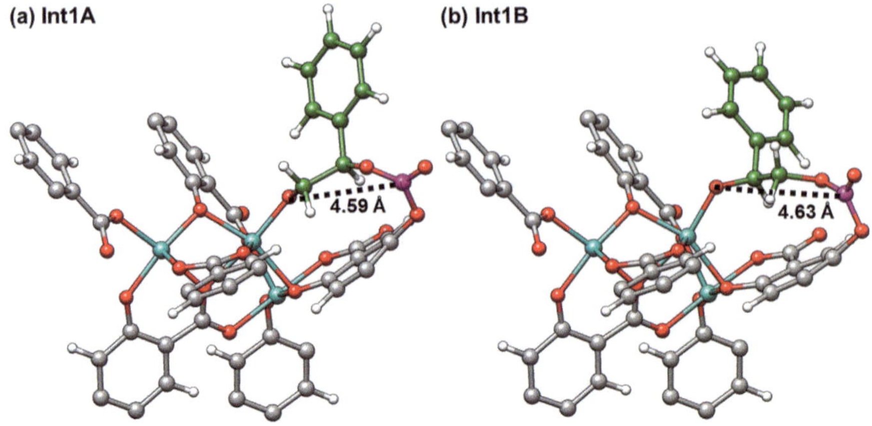

Fig. 18 Intermediates formed after the ring-opening step (Reprinted with permission from [87]. Copyright 2018 American Chemical Society)

the coordination bond between Co(II) and alkoxide or the C–O bond between CO_2 and the Lewis base site; both are strong bonds. In order to allow ring closure to occur, some flexibility should be introduced to the intermediate, which is possible if another molecule of CO_2 binds to the alkoxide oxygen. The addition of a second CO_2 molecule turned out to be a rather facile process (Fig. 19a). Moreover, as a result of the addition of a second CO_2 molecule, the substrate moiety now gains sufficient flexibility to complete the formation of styrene carbonate (Fig. 19b).

QM/MM calculations were also performed on the same reaction catalyzed by Mg-MOF-74, and comparisons were made between the reactions in Co-MOF-74 and Mg-MOF-74 (Fig. 20). The mechanisms of these two MOF-catalyzed reactions are basically similar, and there are three major steps in both cases: (1) ring opening of the epoxide, (2) addition of CO_2 for carbonate formation, and (3) five-membered ring formation (ring closure). However, a few differences were noted. According to the QM/MM results, ring closure is the rate-determining step in the Co-MOF-74 reaction, whereas the ring-opening step has the highest barrier in the Mg-MOF-74 reaction. In addition, Co-MOF-74 was found to have a more rigid coordination environment around the metal center than Mg-MOF-74, probably because of the involvement of d-orbitals in coordination bonds in the former. Thus, Mg-MOF-74 is more flexible, and in this case, there is a structural rearrangement before the final ring-closure step.

5.2 *Force-Field Parameterization for MOF Simulations*

Although QM/MM can essentially be applied to metalloenzymes and MOFs, from a technical perspective, a major difference between the QM/MM calculations on these two different types of systems is the availability of MM parameters. Good force-

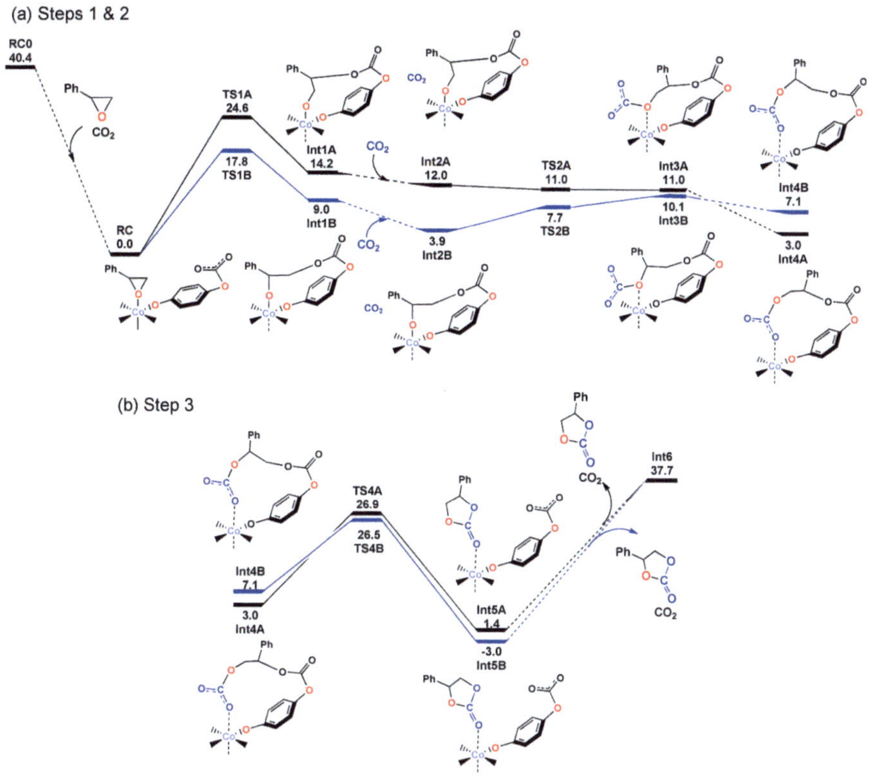

Fig. 19 Reaction energy profile (in kcal/mol) for the cycloaddition of CO_2 to styrene oxide in Co-MOF-74 (Reprinted with permission from [87]. Copyright 2018 American Chemical Society)

field parameters have been developed for proteins, which can be used for QM/MM calculations of enzymes. In addition, even though enzymes have diverse structures and intricate folding patterns, their building blocks are essentially 20 amino acids that have stiff covalent bonds (Fig. 21a). By contrast, MOFs have much more diverse bonding patterns than enzymes do, and coordination bonds of MOFs are essentially softer than the covalent bonds in enzymes. As such, previously developed force-field parameters for proteins or organic/inorganic molecules are not always applicable to MOFs, and the development of a transferable force field for MOFs is more difficult than for proteins.

Therefore, when performing MM or QM simulations of MOFs, one often needs to develop a new parameter set or modify existing parameters for a given MOF system. We have recently developed force-field parameters for Fe-MOF-74 using the MOF-FF method, which uses a genetic algorithm for force-field parameterization [92, 93]. The optimized parameters were first tested for performance in MM geometry optimization of Fe-MOF-74 in the periodic boundary condition, which yielded encouraging results. The parameters were then used for periodic QM/MM calcula-

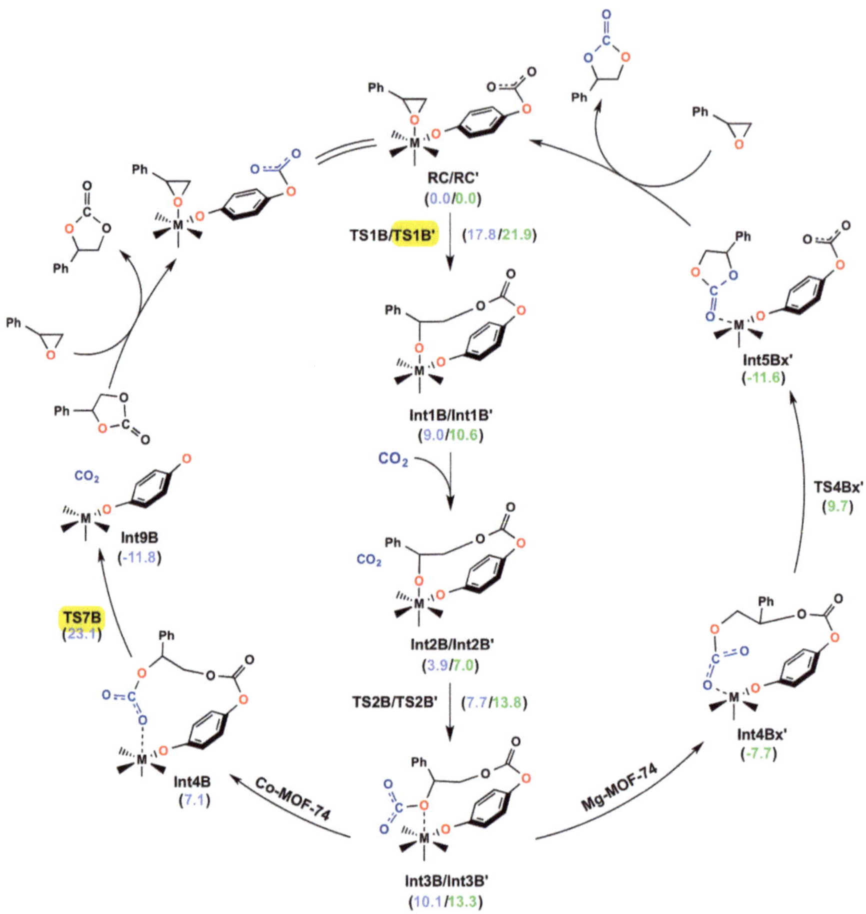

Fig. 20 QM/MM-derived mechanisms for the reactions in Co-MOF-74 and Mg-MOF-74 (Reprinted with permission from [87]. Copyright 2018 American Chemical Society)

tions. The QM/MM calculations were able to show that the side-on geometry is more stable than the end-on geometry for the ferric superoxide species. The experimentally determined binding energy between O_2 and Fe-MOF-74 was also well reproduced.

We have also developed our own methods for rapid force-field parameterization, which we call partial Hessian fitting (PHF), full Hessian fitting (FHF), and internal Hessian fitting (IHF) [94, 95] (Fig. 22). We formulated these methods in a way that minimizes the deviation between MM- and QM-calculated Hessian elements without going through iterative deviation-minimization processes. These Hessian fitting schemes performed very well in determining force-constant parameters for 23 different molecules including three secondary building units of MOFs, but in general, the best performance was obtained with IHF, followed by PHF and then FHF. We

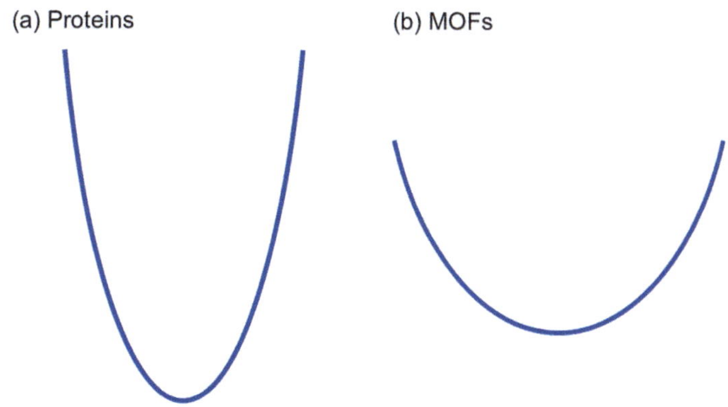

Fig. 21 Schematic illustrations of the bond stiffness of **a** covalent bonds in proteins and **b** coordination bonds in MOFs

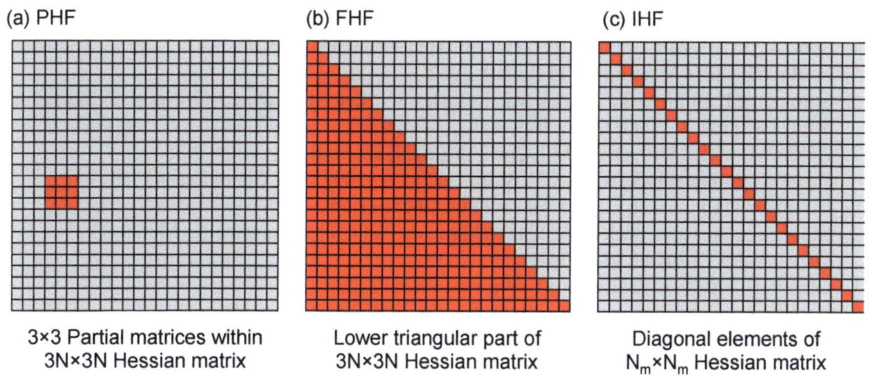

Fig. 22 Schematic illustration of the recently developed three Hessian fitting schemes (Reprinted with permission from [95]. Copyright 2017 Wiley

believe that the development of efficient force-field parameterization schemes will broaden the applicability of MM and QM/MM methods to MOF systems.

6 Conclusions

This chapter has described our recent attempts to understand and predict catalytic mechanisms of various transition-metal-containing systems using computational methods. For such studies, it is essential that quantum mechanical insights be derived. Kohn–Sham DFT methods are often the first choice for describing involved molecules quantum mechanically. However, sole reliance on DFT is precluded by its high computational demands in describing large molecular systems. As realistic solutions to this problem, we have applied DFT, QM/QM', and QM/MM methods to systems of varying sizes, i.e., homogeneous catalytic systems, enzymes, and MOFs. These methods can provide invaluable microscopic and quantum mechanical insights and thus serve as indispensable partners of experimental studies.

Acknowledgements The author thanks collaborators for fruitful collaborations and City University of Hong Kong (grant numbers 7200534 and 9610369) and JST-PRESTO (grant number JPMJPR141B) for generous financial support.

References

1. Ortiz de Montellano PR (2005) CytochromeP450: structure, mechanism, and biochemistry, 3rd edn. Kluwer Academic/Plenum Publishers, New York, NY
2. Meunier B, de Visser SP, Shaik S (2004) Chem Rev 104:3947–3980
3. Groves JT (1985) J Chem Educ 62:928–934
4. Sono M, Roach MP, Coulter ED, Dawson JH (1996) Chem Rev 96:2841–2887
5. Poulos TL (2014) Chem Rev 114:3919–3962
6. Solomon EI, Brunold TC, Davis MI, Kemsley JN, Lee SK, Lehnert N, Neese F, Skulan AJ, Yang YS, Zhou J (2000) Chem Rev 100:235–350
7. Costas M, Mehn MP, Jensen MP, Que L Jr (2004) Chem Rev 104:939–986
8. Krebs C, Fujimori DG, Walsh CT, Bollinger JM Jr (2007) Acc Chem Res 40:484–492
9. Bollinger JM Jr, Krebs C (2007) Curr Opin Chem 11:151–158
10. Mukherjee A, Cranswick MA, Chakrabarti M, Paine TK, Fujisawa K, Münck E, Que L Jr (2010) Inorg Chem 49:3618–3628
11. Oloo WN, Que L Jr (2015) Acc Chem Res 48:2612–2621
12. Nam W, Lee YM, Fukuzumi S (2014) Acc Chem Res 47:1146–1154
13. Nam W (2015) Acc Chem Res 48:2415–2423
14. Cho KB, Hirao H, Shaik S, Nam W (2016) Chem Soc Rev 45:1197–1210
15. Hohenberger J, Ray K, Meyer K (2012) Nature Commun 3: Article 720
16. Shaik S, Kumar D, de Visser SP, Altun A, Thiel W (2005) Chem Rev 105:2279–2328
17. Shaik S, Hirao H, Kumar D (2007) Acc Chem Res 40:532–542
18. Shaik S, Hirao H, Kumar D (2007) Nat Prod Rep 2007(24):533–552
19. Siegbahn PEM, Borowski T (2006) Acc Chem Res 39:729–738
20. Bassan A, Blomberg MRA, Borowski T, Siegbahn PEM (2006) J Inorg Biochem 100:727–743
21. Neese F (2017) Angew Chem Int Ed 56:11003–11010
22. Ye S, Geng CY, Shaik S, Neese F (2013) Phys Chem Chem Phys 15:8017–8030
23. de Visser SP, Rohde JU, Lee YM, Cho J, Nam W (2013) Coord Chem Rev 257:381–393
24. Hirao H, Thellamurege N, Zhang X (2014) Front Chem 2: Article 14
25. Kohn W, Sham LJ (1965) Phys Rev A 140:A1133–A1138
26. Seidl A, Görling A, Vogl P, Majewski JA, Levy M (1996) Phys Rev B 53:3764–3774
27. Ziegler T, Autschbach J (2005) Chem Rev 105:2695–2722
28. Harvey JN (2006) Annu Rep Prog Chem Sect C 102:203–224
29. Cramer CJ, Truhlar DG (2009) Phys Chem Chem Phys 11:10757–10760
30. Liu P, Houk KN (2011) Inorg Chim Acta 369:2–14
31. Sameera WMC, Maseras F (2012) Wiley Interdiscip Rev Comput Mol Sci 2:375–385
32. Blomberg MRA, Borowski T, Himo F, Liao RZ, Siegbahn PEM (2014) Chem Rev 114:3601–3658
33. Sakaki S (2015) Bull Chem Soc Jpn 88:889–938
34. Cheng GJ, Zhang X, Chung LW, Xu L, Wu YD (2015) J Am Chem Soc 137:1706–1725
35. Sperger T, Sanhueza IA, Kalvet I, Schoenebeck F (2015) Chem Rev 115:9532–9586

36. Eisenstein O, Milani J, Perutz RN (2017) Chem Rev 117:8710–8753
37. Hopmann KH (2018) Int J Quantum Chem 115:1232–1249
38. Guan J, Zarić SD, Brothers EN, Hall MB (2018) Int J Quantum Chem 2018(118):e25605
39. Schultz NE, Zhao Y, Truhlar DG (2005) J Phys Chem A 109:11127–11143
40. Bühl M, Kabrede H (2006) J Chem Theory Comput 2:1282–1290
41. Furche F, Perdew JP (2006) J Chem Phys 124:044103
42. Weymuth T, Couzijn EPA, Chen P, Reiher M (2014) J Chem Theory Comput 10:3092–3103
43. Sousa SF, Fernandes PA, Ramos MJ (2007) J Phys Chem A 111:10439–10452
44. Cohen AJ, Mori-Sánchez P, Yang W (2011) Chem Rev 112:289–320
45. Verma P, Varga Z, Klein JEMN, Cramer CJ, Que L Jr, Truhlar DG (2017) Phys Chem Chem Phys 19:13049–13069
46. Grimme S (2011) Wiley Interdiscip Rev Comput Mol Sci 1:211–228
47. Senn HM, Thiel W (2009) Angew Chem Int Ed 48:1198–1229
48. Lonsdale R, Harvey JN, Mulholland AJ (2012) Chem Soc Rev 41:3025–3038
49. Quesne MG, Borowski T, de Visser SP (2015) Chem Eur J 22:2562–2581
50. Chung LW, Hirao H, Li X, Morokuma K (2012) Wiley Interdiscip Rev Comput Mol Sci 2:327–350
51. Chung LW, Sameera WMC, Ramozzi R, Page AJ, Hatanaka M, Petrova GP, Harris TV, Li X, Ke Z, Liu F, Li HB, Ding L, Morokuma K (2015) Chem Rev 115:5678–5796
52. Hirao H, Xu K, Chuanprasit P, Moeljadi AMP, Morokuma K (2017) Key concepts and applications of ONIOM method. In: Tunon I, Moliner V (eds) Simulating enzyme reactivity: computational methods in enzyme catalysis. The Royal Society of Chemistry, Cambridge, pp 245–293
53. Wu C, Yue G, Nielsen CDT, Xu K, Hirao H, Zhou J (2016) J Am Chem Soc 138:742–745
54. Pooi B, Lee J, Choi K, Hirao H, Hong SH (2014) J Org Chem 79:9231–9245
55. Gazi S, Ng WKH, Ganguly R, Moeljadi AMP, Hirao H, Soo HS (2015) Chem Sci 6:7130–7142
56. Gazi S, Đokić M, Moeljadi AMP, Ganguly R, Hirao H, Soo HS (2017) ACS Catal 7:4682–4691
57. Zakzeski J, Bruijnincx PCA, Jongerius AL, Weckhuysen BM (2010) Chem Rev 110:3552–3599
58. Sun Z, Fridrich B, de Santi A, Elangovan S, Barta K (2018) Chem Rev 118:614–678
59. Xu H, Yang P, Chuanprasit P, Hirao H, Zhou JS (2015) Angew Chem Int Ed 54:5112–5116
60. Yang P, Lim LH, Chuanprasit P, Hirao H, Zhou JS (2016) Angew Chem Int Ed 55:12083–12087
61. Hirao H, Cheong ZH, Wang X (2012) J Phys Chem B 116:7787–7794
62. Hirao H, Chuanprasit P, Cheong YY, Wang X (2013) Chem Eur J 19:7361–7369
63. Hirao H, Thellamurege NM, Chuanprasit P, Xu K (2013) Int J Mol Sci 14:24692–24705
64. Hirao H, Chuanprasit P (2015) Chem Phys Lett 621:188–192
65. Chuanprasit P, Goh SH, Hirao H (2015) ACS Catal 5:2952–2960
66. Yoshimoto FK, Guengerich FP (2014) J Am Chem Soc 136:15016–15025
67. Xu K, Wang Y, Hirao H (2015) ACS Catal 5:4175–4179
68. Ghosh D, Griswold J, Erman M, Pangborn W (2009) Nature 457:219–223
69. Hirao H (2011) Chem Lett 40:1179–1181
70. Thellamurege N, Hirao H (2013) Molecules 18:6782–6791
71. Thellamurege N, Hirao H (2014) J Phys Chem B 118:2084–2092
72. Hirao H, Morokuma K (2009) J Am Chem Soc 131:17206–17214
73. Hirao H, Morokuma K (2010) J Am Chem Soc 132:17901–17909
74. Hirao H, Morokuma K (2010) J Phys Chem Lett 1:901–905
75. Hirao H, Morokuma K (2011) J Am Chem Soc 133:14550–14553
76. Hirao H, Morokuma K (2011) Yakugaku Zasshi – J Pharm Soc Jpn 131:1151–1161
77. Hirao H (2011) J Phys Chem B 115:11278–11285
78. Xu K, Hirao H (2018) Phys Chem Chem Phys 20:18938–18948
79. Dobbek H, Gremer L, Kiefersauer R, Huber R, Meyer O (2002) Proc Natl Acad Sci USA 99:15971–15976
80. Siegbahn PEM, Shestakov AF (2005) J Comput Chem 26:888–898
81. Hofmann M, Kassube JK, Graf T (2005) J Biol Inorg Chem 10:490–495
82. Stein BW, Kirk ML (2014) Chem Commun 50:1104–1106

83. Rokhsana D, Large TAG, Dienst MC, Retegan M, Neese F (2016) J Biol Inorg Chem 21:491–499
84. Hirao H, Ng WKH, Moeljadi AMP, Bureekaew S (2015) ACS Catal 5:3287–3291
85. Doitomi K, Xu K, Hirao H (2017) Dalton Trans 46:3470–3481
86. Doitomi K, Hirao H (2017) Tetrahedron Lett 58:2309–2317
87. Xu K, Moeljadi AMP, Mai BK, Hirao H (2018) J Phys Chem C 122:503–514
88. Ghalei B, Sakurai K, Kinoshita Y, Isfahani AP, Song Q, Doitomi K, Furukawa S, Hirao H, Kusuda H, Kitagawa S, Sivaniah E (2017) Nat Energy 2: Article 17086
89. North M, Pasquale R, Young C (2010) Green Chem 12:1514–1539
90. Beyzavi MH, Stephenson CJ, Liu Y, Karagiaridi O, Hupp JT, Farha OK (2015) Front Energy Res 2: Article 63
91. Cho HY, Yang DA, Kim J, Jeong SY, Ahn WS (2012) Catal Today 185:35–40
92. Moeljadi AMP, Schmid R, Hirao H (2016) Can J Chem 94:1144–1150
93. Bureekaew S, Amirjalayer S, Tafipolsky M, Spickermann C, Roy TK, Schmid R (2013) Phys Status Solidi B 250:1128–1141
94. Wang R, Ozhgibesov M, Hirao H (2016) J Comput Chem 37:2349–2359
95. Wang R, Ozhgibesov M, Hirao H (2018) J Comput Chem 39:307–318

How Metal Coordination in the Ca-, Ce-, and Eu-Containing Methanol Dehydrogenase Enzymes Can Influence the Catalysis: A Theoretical Point of View

Tiziana Marino, Mario Prejanò and Nino Russo

Abstract Methanol dehydrogenase (MDH) enzymes are quinoproteins that require calcium or magnesium ion as well as pyrroloquinoline quinone as a cofactor for activity in the oxidation of methanol to formaldehyde. Lately, MDH enzymes containing lanthanide ions in the active site have been isolated in drastic conditions from *Methylacidiphilum fumariolicum* bacterium. The present theoretical study performed in the framework of the density functional theory employing the quantum mechanical cluster approach mainly focused on the catalytic mechanism of cerium containing MDH enzyme. In order to rationalize the effect of the metal ion substitution on the catalytic activity, geometrical and electronic properties of the "Michaelis–Menten" enzyme–methanol complexes of Ce-MDH and Eu-MDH are also discussed as well as the substrate's activation mediated by the metal ion. With the aim to better describe the Lewis acidity of metal ions in the methanol oxidation, the comparison of the catalytic performance between Ce-MDH and Ca-MDH was also made.

1 Introduction

Metal ions are ubiquitous and widely distributed in nature and account extremely important roles in chemistry, geochemistry, biochemistry, material sciences, and medicine. Approximately one-third of the structures in the Protein Data Bank (PDB) contains metal ions.

Metals such as copper (Cu), iron (Fe), lanthanides (Ln), cobalt (Co), zinc (Zn), molybdenum (Mo), and tungsten (W) are particularly important for methanotrophy [1].

In particular, rare earth elements (REE) including lanthanum and lanthanoid (17 elements) as well as scandium and yttrium are found in bacteria, archaebacteria, fungi, vegetables, and animals [2]. They are known to act as calcium analogs in the

T. Marino (✉) · M. Prejanò · N. Russo
Dipartimento di Chimica e Tecnologie Chimiche, Università della Calabria, 87036 Rende, Italy
e-mail: tiziana.marino65@unical.it

E. Broclawik et al. (eds.), *Transition Metals in Coordination Environments*, Challenges and Advances in Computational Chemistry and Physics 29,
https://doi.org/10.1007/978-3-030-11714-6_16

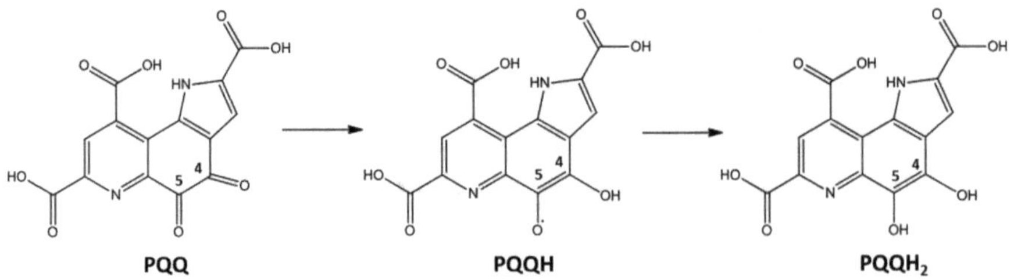

Scheme 1 Three different oxidation states of the pyrroloquinoline quinone during the methanol oxidation by MDH enzyme

biological systems, and in particular, lanthanum ions (III) may replace calcium in many proteins including enzymes [3, 4].

REE were considered for long time to be biologically inert but very recently received a great attention since revealed to be essential metals for activity and expression of a special type of methanol dehydrogenase, XoxF. This gene, first discovered in 1995 [5, 6], is associated with the activity of a novel methanol dehydrogenase (MDH) in a methylotrophic bacterium *Methylobacterium radiotolerans* [7]. XoxF sequences share almost 50% amino acid residues with those of MxaF, the extensively studied and well-characterized two-subunit methanol dehydrogenase, that contains 2,7,9-tricarboxypyrroloquinoline quinone (PQQ) and Ca^{2+} ion as cofactors in its catalytic center [8, 9].

PQQ cofactor accepts formally two electrons and two protons from the alcohol substrate to give rise the corresponding aldehyde and the reduced form $PQQH_2$ form [10, 11]; see Scheme 1. The presence in the REE-MDH active site of an ion with larger size and higher charge compared to calcium ion requires a further amino acid residue, the Asp 301, to balance the excess of positive charge of the metal center [1, 12].

In nature, the ability of methylotrophic microorganisms to use reduced C1 compounds, such as methane and methanol, as an energy source, is a significant biological and geochemical process involved in the global carbon cycle. In the case of methanotrophic species, instead, the methane must be oxidized to methanol by a methane monooxygenase [13–15]. Studies performed on methanotrophic and methylotrophic bacteria demonstrated that they carry both MxaF and XoxF genes, but the environment plays a crucial role in their expression [16].

The XoxF REE-MDH that uses different metal ions can be considered an example of enzyme condition promiscuity where the enzyme can show catalytic activity in reaction conditions different from their natural ones. In fact, REE-MDH from SolV operates at neutral pH, whereas the Ca-MDH displays its highest activity at higher pH values [1, 17, 18].

The isolation of the XoxF-type MDH from the extremophile *Methylacidiphilum fumariolicum* SolV bacterium cultivated with mudpot water in the Solfatara crater in south Italy, allowed to identify in the active center the most abundant lanthanide, the cerium element [18]. Recently, the cultivation and purification of the strictly REE-dependent methanotrophic bacterium SolV with europium have been also reported [19].

Despite the mechanism of methanol oxidation by calcium-containing enzyme has attracted the attention in the last 40 years and different mechanistic proposals have been suggested and discussed, today controversial opinions are present and the debate is still open [20–28]. In the case of REE-MDH, owing to the most recent discoveries [7, 14, 19, 29–31], the discussion is just beginning.

Furthermore, a deeper investigation on the reaction mechanism of cerium containing MDH has been performed by employing the quantum chemical cluster methodology in the framework of density functional theory (DFT).

Having in mind that the coordination chemistry of the lanthanides can generate structural diversity depending on the progressive decrease in ionic radius, in the present work, we present the results obtained by the comparative DFT analysis of the catalytic behavior of the cerium- and europium-containing methanol dehydrogenase. In particular, the step of the enzymatic mechanism where the metal ion plays a critical role has been considered. Both cerium and europium metal ions act as good Lewis acids in the enzyme–substrate complex (ES), and our study is focused on this step and the analysis of the structural, electronic properties and of the charges distribution can contribute to gain a more comprehensive knowledge of this important event. In addition, during the discussion, the catalytic behavior of the cerium-containing enzyme is compared with that of the more usual calcium-containing enzyme.

2 Computational Methods

2.1 Active Site Model

For the cerium enzyme, the ES starting structure has been obtained by modeling the X-ray structure of the Xoxf-type natural cerium-dependent MDH (PDB: 4MAE, 1.6 Å) [18]. It corresponds to the precatalytic enzyme–substrate complex where the polyethylene glycol instead of the natural substrate prevents the occurrence of the reaction. In the case of the Eu-MDH-used structure, the achieved crystals report the enzyme without substrate. (PDB: 6FKW, 1.4 Å) [19].

The QM cluster includes the amino acid residues of the inner coordination shell of Ce^{3+} ion (Glu172, Asn256, Asp299, Asp301) and the cofactor PQQ and Glu55, Arg110, Ser169, Arg326, Asp388 residues of the outer coordination shell that potentially form hydrogen bonds with the cofactor. All amino acids were truncated at the α carbons, and hydrogen atoms were added manually. The methanol substrate is coordinated to the metal center by the hydroxyl group. The overall model consists of 113 atoms, including the substrate, and has a total charge of zero (Fig. 1).

As usual in the cluster approach for modeling enzymatic reactions [23, 32–37], the truncated amino acid residues at the periphery of the active site are fixed to

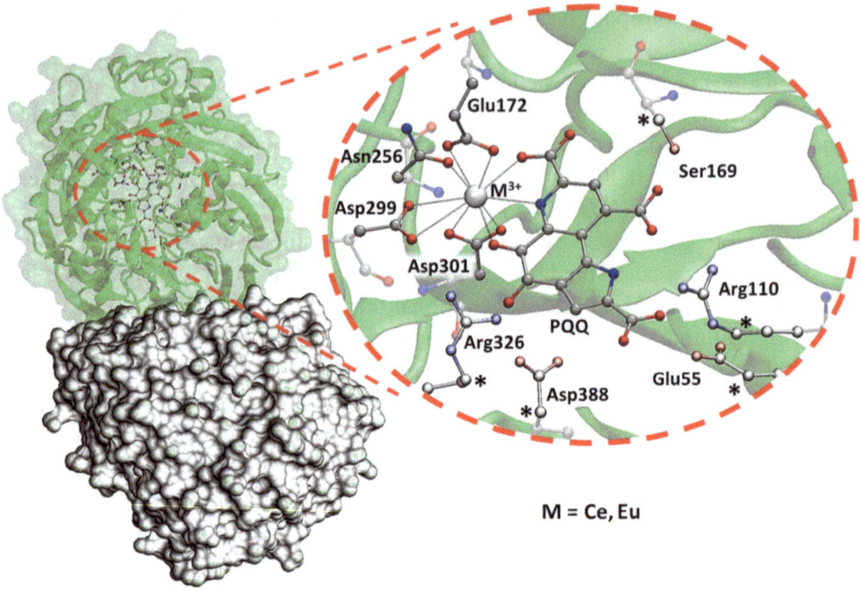

Fig. 1 Overall structure of Ce- and Eu-MDH and close-up view of the active site that shows the model cluster (in ball and stick) representing the active site of. Stars indicate the locked atoms frozen during the calculations

their crystallographic positions during the geometry optimizations. This precludes the artificial expansion of the cluster.

2.1.1 Technical Details

All the computations have been performed employing the Gaussian 09 program [38] with the Becke exchange and the Lee, Yang, Parr correlation (B3LYP) hybrid density functional method [39, 40]. Geometries have been optimized using the 6-31+G(d,p) basis set for the C, N, O, H atoms, and the SDD [41] effective core potential (ECP) coupled with its related basis set for the lanthanides. The D3 dispersions contribution [42] as implemented in the Gaussian package has been considered.

The employed relativistic effective core potential (ECP) [41] based on a number of 30 and 35 valence electrons for cerium and europium, respectively, ensures a better description of the coordination chemistry of these elements. The nature of minima (without imaginary frequency) or maxima (only one imaginary frequency) of every stationary point intercepted along the PES has been determined by frequency calculations. To estimate the energetic effects of the protein environment, single-point calculations by using the polarizable continuum method (C-PCM) coupled to the dielectric constant value $\varepsilon = 4$ have been performed [43, 44]. This value usually represents a good choice of protein surrounding [32, 34, 45–53].

For Ce-MDH, further calculations with $\varepsilon = 24$ and $\varepsilon = 78$ have been also done to test the influence of the more polar environments on its catalytic activity.

More accurate energies with the larger basis set 6-311+G(2d,2p) for all atoms except for the metal ion described by ECP have been obtained by single-point calculations performed on the optimized structures.

The given energies presented are corrected for ZPE, solvation, and dispersion effects. The same computational protocol has been successfully used in the mechanistic studies of other metal proteins [33, 34, 47, 51, 53, 54].

As far as the cerium-containing enzyme is concerned, the doublet spin multiplicity ($2S + 1 = 2$) previously determined [53, 55] has been considered as the ground state in all the calculations by using spin-unrestricted wavefunctions. Almost no spin contaminations have been evidenced due to the $\langle S\hat{}2\rangle$ value equal to 0.77.

Natural bond orbital (NBO) analysis [56] was performed on all the intercepted stationary points of the investigated PESs.

3 Results and Discussion

3.1 Ce-MDH PES

Different mechanisms have been proposed for MDH enzymes over the years [9, 18, 20, 21, 23, 27, 28, 30, 53, 57], but the exact oxidation mechanism of CH_3OH by MDH is still under debate.

In the investigated mechanism shown in Scheme 2, we propose the metal ion, during the first step evidenced in the rectangular box, to facilitate the nucleophile attack of OH group of the coordinated methanol that with its lone pair oriented toward the PQQ is preparing to transfer the proton to the Asp299 base. As a result, the O5 of PQQ coordinated to the metal ion completes the activation role played by the metal ion making C5 more suitable for the nucleophilic addition. Successively, the reaction proceeds with the protonation of the O4 of PQQ with simultaneous CH_2O formation (TS2) and then the H^+ abstraction from Asp299 with the formation of $PQQH_2$ (TS3). Summarizing during the oxidation of methanol to formaldehyde, the PQQ cofactor suffers a reduction in two steps (Scheme 1): from PQQ (starting point) to semiquinone PQQH (intermediate) and the fully reduced $PQQH_2$ form.

During the redox cycling of the PQQ cofactor, it was demonstrated the metal ion to retain the same oxidation state equal to +3 [55].

As reported in our previous works on methanol dehydrogenase enzymes [23, 53], the PESs have been obtained considering the polarization effects caused by the portion of the surrounding enzyme that is not explicitly included in the quantum model by cavity techniques. Other than the environment simulated with $\varepsilon = 4$, in the present investigation, single-point calculations with other dielectric constant values have been performed. In this way, a set of three different dielectric constants ($\varepsilon = 4$, $\varepsilon = 24$ and $\varepsilon = 78$) characterized by a trend of increasing polarity have been tested.

The results shown in Fig. 2 clearly show the absence of relevant effects in both minima and transition states as a function of the dielectric constants. Only in the case of the INT1 species, a major stabilization, by few kcal/mol, can be observed. This behavior can be ascribed to the fact that the INT1 species is the result of a series of rearrangements accomplished by different charge distributions. The absence of large solvent effects is usually related to the "completeness" of the QM selected model that already explicitly includes most of the polarization effects. The rate determining step is localized in the second part of the process (INT1 → TS2_A → INT2_A). The corresponding calculated barrier values are 19.4, 21.2, and 21.5 kcal/mol with $\varepsilon = 4$, $\varepsilon = 24$ and $\varepsilon = 78$, respectively. During this step, the produced formaldehyde is moved outside of the inner coordination shell of the metal ion and is kept close to the cerium by H-bond with Ser169 (2.020 Å).

The next step (INT2_A → TS3_A → EP) describes the restoration of the protonation state of the Asp299 residue acting as acid–base and the consequent protonation of the C5-O5$^{(-)}$ moiety. The barriers for this hydrogen transfer independently of the simulated environments (5.4 kcal/mol for $\varepsilon = 4$, 7.1 kcal/mol for $\varepsilon = 24$ and 6.8 kcal/mol for $\varepsilon = 78$) lie below the ES energy (see Fig. 2). The PQQ reduction to $PQQH_2$ is accomplished. In all cases, a favorable thermodynamics of the process is retained.

The obtained PES can be compared with that of calcium containing MDH previously investigated by our group [23] considering the same mechanism. It is worthy to note that the coordination sphere of the two metal ions is different such as the charge of the metal ion, but the presence of negatively charged residues (mainly Asp and Glu in the inner coordination shell) confirms the hard nature of Lewis acid.

This striking similarity between lanthanides and Ca^{2+} ions, in terms of size, coordination environment, and ligand preferences, is widely reported in the literature [20–28].

In any case, we notice that in the previous investigation on the Ca-MDH [23], the dispersion contributions were not taken into account. Anyway, both enzymes suggest TS2 barrier as rate limiting step owing to 19.0 (Ca-MDH) and 21.4 (Ce-

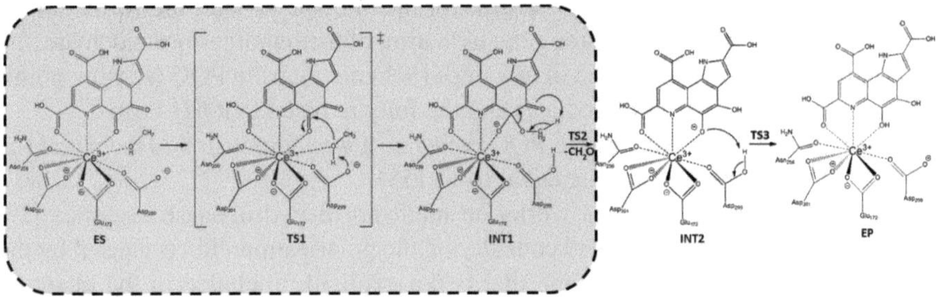

Scheme 2 Mechanism of methanol oxidation catalyzed by the Ce-MDH enzyme. The gray box evidences the mechanism's step where the metal ion is directly involved

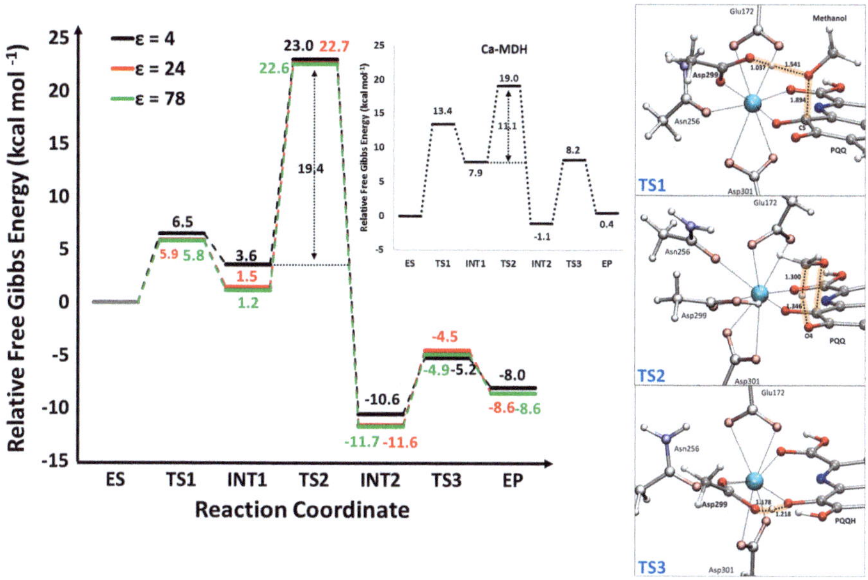

Fig. 2 Calculated energetics of methanol dehydrogenase in the Ce-MDH. The DFT energy profiles, considering three different dielectric constant values, include the electronic energy calculated at the B3LYP-D3/6-311+G(2d,2p)ISDD level of theory, ZPE, dispersion, and solvation corrections in comparison with that of Ca-MDH (ε = 4). In the squared boxes are depicted the optimized geometries at B3LYP-D3/6-31+G(d,p)ISDD of the three intercepted transition states. Distances are in Å

MDH) kcal/mol above the ES. This finding is indicative of a very similar catalytic activity [53].

The better catalytic activity of calcium-containing enzyme with respect to the cerium one cannot be ascribed to the stronger Lewis acidity of the Ce^{3+} since this effect influences the first step of the mechanism where the lanthanide ion makes the C5 more electrophile and intensifies the nucleophilic nature of the OH moiety in the substrate. As a consequence, the first barrier (TS1) is more favorable in Ce-MDH (3.8 kcal/mol without dispersion) than that in Ca-MDH (13.4 kcal/mol) [53].

3.2 *Michaelis–Menten Complex (ES) for Ce-MDH and Eu-MDH*

As the first step, with the aim to predict the catalytic activity of the Eu-MDH, a deep analysis of the catalytic pocket of Ce- and Eu-MDH arising from X-ray structures [18, 19] can give helpful hints about the metal surrounding environment entailed in the catalysis.

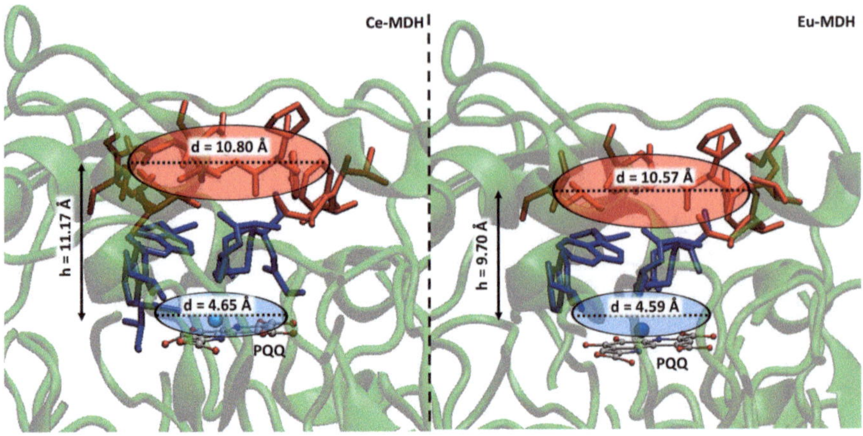

Fig. 3 Comparison of the sizes of catalytic pocket of the Ce- and Eu-MDH. In blue are indicated the hydrophilic residues (Trp260, Trp532, Cys104, Cys105, Glu172), while in red the hydrophobic ones (Eu: Pro424, Phe273, Val426, Gly427, Leu550, Ala103, Val102; Ce: Pro424, Phe273, Val426, Gly427, Leu550, Ala103, Val102; Ala101, Val100, Gly551, Ala552, Ile 535)

The crystal structure of Ce-MDH showed the active site in every monomer located in a large, deep, and conic-shaped cavity (~10.80 Å in diameter at its mouth and 11.2 Å deep). As it can be evinced from Fig. 3, in Eu-MDH, the catalytic pocket shows reduced sizes owing to 9.70 Å deep, 10.57 Å at the top, and 4.59 Å at the bottom. In both enzymes, the binding cavity of the primary CH_3OH alcohol includes a hydrophobic region at the entrance and then a hydrophilic one at the bottom. The twelve amino acid residues included in the hydrophobic region of the cerium with respect to the seven ones of europium justify the different dimensions. The "bipolar" nature of the access channel to the metal center can account for the recognition of the substrate by hydrophobic region, while the hydrophilic counterpart of the cavity in proximity to the active site comes into play during the occurrence of the reaction. This scenario in a simple way reflects the consequences due to the metal ion substitution inside the catalytic pocket owing to the slightly smaller ionic radius of europium (III) relative to cerium (III).

In the ES formation, the substrate binds to the active site, and the "complexation energy" should be an important parameter since our QM model resembles a hypothetic ML_9 species (L = oxygen atoms from the ligands).

$$ML_9 + L' \rightarrow ML_9L'$$

The total interaction energy can be computed according to equation above following the counterpoise method [58, 59] as a sum of the two-body interaction energies.

This energy represents a measure of the bond formation with the methyl alcohol by lanthanide ion increasing of the coordination number from 9 to 10. The obtained energetic values (ΔH of the above reaction) of −32.56 kcal/mol for cerium and

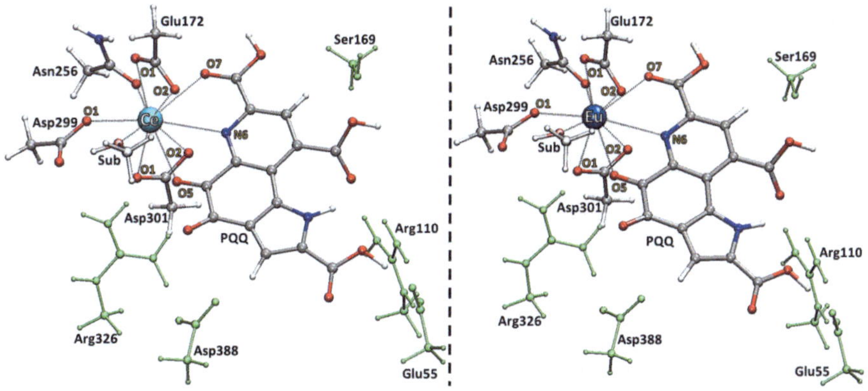

Fig. 4 Optimized geometry of the Michaelis–Menten complexes of Ce- and Eu-MDH. The amino acid residues in green belong to the outer coordination shell of the metal ions. Atoms labeled identify the first coordination shell

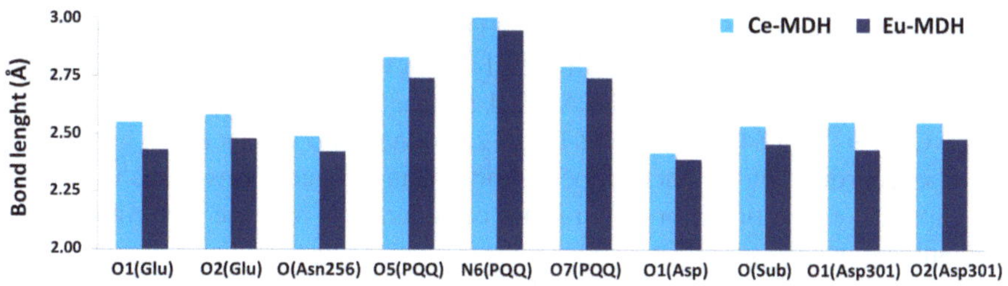

Fig. 5 Distances of the ligand atoms to metal ion in cyan for Ce-MDH and in blue for Eu-MDH. Labels of the atoms are present in Fig. 4

–31.53 kcal/mol for europium reflect as the smaller radius of the central metal the greater the non-bonding inter-ligand repulsive interactions. A possible explanation of this behavior is related to the increase in effective nuclear charge in the cerium experienced by the outer electrons (35 by ECP in Eu^{3+} vs. 30 by ECP in Ce^{3+}), essentially promoted by the incomplete shielding of the 5s and 5p electrons by the 4f ones, with consequent relativistic effects by about 10%. This result can also provide a reason for the reduced affinity constant found in the kinetic study on Eu-MDH [19].

The optimized geometries of the ES complex of Ce- and Eu-MDH are reported in Fig. 4, and the main geometrical parameters (bond distances) related to the atoms directly bonded to the metal ion are collected in Fig. 5.

Both Ce^{3+} (ionic radius 1.196 Å) and Eu^{3+} (ionic radius 1.120 Å) retain all of the amino acids in the inner coordination shell showing a coordination number equal to 9, without considering the substrate because the PQQ factor acts as a bidentate ligand. The nitrogen atom (N6) lies at 2.999 Å from the cerium in the Ce-MDH and 2.946 Å from the europium in Eu-MDH with respect to the shorter distances of M^{3+}-$O5_{PQQ}$ and M^{3+}-$O7_{PQQ}$ (see Fig. 5 and Table 1). Figure 5 clearly shows that

Table 1 Calculated coordination distances (Å) in Ca-, Ce-, and Eu-MDH in ES complexes. For comparison, the corresponding crystallographic values (Å) are reported in parenthesis

M^{n+}	$O1_{Glu172}$	$O2_{Glu172}$	O_{Asn256}	$O5_{PQQ}$	$N6_{PQQ}$	$O7_{PQQ}$	$O1_{Asp299}$	O_{Sub}	$O1_{Asp301}$	$O2_{Asp301}$
Ca^{2+}	2.426 (2.709)	2.500 (2.704)	2.414 (2.861)	2.758 (2.411)	2.787 (2.518)	2.677 (2.729)	2.379 (3.336)	2.513	–	–
Ce^{3+}	2.550 (2.667)	2.581 (2.917)	2.488 (2.715)	2.829 (2.552)	2.999 (2.800)	2.788 (2.697)	2.416 (2.857)	2.535	2.551 (2.780)	2.550 (2.469)
Eu^{3+}	2.431 (2.778)	2.478 (3.110)	2.421 (2.831)	2.740 (2.634)	2.946 (2.834)	2.74 (2.560)	2.387 (2.983)	2.454	2.431 (2.839)	2.478 (2.548)

the variation of the bond lengths in the Ce-MDH and Eu-MDH reflects the expected smooth contraction with increasing atomic number (effect of the periodicity).

A similar trend is observed even if the crystallographic distances are examined (Table 1) although they are in all the cases more elongated. In the same table are reported the coordination bonds values of the Ca-MDH that are close to those of Eu-MDH, except for the distance involving the nitrogen atom (N6) of PQQ.

In our ES optimized structure of Eu-MDH, the metal ion presents Glu172 and Asp301 in bicoordinated fashion differently from that reported in the study of Pol et al. [19]. In fact, in this work the Asp301 residue acts as unidentate ligand owing to the O1–Eu $^{3+}$ (3.842 Å) and O2-Eu^{3+} (2.359 Å) bond distances. In general, all the other coordination distances appear shorter than our ones. This probably reflects the chosen computational procedure that considers few valence electrons for the lanthanide ions [19].

The NBO analysis for the ES complexes, depicted in Fig. 6, demonstrates again a periodic trend giving in general more negative values for the cerium-containing enzyme except for some cases. A more negative charge on the substrate oxygen for the Ce-MDH with respect to the Eu-MDH is obtained. This is important for the catalytic activity because the more polarized O–H-bond of the substrate facilitates the H-transfer and the nucleophilic addition on C5 of PQQ that occurs in the first step of the reaction. In fact, the charge value on O5 atom of the PQQ cofactor retains almost the same amount in both lanthanides such as the value on C5 (0.53 |e|) not reported in the figure. This is a confirmation that the activation of the OH moiety of the methanol, and not of the PQQ cofactor, is subjected to the metal ion Lewis acidity and is crucial for the first step of the catalysis.

A further reasonable basis for evaluating the different behavior of the two Ce- and Eu-dependent methanol dehydrogenases can arise from MO energy diagram depicted in Fig. 7. Here, it is possible to observe that the energetic gap separating the HOMO from LUMO in the ES complex of Ce-MDH is smaller by 0.84 eV than that of the same orbitals in the Eu-MDH. This finding corroborates the better nature of Lewis acid of Ce^{3+} since the HOMO location on the cerium (4f orbital). In both enzymes, the LUMO lies on the PQQ cofactor that must suffer $2e^-$ reduction and is almost isoenergetic.

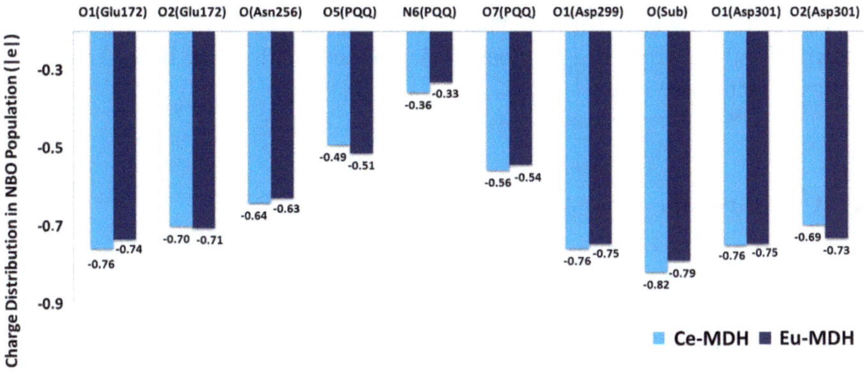

Fig. 6 NBO charges (|e|) of the coordinated atoms for ES complex of Ce-MDH and Eu-MDH

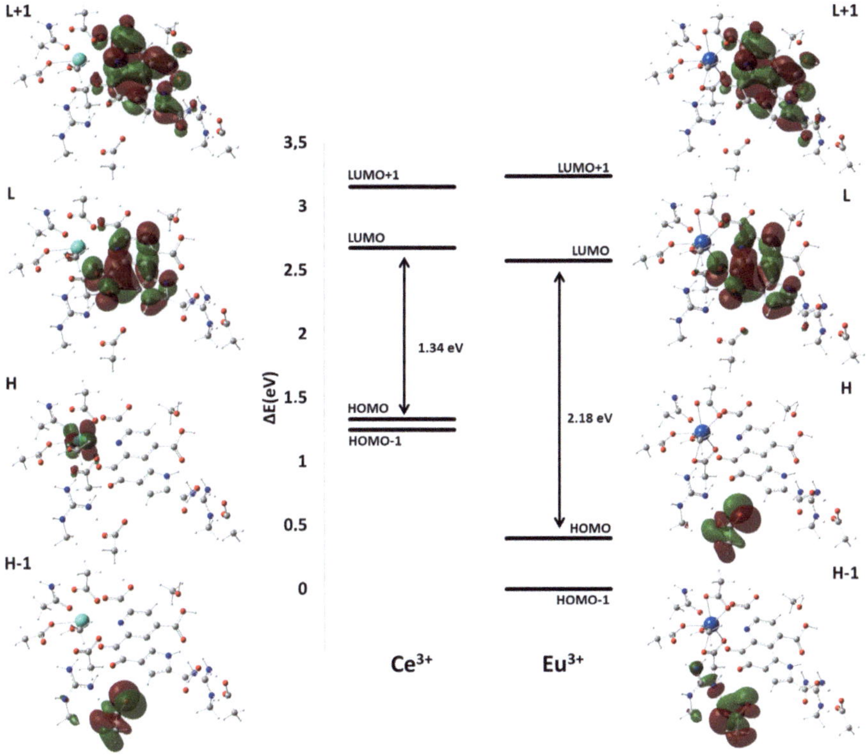

Fig. 7 Energy diagram of frontier molecular orbitals of the enzyme–substrate complexes (ES) for Ce-MDH and Eu-MDH

This behavior is different from that observed in the LUMO comparison between Ca-MDH and Ce-MDH in a previous work [55] and confirms the better Lewis acid nature of lanthanide ions.

4 Conclusions

In the present work, we have performed a theoretical investigation at QM level and using the hybrid B3LYP functional with the dispersion contribution, for giving details about the different coordination chemistry of the two considered lanthanide ions (Ce^{3+} and Eu^{3+}) in their ES complexes. The catalytic behavior of Ce-MDH as compared to Ca-MDH has been also examined. The emerged differences and similarities have been correlated to the periodic properties as atomic radius and relativistic effects (in the case of lanthanide ions).

In particular, results show that:

- PESs calculated for Ce-MDH by taking into account three diverse dielectric constant values do not suffer evident changes.
- Metal ion can play a crucial role in the activation of the methanol substrate (first step of the investigated catalytic mechanism) more than in the PQQ cofactor reduction.
- Cerium ion is a better Lewis acid than calcium and europium ones.
- Cerium more than europium acts as a better biomimetic agent for calcium-containing methanol dehydrogenase.
- Frontier orbital and NBO analysis of the enzyme–substrate complex for Ce- and Eu-MDH support the better catalytic activity of the cerium-containing enzyme.

References

1. Good NM, Vu HN, Suriano CJ, Subuyuj GA, Skovran E, Martinez-Gomez NC (2016) Pyrroloquinoline quinone ethanol dehydrogenase in Methylobacterium extorquens AM1 extends lanthanide-dependent metabolism to multicarbon substrates. J Bacteriol 198:3109–3118. https://doi.org/10.1128/JB.00478-16
2. Panichev AM (2015) Rare Earth Elements: Review of medical and biological properties and their abundance in the rock materials and mineralized spring waters in the context of animal and human geophagia reason evaluation Achiev. Life Sci 9:95–103. https://doi.org/10.1016/j.als.2015.12.001
3. Brittain HG, Richardson FS, Martin RB (1976) Terbium(III) emission as a probe of calcium(II) binding sites in proteins. J Am Chem Soc 98:8255–8260. https://doi.org/10.1021/ja00441a060
4. Martin RB, Richardson F (1979) Lanthanides as probes for calcium in biological systems. Q Rev Biophys 12:181–209. https://doi.org/10.1017/S0033583500002754
5. Chistoserdova L (1996) Metabolism of formaldehyde in M. extorquens AM1. Molecular genetic analysis and mutant characterization. In: Lidstrom ME, Tabita FR (eds) Microbial growth on C1 compounds. Kluwer Academic Publishers, Dordrecht, pp 16–24
6. Harms N, Ras J, Koning S, Reijinders WNM, Stouthamer AH, van Spanning RJM (1996) Genetics of C1 metabolism regulation in Paracoccus denitrificans. In: Lidstrom ME, Tabita

FR (eds) Microbial growth on C1 compounds. Kluwer Academic Publishers, Dordrecht, pp 126–132
7. Hibi Y, Asai K, Arafuka H, Hamajima M, Iwama T, Kawai K (2011) Molecular structure of La^{3+}-induced methanol dehydrogenase-like protein in Methylobacterium radiotolerans. J Biosci Bioeng 111:547–549. https://doi.org/10.1016/j.jbiosc.2010.12.017
8. Chistoserdova L (2016) Lanthanides: new life metals? World J Microbiol Biotechnol 32:138. https://doi.org/10.1007/s11274-016-2088-2
9. Anthony C (2000) Methanol dehydrogenase, a PQQ-containing quinoprotein dehydrogenase. In: Holzenburg A, Scrutton NS (eds) Enzyme-catalyzed electron and radical transfer, vol 35. Springer, New York
10. Anthony C (2004) The quinoprotein dehydrogenases for methanol and glucose. Arch Biochem Biophys 428:2–9. https://doi.org/10.1016/j.abb.2004.03.038
11. Goodwin PM, Anthony C (1998) The biochemistry, physiology and genetics of PQQ and PQQ-containing enzymes. Adv Microb Physiol 40:1–80. https://doi.org/10.1016/S0065-2911(08)60129-0
12. Cotton S (2006) Lanthanide and actinide chemistry. Wiley, Chichester; Evans CH (1990) Biochemistry of the lanthanides. Springer, New York
13. Shiller AM, Chan EW, Joung DJ, Redmond MC, Kessler JD (2017) Light rare earth element depletion during Deepwater Horizon blowout methanotrophy. Sci Rep 7:1–9. https://doi.org/10.1038/s41598-017-11060-z
14. Vu HN, Subuyuj GA, Vijayakumar S, Good NM, Martinez-Gomez NC, Skovran E (2016) Lanthanide-dependent regulation of methanol oxidation systems in Methylobacterium extorquensAM1 and their contribution to methanol growth. J Bacteriol 198:1250–1259. https://doi.org/10.1128/JB.00937-15
15. Ochsner AM, Sonntag F, Buchhaupt M, Schrader J, Vorholt JA (2015) Methylobacterium extorquens: methylotrophy and biotechnological applications. Appl Microbiol Biotechnol 99:517–534. https://doi.org/10.1007/s00253-014-6240-3
16. Semrau JD, Di Spirito AA, Gu W, Yoon S (2018) Metals and methanotrophy. Appl Environ Microbiol 84:e02289–17. https://doi.org/10.1128/AEM.02289-17
17. Pol A, Barends TRM, Dietl A, Khadem AF, Eygensteyn J, Jetten MSM, Op den Camp HJM (2014) Rare earth metals are essential for methanotrophic life in volcanic mudpots. Environ Microbiol 16:255–264. https://doi.org/10.1111/1462-2920.12249
18. Afolabi PR, Mohammed F, Amaratunga K, Majekodunmi O, Dales L, Gill R, Thompson D, Cooper B, Wood P, Goodwin M, Anthony C (2001) Site-directed mutagenesis and X-ray crystallography of the PQQ-containing quinoprotein methanol dehydrogenase and its electron acceptor, cytochrome cL. Biochemistry 40:9799–9809. https://doi.org/10.1021/bi002932l
19. Jahn B, Pol A, Lumpe H, Barends TRM, Dietl A, Hogendoorn C, Op den Camp HJM, Daumann LJ (2018) Similar but not the same: first kinetic and structural analyses of a methanol dehydrogenase containing a europium ion in the active site. ChemBioChem 19:1147–1153. https://doi.org/10.1002/cbic.201800130
20. Zheng YJ, Bruice TC (1997) Conformation of coenzyme pyrroloquinoline quinone and role of Ca^{2+} in the catalytic mechanism of quinoprotein methanol dehydrogenase. PNAS 94:11881–11886. https://doi.org/10.1073/pnas.94.22.11881
21. Zheng YJ, Xia Z, Chen Z, Mathews FS, Bruice TC (2001) Catalytic mechanism of quinoprotein methanol dehydrogenase: a theoretical and X-ray crystallographic investigation. PNAS 98:432–434. https://doi.org/10.1073/pnas.98.2.432
22. Idupulapati NB, Mainardi DS (2010) Quantum chemical modeling of methanol oxidation mechanisms by methanol dehydrogenase enzyme: effect of substitution of calcium by barium in the active site. J Phys Chem A 114:1887–1896. https://doi.org/10.1021/jp9083025
23. Leopoldini M, Russo N, Toscano M (2007) The preferred reaction path for the oxidation of methanol by PQQ-containing methanol dehydrogenase: addition-elimination versus hydride-transfer mechanism. Chem Eur J 13:2109–2117. https://doi.org/10.1002/chem.200601123
24. Hothi P, Sutcliffe MJ, Scrutton NS (2005) Kinetic isotope effects and ligand binding in PQQ-dependent methanol dehydrogenase. Biochem J 388:123–133. https://doi.org/10.1042/BJ20041731

25. Zheng X, Reddy SY, Bruice TC (2007) Mechanism of methanol oxidation by quinoprotein methanol dehydrogenase. PNAS 104:745–749. https://doi.org/10.1073/pnas.0610126104
26. Reddy SY, Bruice TC (2004) Determination of enzyme mechanisms by molecular dynamics: studies on quinoproteins, methanol dehydrogenase, and soluble glucose dehydrogenase. Protein Sci 13:1965–1978 (2004). https://doi.org/10.1110/ps.04673404
27. Idupulapati NB, Mainardi DS (2009) THEOCHEM 901:72–80. https://doi.org/10.1016/j.theochem.2009.01.004
28. McSkimming A, Cheisson T, Carroll PJ, Schelter EJ (2018) Functional synthetic model for the lanthanide-dependent quinoid alcohol dehydrogenase active site. J Am Chem Soc 140:1223–1226. https://doi.org/10.1021/jacs.7b12318
29. Nakagawa T, Mitsui R, Tani A, Sasa K, Tashiro S, Iwama T, Hayakawa T, Kawai K (2012) A catalytic role of XoxF1 as La3+-dependent methanol dehydrogenase in Methylobacterium extorquens strain AM1. PLoS ONE 7:e50480. https://doi.org/10.1371/journal.pone.0050480
30. Chu F, Beck DAC, Lidstrom ME (2016) MxaY regulates the lanthanide-mediated methanol dehydrogenase switch in Methylomicrobium buryatense. PeerJ 4:e2435. https://doi.org/10.7717/peerj.2435
31. Skovran E, Martinez-Gomez NC (2015) Just add lanthanides. Science 348:862–863. https://doi.org/10.1126/science.aaa9091
32. Blomberg RMA, Borowski T, Himo F, Liao R-Z, Siegbahn PEM (2014) Quantum chemical studies of mechanisms for metalloenzymes. Chem Rev 114:3601–3658, and references therein. https://doi.org/10.1021/cr400388t
33. Piazzetta P, Marino T, Russo N (2014) Promiscuous ability of human carbonic anhydrase: QM and QM/MM Investigation of carbon dioxide and carbodiimide hydration. Inorg Chem 53:3488–3493. https://doi.org/10.1021/ic402932y
34. Amata O, Marino T, Russo N (2011) Catalytic activity of a ζ-class zinc and cadmium containing carbonic anhydrase. Compared work mechanisms. Phys Chem Chem Phys 13:3468–3477. https://doi.org/10.1039/C0CP01053G
35. Liao R-Z, Yu G, Himo F (2010) Mechanism of tungsten-dependent acetylene hydratase from quantum chemical calculations. PNAS 107:22523–22527. https://doi.org/10.1073/pnas.1014060108
36. Ramos MJ, Fernandes PA (2008) Computational enzymatic catalysis. Acc Chem Res 41:689–698. https://doi.org/10.1021/ar7001045
37. Marino T, Russo N, Toscano M (2012) Catalytic mechanism of the arylsulfatase promiscuous enzyme from pseudomonas aeruginosa. Chem Eur J 19:2185–2192. https://doi.org/10.1002/chem.201201943
38. Gaussian 09, Revision D.01, Frisch MJ, Trucks GW, Schlegel HB, Scuseria GE, Robb MA, Cheeseman JR, Scalmani G, Barone V, Mennucci B, Petersson GA, Nakatsuji H, Caricato M, Li X, Hratchian HP, Izmaylov AF, Bloino J, Zheng G, Sonnenberg JL, Hada M, Ehara M, Toyota K, Fukuda R, Hasegawa J, Ishida M, Nakajima T, Honda Y, Kitao O, Nakai H, Vreven T, Montgomery Jr JA, Peralta JE, Ogliaro F, Bearpark M, Heyd JJ, Brothers E, Kudin KN, Staroverov VN, Keith T, Kobayashi R, Normand J, Raghavachari K, Rendell A, Burant JC, Iyengar SS, Tomasi J, Cossi M, Rega N, Millam JM, Klene M, Knox JE, Cross JB, Bakken V, Adamo C, Jaramillo J, Gomperts R, Stratmann RE, Yazyev O, Austin AJ, R. Cammi R, Pomelli C, Ochterski JW, Martin RL, Morokuma K, Zakrzewski VG, Voth GA, Salvador P, Dannenberg JJ, Dapprich S, Daniels AD, Farkas O, Foresman JB, Ortiz JV, Cioslowski J, Fox DJ (2013) Gaussian, Inc., Wallingford, CT
39. Lee C, Yang W, Parr RG (1988) Development of the Colle-Salvetti correlation-energy formula into a functional of the electron density. Phys Rev B 37:785–789. https://doi.org/10.1103/PhysRevB.37.785
40. Becke AD (1993) Density-functional thermochemistry. III. The role of exact exchange. J Chem Phys 98:5648–5652. https://doi.org/10.1063/1.464913
41. Andrae D, Haußermann H, Dolg M, Stoll H, Preuß H (1990) Energy-adjustedab initio pseudopotentials for the second and third row transition elements. Theor Chim Acta 77:123–141. https://doi.org/10.1007/BF01114537

42. Grimme S, Ehrlich S, Goerigk L (2011) Effect of the damping function in dispersion corrected density functional theory. J Comput Chem 32:1456–1465. https://doi.org/10.1002/jcc.21759
43. Cossi M, Rega N, Scalamani V, Barone V (2003) Energies, structures, and electronic properties of molecules in solution with the C-PCM solvation model. J Comput Chem 24:669–681. https://doi.org/10.1002/chem.201700381
44. Barone V, Cossi M (1998) Quantum calculation of molecular energies and energy gradients in solution by a conductor solvent model. J Phys Chem A 102:1995–2001. https://doi.org/10.1021/jp9716997
45. Chen W, Fang WH, Himo F (2009) Reaction mechanism of the binuclear zinc enzyme glyoxalase II—a theoretical study. J Inorg Biochem 103:274–281. https://doi.org/10.1016/j.jinorgbio.2008.10.016
46. Ziao RZ, Himo F Yu JG, Liu RZ (2010) Dipeptide hydrolysis by the dinuclear zinc enzyme human renal dipeptidase: mechanistic insights from DFT calculations. J Inorg Biochem 104:37–46. https://doi.org/10.1016/j.jinorgbio.2009.09.025
47. Sousa SF, Fernandes PA, Ramos MJ (2009) The search for the mechanism of the reaction catalyzed by farnesyltransferase. Chem Eur J 15:4243–4247. https://doi.org/10.1002/chem.200802745
48. Alberto ME, Marino T, Ramos MJ, Russo N (2010) Atomistic details of the catalytic mechanism of Fe(III)–Zn(II) purple acid phosphatase. J Chem Theory Comput 6:2424–2433. https://doi.org/10.1021/ct100187c
49. Siegbahn PEM, Eriksson LA, Himo F, Pavlov M (1998) Hydrogen atom transfer in ribonucleotide reductase (RNR). J Phys Chem B 102:10622–10629. https://doi.org/10.1021/jp9827835
50. Warshel A (1991) Computer modeling of chemical reactions in enzymes and solutions. Wiley, New York
51. Piazzetta P, Marino T, Russo N (2014) Insight into the promiscuous activity of human carbonic anhydrase against the cyanic acid substrate from a combined QM and QM/MM investigation. Phys Chem Chem Phys 16:16671–16676. https://doi.org/10.1039/C4CP02363C
52. Ribeiro AJM, Alberto ME, Ramos MJ, Fernandes PA, Russo N (2013) The catalytic mechanism of protein phosphatase 5 established by DFT calculations. Chem Eur J 114:3601–3658. https://doi.org/10.1002/chem.201301565
53. Prejanò M, Marino T, Russo N (2017) How can methanol dehydrogenase from Methylacidiphilum fumariolicum work with the alien CeIII ion in the active center? A theoretical study. Eur Chem J 23:8652–8657. https://doi.org/10.1002/chem.201700381
54. Prejanò M, Marino T, Rizzuto C, Madrid JCM, Russo N, Toscano M (2017) Reaction mechanism of low-spin iron(III)- and cobalt(III)-containing nitrile hydratases: a quantum mechanics investigation. Inorg Chem 56:13390–13400. https://doi.org/10.1021/acs.inorgchem.7b02121
55. Bogart JA, Lewis AJ, Schelter EJ (2015) DFT study of the active site of the XoxF-type natural, cerium-dependent methanol dehydrogenase enzyme. Chem Eur J 21:1743–1748. https://doi.org/10.1002/chem.201405159
56. E. D. Glendening, A. E. Reed, J. E. Carpenter, F. Weinhold, NBO, version 3.1
57. Anthony C (1996) Quinoprotein-catalysed reactions. Biochem J 320:697–711. https://doi.org/10.1042/bj3200697
58. Simon S, Duran M, Dannenberg JJ (1996) How does basis set superposition error change the potential surfaces for hydrogen-bonded dimers? J Phys Chem 105:11024–11031. https://doi.org/10.1063/1.472902
59. Boys SF, Bernardi F (1970) The calculation of small molecular interactions by the differences of separate total energies. Some procedures with reduced errors. Mol Phys 19:553–566. https://doi.org/10.1080/00268977000101561

Challenges in Modelling Metalloenzymes

Tomasz Borowski and Maciej Szaleniec

Abstract Modelling structure and/or reaction mechanism of a metalloenzyme requires constructing reliable computational models. This process usually starts with crystal structure and involves several steps, each potentially influencing the accuracy of the constructed model. In this chapter, we provide an account, from our own works and from the literature, on how one can check the quality of the crystal structure, predict protonation states of titratable residues, including the metal ligands, perform molecular dynamics simulations, chose representative snapshots, and finally, construct a QM cluster model that can be used for mechanistic studies.

1 Introduction

Enzymes catalyse a wide range of reactions making life possible in a variety of environmental conditions. A great majority of enzymes are proteins built from L-α amino acids; however, the range of chemical tools used by nature to assemble efficient catalysts goes significantly beyond residues of amino acids and includes small organic molecules as well as metal ions or metal-containing organic complexes in the role of cofactors or co-enzymes. The latter group of enzymes is referred to as metalloenzymes, which may be one of the first groups of biological catalysts that evolved on Earth, especially in the context of bioenergetic systems [1]. Now metalloenzymes are involved in many biological processes such as nitrogen fixation, respiration, hydrocarbon oxidation, generation of radicals for subsequent catalysis and dismutation of superoxide or oxygenic photosynthesis [2].

The bioinformatic analysis of available protein databases estimates that 30–40% of all proteins require metal ion(s) to carry out their biological functions [3, 4]. These

T. Borowski (✉) · M. Szaleniec (✉)
Jerzy Haber Institute of Catalysis and Surface Chemistry, Polish Academy of Sciences, Niezapominajek 8, 30-239 Kraków, Poland
e-mail: ncborows@cyf-kr.edu.pl

M. Szaleniec
e-mail: ncszalen@cyf-kr.edu.pl

E. Broclawik et al. (eds.), *Transition Metals in Coordination Environments*, Challenges and Advances in Computational Chemistry and Physics 29, https://doi.org/10.1007/978-3-030-11714-6_17

statistics also hold for enzymes, and as much as ca. 40% of the structurally characterized enzymes turn out to be metal-dependent. Furthermore, metalloenzymes can be found among 76% of the EC-based subclasses that were structurally characterized [4]. These facts underline high catalytic versatility and widespread occurrence of metalloenzymes in various forms of life.

Metalloenzymes are often the most efficient catalysts in terms of their activity, selectivity and ability to operate at mild conditions. Additionally, due to the active participation of metal ions in the catalytic process, the scope of the available molecular transformations is greatly enhanced. Furthermore, it is also possible that presence of the metal ions inside the enzymes' active sites significantly contributes to their overall stability and thus increases their operational robustness [5].

The majority of enzymatic reactions tend to be multistep processes proceeding through enzyme-bound intermediates, some of which are very short-lived, and hence rather difficult to characterize directly. Even though in some cases it is possible to trap and structurally characterize such intermediates *in crystallo*, e.g. peracid intermediate in 2-oxoglutarate-dependent dioxygenases [6] or peroxo-bridged intermediate in extradiol dioxygenases [7], in most cases only some indirect experimental evidence, e.g. from spectroscopy or kinetic isotope effect measurements, is available from which one can only infer geometric and electronic structure of the key species. Hence, computational methods that can aid in interpretation of the available indirect data concerning some intermediates and fill in the gaps in the reaction mechanism by describing most likely structures of the remaining intermediates and transition structures are very helpful. For examples on how spectroscopy and computational methods can be efficiently used in tandem for metal containing systems see Chapters "Anisotropic Magnetic Spin Interactions of Transition Metal Complexes and Metalloenzymes from Spectroscopy and Quantum Chemistry", "Computational Versus Experimental Spectroscopy for Transition-Metals" and "Assessing Electronically Excited States of Cobalamins via Absorption Spectroscopy and Time-Dependent Density Functional Theory" in this volume. Chapters "The Quest for Accurate Theoretical Models of Metalloenzymes: An Aid to Experiment", "Applications of Computational Chemistry to Selected Problems of Transition-Metal Catalysis in Biological and Non-biological Systems" and "How Metal Coordination in the Ca-, Ce- and Eu-Containing Methanol Dehydrogenase Enzymes Can Influence the Catalysis: A Theoretical Point of View" provide a good number of examples of computational studies on reaction mechanism of various metalloenzymes. Hence, instead of providing further examples of this kind, in this chapter we decided to limit their number to necessary minimum and instead put more emphasis on: (a) various auxiliary methods that are useful when preparing computational model of metalloenzymes and (b) illustrate some challenges one may face when trying to investigate reaction mechanism of redox-active metalloenzymes.

2 Checking the Structure Quality

Computational studies on metalloenzymes depend heavily on available experimental macromolecular structures, and in most cases, one starts his/her project by downloading a suitable structure from the Protein Data Bank (PDB). Irrespective of if the structure available at PDB is for the actual enzyme of interests or it will be used as a template in homology modelling, it is highly desirable that it is complete and accurate, especially in the vicinity of the active site hosting the metal(s) ion(s). The overall quality of the structure can be inferred from various metrics reported at the PDB web page, such as resolution, R-value free, R-value work, clashscore, Ramachandran outliers and other, yet these global parameters take into account thousands of atoms of the structure, of which only a few are metal ions; hence, their informative value as to the quality of the structure in the immediate vicinity of the metal is rather limited. This is compounded by the fact that force fields commonly used for structure refinement are more accurate for standard amino acids and organic molecules than for inorganic complexes. Thus, it is advisable to invest some time and check the quality of the metal-binding site before undertaking some more demanding modelling tasks. *CheckMyMetal* (CMM, https://csgid.org/metal_sites) is an easy to use metal-binding site validation server that can be readily used for this purpose [8, 9]. For a given macromolecular structure, specified either by its PDB four-letter code or uploaded in the PDB format, CMM computes seven quality parameters derived solely from the geometry of the first coordination shell and two metrics related to the experimental model. Computed values are compared to reference ranges (Acceptable, Borderline and Outlier) that were derived based on benchmark data of 7350 metal-binding sites from 2304 high-resolution crystal structures.

As an illustrative example of using CMM may serve an analysis of a metal-binding site in the structure of α-galactosidase (PDB: 3LRK), even though in this case the metal is most likely not involved in the enzymatic reaction [8, 10]. 3LRK has been solved with a good resolution (1.95 Å), and it also shows rather good values for other metrics reported by the PDB server. In the original structure, one Na^+ cation was modelled to be coordinated by His98, His102, one nitrogen and four oxygen atoms from a bis-tris buffer molecule. Analysis of this site by CMM shows that three parameters (*Ligands*, *Valence* and *Geometry*) fall into the Outlier range, whereas other two (*B factor* and *nVECSUM*) are classified as Borderline. Further inspection of electron density difference maps revealed that sodium cation cannot account for the observed electron density peak, and the site was re-refined with Cu^+ instead of Na^+ and the bis-tris molecule replaced by two glycerol and two water molecules, all four with partial occupancy of 0.5 (Fig. 1).

As can be noted in the figure, the curated model successfully accounts for the observed electron density and in the CMM analysis performed for this model all quality metrics fall either into Acceptable or Borderline ranges. Detailed description of the CMM protocol, its limitations (e.g. the valence model does not cover metal–metal, delocalized and π metal–ligand bonds) and further examples of use can be found in the original CMM papers [8, 9].

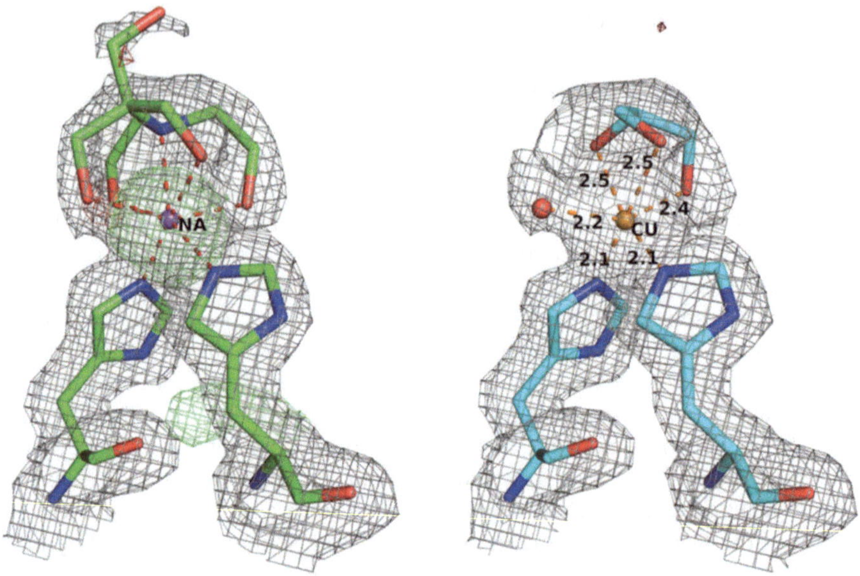

Fig. 1 Example of curating the metal-binding site by CMM. Metal-binding site before re-refinement is shown with green carbon atoms and after re-refinement with carbon atoms coloured cyan. Electron density maps (2Fo-Fc) are shown in grey mesh with 1.0σ cut-off, with difference maps (Fo-Fc) shown in green and red mesh with 4.0σ cut-off. In the re-refined structure, only one of the two alternative glycerol molecules is shown for clarity. Reprinted with permission from [8]. Copyright 2013 Springer Nature

3 Quantum Refinement

As already mentioned above, general and accurate force fields for metal–ligand interactions are not available, yet they are highly desirable during crystal structure refinement of metalloproteins. The reason is that in majority of cases, the exception being very high-resolution data, electron density maps are not accurate enough to allow precise localization of atomic positions and chemical knowledge needs to be used as a restraint when building a structural model. However, metal–ligand interactions are hard to cast in a relatively simple analytical form of classical force field, as the metal–ligand bonds are relatively soft and easily influenced by the presence of other ligands coordinated to the metal ion, e.g. trans effects, changes of spin state on the central metal ion. As a solution to this problem, Ryde and co-workers proposed to use during refinement forces computed with quantum chemistry (DFT) methods for the metal ion(s) and its immediate surroundings within a protein [11]. Just like in other QM/MM methods, the part of the system beyond the quantum region is described with the classical force field used in standard refinement. Such re-refinement focused on the inorganic part of the metalloprotein on the one hand improves the structure around the metal, as shown by comparing re-refined 1.7 Å resolution structure of cytochrome c_{553} to a high-resolution (0.97 Å) one and, on the other, it also allows to discriminate between various possible protonation states on ligands bound to the metal [11–13]. Moreover, in favourable cases, the oxidation state of the metal ion can be determined; however, for higher oxidation states such analysis can be compounded by partial reduction by photoelectrons released during data collection [14]. Knowledge of the oxidation state of the metal and protonation state of its ligands is critical for reliable modelling of metalloenzymes and their catalytic reactions, hence when they are uncertain it may be worthy investing some time into performing quantum re-refinement, as illustrated by a recent work by Ryde and co-workers on nitrogenase [13].

4 Homology Modelling

When experimental structures are not available for a particular enzyme of interest, yet they are known for homologues, then homology modelling may turn out to be very useful. The field of comparative structure modelling is a mature one with many stable software packages and servers, e.g. MODELLER, I-TASSER, ROSETTA or SWISS-MODEL, to name just a few [15–18]. Obviously, the quality of the model will depend heavily on the availability of structures with high sequence identity to the target protein. When the sequence identity between the template and the target is above 60%, one may expect a high-quality model with RMSD to the true structure on the order of 1 Å [16]. Such models should be accurate enough to meaningfully use them in studies on ligand binding or reaction mechanisms. However, when the sequence identity is on the order of 30%, typical RMSD will increase above 3 Å and the model may have serious errors prohibiting accurate modelling at the atomic level. Here as an example may serve comparison of structures of thebaine-6-O-demethylase (T6ODM), a mononuclear non-heme iron enzyme using 2-oxoglutarate as a co-substrate. A homology model of T6ODM was constructed with the use of the I-TASSER prior to the public release of the actual X-ray crystal structure of T6ODM (1.85 Å resolution, PDB: 5O9W) [19]. The I-TASSER server identified and used for modelling several homologous structures, the one with the largest sequence identity of 31% was for anthocyanidin synthase (ANS; PDB:1GP6). Figure 2 shows superimposed structures of the best model generated by the I-TASSER server and the crystal structure. RMSD between the two structures calculated for protein backbone is 3.2 Å, whereas for all heavy atoms it amounts to 3.9 Å, which falls into the 4.4 $\pm$ 2.9 Å range predicted by I-TASSER. As can be noted from the figure, the jelly-roll core of the protein superimposes very neatly, whereas the least precisely modelled fragments are those at the entrance to the active site, which are also responsible for substrate recognition.

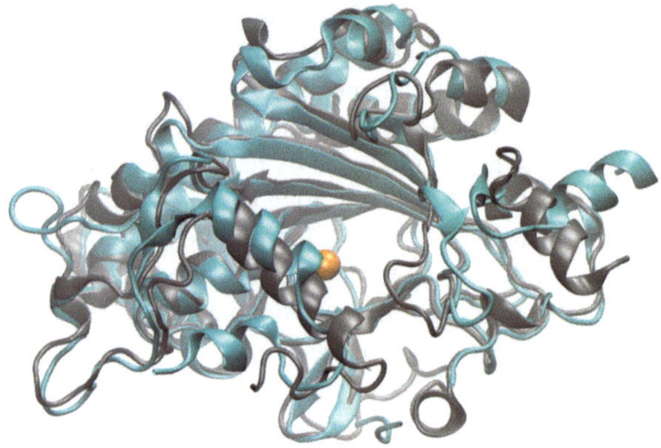

Fig. 2 Superimposed structures of T6ODM. Grey—homology model generated by I-TASSER server, cyan—X-ray crystal structure (PDB: 5O9W). Metal bound in the active site marked as an orange sphere

5 Modelling Loops

Loop regions are enzymes' variable fragments that are often responsible for substrate and/or reaction specificity. However, it is not infrequent that mobile loop regions are not visible in electron density maps derived from XRD data and in consequence such fragments are not present in corresponding crystal structures. Likewise, when comparative modelling based on sequence alignment is done for homologous enzymes, the variable loop regions are often the regions with no sequence overlap, and hence impossible to model based on sequence homology. In such cases, methods of loop structure prediction are very valuable. There are several of such, e.g. loop modelling routine available in MODELLER, which is based on minimization of a pseudo-energy function, [20] fragment assembly method [21] or the DISGRO method using chain growth sequential Monte Carlo sampling strategy and which is particularly effective for longer (10–17) loops [22]. As an example of application of loop modelling techniques to a metalloenzyme may serve the work of Widderich et al., which is devoted to ectoine hydroxylase (EctD), an enzyme from a broad superfamily of α-ketoglutarate-dependent dioxygenases [23]. In the crystal structure of EctD (PDB: 3EMR), there are no coordinates for residues Gly195–Leu211, which form a loop presumably acting as a lid for an entrance to the active site [24]. This is a 17-residue long loop and as such pose a challenge for loop modelling techniques, the authors first tried to build this fragment by homology modelling. However, due to a lack of suitable templates, this procedure failed and they resorted to ab initio loop modelling method implemented in MODELLER. A total of 25 loop models were first generated (this is a relatively small number, probably 500 would allow for substantially better sampling of the conformation space [20]), and then each model was subjected to a thorough structural analysis done with three different methods: Anolea, Dope and PROCHECK [25–27]. Global scores that reflect the structure quality of a given model and which were obtained with the three methods were combined to form a "Composite Ensemble Score"; the model with the lowest value of this score was selected for MD simulations. In the obtained trajectories, the loop did not close the entrance to the active site, nor made direct contacts with the substrate—ectoine. Later on, several other crystal structures were published for EctD (PDB: 4NMI, [28] 4MHR, 4MHU and 4Q5O [23]), yet in neither of them the loop region is resolved. However, crystal structure obtained for EctD - 5-hydroxyectoine complex confirmed the previous modelling study successfully identified the substrate-binding site of EctD.

6 Prediction of pK_a

One of the important steps in preparation of the model of the protein is determination of its protonation state at the selected pH. For modelling of enzymes, the latter will usually be optimal pH for the catalytic reaction. The titration curve of an amino acid that is part of a protein can significantly shift with respect to that known from water solution. This happens due to: (i) desolvation of the residues when transferred from the aqueous solution to less polar interior of the protein, which generally destabilizes the charged state and (ii) electrostatic interactions inside the protein which may either stabilize or destabilize the charged state of a particular residue [29]. It should be underlined that with the change of pH protonation state of the protein residues may change and this in turn affects electrostatic environment of the residue under consideration. As a result, one has to know pH-dependent titration curves of all residues within the protein to determine their probability of protonation. The ionization state of the protein has a crucial role not only for its stability or strength of protein–protein interactions but also for ligand binding and catalysis, which is arguably most important from the enzymology perspective.

Prediction of pK_a values for protein residues and hence their protonation state at a given pH is a necessary step before any further modelling can be conducted, be it MM, MD or QM. Indeed, even in the case when the initial geometry is derived from the crystal structure, the ambiguity of the electron density, especially in the vicinity of the heavier metal sites, makes initial geometry optimization, e.g. by means of MD or QM/MM/MD methods, a necessary step. Hence, the protonation state of all ionisable residues should be determined as accurately as possible.

Although the protonation state of individual residues can be assessed experimentally in solution with the help of NMR, [30] IR or in crystal with neutron diffraction, [31] for computational methods, it is still a challenging task. This is due to the fact that even small differences in free energy of protonation give rise to significant pK_a shifts (1.364 kcal/mol of energy at 298.15 K corresponds to a change of pK_a by a unit) [32].

The majority of computational methods developed for this purpose in the last 20 years concentrate on predicting pK_a shifts of ionisable groups from the known reference state (such as pK_a of isolated amino acid in water or Ala-Ala-X-Ala-Ala pentapeptides) [33]. Presumably, the most accurate and reliable methods to approach this problem are those based on QM or QM/MM computational models. However, they are still too demanding computationally for routine treatment of whole proteins and are more suited for studies of extreme pK_a shifts of individual residues [34]. For the practical purposes, however, fast methods that utilize force field and Poisson–Boltzmann or generalized Born approach, or even empirical methods, are commonly used. It is not our goal to discuss here details of different approaches, but rather we would like to point out some potential pitfalls and dangers associated with use of such tools. For a comprehensive comparison of different methods, see reviews by Awoonor-Williams et al. and Nielsen et al. [35, 36].

As a measure of accuracy of pK_a prediction methods, we can use the isoelectric point (pI), which can be both readily measured, e.g. by means of DLS (Dynamic Light Scattering) [37] or isoelectrofocusing [38], as well as calculated by pK_a-predicting programs. Such comparison seems to be a valid assessment tool as the computational methods are parameterized based on the known data set of pK_a shifts of selected residues (Asp or Glu) and very seldom pI values of proteins are used in parametrization. In our small test, we have used two computational methods to calculate pK_a and pI values: a very popular empirical method of Propka3.1 [39] and a CHARMM-based GB/IMC (Generalized Born/iterative mobile clustering) protocol implemented in Discovery Studio [40]. These have been applied to four proteins for which structures are either known from X-ray diffraction studies (bovine α-chymotrypsin (PDB: 5CHA), human serum albumin HSA (PDB: 1AO6), fibrynogen (PDB: 1M1J)), a homology model of 3-ketosteroid dehydrogenase (KTSD) from *Sterolibacteria denitrificans*, also known as AcmB, as well as two protein models developed using molecular modelling techniques in conjunction with global secondary structural information derived from Raman spectroscopy data: α-casein and β-casein [41, 42]. As can be seen from the data gathered in Table 1, for most of the tested proteins the prediction methods perform relatively well (error below 1 pH unit). In some cases, however, the errors are way beyond the theoretical RMSD for individual pK_a (0.79–1.3 for PROPKA3, 0.5 for GB/IMC) of the training set used to develop these two methods. Large discrepancies are observed for fibrynogen and AcmB with the error for the latter being bigger than 3–4 units. The largest error observed for the AcmB homology model might indicate that some parts of it are less-than-optimally folded. On the other hand, homology models are frequently used to study new enzymes and this case illustrates a potential problem one may encounter. pH reaction optimum of AcmB is around 6.5, and for this range of pH, the computational methods predict the model to be positively charged (with the predicted pI *ca.* 8–9). Meanwhile, the experimental data indicate that in reality this enzyme has pI in the 4.7–5.2 range, and hence, the protein is negatively charged at the optimum pH. As a result, one might obtain qualitatively incorrect charge of the homology model for the optimal pH, which in turn might influence the quality of both MD simulation results and, most importantly, possibly also energy profile of the catalytic reaction. The latter can be expected especially when the active site is close to the protein surface and the reaction involves large charge separation events. Even though the actual source of error in this particular case is still not known (it might be due to failure of the homology modelling protocol, pK_a prediction or both), we advise a healthy dose of scepticism towards results obtained with pK_a predictors

Table 1 Isoelectric points (pI) for selected proteins

Protein	Predicted pI		Experimental pI
	CHARMM GB/IMC	PROPKA3	
α-casein	6.4	6.4	5.5 [44]
β-casein	5.4	5.5	5.5 [44]
α-chymotrypsin	9.4	8.8	8.76 [45]
HSA	6.2	5.9	5.1 [46]
fibrynogen	7.2	6.4	5.8–6.0 [47]
KTSD AcmB	9.3	8.9	4.7–5.2[a]

[a]K. Sofińska, A. M. Wojtkiewicz, M. Guzik, A. Winiarska, P. Waligórski, M. Cieśla, J. Barbasz and M. Szaleniec, Biochim Biophys Acta in press

and, if possible, cross-validation of the obtained results with available experimental data, or at least comparative analysis using different theoretical approaches.

In the context of metalloenzymes, one needs to be aware of the fact that most of the force field-based and empirical methods cannot take into account pK_a perturbation introduced by non-standard residues, such as metal ions or metal complexes. As demonstrated by Ryde and co-workers, one viable approach to tackle this issue can be performing a series of relatively short MD simulations for various plausible microprotonation states of titratable residues in the vicinity of the metal cofactor and comparing the thus obtained structures to experimental X-ray data [43]. When comparing MD results and X-ray structures, the authors used RMSD calculated for a titratable residue, RMSD calculated for the residue and its immediate surroundings, conformation of the residue and/or its H-bonding contacts [13, 43]. This approach works best for titratable residues in protein interior but loses its predictive power for solvent-exposed residues.

Probably, the most challenging is predicting pK_a values for residues that directly coordinate to the metal. The shift in pK_a for such residues can be approximated if the selected computational tools are based on the force field approach and the metal site in question is parameterized. For example, using the GB/IMC method implemented in Discovery Studio, which utilizes the CHARMM force field, one can estimate a pK_a shift of His120, which coordinates heme Fe in cytochrome C (PDB: 1CNG). Assuming the standard partial charges for the heme group from CHARMM, the apparent pK_a of Nε of His120 calculated for the model with and without heme prosthetic group are 2.0 and 6.25, respectively (the protocol does not provide pK_a for Nδ1 but predicts neutral charge in the physiological pH range). Such a result is hardly surprising as close vicinity of the positively charged Fe^{2+} ion influences protonation equilibrium of Nε of His120. Still it demonstrates that even at low pH, His120 will not get protonated which would result in de-coordination from Fe^{2+} ion. However, much less trivial is an increase of pK_a of Asp67 (from 3.13 to 4.1) due to

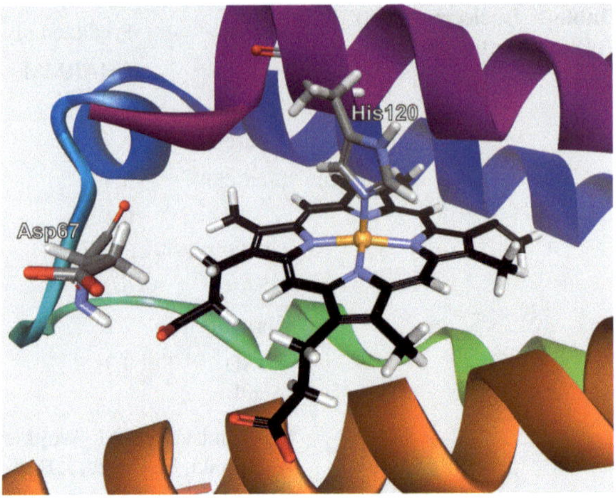

Fig. 3 Residues in cytochrome C (PDB: 1CNG) active site that undergo significant shift of pK_a due to interaction with Fe-heme prosthetic group

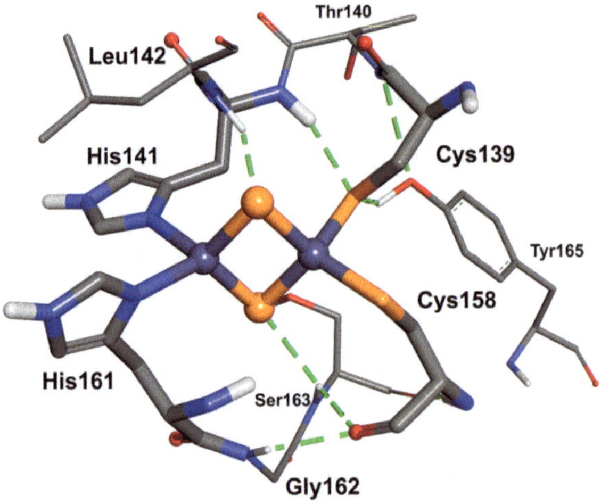

Fig. 4 Rieske-type Fe_2S_2 centre from the iron–sulphur protein of the bovine heart mitochondrial cytochrome bc1-complex (PDB 1RIE). Only polar H atoms are shown. Second coordination shell of residues is presented with thinner sticks [95]

the electrostatic interaction with negatively charged carboxyl groups of the porphyrin ring (Fig. 3).

Unfortunately, in most cases one has to resort to QM-based techniques of pK_a estimation for the crucial residues, especially for redox-active systems where oxidation state of the metal can directly influence the pK_a of its ligands or residues in close vicinity. Several examples of such successfully analyses were described by Ullmann and Noodleman for His coordinated Rieske-type Fe_2S_2 centre in cytochrome *bc1* complex [48], Glu residues in spinach plastocyanin [49], or H_2O/OH ligands in binuclear metal complexes in enzymes [50] (Fig. 4).

As an example of such QM-based pKa prediction approach may serve the work on the Rieske-type Fe_2S_2 by Ullmann et al. [48]. The system was partitioned into three parts: cluster model of the redox active site containing the iron sulphur cluster

($[Fe_2S_2(His)_2(Cys)_2]$) as well as its hydrogen-bonded partners (with dielectric constant $\varepsilon = 1$), rest of the protein ($\varepsilon = 4$) and the solvent ($\varepsilon = 80$). DFT gas-phase calculations for the cluster model allowed establishing geometries and energies of the active site cluster with various numbers of electrons and protons bound in order to account for every combination of protonation and metal oxidation state. In this case, 12 different variants were considered resulting from combination of 4 possible protonation states of two His ligands and three different electronic states of the Fe_2S_2 cluster (fully oxidized, with Fe coordinated by His reduced, with Fe coordinated by Cys reduced). The protein and solvent environment were modelled with a dielectric continuum model with embedded charges representing the protein utilizing macroscopic electrostatics with atomic detail (MEAD) model [51–54]. The calculations were iterative and consisted of the following main steps. First, ESP atomic charges on the cluster were fitted with the CHELPG protocol. Next, the Poisson or Poisson–Boltzmann equation was solved yielding the reaction field. The protein field was computed in the same manner using partial charges from the surrounding protein. Subsequently, the protein and reaction fields were added to the cluster Hamiltonian and the DFT calculations were repeated until convergence was reached. The numerical solution of the Poisson–Boltzmann equation (conducted with the program MEAD [52]) was used to calculate protonation and interaction energies and Monte Carlo simulation for protonation state sampling [55]. Finally, the microscopic protonation equilibrium constants were calculated by methods described by Li et al. [56, 57] using a thermodynamic cycle that represents deprotonation in gas/protein/solvent phase and solvation process. The application of such approach allowed Ullmann et al. to predict apparent pK_a values of the oxidized Rieske centre as 6.9 and 8.8, i.e. values fairly close to those obtained from the experiment (7.5 and 9.2, respectively). However, it turned out that these pK_a values cannot be associated with deprotonations of individual His residues as both of them titrated over a wide range of pH. In accordance with experimental data, the computational results demonstrated that it is the Fe ion coordinated by His that receives electron upon the site reduction and that such redox change results in shift of apparent pK_a to 11.3 and 12.8, i.e. above 10.0, as was also found experimentally. Computational results showed that after reduction of the FeS centre, the His161 exhibits a higher proton affinity than His141. As a result, the His161 residue will be able to transfer a proton from ubiquinol faster than His141. Finally, the authors considered the possibility of coupling between pK_a of the second shell residues and the redox state of the Rieske centre. However, it turned out that for most residues, pK_a are not strongly affected by the redox state of the cluster, with the only exception of Tyr157, whose $pK_{1/2}$ shifted by 0.6 pH unit upon cluster reduction.

7 Molecular Dynamics Simulations

Molecular dynamics simulations are widely recognized as a very powerful tool that allows one to study various processes. For example, it is frequently used to: (a) "relax" the protein structure, i.e. allow it to attain a structure as in solution as opposed to the situation in the crystal, (b) study ligand binding or dissociation, (c) study conformational freedom of a macromolecular system and many more. In the context of metalloenzymes, the key issue is what kind of force field is used for the metal with its ligands and where from its parameters can be obtained. One of the options is to use a so-called non-bonded model for the metal ion, whereby the metal–ligand interactions are described with non-bonded terms of classical force field, i.e. van der Waals and electrostatic terms, the latter possibly also including ion-induced dipole term [58]. The advantage of this approach is that it allows to simulate complexes of a given metal ion with various coordination numbers and compositions of ligands without the need to derive parameters for each particular coordination complex separately. Moreover, as this model does not specify explicit bonds between the metal and the ligands, it permits to simulate the process of ligand binding or dissociation. Up to quite recently, the disadvantage of classical non-bonded models was lack of parameters for transition metal ions, so for example, quite often Mg^{2+} ion was used in simulations in place of Fe^{2+} bound in the enzyme active sites [59, 60]. This situation has changed as parameters for many metal cations were derived by Merz and co-workers [61, 62]. The alternative is a bonded model in which explicit bonds are specified between the central metal ion and its ligands. The major disadvantages of this approach are the need to derive bonded parameters for each specific complex and the fact that bond cleavage/formation steps cannot be studied. There are several different ways in which one can derive such bonded metal–ligand parameters, probably the most straightforward being those projecting the Hessian matrix, calculated at the QM (DFT) level, onto internal coordinates of the complex. This method was originally suggested by Seminario [63, 64], but in its original form it suffers from double counting of 1–4 interactions when the derived bonded parameters are used together with atomic charges. A recent improvement proposed by Hirao and co-workers to a large extent lifts this limitation (see Chapter "Applications of Computational Chemistry to Selected Problems of Transition-Metal Catalysis in Biological and Non-biological Systems"). Recent advances in development of polarizable force fields pave a way to more accurate description of electrostatic interactions in macromolecules including metal ion–ligand interactions. As writing this chapter, the first works devoted to derivation and testing of polarizable force field parameters for metal ions were already published. The preliminary results of molecular simulations employing such force fields let us hope that in the near future such methods will be used more routinely to study structure and dynamics of metalloenzymes [65–67].

A recent work on AsqJ dioxygenase may serve as a case study demonstrating the value of relatively straightforward MD simulations for metalloenzymes [68]. The reaction mechanism for oxidative dehydrogenation (Fig. 5) was studied at the QM/MM (ONIOM) level, yet the starting structures for these investigations were derived from classical MD simulations performed for two reaction intermediates: either for the enzyme–substrate complex, i.e. before O_2 activation and α-ketoglutarate decarboxylation, or for the ferryl species, i.e. the state immediately before C–H bond cleavage (**S** in Fig. 5). Interestingly, even though the QM/MM optimized structures for species **S** looked very similar in these two cases, for the former model a lower

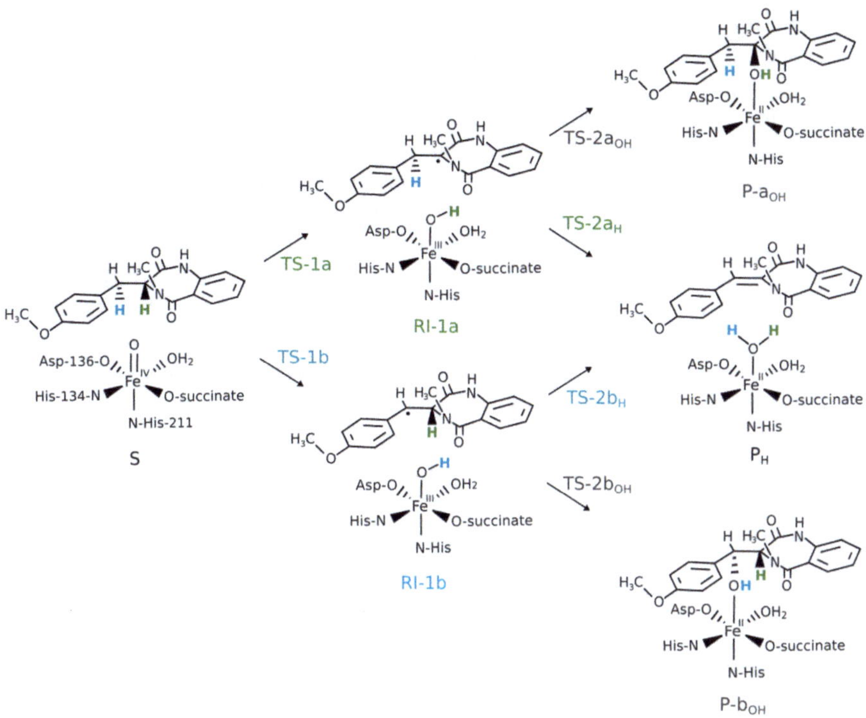

Fig. 5 Possible pathways for AsqJ-catalysed oxidation reactions of 4′-methoxycyclopeptin. C3-bound hydrogen is shown in green; C10-bound hydrogen in blue. Reprinted from [68]

barrier was obtained for the initial C–H activation at the 7-membered ring (H-atom and the reaction pathway marked in green), whereas for the latter the preferred site of the initial C–H cleavage was the benzylic carbon (marked in blue). Initial C–H bond cleavage at the benzylic carbon is consistent with the results of experimental studies where an epimer of the substrate was used as a mechanistic probe [69]. Thus, a simple explanation would be that the enzyme pocket adapts to the changes along the reaction coordinate and the MD simulation for the reactive intermediate **S** reproduces it better than MD for the preceding species (Fe^{2+}/α-ketoglutarate/substrate; not shown) followed by a simple QM/MM minimization.

8 Choosing MD Snapshots for QM or QM/MM Studies

Once the structure of enzyme–substrate (or enzyme–intermediate) complex has been equilibrated through classical molecular dynamics simulations, one needs to select one, or at most a few, configuration(s) that will serve as starting points for QM or QM/MM investigations on the reaction mechanism. This step is still required for open-shell transition metal systems as QM(/MM) dynamics employing a reliable QM model and basis set is computationally too demanding for investigations of reaction path. Note, however, that such studies are perfectly doable with semi-empirical QM methods for systems without transition metals [70]. How such a snapshot(s) should be selected is not obvious, and there are at least several approaches used by the community. One way is to take the last saved snapshot from the simulated trajectory, as this can be viewed as the "final" product of MD simulation. In another approach, one arbitrarily selects specific frames from the saved trajectory. Both of the above approaches may lead to selection of suboptimal structures because at physiological temperatures, simulated with MD, energy of the system fluctuates quite substantially and the trajectory visits various regions of the conformational space, also those characterized by higher free energy values and hence lower probability. On the other hand, reaction mechanisms of most metalloenzymes proceed through very reactive intermediates, e.g. Fe(II)-OO-R, Fe(IV)=O, and for this reason these enzymes usually function according to a "negative catalysis" paradigm [71, 72], which means the enzyme holds the substrate(s) in a well-defined position and conformation, so that only one reaction pathway, i.e. the desired one, is realized. Hence, it seems reasonable to assume that, at least at this stage of the reaction when the reactive species is present, the free energy landscape of the enzyme–substrate (or intermediate) complex will feature a dominating basin of attraction corresponding to the preferred orientation of the reactants. Therefore, another way of selecting snapshots from the trajectory rests on methods capable to identify such basins, i.e. on clustering algorithms. Objective of the latter is to divide a given set of objects, in our case MD snapshots, into groups (clusters) that gather most similar objects under the condition that differences between the clusters are much more pronounced than variance within them. Arguably, the major basin of attraction should correspond to the most populated cluster. An example of cluster analysis comes from our recently published work on acireductone dioxygenase with the key settings and major results presented in Fig. 6 [73]. A total of 500 snapshots saved during the last 10 ns of a stable "production" period of MD were first superimposed minimizing RMSD for selected atoms of residues lying within 10 Å of the metal. The specific atoms used for RMSD calculations are marked with orange spheres in the figure; note that symmetry-equivalent atoms were not taken into account as using them would lead to discrimination between physically indistinguishable conformations. Second, using this RMSD metrics, clustering algorithms that are available in the cpptraj program from the Amber 16 suite were used to cluster the frames. For K-means and hierarchical agglomerative algorithms, the required number of clusters (n) can be explicitly specified and hence, with their use clustering was performed for n ranging from 2 to 6. For the density-based algorithms (DBSCAN and dpeaks), one needs to input either distance cut-off between points for forming a cluster (DBSCAN) or cut-offs determining local density and distance (dpeaks). The choice of the appropriate values of these parameters was aided by visual inspection of a "K-dist" plot (DBSCAN) and the "decision graph" (dpeaks).

The decision graph for the dpeaks algorithm clearly showed 3 or 4 distinctive points characterized by high density and large distance to the point of next highest

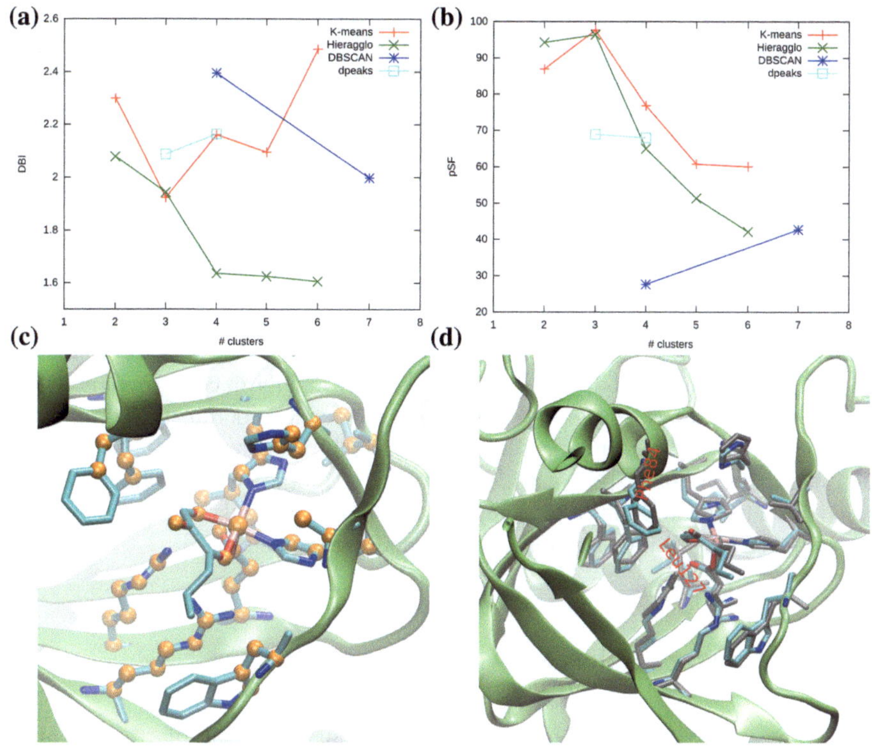

Fig. 6 Clustering MD snapshots based on active site conformation [74]. **a** The Davies–Bouldin Index (DBI) versus number of clusters, **b** the pseudo-F statistic (pSF) versus number of clusters, **c** residues from the metal's first and second coordination sphere, atoms taken for RMSD calculations marked with orange spheres, **d** superposed representative structures for most populated clusters obtained with K-means (sticks in standard colours) and dpeaks (in grey) algorithms for total number of clusters of 3

density. Parameters were read from the graph and used as input for clustering with dpeaks, which yielded 3 or 4 clusters. On the other hand, for the parameters read from the "K-dist" plot a single cluster was obtained, and only by manually changing the parameters it was possible to get results with 4 and 7 clusters.

Two clustering metrics are by default computed by cpptraj: the Davies–Bouldin Index (DBI) and pseudo-F statistic (pSF); plots of their values obtained with the four algorithms versus number of clusters are presented in Fig. 6a and b, respectively. The profiles obtained by the K-means algorithm indicate that the optimal number of clusters is 3 (minimum for DBI, maximum for pSF) and this agrees with the results from dpeaks, which show better statistics for $n = 3$ than for $n = 4$, and with the peak in pSF value at $n = 3$ obtained with the hierarchical agglomerative algorithm. Percentage populations of the obtained clusters for $n = 3$ are reported in Table 2 together with the number identifying a representative frame for a given cluster.

Table 2 Populations [in %] of the clusters obtained for number of clusters $n = 3$. In parentheses frame number of a representative structure for a given cluster

	cluster 1	cluster 2	cluster 3
K-means	38.0 (87)	31.6 (398)	30.4 (164)
Hieragglo	38.2 (87)	31.6 (398)	30.2 (164)
dpeaks	48.2 (312)	38.8 (87)	13.0 (164)

From these values, it immediately follows that K-means and hierarchical agglomerative algorithms gave practically identical results and that the cluster represented by frame 87 has a population of ca. 38–39% irrespective of the algorithm used. Comparison of structures representative for the most populated clusters, i.e. frames 87 and 312, shown in Fig. 6d, reveals that they differ mainly in conformation of two hydrophobic residues (Phe84 and Leu127) in the second coordination shell of the metal. As these residues are most likely not critical for the progress of the reaction and since better DBI and pSF metrics were obtained for the two algorithms that identified the cluster with frame 87 as a representative, this particular frame was selected for further QM/MM investigations on the reaction mechanism [73].

9 QM Cluster Models for Metalloenzymes

Modelling reactions catalysed by enzymes that contain additional cofactors and metal complexes in particular have two advantages. First, the active site can usually be easily localized and, second, it can be studied by spectroscopic means (such as EXAFS, EPR or ORD) also in a non-active state [74–76]. It is also relatively straightforward to select a cluster model of such a metalloprotein, taking into consideration the metal ion(s) and all its ligands from the first coordination shell and some or all ligands from the second one. In order to ensure that the geometry of such a model exhibits structural characteristics of the active site of the actual protein, one has to introduce geometric constraints on selected atoms, whose purpose is to mimic structural influence of the rest of the now absent protein. The usual approach is to freeze Cartesian coordinates of the carbon atom at the cleavage of the C–C bond as well as of the hydrogen atom which was used to end-cap the cleaved bond. It has been argued that most of the metal sites in proteins are in their equilibrium geometry and therefore the cluster model approach is representing their geometry well. Indeed, the seminal work of Siegbahn on structure and reaction mechanism of photosynthetic oxygen evolving complex clearly demonstrated that the structure of this manganese cluster could be predicted using energy as the major guideline [77]. However, there are several cases where both theoretical and experimental evidences suggest the contrary that protein introduces strain into the metal complex destabilizing its reactant state and thus promoting catalysis or electron transfer [78]. This effect, known as "entatic state principle", was proposed as a reason standing behind unexpected spectroscopic properties of blue copper proteins, the Mo-oxidation state-dependent geometry of the molybdenum enzymes as well as structural distortion of metal sites in many other enzymes; for more examples, see recent review by Stanek et al. [79] As a representative example here, we take the case of molybdenum proteins. The strained entatic state was observed especially for DMSO reductase family of molybdenum enzymes that harbour a bis-molybdopterin cofactor (Moco). Such cofactor comprises Mo ion in its centre, two pterins, each providing two dithiolene ligands to Mo, a protein residue of serine, cysteine, aspartate or selenocysteine, depending on the enzyme, as well as one oxo ligand, the latter being sensitive to the Mo-oxidation state. The structurally extended pterin ligands interact with protein surrounding via multiple hydrogen bonds and ionic interactions, and as a result, the strain is introduced to the structure of the Mo complex. This effect can be structurally quantified by dihedral angle between planes of pterin ligands, referred to as an angle ϕ, similar to the Rây–Dutt twist used in studies of arsenate oxidase [80], or by S_1-S_2-S_3-S_4 dihedral, which was used in studies on ethylbenzene dehydrogenase [81, 82].

The oxidized (Mo^{VI}) Moco is six-coordinate and contains the oxo (O) ligand. Despite the presence of six ligands, in most of the enzymes the Mo coordination does not attain a (distorted) octahedral geometry, which is typical for its model complexes [83, 84], but rather a distorted trigonal prism, where each trigonal face is formed by two sulphurs from opposite dithiolenes and one oxygen atom [85]. The S_1-S_2-S_3-S_4 dihedral in oxidized enzymes attains values in the range from $-20°$ to $-30°$ [86–88]. On the other hand, the reduced (Mo^{IV}) Moco is usually five-coordinate (the prior O ligand is already transferred to the reagent) and attains a shape of tetragonal pyramid with sulphur atoms of pterins forming its square base with relatively small twist. Alternatively, the product, which is loosely bound to Mo(IV) by the dative bond, can serve as the sixth ligand. Such geometries were confirmed experimentally by X-ray crystallography results for: (a) the structure of non-active resting state of EBDH [89], in which the reduced Moco features S_1-S_2-S_3-S_4 dihedral of $-4.86°$, (b) the reinterpreted structure of the pentacoordinate Moco of formate dehydrogenase H from *E. coli* (FDH) with S_1-S_2-S_3-S_4 dihedral of $-6.23°$ [90], as well as (c) reduced perchlorate reductase (PChR) with the substrate analogue bound to Moco (hexacoordinate, PDB: 5CHC) or without the substrate (pentacoordinate, PDB: 4YDD), where S_1-S_2-S_3-S_4 dihedrals are in the range from $-6°$ to $-9°$ [91].

The relaxed nature of the almost planar arrangement of pterins in the reduced molybdenum enzymes is confirmed by geometry optimization in vacuum, which yields almost identical conformation as that observed in the EBDH crystal structure (S_1-S_2-S_3-S_4 dihedral of $-5.3°$).

The strain imposed by the protein on Mo coordination geometry is usually modelled by freezing the position of the fringe hydrogen atoms that are saturating bonds at places where the cofactor was truncated. However, in contrast to the amino acid residues, it is much more difficult to select truncation and constrain sites that will allow both to simulate the real flexibility of the cofactor on the one hand and to properly model its electronic characteristics on the other. For example, constraining H atoms (or sometimes carbon atoms of CH_3 groups introduced instead of H) of the $^{-}$S-CH=CH-S^{-} gives a very rigid structure, and as a result, the modelled reaction pathway may be biased towards oxidized or reduced geometry of Moco, depending

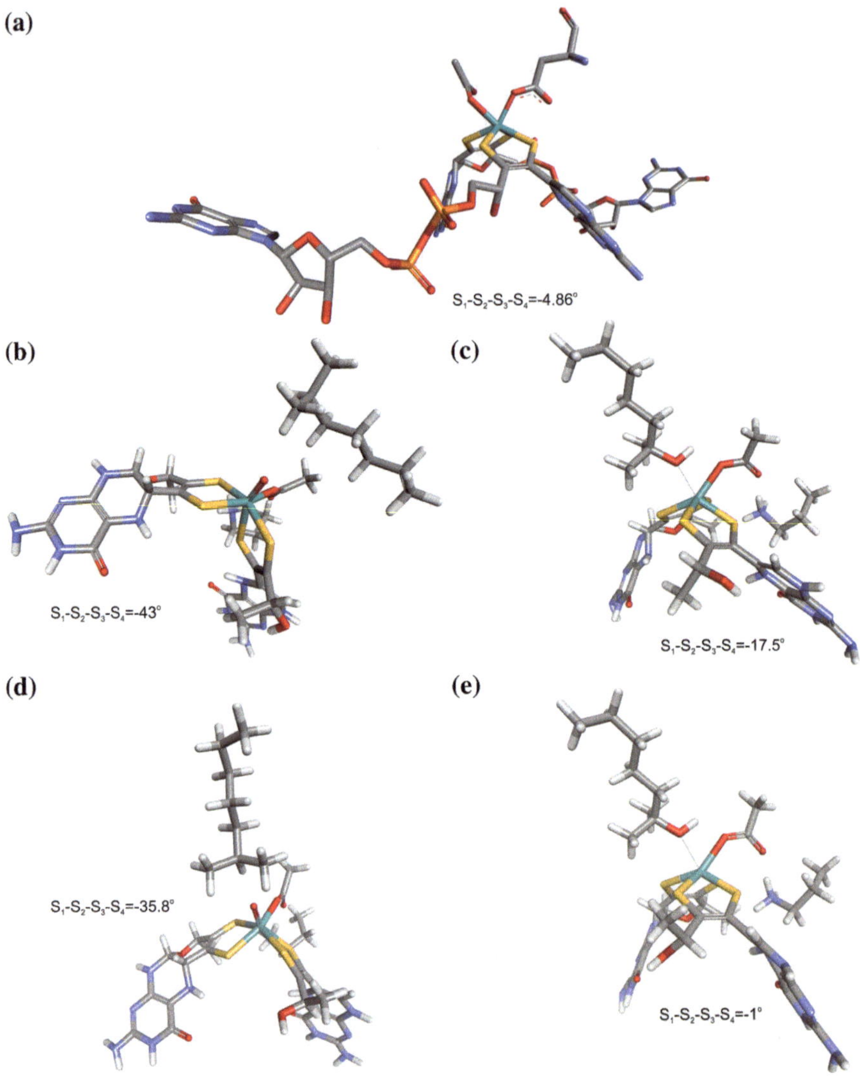

Fig. 7 **a** Structure of six-coordinate reduced Moco in ethylbenzene dehydrogenase; [88] **b** oxidized extended model 1; **c** product-bound reduced model 1; **d** oxidized extended model 2 with constraints imposed on pterins as in (**a**); **e** product-bound reduced extended model 2

on the initial geometry and direction of the process under study [80, 92, 93]. Application of a more extended model of the cofactor (Fig. 7), as proposed in our previous papers, provides more flexibility, which might actually be too large compared with the enzyme active site [81].

The other important issue is for which oxidation state the constraints are introduced to the model. Availability of both reduced and oxidized structures of a particular enzyme is in the case of molybdenum enzyme rather an exception then a rule. Molybdenum enzymes are also often air sensitive, and they are crystalized under anaerobic atmosphere which exhibits reductive properties (5% of H_2 in N_2). As a result, the enzyme can crystalize in the reduced state despite the oxidized state of the protein was prepared. Secondly, the oxidized cofactor can change its oxidation state due to photo-reduction in the very intense synchrotron beam. This leads to additional ambiguity in the actual oxidation state of the cofactor. Therefore, in order to study reaction pathway, one has to carefully construct the working theoretical model, usually from just one geometry (be it reduced or oxidized).

In the case of the EBDH, the reduced and catalytically inactive state was available, and only from that point, the oxidized Moco geometry was inferred by geometry optimization of the whole cofactor with constraints. Such cofactor model was then truncated, constrained at the edges and used for modelling of the reaction pathway. In this case, the S_1-S_2-S_3-S_4 dihedral turned out to be −38.3°, whereas the same angle calculated using a QM/MM approach, which presumably models interactions between Moco and the protein more properly, was only −20.4° [82]. This resulted with the calculated Δ(E+ZPE) for the first transition state of C–H activation higher by 3.0 kcal/mol for the cluster model than calculated with QM/MM.

A slightly different approach was chosen for modelling steroid C25 dehydrogenase (S25DH). More specifically, the oxidized geometry of the enzyme was obtained through homology modelling, MD simulation, which actually yielded pretty decent S_1-S_2-S_3-S_4 dihedral of −21.3°, followed by QM/MM geometry optimization, which gave a significantly more relaxed geometry with dihedral angle of −39.2° [94]. The cluster model cut out from this macromolecular structure and constrained in the analogous way as before relaxed even further yielding the dihedral of −48.6°. As a result, the calculated transition state energies are quite high (approx. 21.2 kcal/mol) and the reduced state exhibits more twisted conformation of the pterin ligand (−17°) than that observed for the EBDH crystal structure.

We decided to illustrate the influence of the way in which the constraints are introduced into the model on the reaction energy profile. To achieve this, we have selected a big model of Moco that contains whole pterin ligands. The constraints were introduced either to the MD-relaxed oxidized model 1 (dihedral −21.3°) or to the reduced homology model 2, which had Moco in geometry as in the reduced EBDH crystal structure (dihedral −4.86°). After optimization, the S_1-S_2-S_3-S_4 dihedral angle for the oxidized Moco in model 1 was −43° and differed from the value obtained for model 2, i.e. −35.8°. This resulted in only slight energy elevation of model 2 ES with respect to model 1; by 0.8 kcal/mol. However, the influence on C–H activation barrier was more pronounced—the barrier was lower by 1.4 kcal/mol for model 2. The largest effect was observed for EP complex where model 2 gave energy lower by 4.0 kcal/mol. Moreover, model 2 not only yielded more favourable reaction energetics but also the geometry of the relaxed reduced cofactor, with the S_1-S_2-S_3-S_4 dihedral of −1°, is much closer to the experimental value of −4.8° than the value observed for EP in model 1 (−17.5°).

In order to further illustrate the degree to which the oxidized Moco is energized by the twist introduced by the enzyme, one can calculate relaxation energy of the ES upon removal of constrains. For EBDH, such difference was estimated to be 14.3 kcal/mol [81], while for S25DH it was 14.9 kcal/mol for the smaller model used by Rugor et al. and 16.8 kcal/mol and 17.7 kcal/mol for extended model 1 and model 2, respectively. Similarly, it has been calculated that in the case of arsenate oxidase, the energy difference between equilibrium and strained geometry is in the range of 12.9 kcal/mol [80].

The above-discussed results clearly illustrate that in certain cases, where the cofactor is fairly rigid and extended, the steric effects imposed by the protein can play non-negligible role for the catalytic process. Hence, the origin of the cluster model, i.e. if it comes from X-ray structure for oxidized or reduced form of the enzyme, its size and the way constraints are introduced into it all will have influence on the energetic profile for the studied reaction.

10 Conclusions

Computational studies on metalloenzymes can provide valuable new insights into structure and reaction mechanisms, yet their credibility depends on multiple factors. In this contribution, we have tried to illustrate the key decisions and steps one has to take in such projects (validating X-ray structures, re-refining metal sites, determining residues protonation state, modelling missing loop regions or whole proteins by comparative modelling, performing MD simulations, choosing representative snapshots and constructing QM models). While this list is by no means exhaustive, we hope that when the content of this chapter will be combined with the material covered in other chapters of this book, it will offer a more comprehensive view on modelling metalloenzymes.

Acknowledgements This research was supported in part by PL-Grid Infrastructure. Computations were performed in the AGH Cyfronet Supercomputer Centre. Authors acknowledge financial support of the statutory research fund of ICSC PAS. M. Sz. acknowledges the input of Ewelina Biała, the undergraduate student conducting modeling of S25DH under supervision of M. Sz.

References

1. Nitschke W, McGlynn SE, Milner-White EJ, Russell MJ (2013) Biochim Biophys Acta Bioenerg 1827:871
2. Lovell T (2003) Coord Chem Rev 238–239:211
3. Andreini C, Bertini I, Rosato A (2009) Acc Chem Res 42:1471
4. Andreini C, Bertini I, Cavallaro G, Holliday GL, Thornton JM (2008) JBIC J Biol Inorg Chem 13:1205
5. Jernigan R, Raghunathan G, Bahar I (1994) Curr Opin Struct Biol 4:256

6. Mitchell AJ, Dunham NP, Martinie RJ, Bergman JA, Pollock CJ, Hu K, Allen BD, Chang WC, Silakov A, Bollinger JM, Krebs C, Boal AK (2017) J Am Chem Soc 139:13830
7. Kovaleva EG, Lipscomb JD (2007) Science 316:453
8. Zheng H, Chordia MD, Cooper DR, Chruszcz M, Müller P, Sheldrick GM, Minor W (2013) Nat Protoc 9:156
9. Zheng H, Cooper DR, Porebski PJ, Shabalin IG, Handing KB, Minor W (2017) Acta Crystallogr Sect D Struct Biol 73:223
10. Fernández-Leiro R, Pereira-Rodríguez Á, Cerdán ME, Becerra M, Sanz-Aparicio J (2010) J Biol Chem 285:28020
11. Ryde U, Nilsson K (2003) J Mol Struct THEOCHEM 632:259
12. Nilsson K, Ryde U (2004) J Inorg Biochem 98:1539
13. Cao L, Caldararu O, Ryde U (2017) J Phys Chem B 121:8242
14. Ryde U(2007) Dalt Trans 607
15. Orry AJW, Abagyan R (eds) (2012) Homology modeling. Humana Press, Totowa, NJ
16. Webb B, Sali A (2016) Current protocols in bioinformatics. Wiley, Hoboken, NJ, USA, pp 5.6.1–5.6.37
17. Yang J, Yan R, Roy A, Xu D, Poisson J, Zhang Y (2014) Nat Methods 12:7
18. Bordoli L, Schwede, T (2011) Methods Mol Biol, 107–136
19. Kluza A, Niedzialkowska E, Kurpiewska K, Wojdyla Z, Quesne M, Kot E, Porebski PJ, Borowski T (2018) J Struct Biol 202:229
20. Fiser A, Do RK, Sali A, Fiser A, Kinh R, Do G, Andrej S (2000) Protein Sci 9:1753
21. Lee J, Lee D, Park H, Coutsias EA, Seok C (2010) Proteins Struct Funct Bioinforma 78:3428
22. Tang K, Zhang J, Liang J (2014) PLoS Comput Biol 10:e1003539
23. Höppner A, Widderich N, Lenders M, Bremer E, Smits SHJ (2014) J Biol Chem 289:29570
24. Reuter K, Pittelkow M, Bursy J, Heine A, Craan T, Bremer E (2010) PLoS ONE 5:e10647
25. Melo F, Feytmans E (1998) J Mol Biol 277:1141
26. Sali A, Sali, A (2006) Protein Sci, 2507
27. Laskowski RA, MacArthur MW, Moss DS, Thornton JM (1993) J Appl Crystallogr 26:283
28. Widderich, N, Hoþpner, A, Pittelkow M, Heider J, Smits SHJ, Bremer E (2014) PLoS One 9
29. Ullmann GM, Bombarda E (2014) Protein model. Springer, Cham, pp 135–163
30. Nielsen, JE (2008) In: Annual reports in computational chemistry. Elsevier, pp 89–106
31. Iwai W, Yagi D, Ishikawa T, Ohnishi Y, Tanaka I, Niimura N (2008) J Synchrotron Radiat 15:312
32. Moser A, Range K, York DM (2010) J Phys Chem B 114:13911
33. Thurlkill RL, Grimsley GR, Scholtz JM, Pace CN (2006) Protein Sci 15:1214
34. Jensen JH, Li H, Robertson AD, Molina PA (2005) J Phys Chem A 109:6634
35. Awoonor-Williams E, Rowley CN (2016) J Chem Theory Comput 12:4662
36. Nielsen, JE, Gunner MR, García-Moreno BE (2011) Proteins Struct Funct Bioinforma 79:3249
37. Wasilewska M, Adamczyk Z, Jachimska B (2009) Langmuir 25:3698
38. Righetti PG (2004) J Chromatogr A 1037:491
39. Olsson MHM, Søndergaard CR, Rostkowski M, Jensen JH (2011) J Chem Theory Comput 7:525
40. Spassov VZ, Yan L (2008) Protein Sci 17:1955
41. Kumosinski TF, Brown EM, Farrell HM (1994) In: Molecular modeling from virtual tools to real problems, pp 368–390
42. Kumosinski TF, Brown EM, Farrell HM (1993) J Dairy Sci 76:931
43. Uranga J, Mikulskis P, Genheden S, Ryde U (2012) Comput. Theor Chem 1000:75
44. Szyk-Warszyńska L, Gergely C, Jarek E, Cuisinier F, Socha RP, Warszyński P (2009) Colloids Surf A Physicochem Eng Asp 343:118
45. Ui N (1971) Biochim Biophys Acta Protein Struct 229:582
46. Jachimska B, Wasilewska M, Adamczyk Z (2008) Langmuir 24:6866
47. Wasilewska M, Adamczyk Z (2011) Langmuir 27:686
48. Ullmann MG, Noodleman L, Case DA (2002) JBIC J Biol Inorg Chem 7:632
49. Bombarda E, Ullmann GM (2010) J Phys Chem B 114:1994

50. Noodleman L, Han W-G (2006) JBIC J Biol Inorg Chem 11:674
51. Bashford D (1991) Curr Opin Struct Biol 1:175
52. Bashford D, Gerwert K (1992) J Mol Biol 224:473
53. Lim C, Bashford D, Karplus M (1991) J Phys Chem 95:5610
54. Bashford D, Case DA, Dalvit C, Tennant L, Wright PE (1993) Biochemistry 32:8045
55. Beroza P, Fredkin DR, Okamura MY, Feher G (1991) Proc Natl Acad Sci 88:5804
56. Li J, Nelson MR, Peng CY, Bashford D, Noodleman L (1998) J Phys Chem A 102:6311
57. Fisher CL, Chen J-L, Li J, Bashford D, Noodleman L (1996) J Phys Chem 100:13498
58. Li P, Merz KM (2014) J Chem Theory Comput 10:289
59. Widderich N, Pittelkow M, Höppner A, Mulnaes D, Buckel W, Gohlke H, Smits SHJ, Bremer E (2014) J Mol Biol 426:586
60. Torabifard H, Cisneros GA (2017) Chem Sci 8:6230
61. Li P, Song LF, Merz KM (2015) J Phys Chem B 119:883
62. Li P, Song LF, Merz KM (2015) J Chem Theory Comput 11:1645
63. Seminario JM (1996) Int J Quantum Chem 60:1271
64. Hu L, Ryde U (2011) J Chem Theory Comput 7:2452
65. Xiang JY, Ponder JW (2013) J Comput Chem 34:739
66. Xiang JY, Ponder JW (2014) J Chem Theory Comput 10:298
67. Jing Z, Qi R, Liu C, Ren P (2017) J Chem Phys 147:161733
68. Wojdyla Z, Borowski T (2018) JBIC J Biol Inorg Chem 23:795
69. Liao H-J, Li J, Huang J-L, Davidson M, Kurnikov I, Lin T-S, Lee JL, Kurnikova M, Guo Y, Chan N-L, Chang W (2018) Angew Chemie Int Ed 57:1831
70. Świderek K, Arafet K, Kohen A, Moliner V (2017) J Chem Theory Comput 13:1375
71. Rétey J (1990) Angew Chemie Int Ed English 29:355
72. Vögeli B, Erb TJ (2018) Curr Opin Chem Biol 47:94
73. Miłaczewska A, Kot E, Amaya JA, Makris TM, Zając M, Korecki J, Chumakov A, Trzewik B, Kędracka-Krok S, Minor W, Chruszcz M, Borowski T (2018) Chem A Eur J 24:5225
74. Vallee BL, Williams RJ (1968) Proc Natl Acad Sci 59:498
75. Pushie MJ, George GN (2011) Coord Chem Rev 255:1055
76. Luthra A, Denisov IG, Sligar SG (2011) Arch Biochem Biophys 507:26
77. Siegbahn PEM (2009) Acc Chem Res 42:1871
78. Williams RJP (1995) Eur J Biochem 234:363
79. Stanek J, Hoffmann A, Herres-Pawlis S (2018) Coord Chem Rev 365:103
80. Warelow TP, Pushie MJ, Cotelesage JJH, Santini JM, George GN (2017) Sci Rep 7:1757
81. Szaleniec M, Borowski T, Schühle K, Witko M, Heider J (2010) J Am Chem Soc 132:6014
82. Szaleniec M, Dudzik A, Kozik B, Borowski T, Heider J, Witko M (2014) J Inorg Biochem 139:9
83. Donahue JP, Goldsmith CR, Nadiminti U, Holm RH (1998) J Am Chem Soc 120:12869
84. Enemark JH, Cooney JJA, Wang J-J, Holm RH (2004) Chem Rev 104:1175
85. Li H-K, Temple C, Rajagopalan KV, Schindelin H (2000) J Am Chem Soc 122:7673
86. Dias JM, Than ME, Humm A, Huber R, Bourenkov GP, Bartunik HD, Bursakov S, Calvete J, Caldeira J, Carneiro C, Moura JJ, Moura I, Romão MJ (1999) Structure 7:65
87. Czjzek M, Dos Santos J-P, Pommier J, Giordano G, Méjean V, Haser R (1998) J Mol Biol 284:435
88. Bray RC, Adams B, Smith AT, Bennett B, Bailey S (2000) Biochemistry 39:11258
89. Kloer DP, Hagel C, Heider J, Schulz GE (2006) Structure 14:1377
90. Raaijmakers HCA, Romão MJ (2006) JBIC J Biol Inorg Chem 11:849
91. Youngblut MD, Tsai C-L, Clark IC, Carlson HK, Maglaqui AP, Gau-Pan PS, Redford SA, Wong A, Tainer JA, Coates JD (2016) J Biol Chem 291:9190
92. Leopoldini M, Russo N, Toscano M, Dulak M, Wesolowski TA (2006) Chem A Eur J 12:2532
93. McNamara JP, Hillier IH, Bhachu TS, Garner CD (2005) Dalt Trans 0:3572
94. Rugor A, Wójcik-Augustyn A, Niedzialkowska E, Mordalski S, Staroń J, Bojarski A, Szaleniec M (2017) J Inorg Biochem 173:28

95. Iwata S, Saynovits M, Link TA, Michel H (1996) Structure of a water soluble fragment of the 'Rieske' iron-sulfur protein of the bovine heart mitochondrial cytochrome bc1 complex determined by MAD phasing at 1.5 A resolution. Structure 4:567–579

Index

E. Broclawik et al. (eds.), *Transition Metals in Coordination Environments*, Challenges and Advances in Computational Chemistry and Physics 29,
https://doi.org/10.1007/978-3-030-11714-6

MIX
Papier aus verantwortungsvollen Quellen
Paper from responsible sources
FSC® C105338

If you have any concerns about our products,
you can contact us on
ProductSafety@springernature.com

In case Publisher is established outside the EU,
the EU authorized representative is:
Springer Nature Customer Service Center GmbH
Europaplatz 3, 69115 Heidelberg, Germany

Printed by Libri Plureos GmbH
in Hamburg, Germany